Chromogenics

Chromogenics

Smart Switchable Optical Materials and Their Applications

Edited by

Carl M. Lampert
Star Science
CA, USA

WILEY

Published by John Wiley & Sons, Inc., Hoboken, New Jersey.

For general information on our other products and services or for technical support, please contact our Customer Care Department within the United States at (800) 762-2974, outside the United States at (317) 572-3993 or fax (317) 572-4002.

Wiley also publishes its books in a variety of electronic formats. Some content that appears in print may not be available in electronic formats. For more information about Wiley products, visit our web site at www.wiley.com.

Library of Congress Cataloging-in-Publication Data applied for:

Hardback ISBN: 9781394159079

Cover Design: Wiley
Cover Image: © mf-guddyx/Getty Images

Set in 9.5/12.5pt STIXTwoText by Straive, Chennai, India

Contents

List of Contributors

Anoop Agrawal
Glass Dyenamics
Tucson, AZ
USA

and

Formerly with Donnelly Corporation
Holland, MI
USA

Wilmert C.S. De Bosscher
Soleras Advanced Coatings
Deinze
Belgium

Melepurath Deepa
Department of Chemistry
Indian Institute of Technology Hyderabad
India

H. Hakemi
Plastic Liquid Crystal Technology
Macherio
Italy

Carl M. Lampert
Star Science
Cotati, CA
USA

and

Formerly, Technical Director
Society of Vacuum Coaters (non-profit)
Albuquerque, NM
USA

and

Staff Scientist University of California
LBNL Energy Technologies, Berkeley
CA, USA

Philippe Lemarchand
Dublin Energy Lab
Technological University Dublin
Dublin
Ireland

and

MaREI – the SFI Research Centre for
Energy, Climate and Marine, Energy
Management
University College Cork
Cork
Ireland

and

Sustainability Intelligence
Technological University Dublin
Dublin
Ireland

Paul Lippens
Vacuum & Thin Film
Technologies/Materials
Marciano (PG)
Italy

José Montero
Department of Materials Science and
Engineering, The Ångström Laboratory
Uppsala University
Uppsala
Sweden

Junichi Nagai
NanoFilm Laboratory
Yokohama
Japan

and

Formerly Principal Researcher
Asahi Glass Co.
Yokohama
Japan

George Neuman
Iridescence Thin Film Consulting
Michigan
USA

Gunnar A. Niklasson
Department of Materials Science and
Engineering, The Ångström Laboratory
Uppsala University
Uppsala
Sweden

Brian Norton
Dublin Energy Lab
Technological University Dublin
Dublin
Ireland

and

International Energy Research Centre
Tyndall National Institute
University College Cork
Cork
Ireland

and

MaREI – the SFI Research Centre for
Energy, Climate and Marine, Energy
Management
University College Cork
Cork
Ireland

Lars Österlund
Department of Materials Science and
Engineering, The Ångström Laboratory
Uppsala University
Uppsala
Sweden

Uwe Posset
Fraunhofer Institute for Silicate Research
ISC
Würzburg
Germany

Marco Schott
Fraunhofer Institute for Silicate Research
ISC
Würzburg
Germany

Preface

I became involved with electrochromics in the mid-1970s when I was studying wavelength selective solar materials at the University of California, Berkeley, in the Departments of Materials Science and Mineral Engineering, and Electrical Engineering and Computer Science. My interest was in developing materials which interact with light. This research involved materials called heat mirrors, now known as low-e (low emittance) coatings. One group of low-e coatings is degenerately doped semiconductors which are utilized as transparent conductors. The development of transparent conductors led to electrically powered displays, such as liquid crystal displays. This in turn led to the concept of large window displays or large area switchable windows which could control the flow of light and heat into a room. The big question was could one find a material with a broad enough wavelength response to effectively switch the solar spectrum. The key was to find a physical process that would be able to switch in the visible and possibly over the entire solar spectrum. This led me to electrochromic materials, such as hydrogen tungsten and molybdenum tungsten bronzes because they had the potential to be a strong switch with dual-ion intercalation and electron injection. Devices could modulate the visible and be modified for use in the near-infrared spectrum. Also, these inorganics showed the necessary stability required for long life in a glazing environment, usually over 10 years. At the time electrochromics were experimental and used in prototype non-transparent information display devices. Following early work on electrochromic displays, I wondered if they could be made transparent.

In the late 1970s, while in the Department of Materials Science and Mineral Engineering at Berkeley, I began a conversation about methods for dynamic light control of building glazing with Dr. Sam Berman at the Lawrence Berkeley National Laboratory (LBNL), Windows and Lighting, which later resulted in being offered a position at the Laboratory. Work at LBNL resulted in over 20 years of materials research on chromogenic and low-e coatings, in partnership with industry. A few select colleagues including Claes Granqvist, Angstrom Institute (Uppsala, Sweden), and I would organize solar and switchable materials conferences for SPIE, The International Optical Engineering Society, The Society of Vacuum Coaters (SVC), and other organizations.

The term "chromogenics" for the field of switchable materials was coined by this writer back in the 1980s during a bus ride with solar materials specialists at the first SPIE European conference on switchable media and solar materials in Hamburg, Germany. Chromogenics comes from Latin for the creation of color. The term chromogenics does exist in other fields. But at the time we needed a broad name to cover all phenomena that changes color or opacity with the application of temperature, light, electricity, electric fields, magnetic fields, or chemicals. So chromogenics was broad enough to be used as a title for future conferences. The term chromogenics covers the entire range of electrochromics, electrophoretics, photochromics, thermochromics and liquid crystals, as well as new technologies yet to be developed.

Colleague Acknowledgements

I wish to thank all the authors for their time and efforts in putting this work together. For those who were invited, but unable to write, I thank you for the information and images you supplied for the book.

I have worked with so many technical colleagues over the last 45 years. It is hard to thank all, so allow me to select a few. I wish to thank posthumously, Profs. Jack Washburn, Marshall Merriam and Alan Searcy for all their work encouraging solar materials research at the University of California, Berkeley. I wish to thank posthumously Mr. Kawahara, former General Manager, Nippon Sheet Glass (NSG) (Itami, Japan), Yuchi Yano, Managing Director of NSG UMU PDLC switchable glazing (Japan), Dr. Junichi Nagai, Head Electrochromics Lab., Asahi Glass (Yokohama, Japan), Dr. Jean-Christophe Giron, VP Sage Electrochromics (Faribault, MN) (former student), and Dr. Philip Yu, Director, PPG (Monroeville, PA) (former student). I wish to thank Dr. Nilgun Ozer (Guest Scientist and Prof. San Francisco State University) for all her work on sol-gel deposited coatings. Special thanks go to Marca Doeff and Steve Visco of the LBNL Energy Technologies for their direction on ionic polymers for electrochromics. This includes polymer work by my former students Yan Ping Ma and Yongxiang He. Another special thanks goes to the team of Prof. Agostino Pennisi and Dr. Francisca Simone (University of Catania, Italy) who were guest electrochromic scientists in my laboratory. I wish to thank my dear colleagues, Profs. Claes Granqvist, and Gunnar Nicklassen, Angstrom Laboratory (Uppsala, Sweden), and Anoop Agrawal, Glass Dyenamics (Tucson, AZ), for discussions about electrochromics starting in the late 1970s. I wish to posthumously thank Prof. Mick Hutchins, Oxford Brooks University, (Oxford, UK) for all his help with optical measurements. It was in Hutchins's lab, where I first met Prof. Xingfang Hu, Chinese Academy of Sciences (Shanghai) working on electrochromics. I wish to thank the late Bob Sax and the current CEO, Joe Harary, Research Frontiers Inc. (Woodbury, NY), for their tireless energy and information about SPD electrophoretic glazing. Also, I thank Dr. Gottfried Haake, American Cyanamid (Stamford, CT), for our discussions about new transparent conductors and electrochromics for displays. In addition, I wish to thank the late John Thornton, Telic Corp. (Santa Monica, CA) and University of Illinois (Urbana-Champaign) for his friendship and advice about post magnetron sputtering in the early days of its development. Also, thanks to John Vossen, RCA Labs (Cambridge, NJ), for our collaborative work on thin-film patents. I wish to thank the late Prof. David Adler, MIT (Cambridge, MA) for his help with the study of thermochromic metal-to-insulator transition materials. Also, I wish to thank Dr. Hulya Demiryont, Eclipse Energy Systems (St. Petersburg, FL), for all our work together on commercial electrochromics. I wish to thank posthumously, Prof. Angus McLeod, Thin Film Center, Tucson, AZ, for his mentoring on thin film optics and the late Michael Andreasen, Guardian Glass and Vacuum Edge (Fairfield, CA), for our work together on industrial PVD coating equipment designs. Also, Walter G. Overacker, Airco Temescal (later BOC Corp., Berkeley, CA), for all his advice about high-volume sputtering and e-beam deposition. Furthermore, I enjoyed working with Rolf Illsley, President, OCLI (Santa Rosa, CA), on solar products. I wish to thank Ric Shimshock, MLD Technologies (Mountain View, CA), for his friendship and practical experience in a wide range of vacuum deposition systems. Also, "Al"

Albany Grubb (BOC Technologies, Fairfield, CA) for his "big vision" of large-scale sputtering plants. I wish to thank posthumously Dr. Stan Ovshinsky, Ovonic Battery Company (Rochester Hills, MI), for his "never give up" attitude in the pursuit of researching the development of amorphous materials. I wish to thank Dr. Harlan Byker, Pleotint (Mt. Olive, MI), and Dr. Martin Preuss, Wiley-VCH publisher, for suggesting I compile this book. Little did I know what I was getting in to. Also, I wish to thank Monica Chandrasekar, Wiley, and her publishing team for their help assembling and editing this work.

Personal Dedication

I wish to dedicate this work to my parents, my mother, Gwen (UCLA '34), for her showing me the importance of deep study and analysis for problem-solving. I thank my father, Gailard, for passing on his practical knowledge in making inventions and his ability to visualize and construct almost anything. Also, I wish to thank my keen-eyed manufacturing engineer and wife, Joyce, for her proofreading.

Organization of the Book

This book is the culmination of several chapters written by some of the best experts in the field of chromogenics. The goal of the work is to give industry and those just getting involved in this field a clear view of chromogenics and its applications in real glazing and other useful products based on this technology. This compilation covers a wide range of technologies. We have seen several books written on electrochromics and research books on photochromics and thermochromics. Some books are very hard to obtain. This work is not intended to be a deep review of ongoing research in the area chromogenics. I will leave that for other academic authors. It does contain some history, especially industry history, as best we know it, that led to several developments in the field. I have collaborated with many companies during my work as a scientist with the University of California, Lawrence Berkeley National Laboratory, as a consultant with Star Science and as Technical Director of the Society of Vacuum Coaters.

This work is organized in two main sections, chromogenic technologies and manufacturing processes, used to make chromogenic products. In the technologies sections, the topics covered are introduction to chromogenics, introduction to measurements and glazing design, inorganic, organic and polymer electrochromics, polymer dispersed liquid crystals, suspended particles, inorganic and organic thermochromics and photochromics. The chapter on emerging technologies includes guest–host liquid crystal windows, electronic ink-based and water-based technologies. Each chapter is designed to be self-contained, so some terms and measurements are duplicated for ease of the reader. In the manufacturing section, the chapters discuss many different and sometimes common manufacturing methods including large-scale sputter, vacuum web coating and pyrolytic coating processes frequently used by the glass and plastic industries. In the manufacturing introduction, I cover manufacturing processes that are not discussed in great detail but are important for the development of products. They include sol-gel, slot die coating and ink-jet deposition processes.

Forward-Looking Statement and Statement of Neutrality

All the product data, costs, future market size and company information are the best of our combined knowledge. Readers should always check with the source of information before using technical data from this work. This is especially true of industry data, as the properties of the devices and materials shown keep changing and improving. Also, companies in this field continue to evolve and change product directions or mixtures of products. Some companies go out of business or have their IP transferred to another company.

We do not intentionally endorse any product or company. The inclusion (or exclusion) of a device or technology in this work is not intended to imply that it is either favored or not favored by the authors. Products have been chosen to illustrate some of the many possible technologies, architecture and manufacturing approaches. In some illustrated cases, companies have given their permission for images to be reproduced here. Our intention is to present information to better inform the reader of the variety of technologies and recommend approaches to improve devices or develop new chromogenic technologies.

October 2025

Dr. Carl M. Lampert
Cotati, CA, USA

Part I

Technologies

1A

Overview of Chromogenics

Carl M. Lampert[1,2,3]

[1] *Star Science, Cotati, CA, USA*
[2] *Formerly, Technical Director, Society of Vacuum Coaters (non-profit), Albuquerque, NM, USA*
[3] *Staff Scientist, University of California, LBNL, Energy Technologies, Berkeley, CA, USA*

Nomenclature

ε_t	thermal emittance
τ	time constant
λ	wavelength
A_λ	absorptance at wavelength λ
AI	artificial intelligence
AIST	National Institute of Advanced Industrial Science & Technology (Nagoya, Japan)
AR	augmented reality
ARPA-E	Advanced Research Projects Agency-Energy (part of USDOE)
ASTM	ASTM International (testing)
CAGR	compound annual growth rate
CAM	color alteration mechanochromism
CES	Consumer Electronics Show (Las Vegas)
ChLCs	Cholesteric liquid crystals
CIE	Commission Internationale de l'Eclairage (Vienna, Austria)
$d\lambda$	incremental wavelength interval.
DGU	Double Glazing Unit (IGU)
DMP	5,10-dimethyl-5,10-dihyrophenazine
EC	electrochromic
EGU	Electrochromic Glazing Unit (double glazing)
ELM	electric light modulator
EM	encryption mechanochromic
EU	European Union
EVA	ethylene vinyl acetate
IGU	insulated glass unit (double glazing) also known as DGU
IP	intellectual property
IR	infrared
L*	Luminance (brightness)

Chromogenics: Smart Switchable Optical Materials and Their Applications, First Edition. Edited by Carl M. Lampert.

LBNL	Lawrence Berkeley National Lab. (Berkeley, CA)
LC	liquid crystal
LCD	liquid crystal device
LED	light-emitting diode
LETC	ligand exchange thermochromic
Low-e	low emittance (coating)
LM	luminescent mechanochromism
MEMS	microelectromechanical system
MIR	middle infrared
MRS	Materials Research Society
nc	nanocrystalline
NCAP	nematic curvilinear aligned phase
NIR	near-infrared
NFRC	National Fenestration Ratings Council
NREL	National Renewable Energy Lab. (Golden, CO) (formerly SERI)
OD	optical density
OLED	organic light-emitting diode
PANI	polyaniline
PC	photochromic
PDLC	polymer-dispersed liquid crystals
PDMS	polydimethylsiloxane (dimethicone)
PE	polyethylene
PEC	photoelectrochemical
PEDOT	poly(3,4-ethylenedioxythiophene)
PET	polyethylene terephthalate
PMMA	polymethyl methacrylate
PVA	polyvinyl alcohol
PVB	polyvinyl butyral
P(VDF-HFP)	poly(vinylidene fluoride-co-hexafluoropropene) (co-polymer)
R_λ	reflectance at wavelength λ
R_{vis}, R_v	visible reflectance (photopic)
R_{sol}	solar reflectance
REM	reversible electrochemical mirror
RoR	rate of return on investment
SDG	Suntutive™ Dynamic Glass
SHGC	solar heat gain coefficient
SID	Society for Information Display
SPD	suspended particle device
T	transmittance
T_a	average visible transmittance (mathematical average, not weighted)
TC	thermochromic
T_c,T_b	transmittance colored, transmittance bleached
T_o,T_∞	transmittance colored, transmittance bleached
TGU	triple glazing unit
TC	transparent conductor
TCO	transparent conductive oxide

$\Delta T(\lambda,\tau)$	change in monochromatic transmittance with time
$\Delta T_{max}(\lambda)$	maximum change in monochromatic transmittance
$T(\lambda)$, T_λ	monochromatic transmittance at wavelength λ
T_{lum}	luminous transmittance (visible)
T_{sol}	solar transmittance
T_{uv}	ultraviolet transmittance
T_{vis}, T_v	visible transmittance (photopic)
TCM	transparency change mechanochromism
TCNQ	tetracyanoquinodimethane
TSER	Total solar energy rejection
TTF	tetrathiafulvalene
UAM	urban air mobility
UAV	unmanned aviation vehicle
USD or US$	United States dollars
USDOE	US Department of Energy
UV	ultraviolet, with spectral regions: UVA, UVB, UVC
UVA	ultraviolet, longwave
VAC	volts alternating current
VDC	volts direct current
VIG	vacuum insulated glass (vacuum window)
Vis,vis	visible light
VLT, VT	visible light transmittance
U	coefficient of heat transfer (U-factor), $Wm^{-2}\ K^{-1}$
U_{cg}	U-factor of the center of glass
UD	ultra definition display 3840 × 2160 pixels
VR	virtual reality
XR	extended reality (includes VR, AR)
x,y	CIE Color coordinates
Y*	lightness index

1A.1 Introduction

The term *Chromogenic* refers to a material that has a switching property usually involving a color change. However, in this work, we use the term in an even broader sense to include physical change that results in a detectable optical, chemical, or structural change. For example, some chromogenic materials do not change color but change in tint, hue, optical density, reflectance, or optical scattering properties. Some materials change from opaque to transparent or the opposite. The material may just have two states, with or without memory. The material may have multiple states or colors. The material may be dynamic with a continuous range of gray scale. Also included are materials that change their properties in a wavelength band outside of the visible region. For example, the material may change its reflectivity or thermal emittance in the infrared. Many chromogenic materials are synthetically made. But natural chromogenesis does occur in nature. For example, when a chameleon's skin changes color, it triggers chromatophores in its skin to adapt to its surroundings or to signal other chameleons [1]. Those changes can be made by chemical coloring or optical diffraction effects, as seen in butterflies. Chromogenic materials have the ability to change color under certain environmental changes (e.g. temperature or light levels) or under applied conditions, such as voltage or electric field, magnetic field, pressure, or change in chemical environment.

The subjects to follow are directed to those who want to understand the world of the chromogenic technology and its business. Chromogenics is a mix of technologies to create mainly large-area optically switchable reversible technologies for applications such as glazing for buildings and vehicles and information display applications. Also, chromogenics can be used for opaque surface alteration, such as change of color or thermal properties. Chromogenics include some heavily commercialized products like electrochromic rearview mirrors as an option in most vehicles, electrochromic glazing now installed in many commercial building windows, and suspended particle devices (SPDs) in automotive sunroofs to control heating and daylighting. Liquid crystal systems that control light or scatter light are used in security or privacy glazing. Some chromogenic devices have been developed for information displays such as the ePaper reader. Photochromic eyeglasses have matured as a widespread product across the major eyeglass companies in the world. Also, some thermochromics (TCs) have matured into building glazing products. Other chromogenic technologies are just emerging as prototypes, and many others are still in research. This book is focused on commercial developments in chromogenics. As a whole, the chromogenic research area is very active. Certain research activities will be covered in the chapters to follow, this book is not intended to be a review of all the research nor all the commercial developments in this field. This book is divided into two main sections: chromogenic technologies and manufacturing. This book covers selected manufacturing processes, since manufacturing is so important to economically producing a commercial item. Manufacturing is emphasized because it is largely overlooked in most books in this field. The major manufacturing processes include pyrolytic spray coating, sputter vacuum deposition and flexible roll coating, slot die coating, heat treatment, photonic curing, and laminating. Total fabrication processes and production plant design are not covered in detail since some steps and formulary represent a larger subject worthy of its own book. Many patents are covered, including some older source patents, which are instructive beyond the open research and commercial literature.

The goal of this introductory chapter is to give the reader an overview of the chromogenic field. This chapter covers classifications of switchable media which is a fairly large expanding area if we include studies under research and development. I anticipate that chromogenics will play a vital role in the emergence of smart materials and systems. There appear to be many commercial applications for switchable media. The largest application groups are automotive mirrors and sunroofs, dynamic and switchable architectural glazing, and information displays. These technologies are mainly produced on glass or flexible plastic substrates, but in some cases, displays can be produced on paper. An example of a shopping center façade with electrochromic glazing is shown in Figure 1A.1. In this chapter, I cover the application side of chromogenics and summarize the chromogenic families of technologies. The final section of this chapter is about the business of chromogenics. Each subsequent chapter covers a selected technology in detail by authors who are specialists in their field.

Figure 1A.1 shows an installation of electrochromic glazing in a shopping center in Sweden.

Figure 1A.1 Grändbystaden Shopping Center (Sweden) shows an electrochromic façade on the curved portion of the building. The panels are fabricated by bonding an electrochromic polymer film to glass. Electrochromic glazing shows both switched and unswitched coloration. The windows were made by ChromoGenics AB, Uppsala, Sweden.

1A.2 The Dynamic Glass Market

According to Grand View Research, the smart glass global market has been estimated to increase from 8.22 billion US$ in 2025 to 13.01 billion US$ in 2030 at a 9.6% compound annual growth rate (CAGR) [2]. This survey includes electrochromics, liquid crystal devices, SPDs, and TCs. Applications encompass automotive mirrors, automotive sunroofs, architectural glazing, and aerospace transparencies. Growth in dynamic glazing will be in automotive, including electric and hybrid cars, aerospace transparencies, and marine glazing. High cost is still a deciding factor. In another study, Markets & Markets predicted the world demand for smart glass will grow from 6.42 billion US$ (2025) to 10.42 billion US$ (2030) at 10.2% CAGR during 2025–2030 [3]. The big unknown will be the effect of government tariffs, which could alter these projections greatly. *Just a word of caution: many market numbers in this book are estimates, numbers vary by the number of companies interviewed and the range of products covered. Some projections include control systems. Many advanced marketing studies give different numbers.*

The future growth in smart glass will be in the sectors of automotive, aerospace, and energy-efficient buildings. The electrochromic (EC) field has over 5,080 patent families with about 38.5% granted and active, 21.3% pending, and 40.2% expired. According to Nowak at Questel [4], the major electrochromic companies and patent activity are shown in Figure 1A.2. Regionally, the largest number of electrochromic patents (53%) has been from mostly Chinese companies over the years 2013–2023. Many of these patents are in the electrochromic display and optics areas. In comparison, the United States company portion of the patents is 17%, with Japan at 9%, Korea at 7%, and Taiwan at 3% of the

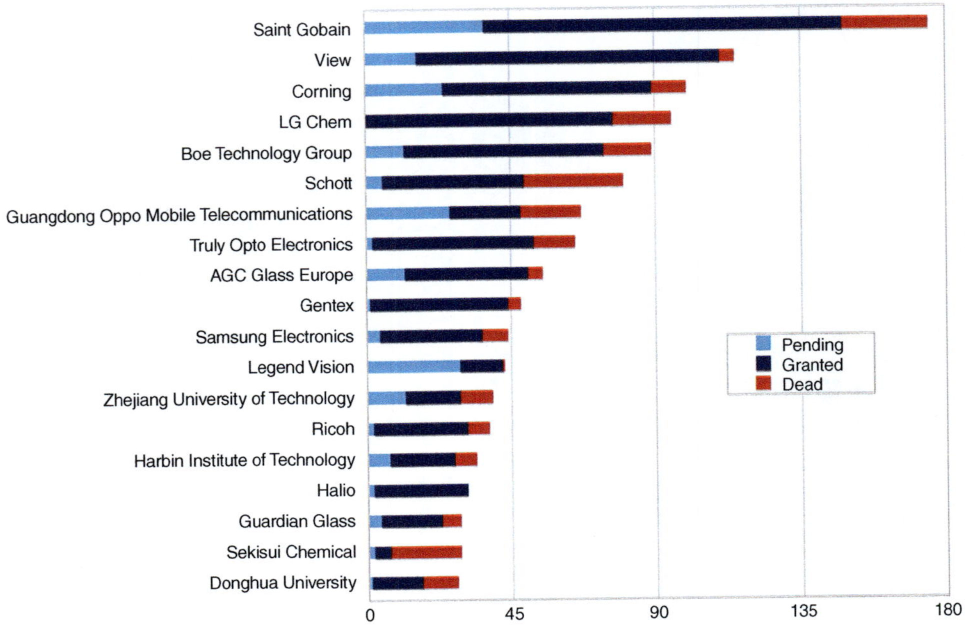

Figure 1A.2 Graph of electrochromic patent activity from 2013 to 2023 showing the major players. Sage electrochromics is included in the Saint-Gobain patents [4]. *Source:* Courtesy of Nicolas Nowak, Questel.

total. The largest company holding patents is Saint-Gobain, owner of SageGlass (175 total, with 148 active patents), followed by View (115 total, with 110 active patents). Another measure of activity in this area is the number of research papers on electrochromism and electrochromics published. There has been a considerable upswing in the number of articles since 1998, with over 800 articles being published in 2020. Further details are discussed in Chapter 2 on Inorganic Electrochromism and Chapters 3 and 4 on Organic and Polymeric Electrochromism, respectively.

1A.2.1 Automotive

The sales of cars in the world were 75.3 million in 2023 (with a value of 2.52 trillion US$), with sales for 2024 of 78 million cars [5]. Sales for 2025 are expected to be around 90 million cars. Manufacturing has picked up after COVID-19 which caused shortages resulting in the lowest sales in 2020 of 63.8 million cars. Toyota leads the world market in sales with a market share of 10.2%. The largest national market is in China with 25.8 million cars sold in 2023. Automotive mirrors represent the major electrochromic market share. Automotive sunroofs, visors, and panels based on chromogenic glass are at their prototype or early production phase. Sunroofs must be manufactured precisely to fit within the compound curved surface of car roof lines. Chromogenic glazing can be used in sunroofs, side windows, sunvisors, and rear windows. The use of electrochromic glazing ensures the reduction of unwanted light and glare and built-up heat inside the vehicle, thereby making them preferable over normal glazing. This improves the vehicle's fuel efficiency or battery efficiency. In the case of less fuel used, less carbon will be emitted. Dynamic glazing systems can be totally automatic or semiautomatic or have manual switching capabilities. The most successful automotive product is the electrochromic anti-dazzle mirror. Gentex (Zeeland, MI) has the majority of the switchable automotive mirror market. Gentex makes an integrated dimming/display/camera mirror system called Full Display Mirror®. Other dimmable mirror makers are Magna International (Holland, MI), Murikami (Japan), Panasonic (Japan), Fudi Technology (China), NanJin AoLian AE&EA Co., Ltd. (China), Ultronix Products (China), Tokai Rika Company (Japan), SMR Automotive (Germany), Triton Intertech (USA), Adayo (China), Sincode (China), Mike Shanghai (China), Guangdong Yangfeng Electronic Technology Co., Ltd. (China), Licon (Germany), MirrorTech (China), Amblilight (US, China), and Chongqing Yimei (China). Switchable mirrors are available for most major makes of cars. As the 86% switchable mirror market leader, Gentex's sales were 2.32 billion US$ in 2024 [6]. ITRI (Taiwan) presented a foldable high-contrast electrochromic car mirror with visible reflectivity, R_{vis} = 80–3%, at the 2024 Society for Information Display (SID) Conference in San Jose, CA.

The demand for switchable sunroof glazing is growing. Leading automotive chromogenic sunroof suppliers are AGC Inc. (nematic curvilinear aligned phase liquid crystals, NCAP) [7], Gauzy (Tel Aviv, Israel) with Research Frontiers Inc. (Woodbury, NY) (SPD), and Gentex electrochromic dimmable smart glass for sunroofs and visors. Car brands showing prototype or production dimmable and graduated sunroofs are Cadillac, BMW, Ioniq, Renault, Mercedes, Porsche, Toyota, and Lexus. ELSTAR Dynamics (Eindhoven, NL) demonstrated prototype sunroofs, including segmented ones, at the SID 2024 Conference in San Jose, CA. The ELSTAR sunroof is a light modulator with a transmittance switch from T_{vis} = 0.2–70% in 20 seconds. The ELSTAR sunroof is made by nanoprinting and electrophoretic inks. At the 2023 Consumer Electronics Show (CES) in Las Vegas, NV, E Ink showed a concept BMW car with full color-switchable body colors, based on their encapsulated electrophoretic inks. Further details are in Chapter 9 – Other Chromogenic Technologies. Also, at CES 2023, Research Frontiers, Inc. (RFI, Woodbury, NY) highlighted their tintable (SPD) side windows on a BMW concept car.

1A.2.2 Eyewear

Eyewear products include eyeglasses, sunglasses, goggles, motorcycle helmets, welding helmets, and face shields. Applications include everyday wear, fashion wear, industry, safety, sports, medicine, and recreational uses. According to the world sales of eyewear were 155.4 billion US$ in 2024, with projected sales for 2025 of 161 billion US$, growing at a rate of 5.85 over the years of 2025 – 2034 (Global Market Insights). Around 85% sales of the eyewear are for non-luxury wear, but the luxury market continues to grow. Sales of photochromic lenses are estimated to be 7.2 billion US$ (2025) with projected growth to 9.65 billion US$ by 2030, with a CAGR of 5.81% [8]. The major upswing in sales is caused by an aging population and increased information display screen time, causing more myopia and other eye problems. Another contributing factor is the increasing exposure to UV and blue radiation. Europe has the largest eyewear market share. Japan has the fastest aging population, with retired people predicted to reach 37.3 million by 2030. The use of polymer lenses instead of glass is on the rise but currently about equal with glass. The major photochromic eyeglass makers are Carl Zeiss Meditec AG (Germany), Corning Incorporated (US), Essilor International (France), HOYA Corporation (Japan), Vision Service Plan (US), Transitions Optical (US), Transitions Optical Limited (Ireland), Optiswiss AG (Switzerland), Tokai Optical Co., Ltd. (Japan), Seiko Optical Products Co., Ltd. (Japan), and Vision Dynamics

LLC (US) [9]. There is an increasing trend in the number of companies offering chromogenic eyewear. The current market is dominated by photochromic eyewear. Notably, Nikon (Japan) introduced electrochromic dimmable eyeglasses back in the late 1980s–early 1990s. Several pairs were sold but at a high price point (~300 US$). They are constructed following the patent US 5187607 (1990). The glasses had a tiny button on each side of the frame: push one button, it darkens; push the other, it bleaches – so it was not automatic. The other issue was it used tiny hearing aid size batteries with limited lifetime. So, the product did not gain enough market sales to be economically viable. At that time, Nikon was developing EC displays/shutters for camera use. Another company, Eyeonics (which the author was working with) (Tualatin, OR), developed prototypes of prescription electrochromic eyewear, but the product never made it to the marketplace due to funding issues.

Ashwin-Uhas has marketed "Fades" sunglasses under the AshChromics company (Marlboro, NJ) [10]. These are organic electrochromic sunglasses and goggles which automatically switch to different shades of darkness in response to ambient light in less than two seconds. Fades have two colors, transparent to light gray or blue-gray to dark blue-black. They work indoors and in automobiles and are very lightweight and priced at 140 US$ per pair (2024). They are competitive with photochromics. Fades are based on organic electrochromics like poly(3,4-ethylenedioxythiophene) (PEDOT). According to AshChromics, the glasses are T_{vis} = 70–7% for "light" electrochromic lenses and T_{vis} = 50–5% for "dark" electrochromic lenses. Switching times are one to two seconds [11]. Also, Nitto Denko (Osaka, Japan) has been granted a world patent (WO 2022/098854A1) for solid-state electrochromic eyewear [12]. Mitsui Chemical (Japan) has sold their electrochromic eyeglass patent US 11294199B2 (2022) to Carl Zeiss Vision International [13].

Other makers of chromogenic eyeglasses include Ampere Technologies, Inc. (Dover, DE) which sells "Dusk" liquid crystal (but called "electrochromic") eyewear. The Ampere company is a crowd- and angel-funded company. For a certain amount of investment, you get a free pair of glasses and a discount toward new purchases, and you own a little bit of the company. They have raised almost 400K US$ so far. Their glasses change from T_{vis} = 4% to 38% tint and cost in the range of 199–399 US$ [14]. They are manufactured in Japan. The glasses were judged "Best of Billboard" at the CES 2023 show. Skugga (Stockholm, Sweden) makes dimmable liquid crystal eyeglasses with wireless charging. Skugga uses LC-Tec liquid crystal lenses. The Skugga glasses sell for about 240 US$ (2024). It is based on the US patent 11487137. ONESEC (Hong Kong, the United States, the United Kingdom, and Taiwan) introduced liquid crystal "shade shift tint" (but called "electrochromic") eyewear in 2023. ONESEC won the Vision Plus Special Award in 2021 and has partnered with McLaren Vision (France) [15, 16]. According to their website, a pair of ONESEC glasses costs between 192 and 265 US$. Another switchable eyewear maker is CTRL (USA) who makes sports and military eyewear. CTRL lenses use "E-Tint" Alpha Micron (Kent, OH) flexible liquid crystal films. Switching time is 0.1 seconds, self-adjusting to the light level with 70 hours of battery life. The standard eyeglass pair sells around 249 US$. It is anticipated that the use of 3D printing of eyeglass frames will revolutionize and reduce frame costs [17, 18].

1A.2.3 Architectural

If we look at some market numbers from Straights Research, the smart window market is projected to be 6.641 billion US$ (2023), growing at 11.2% to 17.265 billion US$ [19a]. If we just look at architectural windows (just switchable, no informational or display windows), the market is 3.5 billion US$ with a projected growth of 12.1% CAGR to 8.6 billion US$ (2030) [19]. Currently, commercial applications are 55%, and residential are 30% of the market share [19b].

According to Markets & Markets [19c], the entire market for smart glass for all uses was projected to be 5.87 billion USD in 2024 and 6.42 billion USD in 2025, growing to 10.42 billion USD by 2030 with a CAGR of 10.2%. The architectural market is about 40% of these values. The projected numbers cover all switchable window technologies, including SPDs, polymer dispersed liquid crystals (PDLCs), and other liquid crystal devices.

Architectural windows are responsible for about 30% of the total heating and cooling loads of a building. In the United States, this is equivalent to 4.3 EJ or 4.1 quads of primary energy used [20]. The leading players in the electrochromic glass market are SageGlass, formerly known as SAGE, Sun Active Glass Electrochromics (Faribault, MN), which is owned by Saint-Gobain (France), and ChromoGenics AB (Uppsala, Sweden). ChromoGenics AB produces electrochromic on flexible plastic for lamination to glass. Another electrochromic glazing manufacturer is View, Inc. (US). View is currently under reorganization from a public to a privately held firm. A new company, Glass Dyenamics (Tucson, AZ), is developing custom electrochromic windows for residential applications. Another company, Furcifer, Inc. (Fremont, CA), is working to make electrochromic to flexible plastic for building windows. Also, MIRU, formerly Click Materials (Vancouver, BC), is developing ink-based electrochromic windows.

Dispersed liquid crystals offer a privacy effect. In the off-state, they appear milky, and in the on-state, they appear clear. Polymer dispersed liquid crystal are manufactured by Polytronix Inc. (Richardson, TX) on plastic film. Other manufacturers of PDLC films include Asahi Glass-AGC Inc. (Yokohama, Japan) and Gauzy (Israel). Originally, PDLCs were developed by Kent State University (OH) [21, 22]. Innovative Glass (Plainview, NY) makes PDLC privacy glass and partitions. A similar product called NCAP was developed by James Fergason and Raychem/Taliq (Mountain View, CA) [23, 24]. NCAP is manufactured on plastic film by Nippon Sheet Glass – (NSG, Japan). There are other makers and fabricators. Smart Films International uses NCAP films to fabricate their windows. A different liquid crystal device is offered by eyrise B.V. (Veldhoven, NL). They make a guest–host liquid crystal glazing. The eyrise product has very low scatter even at high angles compared to other liquid crystal glazing products.

For electrophoretic glazing, Research Frontiers Inc. licenses their SPD technology to Gauzy to manufacture film and to Smartglass International Ltd. (Ireland, SPD) to make glazing, among others. ELSTAR Dynamics has shown electrophoretic prototype windows at the SID-2024 conference in San Jose. The ELSTAR window is an electric light modulator (ELM) based on electrophoretic inks and nanoimprinted spacers.

A thermochromic glazing formulator and manufacturer is Pleotint LLC (West Clive, MI). Pleotint makes ligand-exchange thermochromics (LETC). The LETC product contains polyvinyl butyral (PVB) for glass lamination. Also, there are other companies quietly developing smart glazing products, including combined glazing products. Ghosh et al. reviewed smart glazing for smart cities in 2024 [25]. Khaled and Berardi reviewed future coatings for architectural applications in 2021 [26]. Another good chromogenics review for architectural windows, including simulations of energy savings performance, was done by Tallberg et al. in 2019 [27].

1A.2.4 Aerospace

According to Mordor Intelligence [28], the global aircraft market was estimated to be 333.96 billion US$ in 2024 and predicted to grow to 396.15 billion US$ in 2029 at a 2.87% CAGR. Around 45.7% is fixed-wing aircraft, with much of it for commercial use. During 2023–2024, there was a 20% increase in demand for business jets. Over the next 20 years, the demand is estimated to be 42,595 new aircraft units for commercial service (fixed wing, rotary wing, helicopter, and unmanned aviation vehicle (UAV) or drones. According to Boeing (2023) [29], the growth rate in demand is projected to be 3.5% CAGR over the next 20 years. The major driving forces are fleet modernization, better fuel efficiency, lower carbon emissions and increased globalization. Developments in the aerospace field include new electric aircraft production, amphibious planes, and large growth in unmanned drones. Drones are to be used for freight, search and rescue, fire control, real estate mapping, disaster response, emergency support, power line inspection, and surveillance. Another development is the increased interest in urban air mobility (UAM) which includes air taxis and personal air vehicles. Also, 3D printing and fabrication have revolutionized aircraft component design and weight reduction. The aerospace industry's adoption of electrochromic glass in windows, sun visors, partitions, and doors will drive the aerospace transparency market. The recent trend is to incorporate electronically dimmable windows in aircraft cabins with increased window size. Leading aircraft manufacturers are using dimmable glass in aircraft cabins or have plans to use them in future designs. The use of smart glass in aircraft cabins and windows enables passengers to regulate the tint or amount of light, heat, and glare they would be exposed to.

Several of the major airplane manufacturers, including Boeing (Crystal City, VA) and Airbus Industrie (Blagnac, France), have installed some switchable passenger windows. Boeing has made dynamic windows available as an option in their 777X and 787 Dreamliner aircraft. These windows are made by PPG Aerospace using Gentex electrochromic glazing called Alteos® Interactive Window System. The windows have a 10,000 : 1 contrast ratio with the deeply colored "blackout" visible transmittance less than 0.001% for the Alteos® II window [30, 31]. Both Airbus A380 and A330NEO airplanes have specified electrochromic windows as an option. Saint-Gobain has publicly shown Airbus cabin windows. Vision Systems (Brignais, France, now owned by Gauzy) has installed SPD Nuance™ Films and motorized shading systems on boats, business jets, helicopters, and yachts. Vision Systems has worked with airplane manufacturers including PPG Aerospace (US), Airbus, Dassault (France), Hondajet (US), and Embraer (Brazil) [32]. Vision Systems has facilities in the United States (Melbourne, FL), Canada, and the Middle East and Asia. InspecTech Aero Services (Fort Lauderdale, FL) supplies King Air, along with other aircraft companies, with SPD iShade™ [33]. SPD films have been installed on several helicopters and business jets.

1A.2.5 Information Display

According to the market study made by Markets & Markets [34], the world market for displays was estimated to be about 135.2 billion US$ in 2024, growing to about 173.7 billion US$ by 2029 with a CAGR of 5.1%. The display market is divided into three major technologies: organic light-emitting diodes (OLEDs), liquid crystal displays (LCDs), and inorganic light-emitting diode (LED) displays. The major volume is driven by cell phones, tablets, and large-screen televisions. Transparent micro-LEDs have been steadily improving. Due to their new form factor, LEDs can be integrated into chromogenic windows. There are some more unique displays, such as those used for e-paper. Kindle readers are based on encapsulated electrophoretics made by E Ink (Taiwan). Some electrochromics are used for passive displays (charge indicators, point-of-sale displays, and smart labels). In E Ink displays, electrophoretic inks containing black, white, or colored particles are encapsulated in polymer bubbles and then embedded in a flexible polymer matrix. Depending on the applied field, the particles move up or down in the bubble giving a contrast effect. Typically, E Ink displays operate at 15–20 VDC and have contrast ratios of at least 6 : 1 [35]. Also, they are bistable and exhibit a memory effect. Encapsulated electrophoretics have been used in transparent switchable windows. The E Ink JustTint™ product uses encapsulated black particles which move sideways in an electric field to give higher transmission. Information displays can be part of a glazing in applications where fast switching speed is not required, such as for airport display boards. ACREO (Sweden) has made electrochromic displays printed on paper and plastic laminates for electronic pricing labels. Sony (Japan) has made a flexible, reversible silver deposition display on plastic. Ynvisible (Portugal, Canada) manufactures smart labels, freshness indicators, and pricing displays. LinePro Controls, Pvt. Ltd. (Bhiwandi, India) and Polar Printed Electronics (Munich, Germany) make screen-printed electrochromic labels and wearable displays. For the application of chromogenics in smart packaging, a review by Sedeghi et al. is available [36].

If we look at the smart TV display market, we can see how smart windows fit into wall display systems. A smart TV is one that combines TV, internet, streaming video, and music. Larger displays are rapidly improving their quality and lowering overall cost. The largest LED TV screen, "Extreme Home Cinema" is 8.52 m (325″ diagonal), announced by LG in September 2021. Samsung's is called "The Wall" was shown in 2022. At the SID-2024 show, TCL CSOT (Shenzhen China Star Optoelectronics Technology Co. and Samsung) premiered a 163″ ultra-definition (UD) microLED display with a contrast ratio of 1,000,000 : 1. An image of a TCL wall display is shown in Figure 1A.3. We are just beginning to see a bridge forming between the processing of large-area window glazing and large-area display glass. Transparent displays were shown at the SID-2024 show. AUO (Taiwan) showed a 60″ transparent window display as shown in Figure 1A.4. Several other smaller transparent displays were noted by other makers. On a LG (Korea) refrigerator, a smart door window called InstaView™ illuminated the interior when the window was tapped twice. The smart door image is shown in Figure 1A.5. Also, it can be used as a message center. Another model allows Wi-Fi connectivity for messaging and entertainment. Electrochromic displays for electronics have been reviewed [37].

1A.3 Families of Chromogenic Materials

There are several chromogenic families which switch by a wide range of driving forces. We can divide them into four major families. The first family switches by *collective physical movement*, such as liquid crystal or a suspended particle being aligned with an applied electric field to effect an optical change. The second family is the *chemichromic family*, meaning materials that change color or become colored through some chemical reaction. The third family is the *electrochemichromic* family where materials change color or become colored through some electrochemical reaction. The fourth family switches by *discrete mass movement* involving ions, atoms, and molecules. An example would be ions and electrons

Figure 1A.3 Mock-up of a 163″ microLED display the size of a wall. The display is made by TCL, CSOT (Samsung [Korea] and TCL).

intercalating (moving) into an electrochromic. The families are discussed in their own sections.

There are other *chromic* processes not covered here. They include biochromism – color change with contact to biological sources (detection of viruses or bacteria), tribochromism – color change upon grinding (seen in some pigments), radiochromism – color change with ionizing radiation (radiation detector), chronochromism – color change with time (freshness and UV exposure indicators), and the rare concentratochromism – color change with chemical concentration. An example of concentratochromism is when dithizone changes from yellow to green with increasing concentration of methanol [38, 39]. Another phenomenon is electrofluorochromism and electroluminochromism, where the fluorescent or phosphorescent properties of a material can be controlled or modulated by applied potential [40]. I am sure there are more chromisms yet to be discovered (or uncovered). For now, I will cover the processes most relevant to large-area switching.

1A.3.1 Chromogenics Switching by Collective Physical Movement

The first category of chromogenic technologies is the collective physical switching family. This family includes the processes of electromechanical, electrokinetic, magnetochromic, and mechanochromic, as shown in Figure 1A.6.

Figure 1A.4 Two images of the same AUO 60″ transparent window interactive display shown at the 2024 SID show. The window transmittance is $T_{vis} > 60\%$. With a color gamut of 86%, a resolution of 1820 × 1080 pixels, and brightness of 600 cd m^{-2}. The development of transparent electronics has made this possible. This display gives us a new definition of a smart window.

The electromechanical group has been partly commercialized in the form of large mechanical irises or metal shutters. Microshutters or microelectromechanical systems (MEMS) have been used for cell phone displays. Some MEMS devices were made for evaluation in building glazing. Electromechanical shades for windows have been developed. With further discoveries, one expects to add other technologies to this family. The mechanochromic group changes color with applied pressure or force. The magnetochromic group can show a rainbow of colors with an applied magnetic field, but the materials are still in research. In certain applications, this group is used to visualize magnetic field lines and intensity.

The electrokinetic group switches with an applied electric field. In some cases, the off-state is in the scattering or opaque privacy condition. This group covers many commercialized

(a) (b)

Figure 1A.5 LG refrigerator with an InstaView™ touch-sensitive door window. Image (a) shows the darkened door window. Two touches and the window illuminates so you can see inside without opening the door. The user can view the inside of the refrigerator without opening the door, as shown in (b). Another version of this model has AI vision to tell you what you need to buy at the grocery store, with touch screen for messages and entertainment.

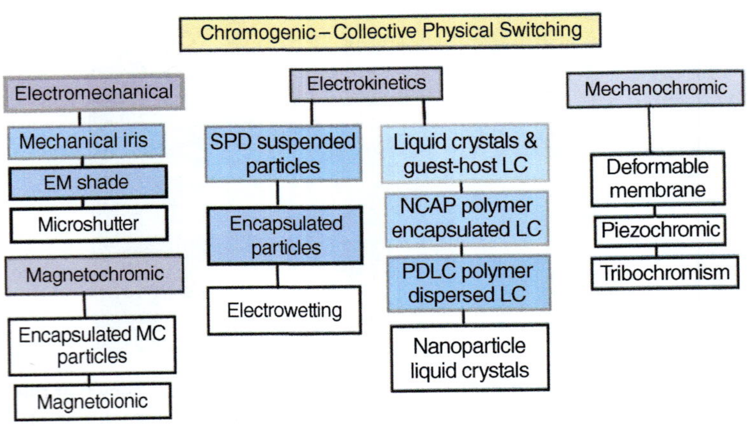

Figure 1A.6 The chromogenic family that switches by collective physical movement. An example would be increasing transmittance by alignment of liquid crystals with an applied field. The blue (shaded) boxes show technologies that have been commercialized. The technologies in the white boxes are not commercialized for large-area applications.

products used for glazing. SPDs are used for both view and skylight glazing. Some haze is seen at high viewing angles. E ink has made encapsulated particle glazing to modulate light. The PDLC and NCAP liquid crystals are used for privacy glazing. Dispersed liquid crystals are typically opaque and scatter light in their off condition. When turned on they switch on electrically, and they become transparent. A new family of guest–host liquid crystal layers is commercialized by eyrise B.V. (owned by Merck) using Merck liquid crystals. Laminated eyrise glazing has been installed in commercial buildings. By using of liquid crystals and dye consisting of dipole particles which have a very large aspect ratio, the liquid crystal remains in an ordered state, reducing scatter. This is unique for liquid crystals since in its off-state it does not appreciably scatter light. Also, it has very low haze at high angles. It is designed to be used in a "view" glazing rather than a privacy glazing. As of 2024, eyrise has 50 building projects, installing a total of 50,000 m^2 of eyrise glazing. More details about liquid crystal eyrise technology are covered in Chapter 9 – Other Chromogenic Technologies.

1A.3.1.1 Electromechanical

The electromechanical group consists of mechanical iris systems and MEMS devices. Viewability, either by visual blocking or distortion, can be an issue with these systems. Sensing and control systems are important to make these devices perform efficiently. MEMS have the ability to create a pixelated blocking image that could shield the viewer from the direct solar image in windows that face the Sun. Also, MEMS may be useful in vehicle visor design when driving directly into the Sun.

Mechanical Iris Systems I mention here mechanical iris and louver systems as they actually go far back into history. One notable system is in Paris at the Institut du Monde Arabe (Arab World Institute; Figure 1A.7). The library building façade has a photosensor which controls a motor that opens or closes the mechanical iris structure for each window, depending on the level of sunlight that hits it, see Figure 1A.8. The iris controls the amount of solar heat and light in the library. The Iris system was designed by Jean Nouvel. The Iris system won the 1981 design competition and the Aga Khan Award for architectural excellence in 1989.

Figure 1A.7 Institut du Monde Arabe, 1 Rue des Fosses Saint-Bernard, Paris, France. The windows are a light-sensitive, motor-controlled iris structure. The iris system was designed by Jean Nouvel, the 1981 design competition winner. The building is situated between the Jussieu Campus of Pierre and Marie Curie University (Paris VI) and the River Seine. The façade is designed to control solar energy and daylight entering the building. *Source:* Fred Romero / Wikimedia Commons / CC BY 2.0.

Figure 1A.8 Interior view of the iris window façade at the Institut du Monde Arabe, Paris. *Source:* ActuaLitté / Wikimedia Commons / CC BY SA 2.0.

Though elegant, one of the drawbacks is the mechanical irises sometimes stick, as they are sliding pieces of metal. In the winter, they can freeze together. The other issue is they are not really a view window, having some difficulty seeing out.

Microshutters Another louver system is based on MEMS. Many people are familiar with the development by Qualcomm (San Diego, CA) of capacitive MEMS displays for cell phones. The MEMS display consists of arrays of very tiny addressable shutters built onto optical cavities [41]. They can be configured as roll-up devices, tilting tiles, or optical cavities to give an optical switching effect or change in coloration. The switching effect is reflective using ambient light. MEMS are made up of small cells between 0.001 and 0.1 mm in size, and MEMS devices generally range in area from 20 μm to more than 1 m^2. The cells are addressable by an applied localized field. Ohmic MEMS are controlled by electrostatically controlled cantilevers or tabs that roll up. Ohmic MEMS switches can fail from metal fatigue of the MEMS actuator and contact wear, since curling can deform over time. A large-area roll-up or microshutter version of MEMS was developed by Boris Lamontagne, patent US7684105B2, NRC Canada, for use as a switchable glazing [42, 43]. It is called the Louver window. When applying a voltage, a 15–25 VDC bipolar square wave forms an electric field between the electrodes; the top electrodes roll down onto the surface, thus blocking the light transmission. Switching time is 40 μs. Microshutter technology has been reviewed by Tong et al. [44]. The louver window is discussed further in Chapter 9 – Other Chromogenic Technologies.

Electric Shade Macro-Shutters Notable electric macro-shutter work has been made by Eliot Schlam's New Visual Media Group, LLC. (Eatontown, NJ) They developed electrostatically driven shades or shutters within IG units. Early work on a similar system was done in the 1970s by C. Kalt of Dielectric Systems International (North Adams, MA) (US3989357A, 1976) [45]. A polymer panel is rolled up or down between retracted and extended positions by applied voltage (100–500 VDC) (US8736938B1, 2014) [46]. The dynamic shutter uses on-glass layers, including a transparent conductor and a dielectric film. The shutter is a resilient polymer with a conducting surface. The polymer may be surface modified to increase reflectance. The shutter can be rolled up or rolled down as desired by the user. Shutters as large as 5 m^2 can be made. The technology has been promoted for architectural glazing and aircraft windows. This shutter technology was acquired by Guardian Glass (Auburn Hills, MI) in 2018. World patents were filed by Guardian in 2020 (e.g. WO2020008436A1) [47]. Other organizations have worked with the shutter concept, including the United States Air Force, Samsung, Fiat, and the University of Kassel (Germany). K. Mori at Ritsumeikan University, Japan, has combined a roller shade with MEMS shutters [48]. The electric window shutter is a viable window concept when an automatic shade is needed, especially for blackout or when fixed opacity conditions are acceptable.

1A.3.1.2 Mechanochromism

Mechanochromic systems include deformable membranes, piezochromics, and tribochromism. Much of this group is under research and development. Tribochromism refers to a change in the color of a material caused by mechanical friction. It is similar to piezochromism, the change in the color of a material caused by applied pressure, a phase change, or a structural change that can be induced, forming a different color. For example, a gold complex emits a blue color when subjected to pressure [49]. Also, colloidal liquid crystal films have been studied, showing mechanochromic color change from red, yellow, and green to purple [50].

Deformable Membranes Deformable membranes deform by mechanical strain placed on them and produce an optical response, such as increased scattering or a change in the index of refraction. Generally, this biaxial strain is greater than 10%, which is too high for a window film. The optical change is usually caused by stretching or wrinkling of a surface coating [51]. There are four types of mechanochromism:

(1) transparency change mechanochromism (TCM),
(2) luminescent mechanochromism (LM),
(3) color alteration mechanochromism (CAM) and
(4) encryption mechanochromism (EM).

For example, cephalopods (squid family) can change the color intensity of their skin by relaxing circular muscles in the skin surrounding pigment sacks. The pigment sacks expand when the muscles relax, causing a CAM color change [51].

Much of the current work on deformable membranes is under investigation. Longevity and cyclic strain are the major issues. I will provide a few examples. A double-layer encryption mechanochromic hydrogel with multidirectional force sensing has been studied [52]. There is work on encryption photochromic/mechanochromism based on silane-substituted spiropyran, which is noted in the photochromics section of this chapter. Another notable system is the wrinkling of polydimethylsiloxane (PDMS), which can change from opaque to transparent with an applied strain of 50%. The film was cycled 100,000 cycles [53]. Another system is based on polyvinyl alcohol (PVA), with orthogonal cracks and folds. It has been cycled 50,000 cycles with no significant change. It switches from $T = 0.85$ to $T = 0.30$ with 50% strain. Wrinkled film technology has been reviewed [54]. Work on multistate mechanochromism smart windows has been reported [55].

Piezochromics Regular piezoelectrics are used to generate electricity from applied pressure. Piezochromic substances change color when pressure is applied to them. The intrinsic color of a piezochromic substance is due to the absorption of light in specific regions of the excitation spectrum as a result of an electronic transition from the ground state to a state of higher energy. When a pressure is applied to the substance, these energy states are perturbed, giving rise to a change in the color of the substance. It has been shown that discontinuous change of color occurs when a crystalline solid undergoes a first-order phase transition from one crystal structure to another. A change in color may also be induced by a change in molecular geometry which makes up the substance. In the case of mesomorphic substances, the piezochromic effect occurs because of a change in the Bragg diffraction angle for a given wavelength, which results from a mechanical force, such as applied pressure or shearing force, and friction, which induces a change in refractive index. Reversible piezochromic materials include derivatives of bianthrone, dixanthylene, helianthrone, imidazole, mesonaphthobianthrone, pyrrole, and xanthylidene anthrone. Piezochromic cholesteric elastomers have been studied by Arno Seeboth, Fraunhofer Institute for Applied Polymer Research, Berlin, Germany. The elastomer changes color from red to green in less than one second when a pressure of 0.4 bars is applied [56]. A schematic of the process is shown in Figure 1A.9.

1A.3.1.3 Magnetochromics and Magnetoionics

Two other technologies come from other fields. Magnetochromics use a magnetic field to modify color. Magnetoionics use injected ions to alter the magnetism of a film. So far magnetochromics have been studied mostly on liquid systems. Colored switches can be made by encapsulating magnetic particles of Fe_3O_4 in SiO_2. Depending on the particle size (124–197 nm), a silver-gray switch to blue, green, or red switch can be achieved when a field is applied [57]. Particles in the form of platelets have been studied to create different optical effects [58]. A commercial magnetochromic plastic sheet has been created so that magnetic fields can be visualized [59]. As the magnetic field strength goes up, the field lines are visualized by optical interference. Depending on the field strength, the colors range from 50 Gauss (orange), 250 Gauss (green) to 1000 Gauss (deep blue). This visualization is very

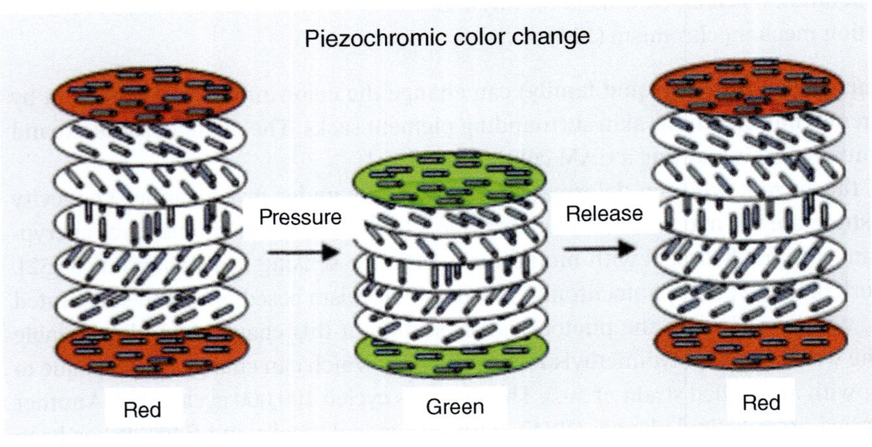

Figure 1A.9 Pressure-dependent color changes of the selective reflection wavelength of a helical cholesteric elastomer. *Source:* Modified from Seeboth et al. [56], courtesy of Scientific & Academic Publishing, image is under Creative Commons CC BY 4.0 license.

useful in the teaching of magnetic fields and how they are altered by materials they come into contact with.

Magnetoionics Magnetoionics uses injected ions to give a voltage-induced magnetic change of a thin magnetic film. This is being studied for electronic memories in the field of spintronics [60]. This magnetoionics is quite interesting since some devices are constructed similarly to electrochromics, using proton injection into GdO_x and CoO_x films [61, 62]. Also, films of Li–$Ti_{1-x}Mn_xO_2$, Li–$Ti_{1-x}Fe_xO_2$, and Li–$Ti_{1-x}Co_xO_2$ have been considered as candidates for spintronics memories [63].

1A.3.1.4 Electrokinetics

Electrokinetics is a broader term used to define mechanical switching to control energy passing through a device or glazing. An example is switching light with electrophoretic dipole particles. The particles can be locally confined, such as in a capsule or a bounded region. Much of this technology comes from applications in the display industry. Some of the technology comes from the electrophoretic image copying technology for copy machines. There are a range of technologies using particles, fluids, and inks to make images and optical effects. For fluids, another terminology is electrowetting technology. Electrowetting technology is being investigated even for something as seemingly simple as a colored fluid display [64]. More information can be found in the SID Digest of Technical Papers, *Journal of SID*, and SID Information Display magazine. A successful electrophoretic company is E Ink with their electronic ink-based e-readers and multi-color displays. E Ink is developing encapsulated pigments to be able to change the color of objects, such as an automotive surface, or change the opacity of transparent articles, such as dynamic windows. Other companies working with electrophoretic glazing are ELSTAR Dynamics and Gauzy, working with Research Frontiers. There are other research groups working in this area, too [65–67]. ELSTAR Dynamics and E Ink's developments are discussed in Chapter 9 – Other Chromogenic Technologies.

Liquid Crystals One technology that has developed significantly into several major products over the last 50 years is liquid crystals (LCs). There are several types of liquid crystals, including nematic, smectic, twisted nematic, cholesteric, guest–host, and ferroelectric types. Twisted nematic liquid crystals are the most commonly used liquid crystal for displays. The mechanism of optical switching in liquid crystals is to change the orientation or twist of liquid crystal molecules interspersed between two conductive electrodes with an applied electric field. The orientation of the liquid crystal can alter the overall optical reflectivity properties of the window or display. A review of liquid crystals for light protection is written by Zhang et al. [68].

Cholesteric liquid crystals (ChLCs) have matured over the years to become very useful for large-area chromogenics. Kent Optronics (Hopewell Junction, NY) has shown a reflective LC mirror panel called e-TransFlector™. Its photopic optical properties in the reflective state are $R_v = 87\%$ (off) and $T_v < 1\%$ (off), in the transparent state, $T_v > 87\%$ (with no AR coating) and $R_v < 10\%$, with 3% haze. The device is powered by 100–260 V, 20 Hz square wave, dissipating 6.6 $mWcm^{-2}$ [69]. According to the Kent Optronics website, a panel up to 0.58 m (23″) diagonal can be fabricated.

Another LC development is at eyrise B.V. (Veldhoven, NL). eyrise has made a guest–host LC device where elongated dichroic dye particles coexist with the liquid crystals in a mixture.

The layer of dye particles and the LCs is sandwiched between two panes of glass coated with a transparent conductive oxide (TCO) layer. The TCO layers have orientation layers applied to them. When a low voltage (~25 V) is applied, the LC molecules alter their horizontal orientation and align along the direction of the electric field. The dye molecules follow the same orientation. The dye is responsible for the system's tint, based on the orientation and changing position of its molecules. The glazing is seen as bright or tinted. When no voltage is applied, the dye molecules are oriented parallel to the incoming light, causing a high absorption of light (dark). When an electric field is applied, the dye molecules reorient themselves to be perpendicular to the incoming light, leading to a low absorption state (bright). This transition between tinted and bright states is possible in less than a second, with continuous control for intermediate shading states. More about eyrise B.V. is in Chapter 9 – Other Chromogenic Technologies.

1A.3.1.5 Dispersed Liquid Crystals (PDLC and NCAP)

One special type of liquid crystal structure is the polymer–liquid crystal emulsions [70]. The types are nematic curvilinear aligned phase (NCAP) from James Ferguson, US4435047 (1984), and dispersed liquid crystal (PDLC) from Kent State, US4671618 (1987). Taliq Corp. (part of Raychem (Sunnyvale, CA) which is now owned by TE Connectivity, Switzerland) went on to perfect the industrial production of NCAP films [7]. Both technologies were heavily licensed to major companies, and several were involved in subsequent patent disputes, as discussed by Hakemi [71]. More about this is in Chapter 5 – PDLC Technology. There are preparation and structural differences between PDLC and NCAP, with similar properties. Here they will be called generically PDLC. How a PDLC functions is shown in Figure 1A.10.

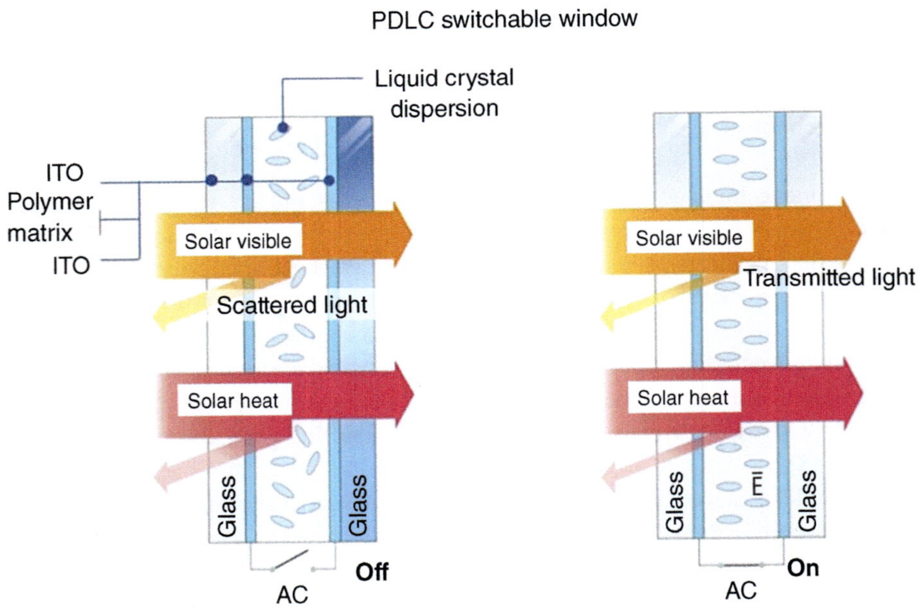

Figure 1A.10 Schematic representations of electrically controlled large-area chromogenic devices. Shown is a polymer-confined liquid crystal dispersion (PDLC or NCAP). The typical voltage is 45–100 VAC. The main effect is change in the amount of scattered light, starting with a milky privacy effect (haze 87%) with voltage off to a specular transmittance with voltage on. Some residual scatter remains after switching (haze 8%). If dipole particles are added, the haze can be considerably reduced. *Source:* Modified from Lee and Song [72], courtesy of Elsevier.

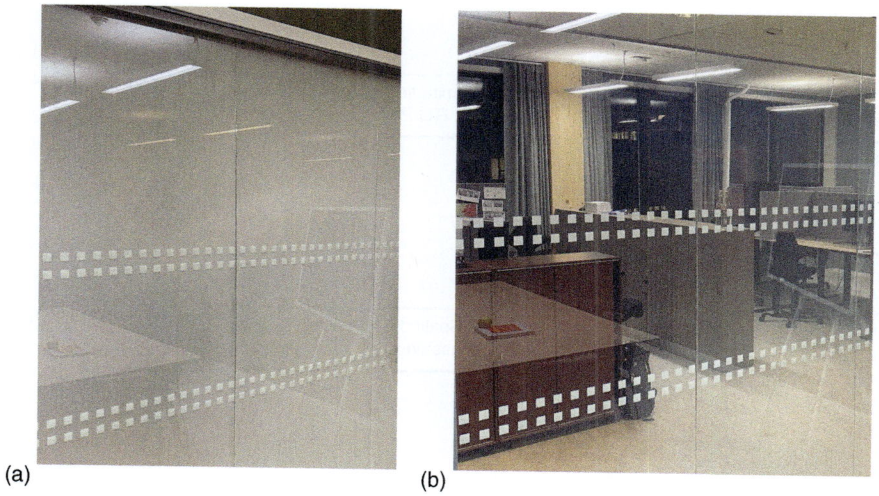

Figure 1A.11 A view of a PDLC interior conference room window in the (a) off (milky) and (b) on (clear) conditions. *Photo Source:* Author's figure.

PDLCs have been commercialized for use in switchable glazing for the control of light and to create a switchable privacy effect. A PDLC glazing used in the interior of a building is shown in Figure 1A.11. The properties of PDLC films for glazing have been studied by Park and Hong [73] and reviewed by Levy and Castellón [74]. Also, PDLCs have been developed for projection display screens. Now that the main patents have expired, we are seeing this technology coming back to the marketplace from several manufacturers. Some of the current producers are Nippon Sheet Glass (NSG, Tokyo, Japan), Polytronix (Richardson, TX & China), Polytron Tech. (Taiwan), DMDisplay (Korea), Innoptec (Italy), and Gauzy (Israel). Dream Glass Group (Spain, Dubai), BestRoom Co. (up to 100 °C use, Korea), PDLC Glass (China), VSOM Glass (China), SNM (China), and Innoglass Technology (China).

NSG produces an NCAP product known as Umu™ for specialty automotive, building, and image display applications. For display applications, the sheet in the off condition serves as an excellent display screen. Large-area LC emulsion glazing is fabricated in 1.42 m × 2.75 m sheets. Open circuit memory is generally not possible with dispersed liquid crystals. However, by adding dipoles to the liquid phase, a memory effect can be achieved. The manufacturing process flows to make UMU rolls and glazing are shown in Figure 1A.12. Some of the issues that remain are long-term UV stability (now controlled by UV cut films), residual haze when activated, and cost, which is about 750–950 US\$ m^{-2} for glazing. For privacy glazing, it is a very good product, and it mimics a rice paper window. An example of an architectural UMU glazing is shown in Figure 1A.13. Another example is shown for a train privacy window in Figure 1A.14. An example of a UMU projection surface is shown in Figure 1A.15. The liquid crystals are encapsulated within an index-matched polymer matrix. The polymer emulsion is fabricated between two sheets of transparent conductor-coated polyester or glass that serve as electrodes.

The switching effect of the UMU glazing spans most of the solar spectrum, up to the absorption edge of glass. In the off-state, the device appears translucent white. When an electric field is applied, the liquid crystal droplets align with the field, and the device becomes transparent. Typically, these devices operate between 60 and 120 VAC but require continuous power to function properly. The NSG UMU glazing power consumption is

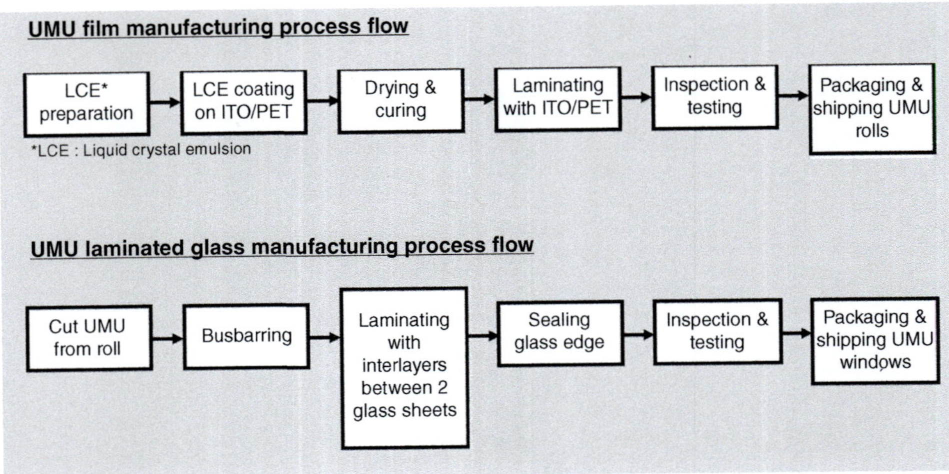

Figure 1A.12 Manufacturing process flows to make UMU (NCAP) rolls and the process to laminate UMU cut sheets to make UMU glazing. *Source:* Redrawn Image, courtesy of Yuchi Yano, Director, NSG-UMU Products, Japan.

(a) (b)

Figure 1A.13 Exterior UMU wall-laminated glazing in the off condition (privacy) (a) and on condition (b). *Photo Source:* Courtesy of Yuichi Yano, NSG-UMU Products, Japan.

(a) (b)

Figure 1A.14 UMU-laminated glazing as a switchable privacy window on a train car, shown off (a) and on (b). *Photo Source:* Courtesy of Yuichi Yano, NSG-UMU Products, Japan.

(a)

(b)

Figure 1A.15 UMU dome glazing as a projection surface at a concert venue, shown under different lighting conditions. *Photo Source:* Courtesy of Yuichi Yano, NSG-UMU Products, Japan.

3.5 W m^{-2}. Other manufacturers of PDLC glazing have power consumption ranging from 3 to 5 W m^{-2}. In general, compared to electrochromics, the power consumption is higher for liquid crystals because of the need for continuous power in the activated state. However, if UMU is used as a privacy glazing and usually off, then the consumption is much lower. The typical integrated normal-hemispherical visible transmission values for a PDLC device are T_v (off-on) = 50–80%. The difference in the numbers is largely due to backscattering. Pleochroic dyes can be added to darken the device in the off-state. The dyed film improves the ΔT range over the visible wavelength region compared to an undyed film.

Dai Nippon Printing (DNP, Japan) showed their "Light Control Film" at the 2024 SID Conference. The film product is a joint project of DNP and AlphaMicron, Inc. (Kent, OH). The film is a neutral-color bistable guest–host LC structure. The flexible film can be used for automotive sunroofs, visors, and display applications. They reported a low 1% haze in both states with a switching range of T_{vis} = 70–20%, with 0.03 s switching time with a driving 7 V square wave. A different film is normally dark with a T_{vis} = <1–40% range, switches at 0.04 seconds using a 30 V square wave [75]. Also, under development are LC systems designed to switch in the near-infrared region of the solar spectrum. The addition of nanoparticles is used to modify the wavelength response of PDLCs [76]. Work on nanocolloids comprising Ag platelets dispersed within a nematic LC has shown that a near-infrared switching of $T_{1000nm} = 80–45\%$ film can be realized. Switching voltage was 10 V [77]. A different approach to solar glare is taken by the researchers at ILEK, University of Stuttgart, Germany. They have constructed glazing that is pixelated to be able to track and block out solar glare as it changes position with the Sun's movement over the window. For the glass panels, they used an array of twisted nematic liquid crystal pixels. Okalux GmbH (Germany) and BMG MIS GmbH (Germany) made glazing modules for testing [78]. The key to the effectiveness of this technology is its detection and control system. Dichroic-doped cholesteric liquid crystal has been studied for smart windows. The liquid crystal pitch can be designed to switch in the UV, visible, or NIR wavelength regions [79].

1A.3.1.6 Suspended Particles

One of the oldest field-effect switchable technologies is suspended particles or electrophoretics, spanning many decades of time. Edwin Land of Polaroid (Minnetonka, MN) in 1934 did some of the earliest work on electrophoretic devices. History does repeat itself, Research Frontiers, Inc. (RFI) has licensed its SPD technology with several refinements to a group of companies. One major maker is Gauzy, who also manufactures PDLC films. An SPD consists of three layers. The active layer has needle-shaped dipole particles (<1 mm long) suspended in a fluid region and dispersed within a polymer matrix. This layer is laminated between two electrical conductors supported by a substrate of polyester or glass, as shown in Figure 1A.16. In the off condition, the particles are random and light-absorbing. When the electric field is applied, the particles align with the field, causing transmission to increase. Typical visible

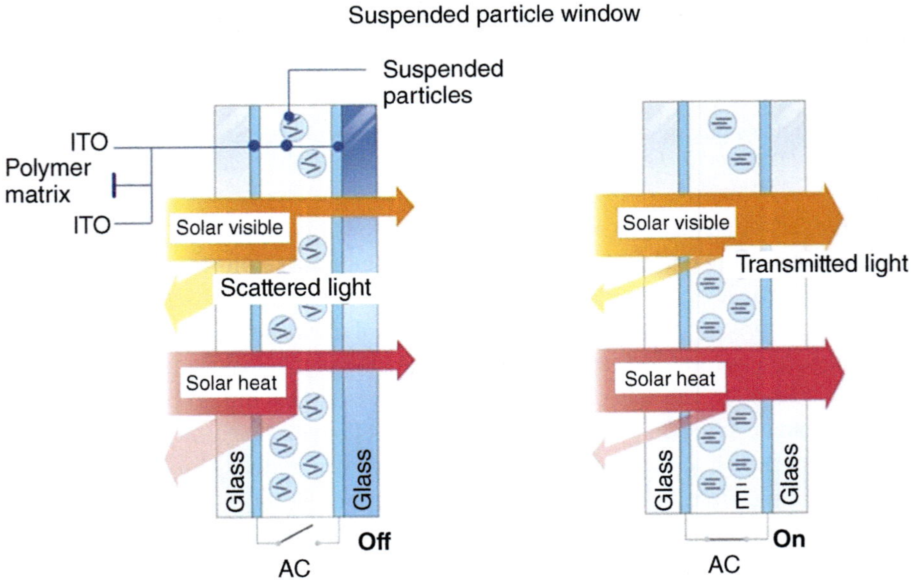

Figure 1A.16 Schematic representations of principal electrically controlled large-area chromogenic devices. Shown is a suspended particle device in its off or scattering condition and in its on condition, showing its direct transmitting effect. The typical operating voltage is 60–100 VAC. *Source:* Modified from Lee and Song [72], courtesy of Elsevier.

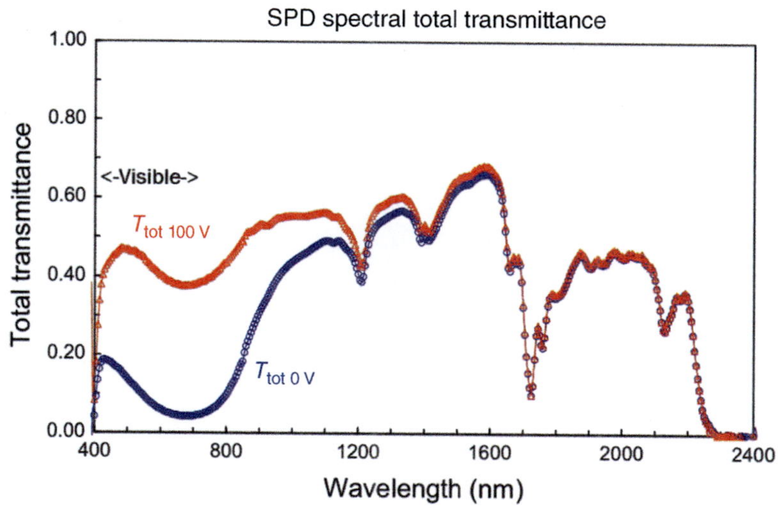

Figure 1A.17 Spectral transmittance of an SPD switchable window for off (darkly colored) and on conditions (lightly colored). Both direct (T_{dir}) and diffuse (T_{diff}) transmittance are shown. There is very little change in reflectance during switching. *Source:* Barrios et al. [80], courtesy of Elsevier.

transmission ranges are 20–60%, 10–50%, and 0.1–10%, with switching speeds of 100–200 ms. The thickness of the particle layer or overall density of the particles determines the transmission range. The voltage required for the device depends on thickness and ranges from about 60 to above 100 VAC. Spectral transmittance (both direct and diffuse) data from Barrios [80] is shown in Figure 1A.17. Solar and luminous transmittance is shown as a function of voltage in Figure 1A.18. A typical SPD color coordinate (x,y) response is shown in Figure 1A.19. The inset in Figure 1A.19 shows the range of coloration with the thickness of the particle layer. Also, CIE color coordinates, color rendering, color fidelity, and color gamut data are available for typical PDLC, SPD, EC, Photochromic (PC), and TC glazing [81].

RFI and their licensees are responsible for the commercial development of SPDs for goggles, eyeglasses, and windows. Hitachi Chemical (Japan) and Dai Nippon Ink have licensed

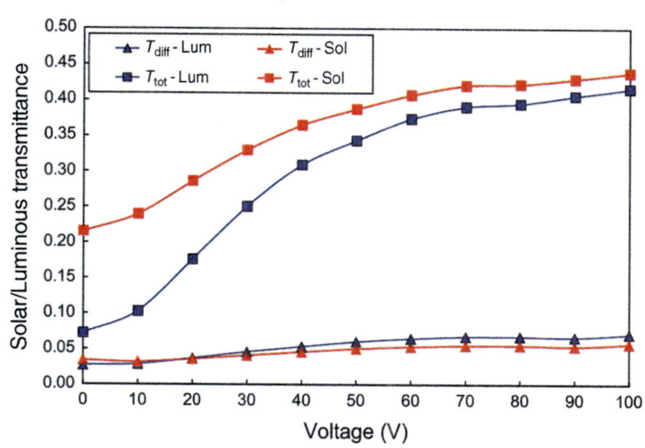

Figure 1A.18 Solar and luminous transmittance of an SPD window shown with increasing voltage. Both diffuse (T_{diff}) and total transmittance (T_{tot}) are shown. *Image Source:* Barrios et al. [80], courtesy of Elsevier.

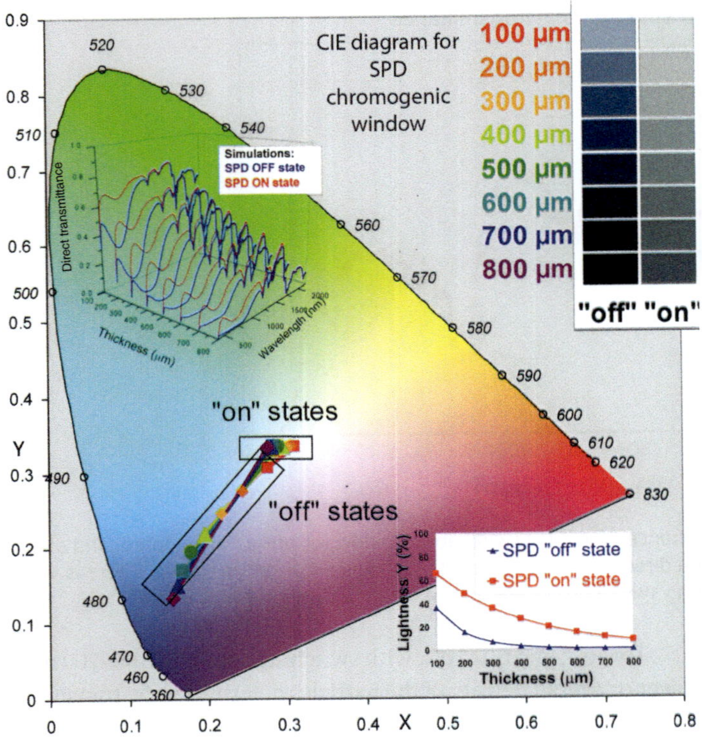

Figure 1A.19 CIE color coordinate (x, y) plot for SPD showing its range of tint either off or on for different thicknesses of SPD particle layer, 100–800 μm thickness. The right top inset shows the range of colors for the different thicknesses. The bottom inset shows lightness Y as a function of SPD thickness. Direct spectral transmittance is shown in the top left inset, shown are on (red) and off (blue, lower line) states. *Image Source:* Barrios et al. [80], courtesy of Elsevier.

the process to make ink and emulsion for devices. Work by RFI and Hitachi Chemical has resulted in a new black particle-based device with $T_L = 10–70\%$ with low haze. Gauzy (Israel) is producing sheet products. Gauzy purchased Hitachi's SPD plant in Japan and has finished Gauzy's new SPD production plant in Germany. Gauzy's product focus is architectural and automotive glazing. They are a supplier to several companies. Smartglass International (Dublin, Ireland) shows some example installations in Figures 1A.20 and 1A.21. Figure 1A.20 shows a panoramic SPD glazing used in a hospital operating room theater in the off condition. Figure 1A.21 shows SPD skylight panels in the dark (off) and light states (on).

More detailed information can be found in Chapter 6 – Suspended Particle Devices. Some work has been done on the energy performance of SPD windows in buildings. It was found that the majority of the energy savings was for the reduction of cooling loads. Also, intelligent control systems are necessary for the balance of daylighting and artificial lighting in a room [83].

1A.3.2 The Chemichromic and Electrochemichromic Families

The chemichromic family, meaning materials that change color or become colored through some chemical reaction, see Figure 1A.22. The chemichromic family includes gasochromic, solvatochromic, hydrochromic, and halochromic materials. Chemichromic dyes will change

Figure 1A.20 Operating room theater with SPD windows shown in the off condition. The SPD glazing reduces the glare of the bright sunlight, seen on the right of the room. *Source:* Courtesy of Michael Kelleher, Director, Smartglass International, Dublin, Ireland [82].

(a) (b)

Figure 1A.21 SPD skylight shown in the off (a) and on (b) conditions. *Image Source:* Courtesy of Michael Kelleher, Director, Smartglass International, Dublin, Ireland [82].

color to indicate a pH change. Chemichromic dyes are sensitive to specific elements or oxidation states. The intensity of color can be used to quantitatively determine elements such as lead, copper, and iron and compounds such as amines. These dyes are incorporated into packaging to determine the freshness of foods and sterilization of medical equipment. Chemichromic dyes can function with packaged articles to indicate whether a gas or a volatile liquid was present inside the packaged article. Chemichromic dyes undergo a

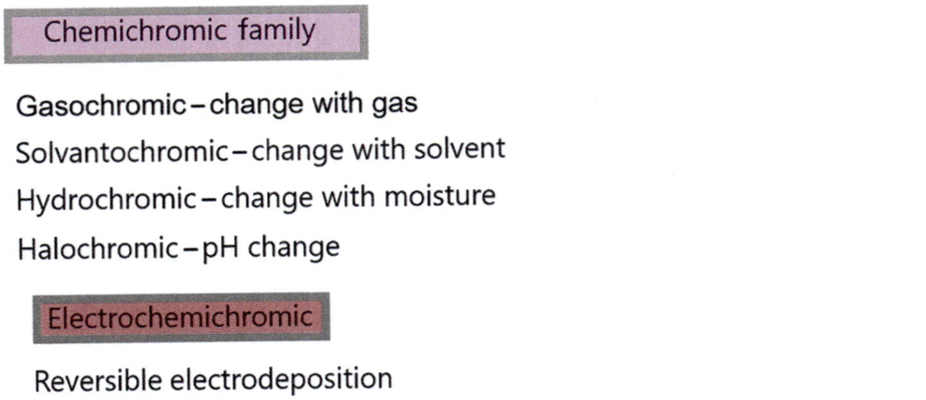

Figure 1A.22 The chemichromic and electrochemichromic families. These materials react with their surroundings to give a color change. Many of these materials are used as indicators or in labels to show a change in environment, such as leaking gas or moisture. Hydrogen detectors are in this category.

change, when exposed to specific chemicals, such as H_2O, CO_2, NO_2, SO_2, or ethylene. The dyes are sensitive to undesirable conditions (such as moisture in moisture-sensitive goods such as electronics). Also, the dyes could detect the degree of product degradation or spoilage of food products (e.g. CO_2, ethylene, SO_2 in food containers). A label, legend, or decal as part of the package could determine if an undesirable product was contained within the package.

The electrochemichromic family consists of a color change when an electrochemical reaction is induced by an applied electrical field or current. These systems can be classified into the following categories: (i) redox reaction resulting in an optical change; (ii) redox reaction coupled with an independent chemical reaction, resulting in a change in color; (iii) redox reaction coupled with a chemical reaction precipitating a colored species; and (iv) redox reaction where a solid film deposited on one electrode is reduced and exhibits a color or reflectance change [84]. One technology that has been studied in detail is reversible electrodeposition. In this work, I will limit the discussion to reversible electrodeposition.

1A.3.2.1 Gasochromic

Gasochromic (or gaschromic) windows have been studied for many years, resulting in some prototype windows. Gasochromic materials change color with exposure to certain gases. A few percent concentration of hydrogen gas in argon is used for switching. The concentration of hydrogen is below the flash point of hydrogen so it is safe to use. A catalyst layer (usually Pd or Pt) ionizes H_2 to $2H^+$ (protons). There are two types of gasochromic windows, one based on WO_3 which colors by adsorption of protons and colors to a deep blue. The second type is based on Mg–Ni and Mg–Y alloys which switch with protons from a mirror-reflective state to a transparent state. Other hydrides, GdMg alloys, certain rare earths, and transition metals have shown a reflective change with hydrogen gas reactions [85].

A gasochromic IGU window is constructed of two glass panes, one coated with a gasochromic film covered with a Pd catalyst. When H_2 gas is fed into the gap region of the IGU, optical switching is achieved. A crosssection of a generic gasochromic window is shown in Figure 1A.23. In the figure, two gasochromic material types are possible, one

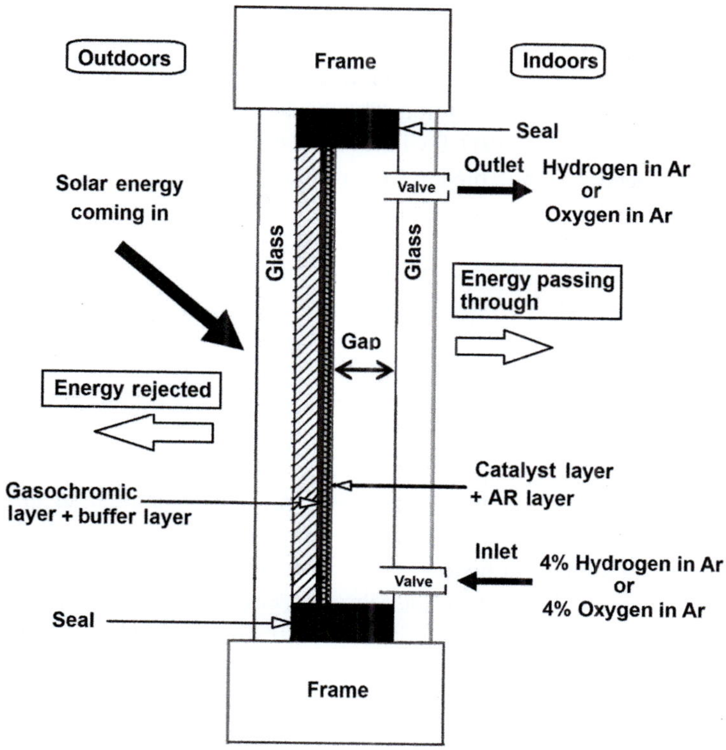

Figure 1A.23 Cross section of a generic gasochromic IGU window construction either with tungsten oxide or a metal hydride. The catalyst layer (Pd or Pt) ionizes H_2 to $2H^+$ (protons). In the first case, the tungsten oxide gasochromic layer darkens when hydrogen is introduced. It lightens when oxygen is introduced. In the second case, a Mg-Y mirror alloy becomes transparent, forming a transparent hydride when it is hydrogenated. When dehydrogenated, it becomes metallic or mirror-like. In some Mg–Y designs, a tantalum buffer layer is used to protect the gasochromic layer. Also, antireflection coatings may be used to increase transmittance or improve color rendition. Solar energy passing through the window can be regulated depending on the flow of hydrogen.

for tungsten oxide that colors with hydrogen and the second for a reflective alloy such as Mg–Y that becomes transparent with hydrogen. In Mg-Y designs, a tantalum buffer layer is used to protect the gasochromic layer. Also, antireflection coatings may be used to increase transmittance or improve color rendition. Work has been done at Wroclaw University, Poland, to enhance WO_3 gasochromic films to increase switching speed and coloration range [86].

Tungsten Oxide Gasochromics Tungsten oxide is the most studied material which reacts by forming a blue color when exposed to ionized H_2S and H_2 gas. WO_3 was used as a hydrogen indicator as early as 1964 by Khoobier (ESSO Research and Engineering, USA) using a Pt/Al_2O_3 as an H_2 catalyst [87]. Considerable work was done on hydrogen sensors [88, 89]. Other materials that react with gases are SnO_2 [90] and also titanium–tin oxide colors when exposed to carbon monoxide. Some of these indicators are used to detect toxic gases in mines, chemical production, and refining plants.

The gasochromic reaction in tungsten oxide involves a proton (H^+) reaction with tungsten oxide to form a deeply colored hydrogen tungsten bronze and a reverse reaction with oxygen to form colorless tungsten oxide. There are three schools of thought on the reaction mechanism. The first describes hydrogen coloration following the electrochromic model with injection of protons and electrons. The second model involves oxygen vacancies in the tungsten oxide to complete the reaction with the proton. The third model involves localized water with the tungsten oxide ($WO_{3-x} \cdot xH_2O$) [91]. I will leave this discussion for the surface scientists to resolve. Using the electrochromic model, we have the following reaction with hydrogen:

$$WO_3 \text{ (colorless)} + xH_2 \rightarrow \text{Catalyst} \rightarrow 2xH^+ + 2e^- \rightarrow H_{2x}WO_3 \text{ (colored)} \tag{1A.1}$$

$$H_{2x}WO_3 \text{ (colored)} + \tfrac{1}{2}\, xO_2 \rightarrow xH_2O + WO_3 \text{ (colorless)} \tag{1A.2}$$

The localized water reaction:

$$WO_3 \text{ (colorless)} + xH_2 \rightarrow \text{Catalyst} \rightarrow WO_{3-x} \cdot xH_2O \text{ (colored)} \tag{1A.3}$$

$$WO_{3-x} \text{ (colored)} \cdot xH_2O + x/2\, O_2 \rightarrow xWO_{3+}\, xH_2O \text{ (colorless)} \tag{1A.4}$$

A suitable catalyst is required to reduce hydrogen to protons. This is similar to an electrochromic reaction with protons and electrons, except with gasochromics, no applied potential is used. Fortunately, the amount of hydrogen gas needed for coloration is below the lower flammability limit in air (under 4% H_2). Gasochromic windows were developed in the early 2000s by Volker Wittwer's Group at The Fraunhofer Institute for Solar Energy Systems (FrG-ISE), Freiburg, Germany [92, 93]. In early gasochromic work, the initial complexity is that it switches with gaseous hydrogen, so a hydrogen storage or generator is needed. To solve this problem, small H_2 generators have been devised to fit into the framing [94]. The disadvantage is that the gasochromic system generates water vapor on the bleach cycle and if not managed properly, can cause condensation. The construction of the window follows a double-pane window construction with a gap between the two panes. On one pane, a film of tungsten oxide is deposited with a very thin layer of catalyst over the tungsten oxide film. H_2/argon gas is fed into the gap causing coloration of the tungsten oxide layer. To switch back, the gap is purged with O_2/argon gas causing the color to disappear. A visible and solar transmittance of $T_v = 75–8\%$ and $T_s = 74–14\%$, respectively, have been measured. The range of solar heat gain coefficient (SHGC) is 0.12–0.46. The transmittance range is better than many electrochromic windows. An image of a prototype window is shown in Figure 1A.24. The Fraunhofer Institute for Solar Energy Systems (Freiburg, Germany), working with Interpane (Germany), developed large prototype windows, 1.5 × 2 m for testing.

Gasochromic WO_3 windows were also studied at AIST (Nagoya, Japan), led by Kazuki Yoshimura. Gasochromic window performance was studied by Wei Feng at Tongji University in Shanghai. The optical and solar heat gain properties of a gasochromic window were measured at Tongji University [95]. The Solar heat gain coefficient versus visible light transmittance for a gasochromic WO_3 IGU compared to a SageGlass electrochromic IGU and other IGU constructions is shown in Figure 1A.25. All data was obtained by simulation using Windows 7.0 software and data base at LBNL, Berkeley.

Switchable Mirror Gasochromics A switchable mirror gasochromic is based on Mg alloy hydrides. Upon switching, they change from reflective metallic to transparent. The work on these alloys is very intriguing. Huibert and Griessen and colleagues in the Netherlands pioneered work on the metallic switching of metal hydrides, YH_3 and LaH_3 [96]. Electrochromic

Figure 1A.24 Gasochromic prototypes shown in the bleached and colored states at the Fraunhofer Institute for Solar Energy Systems. Source image: author.

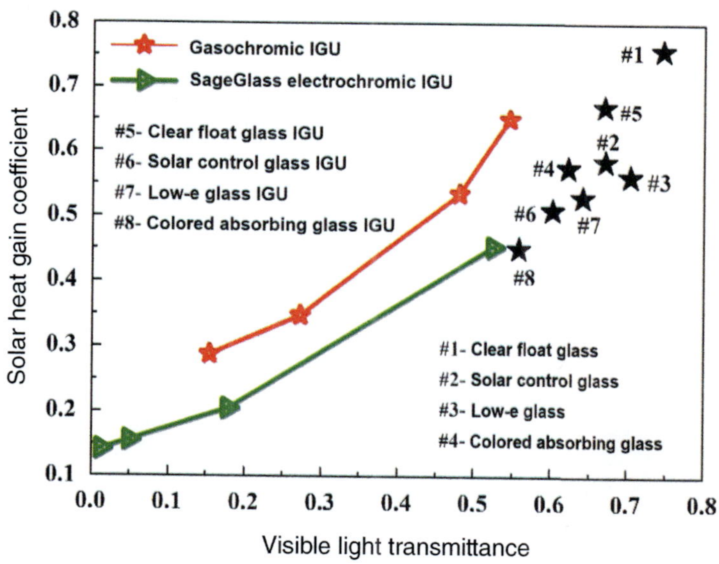

Figure 1A.25 Solar heat gain coefficient versus visible light transmittance for a gasochromic WO_3 IGU compared to a SageGlass electrochromic IGU and other IGU constructions and single-pane windows. All data were obtained by simulation using Windows 7.0 software and database at LBNL, Berkeley. Gasochromic IGU data obtained from test window at Tongji University, Shanghai [95]. *Source:* Modified Image with permission, courtesy of Elsevier.

devices have been made from these hydrides [97, 98]. Work on Mg reflective alloys for windows was done at AIST, led by Kazuki Yoshimura [99–102]. A long-life device has been successfully realized using Mg-Y thin film as a switching layer [101], which would be a good candidate for a switchable energy-reflective glazing. The Mg-Y switchable mirror can be represented by the reactions:

$$\text{Mg (metallic)} + H_2 \leftrightarrow MgH_2\text{ (transparent)} \tag{1A.5}$$

$$\text{Y (metallic)} + 3/2H_2 \rightarrow YH_3\text{ (transparent)} \leftrightarrow YH_2\text{ (metallic)} + 1/2H_2 \tag{1A.6}$$

The device with a wide modulation range of shading SHGC has been developed by the insertion of Ta buffer layer [103] and using TiO_2 anti-reflecting coating to increase the transmittance range [104]. The single-pane values are: T_v = 8–68%, with reflectance on the substrate side, R_{vis} = 39–11%, and on the coated side R_{vis} = 22–6%. The solar values were T_{solar} = 8–55%, with reflectance at the substrate side R_{solar} = 43–17% and the coated side R_{solar} = 28–17. The G-value = 0.24–0.64 and SHGC = 0.24–0.64. They also developed a new gasochromic switching method using a narrow gap of two sheets instead of double pane [105]. Recent work has been on the use of platinum nanoparticles as the catalyst [106]. Prototype Mg-Y IGU windows are shown in Figure 1A.26. A review of research work in the field was made by Tong et al. of Singapore [44]. Beyond gasochromic work, hydrogen fuel, fuel cells, and hydrogen storage for vehicles are being developed in several countries as a clean fuel, including China, Sweden, the United States, South Korea, Spain, Germany, and Japan. The technology for the generation and storage of hydrogen is progressing and will ultimately help move gaschromics forward.

1A.3.2.2 Halochromism

Halochromic materials change color with the change in pH of the chemical environment they are in. Usually, this is a moist or liquid environment. Examples are reversible pH

(a) (b)

Figure 1A.26 Gasochromic Mg–Y mirror glazing showing reflective mirror condition (a) and transparent condition (b). *Source:* With permission from Dr. K. Yoshimura, GIFU University.

indicators and chemical species detectors. For example, halochromic textiles in bandages can be used to monitor patient wound healing. They can be used to monitor bleeding or infections. Depending on the molecular structure, halochromic materials can have prolonged coloring or memory before they reverse back to their original color [107]. One of the newer research areas is halochromic nanoparticle materials [108]. Halochromic substances can detect changes in the acidity of substances, like the detection of corrosion in metals. An example is pH indicator paper. The color change of halochromic substances occurs when the chemical binds to existing hydrogen and hydroxide ions in solution, forming a conjugated molecule. This changes the wavelength of light absorbed, which results in a visible change of color. The various shades result from different concentrations of halochromic molecules with different conjugated systems. But, in general, halochromic substances do not display a full range of colors for a full range of pH because, at certain levels of acidity, the conjugated system will not change.

1A.3.2.3 Solvatochromism

Solvatochromism is a reversible change of the absorption or emission spectrum of a material that is induced by the action of solvents. The color change is the consequence of the absorption peak shift, which occurs due to differences between the solvation energy of the initial and excited state in various solvents [109]. Positive solvatochromism, also known as bathochromic or redshift, occurs when the absorption band is shifted to a longer wavelength. Negative solvatochromism, also known as hypsochromic or blue shift, occurs when the absorption band shifts to a shorter wavelength as the polarity of the solvation environment increases. Examples are pyridinium, merocyanine, and stilbazolium dyes. A positive solvatochromic example is 4,4′-bis(dimethylamino)fuchsone, which is orange color in toluene and red in acetone.

1A.3.2.4 Hydrochromism

Hydrochromic materials change color when exposed to water or moisture. Generally, hydrochromic materials are used in the form of microencapsulated dyes that react in the presence of water or moisture. For example, hydrochromic materials can change from white to a translucent color, allowing you to hide images that only reveal once wet. They can be made to be reversible or nonreversible. Latent indicating hydrochromic labels are used in food safety and can show by color change if a food has been defrosted or allowed to get too hot, or lose water. For example, naphthalenediimide-2D-alkaline earth metal compounds have been studied. These compounds show a yellow-to-red color change with moisture change [110].

1A.3.2.5 Electrochemichromic

As discussed in the introduction of this section, the category of reversible electrodeposition which has been studied and patented for light valves, windows, and displays will only be covered here. Work began in the 1960s and is still continuing today on these interesting chemical systems. Metallic switching is attractive to high-performance window systems since energy is reflected instead of absorbed and reradiated by other technologies. Long-term cyclic stability and incomplete switching have been two of the most difficult problems to solve in reversible electrodeposition.

Reversible Electrodeposition The reversible electrodeposition window is an intriguing electrochemical subject and has been revisited from time to time. It has taken many years to arrive at a good, stable system, and now a reversible electrodeposition appears to have a stable reversible net reaction. Early AgI devices were fabricated as light modulators or display devices in the 1960s–1970s [111, 112]. Bell Labs (Murray Hill, NJ) studied a range of silver halides as potential candidates for display devices [113]. Eastman Kodak (Rochester, NY) patented an electroplating light valve in 1964 [114]. Later, Bernard Eid at Corning Europe (Avon, France) patented reversible deposition for light valves [115, 116]. Polyvision (now Steelcase, Nicholson, GA) patented light modulation and manufacturing processes [117]. One of the major issues was cyclability and residual silver deposits when a device was bleached. A non-silver system was studied by Howard and Ziegler of McDonnell Douglas Technologies (now Boeing, San Diego), researched the Bismuth-Copper system, and achieved a visible reflective switch of $\Delta T_{vis} = 67\%$ [118]. Liquid-based systems, called reversible electrochemical mirrors (REMs), were developed at Rockwell Scientific (now Teledyne, Thousand Oaks, CA) [119, 120]. Also, REM systems have been investigated for tunable antenna applications. REM technology offers a metallic switching effect, a reversible deposition of silver. Rockwell Scientific has developed what promises to be a modulation reflector for windows. It is a very broadband switch covering the visible, near-IR, and IR spectral regions (beyond 20 μm), only limited by the transmittance of the glass substrate, which becomes opaque at about 3 μm. Reversible mirror materials use the reaction on both electrodes of the device to always give a neutral chemical state (Equation. 1.7.)

$$Ag^{+} + e^{-} \leftrightarrow Ag\,(s) \tag{1A.7}$$

A transparent glazing device consists of a metal grid (7 μm line width) on glass/gelled electrolyte/Pt nucleation layer/ITO/glass structure. The electrolyte is gelled in gamma-butyrolactone containing AgI and LiBr. The luminous reflectance properties vary between $R_{Lum} = 5\text{–}80\%$. The switching power is $1.6\,mWcm^{-2}$ and switches between −0.3 and 0.3 V. Since the window is reflective, it does not heat up in the sunlight as absorptive materials do. Windows of 6 × 9 cm with a grid conductor consisting of a line spacing of 144 μm have been fabricated. The McGeehee Group at Stanford University has developed a REM window based on Cu–Pb or Cu–Ag modified with Pt. The window area is $25\,cm^2$. It switches from $T_{600nm} \sim 80\%$ to $T_{600nm} < 5\%$ in three minutes. Switching is between −0.6 and 0.8 V. It has been tested to 5500 cycles with only a minor change in $\Delta T_{600nm} \sim 47\text{–}53\%$ (switching between $T_{600nm} \sim 28$ and 80%) [121]. Recent work has shown a dark-state transmittance of less than 0.1% [122]. A hybrid window using Li^{+} ion insertion and Bi and Cu reversible deposition has been studied at the University of Nevada and Stanford University [123]. A review of transparent-reflective switchable glazing for buildings is compiled by Tong et al. [44]. Tao et al. have reported on REMs for light modulation and thermal management for buildings [124].

1A.4 Chromogenic Switching by Discrete Movement of Ions, Atoms, and Molecules

All the chromogenic families in Figure 1A.27 have materials that span inorganics, organics, and polymers. The color change of these three families is based on discrete or localized mass movement. All of these families involve a temporary, reversible reaction induced by light, temperature, or electric potential. Many chemicals or chemical combinations have been developed into products in these three material groups.

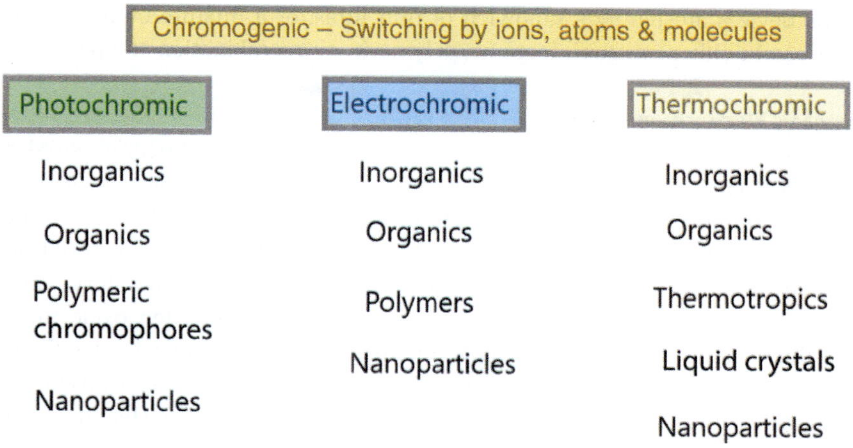

Figure 1A.27 The three most commonly known switching materials groups are photochromic, electrochromic, and thermochromic. They all include the subclassifications of inorganics, organics, and polymers and can involve nanoparticles or nanoengineered materials.

The electrochromic family is the most commercialized for large-area applications. Electrochromics for car mirrors, airplane windows, and architectural windows dominate the applications. Electrochromic materials change by an intercalation reaction involving ions, electrons, and phonons. By utilizing the wide color palette of organic electrochromics, some prototype multicolor displays have been made.

The photochromic family is most commercialized for eyewear. Classic photochromics switch by a UV-induced disassociation reaction, forming color centers which darken the material. Commercialization of both inorganic and organic photochromic materials has been mainly for eyewear and visor applications. It is considered the most mature of the chromogenic fields.

Thermochromics (TCs) are commercialized as temperature indicators and for glazing. TCs require heat to initiate a reaction or phase change. Classical thermochromic inorganic compounds show a semiconductor-to-metallic transition with temperature. Unfortunately, many of the inorganic TCs do not have an acceptable transition temperature for glazing use without chemical modification. Substantial work has been done on the M_yVO_x system (M is a metallic dopant). Also, thermochromic hydrogels show a major structural change with temperature giving a transparent to opaque switching effect (or its reverse). The most successful TCs have utilized the ligand exchange reaction to induce darkening with increased temperature.

One of the newer and most active research areas is nanomaterials used in all families of chromogenics. Nanoparticles are developing in research and moving toward prototypes for testing. Nanoparticles use plasmonics to modify their bulk optical absorption characteristics. The behavior of nanoparticles is just beginning to be understood. Investigating the range of properties is a current research priority. Two-band switchable electrochromics can be made by coupling the properties of plasmonics and polaronics.

1A.4.1 Electrochromics

Electrochromic materials color by the movement of ions and electrons in and out of select materials. Electrochromic materials occur in three forms, inorganic, organic, and polymeric materials. Some devices are purely a sandwich of thin films, or a mixture of chemicals. In practice they can be either inorganic, organic, polymeric, or a combination of types.

1A.4.1.1 Inorganic Electrochromics

The discovery of the tungsten bronzes and their coloration was reported by Friedrich Wöhler in Germany in 1825 [125]. In 1949, S.S. Hsu, working at the Missouri School of Mines, synthesized and reported the coloration of lithium tungsten bronzes [126]. In 1951, work at Linde Air Products (Tonawanda, NY) showed that sodium tungsten bronze could be reversibly colored by ion insertion and extraction [127]. Also, in 1953, laboratory work at Balzers AG (Lichtenstein) reported switching in electrochromic tungsten oxide films [128]. In 1953, RCA was granted a patent for a liquid-based electrochemical color filter, US2632045A [129]. In 1966, Dow Chemical (Midland, MI) patented a color-reversible electrochemical light filter, US 3282157A [130]. In 1967, IBM (Yorktown Heights, NY) patented an electrochromic light valve, US3317266A [131]. In 1969, DuPont (Wilmington, DE) patented an electrochromic device, US3451741A [132], and a compartmented electrochromic device, US3453038A [133]. In the 1970s, American Cyanamid (Woburn, MA) patented an electro-optical device having variable density; this was a field-effect lateral coloring device by Deb and Shaw [134]. This work was followed by an electrochromic seven-segment information display device [135]. From the mid-1970s to the mid-1980s, electrochromic devices for displays were fabricated by numerous investigators working on displays for calculators and watches. These were generally nontransparent displays, meaning the counter-electrode was usually opaque, such as carbon paper. Many companies filed electrochromic display patents. Much of the electrochromic display work stopped when LC displays took the market. However, in current times, companies such as Ynvisible (Portugal & Canada), RISE Research Institutes (Sweden), and ACREO Swedish ICT AB (Kista, Sweden) among others, make electrochromic labels and indicators.

During the 1980s, both inorganic and organic electrochromics were used to make anti-dazzle rearview mirror products, and these mirrors continue as viable products. Very active work on mirrors occurred in the United States, Germany, and Japan. In Germany, Schott Glass made some very durable inorganic solid-state mirrors using WO_3 [136, 137]. In Japan, Nikon made a very stable solid-state WO_3/IrO_2 mirror [138]. However, Gentex (US) was able to stabilize their organic gel-based mirror which was much less expensive to make. Now, Gentex is the dominant manufacturer of electrochromic automotive rearview mirror products.

From the late 1980s to the 2000s, several companies started manufacturing electrochromics for glazing applications [139]. Pilkington (St. Helens, UK) and later their spin-off Flabeg (Furth im Wald, Germany) made some electrochromic commercial window installations. Both architectural and automotive sunroof glazing and visors have been sold commercially. In 1989, Sage, now SageGlass (Faribault, MN), was established to make EC glazing for buildings. ChromoGenics AB (Sweden) was founded in 2003 to make flexible EC film for use in windows and other products like visors. View (Milpitas, CA) evolved from eChromics, and Soladigm (Santa Rosa, Milpitas, CA) and was founded in 2007. View was established to make large-area electrochromic glazing for buildings. Many buildings have been outfitted with View electrochromic windows. Glass Dyenamics (Tucson, AZ) was formed in 2017 to make electrochromic glazing for residential applications.

Even though several companies have been successful at manufacturing EC glazing, others have not. Many organizations have exited the field for a range of reasons, including funding, management change, product competition, product change, corporate change, manufacturing scale-up problems or product order backlog failed to execute. They are Asahi Glass (Yokohama, Japan), Schott Glass (Germany), Pilkington (UK), Flabeg (Germany) which later became EControl-Glas (Plauen, Germany), Ntera (Ireland), Gesimat (Berlin, Germany) [140], Switch Materials (Vancouver, Canada), Heliotrope Technologies (Berkeley, CA), and Halio Inc., formerly known as Kinestral Technologies, Inc. (Hayward, CA) [141]. Also, View has spun-off Neeve AI (San Francisco, CA), an AI-intelligent building, security, and cloud-based building management systems company.

Schematically, the function of an electrochromic glazing is illustrated in Figure 1A.28. Shown is electrochromic glazing constructed between two transparent conductors on glass. The active layers are an electrochromic, electrolyte, or ion conductor and counter-electrode (ion storage layer or complementary electrochromic). For example, in tungsten oxide, when ions and electrons are injected into the material, it colors blue. When the ions and electrons are removed, the material becomes colorless. The device construction is similar to a transparent battery, where the state of charge is shown by coloration. The voltage required is a small 1–5 VDC. Devices can be made with memory (or bipolar), so power is only needed to change color.

Of the companies producing electrochromic IGUs or electrochromic glazing units (EGU)s, the View Company (Milpitas, CA), who has production in Olive Branch, Alabama, is one of the larger volume manufacturers. View was publicly traded but is now privately owned as View Operating Corp. since April 2024. View is going through restructuring, so production has slowed. An example of their EC glazing installed at the Southern Methodist University is shown in Figure 1A.29. SageGlass is another volume maker of EC windows. SageGlass is owned by Saint-Gobain Glass (Paris, France). View and SageGlass produce the most electrochromic windows in quantity and gross area. Both View and SageGlass

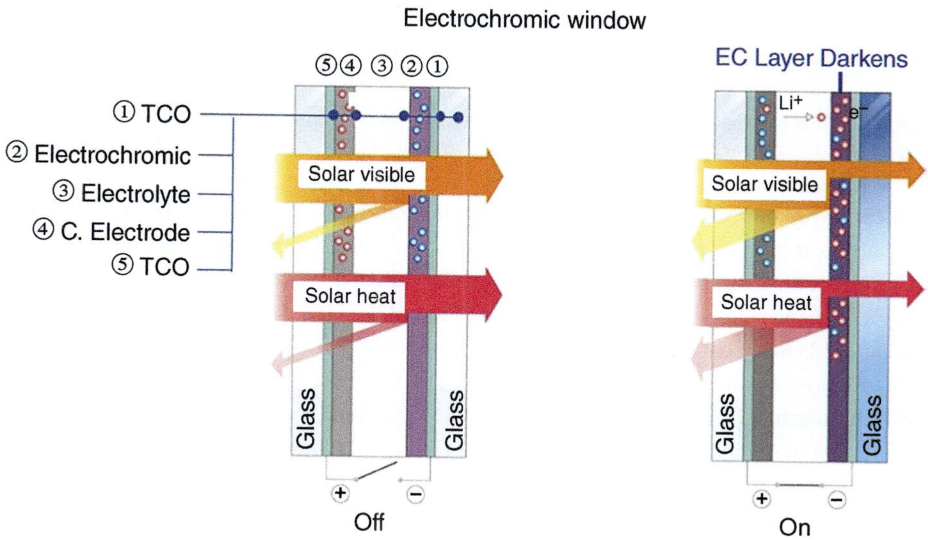

Figure 1A.28 Schematic representation of an electrically controlled chromogenic glass pane. Shown is an electrochromic device constructed between two transparent conductors on glass. The active layers are an electrochromic, electrolyte or ion conductor, and counter-electrode (ion storage layer) or complementary electrochromic. *Source:* Image modified from Lee and Song [72], courtesy of Elsevier.

(a)

(b)

Figure 1A.29 View electrochromic glazing installed at Southern Methodist University. (a) Fully darkened and (b) fully bleached. *Source:* Courtesy of Brian Griedel, View, Inc., Milpitas, CA, USA.

have substantial glazing installations in commercial buildings throughout the world. The following are example properties of EGU windows. The data are not to be used for specifications, as element selection, treatments, and window design can alter these numbers. The following is for 6 mm glass panes. For example, a View IGU (Gen 3) with an argon fill gap of 12.7 mm, and a Low-e coating (type i89) has properties of: $T_v = 57–1\%$, $T_s = 34–1\%$ with an SHGC = 0.42–0.14, U-Factor = 0.23 (BTU h^{-1} ft^2 °F) or 1.31 Wm^{-2} K^{-1}. A View IGU without the Low-e coating has $T_v = 58–1\%$, $T_s = 34–1\%$, with an SHGC = 0.46–0.12, U-Factor = 0.29 (BTU h^{-1} ft^2 °F) or 1.65 Wm^{-2} K^{-1}. Further View IGU design details and data are available [142]. Also, View makes a Gen 4 IGU which is designed to give the most natural color rendition with color rendering index, CRI = 92–86, over the complete tint range.

SageGlass is one of the oldest companies making electrochromic glazing [143]. One of the first products SageGlass made was an EC skylight product manufactured by Velux. Currently, an example SageGlass IGU with an argon fill has a gap of 10.2 mm and a Low-e coating (type 180). The properties are: T_v = 58–1% with an SHGC = 0.40–0.08, U-Factor = 0.25 (BTU h^{-1} ft^2 °F). A SageGlass IGU without the Low-e coating has T_v = 60–1%, with an

SHGC = 0.41–0.09, U-Factor = 0.29 (BTU h^{-1} ft^2 °F), or 1.65 Wm^{-2} K^{-1}. Additional SageGlass IGU design details are available [144].

It is interesting to note that for both SageGlass and View, for windows without Low-e coatings, the SHGC should not be as low as calculated because of a higher amount of transmitted and reradiation of the solar energy into the living space.

A Swedish company, ChromoGenics AB (Uppsala, Sweden), was a spin-off from the IP developed by the Claes Granqvist group at the Angstrom Institute Laboratory (Uppsala, Sweden). ChromoGenics is producing commercial electrochromics on flexible plastic rolls. Flexible electrochromics can be used for lamination to glass or other surfaces. Three key advantages of this material are: (i) it can be easily shipped in rolls, being much lighter weight and not fragile like glass, (ii) the material can be cut to fit a custom or standard glazing, and (iii) the electrochromic film can be laminated to curved surfaces. Glass has always been expensive and fragile to ship long distances. Also, ChromoGenics makes custom EGU windows. More information about ChromoGenics windows is discussed in Chapter 2 on Electrochromics. Many glass and window companies actively continue to monitor developments in chromogenics.

Electrochromic glazing has an interesting attractiveness, but with very subtle complexity. Electrochromic materials are favored for many applications because: (i) electrochromics only require power during switching, (ii) they have a low switching voltage (1–5 V), (iii) they are always specular, (iv) electrochromics exhibit small change in polarization, (v) they have a dynamic gray scale, (vi) many electrochromic multilayer structures have adjustable memory, up to 12–48 hours, and (vii) they can be designed not to have memory, if power goes off, the device fades to clear. Typical electrochromic glazing devices can have upper visible transmission of T_V = 70–50% and fully darkened visible transmittance of T_V = 25–1%. The range of the SHGC for electrochromic glazing ranges from SHGC = 0.35 to 0.50 for the bleached condition and as low as SHGC = 0.09 for the fully darkened condition. The largest drawback to inorganic electrochromics is that their color is less easy to alter. Many devices are based on substoichiometric WO_3 or WO_{3-x}, which shows a strong blue coloration. Certain transition and refractory metal dopants, such as Mo:WO_{3-x}, are used to make a more neutral gray coloration [145]. Also, complimentary coloring electrodes, such as H_yNiO_x and Li_yNiO_x, can be used to make a more neutral coloration studied by the Lampert group at LBNL (Berkeley, CA) [146, 147]. Further information on inorganic electrochromics can be found in Lampert [148, 149] and Granqvist [150, 151] and in Chapter 2 on Electrochromics.

1A.4.1.2 Organic and Polymer Electrochromics

Organic electrochromics are based on the 1,1′-disubstituted-4,4′-bipyridilium salts (viologens), conducting polymers (e.g. PEDOT, polypyrole, and polyaniline), metallopolymers, and metallophthalocyanines. Also, other electrochromic compounds have been studied, including carbazoles, methoxybiphenyls, quinones, pyrazolines, tetracyanoquinodimethane (TCNQ), and tetrathiafulvalene (TTF) [152, 153]. The most commonly used organic electrochromic material is 5,10-dimethyl-5,10-dihyrophenazine (DMP), which is a component in the Gentex switchable mirror [154].

With organic compounds, coloration is achieved by an oxidation–reduction reaction, which may be coupled to a chemical reaction. Originally, organic electrochromics tended to suffer from problems with secondary reactions during switching, but more stable organic systems have been developed. Gentex has developed gelled organic electrochromic materials for automotive mirrors. They have been the most successful company in the field of automotive mirrors. Also, Gentex has made EC sunroofs for cars and aircraft windows with PPG transparencies for aircraft passenger windows. Another company, AshChromics (Marlboro, NJ), markets organic electrochromic eyewear based on propylene-dioxythiophene [155].

Back in the 1980s, Sony (Tokyo, Japan) showed a full-color flat panel concept display based on organic electrochromics. It did have stability issues with certain colors, but it showed a full-color device was possible to make. The John Reynolds research group at the Georgia Institute of Technology (GIT), Chemistry Department (Atlanta, GA), has developed a range of electrochromic polymer materials and flexible films. One of the key group researchers is Anna Österholm. The group has shown a complete color spectrum of electrochromic films. Work on PEDOT has resulted in devices with a change in luminous transmission $\Delta T_{L,580\,nm} = 68\%$. A full range of colors can be made by structural modification of the organic system [156–160]. Also, work at GIT has shown that organic electrochromics can be fabricated into brown tint electrochromic eyeglass lenses [161]. Research in this area of study continues to develop new and more stable compounds. More information is in Chapter 3 – Trends in Organic Electrochromic Materials and their Application and Chapter 4 – Polymeric Electrochromics.

1A.4.2 Light-Induced Switching – Photochromics

Photochromics are covered in detail in a separate chapter. This is a brief summary of its evolution. Photochromic materials change their optical characteristics during illumination of light. These materials are sensitive to both the wavelength and the intensity of light. Photochromic materials are found in inorganic, organic, and biological materials. Photochromism is a classic subject. Fritsche first reported photochromism in 1867 [162]. He observed that tetracene with air and light produced a colorless material that regenerated tetracene upon heating. Phipson observed in 1881 that a gatepost painted with a zinc pigment was black when exposed to sunlight but turned white at night [163]. Phototrophy is a term used to describe photocoloring changes in modern biological systems. However, Markwald coined the term phototrophy in 1899 for organic chemicals [164]. Several major works on photochromism have been written [165–168]. The three most common photochromic products today are photochromic eyewear, photochromic plastic, and paint. Photochromic inorganics include silver halide, TiO_2, ZnS, L_3N, HgCNS, and HgI_2. Many of these compounds are not pure phases and require traces of heavy metals or halogens to be photochromic. The most common photochromic glasses are silver halide-, Ce-, and Eu-doped glasses.

Corning Glass Works (Corning, NY) developed a special photochromic glass that would darken when illuminated by UV light. Corning had been working with opacifying glass and photo-forming images in glass during the 1940s and the 1950s. In 1965, S.S. Stookey, working at Corning, filed a patent on the inorganic photochromic or phototropic eyeglass formulation (Stookey, US 3449103A, 1969) [169]. The lens glass material contains a UV light-sensitive metal halide. When the glass absorbs UVA light from the Sun, a dissociation reaction occurs in the metal halide molecule. This dissociation forms a color center which darkens the glass. When UVA light is removed and there is sufficient thermal energy, the molecule reforms to its rest state, causing the glass to lighten. Roger Araujo of Corning was granted the patent, US3325299A on a non-silver halide phototropic glass in 1967 [170, 171]. The Corning Photosolar™ Photogray™ eyeglass (1.5 mm) can switch from luminous transmittance $T_{Lum} = 85\%$ to $T_{Lum} = 15\%$. Several years ago, Corning made

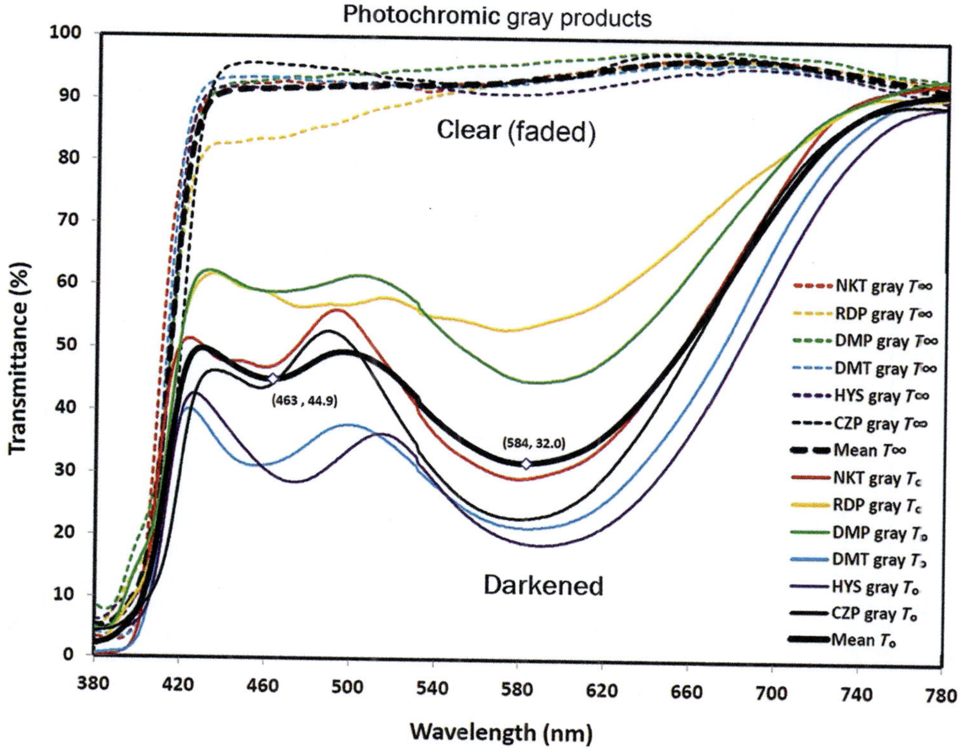

Figure 1A.30 Transmittance for photochromic ophthalmic products from six different manufacturers. Transmittance of gray photochromic lenses at the darkened and clear or faded state. The colored condition is activated with UVA solar radiation and fades with thermal energy [174]. Company codes: NKT, Nikon Lenswear Global (imbibing type); RDP, Rodenstock (coating type), GmbH; DMP, Daemyung Optical (coating type); DMT, Daemyung Optical (imbibing type); HYS, Hoya Corp. (casting); CZP, Carl Zeiss Editec AG (casting). *Image Source:* Moon et al. [174]. This is an open access article distributed under the terms of the Creative Commons Attribution License, which permits unrestricted use, distribution, and reproduction in any medium, provided the original author and source are credited.

prototype photochromic glazing of 1 m^2 for testing as a window glazing [172, 173]. The glass was formed by a casting process. Unfortunately, the glass was considered too expensive to be competitive for architectural windows. The spectral properties of photochromic gray and photochromic brown are shown in Figures 1A.30 and 1A.31, respectively, for samples made by five companies. Two techniques were used to make the lens photochromic, either by coating the lenses or by imbibing a photochromic solution into a plastic lens [174]. Both clear and darkened responses are shown.

Photochromic plastic for eyeglasses is based historically on spiroindolino-naphthoxazines, and multiple dyes have been formulated to give a neutral coloration as developed by Crano and Guglielmetti at PPG Labs (PA) [175]. Currently, diaryl indenonapthopyrans is used in a layer form on the surface of the lens [176]. When a photochromic dye is exposed to UVA radiation, a chemical bond is broken and a structural change is made. The molecule rearranges into a species that absorbs at longer wavelengths in the visible region, causing the lens to darken. When the UVA is removed and the temperature is high enough, the compound reforms back to its colorless form. Improvements in the dyes over the years have

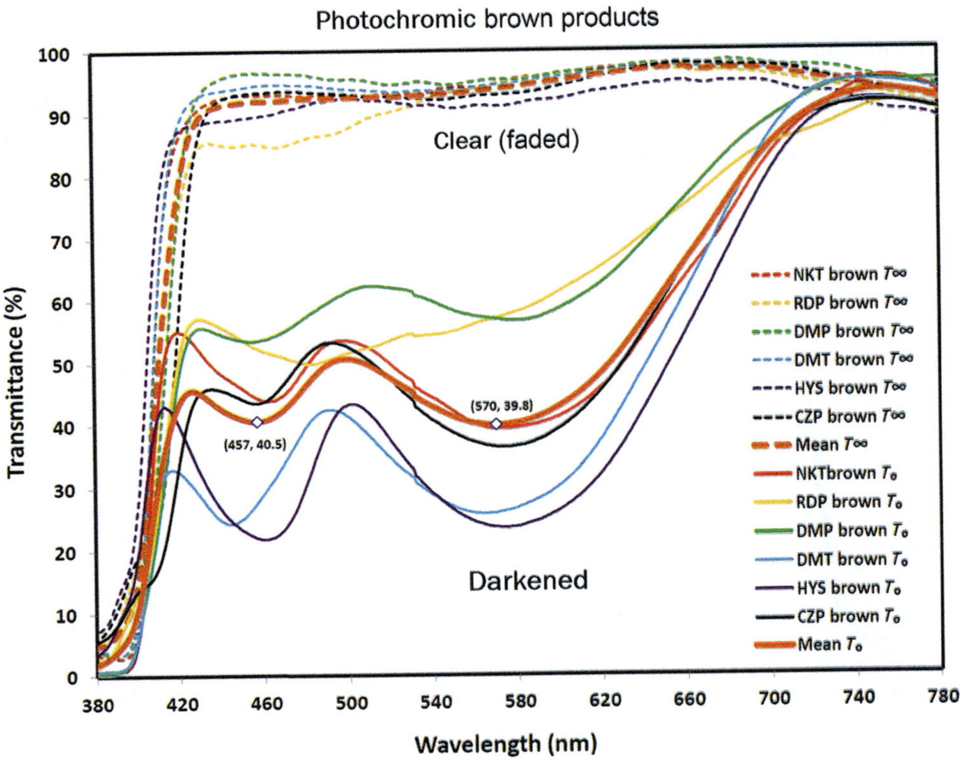

Figure 1A.31 Transmittance for photochromic ophthalmic products from six different manufacturers. Transmittance of brown photochromic lenses at the darkened and clear or faded state at the warm temperature. The colored condition is activated with UV-A solar radiation [174]. Company codes: NKT, Nikon Lenswear Global (imbibing type); RDP, Rodenstock (casting type), GmbH; DMP, Daemyung Optical (casting type); DMT, Daemyung Optical (imbibing type); HYS, Hoya Corp. (coating); CZP, Carl Zeiss Editec AG (coating). *Image Source:* Moon et al. [174]. This is an open access article distributed under the terms of the Creative Commons Attribution License, which permits unrestricted use, distribution, and reproduction in any medium, provided the original author and source are credited.

led to faster coloration and bleaching, as well as longer lifetimes. Photochromic eyeglasses marketed as Transitions™ have been commercially developed by PPG. Transitions™ uses diaryl indenonapthopyrans as an active material. A majority stake in the PPG technology was purchased by Essilor International (Paris, France) in 2013. Essilor is the world's largest manufacturer of ophthalmic lenses. Photochromic plastic in sheet form called SUNICE™ is available online for auto-glazing use. It can be applied as a retro-fit film to existing glazing using ethylene-vinyl acetate (EVA) adhesive. Colors are blue or gray. The properties of a photochromic sheet are visible light transmittance, VLT = 75–25%, R_{IR} = 94.9%, and T_{uv} = 0.1% SHGC = 0.47. Also, for this plastic, the total solar energy rejection, TSER = 44–75%, with a switching time <1 minute. Lifetime is about three to seven years. Photochromic plastic film was evaluated in one study by Khaled and Berardi [177]. To make photochromic plastics typically UV responsive, spiropyrans, spiroxazines, or naphtopyrans are added to polymer surfaces. Another version of the photochromic window, one that combines electrochromics with a photoeffect, is called a photoelectrochromic or PEC device. One PEC IGU style window construction was composed of sol–gel deposited layers onto glass/SiO_2/c-WO_3/TiO_2/dye/electrolyte-LiI//N_2//glass. The N_2 is the fill gas within the gap. The dye is the Ruthenium N719 complex, like the dye used in dye-sensitized solar cells. The performance was $T_{vis} = 0.756$ bleached to $T_{vis} = 0.335$ colored under solar irradiation of 750 Wm^{-2} [178]. Photochromism has been reviewed for inorganic compounds, including combinations of TiO_2, MoO_3, and a-WO_3. Overall, there has been a steady increase in research publications on photochromism over the last 40 years, with 3,215 papers published between 2010 and 2014 and 3,438 papers published between 2015 and 2019 [179].

The growth in the ophthalmic photochromic market is about 7% CAGR, larger than the growth in the overall eyeglass market at 3–4% CAGR. This technology has been reviewed by Ru et al. [180] and Du et al. [181]. The emerging research area for photochromics is the rare earth oxyhydrides. Also, there is work on encryption photochromics based on silane-substituted spiropyran; several compounds have been shown to be photochromic [182].

1A.4.3 Thermal-Induced Switching – Thermochromic and Thermotropics

Thermochromics and thermotropics have been known for a long time. Thermochromic materials with the application of heat show a large change in optical properties of color change, change in opacity, or a reflectivity change. Some TCs switch once permanently indicating a heating event. Other TCs reversibly switch, which is important to large-area glazing. When heated, TCs can go through a chemical reaction or phase transformation, such as a semiconductor-to-metal transition. Some of the first temperature-indicating paints were developed for engines and journal bearings used in railroads and later airplanes. In 1919, H.T. Pinnock described the use of the double salt-mercury-silver iodide, or mercury-copper iodide, as pigments in paints for application to bearings and other machinery parts subject to overheating. The AgI–HgI_2 paint switched from yellow to red at about 100 °F and was reversible [183].

There are many inorganic thermochromic compounds identified by Mott [184] and catalogued by Fritzsche and Adler [185]. The *Mott* Transition Metal Oxides have been studied and used for thermal control on spacecraft by Qui et al. [186]. These materials transform from being poor conductor to good conductors with temperature. A change of eight orders of magnitude in conductivity is seen in V_2O_3. Some thermochromic compounds are: Fe_3O_4, $FeSi_2$, NbO_2, NiS, Ti_2O_3, Ti_4O_7, VO_2, and V_2O_3. Many of these compounds have transition temperatures either higher or lower than room temperature, so they are not useful for terrestrial applications unless they can be chemically modified to have a transition temperature nearer to room temperature. One can modify the transition temperature by metal doping. The challenge is to do it without interfering with the large change in conductivity and optical switching range. Unfortunately, doping usually causes a lower change in conductivity during the transition between semiconducting and metallic states. Some early work was done by Jorgenson and Lee on the modification of $V_{1-x}M_xO_2$ for use in glazing and pigments [187]. K. Yoshimura (now at Gifu University, Nagoya, Japan) studied thermochromic $V_{1-x}W_xO_2$ with Tanamura's group at AIST (Nagoya, Japan) [188]. Also, work was done by the Granqvist and Niklasson groups at Uppsala University in Sweden [189–191]. With further developments of dopants and nanochromic structures, we are seeing more work on binary and ternary thermochromic metal oxides appearing in research. Emittance switchable-stacked structures of Ag/VO_2/Ge/VO_2 have been made for infrared emittance control (5–13 μm) with $\varepsilon_t = 0.10$, switching to $\varepsilon_t = 0.78$ at temperatures above 68 °C [192]. A good review of TCs has been compiled in book by Arno Seeboth and Detlef Lötzsch (FrG-IAP) in 2014 [193]. Also, Arno Seeboth reviews thermochromic materials [194, 195]. Additional works on this topic are found in books by Fukada entitled *Inorganic Chromotropism* [196] and Sone and Fukada, *Inorganic Thermochromism* [197]. These books include a discussion of organometallic thermochromic materials. Classic compilation works on thermochromism have been written by Prof. Jesse Day (Univ Ohio) [198], Greenburg [199], Crano and Guglielmetti [200], and a review by Kamalisarvestani et al. [201]. Studies of the energy efficiency of thermochromic glazing and opaque building envelopes are available too, for example [202, 203]. Further information can be found in Chapter 7 – Inorganic Thermochromics and Photochromics.

1A.4.3.1 Thermal-Induced Switching – Ligand Exchange Thermochromic (LETC)

The organic LETC called Suntuitive™ Dynamic Glass (SDG) was developed by Harlan Byker of Pleotint, LLC (West Olive, MI), US patent 8431045B2. The thermochromic glazing consists of a PVB material containing the ligand exchange compound that can be laminated between glass sheets. The SDG can switch from a clear state to a darkened state when heated. It can be made color neutral following Anderson et al., US Patent 9465239B2. The LETC systems do not switch at a threshold temperature but have continuously variable tinting over a range of temperatures from about 15 °C to about 75 °C. The activation temperature can be formulated from −10 to 50 °C so it is very customizable. Direct sunlight exposure on the SDG causes most of the window heating which results in tinting. The windows only tint on the sides of the building that are exposed to sunlight, which, of course, is the only time when there is excess solar heat gain. At times when windows are not exposed to sunlight, they remain in their highest transmission state to allow for maximum daylighting. The SDG material is stable to 85 °C which satisfies many architectural glass applications. Another advantage of the SDG system is its compatibility with well-established glass lamination techniques using existing equipment. The lamination material (a PVB sheet) can be shipped directly to the glass fabricator. SDG has been incorporated into windows on about 500 buildings throughout the world from 2010 to 2020. This amounts to over 500,000 square feet of windows with SDG installed, in projects ranging from a few hundred square feet to over 30,000 square feet. So far, SDG windows have demonstrated durability for more than 10 years. Example SDG windows are shown in Figure 1A.32.

The LETC systems are incorporated into a polymer layer (interlayer) that is used to laminate two sheets of glass together. Generally, this laminate becomes the outer pane of a double-pane window which involves a third sheet of glass. Almost any commercially available glass, including any color of body tint glass, may be used in the SDG units. A high-performance Low-e coating is provided in the double-pane window structure to ensure most of the solar heat gain accumulated in the thermochromic laminate is rejected to the outside of the building. The LETC thermochromic layer progressively darkens to dark green. The variation of visible transmittance with time of day for a triple-glazed LECT thermochromic window is shown in Figure 1A.33. The data is for a vertical window with window orientation (East, South, or West) for 42.9 N Latitude (West Olive, MI). It is interesting to follow the window response as the Sun rises in the East and sets in the West.

Harlan Byker, CEO of Pleotint, LLC, has recently discovered a very high-performance, lower-cost second-generation LETC system that he hopes to introduce in SDG for dynamic windows in the future. This development may create a whole new cost-performance dynamic in the smart windows arena. Further information can be found in Chapter 8 – Organic Thermochromics.

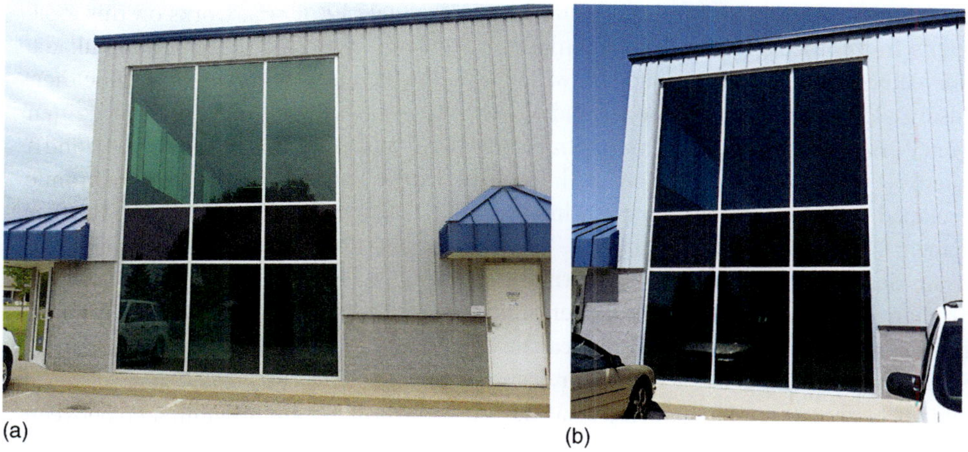

Figure 1A.32 A picture of Suntuitive thermochromic SDG glazing in the lightened state and darkened state. The optical properties are $T_{vis} = 0.60–0.13$, $T_{sol} = 0.25–0.08$, and SHGC = 0.37–0.17 for a temperature change from 25 to 65 °C. *Source:* Courtesy of Harlan Byker, Pleotint.

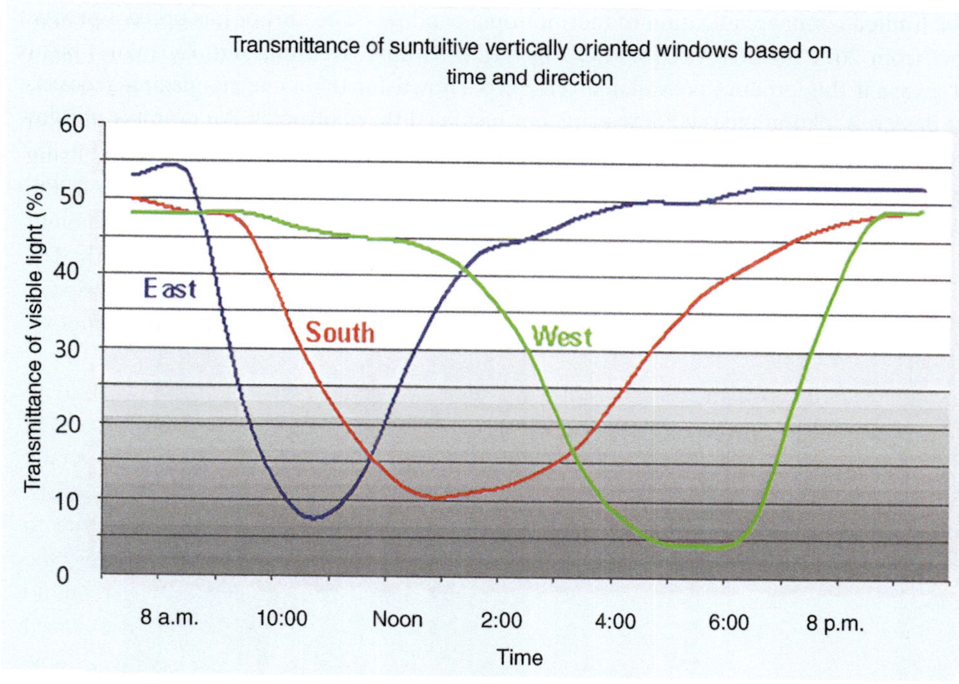

Figure 1A.33 Variation of visible transmittance with time of day for Suntuitive SRT triple glass window with LECT thermochromic. The chart shows the window orientation (East, South, or West) for 42.9 N latitude (West Olive, MI). Modified Image. *Source:* Courtesy of Harlan Byker, Pleotint.

1A.4.3.2 Thermal-Induced Switching – Thermotropics

Thermotropic hydrogels have been studied and developed for glazing. There was a lot of activity in the 1980s–1990s. Day Chahroudi, of SunTek (Tijeras, NM) and later Southwall (Palo Alto, CA), developed Cloud Gel windows following US Patent 4307942. Helen Rose Wilson, along with Werner Platzer and Volker Witwer (as part of International Energy Agency IEA Task 10), studied thermotropics at the Fraunhofer Institute for Solar Energy Systems (Freiburg, Germany) [204, 205]. In Japan, Haruo Watanabe at Affinity Co. (Tokyo) developed large-scale thermotropic windows for building glazing [206] as discussed in Chapter 8 – Organic Thermochromics. These materials exhibit large physical changes at certain design temperatures. Most thermotropics will appear clear at lower temperatures but change to an opaque film at higher temperatures. In the opaque condition, they are highly light scattering. These materials can be used for skylights, inclined glazing, and upper windows where view is not necessary or desired. They can be totally passive, changing with ambient temperature change or by solar absorptive heating. Some designs use a resistive heating layer made from thin metal grids or transparent conductors to artificially heat the switching layer to make it change. A reverse thermotropic is shown in Figure 1A.34. This glazing is a prototype made by Pleotint is not available as a current product (Byker, US patent 6084702). Here it is shown as an example of the thermotropic optical switching effect.

Research continues moving thermotropics forward [207–209]. At this time there appears to be limited commercialization of thermotropic windows. The author has noted a product sheet from 2019 for SOLARDIM®-ECO by TILSE GmbH (Nennhausen, Germany) but is not aware if this product is available. TILSE is known for their marine glazing products. The design is a laminated window using encapsulated thermotropics. An example window has normal-hemispherical values of its highest thermotropic concentration $T_{V(n-h)} = 69–35\%$ and $T_{s(n-h)} = 69–41\%$. This development was part of a government-funded project with Fraunhofer-IAP (Berlin) [210].

(a) (b)

Figure 1A.34 Example of the optical switching of a Pleotint ThermoSEE thermotropic window array. The glazing in the on (a) and in off condition (b). *Source:* With permission, Harlan Byker, Pleotint, Olive, MI.

1A.5 Multiband Switching Windows and Surfaces

The pinnacle of chromogenics is to have independent two-band or even three-band switching over the solar spectrum, as shown in Table 1A.1 revised from [149]. Combination windows are possible with different panes or coatings contributing a switching effect in the visible, NIR, or mid-infrared or possible combination. Different types of chromogenic materials could be used, depending upon the desired result, for daylighting, solar energy heat gain, or even privacy. With this type of switching, one could regulate how much solar visible and near-infrared is allowed to pass through the window. In one case, the solar NIR is allowed to pass through the window. In another case, the solar near-infrared is rejected. The manipulation of the mid-infrared (MIR) is a bit trickier, as one would need to have a selective radiator designed to radiate to the cold troposphere through the atmospheric windows at 3–5 and 8–13 μm. If the coating is covered with a superstrate, then it must be designed with a material that is transparent in the emissive region of interest. Polymers such as polyethylene (PE), polymethyl methacrylate (PMMA), PDMS, and co-polymers of poly(vinylidene fluoride-co-hexafluoropropene) Poly(VDF-HFP) are potential candidates for superstrates [211]. A 300 K (room temperature) black body spectrum spans from 3 to 70 μm with a peak at 10 μm. Passive MIR emitters composed of layers of HfO_2 and SiO_2 have been studied by the Fan group at Stanford University. The emitters have been shown to lower the ambient temperature by 4.9 °C with a radiative cooling power of 40.1 Wm^{-2} [212]. Further work has shown that a transparent system can be achieved using photonic crystal structures [213].

Table 1A.1 Multiband switching matrix for glazing constructions with variable switching elements.

Optical switching materials matrix					
Visible	**Near infrared**	**Mid-infrared**	**Description**	**Example**	**Application**
V	F	F	Tunable visible	Electrochromic, photochromic, SPD, LC	Glazing, eyeglasses, daylighting control
V	V	F	Tunable solar	Electrochromic, PDLC	Variable solar gain glazing
V	V	V	Tunable broadband	Microshutters, LC, REM	Light + energy control, radiative cooling
V	F	V	Tunable solar visible + MIR emitter	a-WO_3 + nc-WO_3, nc-conductors	Tunable lighting, radiative cooling
F	V	V	Tunable infrared	nc-Conductors, thermochromic, polyEC	Solar infrared control, radiative cooling
F	V	F	Tunable NIR	c-WO_3, thermochromic	NIR control, 3–5 μm emitter
F	F	V	Infrared variable	Plasmonic EC, polyEC, thermochromic	Tunable MIR emitter, radiative cooling

Codes used in table: REM is a reversible electrochemical mirror, "nc" are nanocrystalline, "cs" are core–shell nanocrystals, "polyEC" are certain polymeric electrochromics such as polyaniline and PEDOT, LC are liquid crystals, and "SPD" are suspended particle devices.
Source: Revised from Lampert [149].

Table 1A.2 Selective radiator materials matrix for hypothetical building skin construction with variable elements coupled with an adjustable radiative cooling emitter.

Optical switching materials matrix + radiative emitter				
Visible	**Near infrared**	**Infrared**	**Description**	**Application**
F	F	V: 8–13 μm emitter	Variable radiative emitter	Adjustable cooling
V	F	V: 8–13 μm emitter	Variable visible + radiative emitter	Variable daylighting + cooling
V	V	V: 3–5 μm; 8–13 μm emitters	Variable solar + radiative emitter	Variable cooling + heating

Core–shell Au@SiO_2-selective nanoparticle emitters with visible color have been studied for glazing applications [214]. Other MIR emitters include SiO, SiC, Si_3N_4, and SiO_xN_y [215]. Also, polymer electrochromics, such as PEDOT and polyaniline, have been studied for MIR emitters [216, 217]. In addition, MIR switching polymer design rules have been summarized [218]. A two-band switching approach was taken at the Heliotrope (Berkeley, CA) to switch the visible and near-infrared solar visible regions independently. They used a combination of polaronic electrochromic switching in the visible with plasmonic particle switching in the near-infrared. Unfortunately, Heliotrope the developer of this technology went out of business. They had a system where the glazing could be in the "warm," "cool," or "dark mode" depending on the applied voltage. The warm mode passes solar radiation through the glazing. The cool mode passes only visible light, reducing the solar heating effect. The dark mode reduces the transmittance of solar energy over the visible and near-infrared. Another manufacturer will probably pick up this work. The Milliron Group at the University of Michigan, Dept of Engineering, Ann Arbor, MI, Michigan is still working in this area of science. Selective thermal radiators are used to lower the solar cell operating temperature and to improve efficiency. Using radiators, a decrease in operating temperatures of 36 °C has been achieved for concentrating cells [219]. This could also be accomplished with a thermochromic if it does not interfere with the collection of the higher energies. Another technique would be active control using selective infrared (3–5 and 8–13 μm) emitters [220]. With this type of emitter, one would want to maximize the solar visible for photovoltaic energy conversion and reemit unusable energy in the infrared. Movable louvers with different surface emittance are used in space to control spacecraft temperature. Radiative cooling materials have been reviewed by Kang et al. [221]. Microlouvers can be used for radiative cooling, too. They are discussed in Chapter 9 – Other Chromogenic Technologies (Table 1A.2).

1A.6 Running a Chromogenics Business

There are several business and materials hurdles for chromogenic device manufacturing. This is from the authors' experience in working with several start-up chromogenic companies and major, well-experienced glass companies with switchable glazing divisions. It turns out organizational communication and funding resiliency are really more important than optimized materials. Even though, as authors, we are pleased that chromogenics have come a long way to maturity, we want the readers to be aware of some of the pitfalls and opportunities for business in this industry.

1A.6.1 Organizational Viewpoint

(1) Need an outstanding management team fully committed to the product and to bring it from prototype, pilot production, full production and to the marketplace.
(2) Have dedicated long-term investment. Have a solid commercialization plan, especially for government-funded work.
(3) Have stable funding for the project and scale-up. Generally, chromogenic projects take more than three years to enter commercialization. If venture capital is involved, you need to keep the "greed/fear ratio" high. Generally, venture companies have many other technology opportunities to put money into. They can quickly put money together, usually teaming with other firms for a hot product. They can also quickly remove funds too. Also, venture funders may want to add more experienced people to your Board or as advisors. This can happen when the company head is a technologist and less experienced with business.
(4) Angel funders can be your best asset. Many times, they will see "your vision." Sometimes they can help you in times of financial difficulty. For certain situations, "crowd funding" through social media has been successful.
(5) Government money in the form of grants or loans is sometimes available. Many times, funding programs are developed to help create jobs, stimulate the economy, develop technology for the greater good, conserve natural resources, or reduce energy use. For research and development, government bodies usually require a commercialization plan. Commercial readiness is measured on a scale of 1–10, where 1 is in research and 10 is ready for commercialization. Governments value success stories, where they have put in seed money to effect a change.
(6) Make sure your performance schedules are realistic. It is important for you to keep everyone informed and happy.
(7) Know when to stay private instead of going public and selling stock. It is true that stockholders can help you raise money, but it is at a cost, as you give up some control or all control and have to nurture the stock value. Follow SEC rules and keep the stockholders happy.
(8) Have a reasonable scale-up manufacturing and equipment plan. Sometimes, scale-up can involve other business partners with large equipment and with an existing supply chain and customer base.
(9) Have an excellent technical team with manufacturing and scaling experience. One major obstacle is companies with strong R&D, experience difficulty scaling from a quantity of a few custom prototype windows to large sizes or volumes.
(10) Reward your workers for their hard work. Keeping top workers motivated can be difficult unless they share your vision.

1A.6.2 Materials Viewpoint – Chromogenics

(1) Good transparent conductors are needed with high transparency and low haze for electrically connected chromogenics. Is the conductivity sufficient for fast switching? Remember to consider the needs of the application and the size of the final product.

(2) Are the transparent conductors sufficient for the size and switching speed of the product?
(3) The product must have good cyclic durability, as windows are expected to last 10+ years.
(4) Need process to remove/fill pinholes, particles, etc., and other deposition/handling defects.
(5) Need process to make electrical connections on large sheets/glazing.
(6) Laboratory processes should be scalable to large coaters.
(7) Doing experiments on large production equipment is expensive and can use a lot of material.

1A.6.3 Materials Viewpoint – Electrochromics

(1) For electrochromics, the interface between the electrochromic and counter electrode must be non-polarizing and be ohmic so electrons can move easily from the conductor into the active layers.
(2) The electrochromic should possess nanoporosity so ions can easily move from one layer to another under low potential and not become trapped. To make an efficient device, charge balance between the anode and cathode layers must be preserved. The value of interfacial treatments should not be underestimated.
(3) The electrolyte must be stable to light, UV, and temperature and have high ionic and low electronic conductivity, leading to a high-performance device.
(4) Some compounds are photochromic, such as a-WO_3, so one needs to compensate for this effect.

1A.6.4 Manufacturing Viewpoint

(1) Introductory cost will have a premium, but the goal is to be in the mass market, ideally below US\$50/ft^2–US\$100/ft^2 (US\$500/m^2–USS1000/m^2) added cost, to an IGU. Flexible film might be closer to US\$35/ft^2–US\$55/ft^2 (US\$350/m^2–US\$550/m^2).
(2) Deposition processes must be scalable.
(3) Strive to increase product yield and lower defects. Expensive equipment (high capital expenditure, CapEx) with low yields is deadly to a business.
(4) Limit the use of expensive components, such as thin borosilicate glass.
(5) Locate manufacturing close to the development group. Offshoring manufacturing may save labor costs, but overall quality control may suffer.
(6) The device must have excellent and reliable driving and control electronics. Can it be easily integrated, or is it compatible with existing energy management and security electronics? It is actually the control system that makes the window smart!
(7) Have a solid plan for reducing production costs to be appropriate for the application.
(8) Have either a dedicated product manufacturer (e.g. IGU manufacturer)/partner or own your own fabrication facility.
(9) Be able to fit the product into existing product supply chains.
(10) Minimize transportation costs of heavy fragile units/components.
(11) Be able to warranty the switchable window product for 10 years (standard window warranty).
(12) Assure reliable suppliers of materials and components.

(13) Have quality controls in place. The windows need to have consistent performance, color, and appearance (low defects, low manufacturing waste).

1A.6.5 Marketing Perspective

(1) Educate the marketplace and team with potential customers. The average customer understands unwanted glare, hot or cold discomfort, UV damage to interiors and artwork, privacy, and security.
(2) The "wow" factor is important. Do they understand the technology? Probably not.
(3) Understand the needs of your client(s).
(4) Consider value-added design and market your switchable window product with other features such as security, impact resistance, and sound deadening to give it wider market exposure.
(5) Stimulate market awareness and need compared to other chromogenic solutions. Ask why would a consumer want to buy this product? Does it make a space more useful/comfortable/valuable? Does it save energy? Is it Green? Does it get LEED or Green building credits? Do you get a utility/tax rebate rebate/incentive for its use?
(6) Product education. Overall, there is general confusion in the marketplace (and even in the patent and research journal space) as to what is defined as electrochromic. Some customers feel whenever there is a connection to electricity, "it's electrochromic"!
(7) Differentiate your product from other chromogenic products. Why is yours better? Know your competitor's products; they might even become future business associates.
(8) Product demonstrations, YouTube and other social media videos, and trade shows have added value.
(9) Have an interactive website that shows how your technology works, with advice on its use and applications. Showing the performance in actual installations is very important.
(10) Don't over promise the capabilities of your manufacturing plant and its yield (which may be low). Need to balance orders with actual production capabilities.

1A.7 Additional Information about Chromogenics

For further information, there are several review articles that cover a wide range of coatings for glazing, including chromogenics. Historical research can be found in Lampert and Granqvist Large-area Chromogenics, SPIE Optical Engineering Press, 1990 [222]. SPIE hosted a 19-year series of SPIE-focused conferences with proceedings ranging from 1982 (Vol. 324) to 2001 (Vol. 4458) edited by Lampert et al. Conventional and smart glazing were reviewed by Lampert in 2005 and 2007 [223, 224], Jelle et al. [225] and Rezaei et al. [226] in 2017, and Casini in 2016 and 2018 [227, 228]. A review from a chemistry viewpoint on chromism in chemical compounds was made by Sasmal and Pal [229]. A review of research papers covering smart windows is in reference [230]. This reference work does contain information on wrinkled coatings as optical switches. Another variation of the chromogenic glazing is its combination with a grating optical filter which has angle dependency. This means at certain angles or seasons, depending on the position of the Sun relative to the glazing, a spatial darkening effect can be achieved. This adds some complication to making standard products but may be appropriate for custom glazing [231].

Acknowledgments

I wish to thank Yuichi Yano, General Manager, NSG-UMU Products, for his help in obtaining UMU Glazing Photos. I wish to thank, posthumously, Mr. Hideo Kawahara, former General Manager, NSG, for his hospitality and all our discussions about smart windows and vacuum glazing over the many years. I wish to thank Kazuki Yoshimura for his review of the gasochromic section and Gunnar Niklasson and Anoop Agrawal for their review of this chapter.

References

1 O'Day, K.E. (2008). Conspicuous chameleons. *PLoS Biology* 6 (1): e21. https://doi.org/10.1371/journal.pbio.0060021.
2 Grand View Research (2025). Smart Glass Market Size, Share & Trends Analysis Report By Technology (PDLC, SPD, Electrochromic, Thermochromic, Photochromic), By Application, By Control Mode (Rheostats, Remote), By Region, And Segment Forecasts, 2025–2030 Smart Glass Market Analysis – Growth & Forecast 2025 to 2032, https://www.grandviewresearch.com/industry-analysis/smart-glass-market (accessed 1 May 2025).
3 Markets & Markets (2025). Smart glass market size share and trends (Jan 2025). https://www.marketsandmarkets.com/Market-Reports/smart-glass-market-907.html (accessed 30 April 2025).
4 Nowak, N. and Questel (2024). Patent landscape analysis of electrochromics. https://www.questel.com/resourcehub/patent-landscape-analysis-of-electrochromic-glass/ (accessed 6 March 2024).
5 Statista (2025). Number of cars sold worldwide from 2010 to 2024, with a 2025 forecast. https://www.statista.com/statistics/200002/international-car-sales-since-1990// (accessed 30 April, 2025).
6 Gentex (2024). Gentex Annual Report 2024, p. 8, Zeeland, MI. https://ir.gentex.com/financials-and-filings/annual-reports-and-proxy-statements (accessed 30 April 2025).
7 van Konynenburg, P., Marsland, S., and McCoy, S. (1989). Solar radiation control using NCAP liquid crystal technology. *Solar Energy Materials* 19: 27. https://doi.org/10.1016/0165-1633(89)90021-X.
8 Mordor Intelligence (2025). Photochromic Lenses Market Size & Share Analysis – Growth Trends & Forecasts (2025–2030). https://www.mordorintelligence.com/industry-reports/photochromic-lenses-market (accessed 30 April 2025).
9 Transparency market research, J89 (2024). Photochromic lenses market. https://www.transparencymarketresearch.com/photochromic-lenses-market.html (accessed 27 August 2024).
10 AshChromics (2024). Fades sunglasses. https://ashchromics.com/products/fades-sunglasses [https://ashwin-ushas.com/electrochromic-sunglasses-goggles/#performance (accessed 27 August 2024).
11 Chandrasekhar, P., Zay, B.J., Laganis, E.J., Romanov, V.V., and LaRosa, A.J. (2016). Electrochromic eyewear, US 9482880B1, granted: 2016-11-01 assigned: Tri-Power LLC, Ashwin-Uhas.
12 Liping, M., Yongjoon, Y., and Peng W. (2022). Solid state electrochromic eyewear, WO 2022/098854, published: 2022-05-12, Applicant: Nitto Denko Corp., Osaka.

13 Kan, R., Ohto, T., and Matsuda, O. (2022). Eyewear, US 11294199B2, granted; 2022-04-05, Assigned Mitsui Chemical, reassigned 2024: Carl Zeiss Vision Int.

14 Ampere (2024). https://ampere.shop/products/dusk-electrochromic-smart-sunglasses? https://republic.com/ampere (accessed 27 August 2024).

15 ONESEC (2024). Onesec eyewear. https://www.oneseceyewear.net/ (accessed 27 August 2024).

16 Feldman, D. (2024). ONESEC. https://www.opticaljournal.com/onesec-eyewear-introduces-electrochromic-sunglasses/ (accessed 27 August 2024).

17 Tsegay, F., Ghannam, R., Daniel, N., and Butt, H. (2023). 3D printing smart eyeglass frames: a review. *ACS Applied Engineering Materials* 1 (4): 1142–1163. https://doi.org/10.1021/acsaenm.2c00265.

18 Lee, J.H., Kim, H., Hwang, J.-Y., Chung, J., Jang, T.-M., Seo, D.G., Gao, Y., Lee, J., Park, H., Lee, S., Moon, H.C., Cheng, H., Lee, S.-H., and Hwang, S.-H. (2020, 2020). 3D printed, customizable, and multifunctional smart electronic eyeglasses for wearable healthcare systems and human–machine interfaces. *ACS Applied Materials & Interfaces* 12 (19): 21424–21432. https://doi.org/10.1021/acsami.0c03110.

19 (a) Straights Research (2023). Smart Windows Market Trends, Growth & Share by 2032, https://straitsresearch.com/report/smart-windows-market (accessed 07 May 2025). (b) Rishabh (Ray) (2025). Straights Research, personal communication (07 May 2025). (c) Markets & Markets (2025). Smart window market, size, share and trends. https://www.marketsandmarkets.com/Market-Reports/smart-window-market-92955532.html (accessed 20 January 2025).

20 Lee, E.S. (2018). Innovative glazing materials (Chapter 6.3). In: *Handbook of Energy Efficiency in Buildings-A Life Cycle Approach* (ed. U. Desideri and F. Asdrubali). NY: Elsevier. ISBN: 9780128128176.

21 Doane, J.W. and West, J.L. (1991). Liquid crystal display material comprising a liquid crystal dispersion in a birefringent polymeric media, US 4994204A, granted 1991-02-11, assigned Kent State University.

22 West, J.L., Doane, J.W., and Zumer, S. (1987). Liquid crystal display material comprising a liquid crystal dispersion in a thermoplastic resin, US 4685771A, Granted 1987-08-11, assigned Kent State University.

23 Fergason, J.L. (1984). Encapsulated liquid crystal and method, US 4435047, granted 1984-03-06, assigned Manchester R&D partnership.

24 Pearlman, K.N. (1991). Latex entrapped ncap liquid crystal composition, method and apparatus, US 4992201A, granted 1991-02-12, assigned Taliq, reassigned Raychem, Tyco.

25 Ghosh, A., Hafnaoui, R., Mesloub, A., Elkhayat, K., Albaqawy, G., Alnaim, M.M., and Mayhoub, M.S. (2024). Active smart switchable glazing for smart city: a review. *Journal of Building Engineering* 84: 108644. https://doi.org/10.1016/j.jobe.2024.108644.

26 Khaled, K. and Berardi, U. (2021). Current and future coating technologies for architectural glazing applications. *Energy and Buildings* 244: 111022. https://doi.org/10.1016/j.enbuild.2021.111022.

27 Tällberg, R., Jelle, B.P., Loonen, R., Gao, T., and Hamdy, M. (2019). Comparison of the energy saving potential of adaptive and controllable smart windows: a state-of-the-art review and simulation studies of thermochromic, photochromic and electrochromic technologies. *Solar Energy Materials and Solar Cells* 200: 109828. https://doi.org/10.1016/j.solmat.2019.02.041.

28 Mordor Intelligence (2024). Aircraft markets. https://www.mordorintelligence.com/industry-reports/aviation-market (accessed 22 October 2024).

29 Boeing (2024). Boeing Aerospace Annual Report, 2024, Arlington, VA.

30 PPG Aerospace (2024). Alteos® Interactive Window Systems. https://www.ppg.com/en-US/aerospace/products/transparencies/transparencies-technology/alteos (accessed 30 April 2025).

31 Gentex (2024). Electronically dimmable windows. https://www.gentex.com/products-technology/aerospace/electronically-dimmable-windows/ (accessed 22 October 2024).

32 Vision Systems (2024). Shading systems. https://www.vision-systems.fr/solutions/composite-expertise/ (accessed 26 April 2024).

33 Glass on Line (2024). Electronically dimmable windows. https://www.glassonline.com/electronically-dimmable-windows-aircraft-passenger-comfort-at-the-push-of-a-button/ (accessed 27 August 2024).

34 Markets and Markets (2024). Display market size, share and growth analysis, July 2024. https://www.marketsandmarkets.com/Market-Reports/display-market-925.html (accessed 12 August 2024).

35 Cairns, D.R., Broer, D.J., and Crawford, G.P. (2023). *Flexible Flat Panel Displays*, 2e. Wiley. ISBN: 978-1-118-75111-4.

36 Sadeghi, K., Yoon, J.-Y., and Seo, J. (2020). Chromogenic polymers and their packaging applications: a review. *Polymer Reviews* 60 (3): 442–492. https://doi.org/10.1080/15583724.2019.1676775.

37 Gu, C., Jia, A.-B., Zhang, Y.-M., and Zhang, S.X.-A. (2022). Emerging electrochromic materials and devices for future displays. *Chemical Reviews* 122 (18): 14679–14721. https://doi.org/10.1021/acs.chemrev.1c01055.

38 Suhag, N. and Singh, S. (2015). Types of chromism & its applications in fashion & textile designing. *International Journal of Enhanced Research In Science Technology & Engineering* 4 (8): 28–36, ISSN: 2319-7463.

39 Ntoi, L.A., Buitendach, B.E., and von Eschwege, K.G. (2017). Seven chromisms associated with dithizone. *The Journal of Physical Chemistry A* 121 (48): 9243–9251. https://doi.org/10.1021/acs.jpca.7b09490.

40 Corrente, G.A., González, D.A., Aktas, E., Capodilupo, A.L., Ruighi, F., Gianluca Accorsi, G., Imbardelli, D., Cristina Rodriguez-Seco, C., Martinez-Ferrero, E., Palomares, E., and Beneduci, A. (2023). Reversible vis-NIR electrochromic/electrofluorochromic switching in dual-functional devices modulated by different benzothiadiazole-arylamine anodic components. *Journal of Materials Chemistry C* 11: 17115–17127. https://doi.org/10.1039/D3TC03577H.

41 Iannacci, J. (2017). *Introduction to MEMS and RF-MEMS: From the Early Days of Microsystems to Modern RF-MEMS Passives.* IOP Press, https://doi.org/10.1088/978-0-7503-1545-6ch1. ISBN: 978-0-7503-1545-6.

42 Lamontagne, B., Fong, N.R., Song, I.-H., Ma, P., Barrios, P., and Poitras, D. (2019). Review of microshutters for switchable glass. *Journal of Micro/Nanolithography, MEMS and MOEMS* 18 (4): 040901. https://doi.org/10.1117/1.JMM.18.4.040901.

43 Lamontagne, B. (2010). US 7684105B2 Microblinds and a method of fabrication thereof Granted 2010-03-23, Assigned: National Research Council of Canada.

44 Tong, S.W., Goh, W.P., Huang, X., and Jiang, C. (2021). A review of transparent-reflective switchable glass technologies for building facades. *Renewable and Sustainable Energy Reviews* 152: https://doi.org/10.1016/j.rser.2021.111615.

45 Kalt, C.G. (1978). Electrostatic device for gating electromagnetic radiation, patent US 4094590A, granted 1978-06-13, assigned Dielectric Systems International.

46 Schlam, E.M.S., Koskulics, J.P., and Finch, J.S. (2014). US 8736938B1 Electronically controlled insulated glazing unit providing energy savings and privacy, granted: 2014-05-27, assigned New Visual Media Group, LLC, reassigned Guardian Glass, 2018.

47 Blush, J. and Frey, T. (2020). Electric potentially-driven shade including shutter supporting surface-modified conductive coating, methods of making the same and method of operating the same, WO 2020008436A1 Published: 2020-01-29.

48 Mori, K., Misawa, K., Ihida, S., Takahashi, T., Fujita, H., and Toshiyoshi, H. (2016). A MEMS electrostatic roll-up window shade array for house energy management system. *IEEE Photonics Technology Letters* 28 (5): 593–596. doi: 10.1109/LPT.2016.2514299.

49 Lee, Y.-L. and Eisenberg, R. (2003). Luminescence tribochromism and bright emission in gold(I) thiouracilate complexes. *Journal of the American Chemical Society* 125 (26): 7778–7779. doi: 10.1021/ja034560k.

50 Duan, L., You, B., Wu, L., and Chen, M. (2011). Facile fabrication of mechanochromic-responsive colloidal crystal films. *Journal of Colloid and Interface Science* 353 (1): 163–168. https://doi.org/10.1016/j.jcis.2010.09.044.

51 Zeng, S., Zhang, D., Huang, W., Wang, Z., Freire, S.G., Yu, X., Smith, A.T., Huang, E.Y., Nguon, H., and Sun, L. (2016). Bio-inspired sensitive and reversible mechanochromisms via strain-dependent cracks and folds. *Nature Communications* 7: 11802. https://doi.org/10.1038/ncomms11802.

52 Zhu, Q., Van Vliet, K., Holten-Andersen, N., and Miserez, A. (2019). A double-layer mechanochromic hydrogel with multidirectional force sensing and encryption capability. *Advanced Functional Materials* 29: 1808191. https://doi.org/10.1002/adfm.201808191.

53 Kim, P., Hu, Y., Alvarenga, J., Kolle, M., Suo, Z., and Aizenberg, J. (2013). Rational design of mechano-responsive optical materials by fine tuning the evolution of strain-dependent wrinkling patterns. *Advanced Optical Materials* 1 (5): 381–388. https://doi.org/10.1002/adom.201300034.

54 Ke, Y., Chen, G., Wang, S., Zhou, Y., Yin, J., Lee, P.S., and Long, Y. (2019). Smart windows: electro-, thermo-, mechano-, photochromics and beyond. *Advanced Energy Materials* 9: 1902066. https://doi.org/10.1002/aenm.201902066.

55 Kim, H.N., Ge, D., Lee, E., and Yang, S. (2018). Multistate and on-demand smart windows. *Advanced Materials* 30: 1803847. https://doi.org/10.1002/adma.201803847.

56 Seeboth, A., Loetzsch, D., and Ruhmann, R. (2011). Piezochromic polymer materials displaying pressure changes in bar-ranges. *American Journal of Materials Science* 1 (2): 139–142. https://doi.org/10.5923/j.materials.20110102.23.

57 Ge, J., Lee, H., He, L., Kim, J., Lu, Z., Kim, H., Goebl, J., Kwon, S., and Yin, Y. (2009). Magnetochromatic microspheres: rotating photonic crystals. *Journal of the American Chemical Society* 131 (43): 15687–15694. https://doi.org/10.1021/ja903626h.

58 He, L., Janner, M., Lu, Q., Wang, M., Ma, H., and Yin, Y. (2015). Magnetochromatic thin-film microplates. *Advanced Materials* 27 (1): 86–92. https://doi.org/10.1002/adma.201403836.

59 Li, C., Madsen, F., Liu, L., Yu, L., Zhao, W., and Skov, A. (2024). Magnetochromic elastomer with instant color changes: a study of the influence of material composition. *Macromolecular Materials and Engineering* 309: https://doi.org/10.1002/mame.202400034.

60 Hirohata, A., Yamada, K., Nakatani, Y., Prejbeanu, I.-L., Diény, B., Pirro, P., and Hillebrands, B. (2020). Review on spintronics: principles and device applications. *Journal of Magnetism and Magnetic Materials* 509: 166711. https://doi.org/10.1016/j.jmmm.2020.166711.

61 Sheffels, S., Balakrishnan, P.P., Huang, M., Muramoto, S., Borchers, J.A., Dura, J.A., Grutter, A.J., and Beach, G.S.D. (2023). Insight on hydrogen injection and GdO_x/Co interface chemistry from *in operando* neutron reflectometry and secondary ion mass spectrometry. *Applied Physics Letters* 122: 022407. https://doi.org/10.1063/5.0128835.

62 Martins, S., Ma, Z., Solans-Monfort, X., Sodupe, M., Rodriguez-Santiago, L., Menéndez, E., Pellicer, E., and Sort, J. (2023). Enhancing magneto-ionic effects in cobalt oxide films by electrolyte engineering. *Nanoscale Horizons* 8.1: 118–126. https://doi.org/10.1039/D2NH00340F.

63 Park, M.S. and Min, B.I. (2003). Theoretical search for spintronic and electrochromic device materials based on Li-intercalated transition-metal-doped anataseTiO_2. *Physical Review B* 68 (3): 033202. https://doi.org/10.1103/PhysRevB.68.033202.

64 Jin, M., Shen, S., Yi, Z., Zhou, G., and Shui, L. (2018). Optofluid-based reflective displays. *Micromachines* 9: 159. https://doi.org/10.3390/mi9040159.

65 You, H. and Steckl, A.J. (2013). Versatile electrowetting arrays for smart window applications -from small to large pixels on fixed and flexible substrates. *Solar Energy Materials and Solar Cells* 117: 544–548. https://doi.org/10.1016/j.solmat.2013.07.027.

66 Mukherjee, S., Hsieh, W.L., Smith, N., Goulding, M., and Heikenfeld, J. (2015). Electrokinetic pixels with biprimary inks for color displays and color-temperature-tunable smart windows. *Applied Optics* 54 (17): 5603–5609. https://doi.org/10.1021/ja034560k.

67 Shian, S., David, R., and Clarke, D.R. (2016). Electrically tunable window device. *Optics Letters* 41: 1289–1292. https://opg.optica.org/ol/abstract.cfm?URI=ol-41-6-1289.

68 Zhang, R., Zhang, Z., Han, J., Yang, L., Li, J., Song, Z., Wang, T., and Zhu, J. (2023). Advanced liquid crystal-based switchable optical devices for light protection applications: principles and strategies. *Light: Science & Applications* 12: 11. https://doi.org/10.1038/s41377-022-01032-y.

69 Li, L. (2024). Liquid crystal switchable mirror. www.kentoptronics.com (accessed 24 August 2024).

70 Montgomery, G.P. Jr. (1990). *Large-area Chromogenics: Materials and Devices for Transmittance Control*, Proceedings 10304, Optical Engineering (ed. C.M. Lampert and C.G. Granqvist), 577. Bellingham, WA: SPIE. https://doi.org/10.1117/12.2283604.

71 Hakemi, H. (2017). Polymer-dispersed liquid crystal technology -industrial evolution and current market situation. *Liquid Crystal Today* 26 (3): 70–73. https://doi.org/10.1080/1358314X.2017.1359143.

72 Lee, S.-J. and Song, S.-Y. (2023). Energy efficiency, visual comfort, and thermal comfort of suspended particle device smart windows in a residential building: a full-scale experimental study. *Energy and Buildings* 298: 113514. https://doi.org/10.1016/j.enbuild.2023.113514.

73 Park, S., Hong, J., and H. (2007). Polymer dispersed liquid crystal film for variable-transparency glazing. *Thin Solid Films* 517 (10): 3183–3186. https://doi.org/10.1016/j.tsf.2008.11.115.

74 Levy, D. and Castellón, E. (2018). Smart windows based on liquid crystal dispersions (Chapters 5.4). In: *Transparent Conductive Materials: Materials, Synthesis, Characterization, Applications* (ed. D. Levy and E. Castellón). Wiley-VCH. ISBN: 9783527804610.

75 Nakagome, S. (2024). Light Control Film, Dai Nippon Printing Co, Ltd. and Akari, LLC.

76 Shen, W. and Li, G. (2022). Recent progress in liquid crystal-based smart windows: materials, structures, and design. *Laser & Photonics Reviews* 17: 2200207. https://doi.org/10.1002/lpor.202200207.

77 Ghosh, S. and Smalyukh, I. (2022). Electrical switching of nematic plasmonic nanocolloids for infrared solar gain control. *Advanced Optical Materials* 10: https://doi.org/10.1002/adom.202201513.

78 Husser, M., Haase, W., Hoß, P., and Sobek, W. (2016). New possibilities of sun and glare protection with a structured switchable glazing. In: *Challenging Glass 5- Conference on Architectural and Structural Applications of Glass* (ed. J. Belis, F. Bos, and C. Louter). Ghent University. ISBN: 978-90-825-2680-6. https://doi.org/10.7480/cgc.5.2236 (accessed 17 September 2024).

79 Baliyan, V.K., Jeong, K.-U., and Kang, S.-W. (2019). Dichroic-dye-doped short pitch cholesteric liquid crystals for the application of electrically switchable smart windows. *Dyes and Pigments* 166: 403–409. https://doi.org/10.1016/j.dyepig.2019.03.045.

80 Barrios, D., Vergaz, R., Sanchez-Pena, J.M., Granqvist, C.G., and Niklasson, G.A. (2013). Toward a quantitative model for suspended particle devices: optical scattering and absorption coefficients. *Solar Energy Materials and Solar Cells* 111 (2013): 115–122. https://doi.org/10.1016/j.solmat.2012.12.012.

81 Astea, N., Leonfortea, F., and Piccolob, A. (2018). Color rendering performance of smart glazings for building applications. *Solar Energy* 176: 51–61. https://doi.org/10.1016/j.solener.2018.10.026.

82 Kelleher, M., Smartglass International (personal communication), Dublin, Ireland, smartglassinternational.com

83 Carmody, J., Selkowitz, S., Lee, E.S., Arasteh, D., and Willmert, T. (2004). *Window Systems for High Performance Buildings*. NY: W.W. Norton Co. ISBN: 0-393-73121-9.

84 Chang, I.F., Gilbert, B.L., and Sun, T.I. (1975). Electrochemichromic systems for display applications. *Journal of the Electrochemical Society* 122 (7): 955. https://doi.org/10.1149/1.2134377.

85 Notten, P.H.L. (1999). Electrochromic metal hydrides. *Current Opinion in Solid State and Materials Science* 4 (1): 5–10. https://doi.org/10.1016/S1359-0286(99)80004-7.

86 Mazur, M., Weichbrodt, W., Kapuścik, P., Domaradzki, J., and Mazur, P. (2024). Enhancement of gasochromic response to hydrogen of WO_3 thin films by post-process modification and catalyst selection. *International Journal of Hydrogen Energy* 79: 802–813. https://doi.org/10.1016/j.ijhydene.2024.07.112.

87 Khoobiar, S. (1964). Particle to particle migration of hydrogen atoms on platinum – alumina catalysts from particle to neighboring particles. *The Journal of Physical Chemistry* 68 (2): 411–412. https://doi.org/10.1021/j100784a503.

88 Lei, G., Lou, C., Liu, X., Xie, J., Li, Z., Zheng, W., and Zhang, J. (2021). Thin films of tungsten oxide materials for advanced gas sensors. *Sensors and Actuators B: Chemical* 341 (2021): 129996. https://doi.org/10.1016/j.snb.2021.129996.

89 Mirzaei, A., Kim, J.-H., Kim, H.W., and Kim, S.S. (2019). Gasochromic WO_3 nanostructures for the detection of hydrogen gas: an overview. *Applied Sciences* 9 (9): 1775. https://doi.org/10.3390/app9091775.

90 Ihokura, K. and Watson, J. (1994). *The Stannic Oxide Gas Sensor, Principles and Applications*. Boca Raton, FL: CRC Press.

91 Gao, C., Guo, X., Nie, L., Wu, X., Peng, L., and Chen, J. (2023). A review on WO_3 gasochromic film: mechanism, preparation and properties. *International Journal of Hydrogen Energy* 48 (6): 2442–2465. https://doi.org/10.1016/j.ijhydene.2022.10.100.

92 Wittwer, V. (2001). Gaschromic windows. Proceedings of Glass Processing Days, Tampere, Finland.

93 Wittwer, V., Datz, M., Ell, J., Georg, A., Graf, W., and Walze, G. (2004). Gasochromic windows. *Solar Energy Materials and Solar Cells* 84 (1–4): 305–314. https://doi.org/10.1016/j.solmat.2004.01.040.

94 Graf, W. and Rox, R. (2010). System comprising a glazing element and gas supply device, US 7774997B2, granted 2010-08-17, Assigned: Fraunhofer Institute.

95 Feng, W., Zou, L., Gao, G., Wu, G., Shen, J., and Li, W. (2016). Gasochromic smart window: optical and thermal properties, energy simulation and feasibility analysis. *Solar Energy Materials & Solar Cells* 144 (2016): 316–323. https://doi.org/10.1016/j.solmat.2015.09.029.

96 Huiberts, J.N., Griessen, R., Rector, J.H., Wijngaarden, R.J., Dekker, J.P., de Groot, D.G., and Koeman, N.J. (1996). Yttrium and lanthanum hydride films with switchable optical properties. *Nature* 380: 231–234. https://doi.org/10.1038/380231a0.

97 Yoshimura, K. (2015). Metal hydrides for smart window applications (Chapter 8). In: *Electrochromic Materials and Devices* (ed. R.J. Mortimer, D. Rosseinsky, and P.M.S. Monk). Weinheim, Germany: Wiley-VCH.

98 Yoshimura, K. and Yamada, Y. (2024). Development of gasochromic sheets using thin-film switchable mirror coatings. *Solar Energy Materials and Solar Cells* 273: 112949. https://doi.org/10.1016/j.solmat.2024.112949.

99 Yoshimura, K., Yamada, Y., and Okada, M. (2002). Optical switching of Mg-rich Mg–Ni alloy thin films. *Applied Physics Letters* 81: 4709–4711. https://doi.org/10.1063/1.1530378.

100 Tajima, K., Yamada, Y., Bao, S., Okada, M., and Yoshimura, K. (2008). High durability of clear transparency all-solid-state switchable mirror based on magnesium–titanium thin films. *Applied Physics Express* 1: 067007.

101 Yoshimura, K., Yamada, Y., Bao, S., Okada, M. (2009) Preparation and characterization of gasochromic switchable-mirror window with practical size, *Solar Energy Materials and Solar Cells*, 93(12), 2138-2142, https://doi.org/10.1016/j.solmat.2009.03.023.

102 Yamada, Y., Miura, M., Tajima, K., Okada, M., and Yoshimura, K. (2013). Optical switching durability of switchable mirrors based on magnesium-yttrium alloy thin films. *Solar Energy Materials and Solar Cells* 117: 396–399. https://doi.org/10.1016/j.solmat.2013.06.050.

103 Yamada, Y., Miura, M., Tajima, K., Okada, M., and Yoshimura, K. (2014). Influence on optical properties and switching durability by introducing Ta intermediate layer in Mg-Y switchable mirrors. *Solar Energy Materials and Solar Cells* 125: 133–137. https://doi.org/10.1016/j.solmat.2014.02.031.

104 Yamada, Y., Kitamura, S., Miura, M., and Yoshimura, K. (2015). Improving the optical properties of switchable mirrors based on Mg-Y alloy using antireflection coatings. *Solar Energy Materials and Solar Cells* 141: 337–340. https://doi.org/10.1016/j.solmat.2015.06.006.

105 Yamada, Y. and Yoshimura, K. (2020). Gas and humidity analyses in gasochromic processes of switchable mirrors. *Solar Energy Materials and Solar Cells* 233: 111389-1–111389-9. https://doi.org/10.1016/j.solmat.2021.111389.

106 Nishizawa, K., Hu, C.-W., and Yasusei, Y. (2022). Room-temperature fabrication of Pt nanoparticle-dispersed porous WO_3 gasochromic switchable films using oxalic acid. *Solar Energy Materials and Solar Cells* 245: 111891. https://doi.org/10.1016/j.solmat.2022.111891.

107 Gerasimova, T.P., Burganov, T.I., Katsyuba, S.A., Kalinin, A.A., Islamova, L.N., Fazleeva, G.M., Ahmadeev, B.S., Mustafina, A.R., Monari, A., Assfeld, X., and Sinyashin, O.G. (2021). Halochromic luminescent quinoxalinones as a basis for pH-sensing in organic and aqueous solutions. *Dyes and Pigments* 186 (2021): 108958. https://doi.org/10.1016/j.dyepig.2020.108958.

108 Theerasilp, M. and Crespy, D. (2021). Halochromic polymer nanosensors for simple visual detection of local pH in coatings. *Nano Letters* 21 (8): 3604–3610. https://doi.org/10.1021/acs.nanolett.1c00620.

109 Marini, A., Muñoz-Losa, A., Biancardi, A., and Mennucci, B. (2010). What is solvatochromism? *The Journal of Physical Chemistry B* 114 (51): 17128–17135. https://doi.org/10.1021/jp1097487.

110 Huo, R., Zhang, T., Sun, L.X., Zhao, L.N., Xing, Y.H., and Bai, F.Y. (2024). Multiple stimulus responsive of a novel naphthalenediimide-based 2D alkaline-earth metal MOF material: hydrochromism, amine-/acid-induced chromism and ionochromism as well as visual artificial film application. *Chemical Engineering Journal* 488: 150917. https://doi.org/10.1016/j.cej.2024.150917.

111 Zaromb, S. (1962). Theory and design principles of the reversible electroplating light modulator. *Journal of the Electrochemical Society* 109 (10): 903.

112 Chang, I.F. (1975). *Brown Boveri Symposium on Nonemissive Electro-optic Displays* (ed. A.R. Kmetz and F.K. Von Willisen), 155–196. NY: Plenum.

113 Camlibel, I., Singh, S., Stocker, H.J., VanUitert, L.G., and Zydzik, G.J. (1978). An experimental display structure based on reversible electrodeposition. *Applied Physics Letters* 33 (9): 793–794. 10.1063/1.90549.

114 Flanagan, W.C. and Miles, C.L. (1964). Electroplating light valve, US 3153113A, granted: 1964-10-13, assigned: Eastman Kodak.

115 Eid, B., Ghislain, A., and Themont, J.-P. (1994). Electrochrome Device, FR 2681444B1, granted 1994-09-30, Assigned: Corning Inc.

116 Eid, B.A.G. and Marque, P.J.J. (1994). Solid gel electrolytic material for modulating light, US 5332530A, granted: 1994-07-26, Assigned: Corning, Inc.

117 Warszawshi, B. (1991). Material for light modulation and manufacturing process, US 5056899A, granted:1991-01-11, Assigned Polyvision Corp.

118 Howard, B.M. and Ziegler, J.P. (1995). Optical properties of reversible electrodeposition electrochromic materials. *Solar Energy Materials and Solar Cells* 39 (2–4): 309–316. https://doi.org/10.1016/0927-0248(95)00065-8.

119 Tench, D.M. (2001). Reversible electrochemical mirror (REM) state monitoring, US 6301039B1 Granted: 2001-10-09 Assigned: Rockwell, Teledyne Scientific & Imaging LLC.

120 Tench, D.M., Cunningham, M.A., Kuo, J.J., Ruwell, P.V., and Warren, L.F. Jr. (2003). Reversible electrochemical mirror (REM) smart window. *Proceedings – Electrochemical Society* 2003-17: abst. 1294, Paris France.

121 Barile, C.J., Slotcavage, D.J., Hou, J., Strand, M.T., Hernandez, T.S., and McGehee, M.D. (2017). Dynamic windows with neutral color, high contrast, and excellent durability using reversible metal electrodeposition. *Joule* 1: 133–145. https://doi.org/10.1016/j.joule.2017.06.001.

122 Yeang, A.L., Hernandez, T.S., Strand, M.T., Slotcavage, D.J., Abraham, E., Smalyukh, I.I., Barile, C.J., and McGehee, M.D. (2022). Transparent, high-charge capacity metal mesh electrode for reversible metal electrodeposition dynamic windows with dark-state transmission <0.1%. *Advanced Energy Materials* 10.1002/aenm.202200854.

123 Islam, S.M., Hernandez, T.S., McGehee, M.D., and Barile, C.J. (2019). Hybrid dynamic windows using reversible metal electrodeposition and ion insertion. *Nature Energy* 4: 223–229. https://doi.org/10.1038/s41560-019-0332-3.

124 Tao, X.T., Liu, D., Yu, J., and Cheng, H. (2021). Reversible metal electrodeposition devices: an emerging approach to effective light modulation and thermal management. *Advanced Optical Materials* 9 (8): 2001847. https://doi.org/10.1002/adom.202001847.

125 Wöhler, F. (1825). XXXVIII. On tungsten. *Philosophical Magazine* 66 (330): 263–269. https://doi.org/10.1080/14786442508673961.

126 Hsu, S.S. (1949). The lithium tungsten bronzes. Master's Thesis. Missouri School of Mines and Metallurgy, Rolla, Missouri.

127 Brimm, E.O., Brantley, J.C., Lorenz, J.H., and Jellinek, M.H. (1951). Sodium and potassium tungsten bronzes. *Journal of the American Chemical Society* 73 (11): 5427–5432. https://doi.org/10.1021/ja01155a121.

128 Granqvist, C.G. (1995). *Handbook of Inorganic Electrochromic Materials*. Amsterdam, NL: Elsevier. ISBN: 978-0-444-89930-8.

129 Sziklai, G.C. (1953). Electrochemical color filter, US 2632045A, granted 1953-03-17, assigned RCA Corp.

130 Jones, G.D. and Friedrich, R.E. (1966). Color reversible electrochemical light filter utilizing electrolytic solution, US 3282157A, granted 1966-11-01, assigned Dow Chemical co.

131 Heller, W.R., Junji, K., Luzzi, J.J.R., Jr, and Powers, J.C. (1967). Electrochromic light valve, US 3317266A, granted: 1967-05-02, Assigned IBM.

132 Manos, P. (1969). Electrochromic device, US 3451741A, granted 1969-06-24, assigned DuPont.

133 Kissa, E., Manos, P., and Wahlig, C.F. (1969). Compartmented electrochromic device, US 3453038A granted 1969-07-01, Assigned DuPont.

134 Deb, S.K. and Shaw, R.F. (1970). Electro-optical device having variable optical density, US 3521941A, granted: 1970-07-28, Assigned: American Cyanamid, Wyeth Holdings, LLC.

135 Berets, D., Castellion, G., and Haacke, G. (1974). Electrochromic Information displays, US 3839857A, granted: 1974-10-08, Assigned: American Cyanamid, Wyeth Holdings, LLC.

136 Baucke, F.G.K., Krause, D., Metz, B, Paquet, V., and Zauner, J. (1984). Electrochromic mirrors, US 4465339A granted: 1984-08-14, Assigned: Schott AG.

137 Baucke, F.G.K. (1991). Electrochromic applications. *Materials Science and Engineering B* 10 (4): 285–292. https://doi.org/10.1016/0921-5107(91)90104-4.

138 Uchikawa, K. and Niwa, T. (1987). Dispersed iridium based complementary electrochromic device, US 4652090A Granted 1987-03-24, assigned Nikon.

139 Lampert, C.M. (1984). Electrochromic materials and devices for energy efficient windows. *Solar Energy Materials* 11 (1–2): 1–27. https://doi.org/10.1016/0165-1633(84)90024-8.

140 Becker, H. and Wittkopf, H. (1998). Variable solar control glazing-an outstanding application for electrochromics. Proceedings of the International Meeting on

Electrochromics-3, Imperial College of Science, Technology and Medicine, London, UK 8D (7–9 September 1998).

141 Hafemann, J. (2024). Lessons learned from smart glass 1.0. https://www.nextenergytech.com/post/lessons-learned-from-smart-glass-1-0 (accessed 12 October 2024).

142 View (2024). View Gen 4 IGU data sheet, Rev. 4 (April 2024). https://view.com/sites/default/files/documents/View%20Gen%204%20Datasheet%20-%20Standard_Rev4_April.pdf (accessed 27 October 2024).

143 Sbar, N.L., Podbelski, L., Yang, H.M., and Pease, B. (2012). Electrochromic dynamic windows for office buildings. *International Journal of Sustainable Built Environment* 1 (1): 125–139. https://doi.org/10.1016/j.ijsbe.2012.09.001.

144 SageGlass (2024). MKT-382 IGU Product Guide. https://www.sageglass.com/sites/default/files/2024-07/MKT-382_IGU_Product_Guide_US.pdf (accessed 27 October 2024).

145 Pennisi, A., Simone, F., and Lampert, C.M. (1992). Electrochromic properties of molybdenum oxide electrodes. *Solar Energy Materials and Solar Cells* 28: 223–247. https://doi.org/10.1016/0927-0248(92)90032-K.

146 Lampert, C.M., Omstead, T.R., and Yu, P.C. (1986). Chemical and optical properties of electrochromic nickel oxide films. *Solar Energy Materials* 14: 161–174. https://doi.org/10.1016/0165-1633(86)90043-2.

147 Yu, P.C., Nazri, G., and Lampert, C.M. (1987). Spectroscopic and electrochemical studies of electrochromic hydrated nickel oxide films. *Solar Energy Materials* 16: 1. https://doi.org/10.1016/0165-1633(87)90003-7.

148 Lampert, C.M. (1984). Electrochromic materials and devices for energy efficient windows. *Solar Energy Materials* 11 (1–2): 1–27. https://doi.org/10.1016/0165-1633(84)90024-8.

149 Lampert, C.M. (1994/2017). Science and applications of electrochromics and chromogenics. In: *Technologies for Intelligent Vehicle Highways* (ed. J.A. Burman), Proceedings of SPIE, Vol. 10318 (Vol. TTS 2), 1031802 (1994), reprinted 2017. SPIE. https://doi.org/10.1117/12.2284057.

150 Granqvist, C.G. (2014). Oxide-based chromogenic coatings and devices for energy efficient fenestration: brief survey and update on thermochromics and electrochromics. *Journal of Vacuum Science and Technology* B 32: 060801/1–060801/13. https://doi.org/10.1116/1.4896489.

151 Granqvist, C.G. (2014). Electrochromics for smart windows: oxide-based thin films and devices. *Thin Solid Films* 564: 1–38. https://doi.org/10.1016/j.tsf.2014.02.002.

152 Mortimer, R.J., Rosseinsky, D.R., and Monk, P.M.S. (ed.) (2015). *Electrochromic Materials and Devices*. Weinheim, Germany: Wiley-VCH. ISBN: 978-3-527-33610-4.

153 Monk, P.M.S., Mortimer, R.J., and Rosseinsky, D.R. (2007). *Electrochromism and Electrochromic Devices*. Cambridge, UK: Cambridge University Press.

154 Byker, H.J. (2015). Solution-phase electrochromic devices and systems. In: *Electrochromic Materials and Devices* (ed. R.J. Mortimer, D.R. Rosseinsky, and P.M.S. Monk), 401–418. Weinheim, Germany: Wiley-VCH.

155 Chandrasekhar, P., Zay, B.J., Cai, C., Chai, Y., and Lawrence, D. (2014). Matched-dual-polymer electrochromic lenses, using new cathodically coloring conducting polymers, with exceptional performance and incorporated into automated sunglasses. *Journal of Applied Polymer Science* 131: 41043. https://doi.org/10.1002/app.41043.

156 Bulloch, R.H., Kerszulis, J.A., Dyer, A.L., and Reynolds, J.R. (2015). 2015, An electrochromic painter's palette: color mixing via solution co-processing. *ACS Applied Materials & Interfaces* 7 (3): 1406–1412. https://doi.org/10.1021/am507514z.

157 Dyer, A.L., Österholm, A.M., Shen, D.E., Johnson, K.E., and Reynolds, J.R. (2015, 2015). Conjugated electrochromic polymers: structure-driven color and processing control (Chapter 5). In: *Electrochromic Materials and Devices* (ed. R.J. Mortimer, D.R. Rosseinsky, and P.M.S. Monk). Weinheim, Germany: Wiley-VCH. ISBN: 9783527336104.

158 Dyer, A.L., Thompson, E.J., and Reynolds, J.R. (2011). Completing the color palette with spray-processable polymer electrochromics. *ACS Applied Materials & Interfaces* 3 (6): 1787–1795. https://doi.org/10.1021/am200040p.

159 Amb, C.M., Dyer, A.L., and Reynolds, J.R. (2011). Navigating the color palette of solution -processable electrochromic polymers. *Chemistry of Materials* 23 (3): 397–415. https://doi.org/10.1021/cm1021245.

160 Argun, A.A., Aubert, P.H., Thompson, B.C., Schwendeman, I., Gaupp, C.L., Hwang, J., Pinto, N.J., Tanner, D.B., MacDiarmid, A.G., and Reynolds, J.R. (2004). Multi-colored electrochromism in polymers: structures and devices. *Chemistry of Materials* 16: 4401–4412. https://doi.org/10.1021/cm049669l.

161 Österholm, A.M., Shen, D.E., Kerszulis, J.A., Bulloch, R.H., Kuepfert, M., Dyer, A.D., and Reynolds, J.R. (2015). Four shades of brown: tuning of electrochromic polymer blends toward high-contrast eyewear. *ACS Applied Materials & Interfaces* 7 (3): 1413–1421. https://doi.org/10.1021/am507063d.

162 Fritsche, M. (1867). Photochromism in tetracene. *Comptes Rendus* 69: 1035.

163 Phipson, T.L. (1881). On a curious actinic phenomenon. *Chemical News and Journal of Physical Science* 43 (1126): 283, 1881. Available at: https://upload.wikimedia.org/wikipedia/commons/b/b5/The_chemical_news._Volume_44%2C_July_-_December_1881._%28IA_s713id13691460%29.pdf (accessed 5 May 2025).

164 Marckwald, W. (1899). Über phototropie. *Zeitschrift für Physikalische Chemie* 30: 1.

165 Bouas-Laurent, H., Dürr, H. (2001). Organic photochromism (IUPAC Technical Report). *Pure and Applied Chemistry* 73(4):639–665. https://doi.org/10.1351/pac200173040639.

166 Durr, H. and Bouas-Laurant, H. (ed.) (2003). *Photochromism-Molecules and Systems*. NY: Elsevier.

167 Dorian, G. and Wiebe, A. (1970). *Photochromism: Optical and Photographic Applications*. San Diego: Academic Press.

168 Fries, K., Fink-Straube, C., Mennig, M., and Schmidt, H. (2004). Photochromic coatings. In: *Sol-Gel Technologies for Glass Producers and Users* (ed. M.A. Aegerter and M. Mennig). Boston, MA: Springer. https://doi.org/10.1007/978-0-387-88953-5_33 (also available: Research Gate, 2011, accessed July 2023).

169 Stookey, S.D. (1969). Photochromic glass making, US 3449103A, granted 1969-06-10, Assigned Corning Glass Works.

170 Araujo, R.J. (1967). Phototropic glass article and method of making it, US 3325299A Granted: 1967-06-13, Assigned: Corning Glass Works.

171 Araujo, R.J. (1968). Kinetics of bleaching of photochromic glass. *Applied Optics* 7 (5): 781–786. https://doi.org/10.1364/AO.7.000781.

172 Glass sent by R.J. Araujo to author's lab for testing at LBNL – University of California, Berkeley.

173 Compagno, A. (1995). *Intelligent Glass Facades*. Basel, Switzerland: Birkhauser Verlag ISBN-13. 978-3764355470.

174 Moon, B.-Y., Kim, S.-Y., and Yu, D.-S. (2020). Differences in the optical properties of photochromic lenses between cold and warm temperatures. *PLoS One* 15 (5): e0234066. https://doi.org/10.1371/journal.pone.0234066.

175 Crano, J.C. and Guglielmetti, R.J. (1999). *Organic Photochromic and Thermochromic Compounds Volume 1: Main Photochromic Families*, 235–236. New York: Kluwer Academic Publishers. ISBN: 978-0-306-45882-8.

176 What's That Stuff (2009). Self-darkening eyeglasses, the science behind dual-purpose lenses. *Chemical & Engineering News* 87 (15): 54. https://cen.acs.org/articles/87/i15/Self-Darkening-Eyeglasses.html.

177 Khaled, K. and Berardi, U. (2023). A cross-climate assessment of the visual and energy performance of flexible photochromic films for in-situ window retrofits. *Energy and Buildings* 300: 113664. https://doi.org/10.1016/j.enbuild.2023.113664.

178 Hočevar, M. and Krašovec, U.O. (2018). A photochromic single glass pane. *Solar Energy Materials and Solar Cells* 186: 111–114. https://doi.org/10.1016/j.solmat.2018.06.035.

179 Badour, Y., Jubera, V., Andron, I., Frayret, C., and Gaudon, M. (2021). Photochromism in inorganic crystallised compounds. *Optical Materials: X* 12: 100110. https://doi.org/10.1016/j.omx.2021.100110.

180 Ru, Y., Shi, Z., Zhang, J., Wang, J., Chen, B., Huang, R., Liu, G., and Yu, T. (2021). Recent progress of photochromic materials towards photocontrollable devices. *Materials Chemistry Frontiers* 5 (21): 7737–7758. https://doi.org/10.1039/D1QM00790D.

181 Du, J., Yang, Z., Lin, H., and Poelman, D. (2024). Inorganic photochromic materials: recent advances, mechanism, and emerging applications. *Responsive Materials* 2: e20240004. https://doi.org/10.1002/rpm.20240004.

182 Wang, Z., Ding, Z., Yang, Y., Hu, L., Wu, W., Gao, Y., Wei, Y., Zhang, X., and Jiang, G. (2023). Time-resolved encryption via photochromic and mechanochromic system based on silane-substituted spiropyran. *Chemical Engineering Journal* 457: 141293. https://doi.org/10.1016/j.cej.2023.141293.

183 Pinnock, H.T. (1919). Chemical compounds for the detection of overheated machinery and bearings. *Journal of the Society of Chemical Industry* 38.

184 Mott, N.F. (1990). *Metal–Insulator Transition*, 2e. London: CRC Press. ISBN: 9780429094880 10.1201/b12795.

185 Fritzsche, H. and Adler, D. (ed.) (1985). *Localization and Metal-Insulator Transition978-1-4612-9521-1*. NY, https://doi.org/10.1007/978-1-4613-2517-8: Plenum Press.

186 Qiu, J.-W., Feng, Y.-D., and Wu, C.H. (2022). *Thermal Control Thin Films- Spacecraft Technology*. Berlin/Boston: Walter De Gruyter, GmbH. ISBN: 978-3-11-061286-8.

187 Jorgenson, G.V. and Lee, J.C. (1986). Doped Vanadium oxide for optical switching films. *Solar Energy Materials* 14 (3–5): 205–214. https://doi.org/10.1016/0165-1633(86)90047-X.

188 Jin, P., Yoshimura, K., Iwama, S., and Tanemura, S. (1995). Formation of V_{1-x} W_x0_2 thermochromic films by reactive magnetron sputtering with an alloy target. In: *Optical Materials Technology for Energy Efficiency and Solar Energy Conversion XIV*, vol. 2531 (ed. C.M. Lampert, S.K. Deb, and C.G. Granqvist) Proceedings of SPIE, 51–59. SPIE. https://doi.org/10.1117/12.217358.

189 Granqvist, C.G. (2014). Oxide-based chromogenic coatings and devices for energy efficient fenestration: brief survey and update on thermochromics and electrochromics. *Journal of Vacuum Science and Technology B* 32: 060801/1–060801/13. https://doi.org/10.1116/1.4896489.

190 Granqvist, C.G. (2016). Recent progress in thermochromics and electrochromics: a brief survey. *Thin Solid Films* 614: 90–96. https://doi.org/10.1016/j.tsf.2016.02.029.

191 Granqvist, C.G., Pehlivan, İ.B., Ji, Y.-X., Li, S.-Y., and Niklasson, G.A. (2014). Electrochromics and thermochromics for energy efficient fenestration: Functionalities based on nanoparticles of In_2O_3:Sn and VO_2. *Thin Solid Films* 559: 2–8. https://doi.org/10.1016/j.tsf.2013.10.033.

192 Liang, S., Xu, F., Li, W., Yang, W., Cheng, S., Yang, H., Chen, J., Yi, Z., and Jiang, P. (2023). Tunable smart mid infrared thermal control emitter based on phase change material VO_2 thin film. *Applied Thermal Engineering* 232: 121074. https://doi.org/10.1016/j.applthermaleng.2023.121074.

193 Seeboth, A. and Lötzsch, D. (2014). *Thermochromic and Thermotropic Materials.* Singapore: Pan Stanford Publishing. ISBN: 978-981-4411-02-8.

194 Seeboth, A., Ruhmann, R., and Mühling, O. (2010). Thermotropic and thermochromic polymer based materials for adaptive solar control. *Materials (Basel).* 3 (12): 5143–5168. https://doi.org/10.3390/ma3125143.

195 Seeboth, A., Schneider, J., and Patzak, A. (2000). Materials for intelligent sun protecting glazing. *Solar Energy Materials and Solar Cells* 60 (3): 263–277. https://doi.org/10.1016/S0927-0248(99)00087-2.

196 Fukada, Y. (2010). *Inorganic Chromotropism: Basic Concepts and Applications of Colored Materials.* Tokyo: Springer & Kodansha. ISBN: 9783642091414.

197 Sone, S. and Fukuda, Y. (1987). *Inorganic Thermochromism.* Berlin: Springer-Verlag. ISBN: 3-540-17662-4.

198 Day, J. and H. (1963). Thermochromism. *Chemical Reviews* 63 (1): 65–80. https://doi.org/10.1021/cr60221a005.

199 Greenberg, C.B. (2000). Chromogenic materials, thermochromic. In: *Kirk-Othmer Encyclopedia of Chemical Technology.* https://doi.org/10.1002/0471238961.2008051807180505.a01.

200 Crano, J.C. and Guglielmetti, R.J. (ed.) (2002). *Organic Photochromic and Thermochromic Compounds. Topics in Applied Chemistry.* Boston, MA. https://doi.org/10.1007/0-306-46912-X_11: Springer. ISBN: 978-0-306-45883-5.

201 Kamalisarvestani, M., Saidur, R., Mekhilef, S., and Javadi, F.S. (2013). Performance, materials and coating technologies of thermochromic thin films on smart windows. *Renewable and Sustainable Energy Reviews* 26: 353–364. https://doi.org/10.1016/j.rser.2013.05.038.

202 Aburas, M., Soebarto, V., Williamson, T., Liang, R., Ebendorff-Heidepriem, H., and Wu, Y. (2019). Thermochromic smart window technologies for building application: a review. *Applied Energy* 255: 113522. https://doi.org/10.1016/j.apenergy.2019.113522.

203 Hu, C., Ge, L., and Meng, X. (2024). Applications of thermochromic material in opaque building envelopes: a comprehensive review. *Materials Today Sustainability* 28: 101009, 2589-2347, https://doi.org/10.1016/j.mtsust.2024.101009.

204 Wilson, H.R. (1994). Potential of thermotropic layers to prevent overheating: a review. *Proceedings of SPIE* 2255: 214. https://doi.org/10.1117/12.185372.

205 Wilson, H.R., Raicu, A., Nitz, P., and Ferber, J. (1995). Thermotropic glazing. Proceedings of Windows Innovation Conference, Toronto, Canada, 5–6 June 1995, 489. ISBN 0-660-16085-4

206 Watenabe, H. (1998). Intelligent window using a hydrogel layer for energy efficiency. *Solar Energy Materials and Solar Cells* 54: 203. https://doi.org/10.1016/S0927-0248(98)00072-5.

207 Raicu, A., Wilson, H.R., Nitz, P., Platzer, W., Wittwer, V., and Jahns, E. (2002). Facade systems with variable solar control using thermotropic polymer blends. *Solar Energy* 72 (1): 31–42. https://doi.org/10.1016/S0038-092X(01)00093-7.

208 Nitz, P. and Hartwig, H. (2005). Solar control with thermotropic layers. *Solar Energy* 79 (6): 573–582. https://doi.org/10.1016/j.solener.2004.12.009.

209 Inoue, T., Ichinose, M., and Ichikawa, N. (2008). Thermotropic glass with active dimming control for solar shading and daylighting. *Energy and Buildings* 40 (3): 385–393. https://doi.org/10.1016/j.enbuild.2007.03.006.

210 Fraunhofer-IAP /TILSE (2019). SOLARDIM®-ECO Self-regulating solar protection glazing. https//iap.fraunhofer.de>iap>documents.pdf (accessed 20 October 2024).

211 Kang, D.-C., Wang, T.-Y., Lin, D., Cheng, Y.-S., and Huang, C.-W. (2025). PDMS with porous PMMA dual-layer coating for passive daytime radiative cooling. *Solar Energy Materials and Solar Cells* 282: 113380. https://doi.org/10.1016/j.solmat.2024.113380.

212 Raman, A., Anoma, M., Zhu, L., Rephaeli, E., and Fan, S. (2014). Passive radiative cooling below ambient air temperature under direct sunlight. *Nature* 515: 540–544. https://doi.org/10.1038/nature13883.

213 Zhu, L., Raman, A.P., and Fan, S. (2015). Radiative cooling of solar absorbers using a visibly transparent photonic crystal thermal blackbody. *Proceedings. National Academy of Sciences. United States of America* 112 (40): 12282–12287. https://doi.org/10.1073/pnas.1509453112.

214 Chen, Q., Huang, T., Cheng, J., Zhang, J., Huang, X., Xu, H., Lu, Y., and Song, W. (2023). Translucent-Colored radiative coolers based on localized surface plasmon resonances for Energy-Efficient windows. *Solar Energy* 253: 472–479. https://doi.org/10.1016/j.solener.2023.02.046.

215 Zhao, B., Hu, M., Ao, X., Chen, N., and Pei, G. (2019). Radiative cooling: a review of fundamentals, materials, applications, and prospects. *Applied Energy* 236: 489–513. https://doi.org/10.1016/j.apenergy.2018.12.018.

216 Topart, P. and Hourquebie, P. (1999). Infrared switching electroemissive devices based on highly conducting polymers. *Thin Solid Films* 352 (1–2): 243–248. https://doi.org/10.1016/S0040-6090(99)00401-0.

217 Brooke, R., Mitraka, E., Sardar, S., Sandberg, M., Sawatdee, A., Berggren, M., Crispin, X., and Jonsson, M.P. (2017). Infrared electrochromic conducting polymer devices. *Journal of Materials Chemistry C* 5 (23): 5824–5830. https://doi.org/10.1039/C7TC00257B.

218 Shen, D.E., Goins, C.L., Jones, A.L., Österholm, A.M., and Reynolds, J.R. (2024). Design rules for high contrast mid-infrared electrochromism in conjugated polymers. *ACS Materials Letters* 6 (2): 528–534. https://doi.org/10.1021/acsmaterialslett.3c01437.

219 Family, R. and Mengüç, M.P. (2017). Materials for radiative cooling: a review. *Procedia Environmental Sciences* 38: 752–759. https://doi.org/10.1016/j.proenv.2017.03.158.

220 Dong, K. and Wu, J. (2023). Radiative cooling, what's next? *Next Energy* 1 (1): 100003. 10.1016/j.nxener.2023.100003.

221 Liu, R., Wang, S., Zhou, Z., Zhang, K., Wang, G., Chen, C., and Long, Y. (2025). Materials in radiative cooling technologies. *Advanced Materials* 37: 2401577. https://doi.org/10.1002/adma.202401577.

222 Lampert, C.M. and Granqvist, C.G. (ed.) (1990). *Large-Area Chromogenics: Materials and Devices for Transmittance Control*. Bellingham, WA: SPIE Optical Engineering Press. https://doi.org/10.1117/12.2283604.

223 Lampert, C.M. (2005). Advances in smart materials technologies-windows. In: *Proceedings of the Society of Vacuum Coaters*, vol. 48, 675–677. Albuquerque, NM: Society of Vacuum Coaters.

224 Lampert, C.M. (2007). Smart materials-dynamically switchable windows and displays. In: *50 Years of Vacuum Coating Technology and the Growth of the Society of Vacuum Coaters* (ed. D.M. Mattox). Albuquerque, NM: Society of Vacuum Coaters.

225 Jelle, B.P., Hynd, A., Gustavsen, A., Arasteh, D., Goudey, H., and Har, R. (2012). Fenestration of today and tomorrow: a state-of-the-art review and future research opportunities. *Solar Energy Materials and Solar Cells* 96: 1–28. https://doi.org/10.1016/j.solmat.2011.08.010.

226 Rezaei, S.D., Shannigrahi, S., and Ramakrishna, S. (2017). A review of conventional, advanced, and smart glazing technologies and materials for improving indoor environment. *Solar Energy Materials and Solar Cells* 159: 26–51.

227 Casini, M. (2016). *Smart Buildings: Advanced Materials and Nanotechnology to Improve Energy-efficiency and Environmental Performance*. Cambridge: Woodhead Publishing Limited.

228 Casini, M. (2018). Active dynamic windows for buildings. *Renewable Energy* 119: 923–934. https://doi.org/10.1016/j.renene.2017.12.049.

229 Sasmal, A.K. and Pal, T. (2021). Chromism of chemical compounds. *Journal of the Indian Chemical Society* 98 (5): 100073. https://doi.org/10.1016/j.jics.2021.100073.

230 Tällberg, R., Jelle, B.P., Loonen, R., Gao, T., and Hamdy, M. (2019). Comparison of the energy saving potential of adaptive and controllable smart windows: a state-of-the-art review and simulation studies of thermochromic, photochromic and electrochromic technologies. *Solar Energy Materials and Solar Cells* 200: 109828. https://doi.org/10.1016/j.solmat.2019.02.041.

231 Zakirullin, R.S. (2022). Smart window with grating optical filter: comparison with smart windows fully coated with chromogenic layer. *Building and Environment* 219: 109258. https://doi.org/10.1016/j.buildenv.2022.109258.

1B

Introduction to Glazing Design and Measurements

Carl M. Lampert[1,2,3]

[1] *Star Science, Cotati, CA, USA*
[2] *Formerly, Technical Director, Society of Vacuum Coaters (non-profit), Albuquerque, NM, USA*
[3] *Staff Scientist, University of California, LBNL, Energy Technologies, Berkeley, CA, USA*

Nomenclature

AU_λ	absorbance, λ
A_λ	absorptance, λ
a^*	chromaticity coordinates (red-green)
A_{cg}	center of glass area
A_{eg}	edge of glass area
A_f	frame area
AL	air leakage rate
AM	air mass (relative thickness of the Earth's Atmosphere)
A_{pf}	total projected area of fenestration
ASTM	ASTM International (testing), (W. Conshohoken, PA)
BRDF	bidirectional reflectance distribution function
BTDF	bidirectional transmittance distribution function
BSDF	bidirectional scattering distribution function
b^*	chromaticity coordinates (yellow-blue)
C	conversion constant 60 min h^{-1}
CEN	European Commission for Standardization (Brussels, Belgium)
CIE	Commission Internationale de l'Eclairage (Vienna, Austria)
CIELAB	CIE, $L^*a^*b^*$ color space (human perception of color and hue)
C^*	chromaticity
C_p	specific heat of air
D50	luminant at 5000 K
D65	luminant at 6500 K
DGI	daylight glare index
DGP	daylight glare probability
DOE	U.S. Department of Energy (Washington, D.C.)
D/M/D	interference film, dielectric/metal/dielectric
$d\lambda$	incremental wavelength interval
ε_t	thermal emissivity (pure materials)

Chromogenics: Smart Switchable Optical Materials and Their Applications, First Edition. Edited by Carl M. Lampert.

ε_{th}	thermal hemispherical emissivity (pure materials)
E_t	thermal emittance
ε_n	thermal normal emissivity
FOM_H	Haacke figure of merit for a TCO
$FOM_{H\text{-}HR}$	Haacke high-resolution figure of merit for a TCO
G	global solar radiation (direct + diffuse)
H	haze
I_t	incident total solar irradiance, W m^{-2}
I	solar irradiance, Solar constant, W m^{-2} nm^{-1}
IEC	International Electrotechnical Commission (Geneva, Switzerland)
ISE	Fraunhofer Institute for Solar Energy Systems (Freiburg, Germany)
ISO	International Standards Organization (Venier, Switzerland)
IR	infrared
IEA	International Energy Agency
IGU	insulating glass unit (double glazing)
k	thermal conductivity of fill gas W m^{-1} K^{-1}
LBNL	Lawrence Berkeley National Lab. (Berkeley, CA)
L^*	lightness index (CIELAB color space)
NFRC	National Fenestration Ratings Council
Low-e	low emittance (coating)
NIR	near-infrared
NIST	National Institute of Standards and Technology (Bethesda, MD)
NPL	National Physical Laboratory (UK)
NREL	National Renewable Energy Lab. (Golden, CO) (formerly SERI)
OD_λ	optical density, λ
PDLC	polymer dispersed liquid crystal
Q_r	radiative flux
Q	charge (coulombs)
Q_i	instantaneous energy
R	reflectance
R_s	sheet resistance
SHGC	solar heat gain coefficient
SPD	suspended particle device
$\varphi_{sol}(\lambda)$	spectral solar irradiance
$\varphi_{vis}(\lambda)$	spectral solar visible irradiance
$\varphi_p(\lambda)$	photopic spectrum (daytime human eye response)
$\varphi_s(\lambda)$	scotopic spectrum (nighttime human eye response)
ρ	density of air (1.293 kg m^{-3} at 273 K and pressure of 101.325 kPa)
σ	Stefan–Boltzmann constant (5.6697×10^{-8} W m^{-2} K^{-4})
SERI	Solar Energy Research Institute, now NREL
T	transmittance
T_a	average visible transmittance (mathematical average, not weighted)
T_c, T_b	transmittance colored, transmittance bleached
TCO	transparent conductive oxide
t_{out}	exterior air temperature, °C
t_{in}	interior air temperature, °C
$T(\lambda)$, T_λ	monochromatic transmittance at wavelength λ

Te(λ)	average solar transmittance between equal energy intervals
T_{lum}	luminous transmittance (visible)
T_{sol}	solar transmittance
T_{uv}	ultraviolet transmittance
T_{vis}	visible transmittance
U	coefficient of heat transfer (U-factor), W m^{-2} K^{-1}
U_{cg}	U-factor of the center of glass
U_{eg}	U-factor edge of glass
U_{f}	U-factor of frame/edge
U_{overall}	U-factor overall (includes all window elements)
UV	ultraviolet, spectral regions: UVA, UVB, UVC
VIG	vacuum insulated glass (vacuum window)
Vis, vis	visible light
VLT, VT	visible light transmittance
λ	wavelength
x,y,z	color coordinates – tristimulus values
X,Y,Z	nonphysical primaries, color coordinates

1B.1 Common Metrics for Glazing

It is important to discuss measurements on chromogenic materials, since the measurements are so important to evaluate light and energy flow through a glazing and fenestrations in general. The performance of a chromogenic glazing includes the performance of its overall construction and installation. The construction includes multiple panes, types of glass or plastic, gas fills, low-e coatings, sealing, and framing materials. The final assembly dictates the overall performance of a glazing. An energy performance label shows the energy saving properties to the consumer. The best performing glazing can be rated as "Energy Star" (in the U.S.) or similar rating which depends on its optical and thermal measurements [1]. Energy Star has been very successful for a wide range of appliances. Utility and government bodies use such ratings for incentives and rebates on best performing glazing such as chromogenic glazing [2]. Proper measurements are important to determine energy and visual performance. In many cases we are dealing with view windows where color, ease of switching, and durability are extremely important to the customer who will be paying a premium for the glazing. Several common measurements used for solar materials and glazing materials are detailed in this chapter. Also, proper measurements of scatter and haze are important to characterize specific chromogenic windows, such as polymer-dispersed liquid crystals and thermotropics. Further measurement techniques can be found in other works [3]. During the development of electrochromic glazing among other switchable media, the working group of electrochromic researchers formulated testing procedures for the Solar Heating and Cooling (SHC) Program of the International Energy Agency (IEA) Task 10, involving participants from 12 countries [4]. These testing procedures were later published by NREL (SERI) [5]. Beyond standard glazing, energy consumption and switching performance are important to other chromogenic products, such as smart labels and color changing or emittance changing surface treatments. Some measurements specific to electrochromics such as coloration efficiency and switching time are discussed in the Chapters 2–4 on electrochromics.

1B.2 Solar Radiation and the Earth's Atmosphere

Solar radiation can be approximated by a blackbody radiation curve with a spectrum corresponding to the sun's surface temperature (about 5778 K) as shown in Figure 1B.1. Blackbodies represent ideal energy emitters with an energy peak corresponding to its temperature. Fortunately, this radiation is attenuated by ozone (O_3), oxygen (O_2), water vapor (H_2O), carbon dioxide (CO_2), and particulate scattering in our atmosphere. The most significant is the absorption of high-energy ultraviolet (UV) by ozone. The UV spectra is divided into three bands: UVC (100–280 nm), UVB (280–315 nm), and UVA (315–380 nm). Fortunately, only about 1.3% of the solar UV reaches the surface of the Earth. But even with a low percentage, UV damage through a glazing is an important factor in the color fading and embrittlement of interior furnishings and art objects. A double pane of glass reduces overall UV transmission, mostly some UVB but not UVA. The majority of solar radiation extends approximately over the wavelength range 320–2500 nm and has a peak intensity at 550 nm (green light). The solar spectrum is divided into four important spectral regions: UV (100–380 nm), visible (380–750 nm) or luminous, near-infrared (NIR) (750–2500 nm), and infrared (IR) (2500–6000 nm). Visible energy and near-infrared account for about 44% and 55%, respectively, of the solar energy. The actual amounts vary considerably with the weather and atmospheric conditions. Note that in the literature one will see some variation of these wavelength and energy numbers.

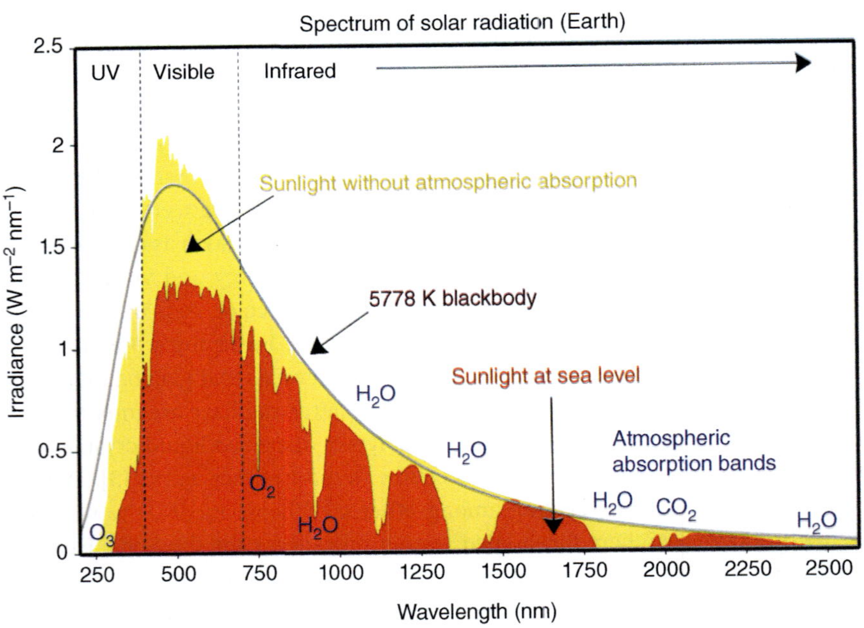

Figure 1B.1 Spectral solar irradiance, $\varphi_{sol}(\lambda)$, showing AM 0 (without atmospheric scattering and absorption) in yellow and AM 1.5D (direct solar energy impinging the Earth at a 48.2° angle to normal) in red. The AM 0 spectra is modeled by a 5778 K blackbody distribution. The spectral range is from 200 to 4000 nm which includes the UV, visible, and infrared spectra. Atmospheric absorptions are shown for a clear day. *Source:* Robert A. Rohde / CC BY 3.0 Unported License.

The amount of energy impinging on a surface can be measured with a pyranometer (total radiation) and a pyrheliometer (direct beam radiation). Many researchers use solar radiation values normal to the sun or at angles from 0° to 60° from normal. For buildings with vertical glazing, much higher angles better typify the intensity of radiation passing through a window. Air mass (AM) is defined as the thickness of the atmosphere that the solar radiation passes through. AM 1.0 is defined as "solar noon" when the sun is directly overhead (at the zenith). Commonly, AM 1.5 and to a lesser degree AM 2.0 are used for energy calculations. Total solar radiation is composed of both direct and diffuse radiation. During cloudy days the direct component of solar is reduced and the diffuse component is increased, as expected. Direct solar is used in many glazing energy calculations. Spectral solar irradiance/nm ($\varphi_{sol}(\lambda)$) and the solar constant (I) data are available from NREL for AM 0 [6] and AM 1.5G (global), 1.5D (direct) [7]. NREL also offers a solar radiation data manual for buildings at typical US locations [8]. Furthermore, NASA has solar spectral data for a range of AMs [9].

The effect of the Earth's AM is shown in Figure 1B.2. At AM 0, the energy above the Earth's atmosphere, using ASTM standard spectra ASTM E490-00a (2019), the integrated solar irradiance or solar constant is $I = 1366.1\ \text{W m}^{-2}$ [11]. AM 1.5G is commonly used to evaluate solar performance (ASTM standard G173-03, and International Electrotechnical Commission, IEC standard 60904). AM 1.5G, (global horizontal radiation from the sun) the integrated solar irradiance for AM 1.5G is standardized to $I = 1000\ \text{W m}^{-2}$ [12]. Solar simulators are classified based on their spectral match to the AM 1.5G spectrum. At AM 1.5 the angle from the Zenith is 48.2°. At 90° from the Zenith, just at sunrise or AM 40, $I = 44\ \text{W m}^{-2}$, see Figure 1B.2. The effect of AM on the attenuation of radiation is shown in Table 1B.1.

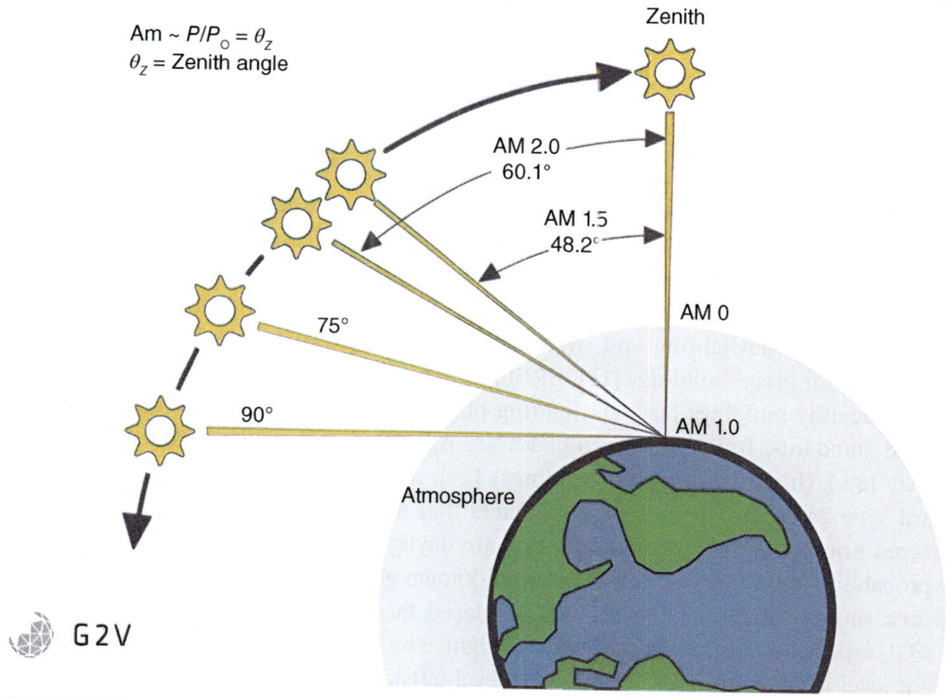

Figure 1B.2 The effect of air mass (atmospheric thickness) on solar radiation. *Source:* [12] / Reprinted with permission of G2V Optics Inc.

Table 1B.1 Calculated solar irradiation (I) at sea level for various air masses (AM) and wavelength intervals, retabulated data after Moon [13].

	$I = \mathrm{W\ m^{-2}}$					
Wavelength (nm)	**AM = 0**	**AM = 1.0**	**AM = 2.0**	**AM = 3.0**	**AM = 4.0**	**AM = 5.0**
290–400 nm	94.6	40.1	19.8	10.0	5.4	2.7
400–700 nm	540.0	419.7	327.8	258.6	205.8	163.7
700–1100 nm	365.4	309.2	267.5	233.4	205.1	181.5
1100–1500 nm	162.5	95.3	70.4	57.0	48.1	40.7
1500–1900 nm	72.8	50.8	45.1	41.0	38.0	35.2
1900-∞[a)]	86.8	12.8	9.2	7.5	6.5	5.8
Total ($\mathrm{W\ m^{-2}}$)	1322	927.9	739.8	607.5	508.9	429.6

a) Effectively the longwave radiation from the sun is very small beyond 3000 nm. In the far-infrared (3,000–100,000 nm), the predominate radiation is from the Earth, clouds, and structures on Earth.

1B.2.1 Clearness Index

The clearness index is a dimensionless quantity and can vary from 0 to 1. The clearness index is a measure of atmospheric clarity. The index is the fraction of total solar radiation on the surface of the Earth during a certain period divided by the theoretical maximum radiation during the same period. There is work on clearness index forecasting [14].

1B.2.2 Daylighting

The use of natural light to provide lighting for a room or structure has been studied in great detail and can be very important in architectural design [15]. Various types of light pipes, fluorescent panels, and other schemes have been used to bring natural light from the outside to indoors. Usually, daylighting results in a productive space, as long as it is not over lit. However, in modern buildings it is sometimes necessary to provide artificial lighting to give the work space the proper illumination. Building designers need to consider several daylighting and overall lighting metrics for building rating systems such as LEED for green buildings [16], BREEAM for sustainability performance [17], and WELL for healthy buildings [18]. Daylighting plays an important role in building energy codes and standards. Important green codes are ASHRAE 90.1 (energy efficient design), ASHRAE 189.1 (high-performance buildings) [19], and the California energy efficiency standard Title 24. There are various parameters that define a daylit space and these will be covered briefly here. Two common terms are daylight glare index (DGI) and daylight glare probability (DGP). DGI and DGP are well-known glare indices that are used to assess large-area sources of daylight glare such as glazed façades. The DGP is more accurate than DGI, especially in bright scenes with daylight. The daylight factor (DF) is the ratio of the light level inside a structure to the light level outside the structure. UDI is the useful daylight illuminance. UDI is the annual occurrence of illuminances that is within a useable range for occupants. Further information can be found in excellent reference texts on daylighting [20–22].

1B.3 Transmittance, Reflectance, and Absorptance Measurements

Before we design a high-performance glazing, it is best to cover the basic optical and thermal measurements of materials. Transmittance (T_λ), reflectance (R_λ), and absorptance (A_λ) measurements are usually made using a spectrophotometer or a scanning fiber-optic spectrometer and are a function of wavelength (λ). For any material at specified wavelength (λ):

$$T_\lambda + R_\lambda + A_\lambda = 1 \tag{1B.1}$$

National Physical Laboratory (NPL) or National Institute of Standards and Technology (NIST) or other national standard organization's materials are used for calibration of a spectrophotometer. The standard spectrophotometer makes spectral measurements of specular materials (for non-scattering media) [23, 24]. Commonly, the measurement samples for chromogenic materials are made either as a material on a substrate or as a sandwich between two sheets of glass or flexible plastic and may include transparent electrodes.

1B.3.1 Solar Transmittance

When spectral transmittance is measured on a material or coating, the solar transmittance (T_{sol}) is determined by a solar spectrum weighting function. The solar spectrum spans principally from 300 to 2500 nm wavelength. Since the solar spectrum is not a constant value over energy or wavelength, the measured transmittance spectrum must be weighted by the solar spectral function to represent an accurate value to characterize the transmission of solar energy. This integration is shown in Eq. (1B.2):

$$T_{\text{sol}} = \frac{\int_{300\text{ nm}}^{2500\text{ nm}} T(\lambda)\, \varphi_{\text{sol}}(\lambda)\, d\lambda}{\int_{300\text{ nm}}^{2500\text{ nm}} \varphi_{\text{sol}}(\lambda)\, d\lambda} \tag{1B.2}$$

where

$T(\lambda)$ is the solar spectral transmittance,
$\varphi_{\text{sol}}(\lambda)$ is the solar irradiance spectrum, AM 1.5 is used frequently for building glazing [11], and
$d\lambda$ is the incremental wavelength interval.

Analogous formulas apply to reflectance (R) and absorptance (A), where $R_\lambda + A_\lambda = 1 - T_\lambda$. An approximation of the integral is the summation using incremental wavelength intervals, $\Delta\lambda$ can also be used [3, 25].

1B.3.2 Luminous, Visible Transmittance, and Visible Light Transmittance (VLT)

In the solar field, we use the visible portion of the solar spectrum as our illuminant and the photopic or daylight human eye response is used for the weighting function. Usually the solar spectrum AM 1.5 direct is used for most calculations. However, AM 1.0 and AM 2.0 are used for specialized calculations. In other fields (such as lighting and display engineering) and for standard color measurement, other illuminates are used. These illuminants include illuminants A, B, C, D, D50, and D65.Illuminate D50 simulates midday daylight with a color temperature of 5000 K. Illuminant D65 simulates average daylight at 6500 K (visible plus UV to 300 nm). In comparison, the sun's temperature is 5772 K. Other optical measurements may be made at a single frequency such as the peak of the visible 550 nm (transmittance is designated as T_{550}). Measurements of T_{vis} or T_{lum} are spectral transmittance values weighted by the standard spectra, such as an illuminant or the human eye response. Another term that is used in the glass industry for visible light transmittance is VLT or just VT and is equivalent to T_{vis} and T_{lum}. Also, the ultraviolet response T_{uv} is important to eye and skin health. It can also control the fading of interior-colored furnishings, such as carpet, paint, upholstery, and art objects. UV transmission is important for aircraft glazing, which are usually designed to block the UV, as it is much more intense at higher altitudes. The intensity of UV energy has been increasing over the airline polar route due to the thinning of the ozone layer. PPG has announced a new airplane transparency called Soloron blue protection UV+ to block about 99% UV. PPG is the vision systems integrator for Boeing transparencies [26].

Visible or luminous radiation extends over the $380 < \lambda < 750$ nm range and has a peak at 550 nm (green light). Quantitative values of luminous transmittance are designated as T_{lum} or alternately visible transmittance, T_{vis}. For the sake of simplicity, we will just use T_{lum}. Luminous transmittance is calculated using Eq. (1B.3):

$$T_{lum} = \frac{\int_{380\text{ nm}}^{750\text{ nm}} T(\lambda)\varphi_{vis}(\lambda)\,\varphi_p(\lambda)\,d\lambda}{\int_{380\text{ nm}}^{750\text{ nm}} \varphi_{vis}(\lambda)\,\varphi_p(\lambda)\,d\lambda} \tag{1B.3}$$

where

$T(\lambda)$ is the spectral transmittance,
$\varphi_p(\lambda)$ is the spectral sensitivity of the human eye, and
$\varphi_{vis}(\lambda)$ is the relative spectral distribution of the sun or an illuminant in the visible range, such as D65.

The photopic spectrum is shown in Figure 1B.3. The other important eye response is the scotopic or human eye nighttime response. It is normally not used, but is important

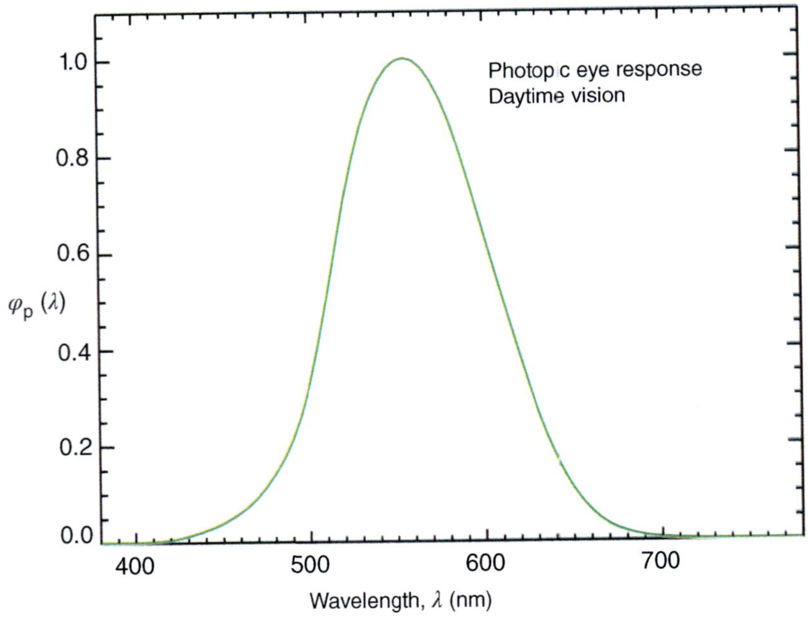

Figure 1B.3 Photopic response, human eye response in daylight. *Source:* Data from Public Domain, https://commons.wikimedia.org/w/index.php?curid=413513.

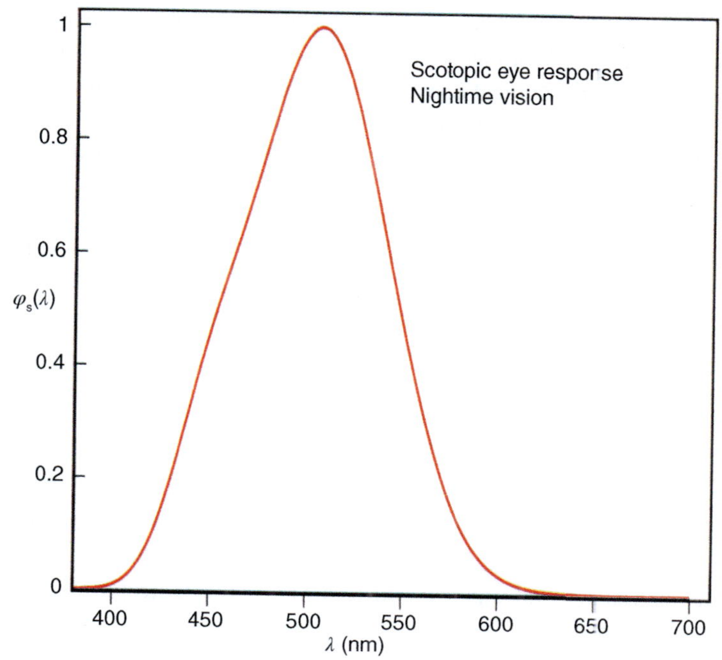

Figure 1B.4 Scotopic response curve, $\varphi_s(\lambda)$. This is the human eye response in dim light. The peak has shifted from 550 nm for the photopic response to 500 nm. This shift affects our color perception at night. *Source:* Data from http://www.cvrl.org/database/text/lum/scvl.htm, Public Domain, https://commons.wikimedia.org/w/index.php?curid=4321290.

for perceived color at night, due to the human eye response shift to shorter wavelengths at night. This is more important for nighttime businesses, signage, and transportation glazing. The nighttime or scotopic response is shown in Figure 1B.4. The scotopic transmittance $T_{\text{lum, scotopic}}$ can be calculated using Eq. (1B.3), substituting $\varphi_s(\lambda)$ for $\varphi_p(\lambda)$.

1B.3.3 Absorbance or Optical Density Measurement

Absorbance (AU_λ) and optical density (OD_λ) are the same measurement. Frequently, absorbance (log scale) is mixed up with absorptance (A_λ) (0–100% scale) measurements. Absorbance measurement arising from the field of chemistry commonly made on fluids to detect optical absorption caused by molecular vibration of chemical species. Measurements may be in the infrared, UV, or visible regions. The data is reported on a log10 or natural log scale or as a percentage. One must be careful when comparing data from different sources. Measurements are commonly made relative to a water-filled cell or empty cell. In this measurement, one is only measuring the switchable media itself not the total glass/quartz containment. In many cases very low or very high absorbance values are obtained, but they do not represent the actual numbers one would get from a functional glazing construction. One must be careful comparing data from the literature. Spectral absorbance is:

$$AU_\lambda = OD_\lambda = \log 10 \left(\frac{1}{T_\lambda}\right) \tag{1B.4}$$

The optical density measurement originates from the photographic and display industries. It is a measure of darkness or opacity. Chromogenic devices/materials change their optical density (ΔOD_λ) under some stimulus. Optical density is wavelength dependent in most cases, so the measurement number must be stated with the wavelength. Frequently, it is measured at 550 nm, the peak of the visible spectrum. The change in spectral optical density is given by:

$$\Delta OD_\lambda = \log 10 \frac{T_{b\,\lambda}}{T_{c\,\lambda}} \tag{1B.5}$$

where

$T_{b\,\lambda}$ is the bleached transmittance at λ, and
$T_{c\,\lambda}$ is the colored transmittance at λ.

1B.3.4 Haze and Scattering Measurements

For materials that are light scattering or highly oriented materials, an integrating sphere or a scanning goniometer must be used to get accurate measurements [27]. This is especially important for materials such as suspended particle devices (SPDs), thermotropic hydrogels and polymer blends, polymer dispersed liquid crystals, and electrophoretics such as suspended particles. With these measurements, one has to be especially careful to collect all the beams of light that can be scattered forward and backward from the sample. Many times, this is not properly measured so the optical values can have significant error. A work that addresses transmissive scattering materials was written by Wilson at Fraunhofer ISE, Freiburg, Germany together with Elstner from AGC Interpane [28]. The team at Fraunhofer ISE has been working to better define the proper measurements of light scattering materials for the glass and solar industry following European Standards EN 410:2011 and EN 14500:2021 and International Standard ISO 9050:2003. Collaboration between LBNL and Fraunhofer ISE resulted in the inclusion of sections within NFRC 300-2023 [E0A1] [29], which specifically address procedures for correct optical measurements of light-scattering (or diffuse) glazing materials. Other techniques for haze measurement have been proposed by researchers working in the plastics industry [30].

For polymer dispersed liquid crystal (PDLC), SPD glazing, and certain thermochromic glazing materials such as hydrogels, haze is an important parameter, especially at high viewing angles, relative to normal. The technical definition of haze is the amount of light subjected to wide angle scattering greater than 2.5° off-center divided by the total light transmission (ASTM D1003). Standards are available to measure system accuracy and obtain a single haze % value for each sample. There are two common standards for the measurement of haze under 30–40%, or light scattering through transmitting materials, ASTM D1003-21 and ISO 14782. Generally, these standards were written for non-switchable plastics, which have light scattering due to their polymer structure and surface texture. ASTM D1003-21, "Standard Test Method for Haze and Luminous Transmittance of Transparent Plastics," provides two procedures for light transmittance and haze or turbidity measurement. One procedure uses a hazemeter [31], and another uses a spectrophotometer with an integrating sphere. A schematic of a classic spectrophotometer with an integrating sphere is shown in Figure 1B.5. Further information is about the use of integrating spheres with spectrophotometers is available by Höpe [33]. In this work the basic techniques of measuring diffuse reflectance and transmission measurements are discussed. This work compares

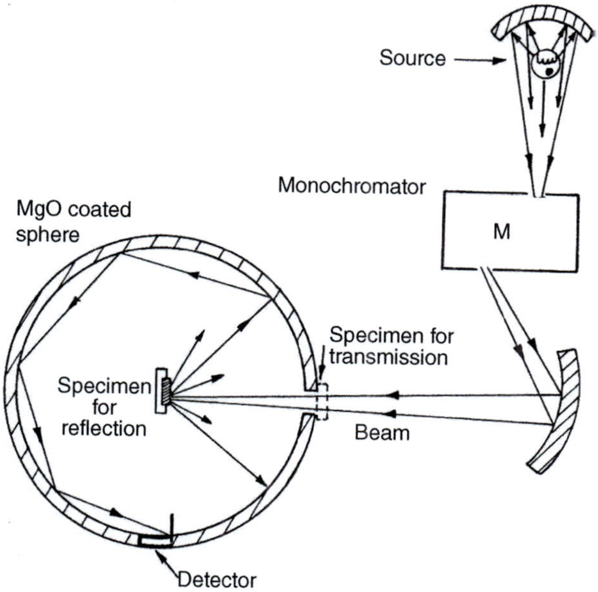

Figure 1B.5 Schematic of spectrophotometer with an integrating sphere for transmission and angular reflectance measurements (0°–80°) [32]. *Source:* Adapted from [33] / Lawrence Berkeley National Laboratory.

standard measurement techniques from the CMC-Database (Calibration and Measurement Capabilities Database) of the BIPM (International Bureau of Weights and Measures) and different NMIs (National Metrology Institutes) for spectral diffuse reflectance and spectral diffuse transmittance.

Hazemeter test values are normally slightly higher and less variable than the spectrophotometer with integrating sphere test results. Typically, regular luminous transmittance is measured by placing a clear specimen at some distance from the entrance port of the integrating sphere. However, when the specimen is hazy, the total hemispherical luminous transmittance must be measured by placing the specimen at the entrance port of the sphere. The measured total hemispherical luminous transmittance will be greater than or equal to the regular luminous transmittance, depending on the optical properties of the sample. With this test method, the specimen is necessarily placed at the entrance port of the sphere in order to measure haze and total hemispherical luminous transmittance. The size of the sphere, the cross-section of the beam, the size of the aperture, the sample thickness, and the exact sample position are all important in determining whether all the scattered light will be captured. The equipment is calibrated using NIST or ISO standard references.

For very strongly scattering materials, showing haze greater than 30% (such as PDLCs in the off state), the ASTM E2387-05 "Standard Practice for Goniometric Optical Scatter Measurements" is used. The ASTM E2387-05 standard describes a goniometric procedure for determining the amount and angular distribution of optical scatter from a surface. This standard focuses on measurement of the bidirectional scattering distribution function (BSDF). BSDF is a means of expressing optical scatter levels. It is often referred to as the bidirectional reflectance distribution function (BRDF) when considering reflective scatter or the bidirectional transmittance distribution function (BTDF) when considering transmissive scatter. The BSDF is a basic description of the appearance of a sample including

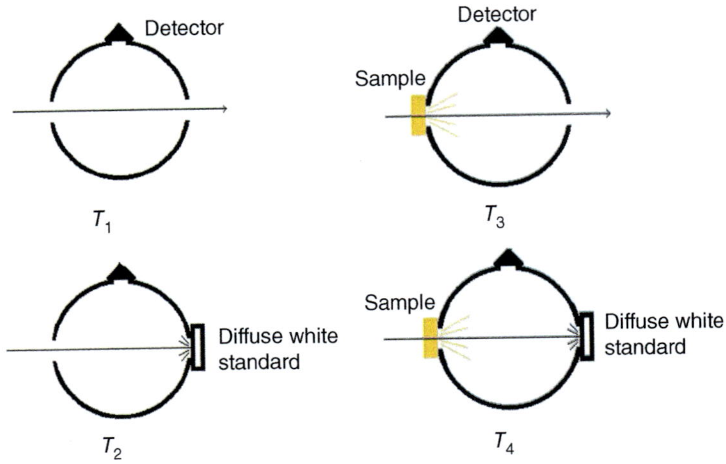

Figure 1B.6 Schematic showing four integrating sphere measurement configurations. T_1 is the transmittance of the equipment with all ports open, no sample or white reference standard. T_2 is the transmittance of the open sample port with the white standard in the exit port. T_3 is the transmittance of the sample in the sample port with no reference standard in the reference port. T_4 is the transmittance of the sample in the sample port with the standard in the exit port. *Source:* Author's image.

characteristics, such as haze, gloss, and color. These characteristics can be represented in terms of integrals of the BSDF over specific geometric and spectral conditions. Also, this technique presents alternative ways of showing angle-resolved optical scatter results, including directional reflectance factor, directional transmittance factor, and differential scattering function. The technique applies to BSDF measurements on transparent, translucent, and opaque samples.

ISO 14782 Standard "Plastics determination of haze for transparent materials," specifies a method for the measurement of transparent materials with haze values less than 40%. In this method, haze or turbidity is determined using an integrating sphere whose efficiency is kept constant by compensation. The integrating sphere and its apertures must be large enough to capture the high angle scattering components. Typically, a Xenon light source is used with a monochromator. The method involves four transmittance measurements to be carried out on a two-port integrating sphere, comprising a sample port and an exit port, each on one side of the sphere, as shown in Figure 1B.6. The measurements are (1) for T_1, all ports open, the beam passes through, (2) measurement T_2 is made with an open sample port and a white reference standard covering the exit port, (3) for T_3, the sample is placed in the sample port, and the exit port is left open, and (4) for T_4, the sample is placed in the sample port with the white standard is placed at the exit port. Haze, H (%), is determined by ratios of transmittance values following equation:

$$H(\%) = \left(\frac{T_3}{T_4} - \frac{T_1}{T_2}\right) \times 100 \tag{1B.6}$$

1B.4 Color Measurements

With all chromogenic materials it is important to consider the color of the glazing and how it affects the color of a living space or the view out of a building or a vehicle. One is especially concerned about the perceived color of objects outside the window. The question generally asked: is it a pleasant view or is it too artificially colored by the window? It is well known that people pay more for a pleasant view. Qualitatively defining the exact appearance and change of a chromogenic coating is a challenge, so color is scientifically defined by the Commission Internationale de l'Eclairage (CIE) color space. The glazing and electronic industries use the CIE standard system [34]. The chromaticity coordinates are a convenient way of quantifying color in color space. These coordinates provide a basis for the description of color. Color is specified by three nonphysical primaries X, Y, and Z. These are derived from the primary colors red, green, and blue. The corresponding tristimulus values are denoted by x, y, and z. Only two of these values are needed to specify color, as the third one can be derived from the other two. Many classic fixed glazing have gray, gold, metallic or bronze, or neutral colors. However, chromogenic glazing can include many other colors such as blue and green. Basic data is derived from spectrophotometry. Spectrophotometry is used to determine transmittance, reflectance, and absorptance over the visible spectral range which is of interest for color measurements. An example of a CIE diagram showing a range of switchable color is shown in Figure 1B.7. This is a special case where a nanochromic electrochromic was used to fabricate a resonant cavity structure, showing that an electrochromic pixel can be made to have the full range of colors, red, blue, and green [35, 36].

A more advanced system of color characterization is CIELAB D65 (6500F). This is commonly used for the human perception of color in natural daylight [37]. CIELAB uses the L^* perceptual lightness index to define pure black to pure white (0–100). The chromaticity coordinates are defined as a^* and b^*. The component a^* represents red as positive and green as negative numbers. The component b^* represents yellow as positive and blue as negative numbers. These represent the unique colors of human vision. By "dark," a common color definition would be (L^*, a^*, b^*) values near (40, 0, 0). Absolute black is at (0, 0, 0). A term called chromaticity, C^*, is defined as $C^* = (a^{*2} + b^{*2})^{0.5}$. This is a useful measure of the quality of color, independent of luminance. An example of this color space is shown in Figure 1B.8.

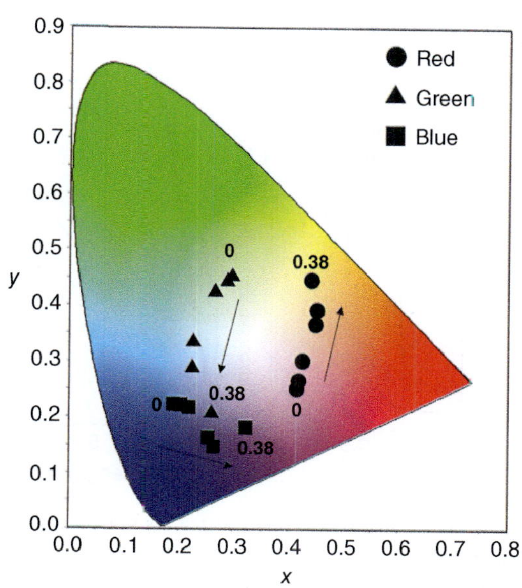

Figure 1B.7 CIE 1931 Diagram showing x,y color coordinates for the switching of nanochromic Li_xWO_3 where the concentration of intercalated xLi^+ ranges from $x = 0$ to $x = 0.38$. Different thicknesses and the resonant cavity effects are used to obtain blue, red, and green colors during switching. Each line represents the color change for a specific resonant cavity (pixel). Color switching is shown by the arrows as the intercalation of lithium increases. *Source:* Reprinted with permission from Lee et al. [35], copyright (2020) American Chemical Society.

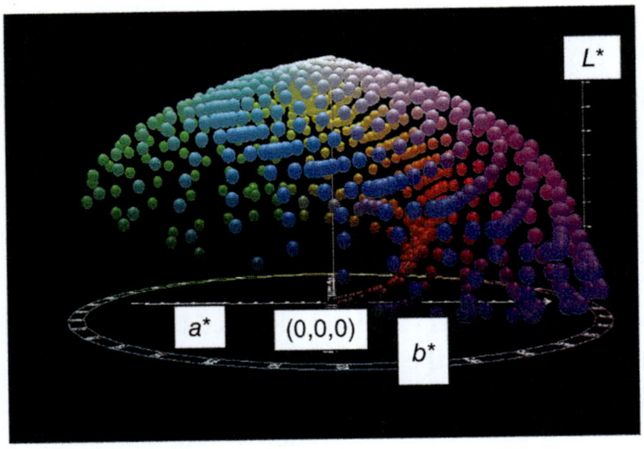

Figure 1B.8 CIELAB color space lightness index, L^*, is shown on the vertical axis ranging from 0 (black) to 100 (white), the chromaticity coordinates, a^* (red-green) and b^* (yellow-blue) are shown in the projected horizontal plane in the circle. At the center of the circle is total black color at (0, 0, 0). For a typical tungsten oxide-based electrochromic, it follows the blue curvature from slightly blue at the top of the figure to a deep blue when colored. *Source:* Adapted from Wiki Media Commons, Holger Everding.

1B.5 Thermal Emissivity and Emittance

Thermal emissivity, ε_t, is important to both architectural windows and to spacecraft surfaces. The radiative properties of pure materials are defined by their thermal emissivity or emittance. Emissivity is the term the National Institute of Standards and Technology (NIST) prefers to use for pure materials (unoxidized), otherwise emittance, ε_t, is used for most engineered materials. A classic reference on the thermophysical properties of materials was compiled by Touloukin and DeWitt in 1975 and serves as a good reference [38]. Ideal emitters are defined by the blackbody spectrum and its surface temperature will dictate the position of the peak of the spectrum. See for example, Figure 1B.9.

Thermal emissivity at a specific temperature is equal to the thermal absorptivity α_t at the same specific temperature. Thermal emittance is a measure of the effectiveness of a

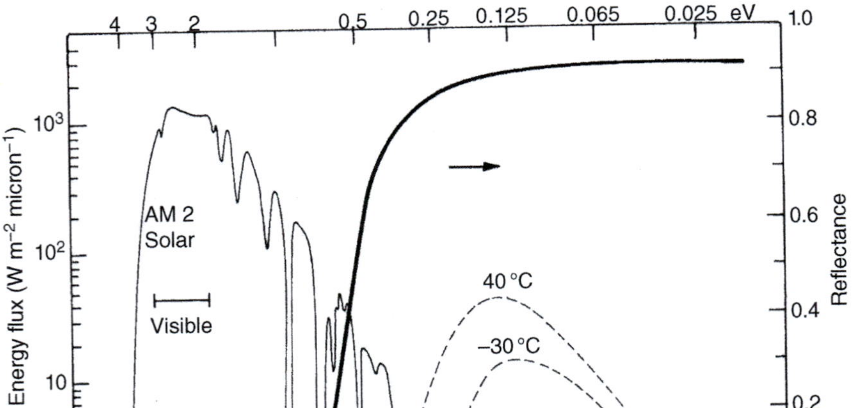

Figure 1B.9 Two blackbody spectra are shown with a doped-tin oxide low-e coating and the AM 2 solar spectrum. Note: as the blackbody gets hotter, the peak shifts to shorter wavelengths (higher energy). *Source:* [39] / Reprinted with permission of Elsevier.

radiator to emit energy over a unit area compared to that of a blackbody, as an ideal emitter. The thermal emittance is important for the control of emitted energy, which is a function of temperature to the fourth power. No other heat transfer component has such a high dependence on temperature. For soda-lime window glass the hemispherical thermal emittance is $\varepsilon_{t,h} = 0.84$. An ideal blackbody emitter has $\varepsilon_{t,h} = 1.0$, so one can see that glass in the infrared behaves close to an ideal emitter. For spacecraft, radiation is the only useful component of heat transfer in space. The ability to emit energy dictates the surface temperature heating or cooling of satellites and spacecraft surfaces. Chromogenic switchable emitters are used in space. The total thermal emittance of a coating or device may be measured hemispherically, using a heated cavity [40]. Also, near-normal emittance measurements can be made using an emissometer [41]. A classic portable emissometer instrument is the Geir-Dunkel DB-100. Alternatively, the infrared reflectance spectrum of an IR-opaque material can be measured by using an IR spectrometer equipped with a gold-coated integrating sphere. The emittance can be obtained by weighting the IR absorptance spectra with the blackbody spectrum for the required temperature. Low-e coatings are used to lower the effective thermal emittance of glass to $\varepsilon_{t,h} = 0.20$–0.03 substantially reducing the intensity of radiative loss of a glazing. For materials with low sheet resistance (R_s), an approximation of the normal emissivity at room temperature (ε_n) has been proposed by Gläser of Von Ardenne [42]:

$$\varepsilon_n = 0.0106 R_s \tag{1B.7}$$

The radiated flux per area Q_r/A is defined for two parallel plates or glazing as [41]:

$$Q_r/A = \sigma \left(T_2^{\,4} - T_1^{\,4}\right) / (1/\varepsilon_1 + 1/\varepsilon_2 - 1) \tag{1B.8}$$

where

σ is the Stefan-Boltzmann constant (5.6697×10^{-8} W m^{-2} K^{-4}),
ε_1 and ε_2 are the emissivity of plates 1 and 2, respectively, and
T_1 and T_2 are the temperatures of plates 1 and 2, respectively.

1B.5.1 Low-e Coatings and Transparent Conductors used in Insulated Glass Units (IGUs)

Low-emittance, or commonly known as low-e, coating is used to reduce the effective emissivity of a glass or plastic surface. In windows this coating is important since it can reduce the radiative portion of a heat transfer. For example, clear window glass (water white) has a hemispherical emissivity of $\varepsilon_{th} = 0.835$, compared to an ideal blackbody with $\varepsilon_{th} = 1.0$. The optical properties of typical soda-lime window glass can be seen in Figure 1B.10.

By the use of a low-e coating on glass, the effective thermal emittance can be reduced to typically below $\varepsilon_t = 0.1$. In an insulating glass unit (IGU) low-e coatings are essential to make a high-performance glazing. Different types of low-e coatings can control how much solar heat passes through a window. There are two major commercial families of low-e, (i) an optical interference type, and a (ii) highly doped transparent conductive oxide, known as a TCO. The first type is a dielectric-metal-dielectric stack (D/M/D) interference film. These are also known as "heat mirrors" in optical thin film science and have been known since the early 1950s [38]. Generally, the dielectric is a metal oxide, MO_x, typically InO_x, SnO_2, TiO_2, ZnO, $AlZnO_x$, $ZnSnO_x$, $SiAlN_x$, Al_2O_3, SiO_2, and Bi_2O_3 [39, 43] with the metal typically Ag, Cu, Zn, In, Bi, Ti, and Zn-Sn. Also, a metal nitride, such as SiN, may be used as a protective layer. Also, the low-e may be a coating of what is known as a "transparent conductor."

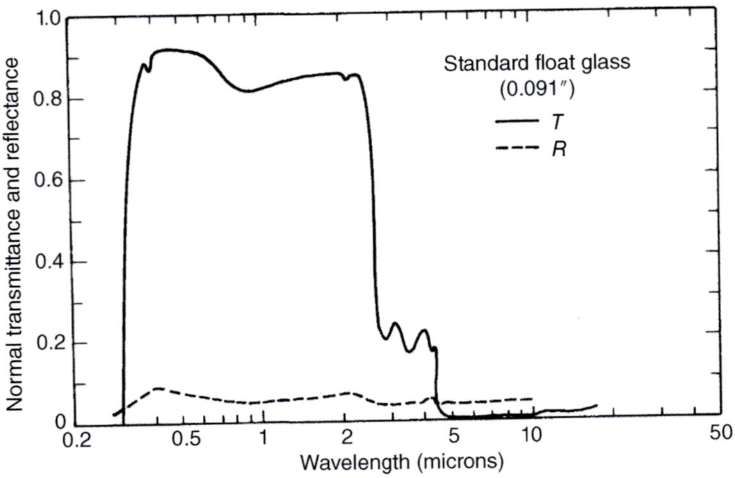

Figure 1B.10 Spectral properties of typical window float glass. In the thermal infrared (>3 microns), the low values of both reflectance and transmittance are the result of high absorptance and emittance [39]. Basically, in the thermal infrared, glass appears similar to an opaque blackbody. *Source:* Author's work.

A D/M/D+ coating can be optically designed to allow only the visible to pass through the glazing or be designed to let a portion of the near-infrared to pass through. For the metal layer silver is preferred. Other metals such as Cu and Cu alloys, Ni, Au, Cr, and Al have been investigated for the metal layer [39]. In Figure 1B.11 a comparison is shown between the optical performance of a single silver layer D/Ag/D (known as low-e (single Ag)) and D/Ag/D/Ag/D (called double low-e (double Ag), also designated as low-e2. Note in the visible range the low-e2 is more selective as to the amount of near-infrared it lets pass through. If a triple low-e stack is made using triple Ag layers, then an even tighter filter can be made,

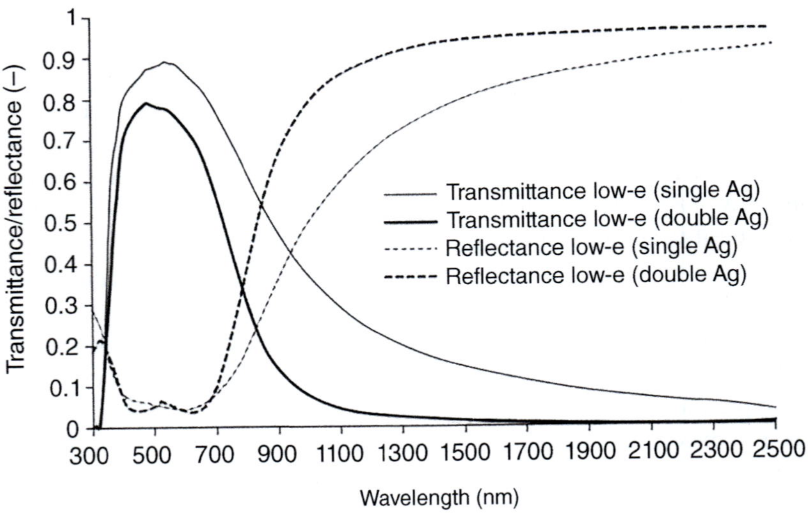

Figure 1B.11 Spectral transmittance and reflectance of a low-e glass with single D/Ag/D and double Ag layer D/Ag/D/Ag/D/ (known as low-e2) (original data from Saint-Gobain). *Source:* Mohelnikova [44], with permission from Elsevier.

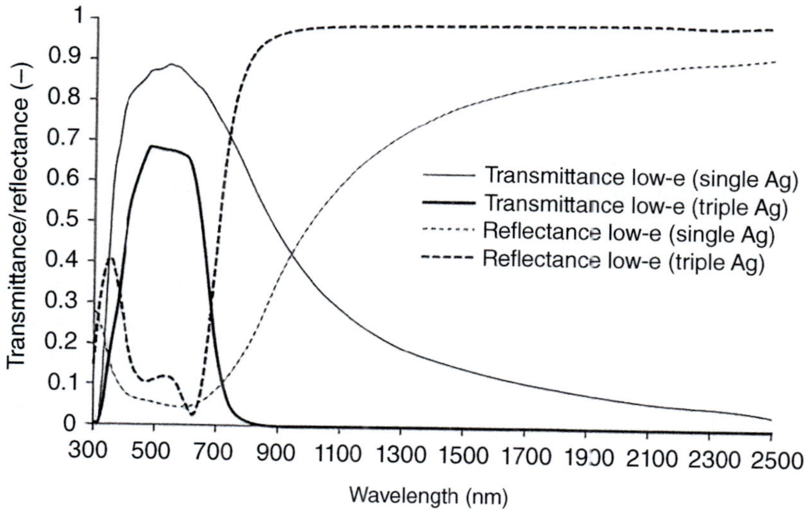

Figure 1B.12 Spectral transmittance and reflectance of a low-e glass with single D/Ag/D and triple Ag layer D/Ag/D/Ag/D/Ag/D (known as low-e3) (original data from Saint-Gobain). *Source:* Mohelnikova [44], with permission from Elsevier.

reflecting much of the near-infrared. A comparison between the optical performance of a single low-e and a triple Ag layer low-e, D/Ag/D/Ag/D/Ag/D (designated as low-e (triple Ag)) or low-e3 is shown in Figure 1B.12.

Low-e3 is generally designated as a D/M/D/M/D/M/D tight bandpass optical filter, which only allows visible light to pass through it, with both UV and solar near-infrared reflected, as shown in Figure 1B.13. In real manufacturing of a D/M/D low-e coating, other layers are used including protective top and barrier coatings, transition, and protective layers of the metal layer. For a seven-layer dielectric/metal optically designed low-e3, it actually takes

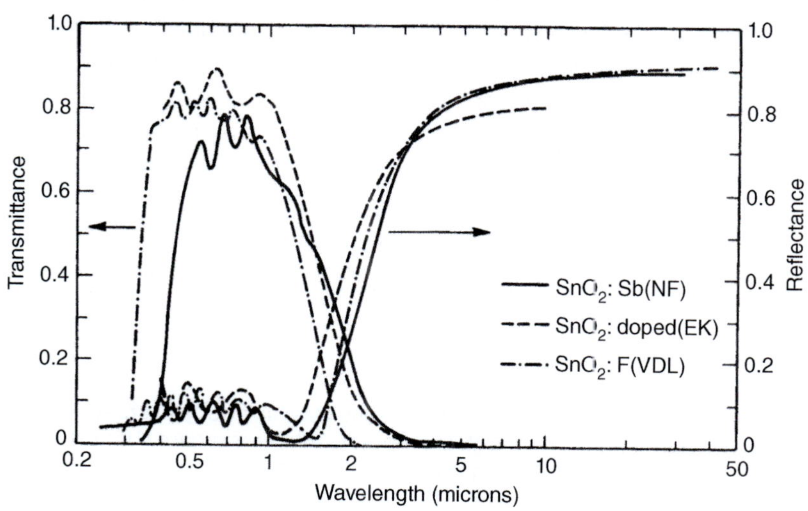

Figure 1B.13 Spectral properties of highly doped tin oxide [38]. Coatings were made by spray pyrolysis of $SnCl_4$ at 460 °C, $e_t = 0.2$ (EK) [45], spray hydrolysis at 500–570 °C, $e_t = 0.15$ (VDL) [46], and physical vapor deposition SnO_2:Sb, $e_t = 0.15$ [NF] [47].

12 distinct layers to make a durable commercial product [48]. For example, a low-e3 stack can be composed of 20 nm ZnO_x dielectric base layer/10 nm Ag reflector/2 nm graded metal oxide protective layer/80 nm $ZnSnO_x$ dielectric layer/10 nm Ag reflector/2 nm graded metal oxide protective layer/80 nm $ZnSnO_x$ dielectric layer/10 nm Ag reflector/2 nm graded metal oxide protective layer/6 nm graded metal oxide protective layer/30 nm $ZnSnO_x$ dielectric layer/4 nm TiO_2 protective top layer. D/M/D+ designs can be found in two excellent reference texts [49, 50].

The second type of low-e coating are the TCOs. Two common examples are heavily doped In_2O_3:Sn or SnO_2:F (both semiconductors are discussed later in chapters on vacuum web coaters and pyrolytic doped tin oxide). These coatings can also serve as an electrical conductor for powered chromogenic materials. Figure 1B.13 shows the spectral properties of doped tin oxide coatings. A figure of merit (FOM_H) for highly doped, highly transparent ($T > 0.85$) semiconductors was developed by Haacke [51]:

$$FOM_H = T_a^{10} R_s^{-1} \tag{1B.9}$$

where T_a is the average visible transmittance (mathematical average, not weighted), and R_s is sheet resistance.

This FOM_H has been further refined to higher resolution $FOM_{H\text{-}HR}$ by [52]:

$$FOM_{H-HR} = T_a R_s^{-1/n} \tag{1B.10}$$

where $n = 10, 11, 12$, for example.

1B.6 Design of an IGU Window System

High-performance windows consist of two or three or four glass panes separated by insulating spacers and installed in insulated frames made of nonconductive wood, fiberglass, or vinyl or composite material. The space between the glass layers is filled with a nontoxic gas like argon or krypton that insulates better than air. The glass panes are coated with a low-e coating that reflects sunlight. A typical double-glazed insulating glass unit is known as an IGU. A cross-section diagram is shown in Figure 1B.14.

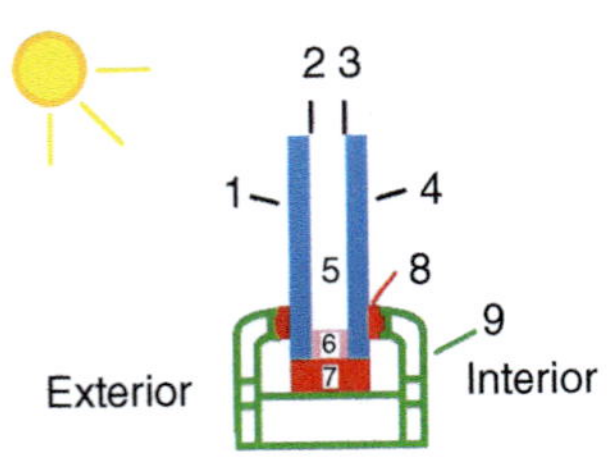

Figure 1B.14 Cross-section diagram of a double-glazed insulating glass unit (IGU) (not to scale), indicating a common numbering convention for the glass surfaces (1–4). Glass surface (1) is the exterior outside, surface (2) is the surface inside the gap of the exterior pane, surface (3) is the surface inside gap of the interior pane, and surface (4) is the outside surface of interior pane. The window gap with low conductivity fill gas, such as argon or krypton (5), a low conductivity (warm seal) spacer (6) and the dual edge seal is (7). The frame seal is (8), and the window frame is (9). Typically surface 3 has a low-e coating applied to it. *Source:* Adapted from NcLean – Own work, CC BY-SA 3.0, https://commons.wikimedia.org/w/index.php?curid=15994521.

1B.7 Energy Flow Mechanisms in Glazing Fenestration

It is best to show the components of an IGU and how they control energy in the static case. By adding different switchable films to the glazing, we can modify the amount of solar visible and near-infrared energy passing through the IGU. As discussed before, the energy from the sun passing through a glazing is roughly split between half in the visible and half in the near-infrared. The types of glass can control the amounts of energy passing through the glazing. Common glazing with glass ranges from clear, to reflective (metallized), and heat absorbing (colored) glass. All have different thermal characteristics. The insulating value or U-factor (or U-value) is a measure of the window's ability to resist heat transfer. Figure 1B.15 shows a cross-section of the IGU with its common heat transfer mechanisms. The three heat transfer mechanisms are conductive, radiative and convective. Conductive heat transfer is a function of the cross-section of the window which is determined by thermal conductivity of the outer glass pane including coatings it may have, air gap, gas fill, conductivity of the inner glass pane including any coatings it may have on it. Although many thermal measurements are taken from the center-of-glass, the effect of thermal conductance of the framing can be significant. This is especially true for smaller windows. In good designs, low-conductivity gap spacers and framing materials are used. Framing materials range from wood, aluminum, vinyl, to other plastics and composites.

The overall energy flow through a fenestration product is a function of three major factors: (1) temperature-driven heat transfer, (2) solar gain, and (3) infiltration of air:

(1) Temperature-driven heat transfer – When there is a temperature difference between inside and outside, heat is lost or gained through the glazing product including frame and glazing by the combined action of conduction, convection, and radiation. This is indicated in terms of the U_{overall}-factor of a glazing assembly. The components of U_{overall} are defined in Section 1B.8.1.

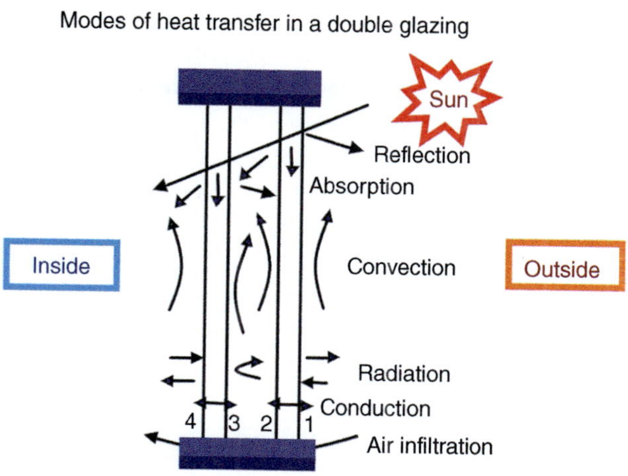

Figure 1B.15 Schematic of a typical double glazing (IGU) showing energy transfer and gain and loss mechanisms. Shown from the top down is first solar transmission and reflection and absorption in the glass panes. Convective losses are shown next, they are a function of surface coefficients and air flow patterns and speed. Conduction losses are a function of window size, gap spacing, fill gas, spacer, and frame conduction. For reference, the IGU glass surfaces are numbered starting from the exterior surface of glass as number (1), the inner gap surfaces (2, 3) and interior surface, number (4), facing the living space. *Source:* Adapted from image courtesy of Dr. Robin Kent, Tangram Technology, Ltd. Consulting Engineers, Hertfordshire, UK, Tangram.co.uk.

(2) Solar gain – Regardless of the outside temperature, heat can be gained through fenestration products by direct transmission of solar radiation and secondary inward heat gain following absorption of solar radiation. The amount of heat gain through the glazing is measured in terms of the solar heat gain coefficient (SHGC).

(3) Air infiltration or air leakage is another factor in the overall performance of a window. Air leakage is driven by wind and thermal gradients through small cracks in the window assembly. In operable windows (ones with sliding panes), air leakage is determined by how well the window seals to the frame when it is shut. Imperfections in seals can result in air leakage. Also, air can leak around frames due to gaps or sealing imperfections between the frame and framing studs. In reality, infiltration varies with wind-driven and temperature-driven pressure changes. Infiltration also contributes to summer cooling loads in some climates by raising the interior humidity level. The heat loss component of air infiltration is usually a function of the seals in movable windows and framing, and any weep holes or gaps. Also, energy loss may be caused by gaps left during framing the window. The NFRC standards for air infiltration are <3 cfm ft^{-2}. The best windows for low air leakage are casement and "tilt and turn" windows with compression seals. As wind pressure goes up the tighter the seals become, minimizing thermal drafts.

1B.7.1 Conductive Heat Transport

Conductive heat transport is temperature driven, for example a warm outdoor on one side of the window and a cooler inside of the window facing the interior of the building. Heat transport is always by conduction along a temperature gradient. If there is no temperature gradient then conduction does not occur. Conduction is the sum of the conductivities of the window frame, glass, edge seal, fill gas, and gap size.

1B.7.2 Convective Heat Transport

Convective heat transfer requires the flow of air or gas for heat transport from surfaces. Convection occurs on both the outside and inside of a window following air flows. Also, within the gap convection can occur even with fill gases. Gap sizes in IGUs typically range from 12 to 20 mm. Only vacuum insulated glass, VIG, has little discernable internal convective heat transport in the gap. Fill gases typically range from argon with thermal conductivities $k = 0.018\ \mathrm{W\ m^{-1}\ K^{-1}}$ and krypton with $k = 0.009\ \mathrm{W\ m^{-1}\ K^{-1}}$, compared to air at $k = 0.026\ \mathrm{W\ m^{-1}\ K^{-1}}$. Further design information is available [47, 53–55]. Convective heat transfer coefficients represent the thermal resistance of the stagnant layer of air or gas on a heat transfer surface, such as a surface of glass [56, 57]. Surface-textured glass would have a higher convective heat transfer coefficient compared to polished glass.

1B.7.3 Radiative Heat Transport

Thermal radiation is a function of surface emissivities of the glass and its coatings. Radiative loss is a function of temperature to the fourth power, T^4, so under hot conditions the radiative loss component can be high. Internally, it is controlled by a low-e coating on typically the third surface of the glass. This was discussed in section 1B.5. The type and design of the

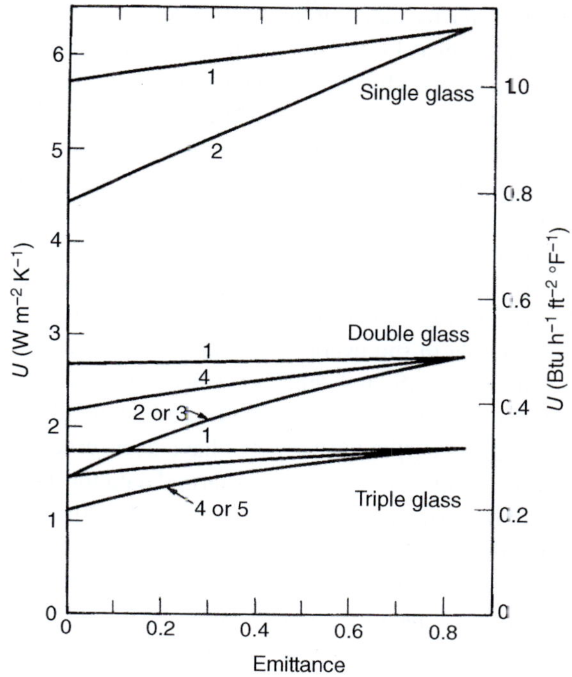

Figure 1B.16 The lowering effect on glazing U-factor by adding a low-e coating with varying emittance to each glazing surface, for single, double, and triple panes [39]. The glazing surfaces are numbered from the exterior (outdoor) pane as (1) to the indoor pane (4), as shown in Figure 1B.15, for a double glazing. For a triple glazing the numbering goes from (1) the exterior (outdoor) surface to (6) the indoor surface of the third pane. The radiative losses are lowest when the low-e coating is applied to the number 2 surface of a single glazing, the number 2 or 3 surface for a double glazing and the number 4 or 5 surface of a triple glazing. By using an NIR chromogenic coating one could regulate the amount of gain or loss by radiative heat transport of the glazing.

low-e coating will determine the radiative properties of the window. Figure 1B.16 shows the lowering effect on glazing U-factor by adding a low-e coating with varying emittance to each glazing surface, for single, double, and triple panes [39]. The glazing surfaces are numbered from the exterior (outdoor) pane as (1) to the indoor pane (4), as shown in Figure 1B.15, for a double glazing. For a triple glazing the numbering goes from (1) the exterior (outdoor) surface to (6) the indoor surface of the third pane. For double glazing the greatest effect is shown for low-e coating on the number 2 or 3 surface. For a triple glazing the greatest effect is shown for the number 4 or 5 surface.

1B.8 Parameters Commonly Used to Characterize Window Glazing

Specific parameters governing the thermal properties, solar heat gain, and air infiltration are used to characterize windows and their assemblies. Various net parameters are used to compare the performance of windows under standard conditions, with conditions set by a standard organization. The standard test conditions from one country to another tend to vary.

1B.8.1 U-Factor

The U-factor, also known as U-value, determines how well a window transfers heat by radiation, convection, and conduction between the indoor and outdoor environments. It is a measure of thermal transport through the glazing. It is one of the key factors in energy rating of a window [58]. Coating one or more of the glass surfaces with low-e coatings reduces heat transfer by radiation and filling the cavity between panes with inert gas rather than air reduces heat losses due to air conduction and convection. The U-factor is the sum of the radiative, conductive, and convective heat transfer mechanisms. It is a measure of total thermal transport through a glazing under standardized conditions. The lower the U-factor, the better it traps in heat. Typical values for windows range from 1.14 to 6.81 W m^{-2} K^{-1} (0.2 to 1.2 Btu h^{-1} ft^{-2} $°F^{-1}$). A low U-factor is especially important in climates with very cold winters. The conversion between units is 1 W m^{-2} K^{-1} = 0.18 (Btu h^{-1} ft^{-2} $°F^{-1}$). The U-Factor is the inverse of the *R*-value. The *R*-value is commonly known as the insulating factor used to characterize thermal insulation. Many times, in the literature one will see the center-of-glass U-factors quoted. In small-area windows the U-factor of the frame plays a major role in the overall U-factor. U-factor can be thermally measured with high accuracy using a calibrated "hot box" (ISO 12567-1:2010). Heat flow meters have been used for installed glazing (ISO 9869-1:2014). Both guarded hot plates (ISO 10291-1994) and infrared thermography (ISO 9869-2:2018) have been used to measure U-factor [59]. Many glazing makers calculate U-factor from thermal data for the individual panes, fill gas, coatings, and overall construction of the window using the Windows 7 glazing program [60], which implements the calculation methods of ISO 15099:2003.

The overall U-factor ($U_{overall}$) includes the whole window area, including edge and frame. The overall U-factor includes separate U-factors for the center of glass U_{cg} for the edge U_{eg} and frame U_f as shown in Eq. (1B.11) [63]:

$$U_{overall} = (U_{cg} A_{cg} + U_{eg} A_{eg} + U_f A_f)/A_{pf} \tag{1B.11}$$

where

U_{cg} is the U-factor of the center of glass,
A_{cg} is the center of glass area, (e.g. m^2),
U_{eg} is the U-factor of edge of glass,
A_{eg} is edge of glass area,
U_f is the U-factor of frame/edge,
A_f is the frame area, and
A_{pf} is the total projected area of fenestration

1B.8.2 Solar Heat Gain Coefficient (SHGC)

The SHGC or "solar factor" is a measure of how well a window blocks solar heat from getting inside. This term is used in North America by the National Ratings and Fenestration Council (NFRC). Fenestration energy performance standards are NFRC 200 and the Canadian standard CSA A440.2. The SHGC takes in account the transmitted, reflected, absorbed and reemitted energy. It considers both inward and outward flows of energy. SHGC can be calculated for the center of glass or for whole window including framing. The SHGC is now more commonly used than the old shading coefficient (SC) term. The SC was an SHGC measure defined relative to the SHGC of a single glazing window. The lower the SGHC, the better the window blocks incident solar radiation. Low SGHC ratings are significant in climates with very hot summers. SHGC is between 0 and 1. Many glazing makers calculate SHGC on the basis of the optical and thermal data for the individual panes (e.g. the transmittance and reflectance spectra for a coated pane) using the Windows 7 program [60].

The calculations use the solar AM1.5 normal for their reference spectra.

1B.8.3 Solar Factor (*g*-Value) or Total Solar Energy Transmittance (TSET)

The g-value or solar factor or total solar energy transmittance (TSET) is closely equivalent to SHGC, especially for high-performance windows using center of glass calculations. The SHGC can be measured calorimetrically or calculated [61]. The *g*-value is defined as the total fraction of solar radiation that is transmitted directly or as transferred heat after absorption by the glazing to the interior. This includes energy transferred by radiation, convection, or conduction. The *g*-value is expressed in % (0–100%) or a value from 0 to 1. The *g*-value is calculated following the European Standard EN 410 standard or ISO 9050. The *g*-value modeling uses the Solar AM 1.0 for the calculation. For whole window calculations, there are some differences between SHGC and *g*-value, depending on different framing data sets [62]. But that discussion is beyond the scope of this work. The *g*-value is used in Europe and other countries.

1B.8.4 Total Solar Transmittance

Total solar transmittance can be calculated for most glazing fenestration using the NFRC 100-2023 Standard [1]. This includes the optical components of solar energy transport through the glazing. The total solar transmittance refers to the total integrated optical transmittance through a window. The optical losses through the window are reflectance and absorptance from the glass and its coatings. Total solar transmittance is the primary heat gain component of the SHGC. Total solar transmittance is always a lower value than the SHGC, as the SHGC includes the secondary solar heat gain components.

Total solar transmittance values can be calculated using the "Windows 7.8.74" modeling program developed by LBNL (Berkeley, CA) [60]. It should not be confused with TSET that was mentioned in Section 1B8.3.

1B.8.5 Instantaneous Heat Flow in a Whole Glazing

The instantaneous heat flow Q_i (W) through glazing fenestration including air infiltration and assuming no humidity difference is described in Eq. (1B.12). The properties of U-factor, SHGC, and air infiltration are major factors that determine the net heat flow through a fenestration product [59].

$$Q_i = U\,A_{pf}(t_{out} - t_{in}) + (\mathrm{SHGC}\,A_{pf}\,I) + C\,(\mathrm{AL})\,A_{pf}\,\rho\,C_p\,(t_{out} - t_{in}) \tag{1B.12}$$

where

Q_i is the instantaneous heat flow, W (Btu h^{-1}),
U is the overall coefficient of heat transfer or U-factor, W m^{-2} K^{-1}, (Btu h^{-1} ft^{-2} °F^{-1}),
t_{out} is the exterior air temperature, °C, (°F),
t_{in} is the interior air temperature, °C, (°F),
A_{pf} is the total projected area of fenestration, m^2 (ft^2),
SHGC is the solar heat gain coefficient, (dimensionless),
I is the incident total irradiance, W m^{-2} (Btu h^{-1} ft^{-2}),
C is the conversion constant 60 min h^{-1},
AL is the air leakage m^3 min^{-1} m^{-2} (cfm ft^{-2}),
ρ is the air density g m^{-3} (lb ft^{-3}), and
C_p is the specific heat of air J g^{-1} °C^{-1} (Btu h^{-1} °F^{-1}).

Acknowledgments

I thank Helen Rose Wlson, Fraunhoffer Institute for Solar Energy Systems, Freiburg, Germany, Gunnar Nicklasson, Uppsala University, Angstrom Laboratory, Uppsala, Sweden, and Anoop Agrawal, Tucson, AZ for their review of this chapter.

References

1 Trząski, A. and Rucińska, J. (2015). Energy labeling of windows – possibilities and limitations, *Solar Energy* 120: 158–174. https://doi.org/10.1016/j.solener.2015.07.029.

2 Dussault, J.-M., Gosselin, L. and Galstian, T. (2012). Integration of smart windows into building design for reduction of yearly overall energy consumption and peak loads. *Solar Energy* 86.11: 3405–3416. https://doi.org/10.1016/j.solener.2012.07.016.

3 Jelle, B.P. (2015). Chapter 15: Electrochromic smart windows for dynamic daylight and solar energy control in buildings. In: *Electrochromic Materials and Devices* (ed. R.J. Mortimer, D.R. Rosseinsky, and P.M.S. Monk). Wiley-VCH.

4 Lampert, C.M. (1994). Glazing materials for solar and architectural applications. Report of the IEA Task group 10, LBL-34436, Sept 1994, https://www.osti.gov/servlets/purl/10124630 (accessed 16 March 2024).

5 Czanderna, A.W. and Lampert, C.M. (1990). Evaluation criteria and test methods for electrochromic windows. SERI/PR-255–3537 July, 1990 Golden, CO-Solar Energy Research Institute. https://www.osti.gov/biblio/6953557 (accessed 24 April 2024).

6 NREL Database (2024). Solar irradiance (AM0). https://www.nrel.gov/grid/solar-resource/spectra-astm-e490.html (accessed 23 March 2024).

7 NREL Database (2024). Solar irradiance (AM1.5). https://www.nrel.gov/grid/solar-resource/spectra-am1.5.html (accessed 23 March 2024).

8 Marion, W. and Wilcox, S. (1995). Solar radiation data manual for buildings. NREL/TP-463-7904, Sept 1995, doi:https://doi.org/10.2172/183223.

9 Mecherikunnel, A.T. and Richmond, J.C. (1980). Spectral distribution of solar radiation. NASA TM 82021, NASA, Goddard Space Flight Center, Greenbelt, MD, https://ntrs.nasa.gov/api/citations/19810016493/downloads/19810016493.pd (accessed 19 Jan 2025).

10 Cromwell, N. (2024). Earth air mass diagram. https://g2voptics.com/air-mass-solar-intensity/, G2V (solar simulators), Canada (accessed 23 March 2024).

11 ASTM International (2019). Standard Solar Constant and Zero Air Mass Solar Spectral Irradiance Tables, ASTM E490-00a, https://www.astm.org/e0490-00ar19.html (updated 27 April 2022).

12 ASTM International (2023). Standard tables for reference solar spectral irradiances: direct normal and hemispherical on 37° tilted surface (AM1.5), ASTM G173-03, ASTM G173-23 (updated 15 Aug 2023).

13 Moon, P. (1940). Proposed standard solar radiation curves for engineering use. *Journal of the Franklin Institute* 230: 607.

14 Martins, S.M. and Giesbrecht, M. (2021). Clearness index forecasting: a comparative study between a stochastic realization method and a machine learning algorithm. *Renewable Energy* 180: 787–805. https://doi.org/10.1016/j.renene.2021.08.094.

15 Button, D. (1974). *Glass in Building*, 2e. Oxford, UK: Architectural Press. ISBN: 9780750621700.

16 US Green Building Council (2024). LEED rating system. usgbc.org/leed (accessed 19 Jan 2025).

17 BREEAM (2024). Sustainable building certification. Breeam.com (accessed 1 July 2024).

18 WELL (2024). WELL certification - health and wellness of people in the built environment. wellcertified.com (accessed 2 July 2024).

19 Kennedy, H. (ed.) (2021). *ASHRAE Handbook—Fundamentals*. Peachtree Corners, GA: ASHRAE. https://www.ashrae.org/technical-resources/ashrae-handbook.

20 Kittler, R., Kocifaj, M., and Darula, S. (2012). *Daylighting Science and Daylighting Technology*. NY: Springer. ISBN: 978-1-4419-8815-7.

21 Ander, G.A. (2003). *Daylighting Performance and Design*, 2e. New York: Wiley. ISBN: 978-0471262992.

22 Ander, G.A. (2016) Daylighting, Whole Building Design Group. wbdg.org/resources/daylighting (accessed 3 July 2024).

23 DeCusatis, C. (1998). *Handbook of Applied Photometry*. Springer-Verlag, NY: Optical Society of America. ISBN: 9781563964169.

24 Germer, T.A., Zwinkels, J.C., and Tsai, B.K. (2014). *Spectrophotometry: Accurate Measurement of Optical Properties of Materials*. Cambridge, MA: Academic Press. ISBN: 9780123860224.

25 Kiedron, P.W. (2001). Spectral irradiance per interval of equal solar flux: convenient spectral grid for atmospheric radiation measurement and modeling. Eleventh ARM Science Team Meeting Proceedings, Atlanta, Georgia (19–23 March 2001).

26 Seidenman, P. and Spanovich, D. (2020). How next-gen aircraft windows have become safer and stronger. *Aviation Week*, 4 Nov 2020. https://aviationweek.com/mro/emerging-technologies/how-next-gen-aircraft-windows-have-become-safer-stronger (accessed 4 July 2024).

27 Rosa, E.B. and Taylor, A.H. (2013). *Theory, Construction, and Use of the Photometric Integrating Sphere: Scientific Papers of the Bureau of Standards*, vol. 75. Gaithersburg, MD: NIST. ISBN: 9781289198282.

28 Wilson, H.R. and Elstner, M. (2018). Determining the light and solar properties of fritted and coated glass challenging glass. In: *CGC6 Proceedings, Conference on Architectural and Structural Applications of Glass* (ed. C. Louter, F. Bos, and J. Belis). Delft, NL: Delft University of Technology, TU Delft Open. https://doi.org/10.7480/cgc.6.

29 National Fenestration Ratings Council (2023). *ANS/NFRC 100-2023 Standards*. Greenbelt, MD, USA: NSFRC. https://nfrccommunity.org/store/viewproduct.aspx?id=1380591 (accessed 24 April 2024).

30 Busato, S. and Perevedentsev, A. (2017). A simple imaging-based technique for quantifying haze and transmittance of materials. *Polymer Engineering & Science* 58 (3): 345–352. https://doi.org/10.1002/pen.24580.

31 Billmeyer Jr., F.W. and Chen, Y. (1985). On the measurement of haze. *Color Research & Application* 10.4: 219–224. https://doi.org/10.1002/col.5080100410.

32 Lampert, C.M. (1979). Chemical, structural and optical characterization of a black chrome solar selective absorber. Lawrence Berkeley National Laboratory. LBNL Report #: LBL-9123. Retrieved from. https://escholarship.org/uc/item/0nd2x713 (accessed 23 July 24).

33 Höpe, A. (2014). Chapter 6: Diffuse Reflectance and Transmittance. In: Germer, T.A., Zwinkels, J.C., Tsai, B.K. (eds) *Experimental Methods in the Physical Sciences*, Academic

Press, Vol. 46, pp. 179–219, ISBN 9780123860224, doi:https://doi.org/10.1016/B978-0-12-386022-4.00006-6.

34 Schanda, J. (2016). CIE 1931 and 1964 Standard Colorimetric Observers: history, data, and recent assessments. In: *Encyclopedia of Color Science and Technology* (ed. M.R. Luo). New York, NY: Springer. https://doi.org/10.1007/978-1-4419-8071-7-323.

35 Lee, Y., Yun, J., Seo, M. Kim, S.-J., Oh, J., Kang, C.M., Sun, H.-J., Chung T.D., and Lee, B. (2020). Full-color-tunable nanophotonic device using electrochromic tungsten trioxide thin film. *Nano Letters* 20 (8): 6084–6090. https://doi.org/10.1021/acs.nanolett.0c02097.

36 Wang, Z., Wang, X., Cong, S., Chen, J., Sun, H., Chen, Z., Song, G., Geng, F., Chen, Q., and Zhao, Z. (2020). Towards full-color tunability of inorganic electrochromic devices using ultracompact fabry-perot nanocavities. *Nature Communications* 11: 302. https://doi.org/10.1038/s41467-019-14194-y.

37 Carter, E.C. (Committee Chair), CIE (2018). *Colorimetry,* 4e. CIE 015:2018, Division 1. ISBN: 978-3-902842-13-8, https://doi.org/10.25039/TR.015.2018.

38 Tonlonkian, Y.S. and DeWitt, D.P. (ed.) (1973). *Thermophysical Properties of Matter, Vol. 7: Thermal Radiative Properties of Metallic Elements; Vol. 8: Non-metallic Solids; Vol. 9: Coatings*, TPRC Data Series. New York: Plenum Press.

39 Lampert, C.M. (1981). Heat mirror coatings for energy conserving windows. *Solar Energy Materials* 6.1: 1–41. https://doi.org/10.1016/0165-1633(81)90047-2 (Also, LBL-11446 report).

40 Honner, M. and Honnerová, P. (2015). Survey of emissivity measurement by radiometric methods. *Applied Optics* 54: 669–683.

41 Duffie, J.A. and Beckman, W.E. (1974). *Solar Thermal Energy Processes*. New York: Wiley.

42 Gläser, H.J. (2000). *Large Area Glass Coating*. Dresden, Germany: Von Ardenne Anlagentechnik GMBH. ISBN: 3-00-004953-3.

43 Cueva, A. and Carretero, E. (2023). Comparison of the optical properties of different dielectric materials (SnO_2, ZnO, AZO, or $SiAlN_x$) used in silver-based low-emissivity coatings. *Coatings* 13 (10): 1709. https://doi.org/10.3390/coatings13101709.

44 Mohelnikova, J. (2011). Chapter 7: Nanocoatings for architectural glass. In: *Nanocoatings and Ultra-Thin Films* (ed. A.S.H. Makhlouf and I. Tiginyanu), 182–202. Woodhead Publishing. ISBN 9781845698126, https://doi.org/10.1533/9780857094902.2.182.

45 Groth, R. and Kauer, E. (1965). Thermal insulation of sodium lamps. *Philips Technical Review* 26: 105–111.

46 van der Leij, M. (1979). Spectral-selective surfaces for the thermal conversion of solar energy. Ph.D. Thesis, Delft University Press, Delft, The Netherlands.

47 Fuschillo, N. (1975). Semi-transparent solar collector window systems. *Solar Energy* 17: 159–165.

48 Timberlake, K. (2014). Investigating Low-e coating technology for glass. https://kierantimberlake.com/updates/investigating-low-e-coating-technology (accessed 27 July 2024).

49 Baumeister, P.W. (2004). *Optical Coating Technology*. Bellingham, WA: SPIE-The International Society for Optical Engineering. ISBN: 0-8194-5313-7.

50 Macleod, H.A. (2017). *Thin Film Optical Filters*, 5e. Boca Raton, FL: CRC Press. ISBN: 9780367781606.

51 Haacke, G. (1976). New figure of merit for transparent conductors. *Journal of Applied Physics* 47: 4086–4089. https://doi.org/10.1063/1.323240.

52 Cisneros-Contreras, I.R., Muñoz-Rosas, A.L., and Rodríguez-Gómez, A. (2019). Resolution improvement in Haacke's figure of merit for transparent conductive films. *Results in Physics* 15: 102695. https://doi.org/10.1016/j.rinp.2019.102695.

53 Haglin, D. (2021). Gas fill 101: gas fill options for IGUs. *Window+Door*, October 19, 2021. https://www.windowanddoor.com/article/gas-fill-101-gas-fill-options-igus (accessed 24 Apr 2024).

54 Sabatiuk, R. (1982). Review of gas-filled windows technology-summary report Oak Ridge National Lab. https://web.ornl.gov/sci/buildings/conf-archive/1982%20B2%20papers/036.pdf (accessed 14 July 2024).

55 Jelle, B.P., Hynd, A., Gustavsen, A., Arasteh, D., Goudey, H., and Hart, R. (2012). Fenestration of today and tomorrow: a state-of-the-art review and future research opportunities. *Solar Energy Materials and Solar Cells* 96: 1–28. https://doi.org/10.1016/j.solmat.2011.08.010.

56 Mirsadeghi, M., Cóstola, D., Blocken, B., and Hensen, J.L.M. (2013). Review of external convective heat transfer coefficient models in building energy simulation programs: Implementation and uncertainty. *Applied Thermal Engineering* 56 (1–2): 134–151. https://doi.org/10.1016/j.applthermaleng.2013.03.003.

57 Kahsay, M.T., Bitsuamlak, G.T., and Tariku, F. (2020). Effect of window configurations on its convective heat transfer rate. *Building and Environment* 182: 107139. https://doi.org/10.1016/j.buildenv.2020.107139.

58 USDOE (2024). Energy performance rating for windows, doors and skylights. Energy Saver, US Department of Energy, Washington, D.C. www.energy.gov

59 Aguilar-Santana, J.L., Velasco-Carraso, M., and Riffat, S. (2020). Thermal transmittance U-value evaluation of innovative window technologies. *Future Cities and Environment* 6 (1): 1–13. https://doi.org/10.5334/fce.99.

60 LBNL (2023). Windows software downloads and users manuals. Windows 7.8.74 (20 Dec 2023). LBNL, Berkeley, CA, windows.lbl.gov/software-tools (accessed 3 March 2025).

61 Kuhn, T.E. (2014). Calorimetric determination of the solar heat gain coefficient g with steady-state laboratory measurements. *Energy and Buildings* 84: 388–402.

62 Eubanks, P.-G. and Richman, R. (2018). Towards harmonizing the NFRC and CEN window performance simulation methods. *Canadian Journal of Civil Engineering* 46 (8): 687–703. https://doi.org/10.1139/cjce-2017-0636.

63 Onatayo, D., Aggarwal R., Srinivasan, R.S., and Shah, B. (2024). A data-driven approach to thermal transmittance (U-factor) calculation of double-glazed windows with or without inert gases between the panes. *Energy and Buildings* 305: 113907. https://doi.org/10.1016/j.enbuild.2024.113907.

2

Electrochromics

Carl M. Lampert[1,2,3], *Anoop Agrawal*[4,5] *and Junichi Nagai*[6,7]

[1] *Star Science, Cotati, CA, USA*
[2] *Formerly, Technical Director, Society of Vacuum Coaters (non-profit), Albuquerque, NM, USA*
[3] *Staff Scientist, University of California, LBNL, Energy Technologies, Berkeley, CA, USA*
[4] *Glass Dyenamics, Tucson, AZ, USA*
[5] *Formerly with Donnelly Corporation, Holland, MI, USA*
[6] *NanoFilm Laboratory, Yokohama, Japan*
[7] *Formerly Principal Researcher, Asahi Glass Co., Yokohama, Japan*

Nomenclature

A_λ	absorptance
a*	chromaticity coordinates (red-green)
ACS	American Chemical Society
AF	Air filtration (through a window)
AGC	Asahi Glass Company
AL	air leakage rate
AM	air mass (thickness of the Earth's atmosphere)
AR	augmented reality
ARPA-E	Advanced Research Projects Agency-Energy (part of USDOE)
ASTM	ASTM International (testing), (W. Conshohocken, PA)
AZO	aluminum zinc oxide
b*	chromaticity coordinates (yellow-blue)
C	coulomb (unit of charge)
C*	chromaticity
CAGR	compound annual growth rate (%)
CE	counter electrode
CE	coloration efficiency
CE_λ	spectral coloration efficiency
CE_{vis}	visible integrated coloration efficiency
CI	condensation index
CIE	Commission Internationale de l'Eclairage (Vienna, Austria)
CIELAB	CIE, L*a*b* color space (human perception of color and hue)
CIGS	copper indium gallium sulfide
CRI	color-rendering index
CVD	chemical vapor deposition

Chromogenics: Smart Switchable Optical Materials and Their Applications, First Edition. Edited by Carl M. Lampert.

D	diffusion coefficient ($cm^2\ s^{-1}$)
DC	direct current
e	elemental electron charge
ε_m	dielectric constant of the surrounding environment
ε_o	vacuum permittivity
e_{ir}	infrared emittance (spacecraft surfaces)
e_r	emittance ratio (for spacecraft surfaces)
EC	electrochromic layer (used only for tables and device constructions)
ECW	electrochromic window
EDG	electrochromic double glazing
EESD	electrochromic energy storage device
EGU	electrochromic glass unit (can be EDG or multiple pane window)
EMI	electromagnetic interference
ETG	electrochromic triple glazing
EU	European Union, Energy Union
γ	bulk collision frequency for free carriers (LSPR) (damping constant)
GZO	gallium doped zinc oxide
IC	ion conductor
ICO	indium cadmium oxide
IEA	International Energy Agency
IGU	insulated glass unit (also DGU, double glazing with gas fill)
IR	infrared
ITO	indium tin oxide
JRC	Joint Research Centre (JRC) Programme
LBNL	Lawrence Berkeley National Lab. (Berkeley, CA)
LC	liquid crystal
LCD	liquid crystal display
Low-e	low emittance (coating)
LSPR	localized surface plasma resonance
m_e*	electron effective mass
MITI	Japan's Ministry of Economic, Trade and Industry
MIR	mid-infrared
N	density of free carriers
nc	nanocrystalline, nanocrystal
NIR	near-infrared
NFC	near-field communication
NFRC	National Fenestration Ratings Council
NREL	National Renewable Energy Lab. (Golden, CO) (formerly SERI)
NTU	National Technical University, Singapore
NW	nanowire
OEM	original equipment manufacturer
OD	optical density
OLED	organic light emitting diode
polyAMPS	poly(2-acrylamido-2-methyl-1-propanesulfonic acid)
PANI	polyaniline
PB	Prussian blue
PDLC	polymer-dispersed liquid crystal
PEDOT	poly(3,4-ethylenedioxythiophene)
PE	polymer electrolyte
PET	polyethylene terephthalate

PEO	polyethylene oxide
PMMA	polymethyl methacrylate
PSS	poly(styrene sulfonic acid)
PVB	polyvinyl butyral
Q	charge (coulombs)
$\varphi_{sol}(\lambda)$	spectral solar irradiance
$\varphi_{vis}(\lambda)$	spectral solar visible irradiance
$\varphi_{p}(\lambda)$	photopic spectrum (daytime human eye response)
$\varphi_{s}(\lambda)$	scotopic spectrum (nighttime human eye response)
R	reflectance
RedOx	in this work- a reversible electron transfer reaction, with change of oxidation states
Rs	sheet resistance (Ω/sq.)
R_{val}	R-value, thermal resistance per unit area
RC	resistor–capacitor (circuit)
SHC	solar heating and cooling
SHGC	solar heat gain coefficient
SNO	doped tin dioxide (transparent conductor)
SPD	suspended particle device
T	transmittance
T_a	average visible transmittance (mathematical average, not weighted)
t	pulse duration time
T_c,T_b	transmittance colored, transmittance bleached
TC	transparent conductor
TCO	transparent conductive oxide
ΔT	change in transmittance
$\Delta T(\lambda,t)$	change in monochromatic transmittance with time
$\Delta T_{max}(\lambda)$	maximum change in monochromatic transmittance
$d\lambda$	incremental wavelength interval.
τ	time constant
$T(\lambda)$, T_λ	monochromatic transmittance at wavelength λ
$Te(\lambda)$	average solar transmittance between equal energy intervals
T_{lum}	luminous transmittance (visible)
T_{sol}	solar transmittance
T_{uv}	ultraviolet transmittance
T_{vis}	visible transmittance
U, U-factor	coefficient of heat transfer (U-factor), $W\ m^{-2}\ K^{-1}$
US	United States of America (USA)
USDOE	US Department of Energy
UNDP	United Nations Development Programme
UV	ultraviolet, spectral regions: UVA, UVB, UVC
VIS, Vis, vis	visible light
VLT, VT	visible light transmittance
VR	virtual reality
ω_p	plasma frequency
ω_{LSRP}	localized surface plasma resonance frequency
λ	wavelength
XR	extended reality
x,y,z	color coordinates – tristimulus values
X,Y,Z	non-physical primaries, color coordinates

2.1 Introduction to the Field of Electrochromics

This chapter focuses on electrochromic key materials, device structures, deposition technologies, and applications. This work covers the evolution of mainly inorganic electrochromics for medium- and large-area applications. In addition, it is important to include the development of organic-based electrochromic products which are significant in the development of the first switchable products. A greater emphasis has been placed on automotive mirrors and building and aircraft windows as these are the most commercially available applications. An electrochromic is a material built like a thin transparent battery where its level of coloration is dependent on the extent of charge. For the case of a cathodic electrochromic (e.g. tungsten oxide), highly charged means deeply colored while discharged means colorless or a slight tint. Electrochromics have a gray scale with intermediate coloration between light to dark. For example, the amount of injected charge determines how dark the electrochromic will get. Some electrochromic devices are designed to have a long-term memory such as for architectural window applications. For car mirrors, they are designed to be completely self-discharging for safety reasons. Electrochromic self-dimming mirrors are a standard product available on the major makes of cars. The majority of all electrochromic product sales are made in this product area. The Gentex Company (Zeeland, MI) holds the largest share of the automotive mirror market. They are followed in market size by Magna International (Aurora, Canada). Prior to the development of electrochromic mirrors, the competing products were mechanical mirrors and liquid crystal mirrors. Several major carmakers have investigated and developed switchable glazing, sun visors, and sunroofs to help control the air conditioner energy load. Chromogenic sunroofs are slow to be adopted, but they are moving forward in the marketplace. They have become more important for electric cars, to help reduce the cabin air-conditioning energy consumption.

Over the years, there have been a lot of developers and manufacturers in the electrochromic arena, ranging from major glass and chemical companies, government and university laboratories to small start-ups and independent inventors. Many government agencies have funded work on electrochromic development and testing. Those agencies include the United States Department of Energy (USDOE), Advanced Research Administration-Energy (ARPA-E, United States), NRC (Canada), INRS (Canada), Energy Union (European Union), Joint Research Centre Programme (European Union), C40 Cities (UK, South Africa, China), China Buildings Programme (China), United Nations Development Programme (UNDP), International Energy Agency (IEA, France), Japan's Ministry of Economic, Trade and Industry (MITI, Japan), Swedish Research Council, Swedish Energy Agency, Industrifonden (Sweden), DRDO (India), Ministry of Science and Technology (India), TUBITAK (Turkey), Russian Science Foundation, IREC (European Union), Fraunhofer Institute (Germany), DIFFER (The Netherlands), SINTEF (Norway), Ministry of Energy (Saudia Arabia), MOHE (Malaysia), Australian Academy of Science, Baesa (Brazil), and ENERCAN (Brazil). These agencies have funded the research, development, and testing of chromogenic materials for building windows. The driving force is to make buildings more energy efficient by reducing carbon footprints, boosting the economy, creating new jobs, and expanding manufacturing. The zero-energy building is under discussion and studied in several regions of the world. It is designed to further reduce carbon and energy demand [1]. The IEA has developed a global roadmap to become carbon neutral by 2050 [2]. Also, the IEA has proposed developing a global energy efficiency workforce in the buildings sector, which over a six-year period could result in 1.3 million new jobs [3]. Zero-energy windows are part of the solution [4].

Dynamic chromogenic windows, combined with highly insulating window components and efficient use of daylight, can help us lower building energy demand [5]. Buildings are responsible for 30% of the global energy needs and 26% of their carbon emissions (IEA) [6]. The annual world consumption of energy is estimated to be 580 million terajoules (2024). If business remains the same, energy consumption is expected to be 740 million terajoules by 2040. China consumes 27% of the world's energy, the United States consumes 16%, and the EU countries consume 10% [7]. In the United States, buildings account for 40% of the energy use in the United States, consume 75% of electricity, and are responsible for 35% of the nation's carbon emissions (NREL) [8a]. In the United States, the peak cooling demand in the summer could be reduced by 25–58% [8b].

By dynamically controlling the light transmission through the windows, space heating needs are decreased in winter by allowing more sunlight through the windows and in summer by limiting it. There are times in the day when judicious control of sunlight through the windows decreases lighting requirements. The energy reduction is dependent on several parameters, which include geographical location of the building, building orientation, building size to window area, and the ratio of window area to wall area. Several physics-based building simulation energy modeling studies using DOE-2 and EnergyPlus have been performed [9–11]. In simulation work done at the University of Southern Florida (Tampa, FL), both the effects of electrochromic windows and thermochromic roofs were studied for buildings in Tampa and Chicago [12]. The United States Department of Energy (USDOE) has developed a window and building envelope road map for emerging technologies, including electrochromic windows [13]. The use of electrochromic windows will decrease building energy cost, by reducing air-conditioning, heating, and lighting consumption, and by reducing demand during peak energy usage. It has been estimated that the overall reduction in building electrical energy will be about 20% and the total building energy use (including heating, cooling and lighting) will be decreased by 10%. In addition, modern architectural designs use and increasing amount of glass, therefore electrochromic technology can provide benefits to control the flow of energy through glazing.

A measure of research activity in the field of electrochromics is shown in Figure 2.1. Figure 2.1 was created using the Web of Science and the search terms "electrochromic" and "electrochromics". After 1998, there was a considerable upswing in research articles with over 800 papers published in 2020 [14]. On the patent side, there are over 5080 patent families in electrochromics with about 38.5% granted and active, 21.3% pending, and 40.2% expired (2024) [15].

2.2 Electrochromic Materials

Electrochromic materials change their optical properties due to the action of an applied electrical potential and can be changed back to the original state by a reverse potential. There are three major categories of electrochromic materials: (i) transition metal oxides such as WO_3 and NiO_x, which can be intercalated with ions forming polarons, or a color indicating reduction-oxidation (RedOx) compound, such as Prussian blue ($Fe_4(Fe(CN)_6)_3$); (ii) organic compounds such as viologens (4-4′bipyridine salts); and (iii) polymeric materials such as poly(3,4-ethylenedioxythiophene):poly(styrene sulfonic acid), also known as PEDOT:PSS. The electrochromic effect occurs in inorganic metal oxides by dual injection (cathodic) or ejection (anodic) of ions and electrons. The ion intercalation involves an ion, and a small

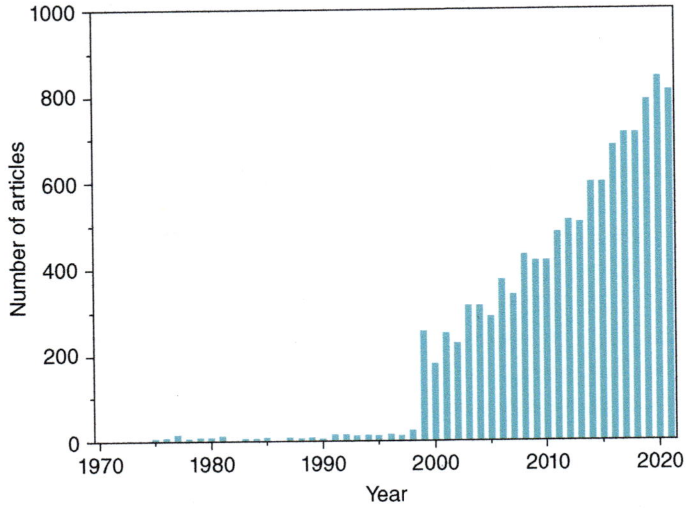

Figure 2.1 The number of electrochromic research articles published by year according to the Web of Science. This is a journal data base, and the chart was created using the search terms, "electrochromic and electrochromism". *Source:* Gu et al. [14], image courtesy of the American Chemical Society, CC BY4.0.

polaron (lattice distortion) with a localized electron. A simplified reversible reaction for a cathodic coloring material (WO_3) using lithium ions (Li^+) and electrons (e^-) as a cathodic coloration ion is:

$$WO_3\ \text{(colorless)} + yLi^+ + ye^- \longleftrightarrow Li_yWO_{3-y}\ \text{(blue)} \tag{2.1}$$

A reversible reaction for an anodic coloring material ($LiNiO_y$) with the extraction of Li^+ and e^- is

$$Li_z\,NiO_{y-z}\ \text{(colorless)} - zLi^+ - ze^- \longleftrightarrow NiO_y\ \text{(bronze)} \tag{2.2}$$

These reactions show that by using two different electrochromic materials, composed of one layer that colors upon intercalation and another that colors on deintercalation, both sides of the device can color at the same time giving high optical density. Depending on the electrochromic various coloration ions can be used, such as H^+, Li^+, Na^+, K^+, Al^{+3}, and Ag^+. However, Na^+ and K^+ ions are so large that they are hard to move and have much smaller diffusion coefficients compared to a proton (H^+) in WO_3. The diffusion coefficients are H^+: $D_{H+} = 4\times10^{-10}$ cm^2 s^{-1} in a-WO_3 and $D_{H+} = 2\times10^{-11}$ cm^2 s^{-1} in c-WO_3 [16], Li^+: $D_{Li+} = 4.76\times10^{-13}$ cm^2 s^{-1} in a-WO_3, and Na^+: $D_{Na+} = 2\times10^{-16}$ cm^2 s^{-1} in $Na_{0.054}WO_3$ [17]. Cathodic coloring inorganic materials that have gained the most interest are WO_3, MoO_3, $Mo_xW_yO_3$ [18], $Ni_xW_yO_3$ [19], TiO_x, VO_x, NbO_x [20, 21], CuO_x [22], and TaO_x [23, 24]. Anodically coloring materials of interest are NiO_x [24–26], V_2O_5, MnO_x [27], IrO_x [28], CoO_x [29], RhO_x [30–32], and Prussian blue (PB) [33]. Depending on the phase, vanadium compounds can be either cathodic or anodic. Phase, physical structure, particle size, particle shape, and composite systems are becoming very important in the tuning and optimization of electrochromic characteristics. This is especially germane to plasmonic electrochromics, which are used for dual-band switching.

Figure 2.2 is the periodic chart showing the EC metal oxides. *Please note, this chart is a summary but is not meant to be exhaustive in covering all the work on oxide electrochromics.*

Amorphous Bi_2O_3 has been added to the table [34]. Also, an alloy of 30% antimony and 70% tin oxide film was noted to be electrochromic [35]. Monsanto studied $Sb{:}SnO_2$ films for printed electrochromics. Cerium oxide has been studied as a non-coloring ion storage material [36]. The transparent conductor (TC), $Sn{:}In_2O_3$ (ITO) was evaluated and determined not to be electrochromic [37]. However, in another study using long-period potentials of ITO in $LiPF_6$ electrolyte and Li metal, a light blue color developed between potentials of 1.5–3.5 V, and $\Delta T_{380\,nm} = 22.5\%$ [38]. Further work needs to be done to determine if the film is being reduced (as reduced $InSnO_2$ is blue) and is stable over long cycling. Indium has not been marked as electrochromic in Figure 2.2. PrO_x was found to have a strong anodic orange electrochromism [39]. Various alloy systems have been studied by the Prof. Claes Granqvist group (Uppsala Univ., Angstrom Lab., Sweden) as potential electrochromic layers, and the examples include $CrNiO_x$, $MgNiO_x$, $MgIrO_x$, and $Ni_{1-x}V_xO_y$ [40]. Rare-earth doping with 2–6% Ce, Sm, Eu, and Gd was studied in porous nanofilms of WO_3. Improved charge capacity and ΔT range were noted for Eu and Sm dopants in nanofilm WO_3 [41]. A research review on sputtered WO_3 coatings covers considerable work done on this popular cathodic electrochromic oxide [42]. Work on one of the most popular anodic electrochromics, NiO_x, includes developments on alloying NiO: with Mg, Al, Si, V, Zr, Nb, Ag, or Ta, to make a clearer bleached color. This alloy study was worked on by Avendaño et al. from the Granqvist group. It was found that the best alloys were Al_xNiO and Mg_yNiO and gave $T_{lum} = 0.85$ in the bleached condition [43].

Another interesting area to look for developments in electrochromic materials is electrochromic energy storage devices (also known as EESD). In this case, the electrochromic can be used as an integrated charge indicator for a battery [44, 45]. These battery devices may cross over to integrated power systems for electrochromic windows (also known as ECWs) or multifunctional systems [46]. There are potentially several metal-ion battery materials with promising optical and chemical properties for EESDs. Metal ion battery materials include oxides, sulfides, phosphides, nitrides and compounds of Li, Na, K, Mg, Cd, Al, Ni, and Zn. With further investigation some of these materials can be developed for EESDs. Also, polymer electrochromics could be developed for EESDs.

Less common electrochromic materials are $Cs_2O_{0.44}W_2O_4$ [47]; chlorides such as $RbPdCl_4$, Rb_2PdCl_6, and Rb_2AuCl_4; and bromides such as Rb_2PdBr_4 and $RbAuBr_4$ [48]. Mg_xNi alloys are used to make electrochromic selective infrared (IR) emitters and visible gasochromic mirrors [49]. Also, $Mg_{1-z}Y_z$ ($z > 0.5$) is being studied to make gasochromic mirrors [50]. Furthermore, $Mg_{24}Y_5$, Mg_2Y, and MgY are being researched for both electrochromics and gasochromics [51]. Also, $GdMgH_5$ has been investigated as an anodic electrochromic with a metallic phase [52]. Co-dopants Nd and Nb have been investigated in SnO_2/a-WO_3 films [53]. Some dopants are used to enhance ion transport in ion conductors (ICs) and electrochromic layers, but are not by themselves electrochromic, such as $La_{0.67-x}Li_{3x}TiO_3$ [54]. A nickel phosphate, $NiHPO_4{\cdot}3H_2O$, was identified as being an anodic electrochromic with a Ni bronze coloration (the phosphate ion is not electrochromic). A device made from the nickel phosphate had a change in transmittance of $\Delta t_{500\,nm} = 73.5\%$ and coloration efficiency (*CE*) of $CE_{500\,nm} = 66.5\,cm^2\,C^{-1}$ [55]. Furthermore, 2D nano platelet structures offer a new range of surface activity to speed up electrochromic reactions, if they can be stabilized and economically scaled it could result in several new electrochromic materials. One application could be the creation of printable electrochromic inks. A broadband (visible and near-infrared [VIS–NIR], 320–1100 nm) electrochromic based on MnO_2 and PB has

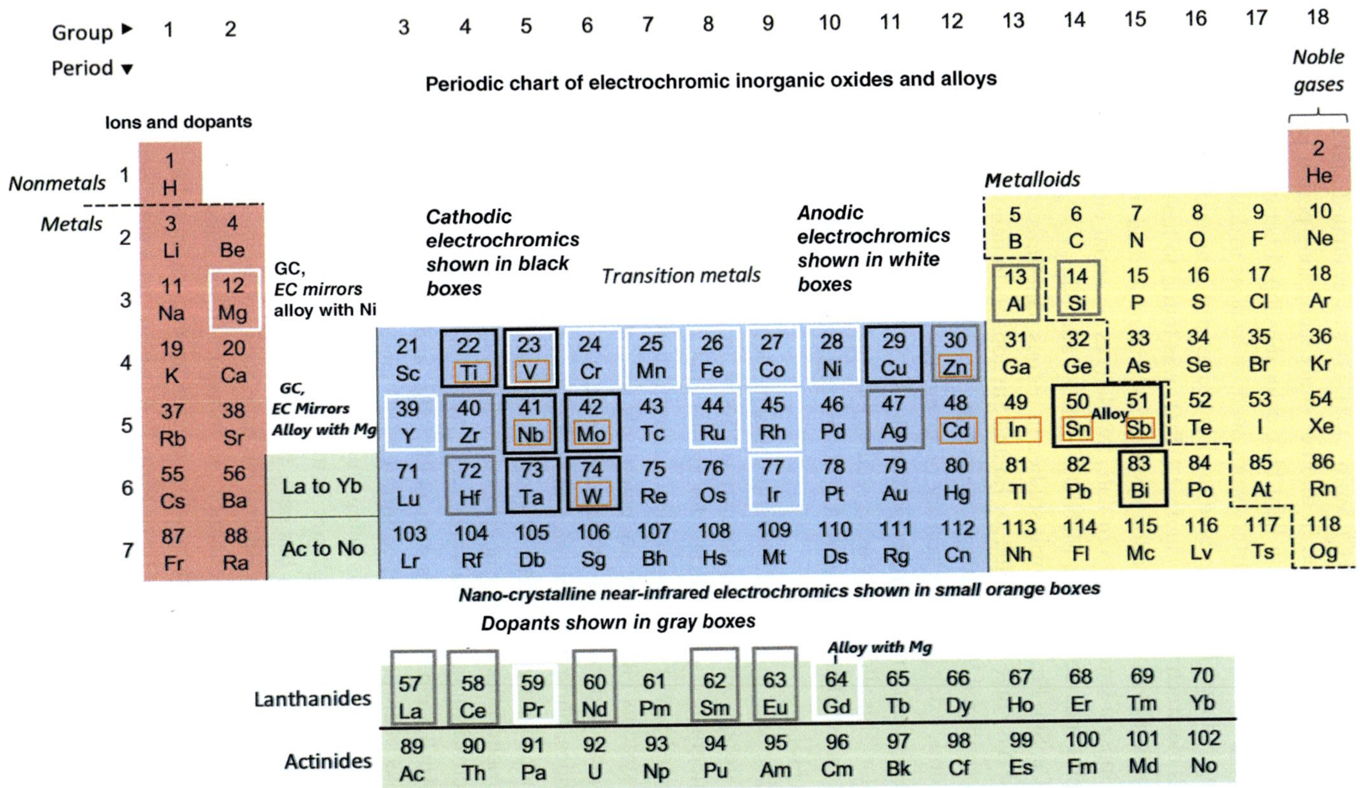

Figure 2.2 Periodic chart showing electrochromic metal oxides. Cathodic electrochromics are in black boxes, anodic electrochromics in white boxes. Dopants are shown in gray boxes. Some dopants are also electrochromics, e.g. Mo_yWO_{3-y}. Localized surface plasma resonance (LSPR) near-infrared electrochromics are nanocrystalline (nc) ITO, AZO, In:CdO_2, $NbTiO_2$, and $Nb_{12}O_{19}$. Also, composite systems show characteristic surface plasma resonance shifts, e.g. $WO_{3-x} + NbO_x$ glass. LSPR oxides are shown in small orange boxes. There are many more compounds yet to be discovered, especially the co-doped nanocrystals, such as F,In:CdO. Visible LSPR metals are Au, Ag, In, and Zn (not marked on chart). Many of the Group 1 and 2 elements are used to open up structures for better ion transport, e.g. Ca_xWO_3 and Gd:WO_3. Alloys of SnSb have shown electrochromism. Alloys of MgY_x and $MgNi_x$ have shown electrochromic mirror characteristics. *Please note, this chart is a summary but not meant to be exhaustive in covering all the work on oxide electrochromics. Source:* Image heavily modified by author (CL) based on periodic chart by Sandbh, CC BY-SA 4.0, https://creativecommons.org/licenses/by-sa/4.0, via Wikimedia Commons.

been reported to block 72% of the solar irradiance, from the Pooi See Lee group, National Technical University (NTU), Singapore [56].

There are probably more oxides, non-oxides, sulfides, nitrides, fluorides, and phosphates and alloy compounds that are electrochromic. Considering this and developments in new materials synthesis, there is a good possibility that new electrochromic compounds will be discovered. That doesn't include all the extensive work in organics and polymer electrochromic – see Chapters 3 and 4 covering organic electrochromics and polymer electrochromics, respectively. A review of multifunctional electrochromic devices is available by Rajesh Kumar's group at IIT Indore reporting in American Chemical Society (ACS) Energy Letters [57]. The work includes collaboration with Prof. Pooi See Lee, NTU Singapore.

2.2.1 Coloration Efficiency

One of the unique characterization parameters for an electrochromic material is its coloration efficiency. In some papers, coloration efficiency (also known as *CE*) is reported as a single number usually measured from the peak wavelength; however, this can be misleading since coloration efficiency is wavelength dependent in several materials. Many tungsten oxide citations will note that the coloration efficiency is given for 900 nm which is in the near infrared (NIR). WO_3 performs better at 900 nm than 550 nm. But if we are speaking of its visual performance, at a minimum the coloration efficiency should be given for the peak of the visible spectrum at 550 nm.

Spectral coloration efficiency (CE_λ) signifies the amount of injected or extracted charge it takes to change the optical density of the electrochromic from light to dark at a specific wavelength (λ). For electrochromics, CE_λ is defined as:

$$CE_\lambda = (\log T_{b\lambda}/T_{c\lambda})/Q \tag{2.3}$$

where

$T_{b\lambda}$ is the bleached transmittance at λ,
$T_{c\lambda}$ is the colored transmittance at λ, and
Q is the applied charge to obtain the coloration change ($cm^2\ C^{-1}$).

Whether the electrochromic is anodically or cathodically coloring will determine if the charge is injected or extracted. The applied charge is measured using a potentiostat in the galvanostatic mode [58, 59]. The higher the coloration efficiency value, the less charge (energy) is needed to gain coloration. Organic electrochromics can have a CE_λ as high as 20,000 $cm^2\ C^{-1}$. The coloration efficiency for inorganics can be as high as $CE_\lambda = 120\ cm^2\ C^{-1}$. It is important to note that the coloration efficiency is not a constant for WO_3, but wavelength dependent, see Figure 2.3. Many investigators make this error in technical papers. It was pointed out in the IEA and solar heating and cooling (IEA–SHC) Task 10 work that integrated spectral coloration efficiency would be a more accurate term [60]. The integrated coloration efficiency for the photopic VIS region is:

$$CE_{vis} = \int_{380\ \text{nm}}^{750\ \text{nm}} CE_\lambda(\lambda)\,\varphi_p(\lambda)\,d\lambda \Big/ \int \varphi_p(\lambda)\,d\lambda \tag{2.4}$$

where $CE_\lambda(\lambda)$ is the spectral *CE* function ($\lambda = 380$–750 nm) and $\varphi_p(\lambda)$ is the photopic energy intensity.

The photopic spectrum along with other optical and thermal performance measurement techniques are discussed in Chapter 1B – Introduction to Measurements.

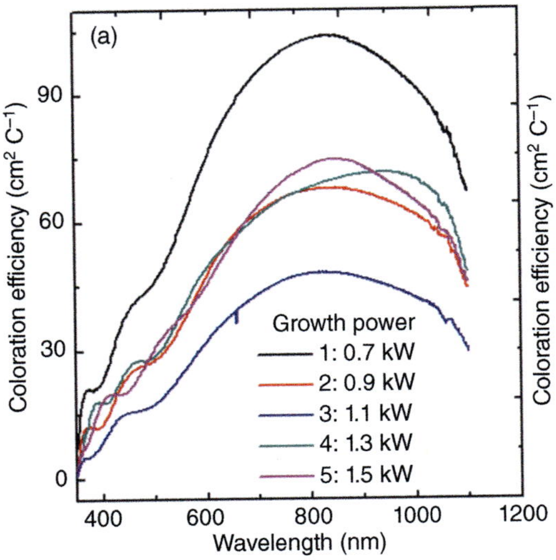

Figure 2.3 Spectral coloration efficiency (CE_λ) given for five applied power conditions using pulsed direct current (DC) sputtering of tungsten/oxygen. The lower coloration efficiencies are caused by densification of the WO_3 film [61]. Note that coloration efficiency is not constant over the entire range of wavelengths. *Source:* Tuna et al. [61] with permission, courtesy of Elsevier.

2.3 Electrochromic Device Design

Electrochromic devices are designed so they shuttle ions and electrons in or out of the electrochromic layer with applied potential. The device colors or bleaches during this process. An electrochromic glazing can be fabricated from as little as three to five layers consisting of two transparent conductors, an ion conductor or electrolyte, counter electrode, and electrochromic layer. Depending on the components used in devices, some of the layers (or chemical components) can be combined to serve multiple functions. Some devices use even more layers depending upon design. Thin interfacial layers can be added to increase performance or control diffusion of detrimental chemical species. The most promising ion conductors are open-structure metal oxides, ionic glasses, or quasi-solid/solvent polymer electrolyte (PE) systems.

There are three types of electrochromic device constructions. Type 1 is the simplest with both the electrochromic and counter electrode (usually a RedOx couple) materials dispersed into the electrolyte. Normally these structures self-discharge when power is removed. Some electrochromic mirrors are based on this design. Type 2 is where one of the electrochromic or counter electrode materials is dispersed into the electrolyte. Type 3 is where all layers are separated into regions in a sandwich structure. Type 3a is a polymer electrolyte (PE)-laminated device as shown in Figure 2.4. This electrolyte laminates the electrochromic layer to the counter electrode layer (also known as the CE layer). Type 3b is a solid-state device using an ion conductor, as shown in Figure 2.5. Commercial products are generally constructed following Type 1 or Type 3a or Type 3b designs.

Electrochromic devices are constructed using generally four types of structures:

(1) substrate/TC or metal/EC + PE + RedOx/TC/superstrate
(2) substrate/TC/EC/PE + RedOx/TC/superstrate
(3a) substrate/TC/EC/PE/CE/TC/superstrate
(3b) substrate/TC/EC/IC/CE/TC/sealing layer or bonded superstrate

where PE is a viscous polymeric electrolyte, usually with sealing properties.

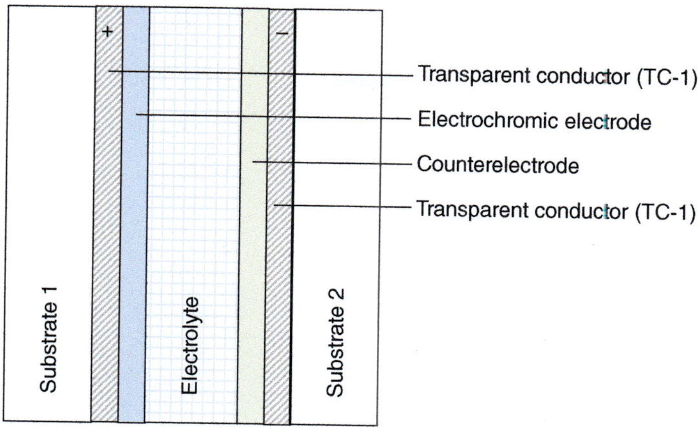

Figure 2.4 Cross section of a laminated electrochromic construction (Type 3a). *Source:* Author (AA).

Figure 2.5 Cross section of a layered electrochromic construction on a single substrate (Type 3b). *Source:* Author (AA).

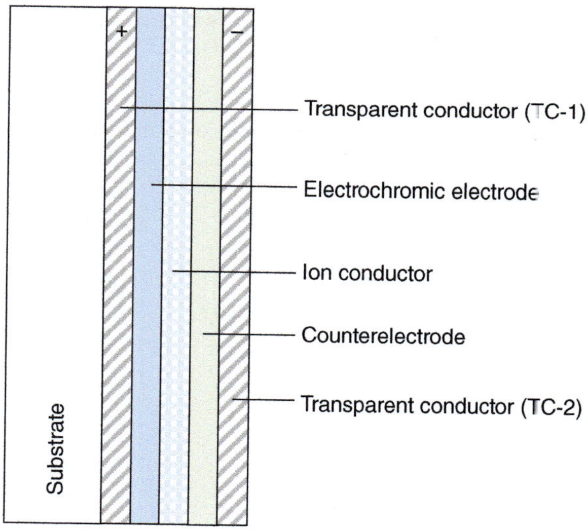

The substrates and superstrates are glass or flexible plastic such as polyethylene terephthalate (PET). The development of electrochromic devices on flexible substrates has been reviewed by Prof. Pooi See Lee's group at Nanyang Technological University, Singapore [62]. Work at the Angstrom Laboratory (Uppsala, Sweden) has been detailed by Prof. Granqvist and Dr. Andris Azens (AGL Technologies SIA, Riga, Latvia) [63–66]. The performance of two very common electrochromic device constructions (Types 3a and 3b) are depicted in Figure 2.6. This diagram shows the movement of ions and electrochromic ions to color the layer. More typical types of materials used in electrochromic devices are shown in Table 2.1.

In Table 2.1, VS is a viologen salt (cathodic), and DDP is dihydro-dimethyl phenazine (anodic), both dissolved in a solvent along with a nonionic polymer and optional salts. Fc is ferrocene (anodic) dissolved in the electrolyte along with a lithium salt and a nonionic polymer; PB is Prussian blue ($C_{18}Fe_7N_{18}$), PANI is polyaniline, PODS is polyorganodisulfide [69], and PEDOT is poly(3,4-ethylenedioxythiophene) is a conducting polymer based on 3,4-ethylenedioxythiophene or EDOT; and PXDOT is poly(3,4-alkylenedioxythiophene) derivatives featuring various structural substitutions. For the anodic counter electrode, MO_x,

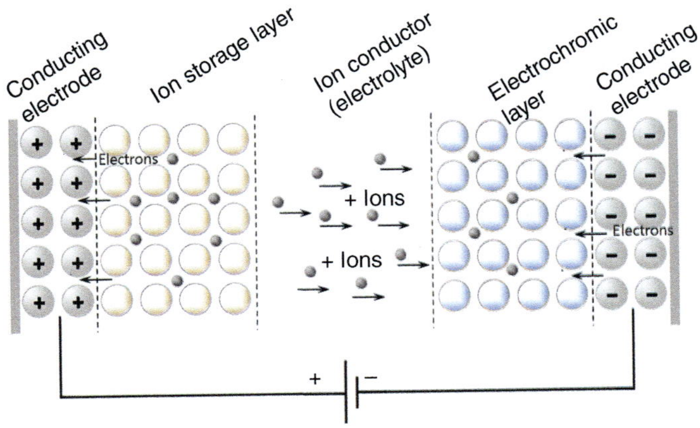

Figure 2.6 Schematic of a cathodic electrochromic device structure. Types 3a and 3b showing the active layers. The ionic and electronic intercalation movement to color the electrochromic layer is shown. In a complementary device, the ion storage layer would be an anodic electrochromic. The anodic electrochromic layer colors when ions and electrons deintercalated from it. *Source:* Modified by CL: By Keyur.tithal - Own work, CC BY-SA 3.0, https://commons.wikimedia.org/w/index.php?curid=40874975.

Table 2.1 Most common types of inorganic and organic electrochromic material combinations.

Cathodic Electrochromic	Ion conductor	Anodic Counter electrode	Ion	Structure Type
In electrolyte	VS + DDP	In electrolyte	Organic	1
WO_3	Li–polymer + Fc	In electrolyte	Li^+	2
WO_3	Li–polymer	TiO_2 + dye	Li^+	3a
WO_3	H–polymer	PB	H^+	3a
WO_3	H–polymer	PANI	H^+	3a
WO_3	Li–polymer	$LiNiO_x$	Li^+	3a
WO_3	Li–PEO	PODS	Li^+	3a
WO_3	$LiSiO_2$	MO_x	Li^+	3b
WO_3	Ta_2O_5	MO_x	H^+, Li^+	3b
$MoWO_x$	Li–Ta_2O_5	MO_x	Li^+	3b
PXDOT	Li–polymer	PEDOT	Li^+	3a

In the table are shown only major components with simplified structures. Other layers seen in commercial devices, such as top protective layers, diffusion barriers, and interfacial layers are not shown in this table [67, 68]. Compiled by author (CL).

M = Ni, Ce, Ti, V, Ir, Rh, Co, Fe, Mn, or Cr. The anodic counter electrode may also be an electrochromic, such as NiO_x, IrO_x, V_2O_5, MnO_x, Mn_2O_3, TiO_2, PrO_x, Li_yNiO_x, and $Li_xAl_yNiO_z$. For cathodic WO_3, there are many variants including Li_yWO_x, Nb_yWO_x, and Ti_yWO_x. One function of the dopant is to open up the a-WO_3 structure to speed up ion transport. Ion conductor compounds include Ta_2O_5, TiO_2–CeO_2, $LiAlF_4$, LiV_2O_5, and $Li_yCe_xTi_zO_2$.

Many devices use classic ionic salt–solvent electrolytes or ionic salt–liquid polymer electrolytes for the study. A classic protonic polymer electrolyte is polyAMPS or poly(2-acrylamido-2-methyl-1-propanesulfonic acid). PolyAMPS was used in electrochromic information display devices in the 1980s. An example is lithium triflate ($LiCF_3SO_3$) in polyethylene oxide (PEO) [69] or lithium perchlorate in liquid propylene carbonate [70, 71]. Other polymer electrolytes include poly(propylene oxide), polyvinyl butyral (PVB), polyacrylonitrile, poly(vinyl chloride), poly(methyl methacrylate), poly(vinylidene fluoride)–hexafluoropropylene, poly(vinylidene chloride), hydroxyethyl methacrylate, N-phenylglycine, and polyurethanes. Lithium salts such as $LiClO_4$, $LiPF_6$, $LiBF_4$, or $LiAsF_6$ are used to prepare polymer electrolytes [72]. Polymer electrolytes have been reviewed by Thakur et al., Nanyang Technical University, Singapore [73].

Also, ionic liquids have been studied to some degree for use in electrochromic devices. The area of ionic liquids is still developing. Ionic liquids are interesting because they have a very wide voltage stability characteristic, without breaking down or generating unwanted byproducts. The ionic liquid, *N*-butyl-*N*-methylpyrrolidinium bis(trifluoromethanesulfonic)imide ($PYR_{14}TFSI$) showed ionic conductivity of $\sigma_{Li+} = 1.5 \times 10^{-3}\ S\,cm^{-1}$ (at room temperature) with a stability window of 5 V [74]. Also in 2022, Zhen at Norwegian University of Science and Technology studied 1-butyl-3-methylimidazolium iron tetrachloride ($BmimFeCl_4$) showed thermal conductivity of $\sigma_{th} = 1.1 \times 10^{-3}\ S\,cm^{-1}$ with a stability window of 3.5 V [75]. A special issue of *Chemical Reviews* is devoted to ionic liquids [76]. Much of this work is being driven by the development of lithium batteries. Many of these liquids are highly colored due to impurities. So, synthesizing a neutral color or colorless form for an electrochromic application is still a synthesis challenge. There has been some success as the compounds, 1-butyl-3-methylimidazolium bromide (BMIMBr), 1-butyl-3-methylimidazolium tetrafluoroborate ($BMIMBF_4$), and 1-hexyl-3-methylimidazolium bromide (HMIMBr) have been shown to be colorless using an "on water" technique to reduce impurities [77].

The conductor layers are transparent conductors made of doped tin oxide or indium tin oxide, thin metals, metal mesh, graphene, nanoparticles, nanowires (also known as NWs), and conductive organics, such as PEDOT. A range of other conductors have been devised but not widely used for transparent electrochromic windows. These materials are discussed in Section 2.3.1. Electrochromic devices use an ion-containing and ion-transporting material (electrolyte) in intimate contact with the electrochromic layer or the transparent conductor for device Types (1) and (2). The conductors must make good ohmic contact with the electrochromic (as a layer or a dispersion) and the polymer electrolyte. The conductors serve as electron injectors or extractors from the active layers. Because the conductor is in contact with the active layers, the conductors must be stable at the applied range of potentials and to the local chemistry of the conductor junction. Over driving voltages or high electrical currents can easily damage these conductors.

A variation of the electrochromic design is the dye-sensitized solar cell structure (also known as the DSSC). This type is fashioned after the DSSC photovoltaic structure. The device relies on a dye-sensitized anatase titanium oxide layer as the counter electrode. Also, it uses a catalyst in the structure. Different structures and their performance have been studied by the Mendes group at University of Portugal-Porto [78a]. Donnelly Corporation (Holland, MI), NREL (Golden, CO), NTera (Dublin, Ireland), and The Grätzel group at École Polytechnique Fédérale de Lausanne (EPFL) (Lausanne, Switzerland) have worked on DSSC electrochromics. There have been other studies on self-powering electrochromics using solar cells, energy storage devices, and triboelectric generators [78b]. Much of this work is still under research.

Electrochromic devices have been successfully used in several commercial applications such as self-dimming rearview automobile mirrors, eyewear, printed displays, windows for transportation, and building windows. Of these applications, the major motivation for the first three is the glare control and visible optical contrast that these devices provide, whereas the major driver for building windows and those for transportation are energy savings, glare control, and privacy, which are impacted both by VIS and the NIR changes. Since the requirements for different applications are varied, several types of electrochromic devices and materials have been developed. A later section will discuss commercial device design.

During the development of electrochromic glazing among other switchable media, the working group of electrochromic researchers formulated testing procedures for the SHC Program of the IEA Task 10, involving participants from 12 countries [60]. These testing procedures were later published by NREL (SERI) [79a] and actual electrochromic window testing results were made by Lampert, et al. [79b]. It has been noted in some Li-WO_3-based electrochromic windows that a reduction of ΔT_{solar} and ΔT_{lum} transmittance are seen after a large number of cycles. This has been shown to be caused by trapping Li ions in the device layers. A method of detrapping using potentiostatic pulsing with UV light has been devised by the Granqvist group [80]. Beyond just glazing, energy consumption and switching performance are important to other chromogenic products, such as smart labels and color changing or emittance changing surface treatments. More details about testing procedures are in Chapter 1B.

2.3.1 Transparent Conductors

It is important to cover transparent conductors since they are a key component to the electronic part of the electrochromic device and must have sufficient transparency for window applications and for top conductors in nontransparent applications. A number of transparent conductors have been developed for a range of applications, including flat panel displays, transparent heating elements for defrosting aircraft windows, electro-magnetic or interference (EMI) shielding, photovoltaics, and low-e coatings for architectural glass [81]. Further information on low-e coatings is covered in Chapter 1B on Introduction to Glazing Design and Measurements. For transparent electrochromic windows and mirrors (top electrode), a transparent conductive oxide (also known as TCO) is preferred. Doped tin oxide (SNO) and indium tin oxide (ITO) are used due to the ease of deposition. The most economic conductor is a form of SNO, SnO:F/glass, which can be produced on the glass float line by pyrolysis. Dopant variants for SnO_2:M are M = Sb, Nb, Ta, or As. See Chapter 13 on pyrolytic tin oxide for further details about the manufacturing of this coating. Roy Gordon of Harvard University has assembled a work on the criteria for choosing TCOs [82]. Claes Granqvist has reviewed transparent conductors for solar applications including chromogenic windows [83]. ITO is generally used on electrochromics with flexible substrates. The price and availability of indium has been a subject of concern, since so much is used in electronics, especially cell phones, touch screens, and copper indium gallium sulfide (CIGS) solar cells. Further information about ITO coatings on plastics can be found in Chapter 12 on vacuum web coating. There are dopant variants of ITO or In_2O_3:M where M = Mo, Ta, W, Zr, F, W, Ge, Nb, Hf, or Mg. The best substitutes for ITO appear to be ZnO:Al, (AZO) and ZnO:Ga (GZO). Also, there are dopant variants of ZnO:M, where M: H, F, In, B, V, Li, Mg, Y, Sc, V, Si, Ge, Ti, Zr, Hf, or As. Other transparent conductor materials include thin-metal films and metal mesh [84],

Ag and Cu NWs and nanotroughs [85], and graphene [86]. For nanowire sheet networks, AgNW has shown T_{vis} = 94.1% with sheet resistance Rs = 12.4 ohms/sq. and Cu NWs have shown T_{vis} = 94.4% and Rs = 60 ohms/sq. A gold nanomesh has shown T_{vis} = 82.5 with Rs = 21 ohms/sq. and Au nanotroughs has shown T_{vis} = 94 and Rs = 7.5 ohms/sq., as reported by Guo and Ren [85]. The cadmium stannates (e.g. Cd_2SnO_4) are excellent TCOs, as studied by Gottfried Haacke, American Cyanamid [87], but environmentally impractical. Chavan et al. has reviewed transparent conductors and their deposition processes for solar cells [88]. Afre et al. (2018) has reviewed TCOs for various applications [89]. Also, Ginley [90] and Gao [91] have reviewed the physics of transparent conductors. A classic study of transparent conductive oxides was written by the well-known Prof. Kasturi Chopra of IIT India [92].

2.3.2 Electrochromic Device Switching Time

The switching time for suspended particle devices (SPDs), polymer-dispersed liquid crystals (PDLCs), and liquid crystals is well defined for on and off conditions in most applications. Since they rely on a physical or structural change, switching is quick and complete. These are fast switching systems (milliseconds) when compared to electrochromics, which can take many seconds to several minutes to switch to color or bleach. Measuring an accurate switching speed of an electrochromic is somewhat difficult and techniques are not standardized in the general literature. Since electrochromic devices rely on the diffusion of ions and electrons, they are inherently slower systems compared to SPDs and liquid crystals. Electrochromic devices are the closest to a battery in behavior with its characteristic capacitive and resistive elements. The steady-state charge and discharge end points are harder to exactly define, but we need a convention to compare different electrochromics. From the mid-1980s to 1990s, the IEA (Task 10) evaluated chromogenic technology for use in glazing. The group consisted of electrochromics researchers from several countries including the Asahi Glass group who supplied electrochromic glazing for testing. Originally, the Task 10 group recommended that time to switch between transmittance of $T = 10\%$ and 90% of the overall $\Delta T_{max} = 100\%$, be defined as the switching time (t) [60]. Later, the electrochromic switching time was modified by NREL (Golden, CO) in conjunction with LBNL (Berkeley, CA) [79a, 79b]. The electrochromic device was modeled by a resistor–capacitor circuit, with the standard time constant $(t) = 1 - e^{-1}$, where steady state is achieved in five-time constants. This was confirmed by impedance spectroscopy. As a result, the switching time (t) is measured between two standard deviations of ΔT_{max}. This amounts to switching time measurements between $T = 84\%$ and 12% of the ΔT_{max} switching range.

Another method to calculate monochromatic electrochromic switching time has been proposed to compare various electrochromic devices. After studying charging characteristics for electrochromic devices, the switching time for both coloration and bleaching is determined by equation (2.5):

$$\Delta T(t) = \Delta T_{max}\left(1 - e^{-\frac{t}{\tau}}\right) \tag{2.5}$$

where ΔT_{max} is the full-switch transmittance contrast obtained for steady state (long driving pulse lengths), t is switching time (or pulse length), and τ is the time constant. So, for example, a device with 90% of coloration range, $t_{90} = \tau \ln 10 = 2.3\tau$, and for 95% of the coloration range, $t_{95} = \tau \ln 20 = 3\tau$, as shown in Figure 2.7. These parameters can describe the switching time for an electrochromic device [93].

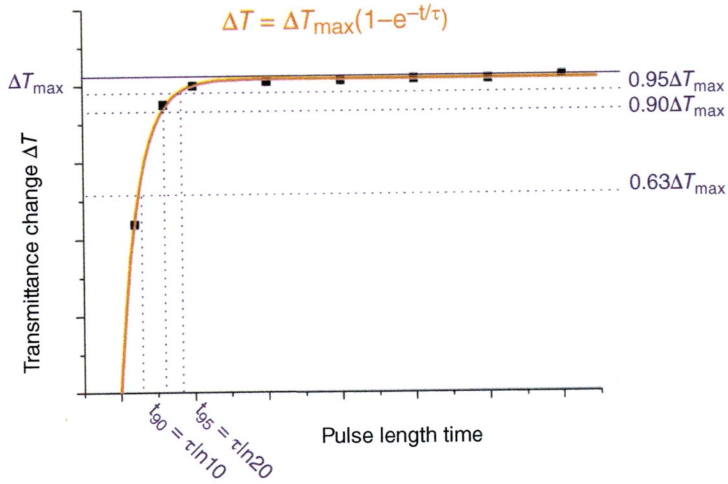

Figure 2.7 Generic electrochromic coloration curve as a function of monochromatic transmittance change versus applied voltage pulse duration. The curve is modeled by an exponential rise where ΔT_{max} is the full-switch transmittance contrast obtained for steady state (long driving pulse lengths), t is switching time (or pulse length), and τ is the time constant. *Source:* Modified from [93], reproduced with permission, courtesy of Elsevier.

2.4 Self-Dimming Automotive Rearview Mirrors

Self-dimming automotive electrochromic mirrors are the most successful commercial application of electrochromics with more than 55 million electrochromic mirrors produced in the year 2024 of which Gentex has 86% of the electrochromic mirror market [94]. These mirrors were first introduced into the market by Gentex Corporation (Zeeland, MI) in 1989 for interior mirrors in limited high-end car models, to replace the prismatic motorized mirrors for glare control. Electrochromic mirrors provide automatic glare reduction to reduce driver distraction at night from the headlights of trailing vehicles which enhances safety. With time, the technology was adapted to the outside flat and curved mirrors and also to an increasing number of vehicles. Figure 2.8 shows an exploded view of an electrochromic mirror construction with an integrated driver monitoring system. The electrochromic element is the first part of the outer frame. Figure 2.9 shows the technology adaptation curve in terms of the number of electrochromic mirrors produced per year. In 2024, the number of Gentex interior electrochromic mirrors produced was 29.9 million compared to 17.8 million outside electrochromic mirrors.

A comprehensive description of electrochromic mirrors technology with automotive test requirements has been discussed in an earlier publication [95]. Many different mirror constructions have been proposed and some of these were commercialized in limited markets, however, the only mirror construction which has succeeded commercially are schematically shown in Figure 2.10 (please note that the layer thicknesses are not to scale). This is known as a Type 1 electrochromic device. An example design for control electronics has been developed by Texas Instruments [96].

The mirror is viewed from the front substrate which is coated with a TC such as ITO. The back substrate is coated with a thin-film stack of materials of the type: dielectric/metal/dielectric (D/M/D) which provides both light reflectivity and electrical conductivity.

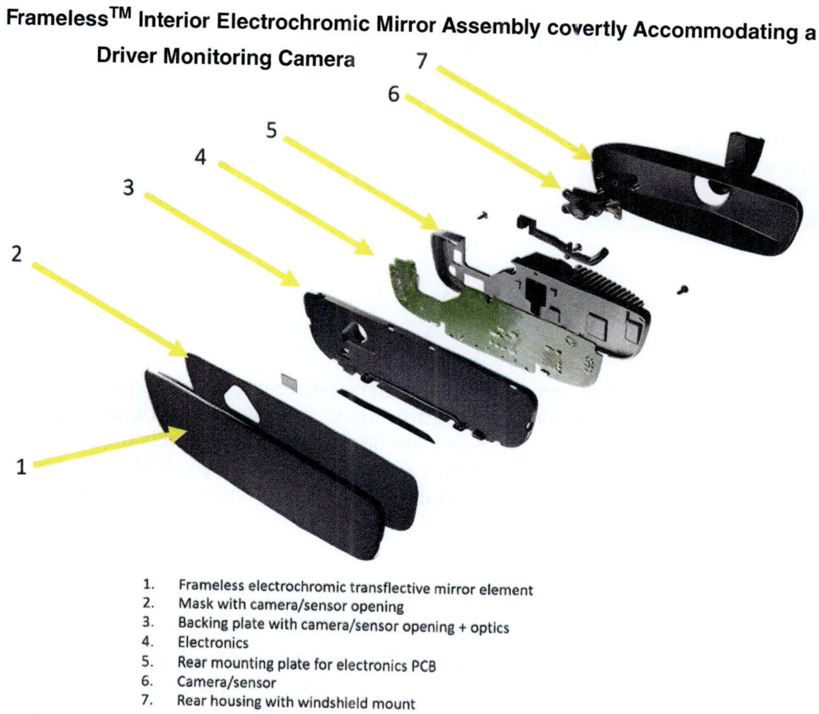

Figure 2.8 Exploded view of Magna Frameless™ interior electrochromic auto mirror assembly. This is a self-dimming mirror with a driver monitoring system. The electrochromic element is the first element. *Source:* Courtesy of Magna International Corp.

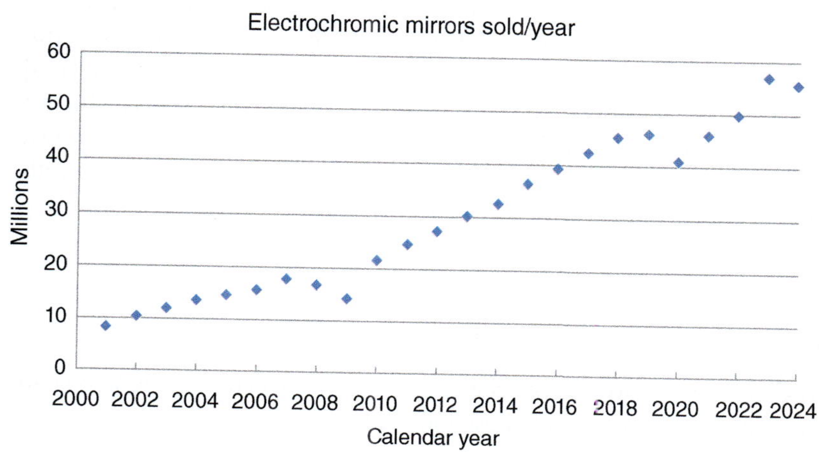

Figure 2.9 Annual number of electrochromic mirrors produced. Data compiled from the annual reports of Gentex, Corp., Zeeland, MI. For 2022, 2023, and 2024, 45.6, 56.9, and 55.4 million EC mirrors were sold, respectively [94]. *Source:* Author AA.

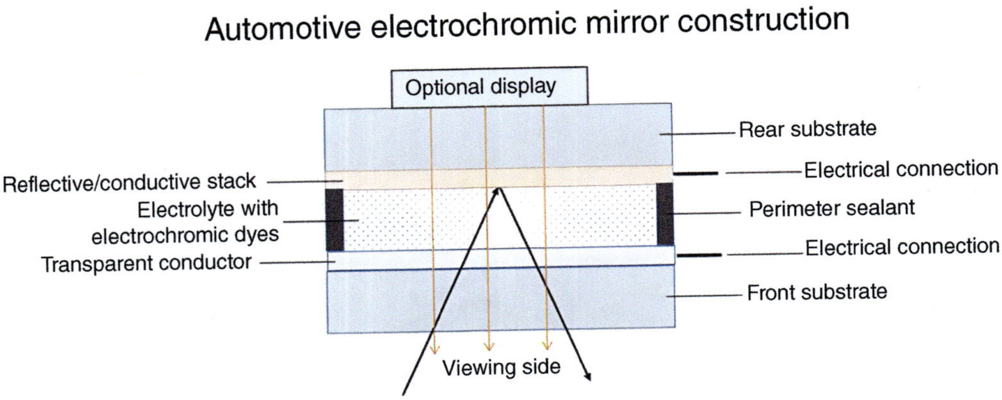

Figure 2.10 Schematic structure of an electrochromic mirror. The reflectivity of the mirror is modified by the charging of the electrolyte layer with electrochromic dyes. An automatic sensing system, which compares the incoming glare with the ambient light, electrically controls the darkness of the reflector. Dielectric/metal/dielectric thin-film stacks are used to enable high visible reflectivity while still allowing good performance of light sensing and camera/display functions (Note: layer thicknesses are not to scale.). *Source:* Author AA.

Depending upon design needs, this multilayer stack may be composed of multiple D/M/D or layer sequences of multiple dielectrics such as M/D1/D2/D1/D2. The reflective materials (M) used are mainly electrochemically stable materials such as silver alloys, ruthenium, and rhodium. For example, the dielectric layers (D) can be made of indium oxide, titanium dioxide, or silicon dioxide [97–99]. In addition, during powering of the mirrors, the metallic layer side is kept cathodic to suppress any metal oxidation. Usually, these stacks with metallic layers have some transparency to allow displays to be placed behind the mirrors. In addition, the layer stack is also optimized to allow the transmission of specific wavelengths/color of the displays to be transmitted effectively. The two coated substrates are then assembled by an adhesive close to the perimeter so that a small uniform gap (typically between 100 and 250 μm) is retained between the two and an electrolyte is introduced in this gap. The electrolyte contains a dye composition, UV stabilizers, and other additives. This is introduced in the gap through a hole in the perimeter sealant via a vacuum backfill process [100] which is then sealed. Electrical leads are attached to the TC and the reflective/conductive stack through which the device is powered for coloration. Generally, the electrolyte has two dyes, an anodic dye (Dye-A) and a cathodic dye (Dye-C) [101]. The coloration reaction with anodic and cathodic dyes may be represented as follows:

$$\text{At the cathode: Dye C (clear)} + e^- \rightarrow \text{Dye C}^- \text{ (colored)} \tag{2.6}$$

$$\text{At the anode: Dye A (clear)} - e^- \rightarrow \text{Dye A}^+ \text{ (colored)} \tag{2.7}$$

The colored dyes then migrate toward the center of the device where they recombine to yield clear (or bleached) species as shown in Reaction (2.8).

$$\text{Dye C}^- + \text{Dye A}^+ \text{ (colored)} \rightarrow \text{Dye C} + \text{Dye A (clear or bleached)} \tag{2.8}$$

Reaction (2.8) is called a back reaction that forces the mirror to self-bleach or go to a clear optical state when the power is removed. This is also a safety mechanism to ensure that the rear objects in the mirror are visible when there is no power. Almost all the commercial

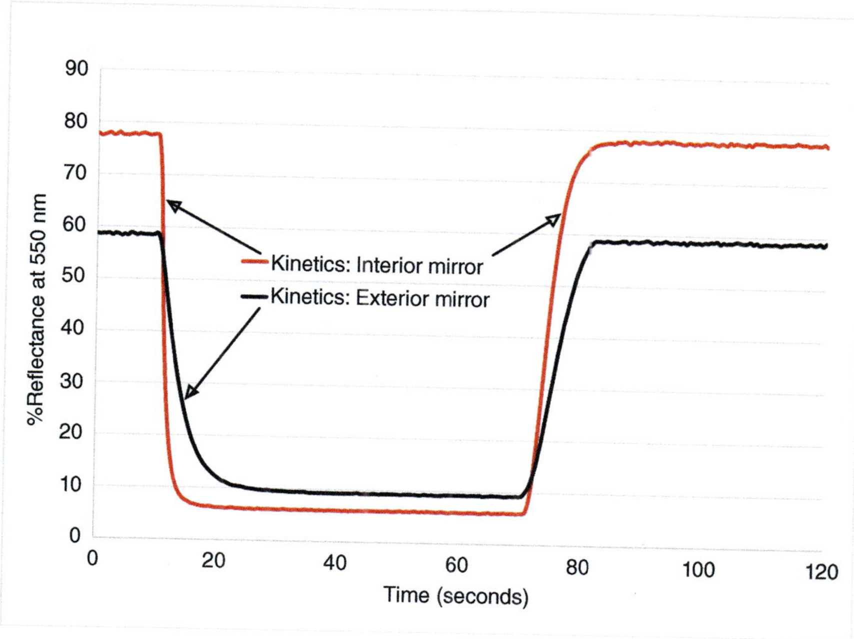

Figure 2.11 Kinetic traces of interior and exterior automotive electrochromic mirrors. *Source:* Author (AA).

EC mirrors produced today use a polymeric gel or a solid electrolyte. These electrolytes contain 4–10% acrylate or urethane monomers (isocyanate and polyol) and a polymerization catalyst [102, 103]. Once the liquid electrolyte is introduced into the mirror cavity, this is polymerized to yield a cross-linked network. Some of the anodic dyes used in the mirrors are ferrocenes ($Fe(C_5H_5)_2$) (with a standard reduction potential of +0.771 V for Fe^{2+}/Fe^{3+}), phenothiazines, and phenazines, and the cathodic dyes are various types of viologen salts [101, 104, 105]. As an alternative to the dual dye system, a single dye molecule where both the anodic and the cathodic dyes are bridged together covalently [106, 107] has been suggested. These bridge dyes have high UV stability, but these are not yet used in commercial mirrors. A good review of small molecule viologens and their applications was done by Shah et al. [108].

Figures 2.11 and 2.12 show the kinetic traces at room temperature and the spectral response of the mirrors in the bleached and the colored states. The data is provided both for an interior mirror and an exterior flat mirror for a sedan (the mirrors were provided by Magna Mirrors of America, Holland, MI). The kinetic traces of the exterior mirrors may vary substantially depending on their size, where the larger mirrors for vans and trucks would have slower kinetics.

Other companies who have done pioneering work on self-dimming mirrors include Nikon Corp. (Tokyo, Japan), Donnelly, now Magna Mirrors of America, Inc. (Holland, MI), OCLI, now Viavi Solutions (Santa Rosa, CA), Schott Glass (Mainz, Germany), VDO Automotive (Germany), Bayer AG (Germany) and Central Glass Products (Japan). Nikon developed the tungsten–iridium electrochromic mirror for exterior mirror automotive use (see patents JP 62270925A, CA 1313562C). The mirror was very durable but expensive to manufacture, due to the cost of iridium. Schott Glass developed a mirror using a proton-conducting reflector layer and tungsten oxide as both the working electrode and counter electrode. This produced

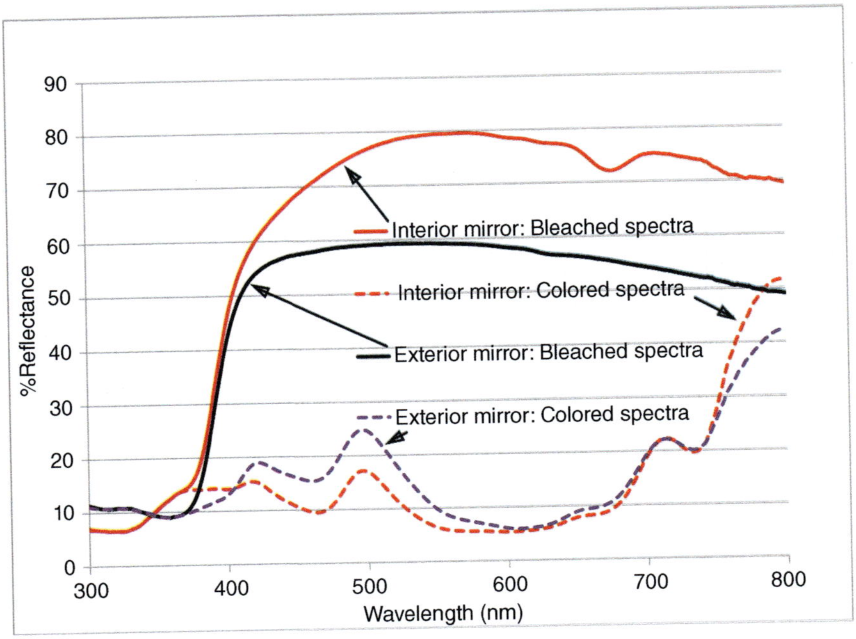

Figure 2.12 Spectral reflectance of electrochromic interior and exterior mirrors in the bleached and colored states. *Source:* Author AA.

the first mirror with completely balanced electrochemistry where the cathodic cycle was identical to the anodic cycle. Other companies in the smart mirror business include Ficosa International SA (Spain), FLABEG Automotive Holding GmbH (Germany), Murakami Corporation (Japan), Samvardhana Motherson Reflectec (India, multinational), Tokai Rika Co., Ltd. (Japan), Valeo SA (France), and Mitsuba Corporation (Japan).

2.5 Early Development of Electrochromic Automotive Sunroofs

It's interesting to look at some of the early development of electrochromic sunroofs. The following is from Dr. Hulya Demiryont, CTO, Eclipse Energy (St. Petersburg, FL). Ford Motor, Glass Division, made some of the first prototype electrochromic sunroofs from 1986 to 1991. Dr. Demiryont was the technical head of the development group. She formerly worked for Şişecam Glass in Turkey. Ford Motor Company, Glass Division, made float glass and it had an online pyrolytic SnO_2:F glass-coating process which made a durable and low-cost conductive coating. The best quality coating was made with a powder spray process. The sheet resistance of the Ford glass was 40–60 ohms/sq., as compared with Pilkington K glass, which was 20 ohms/sq., during that time. The Ford product served as a low-emittance product which could be used as a transparent conductor. The Ford PR&D (Product Research and Development) Laboratory had a pyrolytic deposition facility which duplicated the online production line. This lab coating facility motivated Dr. Demiryont to put together an electrochromic team to develop the individual electrochromic layers and then fabricate a whole system for application of a full-size sunroof for the Lincon Mark VII automobile. Her team made pyrolytic WO_3 and CuO_x. Vacuum deposition processes were also used to make some of the coatings. The electrochromic coating composite layers consisted of a laminated system (Type 3a) of glass/SnO_2:F/$LiWO_3$/Li-PE/V_2O_5/SnO_2:F/glass. The electrochromic electrode layer was WO_3 and it was electrolytically loaded with lithium. The counter electrode was V_2O_5. Both electrodes were laminated with a lithium containing polymer electrolyte. At that time, polymer electrolytes were not well developed, but the system worked well, despite that. The sunroof won the Ford PR&D Eagle Award in 1989. In addition to the electrochromic sunroof, Demiryont's team worked on an electrochromic windshield tint band that was switchable with graded color intensity. This concept was issued patents, US patent 4923289 (1990) and European patent EP 0363045 (1993). A solid-state version of an electromagnetic radiation modulating device was patented EP 0490495 in 1996 [109].

Dr. Demiryont went on to work with Isoclima, SpA (Italy), on all-solid-state electrochromic devices. Isoclima is known for its bulletproof and impact-resistant windows. Their interest in electrochromics was to make side-window privacy glazing for cars and trains. In 2017, Alberto Bertolini, the cofounder of Isoclima patented a laminated electrochromic train, vehicle, and aircraft window (US patent 9778533B2) (2017) [110]. Isoclima could potentially apply their electrochromic technology to building windows. Dr. Demiryont went on to become the CTO of a start-up company, Eyeonics (Tualatin, OR), developing electrochromic eyeglasses. One of the authors (CL) was also a consultant for Isoclima and Eyeonics. Afterward, Dr. Demiryont worked at Leybold (Alzenau, Germany) to make a prototype electrochromic system (Type 3b) on flat glass. This sample was for a Mercedes Benz car sunroof concept project. Unfortunately, the production equipment cost was too expensive for Mercedes to accept. Later Daimler-Benz would adopt SPD glazing for their chromogenic sunroof [111].

Commercial electrochromic sunroofs were introduced at the Geneva Auto Show in 2005 by Saint-Gobain, in a Ferrari Superamerica automobile. It was called the Revocromico sunroof [112]. One of the principal Saint-Gobain researchers was Jean-Christophe Giron, who studied with the author (CL) at LBNL and later went on to become the VP of SageGlass. Dr. Giron's work on tungsten oxide/nickel lithium electrochromics is important work in the development of current electrochromic devices [113]. The electrochromic sunroof was based on a stack of thin films, using a single-pane sunroof following the electrochromic construction types: Types 3a (laminated) and 3b (solid-state). The solid-state structure will have to be protected by a coating or superstrate. Also, the electrochromic technology used for the sunroof can be similar to the technology used for mirrors discussed earlier. This means that the dimming element is also an electrolytic layer containing the two dyes. Instead of a metallic reflector on substrate 2, it has a transparent contact coating. The performance metric for the sunroof is quite different compared to electrochromic mirrors. For example, in a mirror, the coloration and the bleach rates are rapid, typically about two seconds as shown in Figure 2.10, and for a sunroof this may take several minutes. In addition, the contrast ratio of the electrochromic element is less than 100. Prototypes and patents of sunroofs, visors, sensor shrouds, and covers for heads-up-display panels have also been shown and are starting to emerge commercially [94]. Gentex has announced their electrochromic switchable sunroofs for cars [114]. Miru Smart Technologies (Vancouver, Canada) has announced their 1.5×1.6 m prototype EC sunroof. They are working with Glaston Oyj Abp in Tempere, Finland for their scale-up process [115]. The Renault Group (France) has announced their Solarbay PDLC opacifying sunroof in 2024 [116]. The Solarbay sunroof is a joint project with Saint-Gobain Sekurit (Paris, France). With larger glass-area car roofs being the trend in the industry, the automakers are actively looking at smart chromogenic sunroofs.

2.6 Early Electrochromic Windows Developed at Asahi Glass Company

Asahi Glass (Yokohama, Japan) was one of the early fabricators of architectural electrochromic windows. It is instructive to look at the development of the early electrochromic building windows. The following details are from Dr. Junichi Nagai's (Principal Scientist) notes at Asahi Glass in Yokohama. At LBNL (UC Berkeley), the author (cml) worked with Dr. Nagai on the IEA (Task 10) testing program. The Asahi Glass Company (AGC) was one of the first pioneering glass companies working on electrochromic windows for buildings in the 1980s. AGC had a history of developing coatings using vacuum evaporation, sputtering, and chemical vapor deposition (CVD) processes to make functional coatings on glass. The applications were for architectural, automotive glazing, and display applications. Among them, heat-reflective glass (heat mirror or low-e) was favored in terms of conformity and energy savings. A typical film construction was thin silver sandwiched between high refractive index insulators; one construction was $WO_3/Ag/WO_3$/Glass. Tungsten oxide was used because it could be vacuum deposited at a relatively high deposition rate. Parallel to this work, AGC developed electrochromic displays using WO_3 in the late 1970s. The liquid crystal display (LCD) was also worked on, but its visual quality was not as good as the electrochromic display. At this time, the electrochromic coloration/bleaching reaction of WO_3 was known as:

$$xM + WO_3\ (\text{Clear}) + xe^- \Leftrightarrow M_xWO_3\ (\text{Blue}) \tag{2.9}$$

Blue coloration is caused by an optical absorption peak around 1000 nm and was visible and NIR absorptive. A switchable window using WO_3 was thought to be promising as a product candidate as the color of WO_3 can be changed between transparent and blue.

Nagai's group first made the sandwich configuration with a PE (Type 3a):

$$\text{Glass/ITO(+)/PVB/WO}_3\text{/ITO(−)/Glass} \tag{2.10}$$

The configuration shows anodic and cathodic ITO, denoted by ITO (+) and ITO (−), respectively. PVB is commonly used as an adhesive to laminate glass. By applying 100 V DC, coloration occurred but bleaching did not occur completely by changing polarity of the voltage.

In the first experiments, a chelating agent was added to the gelled electrolyte with PVB and alcohol to serve as a redox species on the ITO (+) electrode. Coloration occurred but the beaching process caused metallic deposition onto the ITO (+) electrode, so a clear state could not be obtained. The addition of acid into the gelled electrolyte improved its performance, but cyclic durability was poor (100–1000 cycles) because ITO electrodes were damaged by the acid.

In the next experiments, another redox reagent, such as I^-/I_3^-, ferrocene/ferricenium, or Cu/Cu^{++}, was added into the gelled electrolyte. This time the redox reactions were stable on ITO (+) and good cyclic durability (100,000) was realized in the early 1980s. It was found that alcohol was not a good solvent in terms of weathering, and solvents used for lithium batteries such as propylene carbonate and gamma-butyrolactone were tested and they worked well. In this way, the device configuration became Type 2 structure:

$$\text{Glass/ITO (+)/redox gel electrolyte/WO}_3\text{/ITO(−)/Glass} \tag{2.11}$$

This was established as a product and named commercially, ECW after electrochromic window. Some of the patents that resulted from the work are US 4338000 (1982) [117], US 4671619 (1987) [118], and US 4761061 (1988) [119]. Glass/ITO (+) and a-WO_3/ITO (−) were

Figure 2.13 Seto Bridge Museum with 200 Asahi electrochromic windows installed. The side windows are darkened; the forward windows are becoming bleached (the background is blue). *Source:* Author JN.

laminated using a redox gel electrolyte. In the mid-1980s, electrochromic window products (size 0.4 × 0.4 m) were exhibited at Extension-Renovation Fair in Japan. In 1988, about 200 pieces of ECWs (size 0.4 × 0.4 m) were installed at the Seto Ohashi Bridge Museum (Sakaide, Japan) (see Figure 2.13). Also, electrochromic windows were tested at a private house (Figure 2.14) and in automobile windows. The long-term outdoor durability of ECW was stable if a UV cut filter was used on the outside of the ECW. Also, the filter reduced the photochromic darkening effect of the a-WO_3 layer.

In parallel work, a UV stable electrochromic exterior car mirror was developed at Nikon using IrO-SnO_2 with the Type 3b construction:

$$\text{Glass/Epoxy/Al/}WO_3/Ta_2O_5/IrO_2 - SnO_2\text{/ITO/Glass} \tag{2.12}$$

Figure 2.14 Private home with Asahi electrochromic windows installed bleached (left) and darkened (right). *Source:* Author JN.

The mirror construction was stable to UV but only under 90 °C operating temperature. Some of the Nikon Japanese patents were JPS581130A (1983), JPS6150120A (1986), and JPH0728099A (1995) [120, 121]. Unfortunately, iridium was too expensive to use for a large-area window. Nikon mirrors were used on some of the higher-end production cars as exterior mirrors.

During 1989–1996, AGC received a government grant. One project was to make an all-solid-state window in the Type 3a configuration:

$$\text{Glass/Adhesive (EVA)/ITO(+)/WO}_3\text{/Electrolyte/NiO/ITO(−)/Glass} \quad (2.13)$$

where the electrolyte is either SbO_x, Ta_2O_5, or $LiNbO_3$. The energy performance of the ECWs was simulated for a window in summer in Tokyo. It was determined that the ECW was heat absorptive; but if it is used with low-e coating, the building cooling load is greatly reduced. So, the ECW itself is not heat reflective but the combination with low-e behaves like a switchable heat-reflective glass. Sample windows were evaluated by the IEA Task 10 group [122]. Summarizing the development at AGC, it was concluded that ECW is a good system to realize comfort and energy saving. Today, there are several types of chromogenic windows. Each type has merit and demerit, so there is no one which surpasses the others in terms of all performance tests [123–125].

Asahi Glass and its partner with polymer electrolytes, Eneos (formerly Nippon Oil) (Yokohama, Japan), are not currently active in EC for windows, but are working with non-peeling laminated suspended particle films. For information see AGC patent by Miyasaki and Aoki, EP 3085530A1, entitled "Laminated glass" (2016).

2.7 Designing Materials Systems for Electrochromic Glazing

For architectural applications, the electrochromic glazing is incorporated into an insulated glass unit (IGU) to achieve energy savings. There are mainly two types of electrochromic glazing which are used by the companies mentioned before. Figure 2.4 (Type 3a) shows a different construction where two substrates are used, and each is coated with several layers and the two are combined or laminated by sandwiching a polymeric ion conductor (electrolyte). Figure 2.5 (Type 3b) shows a schematic where several layers are deposited sequentially on a single substrate to complete an electrochromic device. The principle of these devices is very similar, where the electrolyte or the ion conductor binds the two active electrodes (designated EC, CE) together on one side. The two active electrodes are also in contact with transparent conductors on opposite sides. The devices from ChromoGenics AB (Uppsala, Sweden, e.g. European patent EP 3011388B1), Glass Dyenamics (Tucson, AZ, US patent appl. 20210333678A1), Furcifer (Fremont, CA, "Electrochromic system" European patent EP 3518029), and formerly, Helio (Hayward, CA, US Patent 10871695B2, reassigned to Smart Windows, Inc. Hong Kong) resemble the laminated devices (Type 3a) shown in Figure 2.4. Defunct Heliotrope (Alameda, CA, US Patent 9791760B2, modified with UV absorber) and EControl (Germany, US Patent 5859723A) made devices of Type 3a. Also, defunct Gesimat (Germany, European patent EP 1227362B1) made laminated devices of Type 3a. The devices from View (San Jose, CA, e.g. US patents 8300298B2, 9429809B2, and 9958750B2) and SageGlass (Faribault, MN, e.g. US patent 9581875B2) resemble the solid-state layered construction (Type 3b) shown in Figure 2.5. An intermediate design is Type 2 in which the electrochromic or the counter electrode is a film with the counter reaction materials dissolved into the electrolyte. Some of the early Asahi devices used the iodine/iodide (I^{-1}/I^{-3}) redox couple for the counter reaction.

Generally, both electrochromic electrode and counter electrode can have electrochromic properties, but they are complementary. This means that a cathodic electrochromic layer colors by reduction, by incorporation of lithium ions; the electrochromic counter electrode colors by oxidation, by expelling the lithium ions. The most common inorganic electrochromic material for an electrochromic electrode is tungsten oxide, and a common electrochromic material for a counter electrode is nickel oxide. It may also be an alloy containing nickel oxide. For example, the reactions in Reaction (2.6) show that the tungsten atoms in tungsten oxide are hexavalent and are colorless, and when these are reduced to a pentavalent state they change to a blue material. Similarly, a lithium-nickel oxide layer where the nickel is in Ni^{+2} state is colorless, and when it is oxidized to Ni^{+3} state, it colors to a bronze color, shown in Reaction (2.7). Combining the two electrochromic layers results in a device coloring to a deep neutral color as complementary colors from two different electrochromic layers are combined [126, 127]. Devices where only one layer is electrochromic, and the other is only for ion storage have also been fabricated, but their coloration is not deep, and the color is not neutral [128]. Colors such as grey and bronze are favored by the architectural industry.

$$W^{+6}(\text{Bleached}) + Li^{+} + e^{-} \rightarrow W^{+5}Li\ (\text{Blue, Reduced}) \tag{2.14}$$

$$Ni^{+2}(\text{Bleached}) - Li^{+} - e^{-} \rightarrow Ni^{+3}\ (\text{Bronze, Oxidized}) \tag{2.15}$$

In the devices mentioned earlier, the preferred transparent conductor is ITO, and sometimes commercially available $Sb{:}SnO_2$ and SNO can be used instead of ITO. Commercially produced fluorine-doped tin oxide coatings are made by CVD directly on a glass float line, and their surfaces are rough [129]. Sputtering additional layers on such surfaces trap voids since the coating process is not conformal. However, wet-chemical deposition processes can be used since they are conformal and reduce surface roughness [130]. All of the commercially available devices at present for architectural windows use an electrochromic material of tungsten oxide as its main constituent and the counter electrode material containing nickel oxide. As discussed in more detail later, generally all these electrodes contain additional metal oxides for enhancing durability and adjusting electrochemical parameters and their optical properties [131–135]. There are reviews of research on nanostructured electrochromic materials to gain better performance [136–139]. This continues to be an important research area.

Referring to Figure 2.5, each of the layers is deposited sequentially on a substrate. Typically, these are inorganic oxides which are deposited by sputter coating process from metal targets in the presence of a controlled amount of oxygen. The substrate (or the electrochromic device) is not removed from the vacuum until all the layers are deposited. Several of the layers contain lithium, and it is difficult to make mixed-metal sputtering targets with lithium as one of the metals. To overcome this, lithium is incorporated in any layer by sputtering lithium on top of the layer and then heat-treating the assembly to temperatures in a range of 150–500 °C so that it diffuses into the underlying layer(s) before continuing the deposition of the next layer. As an example, after the deposition of tungsten oxide (a cathodic material), a layer of lithium is deposited as a separate layer which then diffuses into the tungsten oxide layer by heat [140–142]. Some publications reveal interesting details on eliminating the use of a separate layer for ion conductor [140–142]. Wang proposed a compositional transition from the cathodic electrochromic layer to the anodic electrochromic layer [140], and Ray proposed that in some embodiments of their invention, a thin-metal oxynitride layer was used [142]. Burdis proposed an ion-conducting layer comprising three layers, where a central buffer layer was selected from a lithium-containing ceramic or a tungsten–nickel oxide composite, and on either side an insulator layer of a metal oxide was deposited [141]. One of the major issues with these devices is metallic particle contamination resulting in electrical damage. Since these can be large devices, a fraction of a square meter to several square meters, and the total thickness of all of the layers is in a range of 1–2 μm, the particles can result in electrical shorts, meaning direct conductive paths may be formed between transparent conductors, TC-1, and TC-2 mostly due to the pinholes caused by particulates. Since the presence of conductive pinholes is only discovered at the end of the process when the devices are being tested, discarding of the devices can be very expensive with low yields. Several methods have been adopted to mitigate low yields; in one method loose particles from the surface are removed at several stages of the layered deposition while also adding insulating layers on the transparent conductors [143]. These insulating layers may have a similar or slightly different composition to the layers that they are in contact with, e.g., the electrochromic electrode or the counter electrode, or these may be deposited under different conditions. In another method, the conductive pinholes are located and the area around them is ablated using lasers to remove the conductive path [144]. These spots are small, so these do not cause a visual distraction when looking through large windows. View has patented techniques to eliminate or obscure defects [145]. Burdis and Sbar (SageGlass) wrote a USDOE report on similar work on their EC system [146]. Also, they have discussed how to make their sol-gel-deposited ion conductor by sputtering [147] and lower manufacturing costs [148].

Figure 2.4 shows an electrochromic device where two substrates are coated with TCs. In the figure, TC-1 is coated with an electrochromic electrode and TC-2 is coated with a counter electrode (called CE). These two substrates are brought together with an organic electrolyte or the ion-conducting medium to combine them. As described earlier, sometimes the ion conductor can be comprised of several layers to provide compatibility with the electrodes [149]. The inorganic electrochromic electrodes as in the layered devices in Figure 2.5 are mostly based on tungsten oxide [150–153] and the counter electrode are chiefly ternary nickel oxide compositions [154]. In one device construction from ChromoGenics AB, the electrochromic device is fabricated using flexible polyester films as substrates, rather than glass [155]. The transparent conductor (ITO) is deposited by sputtering on one of these films followed by a tungsten oxide–containing electrochromic layer deposited by sputtering. The other substrate is coated with ITO and is then coated with a counter electrode containing nickel oxide made by sputtering. These two sheets are assembled in a roll-to-roll process using a liquid polymer electrolyte that is laminated between the two sheets and polymerized [156–158]. These rolls are then shipped to the electrochromic glass fabrication facilities, where they are cut, and contacts are attached and laminated to glass substrates using laminating polymer films [159]. This method of manufacture saves on shipping weight and can be used on curved surfaces. In a Heliotrope device construction, the electrodes are deposited by liquid deposition from suspensions containing nanoparticles of electrode materials. A perimeter sealant is placed on one of the substrates coated with the electrode and then a liquid electrolyte formulation is poured in the central region, and the second electrode is lowered to contain the formulation while it spreads toward the sealant boundary. The lowering of the second electrode also seals the device [160]. The electrolyte formulation is subsequently polymerized. The electrolyte may also contain a sacrificial material so that when an activating voltage is applied, protons/lithium ions from decomposing agents are generated to reduce or intercalate one of the electrodes [161]. In a construction used by Halio, (Hayward, CA) thin glass sheets are laminated so that the total device thickness is 1.2 mm. It is estimated that each substrate thickness is from 300 to 500 μm and the electrolyte is a cross-linked epoxy with a thickness in a range of 100–500 μm [162]. The thickness of the transparent conductor is about 200–400 nm, the tungsten oxide–containing electrochromic electrode about 250–500 nm, and the nickel oxide–containing counter electrode about 200–300 nm. The electrochromic electrode and the counter electrode are deposited from solutions. These contain the precursors which are then heat-treated to form the required oxide, or one or both oxides are formed using preformed nanoparticles which are then used in an ink formulation for deposition [144, 163]. Kinestral Technologies (Hayward, CA) (reorganized as Halio) used a unique approach to color their devices fast and uniformly. The TCs are coated with two insulative materials, one of which is more insulative and is patterned using photolithography so that the electrical resistance from the edge of the device (that is from the bus bars) to different areas on the surface of the devices is similar even though these areas are located at different distances from the busbar. This allows uniform and fast coloration of the device [164]. Halio also stated that their glasses color to a neutral color without changing the color of the transmitted light. In one IGU construction, the color-rendering index (CRI) in the clear or the bleached state was 97.7%, and in the most colored state, the CRI was above 90%.

2.8 Dynamic Building Windows

There are a few companies which are currently offering electrochromic windows in the market. SageGlass, owned by Saint-Gobain and ChromoGenics AB, who manufactures a polymer-laminated electrochromic. SageGlass has developed a gradient coloring electrochromic window called Harmony. View Glass (San Jose, CA) may be still making windows, after their bankruptcy and reorganization as a private company. As of December 2024, View Operating Corp. has been assigned the IP from View Corp. View uses large-scale sputtering to deposit the EC thin-film stack directly onto glass. View has provided a wide range of products including, electrochromic panels, complete windows and control systems. View has many electrochromic window installations in buildings in the United States and abroad. A new company, Miru Smart Technologies (Vancouver, Canada), announced in 2025 that they will be developing ECWs with Glaston OYJ Abp. Glaston is known for its glass processing equipment throughout the world. Miru is an electrochromic startup, using technology based on electrochromic inks. Some Miru staff is from Kinestral and Switch Materials (Vancouver, Canada). Miru showed a large electrochromic sunroof made with Argotech (part of Mativ Holdings, Alpharetta, GA) thermoplastic polyurethane interlayer film. The prototype electrochromic automotive sunroof measured 1.5 × 1.6 m [165]. Another company e-chromic technologies, headed by Loren Burnett (Boulder, CO), began developing "SunStop", a reflective electrochromic switchable window using a thin-metal interlayer [166]. The manufacturing technique is embodied in the Snow, Tenent, and McFadden, US patents 10061176B2 (2011) and US 9880440B2 (2018), and Gittleman and Alie patent application US 20170073805A1 (2017), using high-power impulse magnetron sputtering (HiPIMS) to deposit the films.

Companies that have gone out of business are Halio, formerly Kinestral. Halio constructed electrochromic modules using slot-die coating to make a thin glass laminate that could be fabricated into an IGU. Halio's IP has been transferred to Smart Window Inc. of Hong Kong as of 2025. Heliotrope (Alameda, CA) developed dual-band nanochromics that could regulate heat and light in independent bands of energy (e.g. patent US 9791760B3 (2017)). Heliotrope's IP has been transferred to Hivisq Technology in Spain as of 2023. EControl-Glas GmbH and Co. (Plauen, Germany) with technology developed at Pilkington Glass (St. Helens, UK) and Flabeg (Flachglas AG, Germany), e.g. EP 2342411B1 (2012) and US 5859723A (1999). EControl-Glas made advancements in deposition of large-area glass and control systems for large-area electrochromic glazing, e.g. AU 2009208112B2 (2013) and US 6707590B1 (2004). EControl-Glas has made electrochromic glazing for many building projects in Northern Europe [167]. Two of the key people at EControl were Hartmut Whittkopf and Dirk Jodicke.

During 2021, Switch Materials' IP was sold to Solutia Canada (owned by Eastman Chemicals). A couple of the key people at Switch were Neil Branda and Jeremy Finden. Gesimat (Germany) was liquidated in 2021. A key founder at Gesimat was Alexander Kraft, who is still active in the field [168, 169]. Some of the properties of these windows are in a review by Cuce and Riffat [170].

2.9 Commercial Electrochromic Windows

IGUs with transmissive electrochromic elements have been fabricated by select window companies, such as Viracon working with Halio. Figure 2.15 shows a laminated ChromoGenics ConverLight® electrochromic flexible film bonded to the outboard glass panel or lite. In the IGU, a low-e coating is on the inboard lite.

Figure 2.16 shows a cross-section of an IGU with two glass panels, one of which is an electrochromic panel whose luminous transmission can be changed by applying a low DC

Figure 2.15 ChromoGenics AB (Uppsala, Sweden); two electrochromic window panels. The electrochromic is manufactured on a flexible PET sheet. The sheet is subsequently laminated onto the inboard lite which becomes part of a double glazing. The electrochromic is based on tungsten and nickel oxide as the active films. *Source:* Author's image CL.

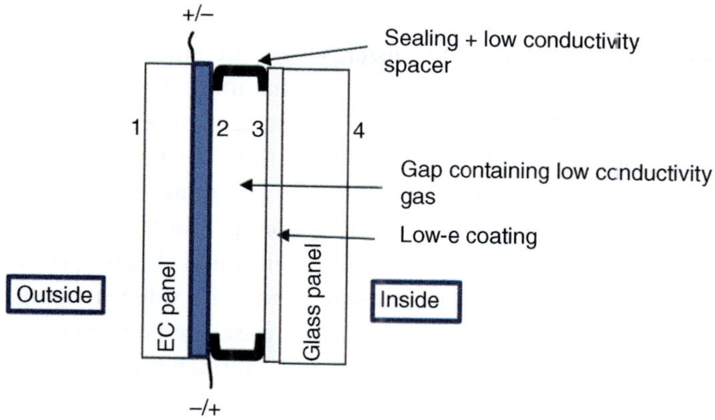

Figure 2.16 Schematic cross section of an insulating glass unit (IGU) window with an electrochromic element (EC panel) is shown (not to scale). Surfaces 1, 2, 3, and 4 are shown progressing from the outside (outdoor) surface 1 to the inside of the building surface 4. The low-e coating is on the 3rd surface, inboard pane. The gap contains low conductivity gas such as argon or krypton. *Source:* Authors image, AA.

voltage, and the other one is a glass panel which is coated with a low emissivity (low-e) coating. These are assembled such that there is a sealed gap (or a cavity) between the two panels. Edge or perimeter spacers are shown sealed in the gap. This gap is typically in the range of 12–20 mm. This gap may have air or an insulating gas such as argon. The configuration of this IGU relative to a building envelope is shown where the electrochromic panel faces outside and the other glass panel faces inside. Since these panels have four surfaces, these are conventionally called surfaces 1, 2, 3, and 4 starting from the outside. If the electrochromic panel is made using multiple panels, each one may be numbered starting from outside or they may be numbered as shown with EC panel being considered as a monolithic unit. The low-e coating is shown on surface 3, but it may also be on surface 2. For electrochromic panel constructions shown in Figure 2.16, the last transparent conductive coating may also act as a low-e surface, and thus it is not necessary to use a low-e coating on surface 3.

The low-e coating is important to control the radiative heat transport of the window. Solar radiation is comprised of ultraviolet (UV), VIS, and IR radiation up to 2500 nm (2.5 μm). From the total energy perspective UV is a small part of the solar radiation (typically less than 5%) and the VIS (46%) and the IR (49%) at Air Mass 1.5, USDOE, NREL [171, 172]. Air Mass is the thickness of the Earth's atmosphere the solar radiation has to traverse to get to the window. Air Mass 1.5 is used generally as a standard for solar building and photovoltaic calculations.

Some makers call the electrochromic IGU an electrochromic double glazing (EDG). A triple-pane IGU (not shown) will have a third glass pane, i.e. the IGU will have two gaps on either side of the middle pane. If the IGU is an electrochromic triple glazing then the term ETG is used.

2.9.1 Window Performance Parameters

Even though measurements and glazing design are discussed in Chapter 1B, we would like to refresh the reader here. There are five important performance parameters which are considered in selection of building windows. The first three parameters are related to energy performance. These three are visible light transmission VLT or VT or "T_{vis}", solar heat gain coefficient (SHGC), known as *U*-factor or *U*-value or just *U*. The lower *U*-factor numbers represent the best performance or lower rate of heat transfer. Where the familiar thermal resistance for building insulation is, R-value = 1/*U*-factor. *U*-factor units in metric are, $W\,m^{-2}\,K^{-1}$ or in British thermal units, $BTU\,h^{-1}\,ft^{-2}\,°F^{-1}$ (or in common engineering terms, $BTU/h\cdot ft^2\cdot°F$ or $W/m^2\cdot K$). For conversion, $1\,W\,m^{-2}\,K^{-1} = 0.1761\,Btu\,h^{-1}ft^{-2}\,°F^{-1}$. The fourth parameter is air infiltration or air leakage rate (AL) which determines the quality of the seal and framing. Common numbers for AL are 0.1–0.3 ft^3 of air/min/ft^2. The fifth parameter is the condensation index (CI). This is important to colder humid climates, such as the northern seaboard. CI is a relative laboratory test measurement with values ranging from 1 to 100. The higher CI numbers show the best condensation resistance performance. Performance rating of glazing is characterized by these parameters. Performance parameters are selected based on the climates where these buildings are located, and the orientation of the windows relative to the sun. The selection of the windows facing different directions may be different for the same building for optimized energy/cost performance. Since windows with electrochromic panels have a variable light transmission, the SHGC and T_{vis} have a range of values.

The visible light transmission (T_{vis}) for static windows for the SHGC (0.2–0.6) range is typically of 30% – 70%, and for electrochromic windows this is adjustable. High visible transmission during the day adds to natural indoor lighting and reduces the lighting energy requirements but may also lead to excessive heat load and glare depending on the task being carried out in that area of the building. Generally, building energy management systems are used to balance T_{vis} and optimize energy use. However, users can override these algorithms depending on the task being conducted and their preference. Privacy is dependent on the lighting conditions inside and outside and the reflectivity of the window, Treado and Bean [173]. At night, it is difficult to achieve total privacy as even 0.1% light-transmitting windows will allow a person, viewing from dark outside conditions, to be able to see inside a well-lit building interior.

The SHGC measures the amount of solar energy entering the window as direct solar radiation or solar radiation that has been converted to heat. SHGC has no units and varies between 0 and 1, where 0 means that no solar radiation enters through the window and 1 means that all solar radiation is let through the window. In North America, depending on the climate and the type of building the SHGC of static windows (IGUs) may vary from about 0.2 to about 0.6. The lower numbers are more suitable for residential warmer climates and the larger numbers for colder climates. For multistoried commercial buildings where large window facades are used, the SHGC tends to be lower for all climates as solar thermal load even in winter may be excessive. Since the windows with electrochromic panels have a range of SHGC values, the properties of these windows can be adjusted based on the time of the day and with changing weather conditions.

The *U*-factor can be for the whole window net performance or just center-of-glass performance. Generally, in the US The National Fenestration Ratings Council (NFRC) *U*-factors are whole window performance values including framing and spacers. The *U*-factor in the window is dependent on the thickness of the glass used, the gap width in the IGU and the type of fill. For example the use of argon instead of air results in a lower *U*-factor. Increasing thickness of glass or increasing the number of panes in an IGU (e.g. going from double to triple panes lowers the *U*-factor. The increasing gap results in decreased *U*-factor but stabilizes around 12 mm beyond which there are no significant benefits in decreasing the *U*-factor, Vitro [174]. These *U*-factors are generally measured at the center of the window.

For the whole window, the numbers also include the thermal properties of the frame. When the electrochromic window is darkened, it changes both visible transmission and SHGC. Darkening has almost no impact on the *U*-factor. Since the SHGC of an electrochromic window can be varied during the day depending on the relative position of the sun to the window and from the winter to summer, it provides energy saving benefits.

2.9.2 Electrochromic Glazing Examples

An example of a View EDG window with an IGU construction is shown in Figure 2.17. Four tint levels are shown as examples of the large dynamic range. The electrochromic window has a continuous grey scale. This window's performance satisfies the California Title 24 compliance [175]. This means that the EDG, *U*-factor is below 0.36 BTU h^{-1} ft^{-2} °F^{-1} (2.04 W m^{-2} K^{-1}), the maximum SHGC is less than 0.25 and the minimum visible transmittance is less than 0.42. For the View electrochromic window: *U*-factor = 0.29 BTU h^{-1} ft^{-2}°F^{-1} (1.65 W m^{-2} K^{-1}), and SHGC = 0.40 (Tint 1) to 0.09 (Tint 4) with transmittance from T_{vis} = 0.52 (Tint 1) to 0.01 (Tint 4). These numbers satisfy the Title 24 compliance.

The spectral properties of each tint level for windows in Figure 2.17 are shown in Figure 2.18. Also, Figure 2.18 overlays the solar AM1.5 irradiance spectrum, showing how the window can control both the solar VIS and NIR passing through the window. The standard AM1.5 spectrum is available from NREL [172]. A more detailed solar AM1.5 spectra is shown in the next section in Figure 2.20. Figure 2.19 shows the visible transmittance and solar heat gain performance of an electrochromic double-glazed window as it changes tint.

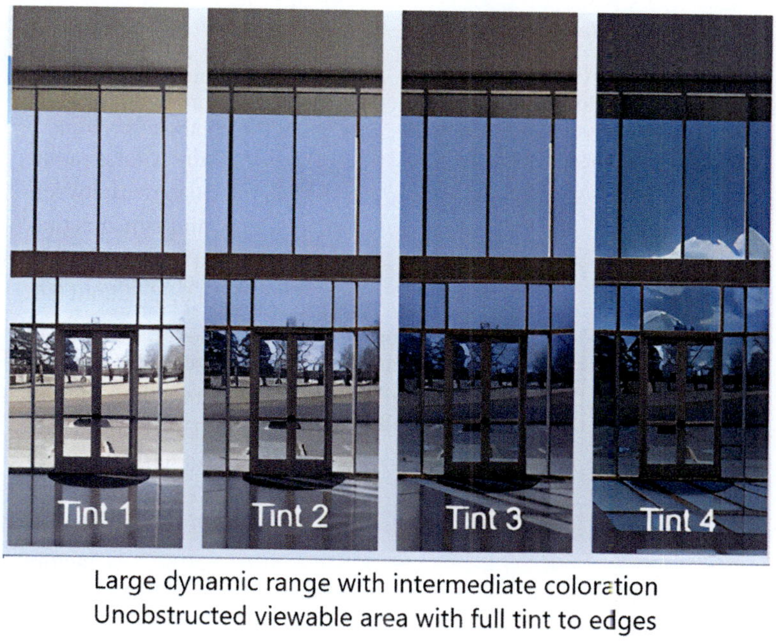

Figure 2.17 View electrochromic double glazing installed in a building. Here the electrochromic glazing shows four tint levels. However, the window is capable of continuous dimming. A large dynamic range is needed to respond to changing environmental conditions, from blocking glare of direct sunlight to letting in plenty of natural light on overcast days. The corresponding spectral characteristics are shown in Figure 2.18. *Source:* Courtesy of Brian Griedel, View, Inc. (Milpitas, CA, USA).

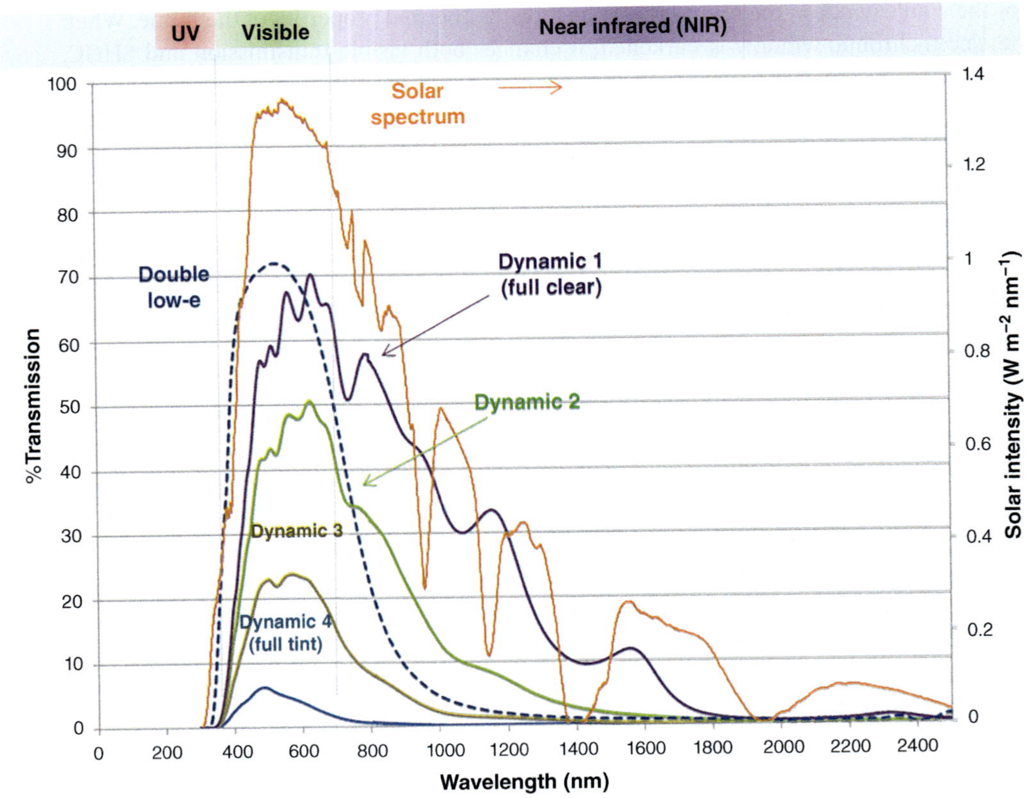

Figure 2.18 Spectral transmittance of a View electrochromic window, showing four dynamic tint levels (as shown in Figure 2.17). The solar terrestrial spectrum (AM1.5) is shown for reference. Also, the dashed line shows the photopic spectrum. Also, the dashed line shows the photopic spectrum. *Source:* Courtesy of Brian Griedel, View, Inc., Milpitas, CA, USA.

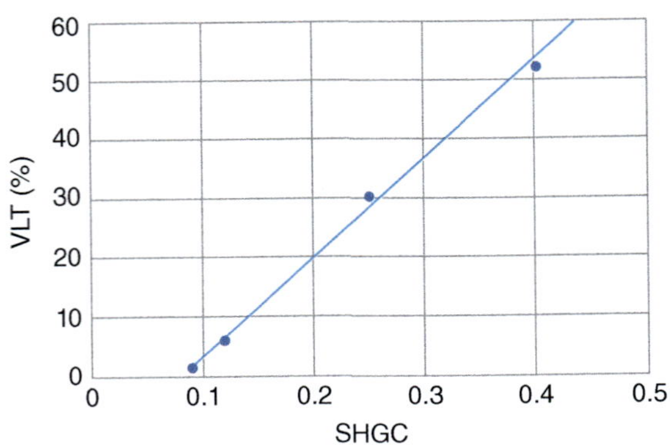

Figure 2.19 Visible light transmittance (VLT) and solar heat gain (SHGC) performance of View's electrochromic double-glazed (EDG) window. *Source:* Authors CL, AA using data from View, Corp.

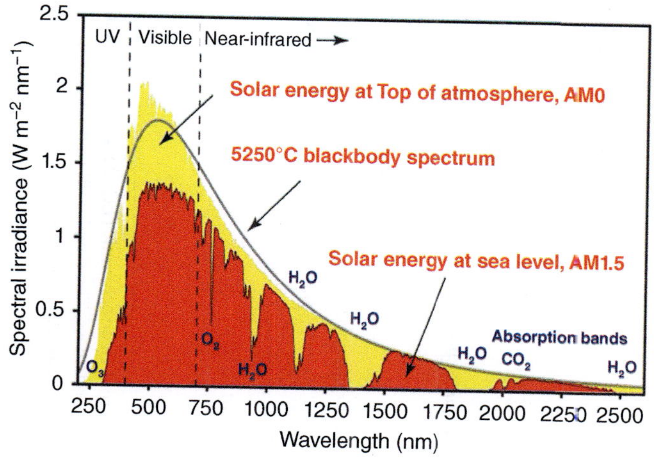

Figure 2.20 Solar spectral irradiance for air mass, AM0, and AM1.5 (terrestrial). The UV, VIS, and NIR regions are shown. The principle atmospheric absorption bands are noted. The data correlates to the NREL (Golden, CO) standard for PV testing. At AM1.5, about 49% of the solar energy is in the NIR range, 46% is in the VIS, and 5% in the UV. The best localized surface plasma NIR coating should switch over the 740–1750 nm region. *Source:* (Modified by CL) Wikimedia Commons, Robert A. Rohde, CL BY 3.0 attribution 3.0 unported. https://en.m.wikipedia.org/wiki/File:Solar_Spectrum.png#file, Accessed 5/20/25.

The average total solar intensity on a summer day at its peak in Arizona, USA (latitude 33°N), could reach about 1000 W m^{-2}. Depending on the orientation of the electrochromic window relative to the sun, the electrochromic window in the dark state can absorb large amounts of solar energy which behaves like a black body and dissipates this energy as heat (or IR radiation in the longer wavelengths of 5–50 μm). Referring back to Figure 2.16, a low-e coating either reflects the longer infrared radiation or does not permit the radiative loss through this surface. In case the low-e coating is on surface 2, the heated window will not radiate this heat inside the building, but it would be radiated to the outside. Similarly, when this coating is on surface 3, it would reflect back the longer IR radiation coming in from the electrochromic panel rather than transmit that to the inside of the building.

For a window with an electrochromic element, the range of SHGC variation and the change in T_{vis} caused by darkening of the electrochromic element can be changed by selecting the static optical properties of the glass used to fabricate these electrochromic elements or the low-e panel. This is illustrated in Table 2.2 which is using data from Halio Inc. Even though Halio is no longer in business, this data serves as a good example of generic electrochromic performance. In reference to a dual-pane IGU shown in Figure 2.16, the outside facing glass electrochromic pane in all cases is a 1.2 mm thick electrochromic device. The device is laminated between two 5 mm thick heat-strengthened clear, low-iron glass sheets (unless specifically mentioned) using two clear PVB sheets, each of which are 1.52 mm in thickness. The gap is 12.7 mm wide filled with gas, and the inside facing, low-e-coated glass is a tempered 6 mm sheet. The total thickness of the double-pane IGU with clear glass including coatings was 33 mm. Table 2.2 illustrates the variation in the performance of these IGUs with the type of low-e glass used. Depending on whether Vitro Sungate 500 or Guardian Sunguard SN68 is used, the color of the outside-facing glass sheet that is laminated to the electrochromic device will be altered. The color can range from clear low-iron glass sheet to Crystal Gray

Table 2.2 Example of optical and thermal performance of tungsten oxide–based electrochromic (EC) IGUs with different gas fills, glass, and low-e types.

	IGU construction				Visible					
Window #	Glass in EC laminate	Fill gas	Low-e glass	EC darkness	%T	%R (front, exterior)	%R (back, interior)	Solar %T	SHGC	*U*
Double-pane IGU with one EC device (glass/PVB/EC/PVB/glass-gap with gas-low-e)										
1	Clear	Ar	Sungate 500	Clear	58	14	16	33	0.43	1.53
				Mid	12	7	12	5	0.13	
				Dark	2	6	13	1	0.09	
2	Clear	Air	Sungate 500	Clear	58	14	16	33	0.44	1.70
				Dark	2	6	13	1	0.1	
3	Crystal Gray	Ar	Sungate 500	Clear	42	9	15	24	0.34	1.48
				Dark	2	5	12	1	0.09	
4	Clear	Ar	Sunguard SN68	Clear	51	13	13	23	0.35	1.36
				Dark	2	6	10	1	0.08	
Double-pane IGU with Two EC devices (glass/PVB/EC/PVB/EC/PVB/glass-argon gap-low-e)										
5	Clear	Ar	Sungate 500	Clear	47	16	17	26	0.36	1.48
				Dark	<1	0.05	6	0.09	0.09	
Triple-pane IGU with One EC device (glass/PVB/EC/PVB/glass-argon gap/low-e/argon gap/low-e)										
6	Clear	Ar	Sungate 500	Clear	50	16	19	26	0.37	0.80
				Dark	2	6	17	2	0.05	
EC device only laminated between two 5mm thick low-iron glasses (not an IGU)										
7	Clear	NA	None	Clear	67	11	12			
				Dark	2	6	7			

Note: U-factor is in $Wm^{-2} K^{-1}$. Sungate 500 is a Vitro product, Sunguard SN68 is a Guardian product. The numbers vary for different electrochromic (EC) device makers and IGU constructions. This table uses example data of a tungsten oxide–based EC on thin glass made by Halio, which is no longer in business. The data is still useful since it shows typical ranges of performance data for a single-EC IGUs (windows #1–4), IGU with two EC panes (window #5), and a triple-pane IGU with a single-EC pane (window #6). Also, a single-pane EC, thin glass–based glazing is shown in window #7.
Source: Author AA using data from Halio.

(from Guardian), in these examples. The thickness of Crystal Gray is 6 mm and the other clear glass used for lamination is also 6 mm.

There are two other IGU variations in this table. One uses two electrochromic panels which are laminated with an additional PVB in the front panel of the double-pane IGU for a total thickness of 37.7 mm. Second is a triple-pane IGU with one panel having a single laminated electrochromic panel as the front panel followed by two argon-filled gaps each being separated by a low-e glass which is Vitro Sungate 500 for a total thickness of 51.6 mm. Various parameters are shown in the fully bleached and fully colored states (although the transmission of the glass may be tuned to different optical densities in between with accompanying SHGC numbers). As a reference, the properties of the electrochromic device panel in the aforementioned IGU (that is the laminated structure with two clear glasses having the 1.2 mm thick electrochromic panel) are given in the last row of Table 2.2.

Electrochromic windows are configurable to multiple tint levels; however, the data in Table 2.2 is only shown for completely clear and completely dark states, except for window #1, where one of the mid-states is included.

For the double-pane IGU, a large variation is obtained in terms of the visible transmission and the SHGC range by changing the cover glass and/or the low-e panel. For example, in northern climates, having a range of SHGC which is high is preferred and for the warmer southern climates a lower high-end SHGC is preferred such as with Crystal Gray glass or with SN68 low-e. Using air instead of argon has more of an impact on *U*-factor only. For window #1, a mid-state of tint is included. The SHGC in the mid-state of Halio is 0.13 very close to the bottom for a double-glazing IGU (see also Table 2.2). This demonstrates that for energy efficiency, it is not necessary to color the windows to an extremely dark state. However, a low transmission of about 1% or lower may be needed to avoid certain glare situations during the day.

As a further comparison, in window #5 where two electrochromic devices are used, the primary impact is on visible and solar transmission and not on any of the factors which influence energy efficiency. The electrochromic in its darkest state is perhaps needed to avoid glare situations when the sun at low angles is directly in the occupant's eyes. However, it is not clear that even this combination will result in total privacy at nighttime with well-lit indoors. A triple-pane IGU which utilized two Vitro Sungate® 500 low-e-coated glasses had a *U*-factor of $0.14\,W\,m^{-2}\,K^{-1}$ and SHGC varied from 0.35 to 0.05 with visible transmission changing from 50% to 2%. The triple glaze results in significantly lower *U*-factor as compared to any of the other windows and lower SHGC without sacrificing the visible transmission. This is generally the case with triple-pane IGU configurations with any electrochromic element, but these come at an added disadvantage of significantly increased thickness and weight. An interesting compromise has been suggested, where the inner pane is a thin glass (usually 1.1 mm thickness or less) having a low-e coating [176a]. Note that Vitro recommends Sungate® 400 as an alternative for Sungate® 500, which is being phased out [176b].

Table 2.3 compares IGU specifications of View, SageGlass, and ChromoGenics AB that are fabricated using clear glass. These can be compared with Window # 1 in Table 2.2 which uses argon as the fill-gas (Halio data).

All of these companies offer their products combined with glass having different optical properties. Also, they offer triple-glass IGUs to achieve better thermal performance. There are some interesting variations among different manufacturers. Some companies offer their products in limited sizes. View offers their products in a variety of sizes and shapes, and they use glass with different optical properties for the second pane to offer a variation of

Table 2.3 Comparative optical and thermal properties for three makers of electrochromic IGUs.

Window	IGU construction			EC darkness	Visible			Solar %T	SHGC	*U*
	Exterior EC panel glass	Fill gas	Interior glass		%T	%R (front, exterior)	%R (back, interior)			
	Double-pane IGU with one EC layer (glass/PVB/EC/PVB/glass/argon gap/glass (low-e)									
View	Clear, 6 mm	Ar, 12.7 mm	Clear, 6 mm	Clear	53	13	18	29	0.41	1.59
				Mid	5	6	16	2	0.1	
				Dark	1	6	16	0	0.09	
SageGlass	Clear, 6 mm + 2.96 mm[a)]	Ar, 12.7 mm	Clear, 6 mm	Clear	60	16	15	33	0.41	1.65
				Mid	6	10	9	2	0.11	
				Dark	1	11	9	0.4	0.09	
ChromoGenics	Clear, 10 mm[b)]	Ar, 16 mm	Clear, low-e 4 mm	Clear	59	14	16	35	0.4	1.0
				Dark	19	8	13	10	0.14	

Note: U-factor is in $Wm^{-2}\,K^{-1}$. This is representative data and will vary depending upon exact IGU configuration and components and construction. Always contact the manufacturer to get more exact and timely numbers.

a) SageGlass: 6 mm clear w/SR2.0, 0.76 mm PVB interlayer, 2.2 mm SageGlass AN, EC2.0, 10.2 mm cavity, 6 mm clear.

b) ChromoGenics: The EC pane is PET film/EC/polymer electrolyte/CE/PET film, laminated between two 4 mm glass sheets.

ChromoGenics IGU: 10 mm EC pane/16 mm argon/4 mm low-e clear.

Source: Data from [177–179].

thermal and solar properties. SageGlass uses different types of glass to modify the color and performance of the IGU properties [178]. Also, SageGlass has made a gradient coloring window called Harmony® (e.g. dark at the top and lighter at the bottom of the window). One of the visual issues that are related to the electrochromic device is their appearance from the outside. When a building has several electrochromic windows, and only some of them are in a colored state, it makes a checkered appearance which has the potential of altering the architectural aesthetics. SageGlass has suggested the use of front glass with coatings which provide reflective properties [180]. Since this reflective glass is located prior to the electrochromic cell as viewed from outside, it masks to a certain degree the optical state of the electrochromic device. Another publication addresses this issue more completely, where a reflective coating optical stack is placed on the substrate prior to the deposition of the transparent conductor. This may also be combined with a certain glass color or color from the reflective coating stack. This reflective stack is placed closer to the outside surface. This has been shown to be an effective mitigator of the variations seen from outside [181]. ChromoGenics electrochromic products have almost as high thermal and solar efficiency as compared to other electrochromic products based on glass, but their visible transmission in the colored state is high and is less able to reduce glare under direct sun. However, ChromoGenics has another approach, unlike other makers where very significant capital is required to manufacture complete EGU panels and include heavy shipping charges to send completed EGU units to the end-customer. ChromoGenics supplies rolls of electrochromic ConverLight product on polyester, which are shipped to fabricators in different geographic regions, where electrochromic sheets are laminated with glass and formed into EGUs [182]. ChromoGenics working with ENERSign, North America (Boulder, CO), has designed a very high-performance triple glazing using ConverLight as one of its panes. It has $T_{vis} = 65\% - 75\%$ at its highest transmittance and can block 80% of solar heat. ENERSign, GmbH (Wittlich, Germany), is known for their high-performance passive solar windows and framing with U-Value = 0.59 W m^{-2} K^{-1}.

There has been considerable work on computer modeling of buildings using a wide range of glazing including dynamic glazing such as electrochromics [183]. Under USDOE and utility funding, several studies have been undertaken showing that by active management of windows, 20–30% energy consumption can be reduced [184]. Electrochromic glazing has been installed in both test buildings and actual buildings for the characterization of energy performance, daylighting, glare reduction, and overall comfort of the occupant.

Even though a significant investment has gone into electrochromic building windows and considerable technical progress has been made, the commercial adoption is quite low. In part, this is because of the high cost, generally in the range of 800–1000 US$ m^{-2} (80–100$ ft^{-2}) with long payback periods for energy savings. Steve Sottile has surveyed the US glazing manufacturers, architects, and energy efficient builders on the driving force for adoption and cost of smart windows. The main attractiveness of smart windows is energy efficiency, daylighting control, occupant wellness, and aesthetics [148, 149]. From a window manufacturer's viewpoint, a 26–50$ ft^{-2} added cost is an 13.5% adoption point and 11–25$ ft^{-2} added cost is a 47.4% adoption point for smart windows [185–187]. Among the building professionals in the US Green Building Council for Leadership in Energy and Environmental Design, a median adoption point for smart windows would be about 50 US$ ft^{-2} added cost [188]. In a 2012 report, the USDOE estimated that based on the energy used and the potential energy savings from the electrochromic windows, the payback for commercial windows will be 57–61 years for commercial buildings and about 26–33 years for residential buildings [189]. In 2020, the USDOE published a zero-energy window roadmap to lower high-performance glazing cost by the years 2030 and 2050. Their goals for dynamic glazing were to reduce the added cost to an IGU to 1.30–6.50 US$ ft^{-2} for residential and 3.5–11.80 US$ ft^{-2} for commercial glazing [190]. Currently, we are far from those goals. In the United States, various tax credits and energy incentives have reduced payback times. The driving force is energy efficiency in heating, cooling and lighting of buildings. Energy savings by electrochromic glazing have been estimated to be 20 - 200 KWh m^{-2} $year^{-1}$, from 10 different studies [191]. Also, smart windows used in the proper space can make that space more comfortable, productive, and valuable.

2.10 Multiband Switching for Glazing

Multiband switching means certain bands of energy can be switched independently from transmitting to reflecting, transmitting to absorptive, or transmitting to scattering in some cases. At air mass AM1.5, the terrestrial solar spectrum has about 49% of solar energy in the NIR range, 46% in the VIS, and 5% in the UV. The bands of interest for glazing are the solar (250–2500 nm), VIS (380–720 nm), and NIR (720–2500 nm), for the control of energy, lighting or privacy through glazing. The UV band for glazing should be blocked unless it is used for the generation of power using transparent photovoltaics. Mid-infrared (MIR) (3–5 and 8–13 μm) is of interest for radiative cooling. A 300 K (room temperature) black body radiates from roughly 3 – 70 μm with a peak at 10 μm. In the chapters of this book, we discuss how to control mainly visible and solar energy using electrochromics, liquid crystals, electrophoretics, thermochromics, and photochromics. However, some technologies like electrochromics can be made broadband covering VIS and NIR. In an IGU design, one could design a window with two different switchable coatings, one specially controlling the visible band and one controlling the NIR band. Currently, we use low-e coating to give fixed control of the NIR and MIR (reradiation). There is a design opportunity to have a single coating that could independently switch at different potentials in the VIS and NIR or both together. The solar spectrum is shown in Figure 2.20 for reference. For a glazing that regulates the NIR, we need to develop technology that switches from 780 to 1750 nm to control most of the energy.

2.10.1 Localized Surface Plasma Resonance (LSPR)

Thus far, we have commercialized visible electrochromic devices which intercalate ions and electrons, forming polarons. The technology is fairly developed for transition metal oxides like WO_3. One technique to make NIR switching material is to utilize the plasmonic properties of small particle physics of conductive metals and doped metal oxides. The plasmonic effect was first observed by Robert Wood in 1902 [192]. The localized surface plasmon was finally detailed by Otto in 1968 [193], which resulted in the development of gas and biological detectors. Incoming light can be coupled to surface plasmons of nanoparticles with metal-like conductivity. A localized surface plasmon is induced when the frequency of the oscillation-excited electron of a metallic particle is matched with the frequency of the incident photon. The local enhancement of the electromagnetic field as a result of the resonance effect leads to the effective light concentration near the surrounding of the metallic particle, mostly in the VIS and NIR regions. The localized electrons in the nanoparticles can be tuned by applying a potential, giving an optical effect in the NIR. Key classic references to plasmonics are Meier [194] and Menguc and Francoeur [195]. The influence of shape and size of metal particles on plasmonic response has been studied by Noguez [196] and Trügler [197]. Boschloo and Fitzmaurice at University College Dublin discovered in 1999 that nc-Sb:SnO_2 exhibited NIR electrochromism [198]. Also, other nanocrystalline degenerately doped nc-Al:ZnO_2, nc-In:CdO, and nc-Sn:InO_x exhibit electrochromism [199]. This has become a developing area of research for energy control of windows since about 2000. The doping of plasmonic nanocrystalline semiconductors has been reviewed by Kriegal et al. [200]. In work on degenerately co-doping of nc-CdO with F and In, it was found that the localized surface plasma resonance (LSPR) peak could be moved from 1.5 to 3.3 mm wavelength, with a change in the concentration of a substitutional cation doping [201]. The design of dual-band NIR electrode materials nc-$Mo_{1-y}W_yO_{3-\delta}$ //nc-$MoCeO_2$ was the subject of a recent thesis by Florian Gillissen (2024) [199].

LSPR is a phenomenon in which nanocrystals can exhibit enhanced light absorption and scattering at certain frequencies. Incident electromagnetic radiation displaces the free electron cloud, which then experiences a restoring force exerted by the positively charged ion cores. At the resonance frequency, scattering and absorption are strongly enhanced. This effect occurs when the incident light frequency matches the natural resonance frequency of collective electron oscillations in the conduction band or of hole oscillations in the valence band of the nanoparticles, as shown in Figure 2.21.

LSPR occurs not only in nanocrystalline metals but also in semiconductor nanocrystals with high free carrier concentration (10^{22}–10^{23} cm^3). The LSPR frequency (ω_{LSPR}) has a positive correlation with the free carrier concentration, and obeys the following expressions:

$$\omega_{LSPR} = \sqrt{\frac{{\omega_p}^2}{1 + 2\varepsilon_m} - \gamma^2} \quad (2.16)$$

$$\omega_p = \sqrt{\frac{Ne^2}{\varepsilon_0 {m_e}^*}} \quad (2.17)$$

Where ω_p is the plasma frequency which is related to the density of free carriers, N, e is the elementary charge, γ is the bulk collision frequency for free carriers (damping constant), ε_o is the vacuum permittivity, ε_m is the dielectric constant of the surrounding environment, ${m_e}^*$ is the electron effective mass, after Gillissen [199]. A selection of LSPR materials is shown in Figure 2.22 [202, 203].

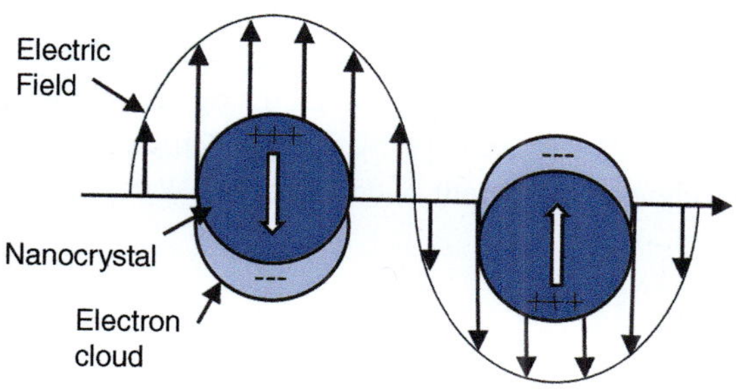

Figure 2.21 Schematic of incident light interacting with the conductive nanocrystals. LSPR is a phenomenon when the electromagnetic field of light induces a collective coherent oscillation of the surface conduction electron in resonance with the frequency of light. Also, the electron can be manipulated with an applied potential. *Source:* Samsuri et al. [192] / EDP Sciences / CC BY 4.0. This is an open-access article distributed under the terms of the Creative Commons Attribution License 4.0.

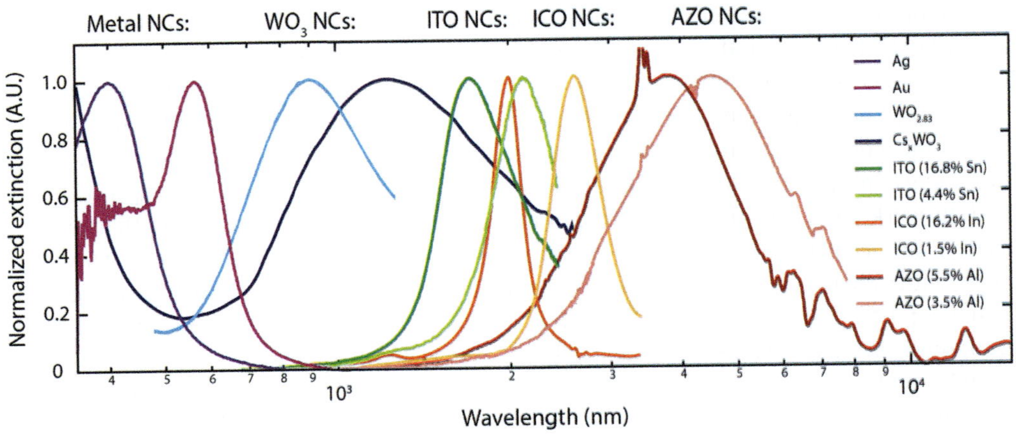

Figure 2.22 Spectral normalized extinction coefficient for LSPR nanocrystal (nc) materials. Shown are a range of conductive metal and oxide nanocrystals, ITO is indium tin oxide, ICO is indium cadmium oxide, and AZO is aluminum zinc oxide. The doping amounts are shown for the conductive oxides. Notice how the doping level shifts the plasma frequency. Particle size, shape and host media can influence the position of the plasma frequency [202–204]. Data is from [205–209]. *Source:* Lounis et al. [204], with permission of American Chemical Society.

To gain a switching effect, the electron cloud can be manipulated with an applied potential, giving a switching effect optically, as shown in Figure 2.23.

2.10.2 Early Commercial Development of LSPR Nanocrystals for Dual-Band Switchable Glazing

Heliotrope Technologies (Alameda, CA) attempted to commercialize this intriguing concept for electrochromic windows in about 2012–2023. This work was a spin-off from the University of California, LBNL (Berkeley, CA). Unfortunately, Heliotrope went out of business. They were working on a system to where the glazing could be in the "warm", "cool," or "dark mode," depending upon applied voltage. One of their first IGU prototypes was warm-to-dark mode, gray-colored EC with T_{vis} = 48–2%, T_{solar} = 27–7%, and SHGC = 0.36–0.12, switching at 1–3 V DC, with power consumption, P = 1.0 W m^{-2}. Further information is found in the Heliotrope patents. There are several patents with a few examples shown here. Work on devices is described by Garcia et al. in US patent 9684219 B2 (2017) [212, 213] and by Mattox et al. in US patent 978503 B2 (2017) [214]. Device manufacturing techniques are described by Holt et al. in US patent 10900277 B1 (2021) [215], and by Wang et al. in US patent 11221534 B2 (2022) [216]. Electrolyte development is described by Wang et al. in US patent 11409176 B2 (2022) [217].

2.10.3 Research in Dual-Band Electrochromics

Tunability of semiconductor nanocrystals has been studied by the Milliron group at the University of Michigan, Engineering Department (Ann Arbor, MI) [218]. By utilizing nanocrystals and plasmonics, the infrared properties are potentially tunable as depicted in Figure 2.24. Three modes are noted, the "bright mode," "cool mode," and "dark mode." When no potential is applied (open-circuit condition), the window is in the bright mode, where the ITO

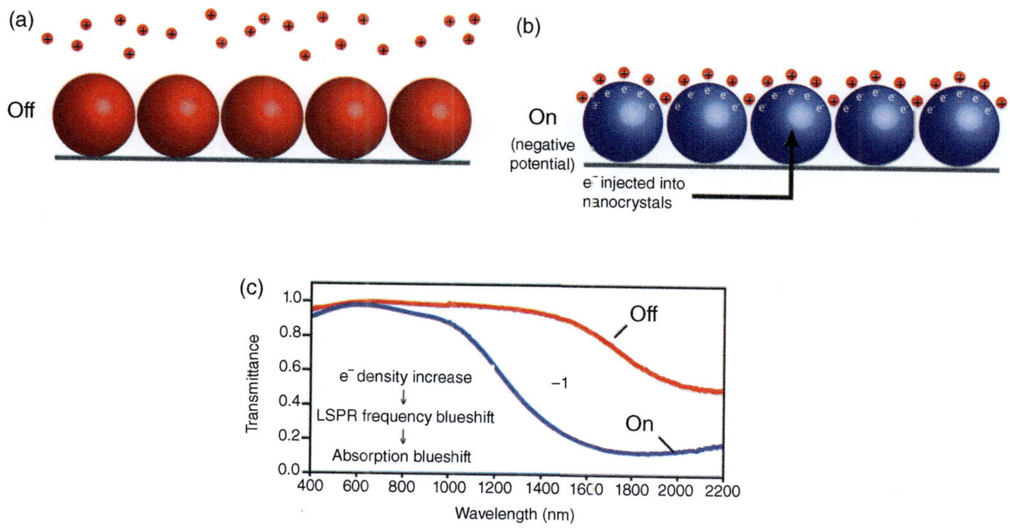

Figure 2.23 Schematic of a nanocrystal-based plasmonic electrochromic coating, along with the associated transmittance change in the NIR with applied voltage (e.g. 3 V). Illustration of a localized surface plasmon in a conductive metallic nanocrystal or semiconductor nanocrystal. (a) In the off condition, a positive potential is applied to the nanocrystals, which are depleted of electrons and lithium ions are repelled. (b) In the on-state, a negative potential is applied to the nanocrystals which injects electrons into the nanocrystals. Lithium ions are attracted to the nanocrystal surface to compensate for the injected charge. (c) Transmittance change of the film when voltage is applied. An increase in carrier density causes a blueshift in the LSPR and absorption. *Source:* Dr. Delia Milliron, University of Michigan, Engineering Dept. (Ann Arbor, MI) [210, 211].

nanocrystals in a NbO_x glass are transparent to solar energy. The bright mode allows heat and natural light to pass through the window. Reducing the voltage to an intermediate value, the window switches to cool mode. The cool mode blocks solar heat (NIR) while allowing natural VIS light to pass through the glazing. At lower potential, the window switches to the dark mode, limiting the amount of solar heat and natural light entering the room. These three modes enable the window to operate under varying temperature and lighting conditions, improving the overall energy efficiency and occupant comfort. One example is indium-doped cadmium oxide studied by the Milliron group [218]. They showed that by using 10 nm diameter particles the plasma resonance could be tuned from 2250 to 1600 nm. The carrier concentration varied from 0.87×10^{21} cm^{-3} to 1.33×10^{21} cm^{-3}. By alteration of the physical size and shape, one can produce properties not seen in bulk materials. The size and shape effect can be modeled by Mie scattering theory [219]. Electrical conductivity can be modeled by Drude free electron theory [220].

ITO nanocrystal composites in NbO_x glass coupled with ITO have shown dual-band switching effects, see Figure 2.25 [221]. The cell structure is TCO/CE/EL/nano-ITO-NbO_x /TCO [222]. Definable modes include a bright mode where both VIS and NIR can pass through glazing, a cool mode is achieved by applying a lower reducing potential (2.3 V vs. Li/Li^+), where the carrier concentration in the ITO nanocrystals increases and the NIR is reflected. A dark mode is achieved by going to an even lower potential (1.5 V vs. Li/Li^+) and further reducing the ITO nanocrystals by insertion of Li^+ ions. In the dark mode, both VIS and NIR are attenuated. The bright mode is obtained by application of a potential (4 V vs. Li/Li^+), whereby Li^+ ions will travel back to the CE.

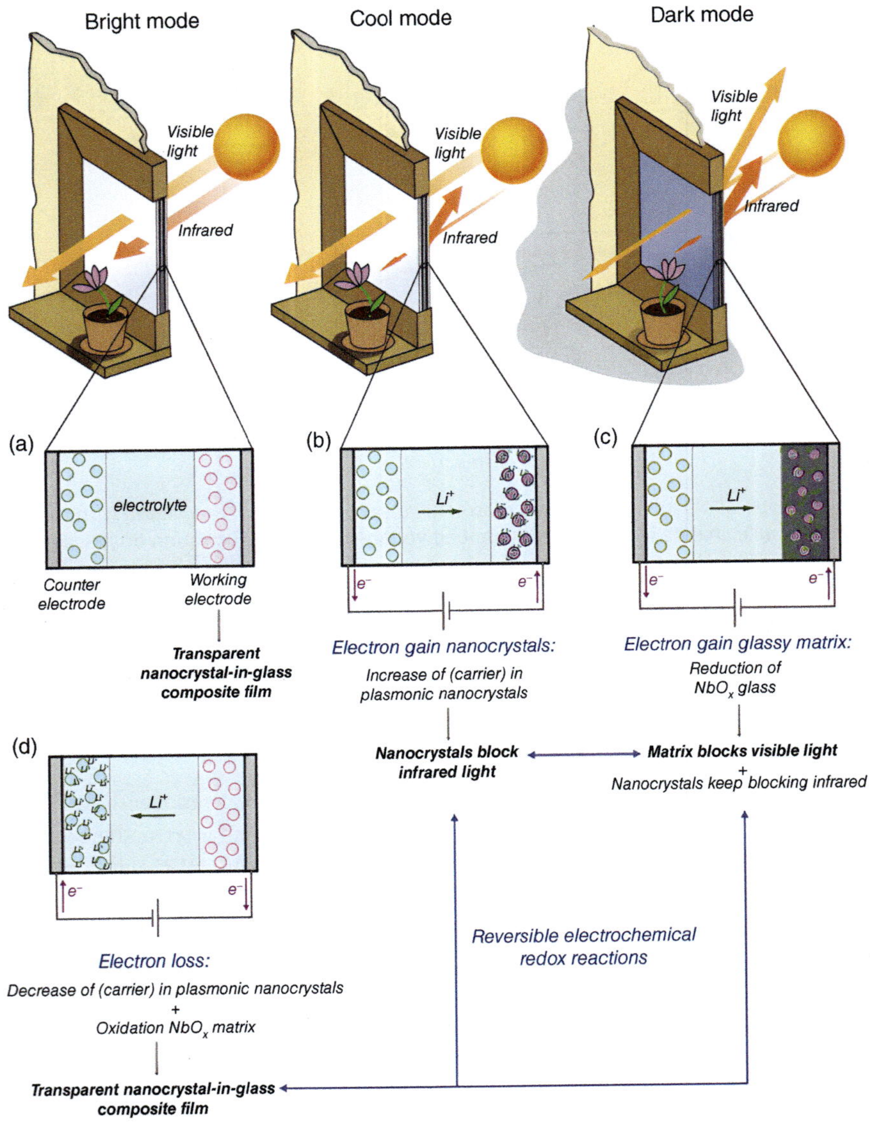

Figure 2.24 Smart window based on dual-band electrochromic concept. The concept is independent modulation of VIS and NIR light transmitted through the window. Both the active and passive films are deposited on transparent conducting electrodes. The ITO nanocrystal-in-NbO_x glass electrochromic film is the working electrode in an electrochemical cell. An optically passive film acts as a CE. When activated, nanocrystals block the passage of infrared light through the glass panes. The panes are separated by a solid electrolyte layer. At open-circuit voltage, the window is in "bright mode", (a) where the glazing is transparent to solar radiation. This allows heat and natural light to enter the room. Reducing the voltage to an intermediate value, the window switches to "cool mode", (b) blocking heat while allowing natural light to enter the room. At lower potentials, the window switches to "dark mode", (c) limiting the amount of heat and natural light that enter the room. These three switching modes enable the window to operate at variable weather conditions, improving thus energy savings and comfort. (d) The reverse potential to go back to the bright mode. *Source:* Adapted from Llordés et al. [221] / with permission of Springer Nature.

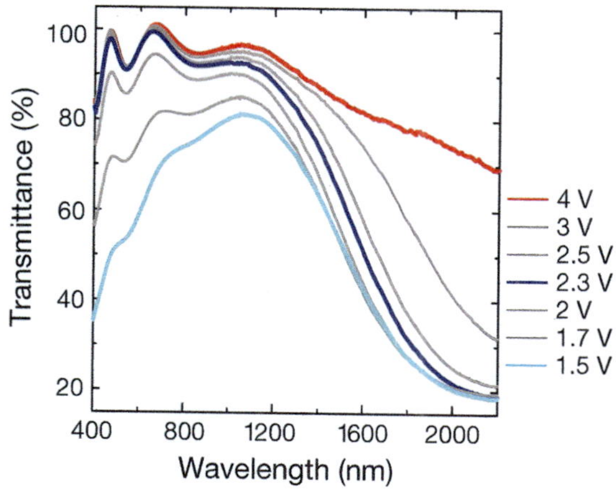

Figure 2.25 Spectral transmittance showing spectral response in the solar range (visible, near-infrared). This is a half-cell electrochemical response. Optical switching response under applied voltage (vs. Li/Li^+) of an ITO-in-NbO_x composite containing 43 vol% of nanocrystals. The ITO-coated glass substrate has been subtracted from the spectra. The carrier concentration in the ITO nanocrystals at 2.3 V was estimated to be 1.9×10^{21} cm^{-3} by applying the Drude free electron theory. The weak, broad NIR absorption feature noticeable at 4 V is not well-understood, but is not plasmon related. The liquid electrolyte was $LiClO_4$ in prop *Source:* Llordés et al. [221] with permission of Springer Nature.

Other materials that exhibit dual-band electrochromism are hydrated $WO_3{\cdot}H_2O$ switching to $Li_{0.12}WO_3{\cdot}H_2O$, with Li^+/e^- injection, giving a visible switching effect. The structural water leads to a 2D layered structure [223].

Anhydrous WO_3, with Li^+/e^- injection, becomes $Li_{0.19}WO_3$ resulting in a broadband switching effect. Intercalation of Al^{+3} in tungsten oxide for dual-band switching has been observed by S. Zhang and group at the National University in Singapore [224].

Research on doped TiO_2 nanocrystals has shown dual-band switching capabilities [225–229]. The Milliron group working on Nb-doped TiO_2 nanocrystalline electrodes has shown excellent dual-band properties. The optical spectral response of Nb-doped TiO_2 showed $\Delta T_{vis} = 72$, $\Delta T_{nir} = 63$ in half-cell devices [225].

The spectral transmittance is shown in Figure 2.26 for different levels of coloration.

Studies on Nb-doped TiO_2 coupled with VO_x and an anode has resulted in a dual-band switching device. The device is shown in Figure 2.27. Cao et al. has reported Ta-doped TiO_2 in half-cells with $\Delta T_{vis} = 92.4$ and $\Delta T_{nir} = 81.4$ [227]. Zhang et al. has reported studies on TiO_{2-x} in half-cell with $\Delta T_{vis} = 95.5$ and $\Delta T_{nir} = 90.5$ [229].

Work continues in this new area of development with LSPR nanocrystals. Having good optical properties is only part of the success, stability, cyclic durability, and manufacturing scale-up cost are key factors for industrial IGU success. Durability of an LSPR IGU is a minimum of 10 years. Ten years is the current warranty period for standard IGU units and electrochromic glass units.

2.11 Electrochromic Windows for Aircraft

Windows used in commercial airplane glazing are known as transparencies. Currently, there are electrochromic passenger windows installed in the upper-class section of selected

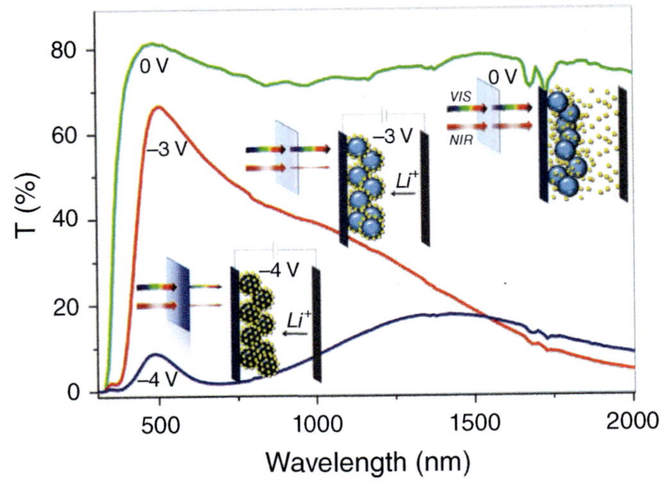

Figure 2.26 Spectral transmittance of 10% Nb-doped TiO_2 nanocrystalline electrochromic device. Three voltage levels are shown: 0 V, completely bleached; −3 V, NIR attenuated; and −4 V, both VIS and NIR attenuated (Barawi et al.) [228]. *Source:* Barawi et al. [228], with permission of American Chemical Society.

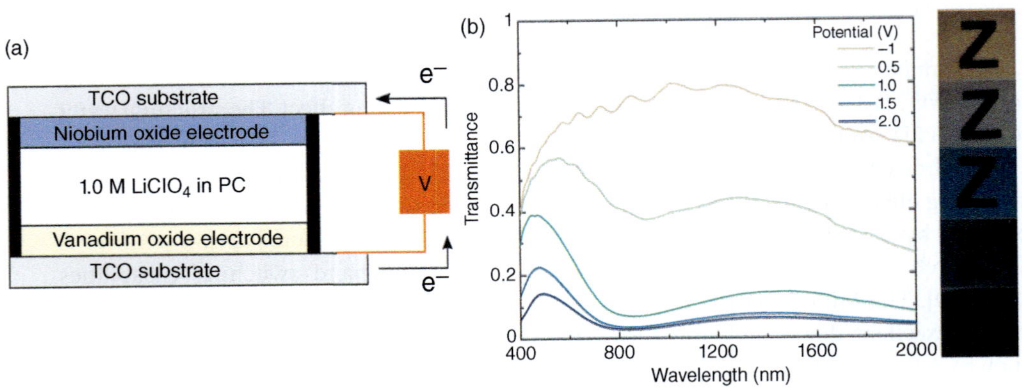

Figure 2.27 (a) Schematic showing the components of a complete electrochromic device incorporating ~500 nm of nanocrystalline niobium oxide ($Nb_{12}O_{29}$) and ~150 nm of vanadium oxide electrodes. (b) Spectral transmittance of a rigid glass electrochromic device at a range of potentials from −1 to +2.0 V DC. Corresponding photographs of coloration in room light with a white background are shown on the right side [225]. *Source:* Zydlewski et al. [225], with permission of American Chemical Society.

Boeing's 787 Dreamliner and the 777X planes. The function of the window is to allow the passenger to adjust the transmittance of the window to the level of lighting and glare reduction the passenger may prefer. Also, it is possible for the flight attendants to dim all the windows for sleep or movies or to brighten all the windows for landing. Figure 2.28 shows an image of a Gentex/PPG aerospace transparency window called ALTEOS™ made for the 787 Boeing Dreamliner. This follows on earlier PPG electrochromic transparency work by Rukavina & Lin, in US Patent 6783099B2 (2004), "electrochromic transparencies for aircraft" [230]. Transparency glazing is shaped, chemically tempered and bent for aircraft glazing. It is an inner glazing behind the structural cabin glazing.

Figure 2.28 ANA Boeing Dreamliner 787-8 with an electrochromic dimming window made by PPG Aerospace with a Gentex electrochromic window element. An intermediate coloration level is shown. The electrochromic is an inner window to the structural cabin glazing. *Source:* Jun Seita / Wikimedia Commons / CC BY 2.0.

A schematic cross section of a Boeing dimmable airplane window can be seen in Figure 2.29 [231]. A three-component window is shown, with a laminated curved exterior window, a middle electronically dimmable window, and an opaque sliding cover on the interior. The exterior window is designed to withstand changes in high pressure and is impact resistant. Daimler-Chrysler Aerospace has made dimmable windows for Airbus Industrie A380 airplanes. Also, the Airbus A330NEO has been specified with dimmable windows. Saint-Gobain has publicly shown gradient electrochromic Airbus cabin windows developed with the SageGlass Harmony® technology [232]. Gentex electrochromic technology used for airplane windows is similar to the technology used in their mirror products. This means that the dimming element is also an electrolytic layer containing the two dyes. Instead of the reflector on the rear substrate, a transparent conductor is used. The performance envelope for electrochromic mirrors is quite different compared to the aircraft windows. For the aircraft windows, the speed of dimming is slower and the depth to which the electrochromic element colors are greater, compared to mirrors. For example, in a mirror, the coloration and the bleach rates are rapid, typically two seconds, and for aircraft windows this could take several seconds to switch. The PPG Alteos® airplane window has a contrast ratio that exceeds 10,000 or more, with the lowest fully colored transmittance of 0.001% [235]. This is achieved by selecting an appropriate mixture of dyes, increasing the electrolyte thickness or the polymeric content to decrease the dye mobility. By decreasing the dye mobility, the back reaction (or self-erasing reaction) can be slowed down, as outlined in Reaction (2.16). The dyes are maintained longer in the colored state. This has a net effect of having a higher concentration of dyes in the colored state.

Other potential electrochromic products include visors which can control glare for pilots and passengers. Gradient dimming pilot windows have been proposed by Saint-Gobain working with SageGlass. These windows are made by PPG Aerospace using Gentex electrochromic glazing [233, 234]. Also, SPDs in the form of a plastic film have been applied

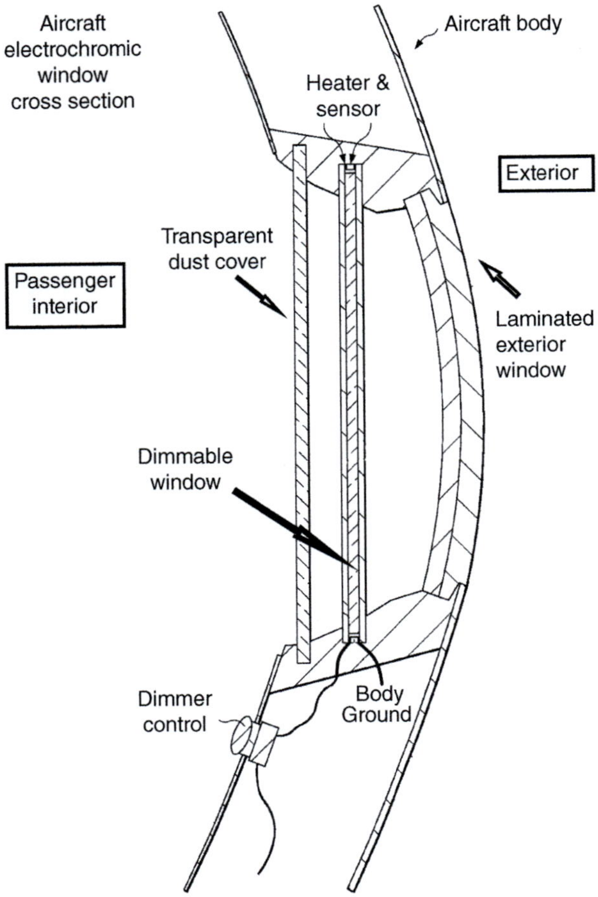

Figure 2.29 A schematic cross section of a dimmable window for a passenger aircraft window. *Source:* Author (CL) modified from Boeing European patent EP 2939919B1 [231].

to inner passenger aircraft windows. Generally, these windows are retrofit and original equipment manufacturer (OEM) for custom orders for executive jets and helicopters. The SPDs can be laminated onto aerospace transparencies [236]. Gauzy (Tel Aviv, Israel) expects to make 240 million US$ over the next 10 years in SPD cockpit shading [237]. Examples of SPDs are shown in Chapter 1 (Introduction) and Chapter 6 (Suspended Particle Devices).

2.12 Electrochromic Eyewear

The early pioneers who showed prototypes of electrochromic eyewear were Nikon (Tokyo, Japan) [238], Eyeonics (Portland, OR) [239–241], and PPG (Pittsburg, PA) [242, 243]. The electrochromic eyewear devices made by Nikon and Eyeonics had a solid-state structure like the one shown in Figure 2.5 (Type 3b) with a protective coating. The PPG eyewear was a laminated structure shown in Figure 2.4 (Type 3a). Nikon did go to market with their electrochromic dark glass eyewear in Japan. The market acceptance was weak probably due to high retail costs and short battery life. Many electrochromic eyewear developers used inorganic solid-state materials for the electrochromic, ion conductor and counter electrode. The electrochromic electrodes in all the early eyewear contained tungsten oxide. For the Nikon [244] and PPG products [245], the counter electrodes contained iridium oxide, a stable but expensive material [244]. Iridium oxide is a complementary electrochromic electrode to tungsten oxide, which means that when tungsten oxide colors by reduction, then at the same time iridium oxide colors by oxidation. Eyeonics used vanadium oxide for their counter electrode [246]. The battery packs for the prototypes from Nikon, Eyeonics, and PPG eyewear were integrated within the eyewear frames. Eyeonics developed a special power system to recycle charge, and prolong battery life [246]. A schematic power and control system is shown in Figure 2.30. Since the battery is limited in size and capacity, any charge that can be saved during bleaching promotes longer battery life. An example of a typical electrochromic eyewear and its components is shown in Figure 2.31 from Nitto Denko (Japan).

AshChromics (Marlboro, New Jersey) has announced electrochromic eyewear "Fades" [248, 249]. The parent company Ashwin-Ushas (Marlboro, New Jersey) was founded in 1992 by Prasanna Chandrasekhar, who has worked with electrochromic organic polymers for many years. The eyewear electrochromic structure is similar to the device shown in

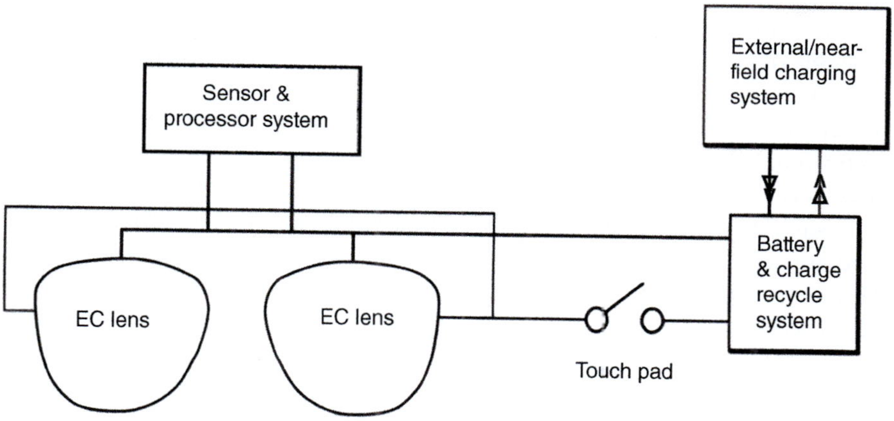

Figure 2.30 Typical electronic components for electrochromic eyewear. This includes a charge recycling system and an external/nearfield charging system which could be integrated into the eyeglass holder, as it is for the Ray-Ban Meta eyewear. *Source:* Modified (CL) from Liping & Youn & Wang, World patent appl. WO2022 / 098854 A1 [247].

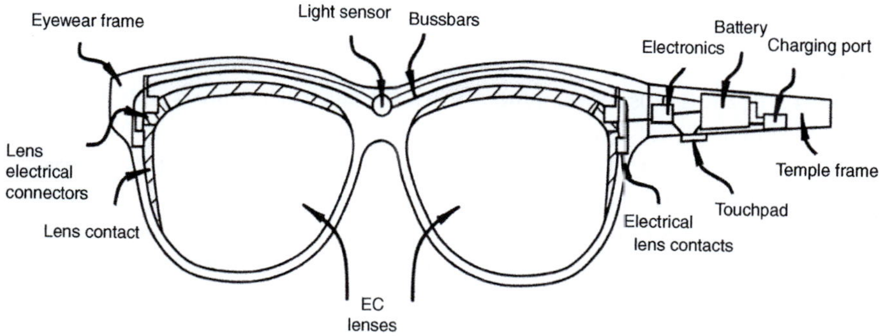

Figure 2.31 The major components of electrochromic eyewear. *Source:* Modified (CL) from Liping & Youn & Wang patent appl. WO2022 / 098854 A1 [247].

the laminated structure (Figure 2.4) where the electrochromic electrodes and the counter electrodes are both complementary coloring conductive polymers. Durability has been tested to 100,000 cycles. The glasses color to a deep blue color with a visible transmission dimmable from T_{vis} = 70–7% or from T_{vis} = 50–5%. Switching is one second to color and two seconds to bleach. Also, a good memory is a feature to keep the glasses in a desired state of transmission while the power is turned off. Further device information can be found in the Ashwin-Ushas patents, US 9482880B1 (2016) [250], USD 834008651 (design patent, 2018) [251], and US 2018025294A1 (2019) [252].

Corning (Corning, NY) and currently Corning SAS (Avon, France) make ophthalmic eyewear. Corning photochromic Photogray® and Photobrown® products have dominated the glass eyewear market since the 1960s. PPG developed organic photochromic materials for their Transitions™ polymer-based eyewear in 1991. This technology was sold to their codeveloper Essilor International (France) in 2013. Overall, the Transitions™ product has dominated the plastic eyewear market. Electrochromics have not made a big impact yet, possibly because prescription eyewear is harder to fabricate with electrochromics, with short battery lifetime and higher expense. Luxexcel (Netherlands) is involved with developing augmented reality (AR) using chromogenic eyeglasses. Luxexcel is a maker of 3D-printed prescription eyeglasses and a developer of smart glasses. Luxexcel was acquired by Meta (owners of Facebook, Instagram) in December 2022. The Ray-Ban Meta eyewear includes a camera, microphone, and earpieces for sound. The AI model lets you ask for information and has virtual Meta platform functions. The Ray-Ban Meta contains laminated electronics throughout the framing structure. The internal layout of the glasses' control and powering components is very interesting in design. YouTube has a few tear-down videos on the glasses (e.g. DigiKey site). The glasses are currently offered with Transitions™ photochromic lenses. There are certainly discussions about a future model with electrochromic lenses. Meta has been buying up extended reality (XR)-related businesses. XR covers AR and virtual reality. The US Federal Trade Commission is watching this area closely as XR has issues of security, safety, and protection of personal information.

On the research side of eyewear, Anna Österholm of the John Reynolds group, Georgia Institute of Technology (Atlanta, GA), has developed electrochromic polymer blends for high-contrast switchable eyewear [253]. Organics and polymer electrochromics offer a full palette of colors. More details can be found in the chapters on organic and polymeric electrochromic.

2.13 Electrochromic Information Displays

From the late 1960s to the 1970s, development work on electrochromic technology was focused on seven-segment digital displays and switchable filters [254–259]. However, with the maturing and stabilization of liquid crystal technology, electrochromics were bypassed except for a few applications where viewing angle and memory were valued. Today, the display market is at 135.2B US$ (2024) predicted to be 173.7B US$ (2029) with a compound annual growth rate (CAGR) of 5.1% [260]. The large growth will be in cell phone, iPad, and portable displays. Currently, the market is dominated by LCDs and organic light-emitting diode (OLED) displays. Most of the future market growth is predicted to be in smart phones and OLED displays. There are several displays with very fast display times, color, and some with foldability for the new foldable cell phones. As power consumption is very important, bistability is favored. Bistability means that after switching, power is not required to maintain the image or coloration. Some displays have a latency of several days, like E Ink electrophoretic displays. Electrochromics can be designed with bistability or direct discharge depending on their construction. Other competing technologies are electrophoretic and dielectrophoretic displays, electrowetting displays, interferometric modulator displays, and photonic crystal displays. Some electrochromic displays behave like batteries, keeping their charge, while others discharge immediately. The niche for electrochromic displays has been largely in small, printed displays for displaying information at slow rates or as indicators (not video displays) [261]. Printed electrochromic e-paper displays are ideal for low-power and cost-sensitive applications. Such applications include digital signage, smart monitoring labels, authenticity of the product and also of the recipient, for controlled and expensive products. Other applications include security (tamper proof) and retail labels and signage for stores and gas stations. These electrochromic labels are usually produced at low cost and may even be disposed of after a single use or several uses. Many of these displays are made of flexible plastic or even paper substrates. An example is an AA battery charge indicator, where you press the indicator and it shows how much charge is left in the battery. Some of the pioneering work in electrochromic displays has been conducted by Ntera (Ireland), Siemens (Germany), RISE Acreo (Norrköping, Sweden), Aveeso (Sweden), AJJER LLC (Tucson, AZ), Chromera (Poway, CA), and Rdot (Sweden), now part of Ynvisible (Vancouver, Canada, and Portugal) [262]. Ntera was focused on electrochromic dye–based technology, where the dyes were anchored to the coatings of inorganic oxides on the electrode surfaces [263, 264]. Ntera has left the field. Aveeso [265], Siemens [266, 267], Acreo [268, 269], and Rdot [270] are developing organic materials including dyes and conductive polymers to create flexible displays which could be made using roll-to-roll printing technology. Acreo's technology was largely adopted by Rdot. Ajjer/Chromera's focus was also on printed displays which could be powered wirelessly to permanently change the display to show the status of a product, for example, if a product passes its expiration date [271]. The best fit for electrochromic e-paper is in intelligent labels for freshness and smart cards. The electrochromic displays are less expensive than electrophoretic e-paper, reflective LCDs, and LEDs. LCD displays fail in sunlight legibility tests and were too bulky, while LED solutions consumed too much power compared to electrochromic displays.

A good review of electrochromic displays is available [272]. Ynvisible started out by incorporating tungsten oxide nanoparticles in their electrochromic layers [273, 274]. However, Ynvisible has largely transformed their technology base to that of Rdot after acquiring them. Of the aforementioned companies listed, Ynvisible is the only one presently in business with several commercial products. Ynvisible is concentrating on making smart electrochromic labels and indicators by a roll-to-roll printing process. Some display samples are shown in Figure 2.32. An authentication card with an electrochromic display is shown. The card is powered by a near field communication (NFC) signal. The card verifies that the product it scanned is authentic and was received by an authorized customer. Also, display products include TempSafe which shows if a package has gone through unacceptable too high or low temperatures. This is important for pharmaceutical packaging.

One of Ynvisible's main products is the seven-segment electrochromic displays which operate at $\pm$1.5 V DC and use 1.67 mW cm^{-2} power, so only very low power is needed for the display [275]. It is possible to power the display with NFC signals [276]. Ynvisible, working with Fortech S.r.l. (Rimini, Italy), has made 25,000 large seven-segment displays for Fortech's

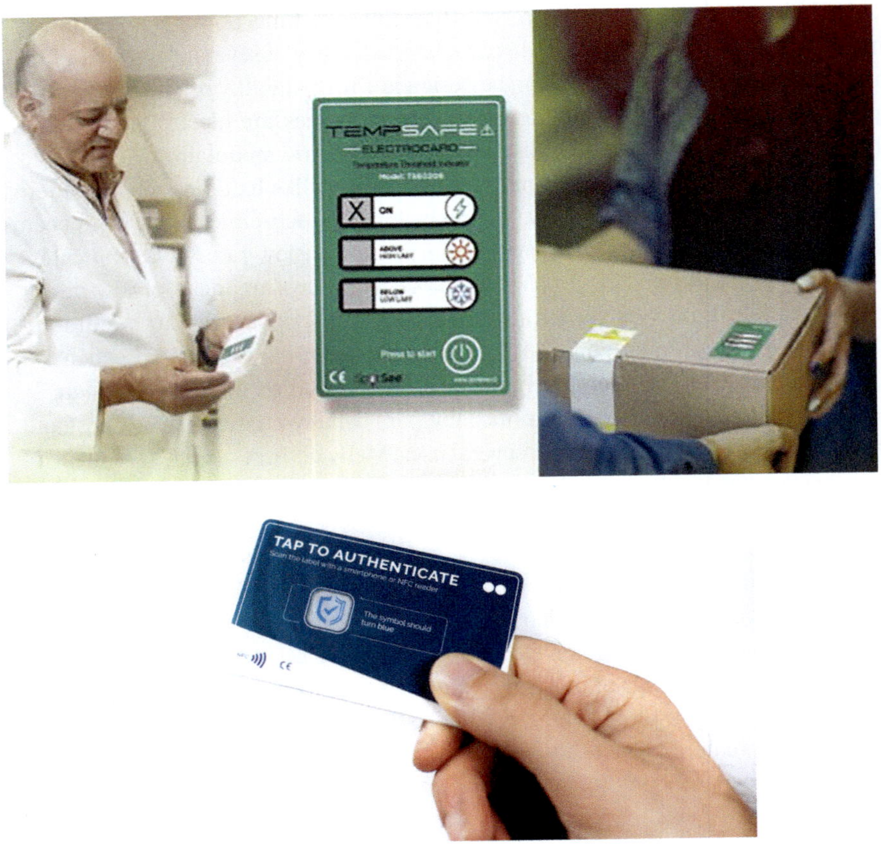

Figure 2.32 A Ynvisible TempSafe card to indicate if a shipment has been exposed to too high or too low temperatures (upper image). The indicator is electrochromic e-paper. This is especially important for drugs, live cultures, and immunization vials. Also, an authentication card is shown with an electrochromic display. This example is powered by a near-field signal. The card verifies that the product it scanned is authentic and received by an authorized customer. *Source:* Courtesy of Carlos Pinheiro, CTO and Marco Alexandre, Ynvisible, Co (Lisbon, Portugal).

(Italy) PriceLink fuel price displays for gas stations. The displays can be battery-operated for wireless convenience, lightweight, and are durable enough to withstand outdoor conditions. The displays have a clear visibility even in direct sunlight. An example fuel display is shown in Figure 2.33.

A prototype erasable electrochromic writing board has been reported using WO_3/SnO_2:F/glass and lithium chloride writing pen [277]. A drawn image shows up as a blue trace, as lithium ions are injected into WO_3 from the writing pen. Color electrochromic organic displays are being developed by the Reynolds group, Georgia Institute of Technology (Atlanta, GA) [278]. More details are in the chapters on organic and polymers electrochromics. Reviews of research on electrochromic displays will give the reader more insight into this topic [272, 279].

2.14 Electrochromic Gradient Filter

An electrochromic gradient filter has been developed for spatial image contrast adjustment to use with cameras [280, 281]. Figure 2.34 shows images of the effect. This work is from

Figure 2.33 Electrochromic ePaper display for fuel pricing, its low power, and light weight. It can be battery powered. *Source:* Courtesy of Carlos Pinheiro, CTO, and Marco Alexandre Ynvisible, Co (Lisbon, Portugal).

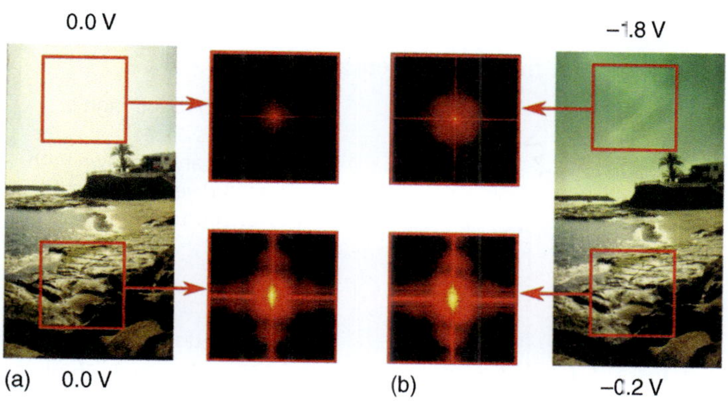

Figure 2.34 Images taken with graduated electrochromic filter placed in front of the camera with no voltage applied (a) transparent, no coloration (left side). The same exposure settings were used to take all the pictures. When different potentials were applied to the top and bottom of the filter, a graduated tinted state (b) was created (right side). The red images are Fourier transforms of the framed areas, which identify the transmitted spatial frequencies. *Source:* Hein et al. [280] / Springer Nature / CC BY 4.0.

the Prof. Oesterschulze group, Dept. of Physics, Physics and Technology of Nanostructures, Technische Universitat Kaiserslautern (Kaiserslautern, Germany). The electrochromic device was ITO/nc-Sb:SnO_2 + TPB/propylene carbonate + $LiClO_4$/nc-TiO_2 + viologen/ITO. TPB is tetrophenylbenzene and with the nanocrystalline Sb:SnO_2 served as the counter electrode. The thickness of all the combined layers without the glass was 55 μm. The filter's spectral transmittance is shown in Figure 2.35. The strongest switch is at 650 nm,

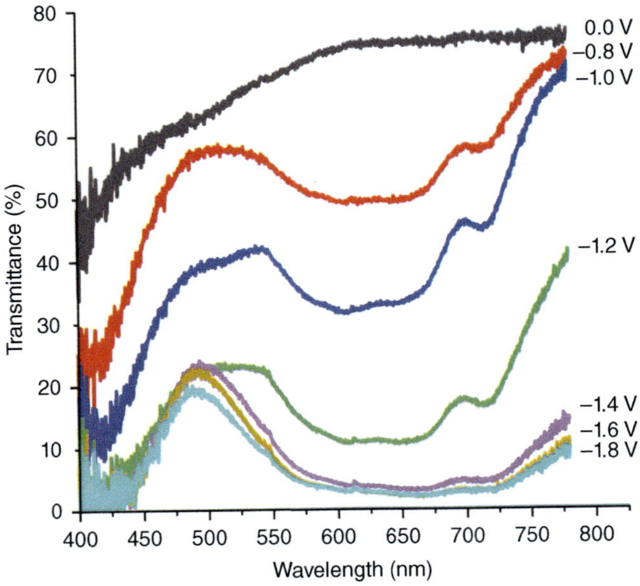

Figure 2.35 Spectral transmittance of an electrochromic filter shown at different potentials. 550 nm is the peak of the visible range. The coloration efficiency was $CE_{690nm} = 280\ mC\ m^{-2}$. *Source:* Hein et al. [280] / Springer Nature / CC BY 4.0.

ΔT_{650} = 68%, and at the peak of the visible at 550 nm, ΔT_{550} = 57%. By using stepped voltage 0.6–1.6 V, average visible transmittance can be changed from T_{ave} = 67% to 5%, as shown in Figure 2.36. A sigmoidal curve can be fit to the optical response. This work looks very promising for gradient electrochromic filters for camera applications.

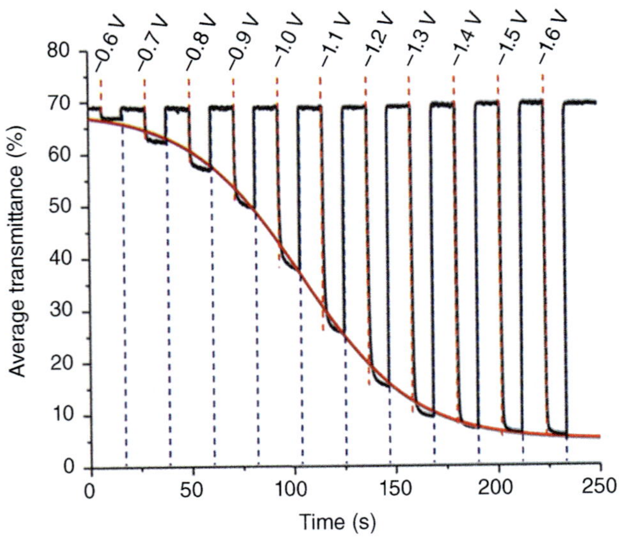

Figure 2.36 Average transmittance of an electrochromic filter using voltage steps from −0.6 to −1.6 V. The transmittance from 400 to 780 nm was the average transmittance. Switching was carried out by applying first the denoted potential (red dotted line) and then +1.5 V (blue dotted line) for 10 seconds each. A sigmoidal red curve was fitted to the transmission values in the colored steady state with 99.9% correlation. *Source:* Hein et al. [280] / Springer Nature / CC BY 4.0.

2.15 Dynamic Thermal Emittance Electrochromics for Spacecraft and Spacesuits

Thermal control is a large issue for both spacecraft and for spacesuits. In vacuum or very thin atmospheres, radiative emission of energy is the only effective method to reduce heat buildup or heat imbalance. The current technology commonly used for temperature control of satellites is the mechanical louver type of radiator [282]. Louver radiators have an emittance $e_{ir} = 0.1$ when open and $e_{ir} = 0.7$ when closed. Their disadvantages are weight, mechanical control, and potential damage. Louvers show a change in infrared emittance $\Delta e_{ir} = 0.6$, with an overall emittance ratio of $e_r = 7$. For space suits, a flexible material must be used [283]. With the development of wearable electronics and flexible electrochromic films now commercialized, there is an interesting potential here for a new type of dynamic space suit [284]. Information about the nature of the space environment on materials has been reviewed [285]. In the space environment, a thermal control surface encounters many potential events as shown in Figure 2.37. Some of these events can be very damaging such as atomic oxygen found in low earth orbit. Micrometeoroids and outgassing can contaminate surfaces, altering their radiative properties.

A few specially designed variable electrochromic modulators have been evaluated experimentally in space. NASA has sponsored research on adaptive radiators devices known as Veri-e and VEC. This application area is still under development and space testing. The special variable electrochromic devices are designed to control the heating of the spacecraft surface by changing the emittance of a surface. This can be done with only a few volts of potential at very low power. Several thermochromic and some electrochromic coatings have been studied for this application and have been reviewed [286]. The modulators are based on the properties of crystalline WO_3 which have been measured by J. Hale et al. working at the J.A. Woolam company (Lincoln, NE) [287, 288]. Hale showed through calculation of emissivity using measured index of refraction, that a strong broad emittance change was achievable in the infrared. From a device of polycrystalline–$WO_3/Ta_2O_5/NiO/Au$, they were able to get $e_{ir} = 0.057$ to switch to $e_{ir} = 0.595$ in the activated state, with an emittance ratio of $e_r = 10.4$. Demiryont et al. group at Eclipse Energy Systems (St. Petersburg, Fl) have made

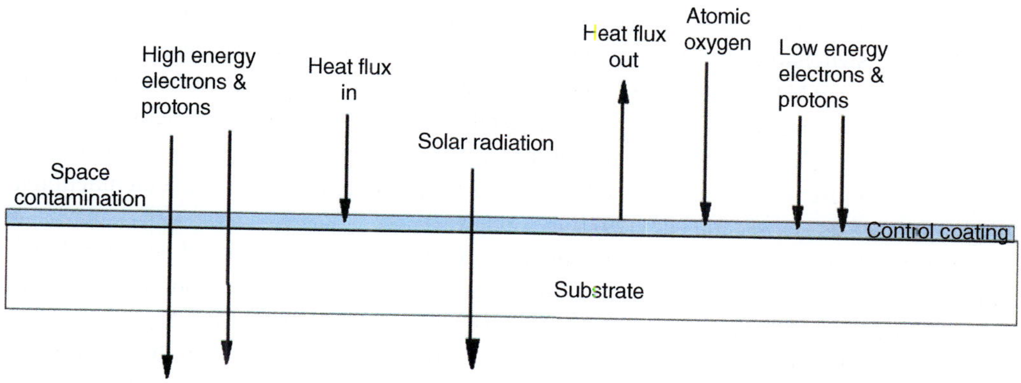

Figure 2.37 Schematic of the impact of the space environment can have on chromogenic optical coatings used in space. *Source:* Author (CL).

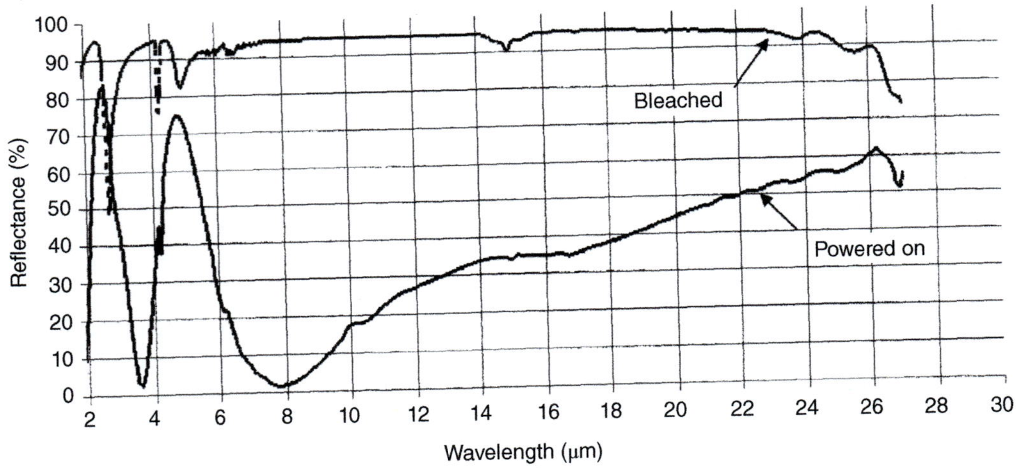

Figure 2.38 Spectral reflectance of an all inorganic electrochromic designed for a variable emissivity modulator for spacecraft application. Special IR transparent conductors were developed for high transmittance over the 0.4–27 μm wavelength range. *Source:* (modified CL) Hulya Demiryont, and US patent 75311239 B2 (2009) [291].

all-inorganic WO_3-based variable emissivity modulators for spacecraft. This modulator has flown on the MidSTAR satellite [289]. The modulator could be switched from $e_{ir} = 0.06$ to $e_{ir} = 0.77$, with $\Delta e_{ir} = 0.71$ and emittance ratio of $e_r = 12.8$ in the thermal infrared [290]. The reflectance spectra are shown in Figure 2.38. The device design is IR conductor/ion storage/solid electrolyte/EC/IR reflector. NASA has validated several switchable thermal control technologies on the ST-5 Satellite. The device is not transparent as it is designed to modulate the reflectance of a metal layer.

Aswin-Ushas Company has made prototype electrochromic variable emittance films for NASA validation [292, 293]. The device construction was PC/CP/gold/microporous membrane/gold/CP/mylar, where CP was the conductive polymer consisting of poly(aniline-co-diphenyl amine) with a long chain dopant polymer. PC was a thin protective coating. Example emittance properties (with protective coating) are $e_{ir} = 0.376$ to $e_{ir} = 0.771$, with $\Delta e_{ir} = 0.401$ [294]. Also, Patent US 20140268283 A1 contains more details of the variable emittance device [295]. A reference book on conductive polymers is available [296].

Another switchable IR emittance device was fabricated based on the intercalation of Li into $Ti_3C_2O_2$, an MXene. A change of emittance of $\Delta e_t = 0.31$ was shown. MXenes are a large family of two-dimensional (2D) transition metal carbides/nitrides with a general structural formula of $Mn_{+1}XnTx$, in which n is from 1 to 3. M represents the early transition metal elements, X represents carbon and/or nitrogen, and T represents surface-terminated functional groups (such as –O, –F, and –Cl) [297].

Thermochromic switchable modulators have been devices with hydrogel polymers [298] using solid state nanophotonic vanadium oxide has been studied by Taylor, et al. at Arizona State University (Tempe, AZ) for NASA. The Arizona group showed a $\Delta e_{ir} = 0.48$ from 20–100 °C. Other work has been done in China on dynamic IR emitters [299, 300]. For those interested in IR windows a review is available [301]. The infrared is a bit trickier, where longer wavelengths raise concern with intrinsic absorptions in your substrate or superstrate materials.

2.16 Other Electrochromic Devices: Photoelectrochromic and Photovoltaic-Electrochromic

Research work has shown some interesting, unusual lab scale devices. Some are called photoelectrochromic devices [302–305]. Other work is developing to combine photovoltaics and electrochromic devices, into essentially a self-powering electrochromic, Liu et al. [306] and Costa et al. [307]. Some designs have two sections of a window, a PV section and an electrochromic section [308]. Other devices that are more innovative combine the two functions together overlaid. One interesting window was made by the group in Slovenia, Hočevar, and Krašovec [309]. They made a nitrogen-filled IGU window with one active pane consisting of glass/SiO_2/c-WO_3/TiO_2/dye/electrolyte. The electrolyte was exposed to the nitrogen fill. They exposed the window to 750 W m^{-2} solar illumination and observed a change of $T_{vis} = 0.756$ switching to $T_{vis} = 0.355$ after 15 minutes of exposure. With the proper design and charge storage techniques, this might make fine devices especially for retrofitting, as running wiring can be expensive and complex with existing framing. Many years ago, photovoltaic devices integrated into the window framing structure were proposed for building integrated photovoltaic (called BIPV) applications. Another concept is to combine a battery with an electrochromic energy storage device (called an EESD) [310]. The EESD includes electrochromic super capacitors and batteries. The electrochromic battery research area has been reviewed by Pathak and Moon [311]. We expect to see more of these combined devices. Overall, the future is bright for electrochromics in expanding range of applications.

Acknowledgments

We wish to thank Brian Griedel, VP of IP, View, Inc., (Milpitas, CA), for supplying images of electrochromic View windows. We wish to thank Niall Lynam, Sr. VP, Magna Mirrors America, (Holland, MI), for his help with editing the electrochromic mirror section and use of the exploded view of an electrochromic mirror construction. Also, we must thank Dr. Hulya Demiryont, Chief Scientist, Eclipse Energy Systems (St. Petersburg, Fl), for her information about the early days of electrochromic sunroof development and spacecraft electrochromic devices. We would like to thank Prof. Delia Milliron, Department of Engineering, University of MI (Ann Arbor), for supplying figures on LSPR nanocrystals for this work. Also, we thank Carlos Pinheiro, CTO, and Marco Alexandre, Senior Designer, Ynvisible, Co. (Lisbon, Portugal), for providing photos of their electrochromic film and smart card prototypes. I (CL) must thank Fredrik Fränding, CEO, Greger Gregard, CTO, and Claes Granqvist, for their hospitality during my visit to ChromoGenics AB (Uppsala, Sweden).

References

1 Wilberforce, T., Olabi, A.G., Sayed, E.T., Elsaid, K., Maghrabie, M.H., and Abdelkareem, M.A. (2023). A review on zero energy buildings – pros and cons. *Energy and Built Environment* 4 (1): 25–38. https://doi.org/10.1016/j.enbenv.2021.06.002.

2 IEA (2024). Developing a global energy efficiency workforce in the buildings sector. https://www.iea.org/reports/developing-a-global-energy-efficiency-workforce-in-the-buildings-sector (accessed 31 Mar 2025).

3 IEA (2021). Net Zero by 2050 – A Roadmap for the Global Energy Sector, 4e. https://www.iea.org/reports/net-zero-by-2050 (accessed 01 Apr 2025).

4 Arasteh, D., Selkowitz, S., Apte, J, LaFrance, M., Zero Energy Windows (2006). ACEEE Summer Study on Energy Efficiency in Buildings, American Council for an Energy-Efficient Economy ACEEE. https://www.aceee.org/files/proceedings/2006/data/papers/SS06_Panel3_Paper01.pdf (accessed 01 Apr 2025).

5 Lee, E.S. (2018). Chapter 6.3 Building envelope. In: *Handbook of Energy Efficiency in Buildings* (ed. F. Asdrubali and U. Desideri). Butterworth-Heinemann, Elsevier Inc. ISBN: 978-0-12-812817-6. https://eta-publications.lbl.gov/sites/default/files/ch6.3-innovative_glazing_lbnl_2001193_eslee.pdf.

6 IEA (2024). Buildings. https://www.iea.org/energy-system/buildings (accessed 11 Aug 2024).

7 University of Michigan, Center for Sustainable Systems (2024). US Energy Fact Sheet. https://css.umich.edu/publications/factsheets/energy/us-energy-system-factsheet (accessed 12 Aug 2024).

8a NREL (2023). NREL-researchers reveal how buildings across the United States do and could use energy. https://www.nrel.gov/news/features/2023/nrel-researchers-reveal-how-buildings-across-the-united-states-do-and-could-use-energy.html (accessed 12 Aug 2024).

8b DOE (2022). Better Windows, Better outcomes: how electrochromics improve health, productivity, and efficiency, December 2022. https://www.energy.gov/sites/default/files/2023-12/bto-electrochromic-window-report-121123.pdf (accessed 25 May 2005).

9 Villa, D.L., Hahn, N.T., Grey, J.K., and Pavich, F. (2024). Futures for electrochromic windows on high performance houses in arid, cold climates. *Energy and Buildings* 315: 114293. https://doi.org/10.1016/j.enbuild.2024.114293.

10 Belzer, D.B. (2010). An exploratory energy analysis of electrochromic windows in small and medium office buildings – simulated results using EnergyPlus. PNNL Report 19637, Richland, WA, August 2010. https://www.pnnl.gov/main/publications/external/technical_reports/PNNL-19637.pdf (accessed 31 Jan 2025).

11 Lee, E.S., Yazdanian, M., and Selkowitz, S.E. (2004). The energy savings potential of electrochromic windows in the US Commercial Buildings Sector, Lawrence Berkeley National Laboratory (LBNL-54966). https://eta-publications.lbl.gov/publications/energy-savings-potential (accessed 31 Jan 2025).

12 Kao, H.-T. (2019). Modeling of buildings with electrochromic windows and thermochromic roofs. M.S. thesis, USF Tampa Graduate Theses and Dissertations. https://digitalcommons.usf.edu/cgi/viewcontent.cgi?article=9236&context=etd (accessed 31 Jan 2025).

13 Sawyer, K. (2014). Windows and Building Envelope Research and Development: Roadmap for Emerging Technologies (energy.gov) US Department of Energy. https://www.energy.gov/sites/prod/files/2014/02/f8/BTO_windows_and_envelope_report_3.pdf (accessed 31 Jan 2025).

14 Gu, C., Jia, A.-B., Zhang, Y.-M., and Zhang, S.X.-A. (2022). Emerging electrochromic materials and devices for future displays. *Chemical Reviews* 122 (18): 14679–14721. https://doi.org/10.1021/acs.chemrev.1c01055.

15 Nowak, N. (2024). Patent landscape analysis of Electrochromics, Questel. https://www.questel.com/resourcehub/patent-landscape-analysis-of-electrochromic-glass/ (accessed 06 March 2024).

16 Randin, J.-P. and Vienret, R. (1982). Proton diffusion in tungsten trioxide thin films. *Journal of the Electrochemical Society* 129 (10): 2349. https://doi.org/10.1021/acselectrochem.5c00137.

17 Xu, G. and Chen, L. (1988). Lithium diffusion in WO_3 films. *Solid State Ionics* 28–30 (Part 2): 1726–1728. https://doi.org/10.1016/0167-2738(88)90450-X.

18 Pennisi, A., Simone, F., and Lampert, C.M. (1992). Electrochromic properties of tungsten-molybdenum oxide electrodes. *Solar Energy Materials and Solar Cells* 28 (3): 233–247. https://doi.org/10.1016/0927-0248(92)90032-K.

19 Green, S.V., Pehlivan, E., Granqvist, C.G., and Niklasson, G.A. (2012). Electrochromism in sputter deposited nickel-containing tungsten oxide films. *Solar Energy Materials and Solar Cells* 99: 339–344. https://doi.org/10.1016/j.solmat.2011.12.025.

20 Özer, N., Rubin, M.D., and Lampert, C.M. (1996). Optical and electrochemical characteristics of niobium oxide films prepared by sol-gel process and magnetron sputtering A comparison. *Solar Energy Materials and Solar Cells* 4 (4): 285–296. https://doi.org/10.1016/0927-0248(95)00147-6.

21 Özer, N., Barreto, T., Büyüklimanli, T., and Lampert, C.M. (1995). Characterization of sol-gel deposited niobium pentoxide films for electrochromic devices. *Solar Energy Materials and Solar Cells* 36 (4): 433–443. https://doi.org/10.1016/0927-0248(94)00197-9.

22 Özer, N. and Tepehan, F. (1993). Structure and optical properties of electrochromic copper oxide films prepared by reactive and conventional evaporation techniques. *Solar Energy Materials and Solar Cells* 30 (1): 13–26. https://doi.org/10.1016/0927-0248(93)90027-Z.

23 Özer, N. and Lampert, C.M. (1998). Electrochromic characterization of sol–gel deposited coatings. *Solar Energy Materials and Solar Cells* 54 (1–4): 147–156. https://doi.org/10.1016/S0927-0248(98)00065-8.

24 Yu, P.C., Nazri, G., and Lampert, C.M. (1987). Spectroscopic and electrochemical studies of electrochromic hydrated nickel oxide films. *Solar Energy Materials* 16 (1–3): 1–17. https://doi.org/10.1016/0165-1633(87)90003-7.

25 Lampert, C.M., Omstead, T.R., and Yu, P.C. (1986). Chemical and optical properties of electrochromic nickel oxide films. *Solar Energy Materials* 14 (3–5): 161–174. https://doi.org/10.1016/0165-1633(86)90043-2.

26 Yu, P.C. and Lampert, C.M. (1989). In-situ spectroscopic studies of electrochromic hydrated nickel oxide films. *Solar Energy Materials* 19 (1–2): 1–16. https://doi.org/10.1016/0165-1633(89)90019-1.

27 Garnich, F., Yu, P.C., and Lampert, C.M. (1990). Hydrated manganese oxide as a counter-electrode material for an electrochromic optical switching device. *Solar Energy Materials* 20 (4): 265–275. https://doi.org/10.1016/0165-1633(90)90059-A.

28 Wen, R.-T., Niklasson, G.A., and Granqvist, C.G. (2014). Electrochromic iridium oxide films: compatibility with propionic acid, potassium hydroxide, and lithium perchlorate in propylene carbonate. *Solar Energy Materials and Solar Cells* 120 (Part A): 151–156. https://doi.org/10.1016/j.solmat.2013.08.035.

29 Özer, N., Chen, D.-G., and Büyüklimanli, T. (1998). Electrochromic characterization of $Co(OH)_2$ thin film prepared by sol–gel process. *Solar Energy Materials and Solar Cells* 52 (3–4): 223–230. https://doi.org/10.1016/S0927-0248(97)00281-X.

30 Lampert, C.M. (1984). Electrochromic materials and devices for energy efficient windows. *Solar Energy Materials* 11 (1–2): 1–27. https://doi.org/10.1016/0165-1633(84)90024-8.

31 Granqvist, C.-G. (2015). *Electrochromic metal oxides: An introduction to materials and devices.* In: *Electrochromic Materials and Devices* (ed. R.J. Mortimer, D.R. Rosseinsky, and M.S. Monk), 3–40. Weinheim, Germany: Wiley-VCH. ISBN: 978-3-527-33610-4.

32 Granqvist, C.G. (1995). *Handbook of Inorganic Electrochromic Materials.* Amsterdam, reprinted 2002: Elsevier. ISBN: 9780444541635.

33 Kraft, A. (2024). Too blue to be good? A critical overview on the electrochromic properties and applications of Prussian blue. *Solar Energy Materials and Solar Cells* 278: 113195. https://doi.org/10.1016/j.solmat.2024.113195.

34 Shimanoe, K., Suetsugu, M., Miura, N., and Yamazoe, N. (1998). Bismuth oxide thin film as new electrochromic material. *Solid State Ionics* 113–115: 415–419. https://doi.org/10.1016/S0167-2738(98)00388-9.

35 Marcel, C., Hegde, M.S., Rougier, A., Maugy, C., Guéry, C., and Tarascon, J.-M. (2001). Electrochromic properties of antimony tin oxide (ATO) thin films synthesized by pulsed laser deposition. *Electrochimica Acta* 46 (13–14): 2097–2104. https://doi.org/10.1016/S0013-4686(01)00378-4.

36 Özer, N. (2001). Optical properties and electrochromic characterization of sol–gel deposited ceria films. *Solar Energy Materials and Solar Cells* 68 (3–4): 391–400. https://doi.org/10.1016/S0927-0248(00)00371-8.

37 Cogan, S.F., Anderson, E.J., Plante, T.D., and Rauh, R.D. (1985). Electrochemical investigation of electrochromism in transparent conductive oxides. *Applied Optics* 24 (15): 2282–2283. https://doi.org/10.1364/AO.24.002282.

38 Zhang, B., Xu, G., Tan, S., and Liu, C. (2020). Effects of indium-doped tin oxide film on electrochromic properties under lithium ion guidance. *Optical Materials* 101: 109756. https://doi.org/10.1016/j.optmat.2020.109756.

39 Kullman, L., Azens, A., and Granqvist, C.G. (1997). Electrochromic praseodymium oxide films. Proceedings of the SPIE 3138, Optical Materials Technology for Energy Efficiency and Solar Energy Conversion XV (14 October 1997), https://doi.org/10.1117/12.290204.

40 Azens, A. (2024). Research conducted on electrochromism in Prof. Granqvist's laboratory between 1992 and 2003 and lessons learned. *Solar Energy Materials and Solar Cells* 273: 112920. https://doi.org/10.1016/j.solmat.2024.112920.

41 Wang, Y., Shen, G., Tang, T., Zeng, J., Sagar, R.U.R., Qi, X., and Liang, T. (2022). Construction of doped-rare earth (Ce, Eu, Sm, Gd) WO_3 porous nanofilm for superior electrochromic and energy storage windows. *Electrochimica Acta* 412: 140099. https://doi.org/10.1016/j.electacta.2022.140099.

42 Mehmood, A., Long, X., Haidry, A.A., and Zhang, X. (2020). Trends in sputter deposited tungsten oxide structures for electrochromic applications: a review. *Ceramics International* 46 (15): 23295–23313. https://doi.org/10.1016/j.ceramint.2020.06.035.

43 Avendaño, E., Azens, A., Niklasson, G.A., and Granqvist, C.G. (2004). Electrochromism in nickel oxide films containing Mg, Al, Si, V, Zr, Nb, Ag, or Ta. *Solar Energy Materials and Solar Cells* 84 (1–4): 337–350. https://doi.org/10.1016/j.solmat.2003.11.032.

44 Chaudhary, A. (2023). A brief status of flexible bi-functional energy storage electrochromic devices. *Materials Today Electronics* 6: 100075. https://doi.org/10.1016/j.mtelec.2023.100075.

45 Yun, T.G., Hwang, B., and Cheong, J.Y. (2024). Recent progress and future research directions for electrochromic zinc-ion batteries. *Journal of Energy Chemistry* 90: 220–232. https://doi.org/10.1016/j.jechem.2023.10.039.

46 Chen, D., Chua, M.N., He, Q., Zhu, Q., Wang, X., Meng, H., Xu, J., and Huang, W. (2025). Multifunctional electrochromic materials and devices recent advances and future potential. *Chemical Engineering Journal* 503: 157820. https://doi.org/10.1016/j.cej.2024.157820.

47 Zhong, Q. and Colbow, K. (1991). Electrochromic properties of cesium tungstate with pyrochlore structure. *Thin Solid Films, v* 205 (1): 85–88. https://doi.org/10.1016/0040-6090(91)90475-D.

48 Loo, B.H., Lee, Y.G., and Emerson, M.T. (1994). Anodically grown electrochromic films of chloro- and bromo-palladates (II) and (IV), and chloro- and bromo-aurates (III) of rubidium: a Raman microprobe study. *Applied Surface Science* 74 (2): 191–195. https://doi.org/10.1016/0169-4332(94)90301-8.

49 Cheng, B., Cheng, H., Jia, Y., Liu, T., and Liu, D. (2023). Infrared electrochromic devices based on thin metal films. *Advanced Materials Interfaces* 10: 2202505. https://doi.org/10.1002/admi.202202505.

50 Yamada, Y., Miura, M., Tajima, K., Okada, M., and Yoshimura, K. (2013). Optical switching durability of switchable mirrors based on magnesium–yttrium alloy thin films. *Solar Energy Materials and Solar Cells* 117: 396–399. https://doi.org/10.1002/admi.202202505.

51 Fu, K., Chen, J., Deng, J., Shuan, L., Gau, Y., Tian, W., and Li, X. (2018). Turning optical switching properties of Mg-Y films in electrochemical process by tailoring composition. *Materials Research Express* 5 (3): 036419. https://doi.org/10.1088/2053-1591/aab6a1.

52 van der Sluis, P. and Mercier, V.M.M. (2001). Solid state Gd–Mg electrochromic devices with ZrO_2H_x electrolyte. *Electrochimica Acta* 46 (13–14): 2167–2171. https://doi.org/10.1016/S0013-4686(01)00375-9.

53 Goei, R., Ong, A.J., Tan, J.H., Loke, J.Y., Lua, S.K., Mandler, D., Magdassi, S., and Tok, A.Y. (2021). Nd–Nb Co-doped SnO_2/α-WO_3 electrochromic materials: enhanced stability and switching properties. *ACS Omega* 6 (40): 26251–26261. https://doi.org/10.1021/acsomega.1c03260.

54 Trong, L., Dinh, N., and Thanh, D. (2016). Preparation and characterization of $La_{0.67-x}$ Li_{3x} TiO_3 solid-state electrolyte used for electrochromic mirrors. *Materials Sciences and Applications* 7: 702–709. https://doi.org/10.4236/msa.2016.711056.

55 Lei, P., Wang, J., Gao, Y., Hu, C., Zhang, S., Tong, X., Wang, Z., Gao, Y., and Cai, G. (2023). An electrochromic nickel phosphate film for large-area smart window with ultra-large optical modulation. *Nano Micro Letters* 15: 34. https://doi.org/10.1007/s40820-022-01002-4.

56 Ma, D., Eh, A.L.-S., Cao, S., Lee, P.S., and Wang, J. (2022). Wide-spectrum modulated electrochromic smart windows based on MnO_2/PB films. *ACS Applied Materials & Interfaces* 14 (1): 1443–1451. https://doi.org/10.1021/acsami.1c20011.

57 Kandpal, S., Ghosh, T., Rani, C., Chaudhary, A., Park, J., Lee, P.S., and Kumar, R. (2023). Multifunctional electrochromic devices for energy applications. *ACS Energy Letters* 8 (4): 1870–1886. https://doi.org/10.1021/acsenergylett.3c00159.

58 von Fraunhofer, J.A. and Banks, C.H. (1972). *Potentiostat and Its Applications*. Butterworths. ISBN: 978-0408702393.

59 Arsalan, M. and Awais, A. (2022). *Electrochemical Workstation-Potentiostat: Working Principle, Instrument Handling and Application in Research*. Saarbrucken, Germany: Lambert Academic Publishing. ISBN: 978-0408702393.

60 Lampert, C.M. (1994). Glazing materials for solar and architectural applications. Report of the IEA Task group 10, LBL-34436, Sept. 1994. https://www.osti.gov/servlets/purl/10124630 (accessed 16 March 2024).

61 Tuna, Ö.A., Sezgin, A., Budakoğlu, R., Türküz, S., and Parlar, H. (2015). Electrochromic properties of tungsten trioxide (WO_3) layers grown on ITO/glass substrates by magnetron sputtering. *Vacuum* 120 (Part B): 28–31. https://doi.org/doi.org/10.1016/j.vacuum.2015.02.036.

62 Eh, A.L.-S., Tan, A.W.M., Cheng, X., Magdassi, S., and Lee, P.S. (2018). Recent advances in flexible electrochromic devices: prerequisites, challenges, and prospects. *Energy Technology* 6: 33. https://doi.org/10.1002/ente.201700705.

63 Granqvist, C.G. (2000). Electrochromic tungsten oxide films: review of progress 1993–1998. *Solar Energy Materials and Solar Cells* 60 (3): 201–262. https://doi.org/10.1016/S0927-0248(99)00088-4.

64 Azens, A. (2024). Research conducted on electrochromism in Prof. Granqvist's laboratory between 1992 and 2003 and lessons learned. *Solar Energy Materials and Solar Cells* 273: 112920. https://doi.org/10.1016/j.solmat.2024.112920.

65 Azens, A., Gustavsson, G., Karmhag, R., and Granqvist, C.G. (2003). Electrochromic devices on polyester foil. *Solid State Ionics* 165 (1–4): 1–5. https://doi.org/10.1016/j.ssi.2003.08.009.

66 Azens, A., Avendaño, E., Backholm, J., Berggren, L., Gustavsson, G., Karmhag, R., Niklasson, G.A., Roos, A., and Granqvist, C.G. (2005). Flexible foils with electrochromic coatings: science, technology and applications. *Materials Science and Engineering B* 119 (3): 214–223. https://doi.org/10.1016/j.mseb.2004.12.085.

67 Demiryont, H., Hu, J., Kallman, W.R., and Schaus, C.F. (1995). Monolithic multi-layer electrochromic composite material with inner buffer layers, AU 2760895A, filed 1995-05-25, granted: 1996-12-11, Assigned: Isoclima SpA, Comdisco Inc, Dominion Ventures Inc.

68 Boulanger, F., Lerbet, F., Papret, C., Perrodo, F., and Thomas, G. (1994). Electrochromic glazing, US patent 5352504A, filed 1991-11-08, granted: 1994-10-04, Assigned: Saint Gobain Glass France SAS.

69 Lampert, C.M., Visco, S.J., Doeff, M.M., Ma, Y.P., He, Y., and Giron, J.-C. (1994). Characteristics of laminated electrochromic devices using polyorganodisulfide electrodes. *Solar Energy Materials and Solar Cells* 33 (1): 91–105. https://doi.org/10.1016/0927-0248(94)90293-3.

70 Lampert, C.M., Ma, Y.-P., Doeff, M.M., and Visco, S.(1995). Electrochromic device using mercaptans and organothiolate compounds, US patent 5442478A, granted: 1995-08-15, Assigned: UC Berkeley.

71 Lampert, C.M. and Visco, S.J. (1992). Electrochromic glazing, US patent 5142406A, granted: 1992-08-25, Assigned: UC Berkeley.

72 Fu, X. (2010). Chapter 12 – Polymer electrolytes for electrochromic devices. In: *Polymer Electrolytes*, Woodhead Publishing Series in Electronic and Optical Materials (ed. C. Sequeira and D. Santos), 471–523. Woodhead Publishing. ISBN: 9781845697723 https://doi.org/doi.org/10.1533/9781845699772.2.471.

73 Thakur, V.K., Ding, G., Ma, J., Lee, P.S., and Lu, X. (2012). Hybrid materials and polymer electrolytes for electrochromic device applications. *Advanced Materials* 24: 4071–4096. https://doi.org/10.1002/adma.201200213.

74 Brazier, A., Appetecchi, G.B., Passerini, S., Vuk, A.S., Orel, B., Donsanti, F., and Decker, F. (2007). Ionic liquids in electrochromic devices. *Electrochimica Acta* 52 (14): 4792–4797. https://doi.org/10.1016/j.electacta.2007.01.025.

75 Zhen, Y. (2022). Electrochromic tungsten oxide nanofilms and ionic liquid based ion conductor for smart windows development: preparation, characterization and applications. Ph.D. thesis, NTNU, Dept. of Civil and Environmental Engineering, Trondheim, Norway. https://ntnuopen.ntnu.no/ntnu-xmlui/handle/11250/3028983 (accessed 31 Feb 2025).

76 Lei, Z., Chen, B., Koo, Y.-M., and MacFarlane, D.R. (2017). Special Issue – Ionic liquids. *Chemical Reviews* 117 (10): 6633–7240. https://pubs.acs.org/doi/10.1021/acs.chemrev.7b00246.

77 Shen, Y., Zhang, Y., Han, D., Wang, Z., Kuehner, D., and Niu, L. (2009). Preparation of colorless ionic liquids "on water" for spectroscopy. *Talanta* 78 (3): 805–808. https://doi.org/10.1016/j.talanta.2008.12.056.

78a Costa, C., Ivanou, D., Pinto, J., Mendes, J., and Mendes, A. (2019). Impact of the architecture of dye sensitized solar cell-powered electrochromic devices on their photovoltaic performance and the ability to color change. *Solar Energy* 182: 22–28. https://doi.org/10.1016/j.solener.2019.02.036.

78b Ke, Y., Chen, G., Wang, S., Zhou, Y., Yin, J., Lee, P.S., and Long, Y. (2019). Smart windows: electro-, thermo-, mechano-, photochromics and beyond. *Advanced Energy Materials* 9: 1902066. https://doi.org/10.1002/aenm.201902066.

79a Czanderna, A.W. and Lampert, C.M. (1990). Evaluation criteria and test methods for electrochromic windows, SERI/PR-255–3537 July, 1990 Golden, CO-Solar Energy Research Institute. https://www.osti.gov/biblio/6953557 (accessed 24 April 2024).

79b Lampert, C.M., Agrawal, A., Baertlien, C., and Nagai, J. (1999). Durability evaluation of electrochromic devices – an industry perspective. *Solar Energy Materials and Solar Cells* 56 (3–4): 449–463. https://doi.org/10.1016/S0927-0248(98)00185-8.

80 Zhou, Q., Shao, P., Zhang, R., Huang, S., Zhang, Y., Zhu, Y., Yin, M., Niklasson, G.A., and Wen, R.-T. (2024). Photo-electrochemical synergistically induced ion detrapping for electrochromic device rejuvenation. *Matter* 101877: https://doi.org/10.1016/j.matt.2024.09.021.

81 Lampert, C.M. (1981). Heat mirror coatings for energy efficient windows. *Solar Energy Materials* 6 (1): 1–41. https://doi.org/10.1016/0165-1633(81)90047-2.

82 Gordon, R.G. (2000). Criteria for choosing transparent conductors. *MRS Bulletin* 25: 52. https://doi.org/10.1557/mrs2000.151.

83 Granqvist, C.G. (2007). Transparent conductors as solar energy materials: a panoramic review. *Solar Energy Materials and Solar Cells* 91: 1529–1598. https://doi.org/10.1016/j.solmat.2007.04.031.

84 Khan, A. (2020). Introduction to transparent conductors. In: *Novel Embedded Metal-mesh Transparent Electrodes*. Singapore: Springer Theses. Springer. https://doi.org/10.1007/978-981-15-2918-4_1.

85 Guo, C.F. and Ren, Z. (2015). Flexible transparent conductors based on metal nanowire networks. *Materials Today* 18 (3): 143–154. https://doi.org/10.1016/j.mattod.2014.08.018.

86 Peleg, R. and Lampert, C.M. (2017). Graphene for coatings. *SVC Bulletin*, Spring 2017 54–57. Society of Vacuum Coaters, Alburqueque, NM, https://www.svc.org/publications/svc-bulletin-back-issues/.

87 Haacke, G. (1977). Transparent conducting coatings. *Annual Review of Materials Science* 7: 73. https://doi.org/doi.org/10.1146/annurev.ms.07.080177.000445.

88 Chavan, G.T., Kim, Y., Khokhar, M.Q., Hussain, S.Q., Cho, E.-C., Yi, J., Ahmad, Z., Rosaiah, P., and Jeon, C.-W.A. (2023). Brief review of transparent conducting oxides (TCO): the influence of different deposition techniques on the efficiency of solar cells. *Nanomaterials* 13: 1226. https://doi.org/10.3390/nano13071226.

89 Afre, R.A., Sharma, N., Sharon, M., and Sharon, M. (2018). Transparent conducting oxide films for various applications: a review. *Reviews on Advanced Materials Science* 53 (1): 79–89. https://doi.org/10.1515/rams-2018-0006.

90 Ginley, D.S. (2010). *Handbook of Transparent Conductors*. New York, NY. https://doi .org/10.1007/978-1-4419-1638-9: Springer. ISBN: 978-1-4419-1637-2.

91 Gao, J., Kempa, K., Giersig, M., Akinoglu, E.M., Han, B., and Li, R. (2016). Physics of transparent conductors. *Advances in Physics* 65 (6): 553–617. https://doi.org/10.1080/ 00018732.2016.1226804. ISBN: 978-981-15-2917-7.

92 Chopra, K.L., Major, S., and Pandya, D.K. (1983). Transparent conductors—a status review. *Thin Solid Films* 102 (1): 1–46. https://doi.org/10.1016/0040-6090(83)90256-0.

93 Hassab, S., Shen, D.E., Österholm, A.M., Da Rocha, M., Song, G., Alesanco, Y., Viñuales, A., Rougier, A., Reynolds, J.R., and Padilla, J. (2018). A new standard method to calculate electrochromic switching time. *Solar Energy Materials and Solar Cells* 185: 54–60.

94 Gentex Corporation (2024). Fifty years, looking back, focused forward. Annual Report 2024, https://ir.gentex.com/static-files/459493a2-35c8-42b6-9c8f-e68fe00e2d0f (accessed 30 Feb 2025).

95 Lynam, N.R. and Agrawal, A. (1990). Automotive applications of chromogenic materials. In: *Large Area Chromogenics: Materials and Devices for Transmittance Control* (ed. C.M. Lampert and C.G. Granqvist), 46–84. Bellingham, WA: SPIE Optical Engineering Press. https://www.spiedigitallibrary.org/conference-proceedings-of-spie.

96 Texas Instruments (2018). TI designs: TIDA-01539 automotive auto-dimming mirror reference design for electrochromic mirrors. https://www.ti.com/lit/pdf/tidudr8.

97 McCabe, I.A., Habibi, H., Varaprasad, D.V., and Lynam, N.R. (2013). Interior Rearview Mirror System for a Vehicle, US patent 8400704B2, granted: 2013-03-19, assigned: Magna Donnelly Corp.

98 Uken, J.T., De Wind, D.P., Foote, K.D., Mambourg, J.R., Blank, R.K., Larson, M.L., and Lynam N.R. (2013). Rearview Mirror Assembly, WO 2013071070 A1, published 2013-05-16, filed by: Magna Mirrors of America.

99 Forgette, J.A., Byker, H.J., Tonar, W.L., Bauer, F.T., Cammenga, D.J., and Roberts, J.K. (2000). Electrochromic Rearview Mirror Incorporating a Third Surface Metal Reflector and a Display/Signal Light, US patent 6111684A, granted: 2000-08-29, assigned: Gentex Corp.

100 Varaprasad, D.V., Lynam, N.R., Habibi, H.R., and Desaraju, P. (1992). High Performance Electrochemichromic Solutions and Solutions and Devices Thereof, US patent 5140455, granted: 1992-08-18, assigned: Magna Donnelly Corp.

101 Byker, H.J. (1990). Single-compartment, self-erasing, solution-phase electrochromic devices, solutions for use therein, and uses thereof, US patent 4902108, granted: 1990-02-20, assigned: Gentex Corp.

102 Varaprasad, D.V., Agrawal, A., Zhao, M., Allemand, P-M, Dornan, C.G., and Lynam, N.R. (1999). Electrochromic polymeric solid films, manufacturing electrochromic devices using such solid films, and processes for making such solid films and devices, US patent 5910854, granted: 1999-06-08, assigned: Magna Donnelly Corp.

103 Kloeppner, L.J., Giri, P., Lin, R., and Theiste, D.A. (2016). Electrochromic polyelectrolyte gel medium having improved creep resistance and associated electrochromic device, US patent 9500927, granted: 2016-11-22, assigned: Gentex Corp.

104 Varaprasad, D.V., Looman, S.D., Zhao, M., Habibi, H., and Lynam N.R. (1993). Electrochromic solutions, processes and preparing and using the same, and devices manufactured with the same, US patent 5239405A, granted: 193-08-24, assigned: Magna Donnelly Corp.

105 Baumann, K.L., Guarr, T.F., Giri, P., Muyskens, M.D., and Franz, S.F. (2008). Electrochromic compounds and associated media and devices, US patent 7428091B2, granted: 2008-09-23, assigned: Gentex Corp.

106 Nishikitani, Y., Uchida, S., Asano, T., Minami, M., Oshima, S., Ikai, K., and Kubo, T. (2008). Photo- and electrochemical properties of linked ferrocene and viologen donor-acceptor-type molecules and their application to electrochromic devices. *Journal of Physical Chemistry C* 112: 4372–4377. https://doi.org/10.1021/jp711103h.

107 Michaelis, A., Berneth, H., Haarer, D., Kostromine, S., Neigl, R., and Schmidt, R. (2001). Electrochromic dye system for smart window applications. *Advanced Materials* 13 (23): 1825. https://doi.org/10.1002/1521-4095(200112)13:23%3c1825::AID-ADMA1825%3e3.0.CO;2-R.

108 Shah, K.W., Wang, S.-X., Soo, D.X.Y., and Xu, J. (2019). Viologen-based electrochromic materials: from small molecules, polymers and composites to their applications. *Polymers* 11 (11): 1839. https://doi.org/10.3390/polym11111839.

109 Demiryont, H. (1996). Electromagnetic radiation modulating device, US patent 5111328A, granted: 1996-12-18, assigned: Visteon Global Tech., Inc.

110 Bertolini, A. (2017). Window construction, US patent 9778533B2, granted: 2017-10-03, assigned: Isoclima SpA.

111 McGrew, J. (2010). Green carpet reports – Mercedes Benz sheds some light on its magic sky control roof. https://www.greencarreports.com/news/1051027_mercedes-benz-sheds-some-light-on-its-magic-sky-control-roof (accessed 1 April 2025).

112 Idzkowski, A. (2005) 2005 Ferrari Superamerica first drive, Car and Driver (1 August 2005). https://www.caranddriver.com/reviews/a18202408/ferrari-superamerica-first-drive-review/ (accessed 1 April 2025).

113 Giron, J.-C. (1995). Couches minces d'hydroxyde de nickel d'oxydes mixtes nickel/lithium et nickel/sodium. Application aux systemes electrochromes, Ph.D., thesis, University of Paris VI, France.

114 Marcus, F. (2020). Electronically dimming glass from Boeing's Dreamliner is headed for your next car, Motor Trend (9 Jan 2020). https://www.motortrend.com/news/electrochromic-glass-gentex-boeing-dreamliner-future-tech (accessed 31 May 2025).

115 Lewis, R. (2025). Miru Unveils one of the largest electrochromic sunroofs for automobiles, Techcouver (18 March 2025). https://techcouver.com/2025/03/18/miru-unveils-one-of-the-largest-electrochromic-sunroofs-for-automobiles/ (accessed 31 May 31 2025).

116 LeBoucher, N. (2024). Solarbay opacifying sunroof: spectacular innovation born from a unique joint-development. *Renault Group Magazine*, 14 Feb 2024. https://www.renaultgroup.com/en/magazine/our-group-news/solarbay-opacifying-sunroof-spectacular-innovation-born-from-a-unique-joint-development/ (accessed 4 June 2025).

117 Kamimori, T. and Mizuhashi, M. (1982). Electrochromic light control panel, US patent 4338000, granted: 6 July 1982, assigned: AGC.

118 Kamimori, T., Nagai, J., and Mizuhashi, M. (1987). Electrooptic device, US patent 4671619, granted: 9 June 1987, assigned: AGC.

119 Nishiyama, H., Mijasaka, S., Kamimori, T., Mizuhashi, M., and Yoshihara, A. (1988). Method for fabrication of Electrochromic device and same fabricated the method, US patent 4761061, granted: 2 Aug. 1988, assigned: AGC.

120 Niwa, T. and Endo, T. (1986). Production of dazzle preventing mirror, Japanese patent JPS6150120A, granted: 1986-03-12, assigned: Nikon.

121 Niwa, T. and Endo, T. (1995). Fully solid-state type electrochromic element and its production, JPH0728099A, granted: 1995-01-31, assigned: Central Glass, Nikon.

122 Lampert, C.M., Agrawal, A., Baertlien, C., and Nagai, J. (1999). Durability evaluation of electrochromic devices – an industry perspective. *Solar Energy Materials and Solar Cells* 56 (3–4): 449–463. https://doi.org/10.1016/S0927-0248(98)00185-8.

123 Nagai, J. (2024). *The Nanofilm Lab*. Japan: Yokohama. https://sites.google.com/site/nanofilmlab (accessed 11 Aug 2024).

124 Nagai, J. (2021). *Evolution and Practical Realization of Smart Windows*. Tokyo, Japan: CMC Research. ISBN: 978-4-904482-98-8, http://cmcre.com/archive/76552 (in Japanese).

125 Nagai, J. (2018). *Fundamentals and Applications of Smart Windows*. Tokyo, Japan: CMC Research. ISBN: 978-4-904482-49-0, http://cmcre.com/archive/34476 (in Japanese).

126 Lampert, C.M. (1984). Electrochromic materials and devices for energy efficient windows. *Solar Energy Materials* 11: 1–27. https://doi.org/10.1016/0165-1633(84)90024-8.

127 Michalak, F., von Rottkay, K., Richardson, T., Slack, K.J., and Rubin, M. (1999). Electrochromic lithium nickel oxide thin films by RF-sputtering from a $LiNiO_2$ target. *Electrochimica Acta* 44 (18): 3085. https://doi.org/10.1016/S0013-4686(99)00024-9.

128 Burdis, M.S., Siddle, J.R., Batchelor, R.A., and Gallego, J.M. (1998). $V_{0.50}Ti_{0.50}O_x$ thin films as counterelectrodes for electrochromic devices. *Solar Energy Materials and Solar Cells* 54: 93. https://doi.org/10.1016/S0927-0248(98)00059-2.

129 Gerhardinger, P.F. and McCurdy, R.J. (1996). Float line deposited transparent conductors-implications for the PV industry. *MRS Online Proceedings Library* 426: 399. https://doi.org/10.1557/PROC-426-399.

130 Cronin, J.C., Torsky, M., and Agrawal, A. (2001). Reduction of haze in tin oxide transparent conductive coatings on glass, US patent 6268059, granted: 2001-07-31, assigned: Magna Donnelly.

131 Gillet, P.A., Fourquet, J.L., and Bohnke, O. (1992, 1992). New electrochromic thin film materials. *Proceedings of SPIE* 1728: 82. https://www.spiedigitallibrary.org/conference-proceedings-of-spie.

132 Allemand, P.-M., Ingle, A., Cronin, J.P., Kennedy, S.R., Yao, Y., Tonazzi, J.C.L., Boulton, J.M., and Agrawal, A. (2001). Electrochromic devices, US patent 6266177B1, granted: 2001-07-24, assigned: Donnelly Corp., Magna Donnelly Corp.

133 Burdis, S.M. and Weir, D.G.J. (2008). Electrochromic devices and methods, US patent 7372610B2, granted: 2008-05-13, assigned: Sage Electrochromics, Inc.

134 Weir, D.G.J., Lin, F., and Engtrakul, C. (2015). Ternary nickel oxide materials for electrochromic devices, US patent 8995041B2, granted: 2015-03-31, assigned: Alliance for Sustainable Energy LLC, Sage Electrochromics Inc.

135 Green, S.V., Granqvist, C.G., and Niklasson, G.A. (2014). Structure and optical properties of electrochromic tungsten-containing nickel oxide films. *Solar Energy Materials and Solar Cells* 126: 248. https://doi.org/10.1016/j.solmat.2013.04.001.

136 Shchegolkov, A.V., Jang, S.-H., Shchegolkov, A.V., Rodionov, Y.V., Sukhova, A.O., and Lipkin, M.S. (2021). A brief overview of electrochromic materials and related devices: a nanostructured materials perspective. *Nanomaterials* 11 (9): 2376. https://doi.org/10.3390/nano11092376.

137 Granqvist, C.G. (2014). Electrochromic for smart windows: oxide based thin films and devices. *Thin Solid Films* 564: 1–38. http://dx.doi.org/10.1016/j.tsf.2014.02.002.

138 Kim, J., Ong, G.K., Wang, Y., LeBlanc, G., Williams, T.E., Mattox, T.M., Helms, B.A., and Milliron, D.J. (2015, 2015). Nanocomposite architecture for rapid, spectrally-selective electrochromic modulation of solar transmittance. *Nano Letters* 15 (8): 5574–5579. https://doi.org/10.1021/acs.nanolett.5b02197.

139 Zhang, W., Li, H., Hopmann, E., and Elezzabi, A. (2021). Nanostructured inorganic electrochromic materials for light applications. *Nanophotonics* 10 (2): 825–850. https://doi.org/10.1515/nanoph-2020-0474.

140 Wang, Z., Pradhan, A., and Rozbicki, R.T. (2012). Electrochromic devices, US patent 8228592B2, granted: 2012-07-24, assigned: View Operating Corp.

141 Burdis, M.S., Greer, B.D., and Weir, D.G.J. (2009). Electrochromic devices having improved ion-conducting layers, US patent 7593154B2, granted: 2009-09-22, assigned: Sage Electrochromics, Inc.

142 Ray, H.L., Schiltz, R.A.S., Yi, J., Mesnger, C., Li, W., Mulder, C.L., and Giron J.-C. (2022). Electrochromic devices and methods of forming the same, US patent 11249366.

143 Rozbicki, R.T. (2021). Fabrication of electrochromic devices, US patent 11079648B2, granted: 2021-08-03, assigned: View Operating Corp.

144 Sbar, N.L. and Kalweit, H. (2013). Apparatus and methods for repair of defects in an electronic energy control or display device, US patent 8500503B2, granted: 2013-08-06, assigned: Sage Electrochromics, Inc.

145 Pradhan, A.A., Rozbicki, R.T., Gillaspie, D., Kailasam, S.K. (2020) Fabrication of low defectivity electrochromic devices, US patent 10663830B2, filed: 2019-02-25, granted: 2020-05-26, assigned: View Operating Corp.

146 Burdis, M. and Sbar, N. (2005). Increasing yields and broadening markets: process innovations in the manufacturing of energy saving glazing, Osti ID: 840353 (01-04-2005C Web. doi:10.2172/840353 (accessed 1 March 25).

147 Burdis, M. and Sbar, N. (2010). Technology advancements to lower costs of electrochromic window glazing, Final Report, (revised) Osti ID: 989107 (02-04-2010), 10.2172/989107 (accessed 01 Jan 2025).

148 Burdis, M., and Sbar, N. (2012). Recovery Act: Electrochromic Glazing Technology: Improved Performance, Lower Price, Final Report, OSTI ID 1111422 (30-06-2012), https://www.osti.gov/servlets/purl/1111422

149 Turner, H., Bergh, H., Bass, J., and Giaquinta, D. (2018). Flexible and multilayer electrochromic devices and methods of making the same, US patent 20180364541A1, granted: 2021-01-26, assigned: Kenistral, reassigned: Smart Window Inc. Ltd.

150 Giaquinta, D., Wiers, B., Bass, J.D., Scott, G., Daniloff, G., and Duncan, H. (2020). Thin film lithium tungsten oxides for electrochromic applications and methods of making the same, US patent 10877347B1, granted: 2020-12-29, assigned: Kinestral, reassigned: Halio, reassigned (2025) Smart Window Inc, Ltd.

151 Karmhag, R., Gregard, G., and Vogt, R. (2018). Electrochromic devices and manufacturing methods thereof, US patent 10012887, granted: 2018-07-03, assigned: ChromoGenics AB.

152 Thomsen, S. and Garcia, G. (2022). Electrochromic device including lithium-rich anti-perovskite material, US patent 11243447, granted: 2022-02-08, assigned: Heliotrope Technologies, Inc., reassigned 2023: Hivisq Technologies, Ltda Soc., Spain.

153 Garcia, G., Basu, S., and Holt, J. (2019). Electrochromic system containing an insulating protective layer and method for controlling photochromic darkening, US patent 20170219905A1, granted: 2019-06-04, assigned: Heliotrope Technologies, Inc, reassigned (2023): Hivisq Technologies, Ltda Soc., Spain.

154 Choi, H.J., Bailey, M., Bass, J.D., von Kugelgen, S.W., Lachman, E., and Turner, H.W. (2020). Electrochromic lithium nickel Group 5 mixed metal oxides, US patent 10845666B2, granted: 2020-11-24, assigned: Kinestral, reassigned Halio, reassigned: Smart Windows Inc. Ltd. Hong Kong.

155 Granqvist, C.G., Lansaker, P.C., Mlyuka, N.R., Niklasson, G.A., and Avendano, E. (2009). Progress in chromogenics: new results for electrochromic and thermochromic materials and devices. *Solar Energy Materials & Solar Cells* 93: 2032. https://doi.org/10.1016/j.solmat.2009.02.026.

156 Karmhag, R., Gustavsson, G., Granqvist, C. G., and Azens, A. (2011). Manufacturing of electrochromic devices, US patent 7952785B2, granted: 2011-05-31, assigned: ChromoGenics Sweden AB.

157 Karmhag, R., Gregard, G., and Vogt, R. (2018) Electrochromic devices and manufacturing methods thereof, US patent 10012887B2, granted: 2018-07-03, assigned: ChromoGenics Sweden AB.

158 Georen, P. and Berenguel, R.M. (2012). Electrolytes for electrochromic devices, US patent 8218225B2, granted: 2012-07-10, assigned: ChromoGenics AB.

159 Granqvist, C.G. (2017). Chromogenic coatings for glazing: toward a new paradigm for buildings. *Society of Vacuum Coaters Bulletin* Spring 2017: 48. https://www.svc.org/publications/svc-bulletin-back-issues/.

160 Wang, X., Davies, E., Gonzalez, P., Garcia, G, Bayati, A., and Bohland, J. (2020). Methods of manufacturing electrochromic devices containing a solid-state electrolyte, US Patent 11650472B2, granted: 2023-05-16, assigned: Hivisq Technologies, Ltda Soc., Spain.

161 Garcia, G., Holt, J., Rosen, E., and Helms, B. (2017). Methods of charging solid state plasmonic electrochromic smart window devices, US patent 9798214B2, granted: 2017-10-24, assigned: Heliotrope Technologies, Inc., reassigned (2023): Hivisq Technologies Ltda Soc., Spain.

162 Timmerman, W., Micklatcher, M., and Kerszulis J., (2017). Electrochromic multi-layer devices with cross-linked ion-conducting polymer, US patent 9720299, granted: 2017-08-01, assigned: Kinestral, (2021) Halio, reassigned (2025): Smart Window Inc., Ltd, Hong Kong.

163 Roudebush, J. H., Zhu, L., Giaquinta, D., Duncan, H., and Turner, H. (2021). Tungsten oxide nanostructure thin films for electrochromic devices, US patent 10558103B2, granted: 2020-02-11, assigned: Kinestral, (2021) Halio, reassigned (2025): Smart Window Inc., Ltd, Hong Kong.

164 Bergh, H., Zieberth, J., and Timmerman, N. (2020). Electrochromic multilayer devices with composite current modulating structure, US patent 10627692B2, granted:

2020-04-21, assigned: Kinestral, (2021) Halio, reassigned (2025): Smart Window Inc., Ltd., Hong Kong.

165 Daswani, V. (2025). Argotec and Miru unveil largest electrochromic sunroof window in automotive industry milestone (18 March 2025). https://mirucorp.com/news/argotec-and-miru-unveil-largest-electrochromic-sunroof-window-in-automotive-industry-milestone/ (accessed 28 March 2025).

166 Hicks, W. (2015). Vision for NREL technology has entrepreneur seeing clearly (3 Sept 2015). https://www.nrel.gov/news/features/2015/19568.html (accessed 18 March 2025).

167 Jödicke, D. and Wittkopf, H. (2007). The 2nd Generation of an Electrochromic Solar Control Glazing - Ready for Projects. 2007 Proceedings of Glass Performance Days, p. 394. ISBN: 9529186754, 9529186746

168 Kraft, A. and Rottmann, M. (2009). Properties, performance and current status of the laminated electrochromic glass of Gesimat. *Solar Energy Materials and Solar Cells* 93 (12): 2088–2092. https://doi.org/10.1016/j.solmat.2009.05.010.

169 Kraft, A. (2021). Interview with Alexander Kraft, Smartglass consultant, Smart Glass World, https://www.smartglassworld.net/alexander-kraft-interview.

170 Cuce, E. and Riffat, S.B. (2015). A state-of-the-art review on innovative glazing technologies. *Renewable and Sustainable Energy Reviews* 41: 695–714. https://doi.org/10.1016/j.rser.2014.08.084.

171 US DOE, Washington, D.C. https://www.energy.gov/energysaver/energy-performance-ratings-windows-doors-and-skylights#:~:text=Solar%20heat%20gain%20coefficient%20SHGC,the%20greater%20its%20shading%20ability (accessed 10 March 2025).

172 View Inc. (2021). California Title 24 compliance with View smart glass. https://view.com/sites/default/files/documents/california_title_24_compliance_with_view_smart_glass_apr2021.pdf (accessed 22 March 2025).

173 NREL Labs (2025). Solar resource spectra. Golden, CO. https://www.nrel.gov/grid/solar-resource/spectra-am1.5.html (accessed 26 April 2025).

174 Treado, S.J. and Bean, J.W. (1992). Optical performance of Commercial Windows, Report# NISTIR 4711, NIST. https://archive.org/details/opticalperforman4711trea?msclkid=b6904218d0b111ec8247c167da5f3c78 (accessed 10 March 2025).

175 Vitro, Inc. (2024). Center of Glass U-Values (factors) for Double and Triple Glazed Insulating Glass Units with Solarban® 60 Low-e Glass with 100% Air, Argon, or Krypton, or Mixtures of These Gases, TD-121. https://www.vitroglazings.com/media/xd4jd0vm/vitro-td-121-cog-u-values-sb60-double-triple-v4.pdf (accessed 6 March 2025).

176 (a) Agrawal, A., Adams, L.L., and Cronin, J.C. (2020). Insulated glass utilizing electrochromic elements, US patent 10824040, granted: 2020-11-03, assigned: Polyceed, Inc. (b) Vitro Architectural Glass (2022). Product data sheet, Sungate® 400 low-e glass. https://www.vitroglazings.com/media/ctjjaehz/vitro_sungate-400_datasheet_2019.pdf (accessed 20 June 2025).

177 (a) SageGlass (2024). IGU Product Guide. https://www.sageglass.com/sites/default/files/2024-07/MKT-382_IGU_Product_Guide_US.pdf (accessed 8 April 2025). (b) View (2024). View gen 4 insulating glass unit (IGU) Data sheet QDM-02-000210. https://view.com/sites/default/files/documents/View%20Gen%204%20Datasheet%20-%20Standard_Rev4_April.pdf (accessed 30 May 2025).

178 SageGlass (2024). SageGlass IGU Product Guide. https://www.sageglass.com/sites/default/files/2024-07/MKT-382_IGU_Product_Guide_US.pdf (accessed 30 May 2025).

179 Ihrner, N. (2025). ChromoGenics AB, Uppsala, Sweden, private communication (2 June 2025).

180 Selles, O., Ripberger, I., Letocart, P., and Orillard, J. (2015). Electrochromic devices with anti-iridescent coating, US patent 9091896, granted: 2015-07-28, assigned: Saint-Gobain Glass France, Sage Electrochromics.

181 Agrawal, A., Castillo, S.J.C., Perananthan, S., and Angelo, C.C. (2022). Variable light transmission structures with improved optical properties, US patent 11287717, granted: 2022-04-29, assigned: Glass Dyenamics.

182 Karmhag, R., Gregard, G., and Vogt, R. (2018). Electrochromic devices and manufacturing methods thereof, US patent 10012887, granted: 2018-07-03, assigned: ChromoGenics AB.

183 Tällberg, R., Jelle, B.P., Loonen, R., Gao, T., and Hamdy, M. (2019). Comparison of the energy saving potential of adaptive and controllable smart windows: a state-of-the-art review and simulation studies of thermochromic, photochromic and electrochromic technologies. *Solar Energy Materials and Solar Cells* 200: 109828. https://doi.org/10.1016/j.solmat.2019.02.041.

184 Carmody, J., Selkowitz, S.E., Lee, E.S., Arasteh, D., and Willmert, T. (2004). *Window Systems for High Performance Buildings*, 94–100. New York, NY: W.W. Norton. ISBN: 0-393-73121-9.

185 Sottile, G.M. (2005). 2004 Survey of United States Architects on the subject of switchable glazing. *Materials Science and Engineering B* 119 (3): 240–245. https://doi.org/10.1016/j.mseb.2004.12.077.

186 Sottile, G.M. (2008). 2008 Study of architecture professionals on the subject of smart glass, daylighting and clean technology. In: *Proceedings of the Society of Vacuum Coaters*, vol. 51, 107–112. Albuquerque, NM. (svc.org): SVC.

187 Sottile, G.M. (2002). Assessments of attitudes and expectations of switchable glass among United States window manufacturers. In: *Proceedings of Society Vacuum Coaters*, vol. 45, 163–169. Albuquerque, NM. (svc.org): SVC.

188 Sottile, G.M. (2007). 2007 Study of United States LEED Accredited professional on the subject of smart glass. In: *Proceedings of the Society of Vacuum Coaters*, vol. 50, 572–575. Albuquerque, NM (svc.org): SVC.

189 Sawyer, K. (2014). Windows and building envelope research and development: roadmap for emerging technology USDOE, Building Technologies Office. https://www.energy.gov/sites/prod/files/2014/02/f8/BTO_windows_and_envelope_report_3.pdf (accessed 07 March 2025).

190 Harris, C. (2020). Pathway to zero energy windows (April 2020). https://www.nrel.gov/docs/fy22osti/80171.pdf (accessed 07 March 2025).

191 Cannavale, A., Ayr, U., Fiorito, F., and Martellotta, F. (2020). Electrochromic windows to enhance energy saving in buildings. *Energies* 13 (6): 1449. https://doi.org/10.3390/en13061449.

192 Samsuri, N.D., Mukhtar, W.M., Rashid, A.R.A., Dasuki, K.A., and Yussuf, A.A.R.H.A. (2017). Synthesis methods of gold nanoparticles for Localized Surface Plasmon Resonance (LSPR) sensor applications. *EPJ Web of Conferences* 162: 01002. https://doi.org/10.1051/epjconf/201716201002.

193 Otto, A. (1968). Excitation of nonradiative surface plasma waves in silver by the method of frustrated total reflection. *Zeitschrift für Physik* 216: 398–410. https://doi.org/10.1007/BF01391532.

194 Maier, S.A. (2007). *Plasmonics: Fundamentals and Applications*. New York, NY: Springer. ISBN: 978-0-387-33150-8.

195 Menguc, M.P. and Francoeur, M. (2023). *Light, Plasmonics and Particles*. New York: Elsevier. ISBN: 9780323999014.

196 Noguez, C. (2007). Surface plasmons on metal nanoparticles: the influence of shape and physical environment. *Journal of Physical Chemistry C* 111: 3806–3819. https://doi.org/10.1021/jp066539m.

197 Trügler, A. (2016). *Optical Properties of Metallic Nanoparticles*. Switzerland: Springer International Publishing, ISBN: 978-3-319-25072-4 (2016), ISBN: 978-3-319-79730-4 (2018), https://doi.org/10.1007/978-3-319-25074-8.

198 Boschloo, G. and Fitzmaurice, D. (1999). Spectroelectrochemistry of highly doped nanostructured tin dioxide electrodes. *The Journal of Physical Chemistry B* 103 (16): 3093–3098. https://doi.org/10.1021/jp9835566.

199 Gillissen, F. (2024). Innovative electrode materials for dual band visible – near infrared electrochromic smart windows. Dr. Science thesis, Department of Chemistry University of Liège, Belgium.

200 Kriegel, I., Scotognella, F., and Manna, L. (2017). Plasmonic doped semiconductor nanocrystals: properties, fabrication, applications and perspectives. *Physics Reports* 674: 1–52. https://doi.org/10.1016/j.physrep.2017.01.003.

201 Ye, X., Fei, J., Diroll, B.T., Paik, T., and Murray, C.B. (2014). Expanding the spectral tunability of plasmonic resonances in doped metal-oxide nanocrystals through cooperative cation–anion codoping. *Journal of the American Chemical Society* 136 (33): 11680–11686. https://doi.org/10.1021/ja5039903.

202 Agrawal, A., Cho, S.H., Zandi, O., Ghosh, S., Johns, R.W., and Milliron, D.J. (2018). Localized surface plasmon resonance in semiconductor nanocrystals. *Chemical Reviews* 118 (6): 3121–3207. https://doi.org/10.1021/acs.chemrev.7b00613.

203 Milliron Research Group, University of Michigan, Dept of Engineering, Ann Arbor, MI, nanocrystal.engin.umich.edu.

204 Lounis, S.D., Runnerstrom, E.L., Llordés, A., and Milliron, D.L. (2014). Defect chemistry and plasmon physics of colloidal metal oxide nanocrystals. *The Journal of Physical Chemistry Letters* 5 (9): 1564–1574. https://doi.org/10.1021/jz500440e.

205 Garcia, G., Buonsanti, R., Runnerstrom, E.L., Mendelsberg, R.J., Llordés, A., Anders, A., Richardson, T.J., and Milliron, D.J. (2011). Dynamically modulating the surface plasmon resonance of doped semiconductor nanocrystals. *Nano Letters* 11: 4415–4420. https://doi.org/10.1021/nl202597n.

206 Manthiram, K. and Alivisatos, A.P. (2012). Tunable localized surface plasmon resonances in tungsten oxide nanocrystals. *Journal of the American Chemical Society* 134: 3995–3998. https://doi.org/10.1021/ja211363w.

207 Mattox, T.M., Bergerud, A., Agrawal, A., and Milliron, D.J. (2014). Influence of shape on the surface plasmon resonance of tungsten bronze nanocrystals. *Chemistry of Materials* 26: 1779–1784. https://doi.org/10.1021/cm4030638.

208 Gordon, T.R., Paik, T., Klein, D.R., Naik, G.V., Caglayan, H., Boltasseva, A., and Murray, C.B. (2013). Shape-dependent plasmonic response and directed self-assembly in a new semiconductor building block indium-doped cadmium oxide (ICO). *Nano Letters* 13: 2857–2863. https://doi.org/10.1021/nl4012003.

209 Buonsanti, R., Llordes, A., Aloni, S., Helms, B.A., and Milliron, D.J. (2011). Tunable infrared absorption and visible transparency of colloidal aluminum-doped zinc oxide nanocrystals. *Nano Letters* 11: 4706–4710. https://doi.org/10.1021/nl203030f.

210 Runnerstrom, E.L., Llordés, A., Lounis, S.D., and Milliron, D.J. (2014). Nanostructured electrochromic smart windows: traditional materials and NIR-selective plasmonic nanocrystals. *Chemical Communications* 50: 10555–10572. https://doi.org/10.1039/C4CC03109A.

211 Garcia, G., Buonsanti, R., Runnerstrom, E.L., Mendelsberg, R.J., Llordés, A., Anders, A., Richardson, T.J., and Milliron, D.J. (2011). Dynamically modulating the surface plasmon resonance of doped semiconductor nanocrystals. *Nano Letters* 11 (10): 4415–4420. https://doi.org/10.1021/nl202597n.

212 Garcia, G. Holt, J., and Thomsen, S. (2017). Electrochromic device containing color-tunable nanostructures, US patent 9684219B, granted: 2017-06-20, assigned: Heliotrope, reassigned 2023: Hivisq Technologies Ltda Soc., Spain.

213 Garcia, G., Buonsanti, R., Llordes, A., Runnerstrom, E.L., Bergerud, A., and Milliron, D.J. (2013). Near-infrared spectrally selective plasmonic electrochromic thin films. *Advanced Optical Materials* 1: 215. https://doi.org/doi.org/10.1002/adom.201200051.

214 Mattox, T.M., Koo, B., Garcia, G., Milliron, D.J., De Trizio, L., and Dahlman, C. (2017). Conductive transition metal oxide nanostructured electrochromic material and optical switching devices constructed thereof, US patent 9785031B2, granted: 2017-10-10, assigned: Heliotrope Technologies, Reassigned 2021: USDOE, Reassigned 2023: Hivisq Technologies Ltda Soc., Spain.

215 Holt, J., Garcia, G., and Thomsen, S. (2016). Monolithic electrochromic window manufacturing methods, US patent 10900277B1, granted: 2021-01-26, assigned: Heliotrope Technologies, Inc., 2023 reassigned: Hivisq Technologies Ltda Soc., Spain.

216 Wang, X., Davies, E., Gonzalez, P. Garcia, G., Bayati, A., and Bohland, J. (2022). Methods of manufacturing electrochromic devices containing a solid-state electrolyte, US Patent 11221534B2, granted: 2022-01-11, assigned: Heliotrope Technologies, Inc., 2023 reassigned: Hivisq Technologies Ltda Soc., Spain.

217 Wang, X., Davies, E., Motter, J., Garcia, G., and Bayati, A. (2022). Gel electrolyte precursor compositions, electrochromic devices including gel electrolytes, and manufacturing methods thereof, US patent 11409176B2, granted: 2022-08-09, assigned: Heliotrope Europe S.L, reassigned 2023: Hivisq Technologies Ltda Soc., Spain.

218 Agrawal, A., Kriegel, I., and Milliron, D.J. (2015). Shape-dependent field enhancement and plasmon resonance of oxide nanocrystals. *Journal of Physical Chemistry C* 119 (11): 6227–6238. https://doi.org/10.1021/acs.jpcc.5b01648.

219 Hergert, W. and Wriedt, T. (2012). *The Mie Theory: Basics and Applications*, Springer Series in Optical Sciences, vol. 169. Berlin: Springer. ISBN: 978-3-642-28737-4.

220 Drude, P. (1900). Zur Elektronentheorie der Metalle. *Annalen der Physik* 306 (3): 566–613. https://doi.org/10.1002/andp.19003060312.

221 Llordés, A., Garcia, G., Gazquez, J., and Milliron, D.J. (2013). Tunable near-infrared and visible-light transmittance in nanocrystal-in-glass composites. *Nature* 500 (7462): 323–326. https://doi.org/10.1038/nature12398.

222 Milliron, D., Llordés, A., Buonsanti, R., and Garcia, G. (2018) Electrochromic nanocomposite films, US patent 9939662B2, granted: 2018-04-10, assigned: University of California, San Diego.

223 Fortunato, J., Zydlewski, B.Z., Lei, M., Holzapfel, N.P., Chagnot, M., Mitchell, J.B., Lu, H.C., Jiang, D.-e., Milliron, D.J., and Augustyn, V. (2023). Dual-band electrochromism in hydrous tungsten oxide. *ACS Photonics* 10 (9): 3409–3418. https://doi.org/10.1021/acsphotonics.3c00921.

224 Zhang, S., Cao, S., Zhang, T., Fisher, A., and Lee, J.Y. (2018). Al^{3+} intercalation/De-intercalation-enabled dual-band electrochromic smart windows with a high optical modulation, quick response and long cycle life. *Energy & Environmental Science* 11 (10): 2884–2892. https://doi.org/10.1039/C8EE01718B.

225 Zydlewski, B.Z. and Milliron, D.J. (2024). Dual-band electrochromic devices utilizing niobium oxide nanocrystals. *ACS Applied Materials & Interfaces* 16 (19): 24920–24928. https://doi.org/10.1021/acsami.4c02997.

226 Liang, Y., Cao, S., Guo, J., Zeng, R., Zhao, J., and Zou, B. (2022). Dual-band electrochromic smart window based on single-component nanocrystals. *ACS Applied Electronic Materials* 4 (11): 5109–5118. https://doi.org/10.1021/acsaelm.2c01225.

227 Cao, S., Zhang, S., Zhang, T., and Lee, J.Y. (2018). Fluoride-assisted synthesis of plasmonic colloidal Ta-doped TiO_2 nanocrystals for near-infrared and visible-light selective electrochromic modulation. *Chemistry of Materials* 30 (14): 4838–4846. https://doi.org/10.1021/acs.chemmater.8b02196.

228 Barawi, M., De Trizio, L., Giannuzzi, R., Veramonti, G., Manna, L., and Manca, M. (2017). Dual band electrochromic devices based on Nb-doped TiO_2 nanocrystalline electrodes. *ACS Nano* 11 (4): 3576–3584. https://doi.org/10.1021/acsnano.6b06664.

229 Zhang, S., Cao, S., Zhang, T., and Lee, J.Y. (2020). Plasmonic oxygen-deficient TiO_{2-x} nanocrystals for dual-band electrochromic smart windows with efficient energy recycling. *Advanced Materials* 32 (43): 2004686. https://doi.org/10.1002/adma.202004686.

230 Rukavina, T.G. and Lin, C.-C. (2004). Switchable electrochromic devices for use in aircraft transparency windows, US patent 6783099B2, granted: 2004-08-31, assigned: PPG Ind. Ohio.

231 Apdalhaliem, S., Meredith, K.D., and Safai, M. (2018). Aircraft dimmable window heating system, European Patent EP 2939919B1, granted: 2018-05-09, assigned: Boeing Corp.

232 Saint-Gobain (2025) Electrochromic Anti-glare windshield. https://www.saint-gobain-aerospace.com/technology/electrochromic-anti-glare-windshield-improved-comfort-windshield (accessed 7 March 2025).

233 Driscoll, D.I., Busscher, B., and Ash, K.L. (2019). Electro-optic assembly, US patent 9146437B2, granted: 2015-09-25, assigned: Gentex Corp.

234 Guarr, T.F., Theiste, D.A., Ash, K.L., and Driscoll, D.I. (2014). Variable transmission electrochromic window and associated aircraft window system, US patent 20130182308A1, granted; 2014-05-27, assigned: Gentex Corp.

235 PPG Aerospace Transparencies (2009). Alteos interactive window systems. https://www.ppg.com/en-US/aerospace/products/transparencies/transparencies-technology/alteos (accessed 19 March 2025).

236 Wang, D. Slovak, S.M., Saxe, R.L., and Chen, X.-P. (2014). SPD films and light valve laminates with improved durability, US patent 8670173B2, granted: 2014-03-11, assigned: RFI.

237 Swisa, B.K. (2024). Gauzy expects to generate an aggregated revenue of $240 million over next 10 years from aeronautic cockpit shading (15 Aug 2024). https://www.gauzy

.com/articles/gauzy-expects-to-generate-an-aggregated-revenue-of-240-million-over-next-10-years-from-aeronautic-cockpit-shading/ (accessed 1 March 2025).

238 Mizuno, M. and Niwa, T. (1991). Eyeglass frame for electrooptical lenses, US patent 4991951, granted: 1991-02-12, assigned: Nikon Corp.

239 Kallman, W.R., Williams, J.D., Ozias, A.E., Dorfman, L.M., Schaus, C.F., Russell, G.A., Wills, S.G., and Keeney, D. (1997). Electrochromic edge isolation interconnect system, process, and device for its manufacture, US patent 5657150, granted: 1997-08-12, assigned: Eyeonics Corp.

240 Kallman, W.R., Roberts, M.B., Baretich, D.F., Dorfman, L.M., and Ahern, J.F. (1995). Electrochromic eyewear, US patent 5455638A, granted: 1995-10-03, assigned: Comdisco.

241 Kallman, W.R., Roberts, M.B., Baretich, D.F., Dorfman, L.M., and Ahern, J.F. (1995). Electrochromic eyewear system, rechargeable eyewear and external charger therefor, US patent 5455637A, granted: 1995-10-03, assigned: Comdisco.

242 Gregory, C.C. (2000). Hinge with wire extending (switchable eyewear), US Patent 6099117, granted: 200-08-08, assigned: PPG Industries Ohio.

243 Coleman, C.R. and Smarto, J.E. (2000). Lens block and method of processing lenses, US patent 6110016, granted: 2000-08-24, assigned: PPG Indust., Ohio.

244 Endo, T. and Mizuno, M. (1990). Electrochromic device with low resistance electrical connections, US patent 5187607, granted: 1990-03-09, assigned: Nikon.

245 Yu, P.C., Backfisch, D.L., and Rukavina, T.G. (1995). Electrochromic device with plastic substrate, US patent 5471338, granted: 1995-11-28, assigned: PPG Industries Ohio.

246 Kallman, B., Roberts, M., and Baretich, D.F. (1999). US patent 5900720A filed 1993/9/10, granted: 1999/05/04, assigned: Eyeonics Corp.

247 Liping, M.A., Youn, Y., and Wang, P. (2022). Solid state electrochromic eyewear. World patent WO2022/098854 A1, Published: 2022-05-12, assigned: Nitto Denko.

248 AshChromics (2024). Electrochromic eyeglasses. https://ashwin-ushas.com/electrochromic-sunglasses-goggles/#performance (accessed 11 May 2022).

249 Chandrasekhar, P., Zay, B.J., Cai, C., Chai, Y., and Lawrence, D. (2014). Matched-dual-polymer electrochromic lenses, using new cathodically coloring conducting polymers with exceptional performance and incorporated into automated sunglasses. *Journal of Applied Polymer Science* 131 (22): 41043. https://doi.org/10.1002/app.41043.

250 Chandrasekhar, P., Zay, B.J, Laganis, E.J., Romanov, V.V., and LaRosa, A.J. (2016). Electrochromic (2016) Electrochromic eyewear, US patent 9482880B1, granted: 2016-11-01, assigned: Tripower Design LLC, Ashwin Ushas.

251 Chandrasekhar, P., Zay, B.J, Laganis, E.J., Romanov, V.V., and LaRosa, A.J. (2018). Electrochromic (2018) Electrochromic eyewear, USD 834008651 design patent, granted: 2018-11-20, assigned: Ashwin Ushas.

252 Chandrasekhar, P., Zay, B.J, Laganis, E.J., Romanov, V.V., and LaRosa, A.J. (2019) Electrochromic eyewear, US patent 2018025294A1, granted: 2019-10-15, assigned: Tripower Design LLC, Ashwin Ushas.

253 Österholm, A.M., Shen, D.E., Kerszulis, J.A., Bulloch, R.H., Kuepfert, M., Dyer, A.L., and Reynolds, J.R. (2015). Four shades of brown: tuning of electrochromic polymer blends toward high-contrast eyewear. *ACS Applied Materials & Interfaces* 7 (3): 1413–1421. https://doi.org/10.1021/am507063d.

254 Heller, W.R., Kumamoto, J., Luzzi, J. Jr, and Powers, J.C. (1967). Electrochromic light valve, US patent 3317266A, granted: 1967-05-02, assigned: IBM, Corp.

255 Deb, S.K. and Shaw, R.F. (1970). Electro-Optical Device having variable optical density, US patent 3521941, granted: 1970-02-07, assigned: American Cyamid, reassigned Wyeth Holdings, LLC.

256 Giglia, R.D. (1975). Self-supporting pigment layers for electrochromic display, US patent 3892472A, granted: 1975-07-01, assigned: American Cyanamid, Co reassigned Wyeth Holdings.

257 Leibowitz, M. (1976). Electrochromic display having improved electrolyte, US patent 3957352 (1976), granted: 1976-05-18, assigned: Timex Corp.

258 Hara, T., Hajimoto, Y., Shirato, Y., and Matsushima, M. (1977). Electrochronic device, US patent 4053209A, granted: 1977-10-11, assigned: Canon, Inc.

259 Champ, R.B. and Shattuck, M.D. (1977). Visual image display device, US patent 4037928A, granted: 1977-07-26, assigned: IBM, Corp.

260 Market and Markets (2024). Display market size, share & growth analysis. https://www.marketsandmarkets.com/Market-Reports/display-market-925.html (accessed 12 May 2025).

261 Marescaux, M., Beunis, F., Strubbe, F., and Neyts, K. (2007). Electrochromic materials for electronic paper. Proceedings of the Doctoral Symposium, Univ. Gent, Belgium. https://www.researchgate.net/publication/237506426 (accessed 12 May 2025).

262 Bush, S. (2021). Updated: Fully printable display offers low power for simple IoT devices (Rdot part of ynvisible). https://www.electronicsweekly.com/news/products/displays-2/fully-printed-flexible-display-offers-low-power-simple-iot-devices-2018-08/ (accessed 11 May 2025).

263 Fitzmaurice, D., Rao, N.S., Cinnsealach, R., and Enright, B. (2005). Electrochromic system, US patent 6861014, granted: 2005-03-01, assigned: Univ. College Dublin.

264 Edwards, M., Hagfeldt, A., Boschloo, G., and Lindqvist, S.-E. (2009). Electrochromic device based on nanocrystalline materials, US patent 7508009, granted: 2009-03-24, assigned: Ntera, Ltd.

265 Vincent, J.B. and Flick, D.W. (2004). EC display device, US patent 6744549, granted: 2004-06-01, assigned: Dow Global Tech. LLC, reassigned: Aveso, Inc.

266 Kanitz, A. and Roth, W. (2012). Electrochromic formulation method for the production thereof, and electrochromic organic component, US patent 8289608, granted: 2012-10-16, assigned: Siemans AG.

267 Kanitz, A. and Roth, W. (2013). Electrochromic organic component, US patent 8373918B2, granted: 2013-02-12, assigned: Siemans AG.

268 Armgarth, M., Chen M.M., Nilsson, D.A., Berggren, R.M., Kugler, T., Remonen, T.M., and Forchheimer, R. (2010). Electrochemical device and methods for producing the same, US patent 7679110, granted: 2010-03-16, assigned: Acreo AB.

269 Ersman, P.A. and Kawahara, J. (2017). Display device, US patent 9709867, granted: 2017-07-18, assigned Lintec Corp., Rise Acreo AB.

270 Xiong, K. and Dahlin, A. (2018). Switchable Plasmonic display device, Published PCT application World patent WO2018135983 A1, published 2018-07-26, pending assignment Rdot AB.

271 Agrawal, A., Cronin, J.P., Adams, L.L., and Tonazzi, J.C.L. (2015). Flexible and printable electrooptic devices, US patent 9030724.

272 Wang, Z., Wang, X., Cong, S., Geng, F., and Zhao, Z. (2020). Fusing electrochromic technology with other advanced technologies: a new roadmap for future development. *Materials Science and Engineering* R140: 100524.

273 Ferreira, I.M.M., Brito Da Costa, C., Fortunato, E.M.C., Ferrao de Paiva Martins, R., and da Silva Henriques, I.D. (2014). Electrochromic device and method for producing same, US patent 8773747.

274 Marques, A.C.L., Baptista, C.A.P., and Araujo, J. (2017). Electrolyte solution, printing method thereof and resulting solid electrolyte, US patent 9625782.

275 Ynvisible (2022). Ynvisible E-paper display, Gen2 (version. 1.2) data sheet. https://cdn.prod.website-files.com/5f7f21e12602d912658dce12/6655aef1a3429585b3e191fb_Ynvisible-E-Paper-Display-G02-V1.2.pdf (accessed 01 June 2025).

276 Ynvisible University (2024). How to power ynvisibles's epaper displays with NFC energy harvesting. https://www.ynvisible.com/news-inspiration/ynvisible-e-paper-displays-nfc-energy-harvesting (accessed 01 June 2025).

277 Fang, H., Zheng, P., Ma, R., Xu, C., Yang, G., Wang, Q., and Wang, H. (2018). Multifunctional hydrogel enables extremely simplified electrochromic devices for smart windows and ionic writing boards. *Materials Horizons* 5 (5): 1000–1007. https://doi.org/doi.org/10.1039/C8MH00856F.

278 Howard, E.L., Osterholm, A.M., Shen, D.E., Panchumarti, P., Pinheiro, C., and Reynolds, J.R. (2021). Cost-effective, flexible, and colorful dynamic displays: removing underlying conducting layers from polymer-based electrochromic devices. *ACS Applied Materials & Interfaces* 13 (14): 16732–16743. https://doi.org/10.1021/acsami.1c00463.

279 Gu, C., Jia, A.-B., Zhang, Y.-M., and Zhang, S.X.-A. (2022). Emerging electrochromic materials and devices for future displays. *Chemical Reviews* 122 (18): 14679–14721. https://doi.org/10.1021/acs.chemrev.1c01055.

280 Hein, A., Kortz, C., and Oesterschulze, E. (2019). Tunable graduated filters based on electrochromic materials for spatial image control. *Scientific Reports* 9: 15822. https://doi.org/10.1038/s41598-019-52080-1.

281 Kortz, C., Hein, A., Ciobanu, M., Walder, L., and Oesterschulze, E. (2019). Complementary hybrid electrodes for high contrast electrochromic devices with fast response. *Nature Communications* 10: 4874. https://doi.org/10.1038/s41467-019-12617-4.

282 Gilmore, D.G. (1994). *Satellite Thermal Control Handbook*, 4–101. El Segundo, CA: The Aerospace Corporation Press. ISBN: 978-1-884989-11-7.

283 Metts, J.G. (2010). Assessing feasibility of electrochromic space suit radiators for reducing extravehicular activity water. Ph.D. thesis. Department of Aerospace Engineering Sciences, University of Colorado.

284 Flex Conference (2025). flex.semi.org and IDTechEX emerging technologies, idtechex.com

285 Fortescue, P., Stark, J., and Swinerd, G. (ed.) (2004). *Spacecraft Systems Engineering*, 3e. Wiley. ISBN: 0-471-61951-5.

286 Qiu, J.-W., Feng, Y.-D., and Wu, C.H. (2022). *Thermal Control Thin Films- Spacecraft Technology*. Berlin/Boston: Walter De Gruyter, Gmbh. ISBN: 978-3-11-061286-8.

287 Hale, J.S., DeVries, M., Dworak, B., and Woollam, A. (1997). Visible and infrared optical constants of electrochromic materials for emissivity modulation applications. *Thin Solid Films* 313–314: 205–209. https://doi.org/doi.org/10.1016/S0040-6090(97)00818-3.

288 Hale, J.S. and Woollam, J.A. (1999). Prospects for IR emissivity control using electrochromic structures. *Presented at the ICMCTF '97 Conference*, San Diego, CA, USA, April 1997.1, published in: *Thin Solid Films* 339(1–2): 174–180. https://doi.org/10.1016/S0040-6090(98)01335-2.

289 Demiryont, H. and Moorehead, D. (2009). Electrochromic emissivity modulator for spacecraft thermal management. *Solar Energy Materials and Solar Cells* 93: 2075–2078. https://doi.org/10.1016/j.solmat.2009.02.025.

290 Demiryont, H. and Moorehead, D. (2009). Electrochromic emissivity modulator for spacecraft thermal management. *Solar Energy Materials and Solar Cells* 93 (12): 2075–2078. https://doi.org/10.1016/j.solmat.2009.02.025.

291 Demiryont, H. Transparent electrode, US patent 7531239B2, granted: 2009-05-12, assigned: Eclipse Energy Systems (FL).

292 Paris, A., Johnson, K., Chandrasekar, P., Zay, B., and McQueeney, T. (2005). Electrochromic radiators for microspacecraft thermal control. Paper VIII-1, 19th Conference on small Satellites AIAA/USU, https://digitalcommons.usu.edu/cgi/viewcontent.cgi?article=1652&context=smallsat.

293 Demiryont, H., Hu, J., Kallman, W.R., and Schaus, C.F. (1996). Monolithic multi-layer electrochromic composite material with inner buffer layers, Australian patent AU 2760895A, 1995-05-25 filed, granted: 1996-12-11.

294 Chandrasekar, P., Zay, B., Lawrence, D., Caldwell, E., Sheth, R., Stephen, R., and Cornwell, J. (2014). Variable-emittance infrared electrochromic skins combining unique conducting polymers, ionic liquid electrolytes, microporous polymer membranes and semiconductor/polymer coatings for spacecraft thermal control. *Journal of Applied Sciences* 131 (19): 40850. https://doi.org/10.1002/app.40850.

295 Chandrasekhar, P. (2015). Variable-emittance electrochromic devices and methods of preparing the same, US patent 20140268283A1, granted: 2015-12-08, assigned: Ashwin-Ushas Corp. (NJ).

296 Chandrasekhar, P. (2018). *Conducting Polymers, Fundamentals and Applications, Including Carbon Nanotubes and Graphene*, 2e. Springer. ISBN: 978-3-319-69376-7. https://doi.org/10.1007/978-3-319-69378-1.

297 Anasori, B., Lukatskaya, M.R., and Gogotsi, Y. (2017). 2D metal carbides and nitrides (MXenes) for energy storage. *Nature Reviews Materials* 2 (2): 16098. https://doi.org/10.1038/natrevmats.2016.98.

298 Taylor, S., Neal Boman, N., Chao, J., and Wang, L. (2021). Cryothermal vacuum measurement of thermochromic variable-emittance coatings with heating/cooling hysteresis for spacecraft thermal management. *Applied Thermal Engineering* 199: 117561. https://doi.org/10.1016/j.applthermaleng.2021.117561.

299 Sun, W., Zhang, X., Chen, M., Deng, J., Zhang, H., Li, Z., Xiao, Y., Hu, L., Hu, Y., Yan, D., and Li, Y. (2023). Infrared-emissivity-regulating electrochromic device with low visible wavelength absorption based on nanolayered $Au/NiO/Ta_2O_5/WO_3/Ge$ photonic structures. *ACS Applied Nano Materials* 6 (24): 23506–23514. https://doi.org/10.1021/acsanm.3c04872.

300 Song, Y., Cheng, B., Cheng, H., Meng, Z., and Liu, D. (2024). All-solid-state infrared electrochromic devices based on thin metal films. *ACS Applied Materials & Interfaces* 16 (27): 35372–35380. https://doi.org/10.1021/acsami.4c07169.

301 Shearer, A., Hauke, B., Montazerian, M., and Mauro, J.C. (2023). A critical review of infrared transparent oxide glasses. *Optical Materials: X* 20: 100258. https://doi.org/10.1016/j.omx.2023.100258.

302 Sarwar, S., Park, S., Dao, T.T., Lee, M.-S., Ullah, A., Hong, S., and Han, C.-H. (2020). Scalable photoelectrochromic glass of high performance powered by ligand attached

TiO_2 photoactive layer. *Solar Energy Materials and Solar Cells* 210: 110498. https://doi.org/10.1016/j.solmat.2017.03.020.

303 Bogati, S., Basnet, R., Graf, W., and Georg, A. (2017). Development of photochromic device with magnetron sputtered titanium dioxide and tungsten trioxide films. *Solar Energy Materials and Solar Cells* 166: 204–211. https://doi.org/10.1016/j.solmat.2017.03.020.

304 Naresh Kumar, P., Narayanan, R., Laha, S., Deepa, M., and Srivastava, A.K. (2016). Photoelectrochromic cell with a CdS quantum dots/graphitic-nanoparticles sensitized anode and a molybdenum oxide cathode. *Solar Energy Materials and Solar Cells* 153: 138–147. https://doi.org/10.1016/j.solmat.2016.04.011.

305 Shen, K., Luo, G., Liu, J., Zheng, J., and Xu, C. (2019). Highly transparent photoelectrochromic device based on carbon quantum dots sensitized photoanode. *Solar Energy Materials and Solar Cells* 193: 372–378. https://doi.org/10.1016/j.solmat.2019.01.004.

306 Liu, Y., Liu, F., Wang, J., Huang, H., Yan, S., Gao, S., Wang, L., Huang, W., and Qin, T. (2021). Tetrakis (N-phenothiazine) spirobifluorene-based hole-transporting material towards high photovoltage perovskite photovoltaics for driving electrochromic devices. *Dyes and Pigments* 188: 109164. https://doi.org/10.1016/j.dyepig.2021.109164.

307 Costa, C., Ivanou, D., Pinto, J., Mendes, J., and Mendes, A. (2019). Impact of the architecture of dye sensitized solar cell-powered electrochromic devices on their photovoltaic performance and the ability to color change. *Solar Energy* 182: 22–28. https://doi.org/10.1016/j.solener.2019.02.036.

308 Syrrokostas, G., Dokouzis, A., Yannopoulos, S.N., and Leftheriotis, G. (2020). Novel photoelectrochromic devices incorporating carbon-based perovskite solar cells. *Nano Energy* 77: 105243. https://doi.org/10.1016/j.nanoen.2020.105243.

309 Hočevar, M. and Krašovec, U.O. (2018). A photochromic single glass pane. *Solar Energy Materials and Solar Cells* 186: 111–114. https://doi.org/10.1016/j.solmat.2018.06.035.

310 Manjakkal, L., Pereira, L., Barimah, E.K., Grey, P., Franco, F.F., Lin, Z., Jose, G., and Hogg, R.A. (2024). Multifunctional flexible and stretchable electrochromic energy storage devices. *Progress in Materials Science* 142: 101244. https://doi.org/10.1016/j.pmatsci.2024.101244.

311 Pathak, D.K. and Moon, H.C. (2020). Recent progress in electrochromic energy storage materials and devices: a minireview. *Materials Horizons* 9 (12): 2949–2975. https://doi.org/10.1039/D2MH00845A.

3

Trends in Organic Electrochromic Materials and Their Applications

Melepurath Deepa[1] *and Anoop Agrawal*[2]

[1] *Department of Chemistry, Indian Institute of Technology Hyderabad, India*
[2] *Glass Dyenamics, Tucson, AZ, USA*

Nomenclature

A	absorptance ($A = 1 - T + R$)
A_U	absorbance (the logarithm of the ratio of incident to transmitted radiant power through a sample)
Bn-*p*CNVio	1-benzyl-1(pcyanophenyl)-4,4-bipyridinium dibromide
CB	conduction band
CE_λ	coloration efficiency at wavelength λ
CE	counter electrode
C_a	areal Capacitance (capacitance/area)
C_s	specific Capacitance (capacitance/mass)
CNT	carbon nanotube
DHV^{2+}	di-heptyl viologen
DBF	dibenzofulvene
EBE	4,7-bis(7-bromo-2,3-dihydrothieno[3,4-*b*][1,4]dioxin-5-yl)benzo[*c*][1,2,5] thiadiazole
EBzE	4,7-bis(7-bromo-2,3-dihydrothieno[3,4-*b*][1,4]dioxin-5-yl)-2-(2-hexyldecyl)-2*H*-benzo[*d*][1,2,3]triazole
EC	electrochromic
ECD	electrochromic device
ESC	electrochromic supercapacitor
ECM	electrochromic mirror
EASC	electrochromic asymmetric SC
EESD	EC energy storage devices
E_d	energy density (energy/volume)
Et-*p*CNVio	1-ethyl-1-(*p*-cyanophenyl)-4,4-bipyridinium dibromide
FD	fluorene-diphenylamine
FTO	fluorine-doped tin oxide (transparent conductor)
IDTT	indacenodithieno[3,2-*b*]thiophene
IL	ionic liquid
IDDT	indacenodithieno[3,2-*b*]thiophene
ITO	indium tin oxide

Chromogenics: Smart Switchable Optical Materials and Their Applications, First Edition. Edited by Carl M. Lampert.

MAP I_3	methylammonium lead iodide
MMA	methyl methacrylate
MV	mixed valance compounds
NDI	naphthalenediimide
NIR	near-infrared
PANI	poly(aniline)
PC	propylene carbonate
PDI	perylenediimide
PECD	photoelectrochromic device
PhCE	photocoloring efficiency (coloration kinetics)
PIDTT	poly[indacenodithieno[3,2-*b*]thiophene]
PMMA	poly(methyl methacrylate)
PPy	polypyrrole
PhV	phenyl viologen
PV	photovoltaic
QD	quantum dot
R	reflectance
SC	super capacitor
SOC	state of charge
T	transmittance
$\Delta T_{\lambda\text{nm}}$	change in transmittance at wavelength λ (nm)
ΔT_{vis}	change in visible transmittance
ΔT_{ir}	change in infrared transmittance
TBT	4,7-bis(5-bromothiophen-2-yl)benzo[*c*][1,2,5]thiadiazole
TBzT	4,7-bis(5-bromothiophen-2-yl)-2-(2-hexyldecyl)-2*H*-benzo[*d*][1,2,3]triazole
TC	transparent conductor
TPOSt	diphenylamine-tetraphenylethene derivative
t_c	switching time for coloration
t_b	switching time for bleaching
VBV	vinyl benzyl viologen
ZPEM	free standing Zn^{2+}-based polymeric electrolyte membrane

3.1 Introduction to Organic Electrochromics

This chapter focuses on new developments in organic electrochromic (EC) materials and resulting devices (ECDs). Advances in research drive the future of commercial switchable large-area devices. As mentioned before, historically, organic ECs have had the most sales in automotive mirrors. Also, other applications such as side windows, sunroofs, and aircraft windows have been made and will continue to be important for the vehicle and aerospace industries.

We provide a summary of recent developments in organic and polymeric electrochrome-based electrochromic devices (ECDs) with high contrast and/or long cycle life and being able to tune color, show multiple colors, and particularly neutral black in the colored state. These ECDs are used in displays, automotive mirrors, and smart windows.

In addition, we see the use of EC materials expanding into other fields, such as supercapacitors and combined photovoltaic/electrochromic (PV/EC) photoelectrochromic devices

(PECDs). These applications will be included in our discussion. It is our intent to show selected examples of developing EC technologies, but not to review all ongoing research work. We will leave that task for the research community to compile.

Organic EC mirrors and displays have attracted considerable interest over the last 50 years [1]. Commercial electrochromic mirrors (ECMs) commonly utilize organic viologens (cathodic EC dyes) and phenazines (anodic EC dyes), enabling them to undergo reversible transitions between deep blue/green and colorless hues when an electric voltage (~1 V) is applied. The most commonly used phenazines are 5,10-dihydro-5,10-dimethyl phenazine and also 5,10-dineopentyl-5,10-dihydrophenazine [2, 3]. These mirrors require a continuous application of voltage to keep them colored, and when voltage is removed (short or open circuit), these mirrors transition to a clear state (self-erasing). This feature is required for automotive safety in case there is a power failure. The electrolyte matrix is further comprised of a plasticizer such as propylene carbonate to dissolve the dyes, along with UV stabilizers, dissociable salts, and monomers, which are polymerized after the formulation is introduced in a preformed cavity having conductive coatings facing inside the cavity [4, 5]. The cavity or the electrolyte thickness for mirrors is generally in a range of about 100–200 μm, where the smaller gaps are used for the interior mirrors as they are smaller in size and are also specified for faster clearing (or bleaching). These cavities are prefabricated with a perimeter sealant, leaving a small hole (~3–4 mm) in the sealant, from which the formulation is vacuum backfilled with the polymerizable liquid composition described above, and then the hole is sealed using a UV curing adhesive [6–9]. A schematic of this construction is shown in Figure 3.1. Typically, there is no offset between the front and rear substrates, which is done by taking the conductive path to the back and using some laser etching, etc. [10]. Displays (small or large or indicators) are almost ubiquitous in mirrors today.

One of these conductive coatings forming this cavity is transparent, and the other coating (usually a stack of coatings) also has both conductive and reflective properties. The metal coatings used in these stacks are electrochemically stable, such as silver alloys, ruthenium, and rhodium. The other coatings in the stack are metal oxides, including conductive metal oxides [11, 12]. This construction is called third surface mirrors as the reflective layer is on

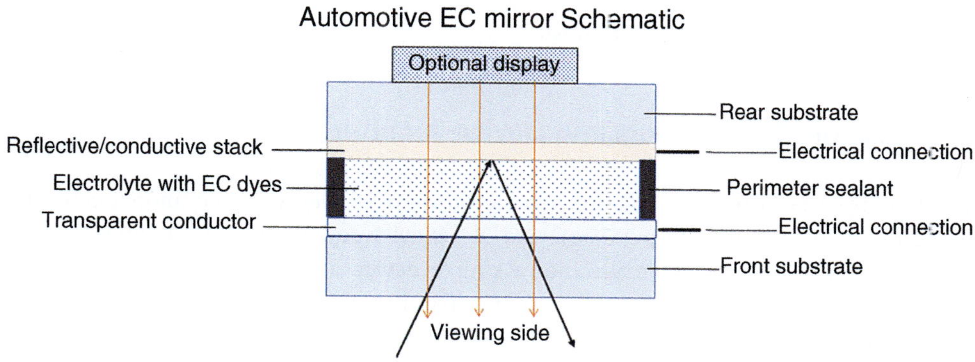

Figure 3.1 Cross-section of an automotive organic mirror showing its principal construction (not to scale). Applying a small voltage across the electrochromic layer causes the mirror to darken. A sensing system (not shown) causes this darkening automatically at night to control mirror dazzling by headlights following from the rear. In many modern mirrors, displays are added to show traffic camera images or other important information about driving conditions.

the third surface as one counts the surface sequentially starting from the outside surface of the first substrate, which faces the observer, and the second surface being the inside surface of the first substrate that faces the cavity, and the third surface is the front surface of the second substrate facing the cavity. The properties of the reflective stacks are optimized to ensure that the color/transmission of integrated displays that are placed or bonded behind the mirror cavities (or cells) is as desired. ECMs, along with sensors and discriminating electronics, automatically dim under glare situations to keep the driver distraction low and promote safety [13]. In this space, Gentex continues to be the global leader in dimmable mirrors for automotive applications. ECMs come in several configurations as interior mirrors and exterior mirrors (flat, convex, and multi-radius, also called aspheric). Aspheric mirrors are used to eliminate the blind spot on the driver's side). ECMs also come with advanced features such as turn indicators, blind spot sensors, and visual cues for exterior mirrors and for interior mirrors. Some of the features include integrated displays, camera systems, intrusion detection, microphones, and speakers that link wirelessly to the phones. When displays are integrated behind the interior mirrors, the reflective coating is partially transmissive, which hides the display when the display is not active as the reflected image is stronger. When the display is activated, its brightness is high, so it overwhelms the reflective image. Another notable supplier of ECMs is Magna International. Almost all the other automotive mirror manufacturers buy EC cells and associated electronics from the above two suppliers and integrate them into their products. Some of these mirror manufacturers include Ficosa International, Samvardhana Motherson Reflectec (SMR) Ichikoh, and Murakami Corporation.

The real challenge is in the processing, formulation chemistry, and its optimization and sealants to achieve long stability for cycling, high temperature, and UV resistance.

Aircraft EC windows such as those introduced by Boeing are also based on chemistry and construction similar to the ECMs but use transparent conductors on both substrates and color to achieve high optical densities of greater than 3, i.e. a contrast of more than 1000 as measured by the ratio of the visible transmission (%) in the clear state divided by the visible transmission (%) in the colored state [14]. EC cells are supplied by Gentex and integrated into window systems by Vitro (a company that acquired the EC assets of PPG Inc). Gentex Corporation has demonstrated EC cells for next-generation aircraft windows that have optical densities greater than 4, implying transmittance of less than 0.001% [15]. EC displays can be built on rigid or flexible substrates such as PET or polyimide plastics. They can also be built on flexible glass and paper substrates.

3.2 Organic EC Materials and Device Architectures

In a study by Alesanco et al. [16], an asymmetric viologen 1-ethyl-1-(*p*-cyanophenyl)-4,4-bipyridinium dibromide (Et-pCNVio) and 1-benzyl-1(pcyanophenyl)-4,4-bipyridinium dibromide (Bn-pCNVio) were made into a simple device architecture. The structure is glass/TC/EC gel/TC/glass, where TC is a transparent conductor. The device exhibited a transmittance change, $\Delta T_{vis} \sim 60\%$, with switching times of $t \leq 15$ seconds and a long cyclic lifetime (~15,000 cycles). The device switched between transparent-yellow and deep-gray states. In another study by Oh et al. [17], flexible substrates and di-heptyl viologen (DHV^{2+}) along with dimethyl ferrocene (dmFc) as an anodic species were used to make an ECD. The ECD showed a very large optical modulation, $\Delta T_{602\,nm} \sim 90\%$, and was colored and bleached in $t_c \sim 17$ seconds (at −0.7 V) and $t_b \sim 32$ seconds (short-circuit condition),

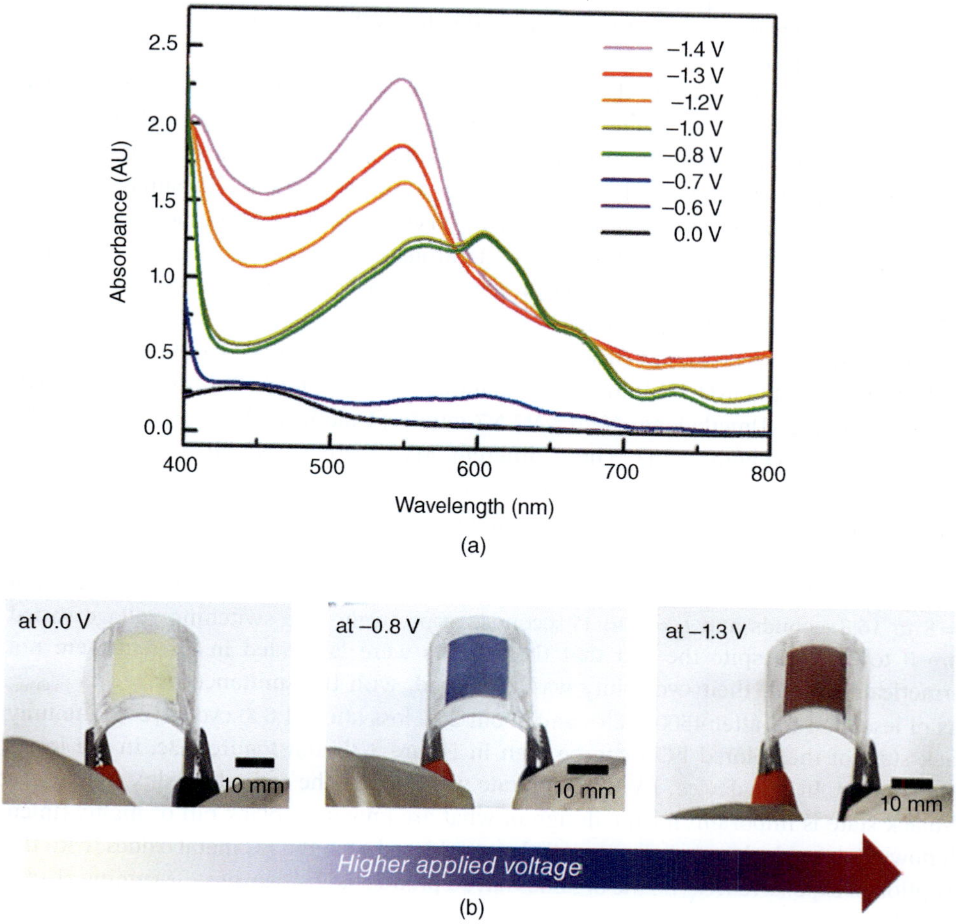

Figure 3.2 Di-heptyl viologen-based ECD in bleached (yellow) and two colored states (blue and red). The device structure is glass/TC/EC gel/TC/glass. The device size is ~20 × 20 mm. *Source:* Reproduced with permission from Oh et al. [17], © 2017 / American Chemical Society.

respectively [17]. The different colors of the flexible ECD can be seen in Figure 3.2. The size of the device is ~20 × 20 mm. A color-tunable window or display device is desirable. EC inorganic oxides generally offer one color, so this is where organic ECs have an advantage.

Another study reported work on a viologen-modified ionic liquid (IL) based on a phosphonium core, which behaved as both the electrolyte and the primary EC material [18]. The colors achieved upon reduction were blue (−0.84 V) and purple (−0.91 V), and the associated coloration efficiency, $CE_{610\,nm} = 10.72\,cm^2\,C^{-1}$, and the switching time was $t \sim 221$ seconds ($0.9\,cm^2$ device area) to reach $A_{610\,nm} = 95\%$ absorptance [17]. Another ECD with an FTO/electrolyte/EC layer/FTO configuration with diphenylamine-tetraphenylethene derivative (TPOSt) as the EC layer, displayed a ΔT of ~50% from 490 to 700 nm wavelength with a color change from a transparent light yellow to a black state. Switching times of 8.1 and 4.2 seconds for coloration and bleaching, respectively, were obtained at 700 nm. The size of the device was 20 × 20 mm. The TPOSt-based ECD demonstrated good EC switching stability over 1500 color-bleach cycles. A transmittance loss was seen in the EC of $\Delta T = 6\%$ after 1500 cycles. The cross-linking of the IL prevented the delamination of the material from the surface of the current collector upon repeated redox cycling and enhanced the long-term cycling stability of the device [19].

The dibenzofulvene (DBF) bridge organic mixed valence compounds (MVs) with arylamine redox centers are an interesting class of small molecules capable of showing NIR (near-infrared) electrochromism. Two-center DBF-MV compounds were synthesized in which the center position was substituted by either a simple linear alkyl chain or a phenyl substituent. Also, in asymmetric 3-center DBF-MVs, a third diarylamine redox center was connected to the exocyclic fulvene bond through an aromatic ring. The synthetic compounds were: *N*2,*N*2,*N*7,*N*7-tetrakis(4-methoxyphenyl)-9-pentylidene-9*H*-fluo-rene-2, 7-diamine (AN2); *N*2,*N*2,*N*7,*N*7-tetrakis(4-methoxyphenyl)-9-(4-methylbenzyli-dene)-9*H*-fluorene-2,7-diamine (PN2); 9-(4-(bis(4-methoxyphenyl)amino)benzylidene)-*N*2,*N*2,*N*7,*N*7-tetrakis(4-methoxyphenyl)-9*H*-fluorene-2,7-diamine (PN3); 9-((5-(Bis(4-methoxyphenyl) amino)thiophen-2-yl)methylene)-*N*2,*N*2,*N*7,*N*7-tetrakis(4-methoxyphenyl)-9*H*-fluorene-2, 7-diamine (TN3) (Figure 3.3a). These different MVs were employed as anodic materials in the construction of the EC devices (size ~ 4 cm^2), while ethyl viologen (EV) diperchlorate was used as the cathodic complementary redox couple. The ECDs exhibited for AN2, $\Delta T_{1390,\,910\,\text{nm}} = 58\%$, 62%, for TN3 $\Delta T_{1390,\,910\,\text{nm}} = 76\%$, 81%, for PN2 + PN3, $\Delta T_{1430,\,920\,\text{nm}} = 88\%$, 95%. The switching times for coloration and bleaching varied from $t_c = 8$ to 16.8 seconds and $t_b = 3$ to 17 seconds, respectively. The switching voltage varied from 0 to 2.4 V. Despite the fact that these ECDs were fabricated in air and were not hermetically sealed, their cyclability was very good, with transmittance range, $\Delta T_{1430\,\text{nm}}$ loss of less than 8% after 4500 cycles and about 25% loss after 11,000 cycles [20]. The fully black state of the colored ECD can be seen in Figure 3.3b and Figure 3.3c, in the lower center image. In this device, EV diperchlorate was used as the cathodic redox couple. An all-black state is important for the design of what are known as black-out or highly tinted windows. A true black state has not really been achieved with the EC metal oxides, with the exception of expensive IrO_x. So the organics have a product advantage here, assuming all else is equal.

Two EC devices were constructed using fluorene-diphenylamine (FD) and its derivative FDOMe as the EC layers. The devices were constructed using a FTO/Li-PMMA-PC electrolyte/EC layer/FTO architecture. FDOMe was prepared by the palladium-catalyzed coupling of 2,7-dibromo-9,9-bis(4-vinylbenzyl)-9*H*-fluorene with 4,4′-dimethoxydiphenylamine (74%). The devices were colorless and transparent in their neutral state. The active area was 2 cm × 1.5 cm. FD changed to yellow and then to dark cyan when the potential was swept from 0 to 2.3 V, whereas its methoxy derivative FDOMe changed from colorless to red and then to blue. The switching times were less than $t \sim 50$ seconds. Spectroelectrochemical studies revealed the two-color EC response of FD (colorless-yellow-dark cyan) and FDOMe (colorless-brick red–blue). The device using FD showed a change of $\Delta T_{485} \sim 44\%$. For the FDOMe device, the optical change was $\Delta T_{500} \sim 63\%$, with stable switching between the neutral and oxidized states for >300 cycles with low switching voltages of <0.9 V for the first oxidation and <1.3 V for the second oxidation [21].

In work by Kao et al. [22], vinyl benzyl viologen (VBV)/poly(methyl methacrylate) (PMMA) in an ECD along with *N*,*N*,*N*′,*N*′-tetramethyl-*p*-phenylenediamine (TMPD) showed $\Delta T > 60\%$ at both $\lambda = 570$ and 615 nm when the device (with an active area of 2 cm × 2 cm) was switched from 0 to 0.5 V. The coloration reaction followed reaction (3.1), right to left:

$$\text{TMPD}^{+}\ (\text{purple}) + \text{PhV}^{+\cdot}(\text{green}) \longleftrightarrow \text{TMPD (colorless)} + \text{PhV}^{2+}\ (\text{colorless}) \quad (3.1)$$

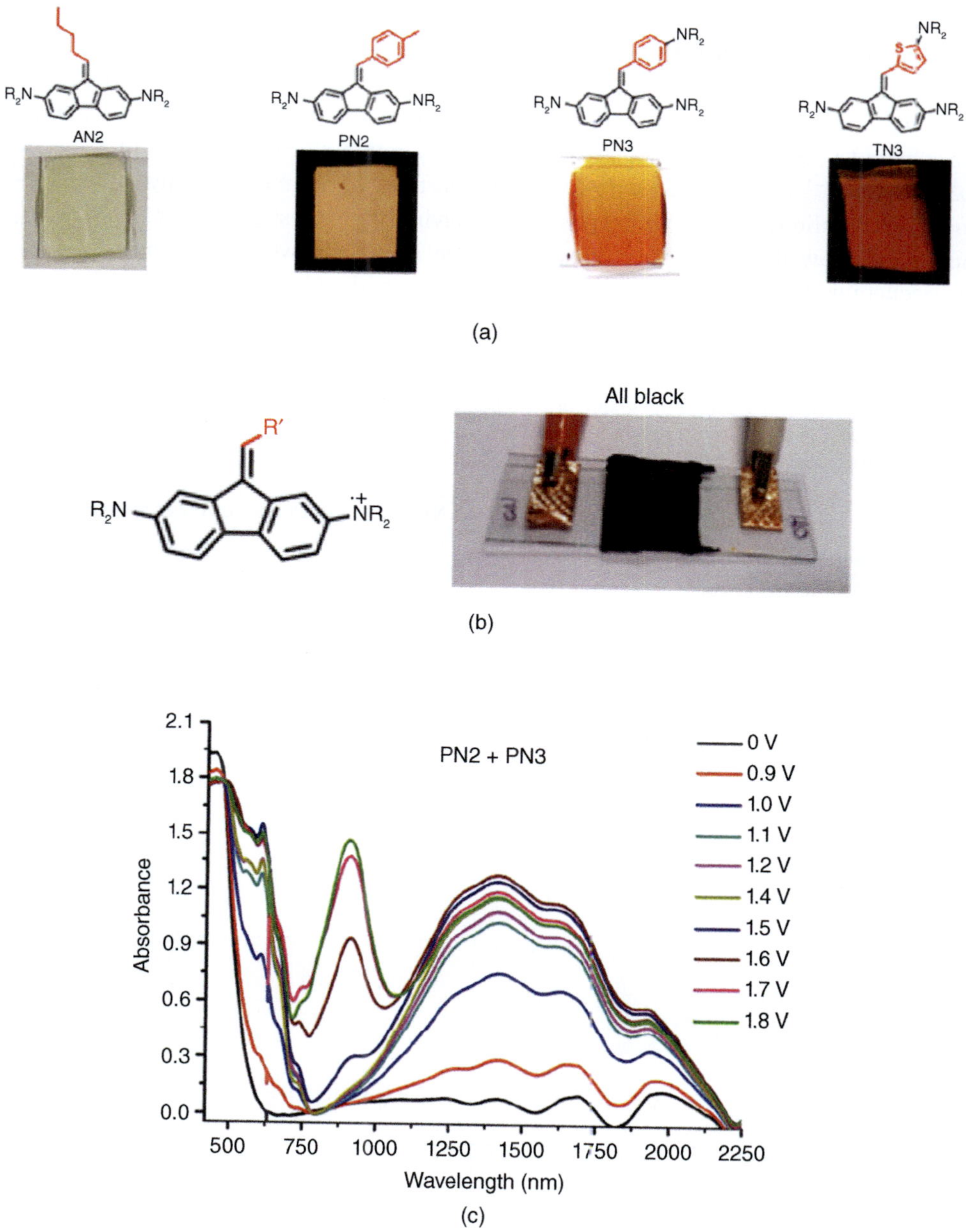

Figure 3.3 (a) Four DBF-based ECDs, showing both the colored and (b) all-black colored states, with a viologen. (c) The near-infrared absorbance spectral response for PN2 + PN3 switched between 0 and 1.8 V. Note the very strong NIR absorbance change. *Source:* Reproduced with permission from Corrente et al. [20], © 2018 / American Chemical Society.

Negligible changes in ΔT at both wavelengths were observed for the thermally cured ECD after being subjected to 1000 cycles of potential. Curing involved polymerization and some cross-linking of the MMA forming PMMA. When devices were made with mixed viologens, VBV and phenyl viologen (PhV), the absorption was in a broader wavelength range and provided more neutral colors. For these, ΔT values were over 60%, with increases in ΔT_{570} of 5.3%

and ΔT_{615} of 6.9%. These values remained stable after 10,000 continuous cycles at bleaching and coloring voltages of 0–0.5 V, respectively.

In a study by Gelinas et al. [23], an ECD was made with an EV and ferrocene-based redox IL. The device turned deep blue at 2 V, with a $\Delta T_{610\,nm} \sim 40\%$. The switching times were t_b: 5.8 seconds, and t_c: 9.5 seconds. A coloration efficiency range of $CE_{610\,nm} = 105–146\,cm^2\,C^{-1}$ was observed with varying proportions of EV and ferrocenylsulfonyl(trifluoromethylsulfonyl)imide, with a 2 : 1 ratio giving the lowest CE. After 1000 cycles, the device showed 94.8% stability in the change of ΔT. In work by Kim et al. [24], a perylenediimide-viologen-based ECD using poly(3-hexylthiophene) as a counter electrode

Table 3.1 Photos of electrodeposited polymer films on indium tin oxide (ITO)/glass made from (i) a π-conjugated donor–acceptor main chain; and (ii) perylenediimide (PDI) or naphthalenediimide (NDI) pendant groups. The films were reversibly switched in a 0.1 mol dm^{-3} Bu_4NPF_6/DCM electrolyte. Shown are the colors obtained.

	n-Doped (reduced states)			Neutral	p-Doped (oxidized states)	
polyM1	 −1.7 V	 −1.6 V	 −1.4	 −0.5 V	 0.6 V	 0.9 V
NDI	No reaction −1.7 V	 −1.6 V	 −1.4	 −0.5 V	No reaction 0.6 V	No reaction 0.9 V
polyM2	No reaction −2 V	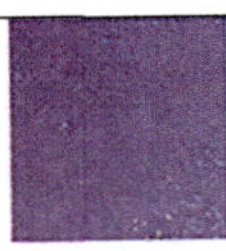 −1.5 V	 −1.4 V	 −0.2 V	 0.5 V	 0.9 V
PDI	No reaction −2 V	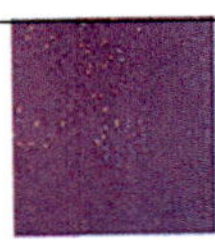 −1.5 V	 −1.2 V	 −0.2 V	No reaction 0.5 V	No reaction 0.9 V
polyM3	 −2 V	No reaction −1.5 V	No reaction −1.4 V	 −0.2 V	 0.6V	 0.9 V

Source: Reproduced with permission from Drewniak et al. [25], © 2019 / American Chemical Society.

(CE) exhibited a strong color change between deep red and dark blue when the applied potential was switched from +2.0 to −2.0 V with $\Delta T_{635\,nm} \sim 33\%$ and switching responses of $t_b = 1.9$ and $t_c = 1.8$ seconds (having an active area of approximately 1.8 cm × 1.5 cm).

In studies by Drewniak et al. [25], conjugated macromolecules comprised of two distinct parts were used to make ECDs. The two parts are : (i) a π-conjugated donor–acceptor main chain; and (ii) perylenediimide (PDI) or naphthalenediimide (NDI) pendant groups. The ECDs showed transmittance changes; for n-doping, $\Delta T_{478\,nm} = 39\%$, $\Delta T_{748\,nm} = 26\%$, depending upon the substituents. For n-type doping, coloration was $t_c = 5.28$ seconds and $t_b = 1.61$ seconds, depending on the substituents. During p-type doping, $\Delta T_{700\,nm} = 18\%$, $t_c = 0.14$ seconds, and $t_b = 0.12$ seconds, much faster than n-doping. The range of colorations is shown in Table 3.1.

In research by Kumar et al. [26], xanthommatin was used as the EC material. ITO was first coated with PEDOT:PSS. By varying the ratio of the two electroactive components, the device's performance was studied. The ratio of electroactive compounds altered the visible color of xanthommatin from different shades of red (−1.5 V) to yellow (+1.5 V), under reducing and oxidizing potentials, respectively. In this ECD, the richest yellow and red films had an optical contrast of $\Delta T_{555\,nm} = 21.5\%$ and $CE_{555\,nm} = 104\,cm^2\,C^{-1}$. The switching response increases from $t_c = 15$ to 25 seconds after 500 cycles. The different colors shown by the ECD are presented in Figure 3.4.

Recent progress has been made on the use of anodic and cathodic dyes for large-area windows, such as for buildings and automotive glazing. In these devices, reactive groups are attached to the dyes covalently so that they can be mixed with compatible polymer-forming monomers, deposited as coatings, and polymerized. The anodic and cathodic coatings containing anodic and cathodic dyes (such as phenazines and viologens, respectively) are deposited on transparent substrates, which are then assembled by sandwiching an electrolyte layer. Typically, the electrolyte layer in these devices does not have any EC dyes. There are several perceived advantages for large-area devices when this construction is used. Since these windows will have superior memory (they do not self-erase rapidly), thus,

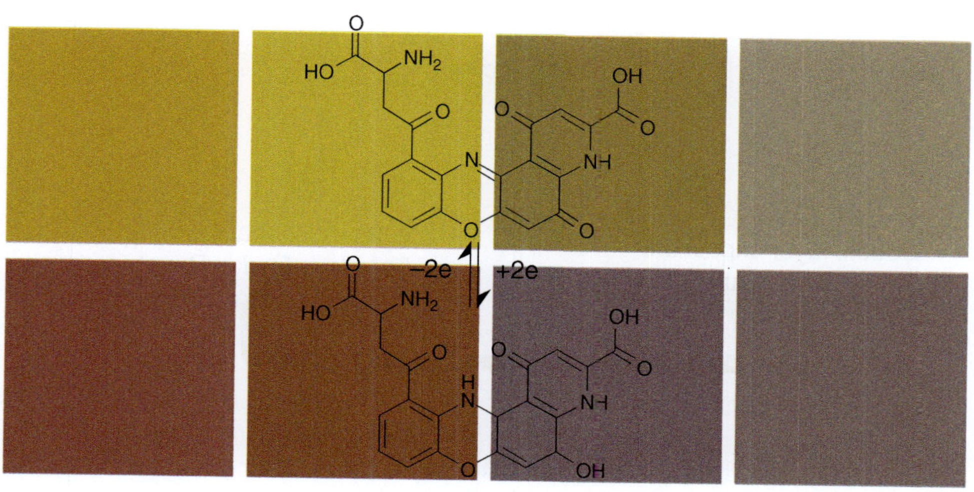

Figure 3.4 Broad spectrum xanthommatin EC materials: showing different shades of yellow, brown, and red. *Source:* Reproduced with permission from Kumar et al. [26]. © 2018 / American Chemical Society.

in larger sizes the color uniformity is superior, and these devices retain optical states longer without the need for continuous electrical power. Furthermore, since many of these dyes have proven to be durable in commercial automotive mirrors, the expectation is that the same success can be repeated here [27–29]. Since these coatings do not use conjugated polymeric backbones in EC coatings, they are less prone to UV degradation. In these references, the devices showed high durability, measured by color appearance change due to degradation of the devices when cycled under UV exposure at 85 °C.

Our survey of organic EC research activities shows that while there are significant studies on high-contrast organic/polymeric EC materials, there is still a lot of room for developing new devices with greater color tunability, even NIR switching, and with durability suitable for practical applications.

3.3 Electrochromic Supercapacitors (ESCs) with at Least One Transparent State

An electrochromic supercapacitor (ESC) is a device that simultaneously demonstrates a visual color change proportional to the energy stored [30, 31]. A supercapacitor is an energy storage system, typically having a capacitance and energy density greater by several orders of magnitude compared to that of a conventional capacitor but a power density only slightly lower than that of a capacitor. Apart from undergoing the optical change, EC materials are also capable of storing and releasing charge. The EC phenomenon involves a redox or a Faradaic process, which is typically the process that a pseudocapacitor relies on for storing and releasing charge. See the supercapacitor family and mechanisms in Figures 3.5 and 3.6.

An ESC consists of an electroactive anode and cathode, where both layers are EC, or at least one layer is. For the single-layer EC, the second electrode serves as an ion-storage layer, and if the ESC is to be used as a window, then this electrode also has to be transparent. This is important when the EC layer is bleached; the device is also in the bleached transparent

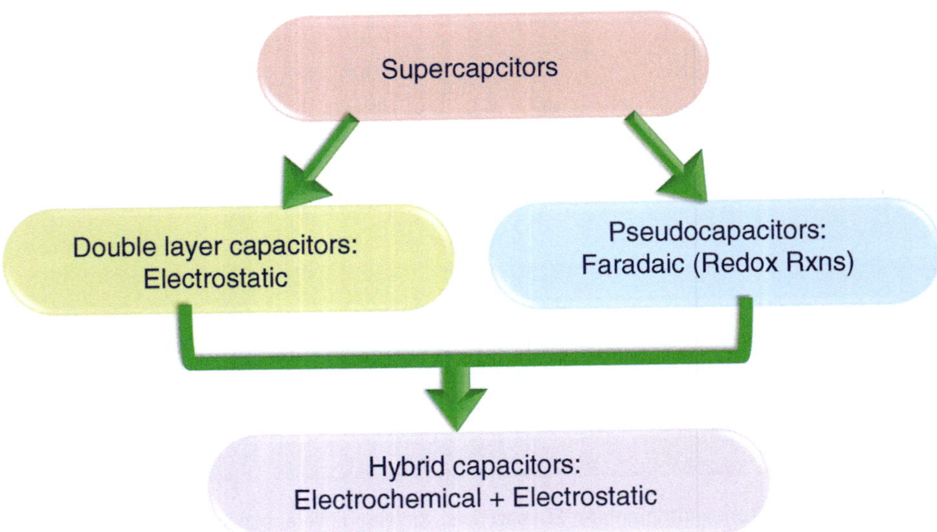

Figure 3.5 Classification of supercapacitors.

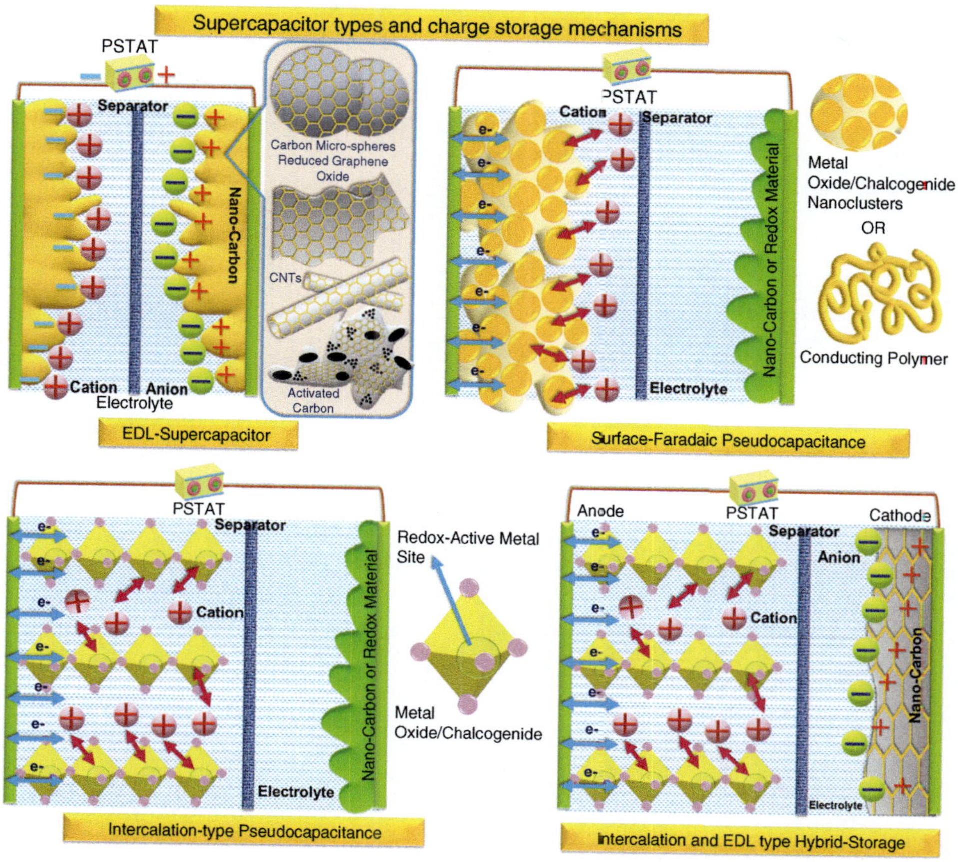

Figure 3.6 The supercapacitor family and the corresponding mechanisms for charge storage. Electrochromic materials behave like pseudocapacitive materials. EDL: Electrical Double Layer and CNTs: Carbon Nanotubes.

state. The state of charge (SOC) in an ESC can be visually realized by the color of the device. As an illustrative example, an ESC can also serve as a smart window, using layers of WO_3 and poly(aniline) (PANI), as shown in Figure 3.7. When this device is connected to a power supply, with an oxidative potential applied to the PANI film, the PANI film is fully oxidized and turns green at given potential and current (e.g. 1 V, 0.2 mA cm^{-2}). Simultaneously, a reductive potential drives electrons from the external circuit to the WO_3 film, and cations migrate from the electrolyte to the oxide, forming the metal tungsten bronze, which is deep blue in color. As a consequence, the ESC is deeply colored and fully charged (as shown in Figure 3.7). Now, when an external load, such as a fan or a light is connected to the ESC, the ESC discharges its stored charge to the load, and the fan runs or the bulb lights up. The capacitance (charge stored in the layers) decays, as shown in Figure 3.7b. As the transmittance of the device increases, and when the device is fully discharged to 0 V, the device is fully bleached. Thus, an ESC will darken during charging and bleach during discharge of its energy to perform work. In this case, an ESC may not function as a switchable window, for example, but can be used to show the SOC of the supercapacitor.

Dual-function ESCs capable of showing real-time charge capacity in color were constructed with WO_3 and Li-doped ion gels containing hydroquinone (HQ). The ESCs

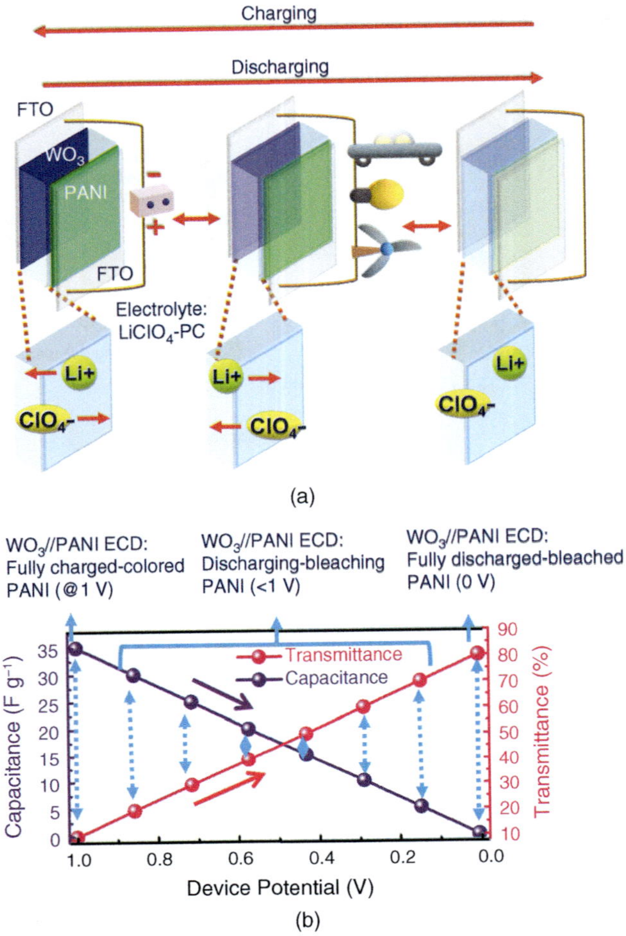

Figure 3.7 Schematic showing the working of a WO_3//PANI electrochromic supercapacitor: (a) change in color as a function of discharging. PANI is polyaniline and FTO is fluorine-doped tin oxide. (b) Change in capacitance and transmittance of the ESC as it discharges (all values are for illustration purposes only).

concurrently served as either ECDs or supercapacitors. The coloration/bleaching and charging/discharging characteristics were measured between 0 and −1.5 V. A large transmittance contrast $\Delta T_{700\,\mathrm{nm}} \sim 91\%$, high coloration efficiency $\mathrm{CE}_{700\,\mathrm{nm}} \sim 61.9\,\mathrm{cm}^2\ \mathrm{C}^{-1}$, high areal capacitance $C_a \sim 13.6\,\mathrm{mF\ cm}^{-2}$, and good charging/discharging cycling stability were achieved. Flexible ESCs were fabricated on plastic substrates by exploiting the elastic characteristics of the gel electrolytes, and they exhibited good bending durability [32]. In work by Murto et al. [33], stable doping of poly[indacenodithieno[3,2-*b*]thiophene] (PIDTT) structures enabled easy color tuning with a significant increase in the ion storage capacity of EC polymers. The IDTT structure was copolymerized with four different donor–acceptor–donor (DAD) units, showing small changes in their electron-donating and electron-acceptor natures. These DAD units were 4,7-bis(5-bromothiophen-2-yl)benzo[*c*][1,2,5]thiadiazole (TBT) and 4,7-bis(7-bromo-2,3-dihydrothieno[3,4-*b*][1,4]dioxin-5-yl)benzo[*c*][1,2,5]thiadiazole (EBE), and 4,7-bis(5-bromothiophen-2-yl)-2-(2-hexyldecyl)-2*H*-benzo[*d*][1,2,3]triazole (TBzT), and 4,7-bis(7-bromo-2,3-dihydrothieno[3,4-*b*][1,4]dioxin-5-yl)-2-

(2-hexyldecyl)-2*H*-benzo[*d*][1,2,3]triazole (EBzE) monomers. The polymers achieved excellent performance requirements for ESCs, with desired switching between −0.4 V (colored) and 1 V (bleached). For example, PIDTT–TBT had a desirable deep gray-black to very light gray with color coordinates: lightness $L^* = 45.9–86.7$ (neutral to oxidized), the red-green balance, $a^* = -3.1$ to -2.2, and the yellow-blue balance $b^* = -4.2$ to -2.7 and chroma $C^*_{ab} = 5.2–3.5$ (color coordinates are discussed in Chapter 1). PIDTT–TBzT showed high optical contrast $\Delta T_{525\,nm} \sim 72\%$, and PIDTT–EBE showed good redox stability $I^{red}/I^{ox} \sim 1.0$. The best overall characteristics were found in poly[indacenodithieno[3,2-*b*]thiophene-2,8-diyl-*alt*-4,7-bis(2,3-dihydrothieno[3,4-*b*][1,4]dioxin-5-yl)-2-(2-hexyldecyl)-2*H*-benzo[*d*][1,2,3]triazole-7,7′-diyl] (PIDTT–EBzE). PIDTT–EBzE showed high contrast, $\Delta T_{550\,nm} = 69\%$, a high doping level, and specific capacitance $C_{s} = 260\ F\ g^{-1}$ [33]. See Figure 3.8.

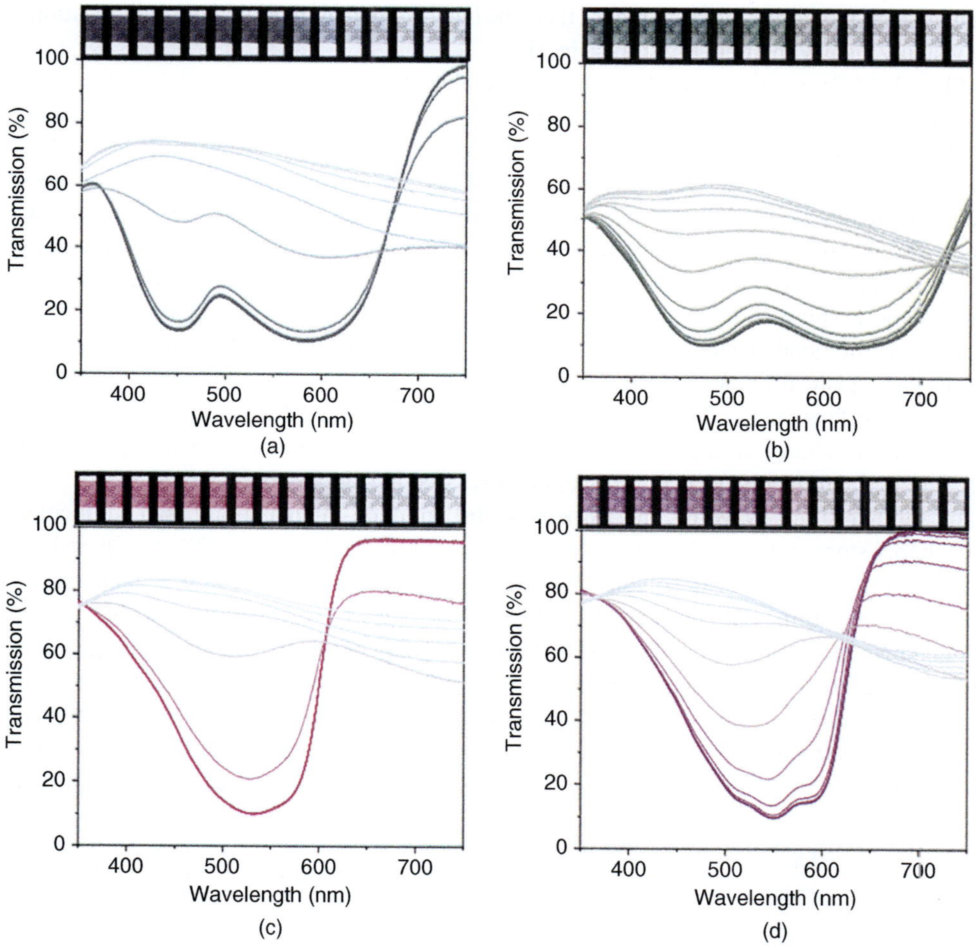

Figure 3.8 Visible spectral transmittance of spin-coated (a) PIDTT-TBT, (b) PIDTT-EBE, (c) PIDTT-TBzT, and (d) PIDTT-EBzE thin films, as measured by increasing the voltage from −0.4 V (colored line) to 1.0 V (bleached line) at intervals of 0.1 V. The colors of the lines represent the pseudo-color of the spectra in RGB color space. The top panels of (a–d) show color photographs of the films in the same potential range, with the chemical structure of IDTT donor depicted in the background. *Source:* Murto et al. [33] / American Chemical Society / CC BY 4.0.

This work introduced IDTT-based polymers as bifunctional electro-optical materials for potential use in self-regulating smart energy systems. These systems can visually show the stored energy levels. If these materials were integrated with PVs, solar energy could be collected and stored.

In work by Zhong et al. [34], a solid-state electrochromic asymmetric supercapacitor (EASC) window was fabricated by using WO_3 and PANI as the negative and positive electrodes, respectively. The two complementary materials contributed to the good electrochemical and EC performance of the fabricated device. With an operating voltage window of 1.4 V and areal capacitance of $C_a = 28.3$ mF cm^{-2}, the EACSs showed a high surface energy density of $E_d = 7.7 \times 10^{-3}$ mWh cm^{-2}. Also, the devices exhibited a reversible color transition between light green (uncharged state) and dark blue (charged state), with an optical transmittance change between $\Delta T_{633\,nm} = 55\%$ and 12%. Hence, the energy storage level of the EASC is directly related to its color and can be determined by the naked eye, as can be seen in Figure 3.9, which means it can be incorporated with other energy applications to visually display their energy storage level.

Flexible ESCs are also currently under considerable investigation as potential smart energy storage components in wearable intelligent electronics. An optical-energy feedback based on EC contrast as a quantitative indicator of the SOC in flexible ESCs was developed by Jiao et al. [35]. The device used a polypyrrole (PPy)-based EC electrode with carbon nanotubes (CNTs)/Au/PPy/poly(ethylene terephthalate) (PET). A linear dependence of real-time EC contrast with the SOC was measured, enabling a quantitative visualization of the energy level. Benefiting from the inherent high color-to-color contrast in PPy during electrochromism, it shows a high detection sensitivity and resolution during charge/discharge. Moreover, the good flexibility, high capacitance of the electrode, and asymmetric device design synergistically benefited the energy storage performance of the flexible ESC, resulting in a high surface energy density of 4.03 μWh cm^{-2} with electrochemical stability. Devices showed a 10% change in the switching range after 2000 deformations at a bending radius of 1.25 cm [35]. More polymeric EC devices are discussed in Chapter 4.

In another study by Wang and associates [36], zinc-ion-intercalation-based multicolor EC energy storage devices (EESDs) were fabricated and studied. The EESD was based on a free-standing Zn^{2+}-based polymeric electrolyte membrane (ZPEM) and a nanocrystal-in-glass V_2O_5 thin film construction. The device showed reversible multi-EC colors of greenish-blue, greenish-yellow, yellowish-green, faint-yellow, yellowish-orange, and reddish-orange colors. These colors were observed in both flexible V_2O_5 thin films and flexible ZPEM/V_2O_5/indium tin oxide (ITO) EESDs. The flexible multicolor EESD exhibited combined electrochromism and energy storage performance with a transmittance modulation of about $\Delta T = 57.9\%$, with good coloration efficiency CE = 36.91 cm^2 C^{-1}, and a high

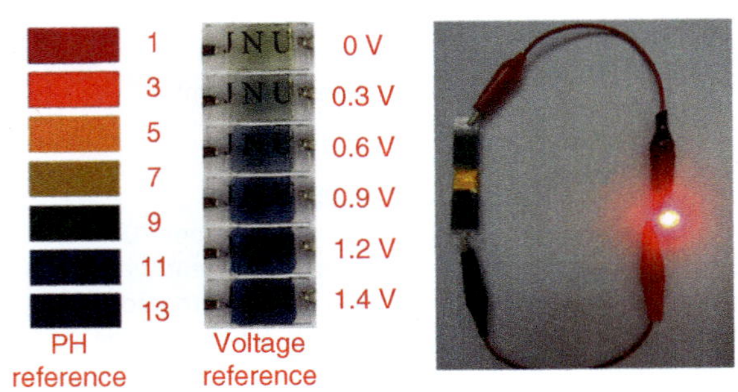

Figure 3.9 A solid-state electrochromic asymmetric supercapacitor (EASC) window shows different colors, representing potential levels and corresponding pH reference colors and numbers. *Source:* Reprinted with permission from Zhong et al. [34]. © 2017 / American Chemical Society.

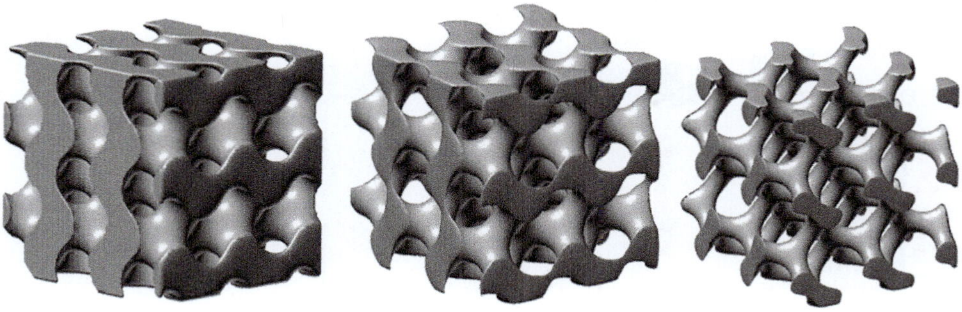

Figure 3.10 Example Gyroid structures. Shown are structures with threshold values $t = 0$, $t = 0.5$, $t = 1$ (left to right). *Source:* Nsiempba et al. [38] / MDPI / CC BY 4.0.

areal capacitance of 51 μF cm^2. In the wearable electronics field, new flexible materials are being developed for flexible circuits that are processable at fairly high temperatures. It is expected that these materials will surpass polyimides and thermoplastic polyurethanes in their properties. This development may offer some new ultra-pliable substrate materials for flexible EC devices.

In a study by Teowee et al., a bicontinuous double-gyroid vanadium pentoxide network was used in an ESC [37]. A freestanding vanadia network was fabricated by electrodeposition into a voided block copolymer template. The template was made by self-assembly in a double-gyroid morphology. Supercapacitors made from such gyroid-structured vanadium oxide electrodes exhibited a high specific capacitance of 155 F g^{-1}. The ESC showed a strong EC color change from green/gray to yellow, indicating the capacitor's charge condition. Examples of gyroid structures are shown in Figure 3.10 [38].

3.4 Integration of PV and ECD–Photoelectrochromic Device

Unlike conventional EC smart windows, which require an external power input (voltage or current) for producing an optical change, PECDs are self-powered and do not require any external electrical stimulus for optical modulation. In some cases, external power is used to reverse the optical change, which is generally done in order to achieve the optical change within a reasonable time period. A PECD typically comprises a PV film or a photoanode (such as an N719 dye or quantum dot [QD] sensitized TiO_2 film), a redox couple-based electrolyte (e.g. I_3^-/I^-), and an EC–CE (e.g. WO_3) [39–41]. The working principle is as follows: upon illumination, the photosensitizer (QD or dye) undergoes electron-hole separation, and the excited electrons are injected from the conduction band (CB) of the dye or QD to the CB of TiO_2. Via the external circuit, these electrons reach the EC–CE and bring about the reduction of the EC film (for example, if it is a WO_3 film, it turns blue). Some of these electrons are consumed in the reduction of oxidized electrolyte species (like I_3^-), and the reduced species (like I^-) then diffuses to the photoanode, where it undergoes oxidation and regenerates the dye or QD. All these processes occur spontaneously and require no other stimulus other than light. The PECD thus colors, and remains in the colored state, as long as solar radiation impinges upon the device. Bleaching is achieved in dark under short-circuit conditions or in light under open-circuit conditions. See Figure 3.11.

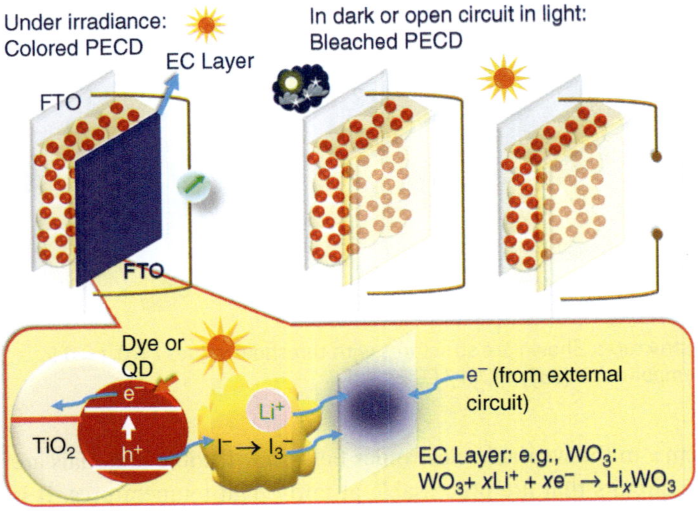

Figure 3.11 Schematic showing the working mechanism of a self-powered photoelectrochromic (PECD). FTO is fluorine-doped tin oxide and QD is quantum dot.

In pioneering works carried out independently by two separate groups simultaneously, Teowee et al. [37, 42] and Bechinger et al. [39, 41] reported PECD configurations. In Bechinger's configuration, a Ru-dye-sensitized nanocrystalline TiO_2 photoanode and WO_3 CE were used as the two primary electrodes. A LiI/4-*tert*-butylpyridine/propylene carbonate electrolyte was sandwiched in between the two electrodes. The resulting PECD showed a transmission modulation of $\Delta T_{788\,nm} = 60\%$ and produced a V_{OC} of 0.6–0.9 V under 75 mW cm^{-2} light irradiance for one minute. The switching kinetics for the PECD are provided in Figure 3.12 [41]. Thus, the device functioned as a self-powered variable transmittance window, where the optical density of the window is controlled by the input light

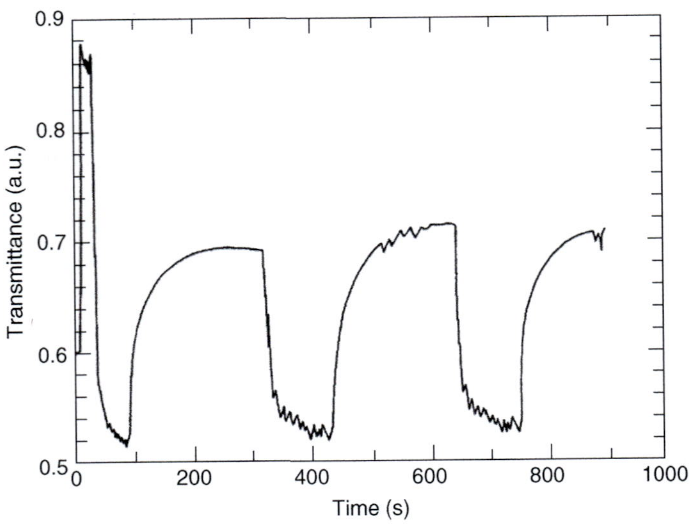

Figure 3.12 Color-bleach switching of a TiO_2: Ru/Li electrolyte/WO_3 PECD under light on-off cycles, $\Delta T_{788\,nm} = 60\%$. *Source:* Adopted from Bechinger et al. [41] / U.S. Department of Energy.

flux. In Teowee's configuration, a colloidal titania solution was synthesized and coated onto a SnO_2:F (TEC10) substrate. This coating was haze-free. The coated electrode was combined with a wet-chemically deposited tungsten oxide coating in a device to make optically clear (haze-free) devices). The best devices changed from $\Delta T_{550\,nm} = 70\%$ transmittance when exposed to sunlight (self-powered) to $\Delta T_{550\,nm} = 3\%$ with the electrodes shorted.

Also, in a study by Syrrokostas et al., they made a user-controllable tandem photochromic device with TiO_2/Ru dye as the photoanode and a WO_3 film as the EC CE [40]. The device showed a transmission modulation of $\Delta T_{vis} \sim 59\%$ after 10 minutes. of applied potential. A carbon-based perovskite solar cell (PSC) "partly covered" the PECD. The device had a three-electrode architecture that decouples the PV and EC operations, thus enabling better control of the device's optical modulation. The PSC unit of the device consisted of FTO/ZnO nanowire/ZrO_2/Carbon stack infiltrated with the methylammonium lead iodide ($MAPbI_3$) perovskite, covering a mere 4% of the device area (see Figure 3.13). The remaining 86% of the device area consists of a redox-type EC in the form of FTO/WO_3/liquid iodine redox electrolyte/Pt/FTO sandwich. The two units, PV and EC, shared a common anode but had different cathodes. The fabricated devices ($3 \times 4\,cm^2$) exhibited fast and reversible transmittance modulation of $\Delta T_{lum} = 49 - 5\%$ (within two minutes) and charge storage of $Q = 20\,mC\,cm^{-2}$, with the PSC having a power conversion efficiency of about 3%. The cells withstood more than 40 continuous coloration-bleaching cycles.

PECDs fabricated with graphene nanoplatelets CE and copper-based redox electrolytes with dye/TiO_2 photoanode exhibited good performance, combining a very fast coloration speed $t_c = 12$ seconds, $t_b = 11$ seconds, with a transmission change of $T_{530} = 67.1 - 5.5\%$. The properties of the PV device were $V_{oc} = 0.83\,V$, $j_{sc} = 6.41\,mA\,cm^{-2}$ with FF = 44%, and PCE = 2.43%. After 400 days (2000 cycles) the PECD showed small changes. The transmission range of switching was $T_{530} = 62.1 - 5.7\%$ after cycling. After cycling, PV values were: $V_{oc} = 0.83\,V$, $j_{sc} = 6.1\,mA\,cm^{-2}$ with FF = 42%, and PCE = 2.13% [43].

In work by Kolay et al. [44], they used a novel PECD architecture. The device coupled a C-fabric/NiO/In_2S_3 photocathode to a TiO_2/CdS/Au/MoO_3 photoanode. The device maximized light absorption by using In_2S_3 nanoflakes on one side and CdS-Au nanoparticles (NPs) on the other. The PCE achieved was 8% for the TiO_2/CdS/Au/MoO_3-nS^{2-}/S_n^{2-}-In_2S_3/NiO/C-fabric tandem device, measured under an air mass of 1.5, light intensity of

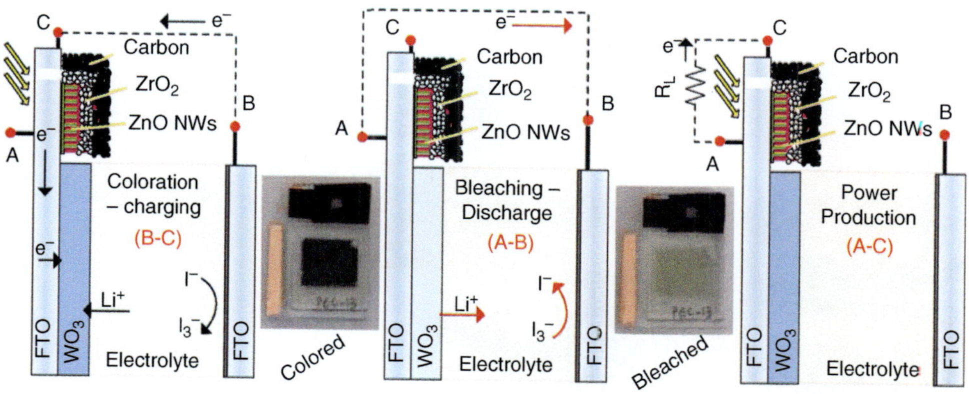

Figure 3.13 PECD tandem schematics and photographs (colored and bleached states). The devices were $3 \times 4\,cm^2$ in size. FTO is fluorine-doped tin oxide and NW are nanowires. *Source:* Reproduced with permission from Syrrokostas et al. [40], © 2020 / ELSEVIER.

100 mW cm^{-2}. The maximum modulation of the transmission was $\Delta T_{550\,nm}$ = 34% upon switching between the bleached state and blue-colored state. The coloration was from the MoO_3 film for the tandem device with optically transparent current collectors (FTO glass), and the photocoloration efficiency (measure of the coloration kinetics), $PhCE_{550\,nm}$ = 16.4 cm^2 min^{-1} W^{-1}.

$$PhCE_\lambda = \Delta OD/I\,t \quad (cm^2\,min^{-1}\,W^{-1})$$

where ΔOD is change in optical density, $\Delta OD = \log(T_b/T_c)$, t = exposure time (minutes), I = incident radiation (W m^{-2}), T_b is bleached state transmittance, and T_c is the colored state transmittance.

In another study, Kolay et al. [45] reported an EC Ag/V_2O_5 electrode exhibiting multi-color switching with good transmittance modulation that was sandwiched with a visible-NIR light-absorbing $TiO_2/CdS/PbS$ photoanode to make a PECD. The direct contact of the plasmonic and conductive metal nanostructures (AgNW) with the layered transition metal oxide V_2O_5 promoted both electron injection and Na^+ ion insertion for coloration. For the PECD, a $\Delta T_{550\,nm}$ = 22.3% was observed with a CE_{655nm} = 45.8 cm^2 C^{-1}, whereas the corresponding value for only V_2O_5 was $\Delta T_{550\,nm}$ = 17.1%, along with a lower $CE_{655\,nm}$ = 30.2 cm^2 C^{-1}. The solar cell efficiency for the PECD filled with the liquid redox electrolyte ($TiO_2/CdS/PbS\text{-}S^{2-}\text{–}Ag/V_2O_5$) was found to PCE = 5.2% measured under a light intensity of 1 sun, which is a substantial improvement over PCE = 3.1% attained with a CE devoid of AgNW and 3.7% in the absence of the PbS QDs co-sensitization.

Reviews by Cannavale et al. [46] and Syrrokostas et al. [47] also give accounts of the evolution and progress in the development of PECDs. Besides, PECDs, photovoltachromic, (PVCD) windows have also been developed by using a bifunctional CE consisting of two physically separated platinum and WO_3 coatings that were juxtaposed in a comb-like pattern (shown in Figure 3.14). When the cell was illuminated, the photovoltage drove electrons into the EC stripes through the PEC circuit and promoted the Li^+ diffusion toward the WO_3 film, which colored and yielded a photocoloration efficiency of $PhCE_{650\,nm}$ = 17 cm^2 min^{-1} W^{-1} under 1 sun and fast kinetics response (coloration time t_c < 2 seconds and bleaching time t_b < 5 seconds). An improved PV performance was made possible due to the independently switchable micro-patterned platinum electrode [48].

Figure 3.14 Photographs of a photovoltachromic device PVCD in the bleached (left) and colored (right) states. *Source:* Reproduced with permission from Cannavale et al. [48], © 2014 / American Chemical Society.

While many novel device architectures have been developed and studied in detail for dual-function devices such as ESCs and PECDs over the last few years, their performance and scale-up still require significant improvements before they can be commercialized. Most of these ESCs, PECDs, and PVCDs have been developed in lab-scale sizes under 25 cm^2. Once perfected and stabilized, they may have usefulness in small-area devices such as self-powering display devices. But for building facades, scaling up to large areas in several square meters is a formidable challenge. Also, long-term durability greater than 10 years is required by industry. Certainly not to be underestimated is the cost of manufacturing the product. Compatibility with existing or modifiable production equipment is a plus for the industry. Energy storage in a window is also very attractive for integration with buildings (for ESCs), but still poses a formidable challenge. Economically attractive manufacturing where compatible processes and materials can be utilized is important.

References

1 Rowley, N.M. and Mortimer, R.J. (2002). New electrochromic materials. *Science Progress* 85: 243–262. https://doi.org/10.3184/003685002783238816.

2 Byker, H.J. (1990). Single-Compartment, Self-Erasing, Solution-Phase Electrochromic Devices, Solutions for use therein, and uses thereof, Gentex Corporation, US Patent 4,902,108A.

3 Baumann, K.L., Guarr, T.F., Giri, P., Muyskens, M.D., and Franz, S.F. (2008). Electrochromic Compounds and Associated Media and Devices, Gentex Corporation, US Patent 7,428,091B2.

4 Varaprasad, D.V., Zhao, M., Dornan, C.A., Agrawal, A., Allemand, P-M., and Lynam, N.R. (1999). Electrochromic Polymeric Solid Films, Manufacturing Electrochromic Devices Using Such Solid Films, and Processes for Making Such Solid Films and Devices, Magna Donnelly Corporation, US Patent 5,910,854A.

5 Kloeppner, L.J., Guarr, T. F., Ash, K. L., and Roberts, K.E. (2003). Electrochromic Medium Having a Self-Healing Cross-Linked Polymer Gel and Associated Electrochromic Device, Gentex Corporation, US Patent 6,635,194B2.

6 Varaprasad, D.V., Lynam, N.R., Habibi, H.R., and Desaraju, P. (1992). High Performance Electrochemichromic Solutions and Devices Thereof, Magna Donnelly Corporation, US Patent 5,140,455A.

7 Varaprasad, D.V., McCabe, I.A., Habibi, H., Lynam, N.R., Zhao, M., and Dornan, C.A. (1998). Electrochromic Mirrors and Devices, Magna Donnelly Corporation, US Patent 5,724,187A.

8 Tonar, W.L. (2005). Electrochromic Device Having a Seal Including an Epoxy Resin Cured with a Cycloaliphatic Amine, Gentex Corporation, US Patent 6,963,439B2.

9 Tonar, W.L., Stray, J.A., Veenman, S.J., Kloeppner, L.J., Poll, D.L., Dozeman, G.J., and Menon, N.C. (2011). Electrochromic Device Having an Improved Fill Port Plug, Gentex Corporation, US Patent 7,884,995B2.

10 Tonar, W.L., Cammenga, D.J., Poll, D.L., Kar, K.B., and Busscher, B.L. (2006). Electrochromic device having no positional offset between substrates, US Patent 7,064,882 B2.

11 Tonar, W.L. and Foote, D.G. (2009). Vehicular Rearview Mirror Elements and Assemblies Incorporating These Elements, Gentex Corporation, US Patent 7,570,413B2.

12 McCabe, I.A., Habibi, H., Varaprasad, D.V., and Lynam, N.R. (2009). Interior Rearview Mirror System for a Vehicle, Magna Donnelly Corporation, US Patent 7,586,666B2.

13 Bechtel, J.H., and Byker, H.J. (1990). Automatic Rearview Mirror System for Automotive Vehicles, Gentex Corporation, US Patent 4,917,477A.

14 Driscoll, D.I., Guarr, T.F., Ash, K.L., Theiste, D.A., and Luten, H.A. (2014). Variable Transmission Electrochromic Window and Associated Aircraft Window System, Gentex Corporation, US patent 8,736,943B2.

15 Driscoll, D. (2017). Electronically dimmable aircraft windows. https://www.mobilityengineeringtech.com/component/content/article/adt/pub/features/articles/27502.

16 Alesanco, Y., Viñuales, A., Palenzuela, J., Odriozola, I., Cabañero, G., Rodriguez, J., and Tena-Zaera, R. (2016). Multicolor electrochromics: rainbow-like devices. *ACS Applied Materials & Interfaces* 8: 14795–14801. https://doi.org/10.1021/acsami.6b01911.

17 Oh, H., Seo, D.G., Yun, T.Y., Kim, C.Y., and Moon, H.C. (2017). Voltage-tunable multi-color, sub-1.5 v, flexible electrochromic devices based on ion gels. *ACS Applied Materials & Interfaces* 9: 7658–7665. https://doi.org/10.1021/acsami.7b00624.

18 Kavanagh, A., Fraser, K.J., Byrne, R., and Diamond, D. (2013). An electrochromic ionic liquid: design, characterization, and performance in a solid-state platform. *ACS Applied Materials & Interfaces* 5: 55–62. https://doi.org/10.1021/am3018948.

19 Abraham, S., Mangalath, S., Sasikumar, D., and Joseph, J. (2017). Transmissive-to-black electrochromic devices based on cross-linkable tetraphenylethene-diphenylamine derivatives. *Chemistry of Materials* 29: 9877–9881. https://doi.org/10.1021/acs.chemmater.7b03319.

20 Corrente, G.A., Fabiano, E., Manni, F., Chidichimo, G., Gigli, G., Beneduci, A., and Capodilupo, A.-L. (2018). Colorless to all-black full-NIR high-contrast switching in solid electrochromic films prepared with organic mixed valence systems based on dibenzo-fulvene derivatives. *Chemistry of Materials* 30: 5610–5620. https://doi.org/10.1021/acs.chemmater.8b01665.

21 Abraham, S., Ganesh, G.P.T., Varghese, S., Deb, B., and Joseph, J. (2015). Cross-linkable fluorene-diphenylamine derivatives for electrochromic applications. *ACS Applied Materials & Interfaces* 7: 25424–25433. https://doi.org/10.1021/acsami.5b08218.

22 Kao, S.-Y., Lu, H.-C., Kung, C.-W., Chen, H.-W., Chang, T-H., and Ho, K.-C. (2016). Thermally cured dual functional viologen-based all-in-one electrochromic devices with panchromatic modulation. *ACS Applied Materials & Interfaces* 8: 4175–4184. https://doi.org/10.1021/acsami.5b11947.

23 Gélinas, B., Das, D., and Rochefort, D. (2017). Air-stable, self-bleaching electrochromic device based on viologen- and ferrocene-containing triflimide redox ionic liquids. *ACS Applied Materials & Interfaces* 9: 28726–28736. https://doi.org/10.1021/acsami.7b04427.

24 Kim, S., Shim, N., Lee, H., and Moon, B. (2012). Synthesis of a perylenediimide-viologen dyad (PDI-2V) and its electrochromism in a layer-by-layer self-assembled multilayer film with PEDOT:PSS. *Journal of Materials Chemistry* 22: 13558–13563. https://doi.org/10.1039/C2JM31407J.

25 Drewniak, A., Tomczyk, M.D., Knop, K., Walczak, K.Z., and Ledwon, P. (2019). Multiple redox states and multielectrochromism of donor–acceptor conjugated polymers with aromatic diimide pendant groups. *Macromolecules* 52: 8453–8465. https://doi.org/10.1021/acs.macromol.9b01069.

26 Kumar, A., Williams, T.L., Martin, C.A., Figueroa-Navedo, A.M., and Deravi, L.F. (2018). Xanthommatin-based electrochromic displays inspired by nature. *ACS Applied Materials & Interfaces* 10: 43177–43183. https://doi.org/10.1021/acsami.8b14123.

27 Agrawal, A. and Tonazzi, J.C.L. (2013). Composite electrode and electrolytes comprising nanoparticles and resulting devices, Ajjer LLC, US Patent 8593714B2.

28 Luten, H.A. (2023). Electro-Optic Element Electrode. US Patent Application 20230266632.

29 Giri, P., Kloeppner, L., and Baumann, K. (2023). Electrochromic Compounds for Memory Electrochromic Devices, Gentex Corporation, US Patent Application 20230266630A1.

30 Wei, D., Scherer, M.R.J., Bower, C., Andrew, P., Ryhänen, T., and Steiner, U. (2012). A nanostructured electrochromic supercapacitor. *Nano Letters* 12 (4): 1857–1862. https://doi.org/10.1021/nl2042112.

31 Reddy, B.N., Kumar, P.N., and Deepa, M. (2015). A poly(3,4-ethylenedioxypyrrole)–Au@WO_3-based electrochromic pseudocapacitor. *ChemPhysChem* 16: 377–389. https://doi.org/10.1002/cphc.201402625.

32 Yun, T.Y., Li, X., Kim, S.H., and Moon, H.C. (2018). Dual-function electrochromic supercapacitors displaying real-time capacity in color. *ACS Applied Materials & Interfaces* 10 (50): 43993–43999. https://doi.org/10.1021/acsami.8b15066.

33 Murto, P., Elmas, S., Méndez-Romero, U.A., Yin, Y., Genene, Z., Mone, M., Andersson, G.G., Andersson, M.R., and Wang, E. (2020). Highly stable indacenodithieno [3,2-*b*]thiophene-based donor–acceptor copolymers for hybrid electrochromic and energy storage applications. *Macromolecules* 53 (24): 11106–11119. https://doi.org/10.1021/acs.macromol.0c02212.

34 Zhong, Y., Chai, Z., Liang, Z., Sun, P., Xie, W., Zhao, C., and Mai, W. (2017). Electrochromic asymmetric supercapacitor windows enable direct determination of energy status by the naked eye. *ACS Applied Materials & Interfaces* 9 (39): 34085–34092. https://doi.org/10.1021/acsami.7b10334.

35 Jiao, X., Li, G., Yuan, Z., and Zhang, C. (2021). High-performance flexible electrochromic supercapacitor with a capability of quantitative visualization of its energy storage status through electrochromic contrast. *ACS Applied Energy Materials* 4 (12): 14155–14168. https://doi.org/10.1021/acsaem.1c02927.

36 Wang, C., Zhang, X., Liu, S., Zhang, H., Wang, Q., Zhang, C., Gao, J., Liang, L., and Cao, H. (2022). Interfacial charge transfer and zinc ion intercalation and deintercalation dynamics in flexible multicolor electrochromic energy storage devices. *ACS Applied Energy Materials* 5 (1): 88–97. https://doi.org/10.1021/acsaem.1c02508.

37 Teowee, G., Gudgel, T., McCarthy, K., Agrawal, A., Allemand, P., and Cronin, J. (1999). User controllable photochromic (UCPC) devices. *Electrochimica Acta* 44: 3017–3026. https://doi.org/10.1016/S0013-4686(99)00015-8.

38 Nsiempba, K.M., Wang, M., and Vlasea, M. (2021). Geometrical degrees of freedom for cellular structures generation: a new classification paradigm. *Applied Sciences* 11: 3845. https://doi.org/10.3390/app11093845.

39 Bechinger, C., Ferrere, S., Zaban, A., Sprague, J., and Gregg, B.A. (1996). Photoelectrochromic windows and displays. *Nature* 383: 608–610. https://doi.org/10.1038/383608a0.

40 Syrrokostas, G., Dokouzis, A., Yannopoulos, S.N., and Leftheriotis, G. (2020). Novel photoelectrochromic devices incorporating carbon-based perovskite solar cells. *Nano Energy* 77: 10524310. https://doi.org/1016/j.nanoen.2020.105243.

41 Bechinger, C.S. and Gregg, B.A. (2002). Self-bleaching Photoelectrochemical Electrochromic Device, US Patent 6,369,934 B1.

42 Teowee, G., Allemand, P.-M., and Cronin, J. P. (1997). Photochromic Devices, US Patent 5,604,626.

43 Lavagna, L., Syrrokostas, G., Fagiolari, L., Amici, J., Francia, C., Bodoardo, S., Leftheriotis, G., and Bella, F. (2021). Platinum-free photoelectrochromic devices working with copper-based electrolytes for ultrastable smart windows. *Journal of Materials Chemistry A* 9: 19687–19691. https://doi.org/10.1039/d1ta03544d.

44 Kolay, A., Potts, N.T.Z., Sardar, K., Gibson, E.A., and Deepa, M. (2019). A dual-function photoelectrochemical solar cell which assimilates light-harvesting, charge-transport and photoelectrochromic nanomaterials in a tandem design. *Sustainable Energy & Fuels* 3: 514–528. https://doi.org/10.1039/C8SE00548F.

45 Kolay, A., Maity, D., Flint, H., Gibson, E.A., and Deepa, M. (2022). Self-switching photo-electrochromic device with low cost, plasmonic and conducting Ag nanowires decorated V_2O_5 and PbS quantum dots. *Solar Energy Materials & Solar Cells* 239: 111674. https://doi.org/10.1016/j.solmat.2022.111674.

46 Cannavale, A., Cossari, P., Eperon, G.E., Colella, S., Fiorito, F., Gigli, G., Snaith, H.J., and Listorti, A. (2016). Forthcoming perspectives of photoelectrochromic devices: a critical review. *Energy & Environmental Science* 9: 2682–2719. https://doi.org/10.1039/c6ee01514j.

47 Syrrokostas, G., Leftheriotis, G., and Yannopoulos, S.N. (2022). Lessons learned from 25 years of development of photoelectrochromic devices: a technical review. *Renewable and Sustainable Energy Reviews* 162: 112462. https://doi.org/10.1016/j.rser.2022.112462.

48 Cannavale, A., Manca, M., De Marco, L., Grisorio, R., Carallo, S., Suranna, G.P., and Gigli, G. (2014). Photovoltachromic device with a micropatterned bifunctional counter electrode. *ACS Applied Materials & Interfaces* 6: 2415–2422. https://doi.org/10.1021/am404800m.

4

Polymeric Electrochromics

Marco Schott and Uwe Posset

Fraunhofer Institute for Silicate Research ISC, Würzburg, Germany

Nomenclature

BHJ	bulk-heterojunction (solar cell)
CMY	cyan–magenta–yellow
CNT	carbon nanotubes
EC	electrochromic
ECP	electrochromic polymer
ECP-Magenta	poly((2,2-bis(2-ethylhexyloxymethyl)-propylene-1,3-dioxy)-3,4-thiophene-2,5-diyl), PProDOT(CH_2OEtHx)$_2$
EDOT	3,4-ethylenedioxythiophene
Fe-MCP	iron metal coordination polymer
FTO	fluorine-doped tin oxide
HOMO	highest occupied molecular orbital
IGU	insulating glass units
IR	infrared
ITO	tin-doped indium oxide
LC	liquid crystals
LED	light-emitting diode
LiTFSI	lithium bis(trifluoromethanesulfonyl)imide
LUMO	lowest unoccupied molecular orbital
MCCP	minimally color-changing polymer
MCP	metal coordination polymer
MLCT	metal-to-ligand charge transfer
NIR	near-infrared
NFC	near-field communication
oCVD	oxidative chemical vapor deposition
OLED	organic light-emitting diodes
OPV	organic photovoltaic
PANI	polyaniline
PANI:PSS	polyaniline:poly(styrene) sulfonate
PB	Prussian blue
PDLC	polymer-dispersed liquid crystals

Chromogenics: Smart Switchable Optical Materials and Their Applications, First Edition. Edited by Carl M. Lampert.

PEDOT	poly(3,4-ethylenedioxythiophene)
PEDOT:PSS	poly(3,4-ethylenedioxythiophene):poly(styrene) sulfonate
PEDOT-EthC6	side-chain-modified poly(3,4-ethylenedioxythiophene)
P3HT	poly(3-hexylthiophene-2,5-diyl)
PET	polyethylene terephthalate
PProDOT-Me_2	poly(2,2-dimethyl-3,4-propylenedioxythiophene)
RGB	red–green–blue
SHGC	solar heat gain coefficient
SPD	suspended particle devices
$TiVO_x$	titanium vanadium oxide
TOLED	transparent organic light-emitting diodes
UV	ultraviolet
VPP	vapor-phase polymerization

4.1 Introduction

This chapter gives an overview of the most relevant research and developments toward industrial applications in the field of polymeric electrochromic (EC) materials. In scientific literature, several reports, book chapters, and review articles on the state-of-the-art of electrochromic polymers (ECPs) have been published. However, these contributions often have a more scientific focus and do not address the scale-up of the best performing materials and large-scale device assembly important for the transfer into industrially relevant processes and applications. EC devices containing inorganic materials (see Chapter 2 – Inorganic and Organic Electrochromics) have been available on the market for many years, in particular deposited on glass for smart windows in architectural applications. In recent years, the demand for alternative flexible EC devices based on polymeric materials is increasing, paving the way for further potential applications in architectural (retrofitting) or automotive glazing, aircraft, marine, eyewear [1], displays [2–4], smart labels and sensors [5, 6], adaptive camouflage [7, 8], etc. The main drivers for the development of polymeric-film-based EC systems are the reduction of production costs (still one of the most relevant obstacles to a wider market penetration of EC windows), light weight, low power consumption, high color variety, and accelerated response. The challenges to be overcome for establishing an industrial-ready all-solid-state polymeric EC cell are manifold, e.g., the assembly of the multi-layer EC cells, proper sealing/encapsulation and contacting as well as integration and testing in relevant environments. To date, most polymeric EC systems lack long-term durability under intense exposure to light and temperature. In this chapter, we focus on the most prominent ECP systems, highlighting their performance, processability, and advances in large-scale manufacturing. Remaining technological challenges and potential strategies for roll-to-roll production of polymeric EC devices are addressed with a view to future industrial applications.

4.2 Electrochromic Polymers

4.2.1 Conjugated Polymers

An important and widely studied class of ECPs reported on in the last decades are conjugated polymers, which offer a large variety of colors [9–15]. The most well-known electroactive conjugated conducting polymers are polyacetylene, polyphenylene, polyaniline, polypyrrole, polyfuran, polycarbazole, polythiophene, and their derivatives [16]. Due to the unique optical and electrochemical properties of conjugated polymers, they can be applied in energy storage devices (supercapacitors, lithium-ion batteries) [17] and various optoelectronic devices, e.g., sensors, transistors, organic photovoltaic (OPV) cells, organic light-emitting diodes (OLEDs), and electrochromic (EC) devices [18–20]. All of these applications benefit from the advantages of polymeric materials, such as low weight, mechanical flexibility and compatibility with large area, cost-effective manufacturing processes. The characteristic EC properties of conjugated polymers result from an extended conjugated π-electron system creating a band-like electronic structure. The band gap and electronic transitions (π–π^* transitions) between the highest occupied molecular orbital (HOMO, valence band) and the lowest unoccupied molecular orbital (LUMO, conduction band) determine the vivid colors. Thus, most conjugated polymers can be adapted in terms of their optical properties by tailoring the structure through synthetic modification. The mechanism for the color change results from oxidation (p-doping) of the neutral (undoped) polymer. Upon doping, polaronic and bipolaronic states are formed that induce changes in the electronic structure and lower the band gap. As a result of the change in the band gap, the spectral absorption shifts to longer wavelengths. A shift within the visible region (380–780 nm) or into the near-infrared (NIR, 780–2500 nm) enables a transition between two colored states and between a colored and colorless state (cathodically coloring), respectively. Anodically coloring polymers are higher band gap materials (>3.0 eV) with a neutral state absorption in the ultraviolet (UV) that color upon oxidation due to a shift of the polaronic/bipolaronic absorption into the visible region of the electromagnetic spectrum [12].

There are many design principles to tailor the optoelectronic properties of conjugated polymers for electrochromics that cannot be covered here. The use of polyheterocycles [21] and the donor–acceptor approach [22, 23] are the most common ones, addressed in a large number of scientific reviews. Understanding the structure–property relationships is of high relevance for improving not only the optical and electrochemical properties, but also the solubility and processability of the materials. In principle, there are three main methods to obtain ECPs and thin films thereof: electrochemical polymerization [24, 25], in situ chemical oxidative polymerization [26, 27], and chemical oxidative polymerization (in bulk) [28]. Electrochemical polymerization is widely used for the lab-scale deposition of polymeric thin films on conductive substrates. However, homogeneous large-area coatings and the use of coating additives to improve the film forming properties are hardly achievable. In contrast, in situ chemical oxidative polymerization has been demonstrated to be a viable technique for fabricating large-area thin-film coatings with high optical quality and good EC properties on roll substrate material [29]. Chemical oxidative bulk polymerization can be used for the preparation of EC inks that are deposited directly in the form of thin films [30, 31].

Since electrochemical polymerization is not suitable for large-area and high-throughput manufacturing, in this chapter we focus on more industrially relevant methods for the fabrication of polymeric thin-film electrodes, i.e., in situ chemical oxidative polymerization and chemical oxidative polymerization. In general, both methods comprise three processing steps (Figure 4.1). The in situ chemical oxidative polymerization requires the deposition of a reactive mixture of the monomer and an oxidizer under controlled conditions. The polymer is formed directly on the substrate followed by a thermal treatment and post-treatment step including the rinsing of the oxidizer and, if appropriate, a further pre-conditioning step after the coating process in order to obtain the undoped (reduced) state of the ECP. The

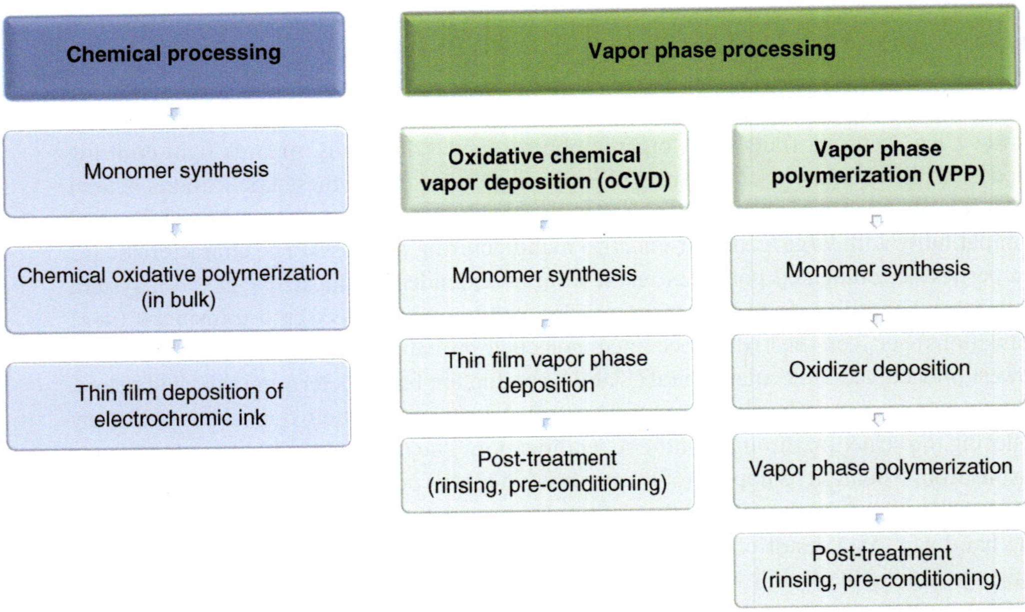

Figure 4.1 Comparison of industrially relevant processing methods for the fabrication of polymeric thin-film electrodes. In situ chemical oxidative polymerization, chemical oxidative polymerization, oxidative chemical vapor deposition (oCVD), and vapor-phase polymerization (VPP).

reproducibility and homogeneity of the polymeric thin films can be optimized by adjusting the coating parameters and employing climate control. The rinsing and pre-conditioning can be accomplished in one process step to save time, resources, and costs. In contrast, during the chemical oxidative polymerization the polymer is formed initially in solution, and then the EC ink is deposited on the substrate. For large-area EC applications, scalable and cost-effective coating methods with reproducible control over thickness and, in particular, homogeneity are of immense importance.

A group of vapor-phase processes is also interesting in this context, namely oxidative chemical vapor deposition (oCVD) [32] and vapor-phase polymerization (VPP) [33]. The processes, which have only been known for roughly a decade, are not yet being used on a large scale or industrially for the production of chromogenic devices, but have great potential to lead to a breakthrough in this area. They precisely provide conformal coatings on complex, even porous substrates and prevent substrate–solvent incompatibility. oCVD is a one-step purely gas-phase process typically yielding electrically conductive or semiconductive films grown via a step-growth polymerization mechanism. The scalability of the oCVD method was demonstrated by fabricating poly(3,4-ethylenedioxythiophene) (PEDOT) thin films on large-area flexible substrates using roll-to-roll deposition [34]. The VPP process still involves a liquid step, but enables the fabrication of highly electronically conductive PEDOT with conductivity exceeding 1000 S cm^{-1} [35]. Figure 4.2 shows an automotive EC rear-view mirror prototype comprising a VPP-processed PEDOT electrode.

In this section, examples of the most promising conjugated polymers that are processable on a large scale by wet chemical deposition, such as roll-to-roll slot-die/microgravure coating [37, 38] or printing techniques (e.g., inkjet [39, 40] or screen printing [41]), are shown. Slot-die and microgravure are commonly used for low-viscosity EC inks (<50 mPa s). In contrast to printing methods, which allow additive processing and electrode patterning, coating

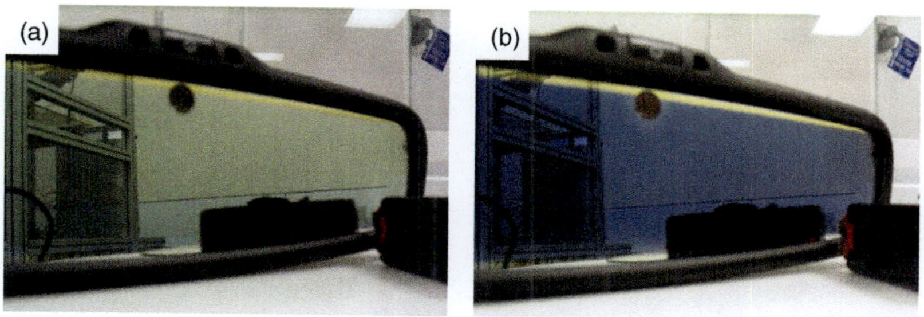

Figure 4.2 Prototype EC device configured as a rear-view mirror in: (a) bleached state and (b) darkened state. The color change is performed by a vapor-phase polymerization (VPP)-processed poly(3,4-ethylenedioxythiophene (PEDOT) electrode. *Source:* Reproduced with permission from [36]. © 2014 ELSEVIER.

techniques are large-area deposition methods without specific structuring options [42]. The ability to use environmentally friendly solvents is essential to making polymeric EC devices commercially viable by reducing environmental impact and cost, and increasing sustainability and safety in the coating and printing processes [43–45]. Collier et al. developed a design strategy to synthesize ECPs with multipolar functionalities that can be processed from environmentally benign solvents and are compatible with aqueous electrolytes [46].

A well-known and widely studied ECP is PEDOT. Its availability as an aqueous dispersion poly(3,4-ethylenedioxythiophene):poly(styrene) sulfonate (PEDOT:PSS) allows for many facile and inexpensive ways of solution processing [47]. PEDOT:PSS is commercially available from Heraeus (Clevios™). Due to its high electrical conductivity and transparency, it is also applied as an alternative material to commonly used transparent conducting oxides, e.g., for solar cells, displays, touch screens, sensors, and electrochromics [48, 49]. A drawback of most ECPs is the lack of a completely colorless or color-neutral clear state, which theoretically would result when the polaronic and bipolaronic absorption lies in the IR region and the bands do not show extensive tailing into the visible region. Very few polymers are characterized by this property. For example, in the oxidized state, PEDOT shows a light blue hue [50], which can be acceptable in some use cases, but usually not for smart window products. Only very thin PEDOT films appear almost colorless in the clear state. However, such thin films in EC devices on the other hand would result in insufficient darkening, which again appear counterproductive, at least for window applications.

The roll-to-roll processing of electrically conductive conjugated polymers was first explored in the context of the development of bulk heterojunction polymer solar cells, which since the early 2000s mostly used poly(3-hexylthiophene-2,5-diyl) (P3HT) as an electron donor polymer in the active layer. The hexyl groups of P3HT confer good solubility of the polymer in organic solvents. Krebs et al. 2009 revised the fabrication process for a bulk-heterojunction (BHJ) solar cell with inverted geometry. This work was performed by computer modeling to develop a new process for complete fabrication of roll-to-roll processed polymer solar cells, starting from a tin-doped indium-oxide (ITO)-coated polyethylene terephthalate (PET) substrate [51]. The aim of all these developments was to establish a process that would enable the transfer of laboratory polymer solar cells into a large-scale industrial polymer solar cell product. Krebs' working group continued the development for several years and pursued new ideas around upscaling the fabrication of polymer solar cells with a focus on cost analysis and practical fabrication approaches [52–55]. Although this work was exclusively about polymer solar cells, it shows impressively that conductive polymers can be roll-to-roll processed into functional large-area optoelectronic devices under ambient conditions. P3HT is an ECP

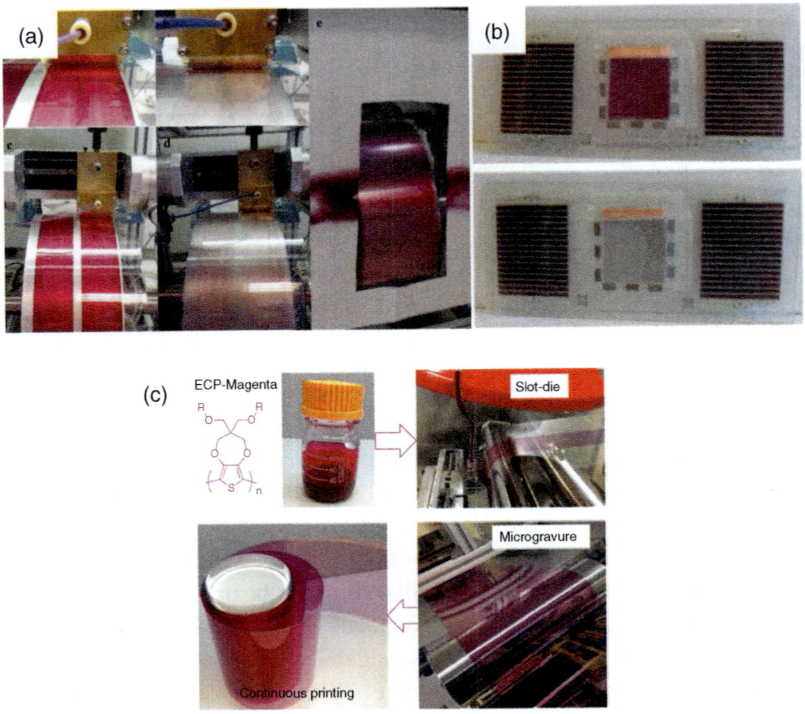

Figure 4.3 (a) Pictures of slot-die and spray coating of electrochromic polymers (ECPs) on a single roll system. (b) Integrated ECP/OPV (organic photovoltaic) demonstrator. (c) Fabrication of ECP-Magenta thin films on flexible substrates using slot-die or microgravure coating. *Source:* (a), (b) Reproduced with permission from [41]. © 2013 John Wiley & Sons. (c) Reproduced with permission from [56]. © 2019 The Royal Society of Chemistry.

with an interesting color change from red-orange to blue, but was never considered for any application other than smart textiles. This is attributed to the low acceptance for reddish and yellowish hues in smart windows because they are commonly associated with yellowing and degradation. However, the use of P3HT is conceivable for textiles and decorative applications as well as warning devices and indicators.

A remarkable development based on the above work is the first demonstration of a magenta-to-colorless EC device prepared on ITO/PET by roll-to-roll slot-die coating (or spray coating) by Jensen et al. (Figure 4.3a). The device consists of the electrochromic polymer poly((2,2-bis(2-ethylhexyloxymethyl)-propylene-1,3-dioxy)-3,4-thiophene-2,5-diyl) (PProDOT(CH_2OEtHx)$_2$, ECP-Magenta) [57] as the vibrantly colored active material and the minimally color-changing polymer (MCCP) poly(*N*-octadecyl-(propylene-1,3-dioxy)-3,4-pyrrole-2,5-diyl) as the charge storage layer. Two electrolyte systems were tested that allow the fabrication of fully printable and laminated EC cells [58]. The EC devices with a size of 10 cm^2 exhibit a transmittance change from approx. 10% (colored) to 54% (bleached) and a switching speed of <10 s at ±2 V.

Due to the low-energy requirements of polymeric EC cells (<100 μA cm^{-2} maximum current density), they can be combined with printed organic photovoltaic devices, allowing the fabrication of a self-powered EC/OPV module (Figure 4.3b). The devices were not encapsulated to avoid the ingress of water and oxygen. Thus, a loss of performance was observed during light exposure. The small-scale approach was scaled up to a full roll-to-roll process of ITO-free EC devices (ECP-Magenta/MCCP, 18 × 18 cm^2) printed directly onto PET using flexographic printing of highly conductive silver grids and slot-die deposition of the ECPs [59]. Recently, Li et al. also showed the roll-to-roll processing of high-quality ECP-Magenta thin films on flexible substrates by slot-die and microgravure coating (Figure 4.3c).

Magenta-to-clear switchable polymeric EC devices based on transparent wood substrates and solution-processed PEDOT:PSS as the transparent conducting layer were demonstrated by Lang et al. using ECP-Magenta. Due to their high strength and toughness, low thermal conductivity, and diffuse transmittance, transparent wood composites offer a new route to replace glass for diffusely transmitting windows [60].

In order to be able to process conjugated polymers wet-chemically on large scale by means of continuous methods, the precursor material must be dissolved or suspended in a solvent. Consequently, most efforts have focused on improving the solubility of the polymers (e.g., by introducing bulky or polar side chains) or controlling their dispersibility by using adjuvants such as dispersants or polyelectrolytes. However, both may affect the electrooptical properties and have further implications such as a very complex multi-step synthesis. The relatively new, alternative in situ chemical polymerization process (see above) also enables access to polymers that are difficult or impossible to produce using conventional methods. An interesting example in this context is a material with unique EC properties, i.e., the side-chain-modified PEDOT derivative (PEDOT-EthC6) [37]. This polymer can be deposited as thin layer on ITO/PET or other transparent conducting films using roll-to-roll slot-die coating. After deposition of the coating solution (polymerization mixture), which contains the EDOT-EthC6 monomers dissolved in butanol, an oxidizing agent and a moderator base, the polymer is formed directly on the substrate resulting in a yellowish colored wet film. The film thickness can be adjusted by the web speed and the delivery volume of the coating solution. After coating and drying, the thin-film electrodes have to be rinsed and pre-conditioned in a second process in order to obtain the deep blue colored undoped (reduced) state of PEDOT-EthC6. Rinsing and pre-conditioning can be implemented in one process step to save time, resources, and costs. High-throughput and large-area processing was demonstrated on a customized roll-to-roll pilot line with a coating width up to 500 mm (Figure 4.4a–d).

Compared to PEDOT:PSS or electrochemically polymerized PEDOT and PEDOT derivatives, the in situ chemically polymerized PEDOT-EthC6 films show improved EC properties in terms of visible light transmittance change (dark: 4% ↔ clear: 63%), contrast ratio (CR: 15.8), coloration efficiency ($\eta = 530\,\text{cm}^2\,\text{C}^{-1}$), color neutrality in the clear state ($L^* = 83.8$, $a^* = -4.3$, $b^* = -4.1$), and switching speed (approx. 45 s for 45 × 65 cm^2) [29]. A high charge retention over 10,000 switching cycles (>91%) was achieved under an inert atmosphere. As PEDOT:PSS and PEDOT-EthC6 are cathodically coloring, they can be combined with various anodically coloring (e.g., Prussian blue, PB) or optically inactive (non-electrochromic) ion storage electrodes in EC full cells. The device structure and EC characteristics will be discussed in Section 4.3.

Efforts to avoid the intermediate steps necessary in the in situ polymerization process and to reduce solvent consumption led to the development of an ECP ink. Bulk oxidative polymerization of the monomer EDOT-EthC6 enables efficient film preparation by via roll-to-roll slot-die coating or inkjet printing [31]. This novel approach combines the advantages of in situ polymerization and standard cross-coupling polymerization and is particularly useful for side-chain-modified EDOT derivatives that cannot be polymerized otherwise. Compared to in situ chemical polymerization with two process steps (coating and rinsing), this method

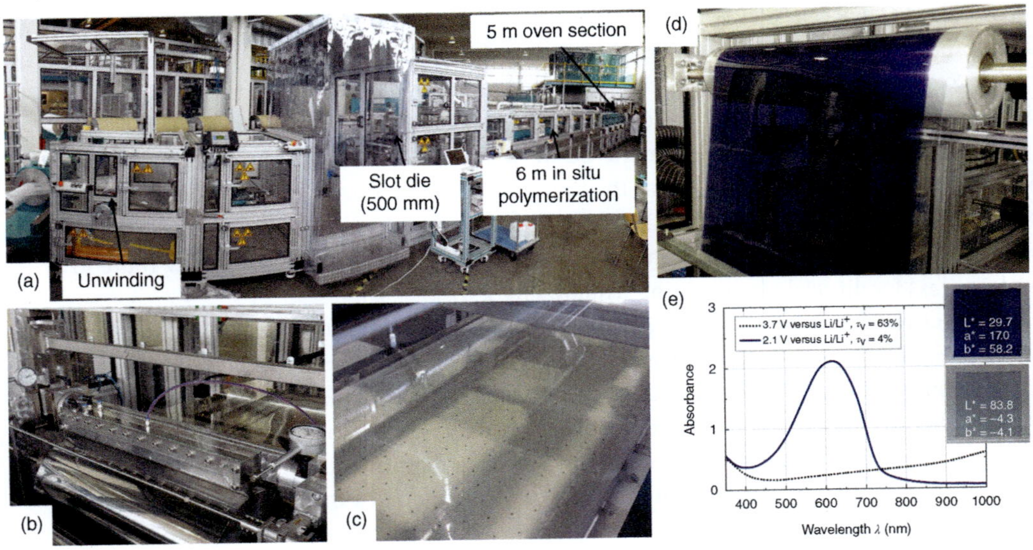

Figure 4.4 Large-scale deposition of side-chain-modified poly(3,4-ethylene dioxythiophene (PEDOT-EthC6) thin films on ITO/PET. (a) Modular roll-to-roll coating line with unwinding unit, slot die, in situ polymerization zone, and oven section. (b) Slot-die deposition. (c) In situ chemical polymerization showing a yellowish colored wet film. (d) Reproduced film after post-treatment. (f) EC switching behavior of the PEDOT-EthC6 film. *Source:* (a–c) Reproduced with permission from [37]. © 2020 John Wiley & Sons.

only requires one process step. The straightforward deposition, followed by a thermal treatment, directly yields homogeneous thin films with good EC performance.

The use of inkjet printing of ECPs is suitable for creating patterned EC devices. This is well illustrated in a report by Shim et al., who showed that printing composite dispersions of polyaniline- or PEDOT-covered silica nanoparticles onto ITO/PET enables the fabrication of EC devices with high resolution [40]. Utilizing photolithographic methods is another approach to create patterned EC surfaces with millimeter-sized dimensions [61] (see Figure 4.5).

There are further studies and reports, e.g., from Argun et al. [62, 63] or Ersman et al. [2, 64–66], demonstrating the preparation of lateral EC devices structured with various patterning methods (e.g., cross/line patterning or additive patterning using screen printing) (see Figure 4.6). PEDOT:PSS is mostly used as the color-changing EC material and/or the transparent conductive layer. Solution-processable ECPs that can be printed and patterned are suitable for applications such as information displays and smart labels [67].

Several studies have been performed on strongly desired "color-neutral" solutions, i.e., black-to-colorless ECPs and devices made therefrom [68–72]. However, these materials/systems often lack processability or durability. In general, there are two implementation strategies for neutral-tint solutions: the donor–acceptor approach and physical color mixing (according to the RGB and CMY color spaces). The main challenges to be overcome for the commercial application of black-to-colorless switching polymers are (i) the synthetic complexity and low reproducibility for large-scale production by using random copolymerization of multiple monomers and (ii) the fact that black ECPs usually show a residual coloration in the bleached state [73]. Black-to-transmissive switching homopolymers represent a groundbreaking approach for future developments [74].

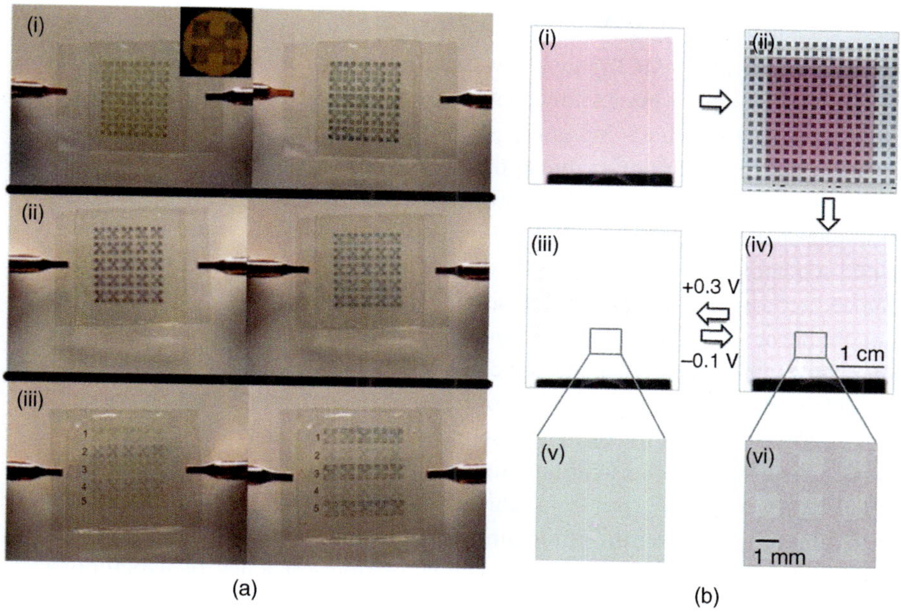

Figure 4.5 (a) Examples of inkjet-printed polyaniline or PEDOT-based EC devices with micrometer-sized patterns. (b) Direct photopatterning of methacrylate-substituted electrochromic polymer (ECP-Magenta). *Source:* (a) Reproduced with permission from [40]. © 2008 The Royal Society of Chemistry. (b) Reproduced with permission from [61]. © 2013 John Wiley © Sons.

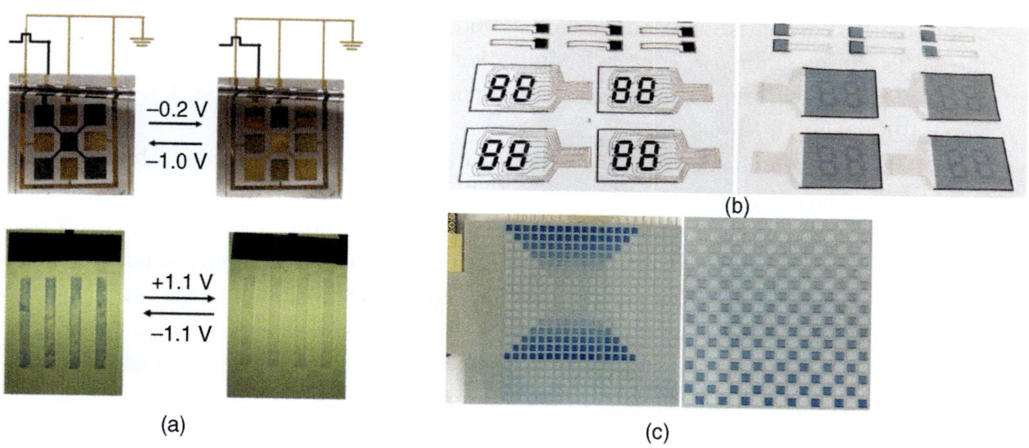

Figure 4.6 (a) Cross and line patterned poly(3,4-ethylenedioxythiophene):poly(styrene) sulfonate (PEDOT:PSS)-based EC devices. (b) Screen-printed EC displays with the white-black side (left) viewed from the printing side and the blue-gray side (right) viewed through the transparent substrate. (c) Screen-printed passive matrix addressed EC display. *Source:* (a) Reproduced with permission from [63] © 2005 The Royal Society of Chemistry. (b) Reproduced with permission from [64] © 2021 IOP Publishing. (c) Reproduced with permission from [2] © 2022 Springer Nature.

4.2.2 Metal Coordination Polymers

Metal coordination polymers (MCPs) are a new and promising class of hybrid EC materials that are generating emerging interest and potential in the field of electrochromics. They combine the advantages of both inorganic and organic EC materials. In recent years, MCPs have attracted special attention due to their outstanding optoelectronic properties, in particular for those of transition metal ions with terpyridine ligands [75, 76]. The optical and electronic properties of MCPs are determined by intramolecular metal–ligand and/or metal–metal interactions in the polymer chain. Linear or three-dimensional MCPs can be synthesized via self-assembly of divalent transition metal ions M^{2+} (M = Fe, Ru, Os, Co, Cu, Ni, Zn) and ditopic or tritopic terpyridine ligands (Figure 4.7) [77, 78].

In comparison to conjugated polymers, the chain length of MCPs changes in solution due to the reversible formation and cleavage of coordinate covalent bonds based on the equilibrium reactions. Thus, high complexation constants result in long polymer chains [79]. The polymer structure influences the solubility and viscosity of MCPs [80]. Common solvents for MCPs are water, aqueous acetic acid, and polar solvents (e.g., short-chain alcohols). Due to their amorphous structure, three-dimensional MCPs exhibit higher mechanical strength and better ionic conductivity [81]. The coloration efficiency of hyperbranched three-dimensional MCPs (color change from magenta to pale green) can also exceed that of linear (one-dimensional) ones [82]. It is assumed that the three-dimensional structure of the MCPs weakens the ionic stacking of the polymers with the anions trapped. The resulting porous structure contributes to a smoother transfer of the anions into the MCP film and enhances the coloration efficiency.

MCPs show a strong absorption in the visible region, which is attributed to a metal-to-ligand charge transfer (MLCT) transition, and hence responsible for the intensive coloration [83]. By variation of the metal ion and/or the design of the ligands the color of the MCPs can be tailored [77, 84–87]. Multicolored MCPs can be synthesized by stepwise complexation of two or more different transition metal ions, e.g., Fe/Ru [88, 89], Fe/Cu [90], Fe/Os [91, 92], or Fe/Ru/Os [93].

MCP thin films can be deposited on various substrates by common solution-based methods including layer-by-layer assembly [94, 95], spin-/dip-/spray-/slot-die coating, and inkjet printing [96, 97]. The layer thickness, and thus the color intensity, can be adjusted via the viscosity and the coating parameters. In order to achieve high reproducibility, which is important for industrial coating processes, a precise control of the viscosity of the MCP coating solutions must be ensured.

In addition, MCP films can also be obtained by oxidative electrochemical polymerization of metal ion complexes with functionalized terpyridine ligands [98, 99]. As with conjugated polymers this method is more suitable for small-scale EC devices, because it allows a convenient and controllable film fabrication.

In particular, iron-based metal coordination polymers (Fe-MCPs) have been extensively studied in recent years and show good processability and excellent EC properties in terms of light transmittance change, coloration efficiency, switching speed, low operating voltages, and cycling stability. Most Fe-MCPs are cathodically coloring, reversibly switching from a blue colored (reduced, Fe^{2+}) state to a highly transmissive (oxidized, Fe^{3+}) state (Figure 4.7c) [100, 101]. Due to the metal–organic structure the Fe-MCPs exhibit high thermal stability. Above 100 °C, irreversible thermochromic effects induced by structural changes in the octahedral coordination geometry can occur [102].

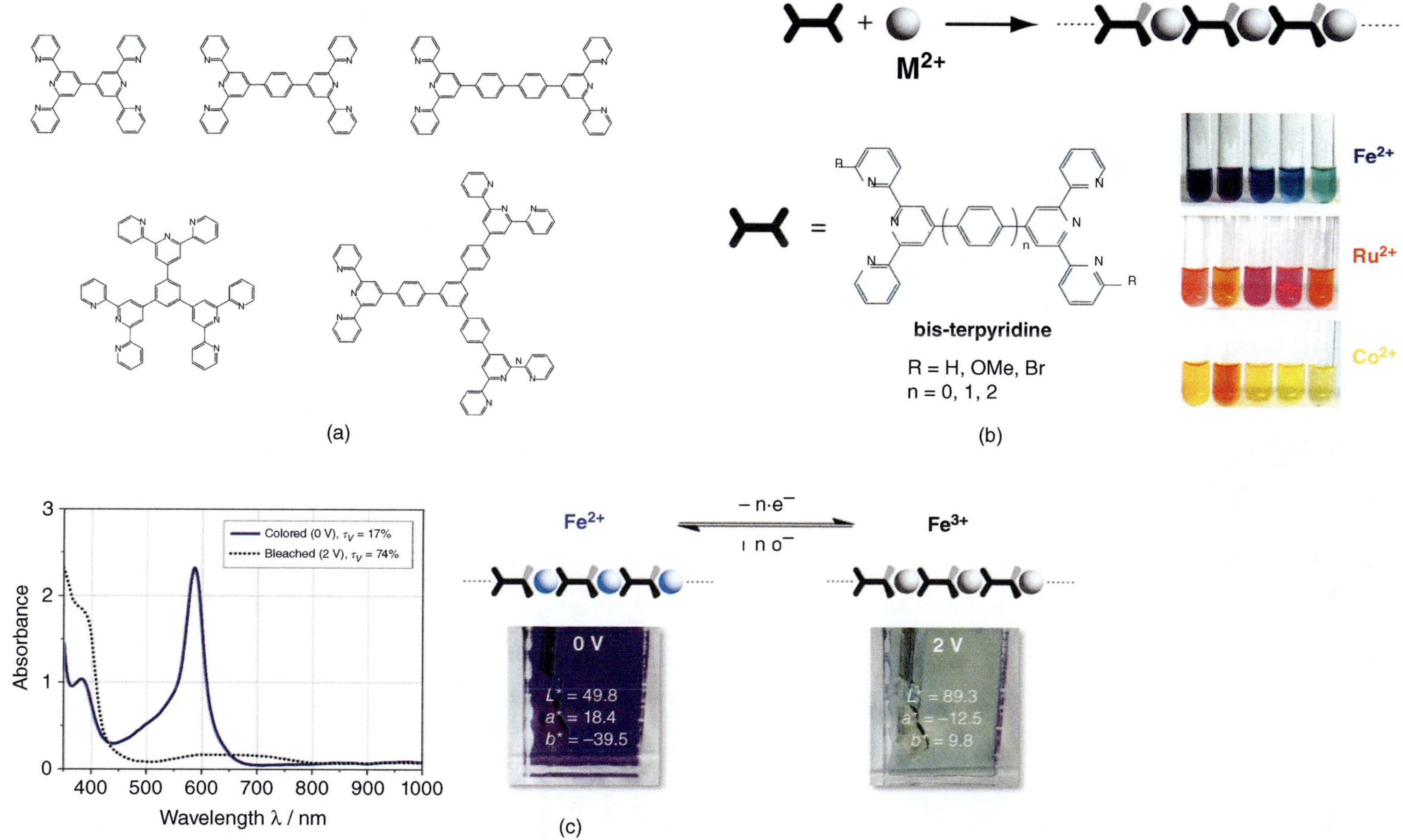

Figure 4.7 (a) Terpyridine ligands, (b) self-assembly of metal coordination polymers with divalent transition metal ions (counter ions omitted), and (c) EC switching behavior of iron metal coordination polymer (Fe-MCP). *Source:* (b) Reproduced with permission from [77]. © 2008 American Chemical Society.

Figure 4.8 Fe-MCP thin films (a) dip-coated on fluorine-doped tin oxide (FTO) glass ($1 \times 1\ m^2$) and (b) roll-to-roll slot-die coated on ITO/PET (coating width: 500 mm).

Large-area metallopolymeric thin-film electrodes with high optical quality can be fabricated by means of dip-coating or roll-to-roll slot-die coating on glass panes or films, respectively (Figure 4.8).

Furthermore, the deposition and EC device fabrication on ultra-thin ITO glass (100 μm) was demonstrated, producing a roll-to-roll-processable transparent conducting electrode with high optical quality that can be an alternative to commonly used ITO/PET [103].

Novel EC device configurations based on Fe- and Ru-MCPs and different ion storage electrode materials, e.g., (mixed) metal oxides [104–108], Prussian blue (PB) [109], or Prussian blue analogues (PBAs) [110, 111], were investigated on lab scale with a focus on color variation (toward neutral tint), optical contrast optimization, thermal and cycling stability. Mondal et al. reported a dual-redox system of MCPs for visible-to-near-infrared switchable EC devices [112]. A combination of electrochromism and photoluminescence can be achieved by MCPs containing Eu^{3+} ions [113].

MCPs are promising materials for next-generation EC devices. However, depending on the application, they need to be tested and validated under real use conditions to prove their maturity and qualify them for commercial products.

4.3 Device Manufacturing

While the properties and performance of ECPs and small-scale devices made thereof are well developed, the major challenges lie in the efficient manufacturing of high-quality coatings in industrial processes and device engineering. In order to develop and manufacture all-solid-state devices for practical use, the EC materials have to be deposited as thin films

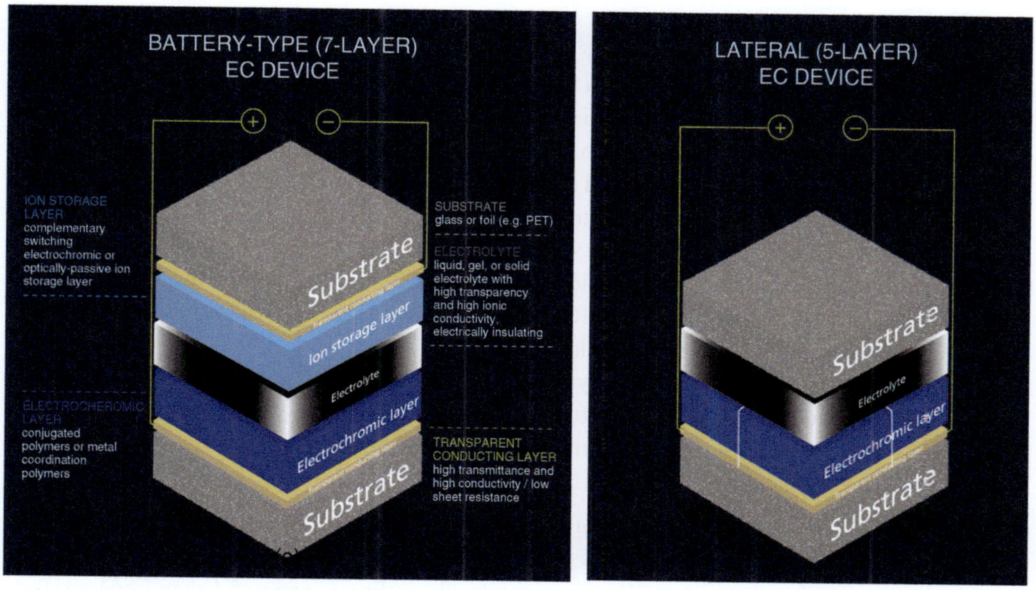

Figure 4.9 Graphical illustration of (a) vertical (battery-type) and (b) lateral (patterned) EC devices.

onto electrodes. Two device structures are conceivable for polymeric EC devices: a vertical (battery-type) design and a lateral design [114] as illustrated in Figure 4.9.

The seven-layer battery-type design, with the electrolyte sandwiched between the two electrodes, is mostly employed for large-area EC devices to achieve maximum optical contrast and switching speed. Mechanical flexibility can be achieved by using flexible electrodes, such as ITO-coated PET with high transmittance and low sheet resistance. Although the manufacturing process is complex and the costs are high, ITO/PET is still widely used as transparent conducting film. Several materials have been developed as alternatives to ITO, e.g., metal grids/nanowires, graphene, carbon nanotubes (CNTs), or conjugated polymers. However, there is always a trade-off between transmittance, sheet resistance, and mechanical ruggedness and flexibility.

In a lateral design, the electrodes are placed side by side and the electrolyte is applied on top of both electrodes. Compared to the layered design, the lateral design is only suitable for small-scale devices due to the higher resistance and longer ionic diffusion paths resulting in lower switching speeds. However, the risk for short-circuits during bending and mechanical deformation is lower, making the lateral design more suitable for smart labels or wearables. In the layered design, short circuits and electrode separation from the electrolyte under mechanical stress are to be avoided by using appropriate insulating materials as spacer between the two electrodes.

An important aspect of device assembly is the charge balancing of both electrodes in order to achieve maximum optical contrast, low cell voltages, fast switching speed, and high cycling stability [115]. An unbalanced configuration, in which one electrode has a higher charge capacity, is possible if this electrode is transparent and/or optically inactive [107, 116–118].

The characteristic features of (rigid or flexible) polymeric or metallopolymeric EC devices are:

- Broad color variety and high optical contrast
- Selective blocking of either visible light or NIR radiation [101, 119]

- High coloration efficiencies and low charge capacities
- Low operating voltage (<±2 V)
- High switching speed (depending on the transparent conducting layer, size/geometry, temperature, operating voltage, interfacial resistance, etc.)
- Mechanically flexible/bendable (lamination on curved surfaces or between curved glass)
- Ability to cut to any desired size and shape ("free-form" concept)
- Usability with common industrial lamination methods
- Lower bi-stability (memory) of the colored and clear states compared to inorganic EC devices based on sputtered metal oxides (generally attributed to faster ionic diffusion)
- Lower solar heat gain coefficient (SHGC) modulation than state-of-the-art EC metal oxides (generally due to lower NIR absorption)
- Long-term durability of metallopolymeric EC materials and devices still not completely investigated/addressed in the literature
- Materials with black or neutral-tint states still in the development phase (academic research).

Large-scale EC devices can be manufactured by either sheet-to-sheet (batch process) or roll-to-roll processes (continuous process). Sheet-to-sheet assembly of EC devices is more time consuming in general, but also offers greater flexibility in terms of design freedom. For roll-to-roll processing of flexible EC films the final design must be defined prior to coating or printing the active materials. Most smart window manufacturer (e.g., SageGlass, View, Halio) use batch processes to produce EC cells and the final glass laminates, with the exception of ChromoGenics. Their dynamic glass product *ConverLight® Dynamic* consists of a completely roll-to-roll processed EC film including sputtering of the EC materials (tungsten oxide and nickel oxide), slot-die deposition of the electrolyte, and lamination [120].

The transfer of flexible EC devices based on PEDOT-EthC6 on ITO/PET as the cathodically coloring electrode, PB on ITO/PET as the anodically coloring electrode, and a solid polymer electrolyte from small scale (sheet-to-sheet) to pilot scale (roll-to-roll processing, 500 mm coating width) has been demonstrated, showing the potential toward industrialization ("from lab to pilot to fab"). Both the ECP and the PB thin films were deposited by means of roll-to-roll slot-die coating using in situ polymerization and a nanoparticle PB ink, respectively.

The polymeric EC cells show improved properties in terms of coloration efficiency, optical contrast, transmittance and color neutrality in the bleached state, as well as cycling stability. The sheet-to-sheet devices were fabricated in a customized lamination process in a dry room up to a size of $45 \times 65\,cm^2$ (active area), whereas the roll-to-roll assembled devices had a maximum size of $50 \times 100\,cm^2$ (Figure 4.10). The devices are bendable to a certain extent, which allows further processing by common lamination processes.

The roll-to-roll lamination of EC films contains four main process steps (shown in Figure 4.11): electrolyte application (e.g., by slot-die coating), nip lamination of both electrodes, electrolyte curing (by UV or thermal heating), cutting and packaging. Depending on how moisture-sensitive the electrode materials and the electrolyte are, the lamination process is ideally performed under inert atmosphere or at low humidity levels, by implementing housing or laminar flow arrangements.

The slot-die coating of the polymer electrolyte, the precise lamination using registration control, and the final UV curing under (ideally) inert conditions are crucial steps in order to obtain functional and stable EC laminates. By avoiding water and oxygen in the final EC cells, faster response and better long-term durability can be expected by minimization of side

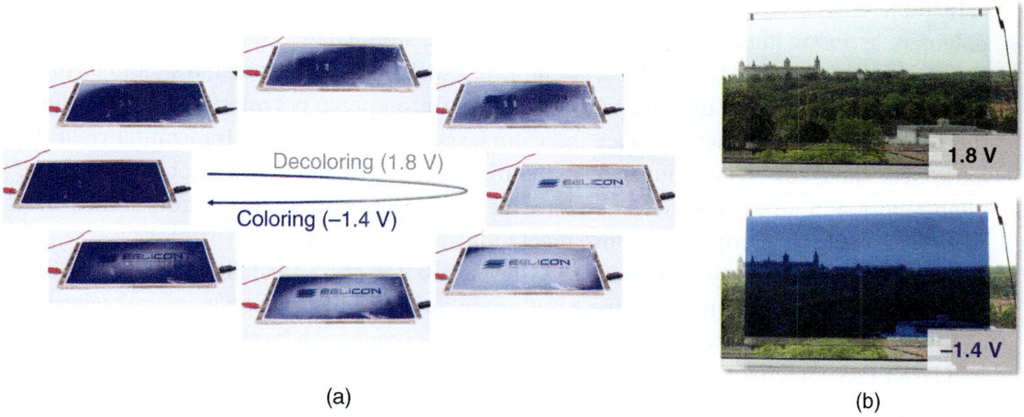

Figure 4.10 Photographic images of flexible EC devices. (a) Sheet-to-sheet assembled (size: $25 \times 40\,\text{cm}^2$) and (b) all roll-to-roll processed ($50 \times 100\,\text{cm}^2$). *Source:* Reproduced with permission from [29]. © 2020 John Wiley & Sons.

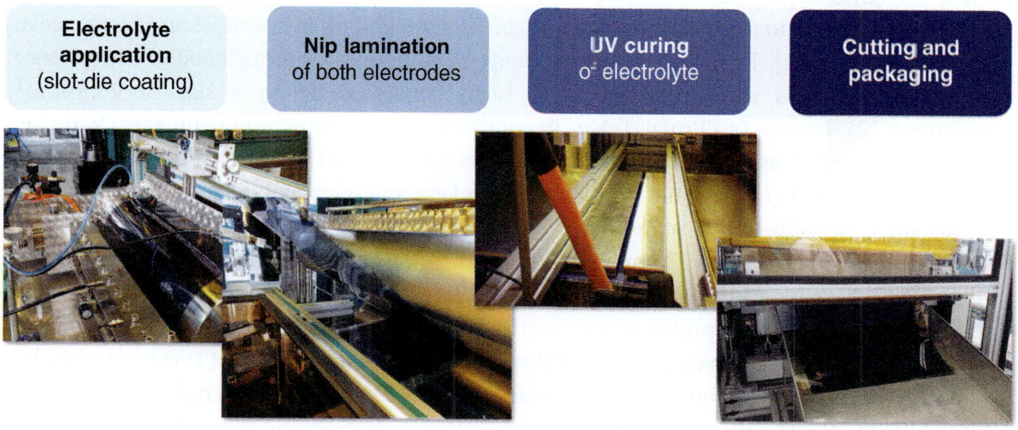

Figure 4.11 Roll-to-roll lamination of flexible EC devices. Shown are the sequential processes (left to right) of electrolyte application using a slot-die, nip lamination of two electrodes - pushing air out, UV curing of a bonding polymer electrolyte and sheet cutting for packaging.

reactions and photo-oxidation [121, 122]. After lamination and UV curing of the electrolyte, the laminates can be cut to a specific length with an integrated knife or laser cutter. The ideal method for contacting EC devices using roll-to-roll processing is still an open task and has to be further elaborated on future developments. An alternative to contacting the devices after or during the lamination process (e.g., with copper tape) is the use of electrodes with pre-fabricated metallic bus bars, e.g., applied by sputter-deposition, evaporation, or printing methods on the conductive electrodes [123–125]. The switching speed depends on the conductivity of the transparent conducting layer, among other factors, and decreases as the size of the EC device increases. To counteract this effect by minimizing the resistive voltage drop over large areas, additional conductor lines can be used (see *SageGlass LightZone*®). At low temperatures, ionic diffusion and thus the coloration/decoloration is also slowed down.

In Table 4.1 device configurations containing conjugated polymers or metal coordination polymers as EC material or proton source [130] are summarized. Only devices that have been processed in larger-scale (>$10 \times 10\,cm^2$) are listed here. At this point, it should be mentioned that it is relatively difficult to compare literature-known polymeric EC devices since the most important key performance indicators, such as (visible light) transmittance of the dark/clear state, coloration efficiency, switching speed [131], and cycling stability [132], are often reported in different ways. Therefore, there is a great need for standardized measurements and evaluation criteria that also meet industrial requirements.

In an outstanding study, a large-area EC device ($50 \times 50\,cm^2$) was fabricated on PET film by a roll-to-roll solution process using a multifunctional conducting polymer synthesized by chemical oxidative polymerization [126]. The device comprises a layer stack of PEDOT:PSS and PANI:PSS, which the authors claim can serve as the electrode and EC layer alone without the need for other electrode materials (e.g., ITO or FTO). The flexible EC device showed excellent optical characteristics with high coloration efficiency ($1873\,cm^2\,C^{-1}$ at 600 nm), low power consumption, and good cycling stability owing to the balanced surface charge capacity.

The fabrication of single-layer ECP devices using an in situ electrochemical polymerization method was demonstrated in further studies [128, 129]. Here, the monomer is dissolved in a UV-curable polymer electrolyte and polymerized directly in the assembled cell by applying a voltage to form a solid-state device. This method offers a simple and inexpensive device design [133]. EC devices utilizing poly(2,2-dimethyl-3,4-propylenedioxythiophene) (PProDOT-Me_2) as active material have been fabricated up to a size of $9 \times 19\,cm^2$ (Figure 4.12). The use of PProDOT-Me_2 in combination with vanadium pentoxide (V_2O_5) as ion-storage layer was also demonstrated in EC sunglasses [134]. However, as mentioned in Section 4.2.1, further upscaling is difficult using electrochemical polymerization.

Recently, all-solid-state proton-based tandem structures were investigated for use in fast-switching EC devices [130]. Compared to the standard 7- or 5-layer structures of polymeric EC devices, in which PEDOT:PSS serves either as EC or electrode material, PEDOT:PSS is used here as a solid-state proton source for tungsten oxide (WO_3) electrodes (Figure 4.13).

Large-area EC devices based on Fe-MCP and titanium vanadium oxide ($TiVO_x$) or complementary switching PB as ion storage electrodes are shown in Figure 4.14. The blue-purple and dark blue EC glass laminates exhibit high transmittance changes and fast response. In particular, the combination of Fe-MCP with PB is promising and characterized by high visible light transmittance changes from 11% to 64% at 2.5 V resulting in coloration efficiencies of approx. $500\,cm^2\,C^{-1}$. The two examples demonstrate the great potential of such novel hybrid EC device configurations combining MCPs with (mixed) metal oxides or PB for next-generation smart window applications.

The combination of organic (e.g., conjugated ECPs or MCPs) and inorganic EC materials can be a viable approach toward novel electrode and device configurations with interesting EC properties. This approach aims to take advantage of the merits of the individual components/materials: the color variety and fast switching of the organic EC materials and the electrochemical, thermal and UV stability of the inorganic ones. In recent years, the combination of ECPs with nanostructured metal oxides as composite thin-film materials has been widely studied [135]. However, most of the hybrid EC materials and devices based on PEDOT, PEDOT derivatives, or PANI are demonstrated on lab scale [10, 136]. Hybrid EC cells (PEDOT:PSS versus NiO or CeO_2/TiO_2) with a size up to $30 \times 30\,cm^2$ combining ECPs and metal oxides as ion storage electrode were reported by Heuer et al. [127]. Rougier et al.

Table 4.1 Overview and key performance indicators of selected polymeric and metallopolymeric EC devices (size > $10 \times 10\ cm^2$), most of them reported in the literature. The table is not a complete list of all polymeric and metallopolymeric EC devices, but an overview of studies addressing large-scale processing.

Cell configuration (size)	Operating voltages (V)	Max. absorption wavelength λ_{max} (nm)	Transmittance at λ_{max} T_{dark}/T_{clear} (%)	Coloration efficiency[a)] at λ_{max} ($cm^2\ C^{-1}$)	Visible light transmittance dark/clear (%)	Cycling stability (tested)	Color	References
PEDOT/PB ($12 \times 15\ cm^2$)	−1.0 1.0	640	n.a.	270	n.a.	21,000	dark blue light blue	[26]
ECP-magenta/MCCP ($18 \times 18\ cm^2$)	−2.0 2.0	n.a.	n.a.	n.a.	10 55 (at 550 nm)	n.a.	magenta colorless	[58, 59]
PEDOT-EthC6/PB ($50 \times 100\ cm^2$)	−1.4 1.8	630	0.5 46.7	563	4 53	10,000	dark blue color-neutral	[29]
PEDOT:PSS/PANI:PSS ($50 \times 50\ cm^2$)	−1.8 0.4	600	10.5 60	1873	n.a.	1,000	dark blue light blue	[126]
PEDOT:PSS/NiO ($30 \times 30\ cm^2$)	−2.5 2.5	565	20 65	n.a.	n.a.	n.a.	dark blue light blue	[127]
PProDOT-Me_2 ($9 \times 19\ cm^2$) ($105\ cm^2$) single-layer EC device	−2.0 2.0	575	n.a.	n.a.	$\Delta = 48$ $\Delta = 46$	n.a.	blue-purple colorless	[128] [129]
WO_3/PEDOT:PSS as proton source ($30 \times 40\ cm^2$) tandem structure	0.0 2.4	670	90 20	109	n.a.	3,000	dark blue colorless	[130]
Fe-MCP/$TiVO_x$ ($40 \times 60\ cm^2$)	0.0 2.0	585	0.8 58	459	16 59	>1,000	blue-purple light yellow	This work
Fe-MCP/PB ($20 \times 30\ cm^2$)	0.0 2.5	584	0.5 60	500	11 64	>1,000	dark blue colorless	This work

a) Coloration efficiency = $\log(T_{clear}/T_{dark})/q$ (with q – charge capacity).

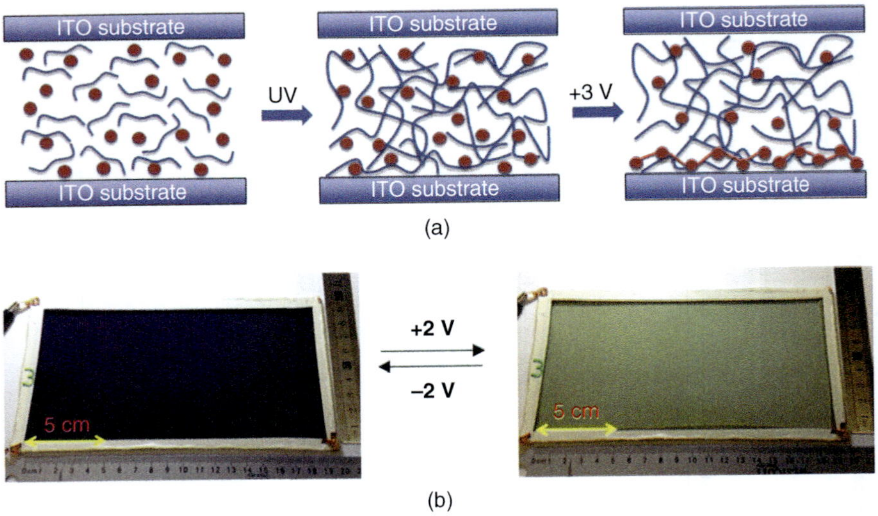

Figure 4.12 (a) Schematic diagram for in situ electrochemical polymerization: (1) liquid monomer electrolyte sandwiched between two ITO-coated substrates, (2) curing under UV light to produce a solid electrolyte gel, (3) ECP/electrolyte blend generated by converting monomers at a constant voltage of +3 V. (b) Colored/bleached state of a single-layer EC device ($9 \times 19\,cm^2$) based on poly (2,2-dimethyl-3,4-propylenedioxythiophene) (PProDOT-Me_2) and a lithium bis(trifluoromethanesulfonyl)imide (LiTFSI) gel electrolyte. *Source:* Reproduced with permission from [128] © 2014 ELSEVIER.

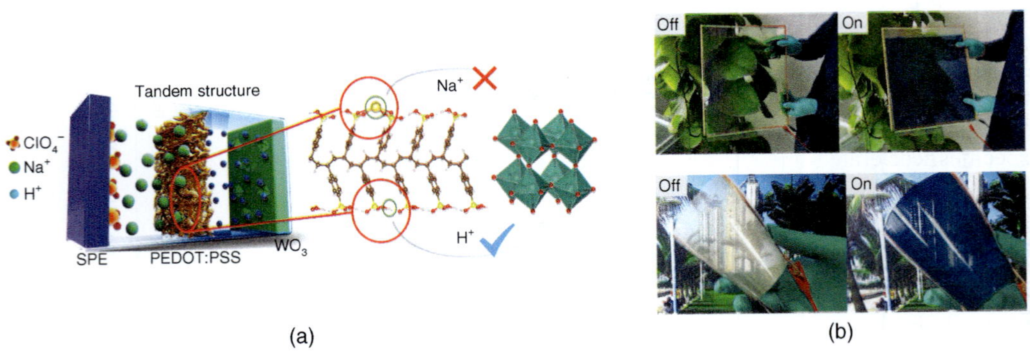

Figure 4.13 Schematic illustration of tandem-structure ECDs and the key step with Na^+ from the solid polymer electrolyte layer pumping out a proton from PEDOT:PSS to WO_3. (b) Photographs of EC devices on glass ($30 \times 40\,cm^2$) and flexible EC devices ($10 \times 10\,cm^2$). *Source:* Reproduced with permission from [130] © 2022 Springer Nature.

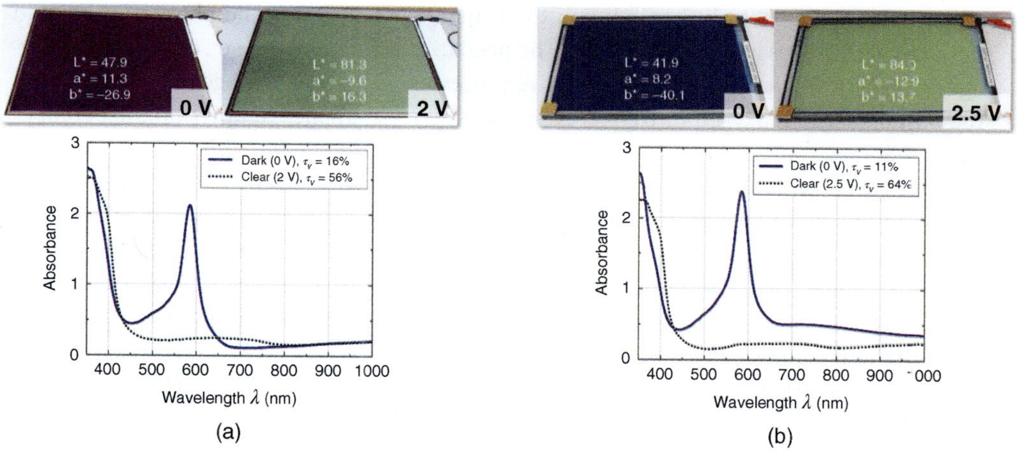

Figure 4.14 Large-area EC glass laminates based on Fe-MCP and (a) $TiVO_x$ ($40 \times 60\ cm^2$) or (b) PB ($20 \times 30\ cm^2$).

investigated novel hybrid EC thin-film electrodes comprising PEDOT:PSS and Fe_2O_3 [137], Nb_2O_5 [138], or V_2O_5 [139] with promising EC properties.

4.4 Industrial Applications of Polymeric Electrochromic Devices

The performance requirements for EC devices strongly depend on the application case. As for outdoor applications, the main requirements are related to the long-term durability of EC products under exposure to light, temperature, and moisture. So far, there have been standards for EC glazing, assembled as insulating glass units (IGU), laminated glass, or a combination of both, only for use in buildings (ISO 18543 and ASTM E2141). This standard specifies accelerated aging tests (including UV and thermal stress tests) and the requirements for EC glazing (based on inorganic or organic materials). Unlike inorganic EC materials used in state-of-the-art dynamic glazing, organic/hybrid EC materials and devices made therefrom have not been fully investigated and evaluated for durability. However, based on bond theory considerations it can be assumed that organic polymers are fundamentally more susceptible to photolytic and photooxidative processes than inorganic oxidic materials (whereas the latter often show photocatalytic effects limiting durability as well). This is also shown by systematic studies on the photostability of polymeric ECDs under different environmental conditions [122, 140].

Because durability is crucial for products exposed to outdoor conditions, in particular in the construction, automotive and transportation sector, UV protection and encapsulation are highly important. Polymeric electrochromic devices should ideally be fabricated and sealed/encapsulated under an inert atmosphere to prevent oxygen and moisture ingress [121], thus ensuring high cyclability and photostability [141]. Common sealant materials used for organic electronics are also applicable to EC devices to improve shelf-life and prevent electrolyte degradation, including from solvent evaporation [142].

It should be pointed out here that polymers are not only used as active materials in EC devices. Rather, polymers are used in commercial and proposed products as everything from the EC materials themselves to solution thickeners, the matrix or solid phase of gels, solid polymer electrolytes and sealants [143]. The need to avoid photolytic (i.e., direct bond cleavage after light absorption) or photooxidative processes (i.e., bond cleavage or oxidation with the participation of oxygen or oxygen-based radicals) also applies to other product components that may consist of organic polymers.

Potential or proposed industrial applications of polymeric electrochromic devices are shown in Figure 4.15. In most of these applications, a fast color change is mandatory (e.g., automotive glazing, goggles/sunglasses, displays), while in architectural glazing longer switching times with a uniform color change are tolerated or even desired. Due to the inherently faster response of polymeric EC devices compared to inorganic ones they are able to fulfill these requirements.

The largest market for electrochromics is the construction sector, followed by the automotive industry. The benefits resulting from the use of smart glass or films consist of improved energy efficiency, active daylight control, glare reduction, and innovative design solutions. For these applications the EC devices are often laminated between glass to ensure durability, safety, and comfort.

For the application of EC functionality in adaptive camouflage and multifunctional textiles (Figure 4.16), in addition to high flexibility, stretchability is also required, which can be achieved through the use of polymeric EC materials [144].

Multi-colored polymers with high coloration efficiency, vivid colors, and high switching speed could open up new possibilities for the military use of electrochromics, e.g., soldier protection or camouflage vehicles. Although there are many concepts or design studies, to date there is no commercial or military product, and most developments lack technological maturity for these applications. In addition, many developments presumably take place under secrecy, which makes it difficult to assess the current state-of-the-art in this field.

So far, there are only a few companies on the market that offer products based on ECPs.

Alphachromics Inc., a start-up company founded in 2010 with the aim to commercialize EC innovations. They claimed to be using low-cost color-changing conductive polymers

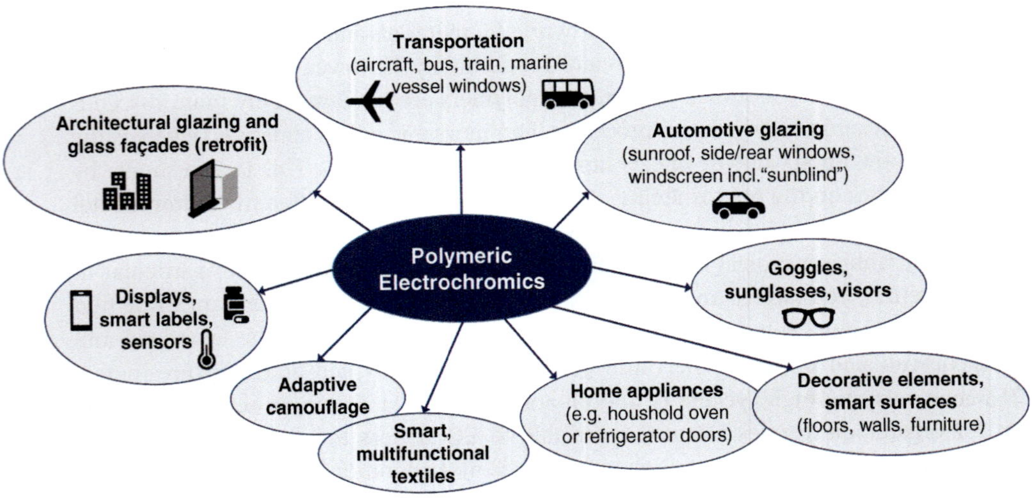

Figure 4.15 Potential or proposed industrial applications of polymeric electrochromics.

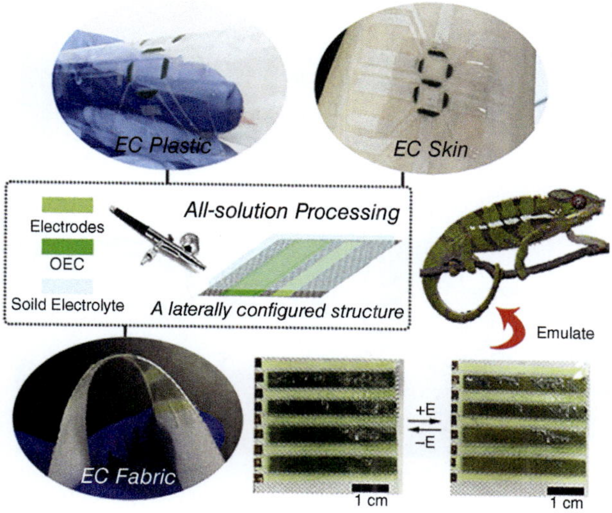

Figure 4.16 All-solution processed EC fabric for adaptive camouflage. *Source:* Reproduced with permission from [144]. © 2019 ELSEVIER.

developed by Sotzing et al. that can be applied to products as diverse as energy-efficient windows, tunable eyewear (goggles and sunglasses), and high-performance switchable fabrics. However, it is unclear if the company is still active in the field of electrochromics.

Ambilight Inc., established in Silicon Valley, Shenzhen, and Indiana in 2017, is dedicated to the commercialization of roll-to-roll manufactured thin-film electrochromics for smart windows in architecture and automotive sectors as well as for consumer electronics [145]. Ambilight's flexible and multi-color EC thin film technology is said to be based on solid polymer materials (π-conjugated chromophores) including black-to-transmissive ECPs [146]. In 2020 Ambilight introduced the world's first mobile phone with an EC rear cover [147].

Saralon GmbH, founded in Chemnitz, Germany, in 2015, focuses on printed electronics applications, offering inks and fully printed EC displays based on PET as a substrate, without a conductive ITO layer [148]. Four different inks are required to print the EC displays, Saral Conducting Polymer 700 is a water-based ECP ink with high conductivity that is screen-printable on plastic and paper. Saralon's EC displays are highly flexible, very thin (approx. 100 μm), and show a color change from light blue to dark blue. Due to their customizable design the EC displays can be integrated in various applications, such as interactive displays (e.g., cards or labels), security tags, and sensors powered by near-field communication (NFC) or printed batteries.

Ynvisible Interactive Inc., a printed e-paper display manufacturer founded in Lisbon, Portugal in 2010, produces customized segmented EC displays for a variety of applications, such as digital signage, smart monitoring labels, authenticity and security tags, as well as retail signage and labels (Figure 4.17). Ynvisible's EC displays are manufactured using cost-effective and quality-controlled roll-to-roll printing [149]. Therefore, suitable ECPs (e.g., PEDOT:PSS) and ink formulations are required. In addition, a transparent solid electrolyte was developed by Ynvisible that is printable and UV-curable [150]. Besides printed e-paper displays, Ynvisible also offers sensors, energy harvesting, and wireless communication components.

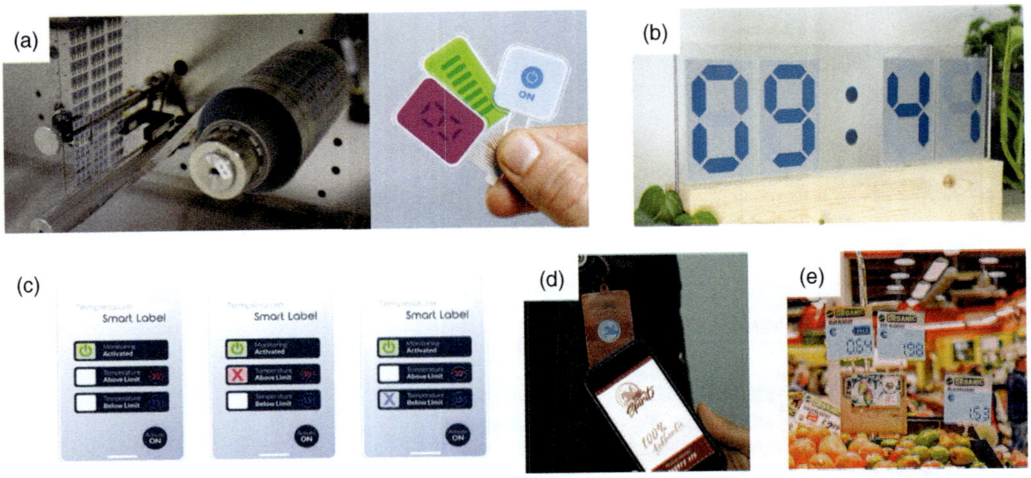

Figure 4.17 (a) Roll-to-roll printing of EC displays for use as (b) digital signage, (c) temperature labels, (d) brand protection, and (e) retail signage and labels. *Source:* (a), (b), (d), (e) © 2022 Ynvisible Interactive Inc. (c) Reproduced with permission from [3]. © 2021 American Chemical Society.

In addition to the above-mentioned companies developing and offering EC products, there are a few companies providing ECPs.

Aqueous PEDOT:PSS dispersions are commercially available from **Heraeus Epurio** under the trade name Clevios™. The company offers a number of product forms differing in terms of solids content, additives, viscosity, particle size, and conductivity [151]. There are numerous resellers and suppliers offering PEDOT:PSS products that are presumably based on (or very similar to) the Clevios™ products and are therefore not listed separately here.

A Germany- and India-based company selling polyaniline formulations in different variants is **Ormecon Pvt. Ltd.**, formerly known and founded as Ormecon Chemie GmbH & Co in 1996. The company manufactures and markets an electrically conductive polymer for industrial applications such as anticorrosive additives for paints, electromagnetic interference shielding, antistatic, radar-absorbing material, electroluminescence, and solid electrolyte capacitors [152]. Ormecon's mentor and founder, Dr. Wessling, can be considered a pioneer in showing how to process insoluble conductive polymers by dispersion [153].

4.5 Conclusion and Outlook

Despite extensive research on polymeric electrochromics over the past few decades, few of them have proven applicable to large-scale production. Further research efforts are needed to bring the most promising polymeric EC materials (including conjugated polymers as well as metal coordination polymers) toward a higher technology readiness level and validate them in devices and products in application-oriented scenarios.

For most industrial applications the color of EC devices is most relevant when it comes to the design and visual appearance. Here, the use of polymeric electrochromics can be beneficial due to the fact that these materials are able to offer a large variety of colors including highly desired "color-neutral" systems switching from gray/black to completely colorless. However, polymeric EC materials with black or neutral-tint states are still in the development phase and lack processability or durability. Thus, for the implementation of black-to-colorless polymeric EC devices in commercial products these two main challenges need to be addressed in future developments. A viable approach could be the use of black-to-transmissive switching homopolymers.

Improving the long-term durability of polymeric EC devices is one of the key issues to be addressed in parallel with the on-going research and development of EC materials, electrolytes, devices, and robust large-area and high-throughput manufacturing processes. The high production cost of inorganic EC materials and devices is still one of the main drawbacks limiting broad large-area applications to custom windows. Also, competition with other lower cost-switchable technologies, such as thermochromics, is challenging for electrochromics. Other technologies such as (polymer-dispersed) liquid crystal (PDLC/LC) devices or suspended particle devices (SPD) are competitive with inorganic electrochromics, but all need to be less expensive. The polymeric electrochromics have a potential manufacturing cost advantage and custom coloration capability which can help increase their market share of switchable materials. If we look at the development of organic electrochromics in

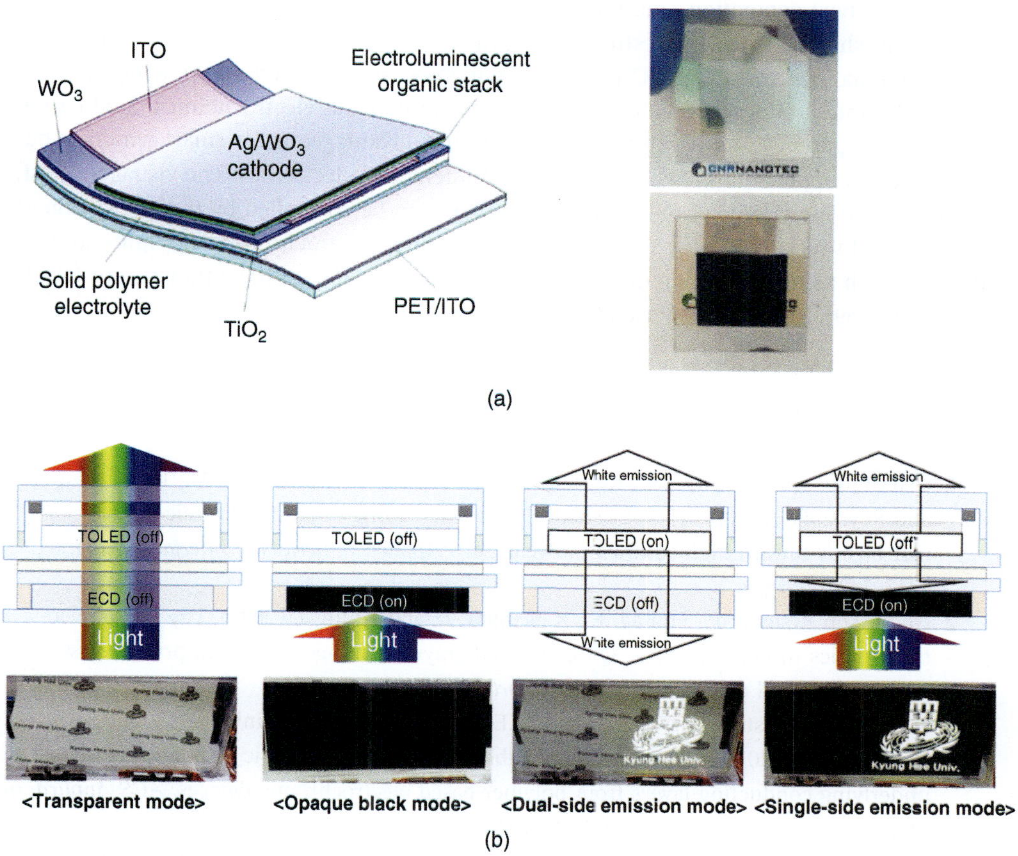

Figure 4.18 (a) Flexible electrochromic/organic light emitting diode (EC/OLED): Device structure and photographs of the bleached/colored state. (b) EC/TOLED (transparent organic light-emitting diode) smart window display with four different operating modes: transparent (EC, TOLED: off), opaque black (EC: on, TOLED: off), dual-side emission (EC: off, TOLED: on), single-side emission (EC: on, TOLED: off). *Source:* (a) Reproduced with permission from [157] © 2021 John Wiley & Sons. (b) Reproduced with permission from [159] © 2018 OSA Publishing.

the dimmable rear-view mirror area, we know the potential is there for durable large-area glazing and further applications.

The demand for more energy-efficient buildings on the global market is growing rapidly, also in connection with the development and legislation of switchable technologies. A key factor with high-energy-saving potential will be the renovation/retrofitting of the existing building stock through the use of new intelligent materials and technologies. Therefore, the overall technical and operational performance of existing polymeric materials needs to be further enhanced in order to provide durable EC systems with superior performance compared to the state-of-the-art.

Opportunities for future developments are application scenarios in which energy-harvesting technologies (e.g., photovoltaics) are combined with energy-saving technologies (e.g., batteries, EC devices). Especially for the application in architectural or automotive glazing, the combination of (O)PV and EC has major advantages for increasing the energy efficiency by using sustainable power resources. The smart windows are powered by the integrated solar cells. Besides the sunlight/glare control and self-powered functionality of next-generation smart windows tend to integrate even display functionality. Kim et al. showed some interesting developments of polymeric EC devices with multiple functionalities including EC capacitive windows, self-powered smart windows, and electrochromic-luminescent windows [154]. The simultaneous electrochemical modulation of both transparency and light emission is not only advantageous for smart windows but also for other applications such as displays and interactive electronics. The visual contrast, intensity, and directionality of transparent displays can be controlled by the integration of an EC cell. In recent years several examples combining electrochromic and light-emitting devices, such as light-emitting diode (LED) [155], OLED [156–158], or transparent OLED (TOLED) [159], have been reported (see Figure 4.18).

References

1 Österholm, A.M., Shen, D.E., Kerszulis, J.A., Bulloch, R.H., Kuepfert, M., Dyer, A.L., and Reynolds, J.R. (2015). Four shades of brown: tuning of electrochromic polymer blends toward high-contrast eyewear. *ACS Applied Materials & Interfaces* 7 (3): 1413–1421.

2 Andersson Ersman, P., Freitag, K., Kawahara, J., and Åhlin, J. (2022). The rise of electrochromics through dynamic QR codes and grayscale images in screen printed passive matrix addressed displays. *Scientific Reports* 12 (1): 10959.

3 Howard, E.L., Österholm, A.M., Shen, D.E., Panchumarti, L.P., Pinheiro, C., and Reynolds, J.R. (2021). Cost-effective, flexible, and colorful dynamic displays: removing underlying conducting layers from polymer-based electrochromic devices. *ACS Applied Materials & Interfaces* 13 (14): 16732–16743.

4 Brooke, R., Petsagkourakis, I., Wijeratne, K., and Andersson Ersman, P. (2022). Electrochromic displays manufactured by a combination of vapor phase polymerization and screen printing. *Advanced Materials Technologies* 2200054.

5 Celiesiute, R., Ramanaviciene, A., Gicevicius, M., and Ramanavicius, A. (2019). Electrochromic sensors based on conducting polymers, metal oxides, and coordination complexes. *Critical Reviews in Analytical Chemistry* 49 (3): 195–208.

6 Aller Pellitero, M. and del Campo, F.J. (2019). Electrochromic sensors: Innovative devices enabled by spectroelectrochemical methods. *Current Opinion in Electrochemistry* 15: 66–72.

7 Yu, H., Shao, S., Yan, L., Meng, H., He, Y., Yao, C., Xu, P., Zhang, X., Hu, W., and Huang, W. (2016). Side-chain engineering of green color electrochromic polymer materials: toward adaptive camouflage application. *Journal of Materials Chemistry C* 4 (12): 2269–2273.

8 Brooke, R., Fabretto, M., Pering, S., Switalska, E., Reeks, L., Evans, D., and Murphy, P. (2015). Diffuse color patterning using blended electrochromic polymers for proof-of-concept adaptive camouflage plaques. *Journal of Applied Polymer Science* 132 (26).

9 Meng, H. (2021). *Organic Electronics for Electrochromic Materials and Devices.* Weinheim: Wiley-VCH.

10 Mortimer, R.J., Rosseinsky, D.R., and Monk, P.M.S. (2013). *Electrochromic Materials and Devices.* KGaA: Wiley-VCH Verlag GmbH & Co.

11 Amb, C.M., Dyer, A.L., and Reynolds, J.R. (2011). Navigating the color palette of solution-processable electrochromic polymers. *Chemistry of Materials* 23 (3): 397–415.

12 Beaujuge, P.M. and Reynolds, J.R. (2010). Color control in pi-conjugated organic polymers for use in electrochromic devices. *Chemical Reviews* 110 (1): 268–320.

13 Kerszulis, J.A., Johnson, K.E., Kuepfert, M., Khoshabo, D., Dyer, A.L., and Reynolds, J.R. (2015). Tuning the painter's palette: subtle steric effects on spectra and colour in conjugated electrochromic polymers. *Journal of Materials Chemistry C* 3 (13): 3211–3218.

14 Rosseinsky, D.R., Monk, P.M.S., and Mortimer, R.J. (ed.) (2015). *Electrochromic Materials and Devices.* Wiley-VCH.

15 Dyer, A.L., Thompson, E.J., and Reynolds, J.R. (2011). Completing the color palette with spray-processable polymer electrochromics. *ACS Applied Materials & Interfaces* 3 (6): 1787–1795.

16 Mortimer, R.J. (2011). Electrochromic Materials. *Annual Review of Materials Research* 41 (1): 241–268.

17 Zhang, Z., Liao, M., Lou, H., Hu, Y., Sun, X., and Peng, H. (2018). Conjugated Polymers for Flexible Energy Harvesting and Storage. *Advanced Materials* 30 (13): e1704261.

18 Reynolds, J.R., Thompson, B.C., and Skotheim, T.A. (2019). *Conjugated Polymers: Properties, Processing, and Applications.* CRC Press.

19 Gross, M., Nothofer, S., Neher, B., and Meerholz, K. (2000). Improving the performance of doped pi-conjugated polymers for use in organic light-emitting diodes. *Nature* 405 (6787): 661–665.

20 Cheng, Y.-J., Yang, S.-H., and Hsu, C.-S. (2009). Synthesis of conjugated polymers for organic solar cell applications. *Chemical Reviews* 109 (11): 5868–5923.

21 Gunbas, G. and Toppare, L. (2012). Electrochromic conjugated polyheterocycles and derivatives–highlights from the last decade towards realization of long lived aspirations. *Chemical Communications* 48 (8): 1083–1101.

22 Amb, C.M., Beaujuge, P.M., and Reynolds, J.R. (2010). Spray-processable blue-to-highly transmissive switching polymer electrochromes via the donor-acceptor approach. *Advanced Materials* 22 (6): 724–728.

23 Beaujuge, P.M., Vasilyeva, S.V., Liu, D.Y., Ellinger, S., McCarley, T.D., and Reynolds, J.R. (2012). Structure-performance correlations in spray-processable green

dioxythiophene-benzothiadiazole donor–acceptor polymer electrochromes. *Chemistry of Materials* 24 (2): 255–268.

24 Poverenov, E., Li, M., Bitler, A., and Bendikov, M. (2010). Major effect of electropolymerization solvent on morphology and electrochromic properties of PEDOT films. *Chemistry of Materials* 22 (13): 4019–4025.

25 Sotzing, G.A., Reynolds, J.R., and Steel, P.J. (1996). Electrochromic conducting polymers via electrochemical polymerization of bis(2-(3,4-ethylenedioxy)thienyl) monomers. *Chemistry of Materials* 8 (4): 882–889.

26 Duluard, S., Celik-Cochet, A., Saadeddin, I., Labouret, A., Campet, G., Schottner, G., Posset, U., and Delville, M.-H. (2011). Electrochromic devices based on in situ polymerised EDOT and Prussian Blue: influence of transparent conducting oxide and electrolyte composition—towards up-scaling. *New Journal of Chemistry* 35 (10): 2314.

27 Ha, Y.-H., Nikolov, N., Pollack, S.K., Mastrangelo, J., Martin, B.D., and Shashidhar, R. (2004). Towards a Transparent, Highly Conductive Poly(3,4-ethylenedioxythiophene). *Advanced Functional Materials* 14 (6): 615–622.

28 Kumar, A. and Reynolds, J.R. (1996). Soluble alkyl-substituted poly(ethylenedioxythiophenes) as electrochromic materials. *Macromolecules* 29 (23): 7629–7630.

29 Macher, S., Schott, M., Dontigny, M., Guerfi, A., Zaghib, K., Posset, U., and Löbmann, P. (2021). Large-area electrochromic devices on flexible polymer substrates with high optical contrast and enhanced cycling stability. *Advanced Materials Technologies* 6 (2): 2000836.

30 Österholm, A.M., Shen, D.E., Gottfried, D.S., and Reynolds, J.R. (2016). Full color control and high-resolution patterning from inkjet printable cyan/magenta/yellow colored-to-colorless electrochromic polymer inks. *Advanced Materials Technologies* 1 (4): 1600063.

31 Macher, S., Sassi, M., Beverina, L., Posset, U., Schott, M., Giffin, G.A., and Löbmann, P. (2021). Electrochromic polymer ink derived from a sidechain-modified EDOT for electrochromic devices with colorless bright state. *ChemElectroChem* 8 (4): 726–734.

32 Gleason, K.K. (2020). Nanoscale control by chemically vapour-deposited polymers. *Nature Reviews Physics* 2 (7): 347–364.

33 Brooke, R., Fabretto, M., Hojati-Talemi, P., Murphy, P., and Evans, D. (2014). Evidence for 'bottom up' growth during vapor phase polymerization of conducting polymers. *Polymer* 55 (16): 3458–3460.

34 Kovacik, P., Hierro, G.d., Livernois, W., and Gleason, K.K. (2015). Scale-up of oCVD: large-area conductive polymer thin films for next-generation electronics. *Materials Horizons* 2 (2): 221–227.

35 Evans, D., Fabretto, M., Mueller, M., Zuber, K., Short, R., and Murphy, P. (2012). Structure-directed growth of high conductivity PEDOT from liquid-like oxidant layers during vacuum vapor phase polymerization. *Journal of Materials Chemistry* 22 (30): 14889.

36 Fabretto, M., Autere, J.-P., Hoglinger, D., Field, S., and Murphy, P. (2011). Vacuum vapour phase polymerised poly(3,4-ethyelendioxythiophene) thin films for use in large-scale electrochromic devices. *Thin Solid Films* 519 (8): 2544–2549.

37 Macher, S., Schott, M., Sassi, M., Facchinetti, I., Ruffo, R., Patriarca, G., Beverina, L., Posset, U., Giffin, G.A., and Löbmann, P. (2020). New roll-to-roll processable PEDOT-based polymer with colorless bleached state for flexible electrochromic devices. *Advanced Functional Materials* 30 (6): 1906254.

38 Jensen, J. and Krebs, F.C. (2014). From the bottom up – flexible solid state electrochromic devices. *Advanced Materials* 26 (42): 7231–7234.

39 Moreira, T., Maia, M., Parola, A.J., Zangoli, M., Di Maria, F., and Laia, C.A.T. (2021). Chapter 12 – Ink-jet-printed semiconductor electrochromic nanoparticles: development and applications in electrochromism. In: *Chemical Solution Synthesis for Materials Design and Thin Film Device Applications* (ed. S. Das and S. Dhara), 407–437. Elsevier.

40 Shim, G.H., Han, M.G., Sharp-Norton, J.C., Creager, S.E., and Foulger, S.H. (2008). Inkjet-printed electrochromic devices utilizing polyaniline–silica and poly(3,4-ethylenedioxythiophene)–silica colloidal composite particles. *Journal of Materials Chemistry* 18 (5): 594–601.

41 Søndergaard, R.R., Hösel, M., and Krebs, F.C. (2013). Roll-to-roll fabrication of large area functional organic materials. *Journal of Polymer Science Part B: Polymer Physics* 51 (1): 16–34.

42 Jensen, J., Hösel, M., Dyer, A.L., and Krebs, F.C. (2015). Development and manufacture of polymer-based electrochromic devices. *Advanced Functional Materials* 25 (14): 2073–2090.

43 Collier, G.S., Pelse, I., Österholm, A.M., and Reynolds, J.R. (2018). Electrochromic polymers processed from environmentally benign solvents. *Chemistry of Materials* 30 (15): 5161–5168.

44 Ponder, J.F., Österholm, A.M., and Reynolds, J.R. (2017). Conjugated polyelectrolytes as water processable precursors to aqueous compatible redox active polymers for diverse applications: electrochromism, charge storage, and biocompatible organic electronics. *Chemistry of Materials* 29 (10): 4385–4392.

45 Schmatz, B., Yuan, Z., Lang, A.W., Hernandez, J.L., Reichmanis, E., and Reynolds, J.R. (2017). Aqueous processing for printed organic electronics: conjugated polymers with multistage cleavable side chains. *ACS Central Science* 3 (9): 961–967.

46 Collier, G.S., Pelse, I., and Reynolds, J.R. (2018). Aqueous electrolyte compatible electrochromic polymers processed from an environmentally sustainable solvent. *ACS Macro Letters* 7 (10): 1208–1214.

47 Wen, Y. and Xu, J. (2017). Scientific importance of water-processable PEDOT-PSS and preparation, challenge and new application in sensors of its film electrode: a review. *Journal of Polymer Science, Part A: Polymer Chemistry* 55 (7): 1121–1150.

48 Singh, R., Tharion, J., Murugan, S., and Kumar, A. (2017). ITO-free solution-processed flexible electrochromic devices based on PEDOT:PSS as transparent conducting electrode. *ACS Applied Materials & Interfaces* 9 (23): 19427–19435.

49 Kousseff, C.J., Taifakou, F.E., Neal, W.G., Palma, M., and Nielsen, C.B. (2022). Controlling morphology, adhesion, and electrochromic behavior of PEDOT films through molecular design and processing. *Journal of Polymer Science* 60 (3): 504–516.

50 Graßmann, C., Mann, M., van Langenhove, L., and Schwarz-Pfeiffer, A. (2020). Textile based electrochromic cells prepared with PEDOT: PSS and gelled electrolyte. *Sensors* 20 (19): 5691.

51 Krebs, F.C., Gevorgyan, S.A., and Alstrup, J. (2009). A roll-to-roll process to flexible polymer solar cells: model studies, manufacture and operational stability studies. *Journal of Materials Chemistry* 19 (30): 5442.

52 Helgesen, M., Søndergaard, R., and Krebs, F.C. (2010). Advanced materials and processes for polymer solar cell devices. *Journal of Materials Chemistry* 20 (1): 36–60.

53 Krebs, F.C., Tromholt, T., and Jørgensen, M. (2010). Upscaling of polymer solar cell fabrication using full roll-to-roll processing. *Nanoscale* 2 (6): 873–886.

54 Krebs, F.C., Fyenbo, J., and Jørgensen, M. (2010). Product integration of compact roll-to-roll processed polymer solar cell modules: methods and manufacture using flexographic printing, slot-die coating and rotary screen printing. *Journal of Materials Chemistry* 20 (41): 8994.

55 Helgesen, M., Carlé, J.E., Andreasen, B., Hösel, M., Norrman, K., Søndergaard, R., and Krebs, F.C. (2012). Rapid flash annealing of thermally reactive copolymers in a roll-to-roll process for polymer solar cells. *Polymer Chemistry* 3 (9): 2649.

56 Li, X., Perera, K., He, J., Gumyusenge, A., and Mei, J. (2019). Solution-processable electrochromic materials and devices: roadblocks and strategies towards large-scale applications. *Journal of Materials Chemistry C* 7 (41): 12761–12789.

57 Reeves, B.D., Grenier, C.R.G., Argun, A.A., Cirpan, A., McCarley, T.D., and Reynolds, J.R. (2004). Spray coatable electrochromic dioxythiophene polymers with high coloration efficiencies. *Macromolecules* 37 (20): 7559–7569.

58 Jensen, J., Dam, H.F., Reynolds, J.R., Dyer, A.L., and Krebs, F.C. (2012). Manufacture and demonstration of organic photovoltaic-powered electrochromic displays using roll coating methods and printable electrolytes. *Journal of Polymer Science Part B: Polymer Physics* 50 (8): 536–545.

59 Søndergaard, R.R., Hösel, M., Jørgensen, M., and Krebs, F.C. (2013). Fast printing of thin, large area, ITO free electrochromics on flexible barrier foil. *Journal of Polymer Science Part B: Polymer Physics* 51 (2): 132–136.

60 Lang, A.W., Li, Y., de Keersmaecker, M., de, Shen, D.E., Österholm, A.M., Berglund, L., and Reynolds, J.R. (2018). Transparent wood smart windows: polymer electrochromic devices based on poly(3,4-ethylenedioxythiophene):poly(styrene sulfonate) electrodes. *ChemSusChem* 11 (5): 854–863.

61 Jensen, J., Dyer, A.L., Shen, D.E., Krebs, F.C., and Reynolds, J.R. (2013). Direct photopatterning of electrochromic polymers. *Advanced Functional Materials* 23 (30): 3728–3737.

62 Argun, A.A., Aubert, P.-H., Thompson, B.C., Schwendeman, I., Gaupp, C.L., Hwang, J., Pinto, N.J., Tanner, D.B., MacDiarmid, A.G., and Reynolds, J.R. (2004). Multicolored electrochromism in polymers: structures and devices. *Chemistry of Materials* 16 (23): 4401–4412.

63 Argun, A.A. and Reynolds, J.R. (2005). Line patterning for flexible and laterally configured electrochromic devices. *Journal of Materials Chemistry* 15 (18): 1793.

64 Linderhed, U., Petsagkourakis, I., Ersman, P.A., Beni, V., and Tybrandt, K. (2021). Fully screen printed stretchable electrochromic displays. *Flexible and Printed Electronics* 6 (4): 45014.

65 Brooke, R., Petsagkourakis, I., Majee, S., Olsson, O., Dahlin, A., and Andersson Ersman, P. (2022). All-printed multilayers and blends of poly(dioxythiophene) derivatives patterned into flexible electrochromic displays. *Macromolecular Materials and Engineering* 308 (2): 2200453.

66 Andersson Ersman, P., Freitag, K., Nilsson, M., Åhlin, J., Brooke, R., Nordgren, N., Aulin, C., Fall, A., Nevo, Y., and Beni, V. (2022). Electrochromic displays screen printed on transparent nanocellulose-based substrates. *Advanced Photonics Research* 4 (1): 2200012.

67 Gu, C., Jia, A.-B., Zhang, Y.-M., and Zhang, S.X.-A. (2022). Emerging electrochromic materials and devices for future displays. *Chemical Reviews* 122 (18): 14679–14721.

68 Beaujuge, P.M., Ellinger, S., and Reynolds, J.R. (2008). The donor-acceptor approach allows a black-to-transmissive switching polymeric electrochrome. *Nature Materials* 7 (10): 795–799.

69 Sassi, M., Salamone, M.M., Ruffo, R., Patriarca, G.E., Mari, C.M., Pagani, G.A., Posset, U., and Beverina, L. (2016). State-of-the-art neutral tint multichromophoric polymers for high-contrast see-through electrochromic devices. *Advanced Functional Materials* 26 (29): 5240–5246.

70 Xu, Z., Yue, H., Wang, B., Zhao, J., Wang, M., Zhang, Y., and Xie, Y. (2020). Color tuning for black-to-transmissive conjugated copolymer with excellent electrochromic properties via electrochemical copolymerization of two donor-acceptor type monomers. *Materials & Design* 194: 108903.

71 Savagian, L.R., Österholm, A.M., Shen, D.E., Christiansen, D.T., Kuepfert, M., and Reynolds, J.R. (2018). Conjugated polymer blends for high contrast black-to-transmissive electrochromism. *Advanced Optical Materials* 6 (19): 1800594.

72 Jarosz, T., Gebka, K., Stolarczyk, A., and Domagala, W. (2019). Transparent to black electrochromism – the "Holy Grail" of organic optoelectronics. *Polymers* 11 (2): 273.

73 Shi, P., Amb, C.M., Knott, E.P., Thompson, E.J., Liu, D.Y., Mei, J., Dyer, A.L., and Reynolds, J.R. (2010). Broadly absorbing black to transmissive switching electrochromic polymers. *Advanced Materials* 22 (44): 4949–4953.

74 Yan, S., Zhang, L., Lv, X., Sun, J., Zhang, Y., and Zhang, C. (2022). Black-to-transmissive electrochromism in π-conjugated polymer-based materials and devices. *Advanced Photonics Research* 3 (1): 2000199.

75 Schubert, U.S., Hofmeier, H., and Newkome, G.R. (2006). *Modern Terpyridine Chemistry*. Wiley-VCH.

76 Schubert, U.S., Winter, A., and Newkome, G.R. (2011). *Terpyridine-based materials: for catalytic, optoelectronic and life science applications*. Wiley-VCH.

77 Han, F.S., Higuchi, M., and Kurth, D.G. (2008). Metallosupramolecular polyelectrolytes self-assembled from various pyridine ring-substituted bisterpyridines and metal ions: photophysical, electrochemical, and electrochromic properties. *Journal of the American Chemical Society* 130 (6): 2073–2081.

78 Hu, C.-W., Sato, T., Zhang, J., Moriyama, S., and Higuchi, M. (2014). Three-dimensional Fe(II)-based metallo-supramolecular polymers with electrochromic properties of quick switching, large contrast, and high coloration efficiency. *ACS Applied Materials & Interfaces* 6 (12): 9118–9125.

79 Higuchi, M. (2019). *Metallo-Supramolecular Polymers: Synthesis, Properties, and Device Applications*. Tokyo: Springer Japan.

80 Schwarz, G., Sievers, T.K., Bodenthin, Y., Hasslauer, I., Geue, T., Koetz, J., and Kurth, D.G. (2010). The structure of metallo-supramolecular polyelectrolytes in solution and on surfaces. *Journal of Materials Chemistry* 20 (20): 4142.

81 Pandey, R.K., Hossain, M.D., Sato, T., Rana, U., Moriyama, S., and Higuchi, M. (2015). Effect of a three-dimensional hyperbranched structure on the ionic conduction of metallo-supramolecular polymers. *RSC Advances* 5 (61): 49224–49230.

82 Narayana, Y.S.L.V., Mondal, S., Rana, U., Ninomiya, Y., Santra, D.C., Yoshida, T., and Higuchi, M. (2021). One-step synthesis of a three-dimensionally hyperbranched Fe(II)-based metallo-supramolecular polymer using an asymmetrical ditopic ligand for durable electrochromic films with wide absorption, large optical contrast, and high coloration efficiency. *ACS Applied Electronic Materials* 3 (5): 2044–2055.

83 Higuchi, M. (2014). Stimuli-responsive metallo-supramolecular polymer films: design, synthesis and device fabrication. *Journal of Materials Chemistry C* 2 (44): 9331–9341.

84 Pai, S., Schott, M., Niklaus, L., Posset, U., and Kurth, D.G. (2018). A study of the effect of pyridine linkers on the viscosity and electrochromic properties of metallo-supramolecular coordination polymers. *Journal of Materials Chemistry C* 6 (13): 3310–3321.

85 Hossain, M.D., Sato, T., and Higuchi, M. (2013). A green copper-based metallo-supramolecular polymer: synthesis, structure, and electrochromic properties. *Chemistry, an Asian Journal* 8 (1): 76–79.

86 Han, F.S., Higuchi, M., and Kurth, D.G. (2007). Metallo-supramolecular polymers based on functionalized bis-terpyridines as novel electrochromic materials. *Advanced Materials* 19 (22): 3928–3931.

87 Pai, S., Moos, M., Schreck, M.H., Lambert, C., and Kurth, D.G. (2017). Green-to-red electrochromic Fe(II) metallo-supramolecular polyelectrolytes self-assembled from fluorescent 2,6-bis(2-pyridyl)pyrimidine bithiophene. *Inorganic Chemistry* 56 (3): 1418–1432.

88 Hu, C.-W., Sato, T., Zhang, J., Moriyama, S., and Higuchi, M. (2013). Multi-colour electrochromic properties of Fe/Ru-based bimetallo-supramolecular polymers. *Journal of Materials Chemistry C* 1 (21): 3408.

89 Bera, M.K., Ninomiya, Y., and Higuchi, M. (2020). Synthesis of an alternated heterobimetallic supramolecular polymer based on Ru(II) and Fe(II). *Molecules* 25 (22): 5261.

90 Hossain, M.D., Zhang, J., Pandey, R.K., Sato, T., and Higuchi, M. (2014). A Heterometallo-Supramolecular Polymer with Cu I and Fe II Ions Introduced Alternately. *European Journal of Inorganic Chemistry* 2014 (23): 3763–3770.

91 Bera, M.K., Ninomiya, Y., Yoshida, T., and Higuchi, M. (2020). Precise synthesis of alternate Fe(II)/Os(II)-based bimetallic metallo-supramolecular polymer. *Macromolecular Rapid Communications* 41 (1): e1900384.

92 Bera, M.K., Ninomiya, Y., and Higuchi, M. (2020). Constructing alternated heterobimetallic Fe(II)/Os(II) supramolecular polymers with diverse solubility for facile fabrication of voltage-tunable multicolor electrochromic devices. *ACS Applied Materials & Interfaces* 12 (12): 14376–14385.

93 Bera, M.K., Ninomiya, Y., and Higuchi, M. (2021). Stepwise introduction of three different transition metals in metallo-supramolecular polymer for quad-color electrochromism. *Communications Chemistry* 4 (1): 56.

94 Schott, M., Szczerba, W., and Kurth, D.G. (2014). Detailed study of layer-by-layer self-assembled and dip-coated electrochromic thin films based on metallo-supramolecular polymers. *Langmuir* 30 (35): 10721–10727.

95 Schott, M., Beck, M., Winkler, F., Lorrmann, H., and Kurth, D.G. (2015). Fabricating electrochromic thin films based on metallo-polymers using layer-by-layer self-assembly: an attractive laboratory experiment. *Journal of Chemical Education* 92 (2): 364–367.

96 Chen, B.-H., Kao, S.-Y., Hu, C.-W., Higuchi, M., Ho, K.-C., and Liao, Y.-C. (2015). Printed multicolor high-contrast electrochromic devices. *ACS Applied Materials & Interfaces* 7 (45): 25069–25076.

97 Xu, L.-J., Li, B., Gao, G.-L., and Jiang, Z. (2022). A metallosupramolecular polymer deposited via inkjet printing for fast-switching pixelated electrochromic devices. *Journal of Materials Chemistry C* 10 (9): 3353–3359.

98 Bao, X., Zhao, Q., Wang, H., Liu, K., and Qiu, D. (2013). Metallopolymer electrochromic film prepared by oxidative electropolymerization of a Fe(II) complex with arylamine functionalized terpyridine ligand. *Inorganic Chemistry Communications* 38: 88–91.

99 Liang, Y., Strohecker, D., Lynch, V., Holliday, B.J., and Jones, R.A. (2016). A thiophene-containing conductive metallopolymer using an Fe(II) bis(terpyridine) core for electrochromic materials. *ACS Applied Materials & Interfaces* 8 (50): 34568–34580.

100 Schott, M., Lorrmann, H., Szczerba, W., Beck, M., and Kurth, D.G. (2014). State-of-the-art electrochromic materials based on metallo-supramolecular polymers. *Solar Energy Materials and Solar Cells* 126: 68–73.

101 Schott, M., Niklaus, L., Clade, J., and Posset, U. (2019). Electrochromic metallo-supramolecular polymers showing visible and near-infrared light transmittance modulation. *Solar Energy Materials and Solar Cells* 200: 110001.

102 Szczerba, W., Schott, M., Riesemeier, H., Thünemann, A.F., Kurth, D.G. (2014). Thermally induced structural rearrangement of the Fe(II) coordination geometry in metallo-supramolecular polyelectrolytes. *Physical Chemistry Chemical Physics* 16 (36): 19694–19701.

103 Schott, M., Niklaus, L., Müller, C., Bozkaya, B., and Giffin, G.A. (2021). Flexible electrochromic devices prepared on ultra-thin ITO glass. *Materals Advances* 2 (14): 4659–4666.

104 Niklaus, L., Schott, M., Mihelčič, M., Jerman, I., Posset, U., and Sextl, G. (2019). Metallopolymers and non-stoichiometric nickel oxide: towards neutral tint large-area electrochromic devices. *Solar Energy Materials and Solar Cells* 200: 110002.

105 Niklaus, L., Schott, M., Posset, U., and Giffin, G.A. (2020). Redox electrolytes for hybrid type II electrochromic devices with Fe–MEPE or $Ni_{1-x}O$ as electrode materials. *ChemElectroChem* 7 (15): 3274–3283.

106 Niklaus, L., Schott, M., Posset, U., Mihelčič, M., Jerman, I., and Giffin, G.A. (2021). Charge balancing and optical contrast optimization in Fe-MEPE/$Ni_{1-x}O$ electrochromic devices containing a Li reference electrode. *Solar Energy Materials and Solar Cells* 227: 111080.

107 Niklaus, L., Schott, M., Subel, J., Ulrich, S., Reichert, D., Posset, U., and Giffin, G.A. (2021). Mixed metal oxides as optically-passive ion storage layers in electrochromic devices based on metallopolymers. *Solar Energy Materials and Solar Cells* 223: 110950.

108 Schott, M., Szczerba, W., Posset, U., Šurca Vuk, A., Beck, M., Riesemeier, H., Thünemann, A.F., and Kurth, D.G. (2016). In operando XAFS experiments on flexible electrochromic devices based on Fe(II)-metallo-supramolecular polyelectrolytes and vanadium oxide. *Solar Energy Materials and Solar Cells* 147: 61–67.

109 Mondal, S., Yoshida, T., Rana, U., Bera, M.K., and Higuchi, M. (2019). Thermally stable electrochromic devices using Fe(II)-based metallo-supramolecular polymer. *Solar Energy Materials and Solar Cells* 200: 110000.

110 Lu, H.-C., Hsiao, L.-Y., Kao, S.-Y., Seino, Y., Santra, D.C., Ho, K.-C., and Higuchi, M. (2021). Durable electrochromic devices driven at 0.8 V by complementary chromic combination of metallo-supramolecular polymer and prussian blue analogues for smart windows with low-energy consumption. *ACS Applied Electronic Materials* 3 (5): 2123–2135.

111 Santra, D.C., Mondal, S., Yoshida, T., Ninomiya, Y., and Higuchi, M. (2021). Ru(II)-based metallo-supramolecular polymer with tetrakis(*N*-methylbenzimidazolyl)bipyridine for a durable, nonvolatile, and electrochromic device driven at 0.6 V. *ACS Applied Materials & Interfaces* 13 (26): 31153–31162.

112 Mondal, S., Chandra Santra, D., Ninomiya, Y., Yoshida, T., and Higuchi, M. (2020). Dual-redox system of Metallo-supramolecular polymers for visible-to-near-IR modulable electrochromism and durable device fabrication. *ACS Applied Materials & Interfaces* 12 (52): 58277–58286.

113 Sato, T. and Higuchi, M. (2013). An alternately introduced heterometallo-supramolecular polymer: synthesis and solid-state emission switching by electrochemical redox. *Chemical Communications* 49 (46): 5256–5258.

114 Wang, Z., Wang, X., Cong, S., Geng, F., and Zhao, Z. (2020). Fusing electrochromic technology with other advanced technologies: a new roadmap for future development. *Materials Science & Engineering R: Reports* 140: 100524.

115 Eric Shen, D., Österholm, A.M., and Reynolds, J.R. (2015). Out of sight but not out of mind: the role of counter electrodes in polymer-based solid-state electrochromic devices. *Journal of Materials Chemistry C* 3 (37): 9715–9725.

116 Hassab, S., Shen, D.E., Österholm, A.M., Reynolds, J.R., and Padilla, J. (2018). Exploring unbalanced electrode configurations for electrochromic devices. *Journal of Materials Chemistry C* 6 (2): 393–400.

117 Knott, E.P., Craig, M.R., Liu, D.Y., Babiarz, J.E., Dyer, A.L., and Reynolds, J.R. (2012). A minimally coloured dioxypyrrole polymer as a counter electrode material in polymeric electrochromic window devices. *Journal of Materials Chemistry* 22 (11): 4953.

118 Hassab, S. and Padilla, J. (2016). Using WO_3 as a transparent, optically-passive counter electrode in an unbalanced electrochromic configuration. *Electrochemistry Communications* 72: 87–90.

119 Brooke, R., Mitraka, E., Sardar, S., Sandberg, M., Sawatdee, A., Berggren, M., Crispin, X., and Jonsson, M.P. (2017). Infrared electrochromic conducting polymer devices. *Journal of Materials Chemistry C* 5 (23): 5824–5830.

120 Granqvist, C.G., Bayrak Pehlivan, İ., and Niklasson, G.A. (2018). Electrochromics on a roll: web-coating and lamination for smart windows. *Surface and Coatings Technology* 336: 133–138.

121 Macher, S., Rumpel, M., Schott, M., Posset, U., Giffin, G.A., and Löbmann, P. (2020). Avoiding voltage-induced degradation sin PET-ITO-based flexible electrochromic devices. *ACS Applied Materials & Interfaces* 12 (32): 36695–36705.

122 Jensen, J., Madsen, M.V., and Krebs, F.C. (2013). Photochemical stability of electrochromic polymers and devices. *Journal of Materials Chemistry C* 1 (32): 4826.

123 Woodard, F.E. and Furler, R. (2001). Vacuum deposition of bus bars onto conductive transparent films. US6204480B1.

124 Stuart, B.W., Tao, X., Gregory, D., and Assender, H.E. (2020). Roll-to-roll patterning of Al/Cu/Ag electrodes on flexible poly(ethylene terephthalate) by oil masking: a comparison of thermal evaporation and magnetron sputtering. *Applied Surface Science* 505: 144294.

125 Välimäki, M.K., Jansson, E., von Morgen, V.J.J., Ylikunnari, M., Väisänen, K.-L., Ontero, P., Kehusmaa, M., Korhonen, P., and Kraft, T.M. (2022). Accuracy control for roll and sheet processed printed electronics on flexible plastic substrates. *The International Journal of Advanced Manufacturing Technology* 119 (9-10): 6255–6273.

126 Park, C., Kim, J.M., Kim, Y., Bae, S., Do, M., Im, S., Yoo, S., and Kim, J.H. (2021). High-coloration efficiency and low-power consumption electrochromic film based on multifunctional conducting polymer for large scale smart windows. *ACS Applied Electronic Materials* 3 (11): 4781–4792.

127 Heuer, H.W., Wehrmann, R., and Kirchmeyer, S. (2002). Electrochromic window based on conducting poly(3,4-ethylenedioxythiophene)-poly(styrene sulfonate). *Advanced Functional Materials* 12 (2): 89–94.

128 Zhu, Y., Otley, M.T., Alamer, F.A., Kumar, A., Zhang, X., Mamangun, D.M.D., Li, M., Arden, B.G., and Sotzing, G.A. (2014). Electrochromic properties as a function of electrolyte on the performance of electrochromic devices consisting of a single-layer polymer. *Organic Electronics* 15 (7): 1378–1386.

129 Kumar, A., Otley, M.T., Alamar, F.A., Zhu, Y., Arden, B.G., and Sotzing, G.A. (2014). Solid-state electrochromic devices: relationship of contrast as a function of device preparation parameters. *Journal of Materials Chemistry C* 2 (14): 2510–2516.

130 Shao, Z., Huang, A., Ming, C., Bell, J., Yu, P., Sun, Y.-Y., Jin, L., Ma, L., Luo, H., Jin, P., and Cao, X. (2022). All-solid-state proton-based tandem structures for fast-switching electrochromic devices. *Nature Electronics* 5 (1): 45–52.

131 Hassab, S., Shen, E., Österholm, A., Rocha, M., Song, G., Alesanco, Y., Viñuales, A., Rougier, A., Reynolds, J., and Padilla, J. (2018). A new standard method to calculate electrochromic switching time. *Solar Energy Materials and Solar Cells* 185: 54–60. https://doi.org/10.1016/j.solmat.2018.04.031.

132 Padilla, J., Niklaus, L., Schott, M., Posset, U., Faceira, B., Mjejri, I., Rougier, A., Alesanco, Y., Viñuales, A., Shen, E., Österholm, A, and Reynolds, J. (2023). Quantitative assessment of the cycling stability of different electrochromic materials and devices. *ACS Applied Optical Materials* 1 (6): 1174–83. https://doi.org/10.1021/acsaom.3c00087.

133 Ding, Y., Invernale, M.A., Mamangun, D.M.D., Kumar, A., and Sotzing, G.A. (2011). A simple, low waste and versatile procedure to make polymer electrochromic devices. *Journal of Materials Chemistry* 21 (32): 11873.

134 Ma, C., Taya, M., and Xu, C. (2008). Smart sunglasses based on electrochromic polymers. *Polymer Engineering and Science* 48 (11): 2224–2228.

135 Neo, W.T., Ye, Q., Chua, S.-J., and Xu, J. (2016). Conjugated polymer-based electrochromics: materials, device fabrication and application prospects. *Journal of Materials Chemistry C* 4 (31): 7364–7376.

136 Thakur, V.K., Ding, G., Ma, J., Lee, P.S., and Lu, X. (2012). Hybrid materials and polymer electrolytes for electrochromic device applications. *Advanced Materials* 24 (30): 4071–4096.

137 Levasseur, D., Mjejri, I., Rolland, T., and Rougier, A. (2019). Color tuning by oxide addition in PEDOT:PSS-based electrochromic devices. *Polymers* 11 (1): 179.

138 Mjejri, I., Grocassan, R., and Rougier, A. (2018). Enhanced coloration for hybrid niobium-based electrochromic devices. *ACS Applied Energy Materials* 1 (8): 4359–4366.

139 Futsch, R., Mjejri, I., Rakotozafy, H., and Rougier, A. (2020). PEDOT:PSS-V2O5 hybrid for color adjustement in electrochromic systems. *Frontiers in Materials* 7 (78).

140 Lang, A.W., Ji, Y., Dillon, A.C., Satam, C.C., Meredith, J.C., and Reynolds, J.R. (2021). Photostability of ambient-processed, conjugated polymer electrochromic devices encapsulated by bioderived barrier films. *ACS Sustainable Chemistry & Engineering* 9 (7): 2937–2945.

141 Bulloch, R.H. and Reynolds, J.R. (2016). Photostability in dioxyheterocycle electrochromic polymers. *Journal of Materials Chemistry C* 4 (3): 603–610.

142 Kim, Y., Kim, H., Graham, S., Dyer, A., and Reynolds, J.R. (2012). Durable polyisobutylene edge sealants for organic electronics and electrochemical devices. *Solar Energy Materials and Solar Cells* 100: 120–125.

143 Byker, H.J. (2001). Electrochromics and polymers. *Electrochimica Acta* 46 (13-14): 2015–2022.

144 Yu, H., Qi, M., Wang, J., Yin, Y., He, Y., Meng, H., and Huang, W. (2019). A feasible strategy for the fabrication of camouflage electrochromic fabric and unconventional devices. *Electrochemistry Communications* 102: 31–36.

145 Perera, K., Yi, Z., You, L., Ke, Z., and Mei, J. (2020). Conjugated electrochromic polymers with amide-containing side chains enabling aqueous electrolyte compatibility. *Polymer Chemistry* 11 (2): 508–516.

146 You, L. and Mei, J. (2022). All-donor black electrochromic polymer and method for preparing same. US20220017689A1.

147 https://www.ambilightinc.com/ (accessed 5 October 2022).

148 https://www.saralon.com/de/blog/product-updates-print-electrochromic-flexible-displays-through-saral-inks/ (accessed 5 October 2022).

149 https://www.ynvisible.com/ (accessed 5 October 2022).

150 Marques, A.C.L., Baptista, C.A.P., and Araújo, J. (2012) Electrolyte solution, printing method thereof and resulting solid electrolyte. EP2821458B1; (2018) US9625782B2 (2017).

151 https://www.heraeus.com/en/hep/products_hep/clevios/clevios_prod/clevios_1.html (accessed 4 October 2022).

152 https://www.ormecon.co/ (accessed 5 October 2022).

153 Wessling, B. and Volk, H. (1986). Post-polymerization processing of conductive polymers: a way of converting conductive polymers to conductive materials? *Synthetic Metals* 15 (2-3): 183–193.

154 Kim, J., Rémond, M., Kim, D., Jang, H., and Kim, E. (2020). Electrochromic conjugated polymers for multifunctional smart windows with integrative functionalities. *Advanced Materials Technologies* 5 (6): 1900890.

155 Zhang, H., Tu, S., Li, L., Chen, X., Zhao, Y., Wu, M., Zhang, X., Zhang, S., and Chen, L. (2022). Large scale transparency-adjustable mini-LED display with recoverable color gamut by a highly transparent electrochromic shutter. *Optics Express* 30 (22): 39904–39910.

156 Lu, X., Liu, G.-J., Li, Y.-H., Ma, Y.-Y., Fan, J., and Fung, M.-K. (2018). Smart OLED Lighting on Electrochromic Glass. *Physica Status Solidi A: Applications and Materials Science* 215 (21): 1800102.

157 Cossari, P., Pugliese, M., Gigli, G., and Maiorano, V. (2021). All solid-state flexible electrochromic-organic light-emitting diode devices on single-plastic substrate for see-through display technologies. *Advanced Materials Technologies* 6 (11): 2100289.

158 Cossari, P., Pugliese, M., Gambino, S., Cannavale, A., Maiorano, V., Gigli, G., and Mazzeo, M. (2018). Fully integrated electrochromic-OLED devices for highly transparent smart glasses. *Journal of Materials Chemistry C* 6 (27): 7274–7284.

159 Kim, G.W., Lampande, R., Choe, D.C., Ko, I.J., Park, J.H., Pode, R., and Kwon, J.H. (2018). Next generation smart window display using transparent organic display and light blocking screen. *Optics Express* 26 (7): 8493–8502.

5

Evolution of Industrial Polymer Dispersed Liquid Crystal (PDLC) Technology in Europe: A Review of Research, Development, Manufacturing, and Potential Emerging Technologies

H. Hakemi

Plastic Liquid Crystal Technology, Macherio, Italy

Nomenclature

C&L	coating and lamination
EHDI	electro-hydro-dynamic Instability
ITO	indium tin oxide
KSU	Kent State University
LC	liquid crystal
LCD	liquid crystal device
PR	process rollers
ME	micro-emulsion
NCAP	nematic curvilinear aligned phase
PET	polyethylene terephthalate
PS	phase separation
PIPS	polymer induced phase separation
PDLC	polymer dispersed liquid crystal
PVA	polyvinyl alcohol
SIPS	solvent induced phase separation
SNR	Snia-Ricerche, S.p.A.
TT	thermal treatment
TIPS	thermal induced phase separation
UV	ultraviolet light

5.1 Introduction to PDLC Technology

Polymer dispersed liquid crystal (PDLC) technology is the generic name of a composite film device consisting of dispersions of liquid crystal in a polymer matrix with various morphologies and process techniques. From the application perspective, PDLC belongs to a class of electrically active film and glass products referred to as "Smart Glass." Industrial development of PDLCs has been made by Gauzy (Israel), NSG (Japan), Polytronix/Polytron (USA, TX), DMDisplays (S. Korea), ABTech (S. Korea), Innoptec (Italy), DreamGlass (Spain), Vanlong (China), CNCNext (China), Scienstry (USA, TX), etc. Also, Smart Glass technologies

Chromogenics: Smart Switchable Optical Materials and Their Applications, First Edition. Edited by Carl M. Lampert.

include the *suspended particle device* (SPD) developed by RFI-Research Frontiers, Inc (USA, NY) with manufacturing made by Hitachi Chemical (Japan), Gauzy, Vision Systems (France), Isoclima (Italy), NSG, SPD CSC (China), PPG (USA, PA), and SGI-Smart Glass, Int. (Germany); ***Electrochromic* (EC)** developed by Saint-Gobain/SageGlass (USA, MI), View (USA, CA), Gentex (USA, MI), Econtrol Glass (Germany, defunct), Gesimet (Germany, defunct), and Tilse-FormGlass (Germany); ***Thermochromic* (TC)** developed by RavenWindow (USA, CO) and Pleotint (USA, MI); ***Thermotropic* (TT)** developed by FormGlas-Solardim ECO (Germany); and ***Photochromic* (PC)** developed by Corning (USA, NY), PPG, SMT(China), SYS Optical (China), HDCrafter (China), and Porters. Their applications include large-area switchable films, screens, and windows for architectural, construction, automotive, entertainment, security, advertising, and other high-end glass markets.

The original interest in the research and industrial development of PDLC technology originated on the premise of stabilizing liquid crystals (LC) through dispersions in a polymer matrix for manufacturing large-area flexible liquid crystal display (LCD) devices by simple and low-cost roll-to-roll plastic converting techniques. The premises of such alternative technology included:

(a) Possibility of a roll-to-roll production of polymer-stabilized liquid crystal flat panel films on flexible plastic substrates for large-area applications.
(b) Simplifying the fabrication techniques and cost reduction of conventional LCD devices.
(c) Improving the electro-optical and other performance criteria of these flexible devices with respect to those of commercial LCD products.
(d) Expanding the LCD market to new areas of applications and global markets.

However, due to a number of technical problems in the development of flexible LCD, the PDLC technology has found its way in new field of application as large-area switchable films, shutters, and windows for architectural, construction, automotive, marine, entertainment, information, security, advertising, and other markets.

The past decades have witnessed a parallel surge of academic research and industrial activities on PDLC technology. However, at industrial scale, the PDLC films have been manufactured at large scale on flexible substrates with different materials, formulations, and processes by continuous roll-to-roll techniques. It should be noted that the majority of PDLC literature has been carried out by academic institutes, because the PDLC industries, due to confidentiality against competition, rarely publish their manufacturing techniques in detail. Some information can be gleaned through the study of patents.

A flexible PDLC device is an electrically switchable composite device consisting of a thin layer (10–30 μm) containing LC microdroplets or nano-droplets in a solid polymer matrix. The PDLC layer is prepared as a thin film which is sandwiched between two transparent conductive indium tin oxide (ITO)-coated glass sheets or polyethylene terephthalate (PET) substrates by various methods. The operational principle of a standard PDLC film is that in the absence of electric field (off-state), the PDLC film is highly scattering, due to random orientations of LC molecules inside droplets, and the PDLC device appears "opaque" or "translucent." By application of an electric field (on-state), the LC molecules inside droplets are aligned with the field and the device becomes "transparent." Such transparent condition is achieved only when the refractive index of matrix is matched with ordinary refractive index of LC. The fast millisecond switching response of LC droplets to the electric field causes a rapid interchange from "opacity" to "transparency," a mechanism that could be used for many far-reaching applications. The basic construction, operational principle, and conventional applications of typical PDLC device are presented in Figures 5.1a–c.

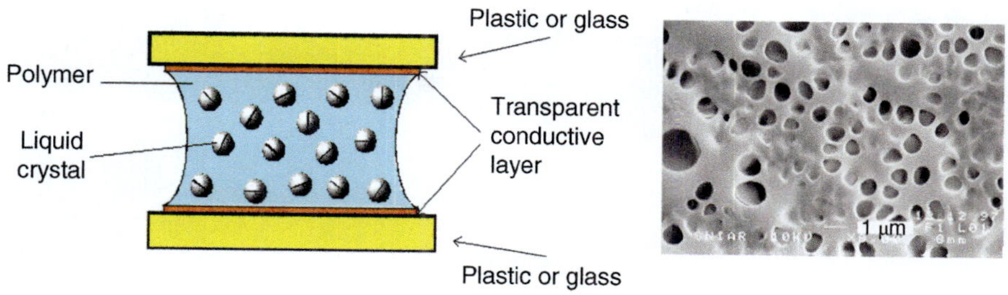

Figure 5.1a On the left is shown how the PDLC device is constructed with transparent conductors and supporting sheets of glass or plastic (not to scale). On the right is a scanning electron micrograph showing typical micron-size liquid crystal droplets.

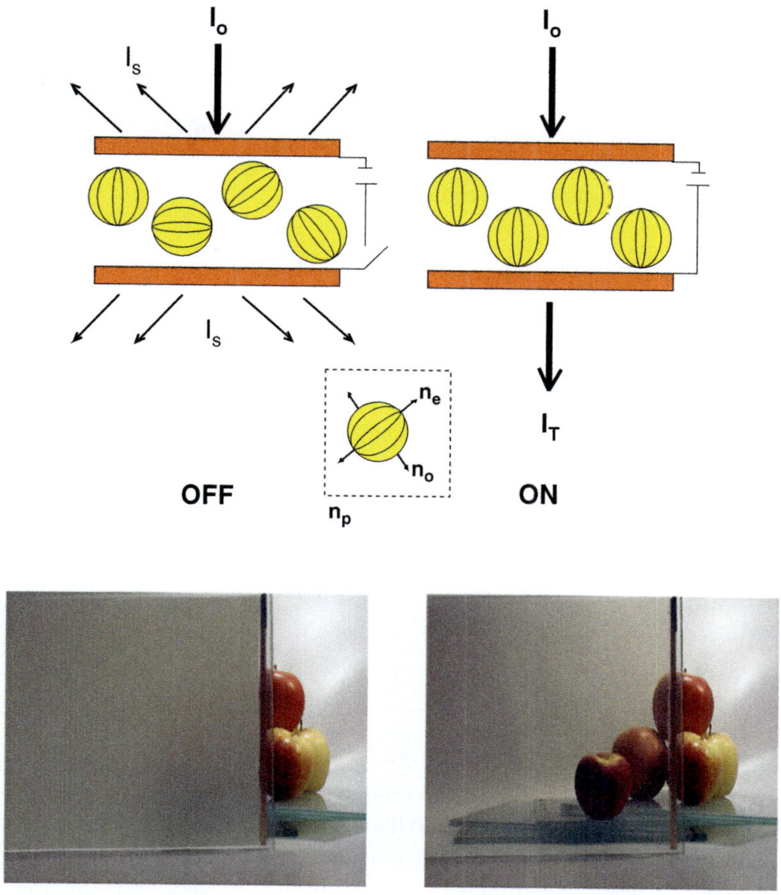

Figure 5.1b In the top figure, a schematic diagram shows how the liquid crystal directors inside microdroplets are randomly aligned with no electric field causing scattering mode and become transparent due to alignment of liquid crystal directors with the applied electric field. This result is shown in the lower images of an actual PDLC window.

Figure 5.1c Examples of PDLC privacy window applications.

The main electro-optical properties of commercial PDLC products are the typical *transmittance–voltage* (T–V) plot, *intensity–time* (I–t) plot, and LC morphology. The T–V plot provides the off-state (opacity) and on-state (transparency) transmittances, which are measured through variations in light transmission intensity as a function of VAC applied electric voltage within 25–50 Hz frequency range. The typical electro-optical parameters of PDLC are *opacity* (T_{off}), *transparency* (T_{on}), *threshold voltage* (V_{10}), and *saturation voltage* (V_{90}). A typical intensity-time plot provides the switching parameters of *rise time* (τ_{on}) and *decay time* (τ_{off}) through application of short electric pulses to PDLC and measurement of their characteristic responses. Further descriptions of electro-optical properties of PDLC have been published in many academic and industrial literature, c.f. [1].

5.2 The Original PDLC Inventions

Thse PDLC technologies were originally invented and patented in the United States during the 1980s by two independent industrial and academic institutes, Raychem, Redwood City, CA, USA, and Kent State University, Ohio, USA. Since then, there has been a surge of academic and industrial research and developments which have resulted in a large number of academic studies and many industrial developments and commercialization of switchable PDLC film and glass products for various applications and markets worldwide. During the industrial development period, as a result of patented invention, issues will be discussed in the following section. Many patented inventions use PDLC as a trademark or common terminology. The early technological inventions were classified as micro-emulsions (ME) and phase separation (PS).

5.2.1 Micro-Emulsion (ME) Invention

The micro-emulsion (ME) method was invented and patented by J. L. Fergason in 1984, following the out-licensing of ME technology by Fergason to Raychem, other patents published under Raychem's ownership [2]. During the industrial development, the ME method was referred to as nematic curvilinear aligned phase (NCAP). The ME is a mechanical process based on encapsulation of liquid crystal (LC) as droplets in a water-based polyvinyl-alcohol (PVA) matrix. The encapsulation of LC is carried out by ultrasonic mixing in immiscible water-based PVA and LC solution, coating of the encapsulated mix on a first metalized glass or film support, vaporization of water, and laminating of the remaining paste with a second metalized glass or film support.

5.2.2 Phase Separation (PS) Invention

The phase separation (PS) technique, referred to as PDLC, is a thermodynamic approach and was originally invented and first patented in 1987 along with later patents by Kent State University (KSU) [3]. The PS techniques are classified into three categories according to the type and nature of materials, curing source, and curing process.

5.2.2.1 Polymer Induced Phase Separation (PIPS)

The PIPS technique generally utilizes polymer matrices that are formed by either a step/growth reaction or a free-radical chain reaction to produce the desired morphology of LC droplets. The basic difference between the PIPS and the other two techniques is that with PIPS, the starting binding material is not a polymer but rather a prepolymer, polymer precursor, or oligomer. These materials are either of "thermoset" (epoxy-based) resins or a "radiation curable" (UV-curable) prepolymers. The first PDLC processing at industrial scale was carried out with epoxy-based (TS-PIPS) method. The main advantage of PIPS is its rapid and complete curing, as well as it its simpler and more efficient roll-to-roll processing compared to the other PS techniques. Although the PS and curing kinetics in UV-PIPS is similar to that of TS-PIPS method, there are some differences. One of the major differences at industrial scale is that with the much faster UV-PIPS method, around 80–90% of the matrix is cured within the first few seconds of the UV exposure. Whereas the degree of curing with TS-PIPS method is around 50–60% during the first heat exposure and the PDLC usually requires a lengthy post-curing procedure. The phase separation with UV curing (UV-PIPS) method became the most preferred PDLC processing technique in both academic research and industrial development.

5.2.2.2 Solvent Induced Phase Separation (SIPS)

The SIPS method requires a common solvent to dissolve the LC material into the thermoplastic polymer matrix. The SIPS process, similar to ME method, includes the coating of a homogeneous solution of LC and polymer on a first glass or plastic substrate, thermal evaporation of solvent and PS of LC droplets from polymer, subsequent solidification of matrix, and final lamination with a second glass or plastic substrate. Whereas the thermoplastic materials in SIPS technique are structurally not tractable polymers and do not melt below their degradation temperatures.

5.2.2.3 Thermal Induced Phase Separation (TIPS)

The third phase separation technique is TIPS, whose PDLC process results from coating of a "hot melt" (solvent-less) homogeneous mixture of LC and thermoplastic in cosolvent directly between two glass or plastic substrates and creation of PDLC morphology by cooling at a desirable rate. Whereas the thermoplastic materials in TIPS technique are structurally similar to those used in the SIPS, they are tractable polymers and melt below their degradation temperatures. It is worth noting that, since the polymer matrix in both TIPS and SIPS methods are thermoplastic materials, it is also possible to "reset" the PDLC morphology by using a combined SIPS/TIPS process, where the initial PS is carried out by the SIPS method followed by heating and a subsequent cooling and resetting of morphology by the TIPS method. In this approach, the use of a slow-reacting cross-linking agent can permanently lock in the morphology of PDLC after cooling.

In general, the PDLC film process by PIPS and TIPS techniques provides composites based on thermodynamic phenomenon that mainly consists of liquid crystal and polymer components, whose description can be presented by a schematic phase diagram and morphologies presented in Figure 5.2. Depending on the nature and miscibility of raw materials, up to certain low liquid crystal concentration of around 30%, the system becomes a one-phase composite. In this type of composite, the liquid crystal is totally dissolved in the matrix as plasticizer and the device has no electrooptical value.

According to Figure 5.2, the composite within the liquid crystal concentration of around 30–80% exhibits a two-phase composite, where based on material miscibility and curing conditions, the phase-separated portion of liquid crystal as droplets (nano, micro, or macro) becomes a discontinuous phase in PDLC or becomes a continuous phase in liquid crystal dispersed polymer, LCDP morphologies. Interestingly in traditional literature, the PDLC morphology is also referred to as Swiss Cheese and LCDP morphology is referred to as "Polymer Balls." It should be noted that, there is no clear distinction between PDLC and LCDP concentrations, where the latter morphology could also occur at lower concentrations by properly selected materials and curing conditions. At liquid crystal concentrations above around 80%, the morphology of the composite becomes a discontinuous polymer network within the continuous liquid crystal phase and such gel network morphology is commonly

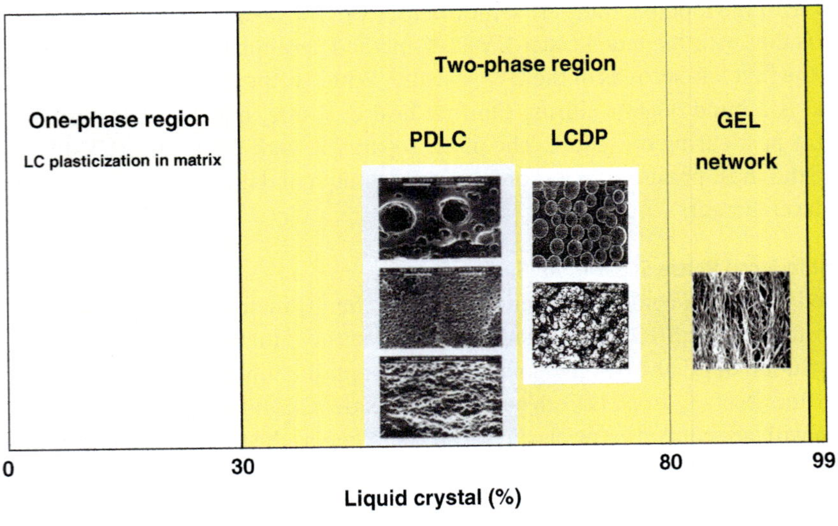

Figure 5.2 Phase diagram of liquid crystal (LC) concentration in polymer composites. Three major regions are shown: one phase, two phase, and gel networks. The two-phase region shows the PDLC where liquid crystals are dispersed in a continuous polymer matrix. Also, this region shows the LCDP region, where polymer is dispersed in a continuous liquid crystal matrix.

referred to as polymer-stabilized liquid crystal (PSLC), polymer network liquid crystal (PNLC), or polymer-stabilized cholesteric texture (PSCT).

By now, there exists a vast academic and some industrial literature on material, process electro-optics, and applications of PDLC device technology, and there are too many to refer to in the present work. Therefore, here we mention some selected academic and industrial literature on UV-PIPS, TS-PIPS, and SIPS by phase separation techniques published after 2000 [4–43].

5.3 Manufacturing Methods of PDLC Film

The coating and lamination (C&L) of plastics by roll-to-roll, known as converting technique, is a well-known manufacturing method in many plastic, film, and printing industries. Such technique is an essential and integral part of PDLC film manufacturing. However, in addition to a conventional converting method, the roll-to-roll manufacturing of PDLC film also requires a complicated and careful integration with phase separation and polymer curing kinetics, as well as the presence of process parameters, such as material formulation, radiant energy and heating rates, curing time and temperature, line speed, and thickness homogeneity.

The quest for development of flexible plastic liquid crystal devices began in the early 1980s on the premise of roll-to-roll manufacturing of large-area polymer-stabilized LCD devices. However, among the promising approaches, the PDLC technology had found its way into industrial development, manufacturing, and commercial applications of large-area electrically switchable films and windows. The interest and activities in both fundamental research and industrial development of PDLC and related technologies are expected to continue to grow, as functional switching and display devices, but include electro-optical devices and photonic systems, such as optical waveguides and fiber optics.

In contrast to conventional converting methods, such as the roll-to-roll coat and laminate, C&L manufacturing of PDLC film requires different approach, among which are the use of very expensive raw materials (i.e., LC and doped ITO on polyethylene terephthalate [ITO-PET] film) and a rather slow processing speed (0.5–1.0 m min^{-1}). These two factors alone are sufficient to add many complications in design, engineering, processing quality, and efficiency, as well as the manufacturing yield and cost.

Conventionally, the roll-to-roll converting technique is carried out by both dry/C&L (solvent-less) and wet/C&L (solvent-based) techniques. In PDLC film technology, the choice of wet or dry technique depends on the nature of PDLC technology. The industrial production of PDLC film with wet/C&L method is utilized by micro-emulsion and phase separation SIPS methods via "slot-die" or "doctor blade" by converting techniques. Whereas the dry/C&L approach is utilized in manufacturing of PDLC film by PIPS and TIPS phase separation techniques via "in situ," "slot-die," and "doctor blade" and "dip coating" converting techniques. Eventually, due to the limitations of material and process, as well as the manufacturing complexity and cost of ME technique, the PS approach has become the preferred technique in both academic research and industrial production of PDLC devices. Currently, the majority of industrial manufacturing of PDLC films are mainly carried out by UV-PIPS technique.

The industrial NCAP process by ME method is carried out by coating of a solvent-based mixture of LC in water-based polymer, using a cosolvent. With ultrasonic mixing, this combination creates LC microdroplets. The nonhomogenous mixture is coated onto a first plastic

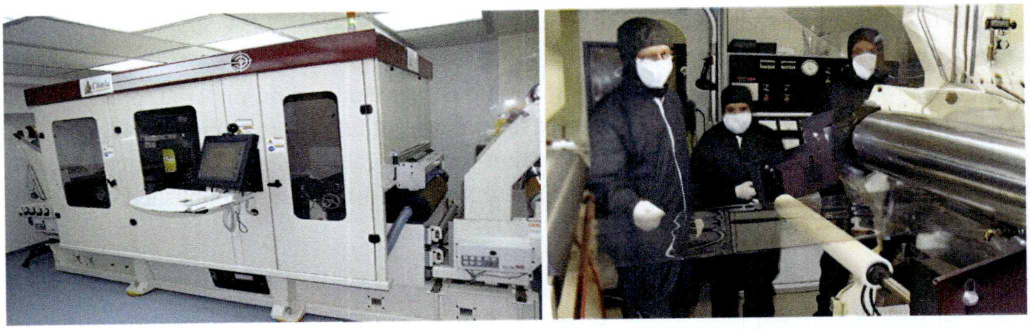

Figure 5.3a Manufacturing of NCAP film. *Source:* Raychem RPG Private Limited.

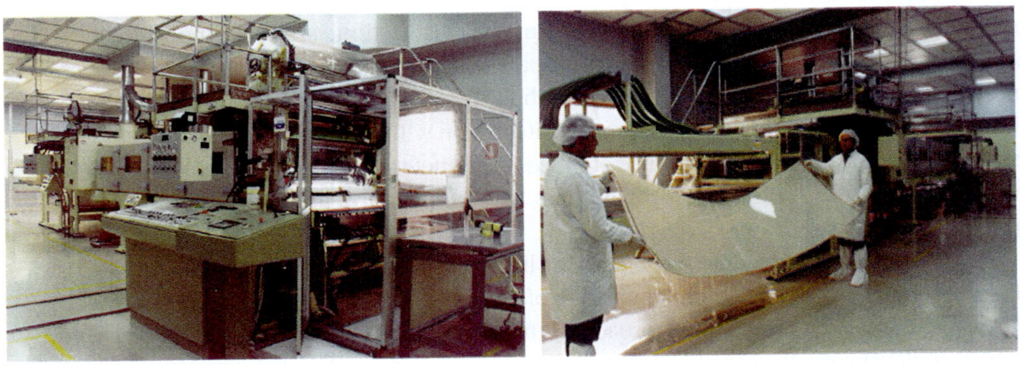

Figure 5.3b Manufacturing of PDLC film. *Source:* Snia PD/Fiat.

ITO-PET. Solvent evaporation creates a dense paste-like NCAP layer. A subsequent process laminates a second plastic ITO-PET sealing layer. This is followed by inspection and testing as necessary. The final NCAP product can be in the form of sheets or rolls. An example of the manufacturing process of NCAP film by wet/C&L ("slot-die" or "doctor blade") method is presented in Figure 5.3a. Due to this type of process, the NCAP manufacturing is rather slow and requires special and costly facilities.

The roll-to-roll process of PDLC by PS approach consists of a simultaneous C&L of homogenous LC/polymer mixture between two plastic ITO-PET supports. The uncured PDLC sandwiched is then phase separated and cured by dry/C&L method with either UV radiation (UV-PIPS) or heat (TS-PIPS) and, subsequently, the PDLC film is inspected and collected as sheets or rolls. It should be noted that, as the PDLC morphology is created after C&L of already sandwiched PDLC layer, its manufacturing is rather fast and does not require special and costly facilities. An example of industrial production of PDLC film with phase separation PIPS method by "in situ" dry coating technique is shown in Figure 5.3b.

5.4 Historical Evolution of Industrial PDLC Technology

Since the invention in the 1980s, the industrial development of PDLC technology has been subject of worldwide cycles of progress and decline. There were a number of factors for these historical evolutions, among which patent opposition was the most influential factor. According to the few patent oppositions issued by KSU against Raychem's patents in the United

States, Europe, and Japan during 1989–1991 period, we decided to classify the industrial PDLC evolution accordingly in the following three periods:

5.4.1 The Early Period (<1995)

Following the invention of ME technology, J. L. Fergason provided its exclusive license and patent rights to Raychem Corporation (USA), where the startup company Taliq was established in 1987 by Raychem in Sunnyvale, CA (USA), and began the first manufacturing of ME-based technology under the NCAP trademark. The NCAP products included the switchable flexible film and display. During the 1989–1990 period, Raychem out-licensed the production of NCAP technology to 3M Corporation in the United States, Nippon Sheet Glass in Japan, as well as to Saint-Gobain Vitrage (SGV) in Europe in 1995. Also, during the 1989–1990 period, KSU out-licensed the industrial development of PDLC technology to Polytronix and 3M Corporation in the United States, Asahi Glass and Ajinomoto in Japan, and Snia BPD (FIAT Group) in Europe. During this period, although General Motors (USA) also obtained a PDLC development license from KSU, they did not commercialize the technology. In addition, other research groups including Hughes Research Labs (USA), Dai Nippon Ink (Japan), the University of Ljubljana (Slovenia), Case Western Reserve (USA), and Wright-Patterson Air Force (USA) had worked on development of phase PDLC by separation PIPS techniques. Finally, Peregrine (USA) was the first startup company who developed PDLC film by the phase separation SIPS method. A list of NCAP and PDLC licensees, trademarks, and products in this early industrial period is tabulated in Table 5.1.

The industrial manufacturing of Raychem's NCAP and KSU's PDLC films and glass products by these licensees were mainly on commercialization of products for switchable privacy window markets. It is worth mentioning that, in this period except for Polytronix, which was a small independent startup, the major licensees of NCAP and PDLC technologies were mainly large glass, plastic, and chemical industries (see Table 5.1). Although in 1995, the French company SGV also obtained the NCAP license from Raychem (USA), the SGV decided to not manufacture PDLC film technology and began purchasing PDLC film products from Taliq/Raychem, Nippon Sheet Glass and 3M. In addition, in 1998, SGV began negotiating with Snia BPD for evaluation and eventual purchasing of PDLC film as alternative supplier in Europe through its PDLC glass production site in St. Rock (BL). The global PDLC market values for film and window were around \$10 and \$30 million (US) and

Table 5.1 Worldwide industrial licensees of NCAP and PDLC in the Early Period.

Company	Licensed technology	Trademark	Product
Taliq/Raychem(USA)	NCAP	"Varylite," "NCAP"	Film & Display
Nippon Sheet Glass (JAP)	NCAP	"UMU"	Film & Glass
St. Rock/Saint-Gobain (EU)	NCAP	"Privalite"	Glass
3M/Marvin Windows (USA)	NCAP & PDLC	"PrivaView"	Film & Glass
Asahi Glass (JAP)	PDLC	"ViewTech"	Film & Glass
Polytronix (USA)	PDLC	"PolyVision"	Film
Ajinomoto (JAP)	PDLC	"ACT"	Film
Snia BPD/Fiat (EU)	PDLC	R&D	Film

the average PDLC film and glass prices were still at high values around $350 and $1500 (US) per square meter, respectively.

5.4.2 The Setback Period (1995–2005)

During almost one decade (1995–2005), it has been a period in which the industrial development of technology was subjected to a setback, mainly due to major patent litigations between KSU and Raychem in the United States, Japan, and Europe. The first patent opposition case happened in 1989, when Raychem, consolidating its patent position contested by KSU in the United States, had requested the US patent office to reexamine and revalidate its patent profile. Such revalidation in late 1989 had confirmed the Raychem's patent position in the United States. In fact, the reason that 3M obtained both Raychem and KSU licenses (see Table 5.1) was to avoid any risk of eventual future patent litigation between Raychem and KSU in the United States.

The second PDLC patent oppositions unfolded in 1991, when KSU opposed Raychem's filed patents in Japan and Europe. The real setback occurred immediately after 1995, when the Raychem's patents were granted by both Japanese and European patent offices, which had confirmed the dominance of Raychem patents over KSU patents, not only in the United States, but also in Japan and Europe. Consequently, in order to avoid the risk of patent litigations by Raychem, Asahi Glass and Ajinomoto were forced to abandon their manufacturing and commercial activities. The suspension of 3M/Marvin Windows manufacturing in the United States in 1998 was not due to patent litigation issue, but was due to unrealized market forecast and production cost issues. During 1995–2005 Setback Period, to avoid any opposition from Raychem, Polytronix began to establish its Polytron subsidiary in Taiwan and also Raychem sold Taliq NCAP film and display production site to Xymox (Tyco Group). Perhaps another important issue resulting in the drawbacks of PDLC technology was the incompatible technology push market (rather than market pull) strategy by major licensees. While the worldwide interest in PDLC film and glass products began to increase by the 2000s, there were not sufficient suppliers to exploit potential markets.

By the end of Setback Period, a few small companies, such as DMDisplay (Korea) and Polytron (Taiwan), began production of PDLC film products in these countries, which were outside of the Raychem–KSU patent litigation areas. Meanwhile in Europe, Snia BPD continued its industrial R&D program and began collaboration with St. Rock (a SGV subsidiary) as an eventual European PDLC film supplier. It is worth mentioning that, since this period the term PDLC has become the popular terminology to describe all products by both PS and EM technologies.

The worldwide market of PDLC in this period did not change much with respect to the Early Period, where the global PDLC market values of film and window were around $15 and $40 million and the average PDLC film and glass prices were still at high values around $300 and $1200 per square meter, respectively.

5.4.3 The Revival Period (>2005)

The resurgence of PDLC industrial, commercial, and market activities and interest in global demands first began after the expiration of original Raychem patents in 2002 and main KSU patents in 2005. Since the expiration of PDLC patents, the field had witnessed rapid progress in industrial developments based on growing global demands. In 2007, Xymox sold its NCAP production assets to Citala (USA), who began commercialization of NCAP films for privacy window market until its closure in 2012. In 2015, Citala's manufacturing facilities were purchased by Glass Apps (a US subsidiary of DMDisplay), and since then, the NCAP production was phased out in the United States. Following the expiration of original PDLC patents, many film manufacturers and glass processors appeared entering the global markets. It is noteworthy to mention that, following decades of industrial progress and decline, the Revival Period witnessed the global disappearance of large industries and dominance by many small startups. The majority of the current companies are manufacturers of PDLC film, but few also supply PDLC glass products. In Table 5.2, we tabulated a list of the most

Table 5.2 Current worldwide manufacturers of PDLC film products.

Company	Location	Trademark
	ASIA-PACIFIC	
Toppan Electronics	Japan	LC-Magic
Nippon Sheet Glass	Japan	UMU
DMDisplay	Korea	Magic Screen
QSys	Korea	PDLC Film
Hanwha	Korea	PolyDreamer
Bestroom	Korea	Smart UV Film
Ouyi Materials	China	OYPDLC
CNCnext	China	CNC-Smart
Magic Film	China	Smart PDLC Film
Vanlong	China	Smart PDLC Film
Beijing All Brilliant Tech	China	Abtech
InnoGlass	China	InnoGlass
Singyes New Materials	China	PDLC Smart Film
UniteGlass	China	Smart PDLC Film
T-Photon Technology	China	PDLC Film
Noyark	China	Magic Film
Polytron	Taiwan-China	Switchable Privacy Film
Iris Film	Taiwan-China	Smart Film
YuGuang	Taiwan	Smart Film
Chiefway	Malaysia	PDLC Film
iGlass	Australia	iGlass Privacy
	USA	
Polytronix	Texas	Polyvision
Scienstry	Texas	3G Film
	EUROPE & MIDDLE-EAST	
Innoptec	Italy	LPF
Dream Glass	Spain	Privacy Film
Gauzy	Israel	LCG

current global PDLC film manufacturers, excluding the most independent PDLC glass processors who themselves do not produce the PDLC films.

Following the expiration of KSU's original patent in 2005, the industrial activities and global demands for PDLC film and glass products began to grow rapidly worldwide. Such growth accelerated the reduction of prices due to the competitive appearances of a growing number of manufacturers of raw materials (LC, ITO-PET), as well as PDLC film producers and glass processors worldwide. Currently, the global players of PDLC technology are over 15 film producers (see Table 5.2) and more than 100 glass processors.

In the past decades, increase in competition, gradual reduction of production costs and product prices, increased expertise in glazing techniques, as well as extensive market promotion campaigns have resulted in a surge of market demands and continuous growth of PDLC business worldwide. The current globalization of the technology is mainly due to the industrial awareness that R&D, product application, and business diversifications, which are the realistic strategies for market acceptance and growth of PDLC technology. Currently, over 85% of PDLC film production is in the Asia-Pacific region and the rest is in North America and Europe. Also, with respect to the main raw materials, whereas all suppliers of liquid crystal mixtures are in the Asia-Pacific rim, the ITO-PET film producers are mainly in the United States, China, and Europe.

The current worldwide PDLC industrial activities and market demand have resulted in increased high competition, gradual reduction of production cost and product price, increase in film and glazing manufacturing efficiency, and strong market promotion campaigns resulting in significant business growth worldwide. Such ongoing globalization of PDLC technology has been mainly due to the industrial awareness in research, development, product, application, and business diversifications, as well as the realistic strategies for market acceptance and growth of technology.

The current worldwide manufacturers of PDLC products consist of over 15 PDLC film producers and over 100 PDLC window glass processors. During this industrial development period, the overall worldwide PDLC market value has grown dramatically and the commercial prices are continuously falling below US$50/m^2 for PDLC film and US$500/m^2 for PDLC glass products, respectively. Further accounts of industrial evolution of PDLC technology are reported in a few selected recent literature [1, 44–47].

5.5 PDLC Industrial Development in Italy

The scope of this section is to make a brief review of the industrial development of PDLC technology in Europe, which has been initiated since 1990 in Italy at Snia-Ricerche, S.p.A., Pisticci Scalo (Matera), Italy (hereafter referred to as SNR) – the research center of Snia BPD. This first European industrial initiation at SNR had been spearheaded by the author, which had resulted in comprehensive research, development, and manufacturing programs of PDLC technology. The outcome is the technology transfer to a few startup companies in Europe since 2001. A brief account of these industrial developments is as follows.

5.5.1 SNR (Italy) PDLC License

The initiation and development of PDLC technology in Europe began in early 1989 by Snia BPD – the chemical division of FIAT Group. This involvement originated with Isoclima

Table 5.3 The SNR–KSU 1990 example license agreement on PDLC film technology.

The SNR–KSU License Agreement
Licensor: Kent Research Corporation (Kent State University, Ohio, USA).
Licensee: Snia BPD/FIN.IND./Inglen (*Snia Research* since 1/2/1995).
Signed: October 17, 1990 (Revised in 1/2/1995).
Termination: October 17, 2008 (or last to expire the licensed patents).
Licensed Technology: Polymer Dispersed Liquid Crystals (PDLC).
European counterpart patents *(EP0238626B1-3/5/1995)* of:
US4,671,618 (9/6/1987); US4,673,255 (16/6/1987); US4,685,771(11/8/1987); US4,688,900 (28/8/1987); US4,890,902 (2/1/1990).
Licensed Products: – PDLC films for architectural & vehicle glazing. – Segmented low-resolution displays. – Optical Shutters devices.
License Territory:
Italy, France, Germany, Belgium, Luxembourg, Netherlands, Switzerland, Spain, United Kingdom, Denmark, Greece, Ireland, Portugal, Austria, Former Yugoslavia, Former USSR, Former Czechoslovakia, Poland, Hungry, Romania, Bulgaria, Norway, Sweden, Finland, Brazil, Argentina, Mexico.
* The countries in bold letters are not subject to PDLC patent opposition
Potential Infringement: Corresponding to Taliq/Raychem US patents:
1- US4,435,047 (6/3/1984), 2- US4,579,423(1/4/1986),
3-US4,616,903 (14/10/1986), 4- US4,605,284 (12/8/1986).

(Italian Automotive Glass company), which had been involved in evaluation of Taliq/Raychem NCAP technology for Fiat Research Center (Centro Ricerche Fiat) in the late 1980s on development of automotive switchable glass. In this period, Isoclima approached Snia BPD to obtain the KSU PDLC license for manufacturing of PDLC film in Italy. The outcome of this initiative had been a series of negotiations between Snia BPD and KSU, resulting in a joint Snia BPD/FIN.IND-Inglen (the financial sponsor of Isoclima) license in 1990. The main aspects of this PDLC license agreement are tabulated in Table 5.3.

In 1995, the KSU's PDLC license agreement was revised and granted to Snia BPD's Research Center SNR. The revised license (see Table 5.3) included a wide range of territories in Western and Eastern Europe, former USSR, as well as Argentina and Mexico. In addition, in order to defer the risk of Taliq/Raychem patent infringement, which had been discussed in the previous section, KSU had committed to compensate SNR for the expenses of an eventual patent infringement.

The initial industrial research, development, and manufacturing of PDLC technology after the granting of a KSU license led by the author began in 1991 with the formation of a new Electro-Optical Film Group (EOF GROUP) at SNR center in Pisticci (MT) Italy. This initiative began through a comprehensive and long-term industrial R&D program, laboratory, scale-up, and pilot plant levels, as well as collaborations with academic and industrial partners in Italy and Europe.

During the 1991–2000 period, the EOF GROUP scientists had pursued a comprehensive industrial program on materials, formulations, characterization, process development, prototyping, pilot plant scale-up, quality control, and manufacturing of various PDLC film products. These activities resulted in the development of novel industrial knowledge, design and construction of a unique multipurpose production facility with capabilities of coating and processing diverse PDLC film products by all phase separation (UV-PIPS, TS-PIPS, SIPS, and TIPS) techniques.

The professionally trained scientists and technical staff of EOF GROUP have extensive experience in materials, preparation, formulation, coating and lamination, processing, scale-up, and manufacturing of large-area PDLC film products. Also in this period, the EOF GROUP initiated an industrial program targeted at full-scale production and commercialization of PDLC film technology for applications such as switchable privacy windows and projection screens for architectural and automotive glazing markets. The industrial facility included the innovative capability of formulating and processing flexible PDLC products with PS techniques.

In late 1997, SNR achieved a preproduction status of plastic PDLC and LC-gel network technologies, and subsequently the management decided, in addition to commercial promotion of PDLC technology, to seek for external contract research, development, and production services. SNR favors companies who intend to develop large-area flexible optical, electro-optical, as well as non-display devices for photonics, displays, holographic, polarizing, and reflective filter applications.

5.5.2 SNR Industrial R&D Program

The initial lab-scale R&D program began in late 1991, after the formation and training of the core scientific team and establishment of the main laboratory equipment and facilities. The first lab-scale activities carried out on PDLC material were the formulation and hand-coating process developments and characterization on metallized glass substrates by UV-PIPS and TS-PIPS phase separation methods. The process included the effect of curing UV radiation intensity and temperature on the PDLC morphology at various film thicknesses. The characterization and improvements consisted of testing the electro-optical properties of the samples, such as off-state and on-state visible transmittance, operating voltages, and response times. The second lab-scale activities were done in late 1992. This was performed by transferring all formulations from discontinuous roll-to-roll processes to continuous processes, characterizations, aging, quality control, and improvements on flexible PDLC prototypes. Also, all lab-scale and prototyping activities continued well beyond this manufacturing period.

The R&D equipment and facilities included the chemistry laboratory, specially constructed equipment, such as wet and dry plastic C&L, dynamic UV and heat curing, electro-optical testing, optical and electron microscopy, DSC, UV-visible and FTIR spectroscopy, accelerated aging system, adhesion, post-curing oven, and other equipment. Examples of these equipment and facilities are presented in Figure 5.4.

The SNR's R&D program on PDLC technology also consisted of a series of external collaborations on training, as well as material and process developments with academic and research institutes in Italy, Europe, the United States, and China. These collaborations included the Universities of Calabria, Naples, Basilicata, Fiat Research Center, Isoclima, Magneti Marelli, and Vacuum Research Technology Center (Italy); University of Sheffield (UK), Merck (Germany), and Technology Partnership (UK); Optrel (Switzerland); Liquid Crystal Institute-KSU, and Peregrine Optifilm (USA); Valiant Fine Chemicals (China).

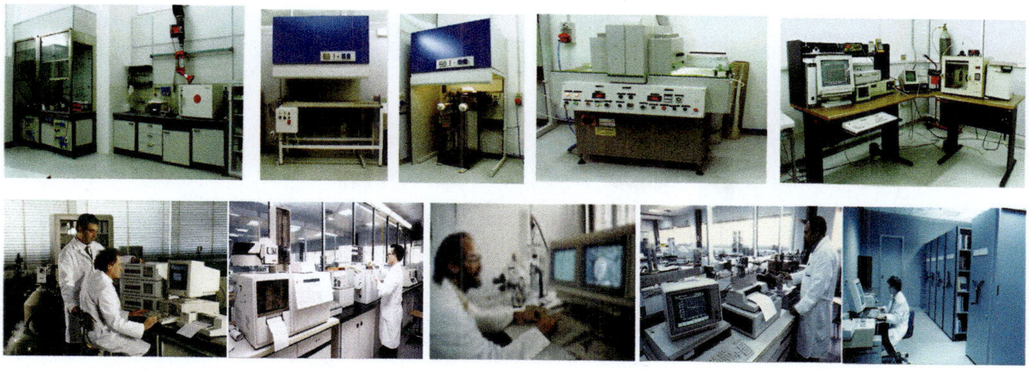

Figure 5.4 Some of the major SNR's R&D facilities.

5.5.3 SNR Production Program

In contrast to the conventional roll-to-roll converting technique, the production of PDLC film requires rather different processing conditions (slow production line speed), expensive raw materials (LC and ITO-PET film) as well as the integration of curing and PS kinetics with plastic C&L technique. These unconventional factors alone are sufficient to introduce further complications in design, engineering, process control, and manufacturing cost of flexible PDLC products. In roll-to-roll production of flexible PDLC product, the choice of dry (solvent-less) or wet (solvent-based) method depends on the type of PS technique utilized. The dry C&L is utilized in production of PDLC by PIPS or TIPS method, whereas the wet C&L is used in the case of SIPS method. This is critical to the manufacturing reliability and product quality through a complex design and engineering of PDLC pilot plant. Consequently, the manufacturing of PDLC film is the result of in-depth understanding and complex integration between the PDLC processing and roll-to-roll coating–laminating technique. In actual industrial practice, SNR acquired extensive knowledge and experience as the cornerstone for successful manufacturing of PDLC film products.

In 1992, SNR commissioned Tecnicol S.p.A. (Italy) for joint design and construction a multipurpose roll-to-roll PDLC pilot plant system with both dry and wet web methods and up to few hundred meters long production capability. The pilot plant was constructed and set up at SNR by the end of 1993. The pilot plant was placed in a special class 100,000 (ISO 8) clean room laboratory while the C&L section was housed in a class 10 (ISO 4) cabinet during the manufacturing with variable converting technologies. The SNR's unique multipurpose PDLC film pilot plant became operational since 1999 with a unique potential of manufacturing PDLC films by both dry and wet C&L methods with all phase separation (UV-PIPS, TS-PIPS, SIPS, and TIPS) techniques. As it will be discussed in the next section, during the 2002–2005 period, the pilot plant had been utilized to manufacture PDLC film products for Innoptec S.p.A. (an Ergom/Isoclima startup company) in Rovereto (Trentino), Italy.

In Figure 5.5, we present the configuration layout and snap-shot photographs of SNR's roll-to-roll pilot plant, capable of manufacturing large-area PDLC films by PS dry and wet C&L techniques. The SNR pilot plant was a custom-made continuous roll-to-roll hardware system designed to perform at low line speed (0.25–4.0 m min^{-1}) PDLC processing. The machine had a high degree of process variability in coating head technologies, thermal and radiation curing capabilities, solvent evaporation, and lamination technique. The overall features of this pilot plant are shown in the following drawing and photos.

Figure 5.5 Three (a–c) photos show the SNR PDLC roll-to-roll pilot plant in operation. The photo (c) shows a PDLC/PET product sheet.

To the best of our knowledge, the SNR's pilot plant had been the only multipurpose hardware system of its kind in the world. SNR is the only industry in the world that has incorporated both dry/C&L and wet/C&L techniques into a unique multifunctional PDLC pilot plant facility, being capable of processing PDLC films using all three PS methods. Such a "multipurpose" production facility provides ample versatility in development of novel processing methods, such as dry/dry (PIPS/PIPS), dry/wet (PIPS/SIPS), as well as new products, which are the subject of current progress. In Figure 5.6, we present the layouts of dry/C&L and wet/C&L manufacturing of SNR pilot plant and discuss the details as follows:

5.5.3.1 Dry/Coating & Lamination Technique

An example of industrial production of PDLC film by PIPS and TIPS dry/C&L technique is presented in Figure 5.6a. This roll-to-roll process consists of a simultaneous C&L of homogenous LC/polymer mixture between two plastic ITO-PET supports, by "dip-coating" (top), "slot die" (middle), or "in-situ" (bottom) method. The uncured PDLC sandwich is then phase separated and cured by UV radiation (UV-PIPS) or heat (TS-PIPS) and subsequently the PDLC film product is inspected and collected as sheets or rolls. The dry/C&L processing is basically similar in both PIPS and TIPS methods, where the latter only requires a controlled cooling of PDLC film.

5.5.3.2 Wet/Coating & Lamination Technique

An example of industrial production of PDLC film by SIPS wet/C&L technique is presented in Figure 5.6b. This process is carried out by "dip-coating" (top) or "slot die" (bottom) method, which consists of first coating of homogeneous mixture of LC/polymer in cosolvent onto a first plastic support, PS of LC droplets by solvent evaporation, lamination of the PDLC layer with a second plastic support, final inspection, quality control, and collection of the PDLC film product. It should be noted that, in addition to SIPS, the pilot plant was also capable of manufacturing the NCAP product by solvent-based ME technique.

In general, the SNR pilot plant hardware was equipped with the state-of-the-art solvent evaporation and thermal ovens, as well as UV curing lamps that could be positioned at different points in the downstream line of the coating head. The lamination with a second substrate could be made either before or after the curing and/or drying processes. The processing speed of the roll-to-roll pilot plant line could vary between 0.25 and 4.0 m min^{-1}, where the film product could be either cut as sheets or collected on rolls.

The outcome of integration of PDLC phase separation and polymer curing with continuous roll-to-roll technique provided a new multipurpose industrial system. These processing conditions, which are not dealt with in the academic literature, have been well developed

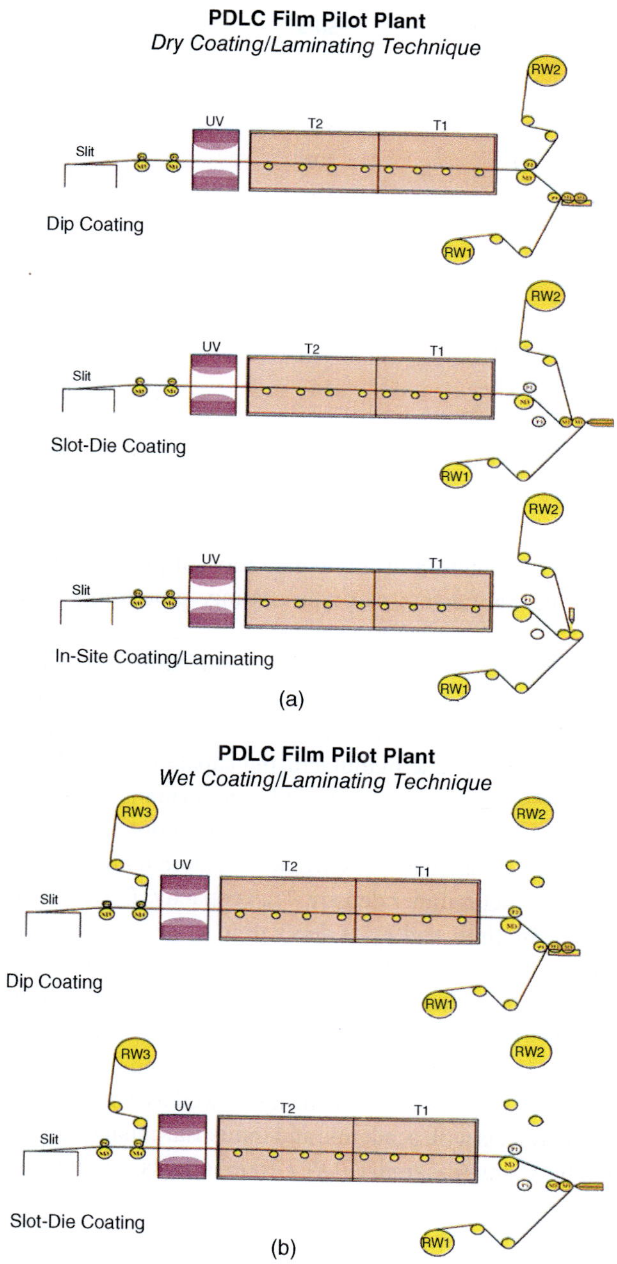

Figure 5.6 The SNR's coating and laminating pilot plants showing five processes: (a) dry/C&L and (b) wet/C&L techniques. RW are winding rollers for plastic film, M are process rollers, T is thermal treatment, and UV signifies UV curing stage. Note that to the right in each diagram is where the PDLC emulsion is applied.

Table 5.4 Comparison of pilot-scale PDLC film performance made by three phase separation techniques.

Parameter	UV-PIPS	TS-SIPS	SIPS
d (μm)	15	15	17
ρ (10^{11} Ω m)	1.4	1.2	4.9
T_{ni} (°C)	101	99	102
(μm)	1.2	2.3	1.4
<N> (μm^{-2})	0.47	0.13	0.28
T_{off} (%)	1.2	0.5	0.1
$T_{on}^{0°}$ (%)	84.3	85.6	94.2
$T_{on}^{60°}$ (%)	60.7	52.0	59.1
V_{90} (Volts)	42	16	55
τ_{on} (ms)	2	3	2
τ_{off} (ms)	13	68	26

and safeguarded by industry through development and implementation of precise quantitative correlations of the corresponding parameters at preprocessing, processing, and post-processing stages.

In actual industrial practice, a direct correlation between *independent parameters* (i.e., material chemical and physical properties and process conditions) and *dependent parameters* (i.e., morphology and electro-optical performances) is the key to successful manufacturing of PDLC products. In order to provide the diverse manufacturing capabilities of SNR's roll-to-roll pilot plant, in Table 5.4, we provide examples of PDLC film performance by three phase separation methods, UV-PIPS, TS-PIPS, and SIPS, produced with this pilot plant.

In Table 5.4, d is thickness, ρ is resistivity, T_{ni} is nematic-isotropic transition temperature, <D> is average droplet size, <N> is the droplet number density, T_{off} is opacity, T_{on} is transmittance, V_{90} is saturation voltage, τ_{on} is rise time, and τ_{off} is decay time.

In 2000, SNR was purchased by Ergom Group (a FIAT-affiliated automotive plastic company) in Torino, Italy, and until 2007 continued the industrial development of PDLC film technology. During the 1991–2007 period, SNR became the only industrial developer of PDLC technology in Europe and had succeeded in developing industrial know-how and innovative products and processes, a strong IP portfolio patents, manufacturing, as well as a number of national and European funded research projects with academic and industrial collaborations in Europe. In 2007, SNR was acquired by Magneti Marelli (FIAT Group) and its activity was subsequently closed.

5.5.4 SNR Intellectual Property

The industrial activities of SNR under Snia BPD and Ergom groups had resulted in a range of innovations, including material, processing, and manufacturing techniques. The company's strong intellectual property (IP) profile consisted of 10 published patent filings covering a range of inventions on formulations, processes, manufacturing, and applications of plastic PDLC, PSLC, and LCDP technologies [48–57].

5.6 Important Industrial Development Issues

There exist a number of important issues in industrial development of PDLC, from R&D to manufacturing, which are not dealt with in academic research and literature on PDLC. In order to clarify their importance for the readers, we only discuss two important issues of significance of scale and time accordingly.

5.6.1 The Significance of Scale

In academic research, the preparation of PDLC devices is carried out at cm^2-scale on glass supports. Industrial development requires a systematic transformation of materials, formulations, and processing from laboratory scale (cm^2-scale) to prototyping (dm^2-scale) and manufacturing (m^2-scale) levels on plastic supports. These industrial processes include both dry and wet methods by PS and ME techniques.

At laboratory scale, like in academic studies, the PDLC materials and formulations are prepared at gram-scale and cured by discontinuous hand coating process on cm^2-scale metallized glass supports. At prototyping scale, the PDLC material and formulations are carried out at decigram-scale and cured by discontinuous roll-to-roll process on dm^2-scale metallized plastic supports. Finally, at manufacturing scale, following the transfer from prototyping scale, the PDLC product is produced at kg-scale by continuous roll-to-roll process on m^2-scale metallized plastic supports.

At all industrial development levels, the PDLC devices are characterized for their chemical, physical, electrical, and electro-optical performances, whereas only at prototyping and manufacturing scales, do they also require reproducibility, uniformity, adhesion, aging, and quality control tests.

5.6.2 The Significance of Time

Another key factor in the industrial development of PDLC is the parameter of time, some of which from milliseconds (i.e., electro-optics response times) to seconds and minutes (i.e., phase separation and curing) are also of concern in academic studies. However, other time scales become crucial in industrial development, including minutes (line speed); hours (preparation, production, post-curing, and quality control); days (post-curing, accelerated aging, and transient phenomenon); months (natural aging, transient phenomenon, and product performance); and years (product life time and warranty).

Consequently, the significances of scale and time are at least two prominent issues that with other issues, such as production yield, product reliability, consistent electrical and optical properties, uniformity, and quality control, are among the crucial manufacturing parameters of PDLC at industrial scale. Such issues are not of concern in academic research.

5.7 Industrial Development of PDLC in Europe

All industrial development of PDLC technology in Europe, which originated by SNR in Italy, was transferred to three startup companies in Italy, Spain, and Israel during the 2008–2017 period. Due to confidential issues, we provide only a brief description of these industrial developments in the following sections.

5.7.1 Innoptec S.p.A. (Italy)

After the acquisition of SNR by Ergom Group in 2000, upon the decision of the company and the author's initiative, Ergom and Isoclima jointly founded as startup company – Innoptec S.p.A. [www.innoptec.it] in Rovereto (TN) Italy – in 2001 in order to initiate the first European manufacturing of PDLC film in Europe. The Innoptec R&D and production facilities were initially based on and supplied by SNR's industrial expertise and assets. Innoptec had been using SNR's pilot plant for manufacturing of PDLC film until 2005 when the company constructed and tested its own new state-of-the-art manufacturing line and began full commercialization of PDLC film products for privacy glass markets in Europe and elsewhere. Currently, Innoptec is among the three PDLC film manufacturers in Europe.

5.7.2 Dream Glass S.L. (Spain)

Following the collaborations of SNR and the Department of Engineering of Cambridge University (UK) on a few European Network R&D projects, upon the request of Cambridge Integrated Knowledge Centre (CIKC) [www.cikc.eng.cam.ac.uk] in early 2007, the author set up lab-scale plastic liquid crystal coating equipment and facilities at the Centre for Applied Photonics & Electronics (CAPE) laboratories of Cambridge University and began initial research activities. During these activities, CIKC management had been searching for a European industrial sponsor for collaboration on industrial development of plastic PDLC technology.

Also in late 2007, the Spanish glass company Dream Glass S.L. [www.dreamglassgroup.com] had been looking for industrial manufacturing and commercialization of PDLC technology. Accordingly, after an agreement between Dream Glass and the author in early 2008, the company had been introduced to CIKC as industrial sponsor for a joint venture on industrial development of plastic PDLC technology. This collaboration resulted in an industrial program under *Placord* project. The *Placord* project initiated by a CIKC–Dream Glass team at CAPE laboratories and accomplished new plastic PLDC prototypes in late 2010 through development of materials and processes. Subsequently, the acquired techniques from the *Placord* project were transferred to the laboratory and production facilities of Dream Glass, which had been under development and setup since 2010. In this respect, Dream Glass became the second producer of made-in-Italy PDLC film technology by manufacturing and commercializing PDLC film and glass products for European and other markets.

5.7.3 Gauzy Ltd. (Israel)

During the 2014–2018 period, the startup company Gauzy [www.gauzy.com] commissioned the author to set up an R&D team, program, and facilities, as well as the design and construction of a new manufacturing line for production and commercialization of diversified PDLC film products for various market applications. Through its diversified R&D team and program, Gauzy established new material and device innovations. Also, they made diverse products and applications, as well as creating a strong patent portfolio for a worldwide market, including the construction, automotive, multimedia, and energy markets. These industrial developments are based on the Gauzy new corporate strategy and business model [47], which resulted in a series of filed and granted patents [58–63] and scientific publications on PDLC technology [64, 65]. Currently, Gauzy as the third industrial developer of SNR technology is in the forefront of market development and commercialization of a range of PDLC and SPD film products at worldwide markets.

5.8 Potential Emerging Industrial PDLC Technologies

Originally, the PDLC technology had been mainly developed for large-area electrically switchable flat-panel LC displays and windows. However, with regards to the emerging PDLC products, there have been rather different views between academia and industries. On one hand, academic literature reports on innovations such as reflective displays [66]; holographic light shutters, beam splitters, switchable lasers, tunable photonic crystals [67, 68]; low-voltage devices [69]; spatial beam splitters [70]; holographic optical switching, gratings, and Fresnel lens [71–73]; optical rewritable display [74]; nanoparticle and dye-doped films along with other applications [75, 76].

On the other hand, the industrial PDLC makers are mainly interested in developing technology and product innovations that could be realized by large-area roll-to-roll manufacturing and commercializing cost-effective and marketable products. With such a difference in academic research, few innovations have been commercialized. In the past, reflective displays were commercialized by Taliq and Xymox Technologies (USA) using the NCAP method, but due to more efficient and competitive display technologies, the NCAP display production was phased out. Regarding the commercial feasibility of emerging PDLC technologies, aside from the current privacy windows, projection screens, self-adhesive, and dye-doped products, there are other emerging PDLC technologies which are in the course of development by some PDLC film and glass suppliers for large-area production and market applications. In the following sections, we provide a brief description of a few selected emerging PDLC technologies. It should be noticed that due to industrial confidentiality, we are not able to provide these developments in detail. We selected the following four emerging technologies.

5.8.1 Direct PDLC Glazing

The so-called Direct Glazing innovation originates from the necessity of reducing the production cost and improving the performance of PDLC glass products. In Figure 5.7a, we present the evolution of Direct Glazing. The standard double PDLC glazing is a seven-layer product consisting of a three-layer PDLC film, two interlayer adhesives laminated between two glass supports. Also, the standard self-adhesive PDLC glass consists of a three-layer PDLC film, one interlayer adhesive laminated onto one glass support (see Figure 5.7a).

Now, both *single* and *double* Direct Glazing products (first and second right in Figure 5.7a) are based on utilizing a new highly adhesive PDLC layer to eliminate the use of the ITO PET layer and interlayer adhesives. In this respect, the *single Direct Glazing* product will consist of only three layers, namely one ITO PET, one PDLC layer, and one metalized glass. Instead, *double Direct Glazing* will be three layers, consisting of one PDLC layer and two metallized glass supports (see Figure 5.8). The first phase of Direct Glazing production is to process a highly adhesive PDLC preform on one or two release sheets and, after peeling release sheets, laminate the free-standing PDLC layer onto metalized glass supports. The main industrial challenges of this invention are: (a) material and process development of highly adhesive PDLC layer, (b) roll-to-roll process on release sheets, and (c) lamination onto metallized glass supports. The Direct Glazing invention was first developed and patented by SNR (57) and recently developed further and patented by Gauzy Ltd. [51].

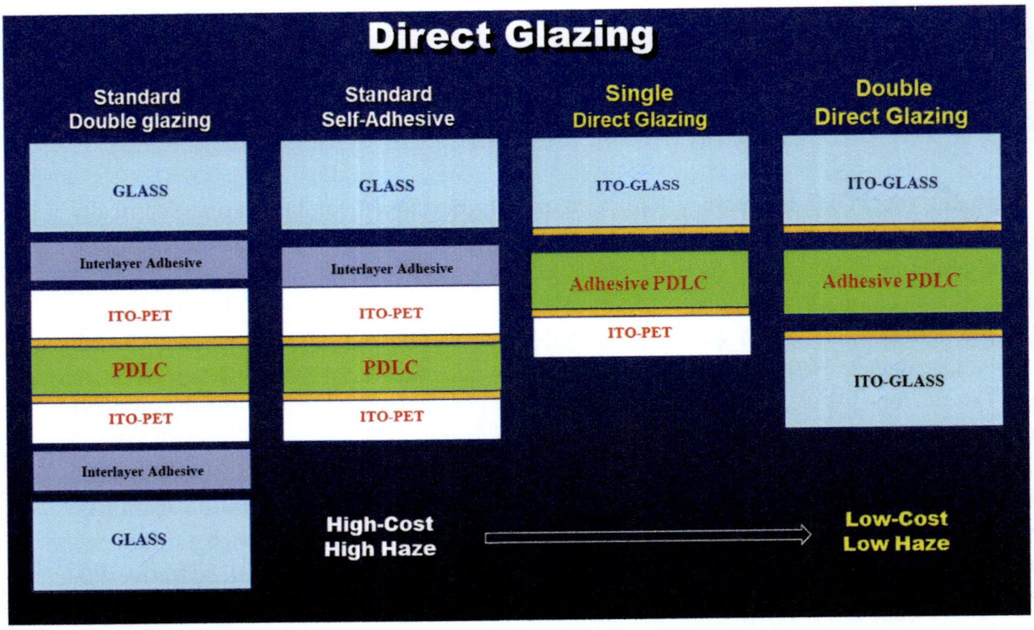

Figure 5.7a Comparison of standard PDLC and Direct Glazing PDLC.

5.8.2 Bistable PDLC

The idea of utilizing smectic-A liquid crystal in bistable display devices is not a new subject on its own. This technology enables the development of low-power, paper-like displays because of its unique concurrent properties, namely bi-stability, wide viewing angles, better sunlight readability, and simpler construction methods. The principle of smectic-A display operation is field-controlled light scattering, where it can be switched between stable transparent state (homeotropic) and opaque state (focal conic) textures by applying an electric field at different frequencies. As presented in Figure 5.7b, by application of a low-frequency (<1 kHz) electric field to smectic-A molecules having positive dielectric anisotropy, focal conic scattering textures (opacity) result from electro-hydro-dynamic instability (EHDI) mechanism. However, at higher frequencies (>1 kHz), the homeotropic texture (transparency) forms due to the dielectric reorientation of smectic-A molecules. The functions and features have been briefly presented in Figure 5.7b. In the academic studies, the bistable smectic-A innovation has reported on the development of lab-scale reflective LCD and flexible PDLC devices, but these innovations have not yet been industrially scaled up or commercialized [77–84].

To the best of our knowledge, only a few industries have been involved in industrial development of smectic-A films and glass products. Among them is Poly Display (Norway), which has commercialized small-size bistable signage, as well as Dow-Corning (USA), which has developed smectic-A (Si-SmA) materials and small-scale bistable dynamic window blinds through a glass partner. Also recently, Gauzy Ltd (Israel) has begun a joint development of flexible bistable PDLC films by roll-to-roll technique to commercialize large-scale privacy glass products in a joint venture with smectic-A materials supplied by Dow-Corning [85]. Gauzy has conducted development of flexible bistable PDLC for industrial manufacturing

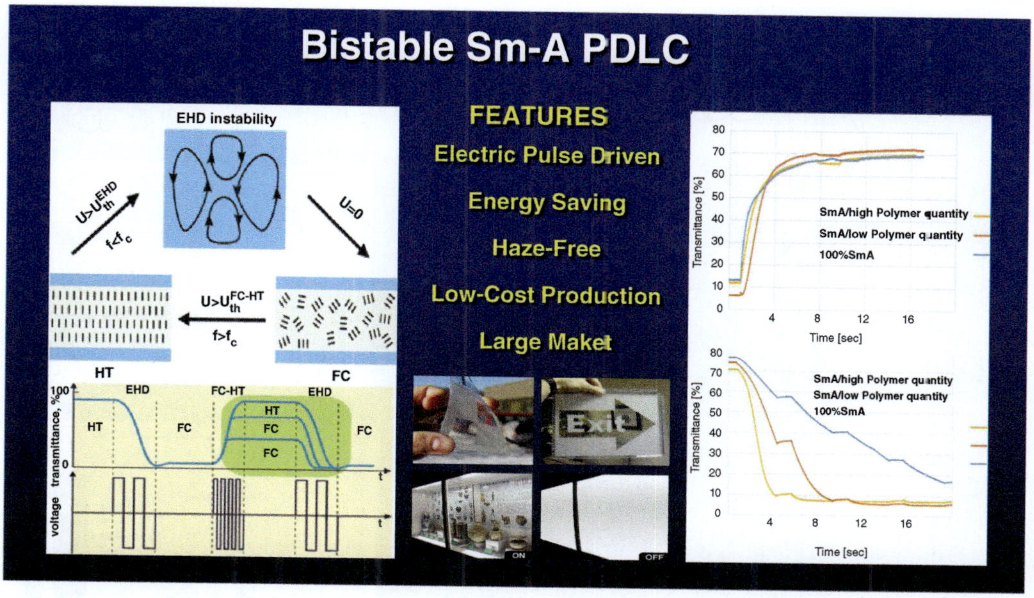

Figure 5.7b The functions and features of bistable smectic-A devices.

and subsequently filed an application patent [60]. The manufacturing and commercialization of large-area bistable PDLC products is expected to significantly expand the PDLC glass businesses and markets worldwide.

5.8.3 Solar-Control PDLC

As the standard PDLC product is not capable of reflecting or absorbing most of solar heat energy belonging to infrared light spectra, there have been few industrial attempts to control solar energy. This type of emerging technology, which is being developed mainly by PDLC industries, consists of two categories of *static* and *dynamic* solar-control PDLC films and glazing. The *static* approach, which is being produced by some PDLC manufacturers, is to utilize solar-reflective and absorptive metalized PET supports during the production of PDLC films. On the other hand, the *dynamic* approach is based on the use of tunable cholesteric liquid crystal materials in PDLC formulation and curing. In Figure 5.7c, we present examples of functions and features of static (top right) and dynamic (bottom right) solar-control PDLC device. The industrial development of both types of solar-control PDLC has been carried out by Gauzy Ltd. through early R&D and a recent patent filing [59].

5.8.4 Dynamic PDLC Signage

The dynamic PDLC signage is one of the promising emerging industrial PDLC technologies that can be realized at online roll-to-roll manufacturing of PDLC film only by UV-PIPS curing process. As presented in Figure 5.7d, this process includes the attachment of a logo or patterned plastic mask to the outside of uncured PDLC film on the side of UV source. This online UV curing process will create various PDLC morphologies due to various UV intensities passing through the different patterned mask.

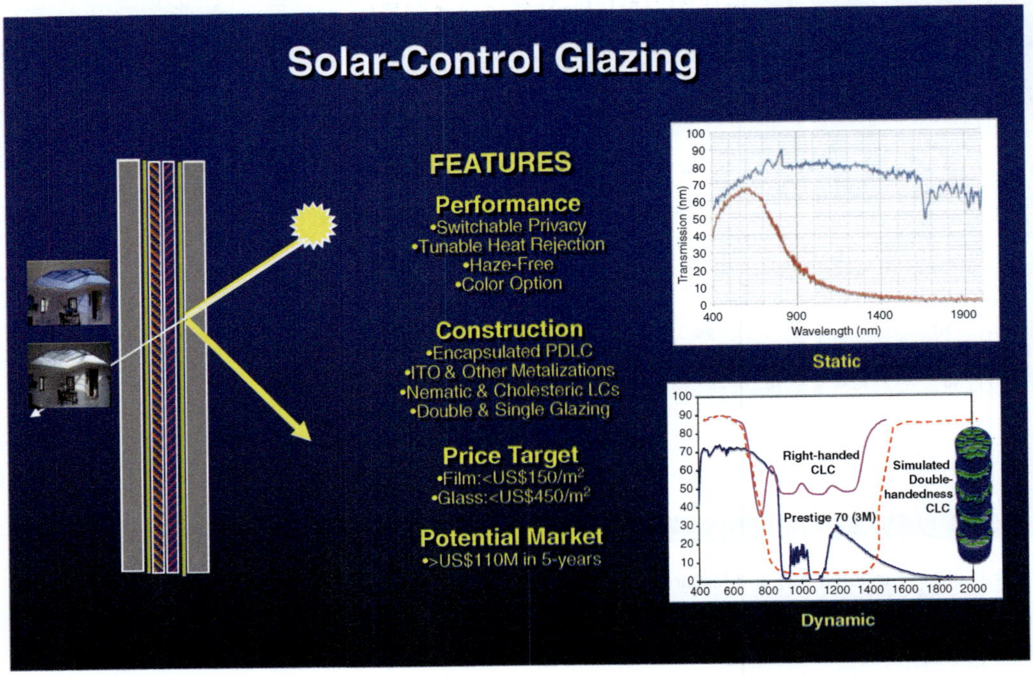

Figure 5.7c The functions and features of static and dynamic solar-control window.

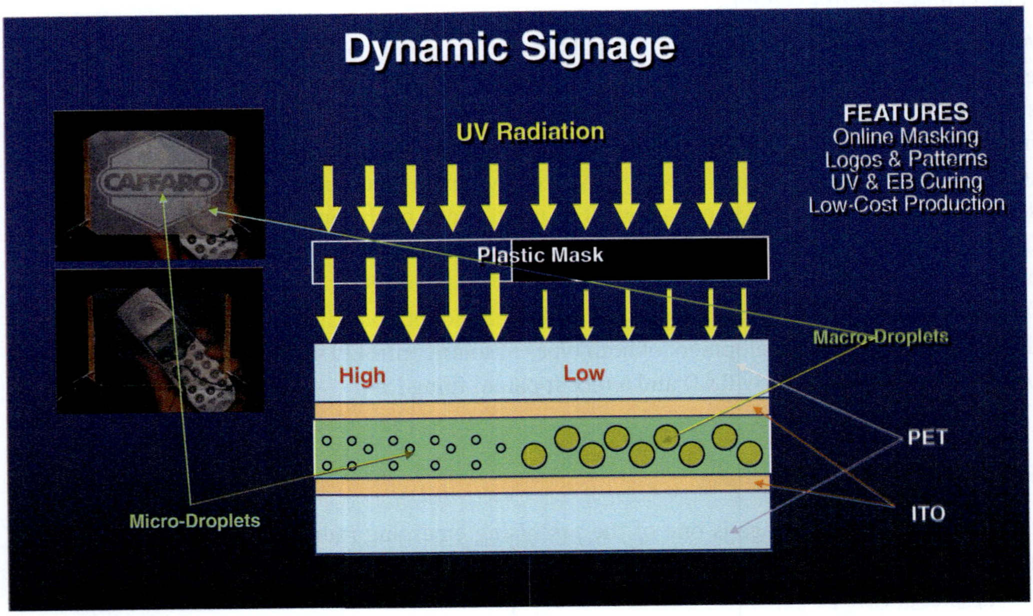

Figure 5.7d The process and features of dynamic PDLC signage.

To the best of our knowledge, the online PDLC signage technology has not been the subject of academic research and the industrial work has been first developed at SNR and continued R&D resulting in a recent patent filing [62].

5.9 The PDLC Market Situation

According to some recent market studies [86–88], the worldwide market value of total "Smart Glass," including PDLC, SPD, EC, TC, TT, and PC for industrial, commercial, residential, architectural, automotive, electronics, and other end-use applications reached around US $4.5 billion in 2019 and is expected to reach near $9.8 billion by 2024. The Compound Annual Growth Rate values in these forecasts are widely variable within 9–21% range, which provides an overall confidence margin of around 80%. Excluding the PDLC manufacturers, the prominent global players of Smart Glass production are SAGE Electrochromics, View Inc., AGC Ltd., PPG Industries, Gentex Corp., Hitachi Chemicals Co., Pleotint, LLC, RavenWindow LLC, Corning Corp., Research Frontiers Inc., and SPD Control System Corp.

With respect to the PDLC market which covers around 15% of total global Smart Glass market, in addition to our reports on the market size of PDLC film [35, 38], there are more recent reports on the global markets of only PDLC film and glass by other independent surveyors [89–98]. The overall forecast of these studies indicates that the global PDLC film and glass market had a revenue of around US$500–600 million in the year 2020 and is forecasted that by the end of 2030 will garner a revenue of around US$1.5–2.0 billion with a CAGR of 12.5%. However, it should be noted that these market surveys are dealing only with few major PDLC film and glass producers, such as Polytronix, DMDisplay, Qingdao, IRISFilm, Glass Apps, Scienstry, Smart Glass International, Gauzy, and Innoptec, and not including other PDLC film suppliers mentioned in Table 5.2. It should also be mentioned that not all the suppliers mentioned in these surveys are PDLC film manufacturers. For example, Qingdao is a supplier of liquid crystal mixture for PDLC, whereas Glass Apps (a US subsidiary of DMDisplay) and Smart Glass International do not produce either PDLC film or glass products. To our understanding, in the absence of a more dedicated market study to include at least the majority of PDLC suppliers, one has to assume that the available market surveys only provide a partial perspective of the global market value of PDLC technology.

References

1 Drzaic, P.S. (1995). *Liquid Crystal Dispersions*. Singapore: World Scientific.

2 Fergason, J.L. (1984). Encapsulated liquid crystal and method, US Patent 4,435,047 (granted 06/03/1984); Encapsulated liquid crystal and method, US4,579,423A (4/1/1986); US4,616,903A (14/10/1986); US4,605,284A (12/8/1986). Assignee: Manchester R&D Partnership.

3 Wu, G.-W. and Doane, J.W. (1987). Liquid crystalline-plastic material having submillisecond switch times and extended memory, US4,671,618 (granted 9/6/1987); US4,673,255 (16/6/1987). West, J., Golemme A., and Doane, J.W. (1987). Method of controlling microdroplet growth in polymeric dispersed liquid crystal, US4,685,771 (granted 11/8/1987). West, J.L., Doane, J.W., and Zumer, S. (1987). Liquid crystal display material comprising a liquid crystal dispersion in a thermoplastic resin, US4,688,900 (granted 28/8/1987). Doane, J.W., Wenninger, E.P., West, J., and Wu, B.-G. (1987). Liquid crystal light modulating materials with selectable viewing angles, US4,890,902 (2/1/1990) Assignee: Kent State University.

4 Luther, B.J., Springer, G.H., and Higgins, D.A. (2001). Templated droplets and ordered arrays in polymer-dispersed liquid-crystal films. *Chemistry of Materials* 13: 2281.

5 Gorkhali, S.P., Cairns, D.R., Esmailzadeh, S., Vedrine, J., and Crawford, G.P. (2002). Conformable-polymer dispersed liquid crystals (C-PDLC) displays with indefinitely captured form (published July 2012). *SID Digest* 1004. https://doi.org/10.1889/1.1830111.

6 Klosowicz, S.J. and Aleksander, M. (2004). Effect of polymer-dispersed liquid crystal morphology on its optical performance. *Opto-Electronics Review* 12 (3): 305–312.

7 Abdoune, F.Z., Benkhaled, L., Coqueret, X., and Mechernene, L. (2004). Effects of ultraviolet-curing conditions on the electro-optical behavior of polymer dispersed liquid crystal films. *Molecular Crystals and Liquid Crystals* 422 (1): 163–172.

8 Tortora, L., Chidichimo, G., Kreminska, L., Golovin, A., and Lavrentovich, O.D. (2005). Electro-optical properties of UV stabilized sub-micron dispersion with low liquid crystal content. *Molecular Crystals and Liquid Crystals* 435: 173–183. https://doi.org/10.1080/15421400590954830.

9 Seok, J.W., Han, Y.S., Kwon, Y., and Park, L.S. (2006). Structural effect of photoinitiators on electro-optical properties of polymer-dispersed liquid crystal composite films. *Journal of Applied Polymer Science* 99 (1): 162–169.

10 Scherschener, C.D., Perciante, E.A., Dalchiele, E.M., Frins, M., Korn, J.A., and Ferrar, A. (2006). Polymer-dispersed liquid-crystal voltage sensor. *Applied Optics* 45 (15): 3482.

11 Ding, H.L., Zhao, T., Cheng, Y. Pang, Y., Xu, H., Shi, G., and Jin, L. (2007). An investigation on a novel PDLC film's fabrication and its electro-optical properties. *Science in China, Series B Chemistry* 50 (N3): 358.

12 White, T.J., Natarajan, L.V., Bunning, T.J., and Guymon, C.A. (2007). Contribution of monomer functionality and additives to polymerization kinetics and liquid crystal phase separation in acrylate-based polymer-dispersed liquid crystals (PDLCs). *Liquid Crystals* 34 (12): 1377–1385. http://dx.doi.org/10.1080/02678290701663936.

13 Koo, J.J., No, Y.-S., Jeon, C.W., and Kim, J.H. (2008). Improvement of electro-optic properties in PDLC device by using new cross-linker for the control of the contrast ratio, response time and driving voltage. *Molecular Crystals and Liquid Crystals* 491 (1): 58–66. http://dx.doi.org/10.1080/15421400802328899.

14 Jin, F.-L. and Park, S.-J. (2008). Effect of fluorine group on electro-optical properties of PDLC films prepared with water-soluble polymers. *Bulletin of the Korean Chemical Society* 29 (12): 2521. http://dx.doi.org/10.5012/bkcs.2008.29.12.2521.

15 Folkes, R.L. (2008). *Controlling the morphological and electro-optical properties of polymer dispersed Liquid crystals*, Ph.D. Thesis Durham University. Durham, UK.

16 Deshmukh, R.R. and Malik, M.K. (2008). Effect of temperature on the optical and electro-optical properties of poly(methyl methacrylate)/E7 polymer-dispersed liquid crystal composites. *Journal of Applied Polymer Science* 109 (1): 627–637. https://doi.org/10.1002/app.27933.

17 Dong, S., Yan, B., Wan, X., Zhang, C., and Wang, Y. (2009). *Device Technology & Physics* 24 (1): 48.

18 Jeong, E.H., Sun, K.R., Kang, M.C., Jeong, H.M., and Kim, B.K. (2010). Memory effect of polymer dispersed liquid crystal by hybridization with nanoclay. *Express Polymer Letters* 4 (1): 39–46. https://doi.org/10.3144/expresspolymlett.2010.7.

19 Meng, Q., Cao, H., Kashima, M., Liu, H., and Yang, H. (2010). Effects of the structures of epoxy monomers on the electro-optical properties of heat-cured polymer-dispersed liquid crystal films. *Liquid Crystals* 37 (2): 189–193.

20 Li, M.-S., Huang, S.-Y., Wu, S.-T., Lin, H-C., and Fuh, A.Y-G. (2010). Optical and electro-optical properties of photonic crystals based on polymer-dispersed liquid crystals. *Applied Physics B: Lasers and Optics* 101 (1–2): 245–252. https://doi.org/10.1007/s00340-010-4110-y.

21 Kashima, M., Cao, H., Meng, Q., Liu, H., Wang, D., Li, F., and Yang, H. (2010). The influence of crosslinking agents on the morphology and electro-optical performances of PDLC films. *Journal of Applied Polymer Science* 117 (6): 3434–3440. https://doi.org/10.1002/app.32101.

22 Dzhons, M.M., Bulgakova, S.A., Pantyukhina, I.A., and Kazantzeva, I.A. (2011). Effects of chemical structure and composition of the polymer matrix on the morphology and electro-optical performance of polymer-dispersed liquid crystal films. *Liquid Crystals* 38 (10): 1263–1268. https://doi.org/10.1080/02678292.2011.60591405914.

23 Ahmad, F., Jamil, M., Jeon, Y.J., Woo, L.J., Jung, J.E., Jang, J.E., Lee, G.H., and Park, J. (2011). Comparative study on the electrooptical properties of polymer-dispersed liquid crystal films with different mixtures of monomers and liquid crystals. *Journal of Applied Polymer Science* 121 (3): 1424–1430. https://doi.org/10.1002/app.3364.

24 Ahmad, F., Jamil, M., Jeon, Y.J., Woo, L.J., Jung, J.E., and Jang, J.E. (2012). Investigation of nonionic diazo dye-doped polymer dispersed liquid crystal film. *Bulletin of Materials Science* 35 (2): 221–231. https://doi.org/10.1007/s12034-012-0286-6.

25 Song, P., Yu, L., Cao, H., Wang, F., Liu, F., Zhang, C., and Yang, H. (2012). Studies on electro-optical properties of polymer dispersed liquid crystal films based on epoxy resins prepared by UV-initiated cationic polymerization. *Liquid Crystals* 39 (3): 313–321. https://doi.org/10.1080/02678292.2011.643247.

26 Malik, M.K., Bhatia, P.G., and Deshmukh, R.R. (2012). Effect of nematic liquid crystals on optical properties of solvent induced phase separated PDLC composite films. *Indian Journal of Science and Technology* 5 (10): 1–13. https://doi.org/10.17485/ijst/2012/v5i10.11.

27 Shanks, R.A. and Staszczyk, D. (2012). Thermal and optical characterization of polymer-dispersed liquid crystals. *International Journal of Polymer Science* 2012: 767581. https://doi.org/10.1155/2012/767581.

28 Song, S., Jeong, J., Chung, S.H., Jeong, S.M., and Choi, B. (2012). Electroluminescent devices with function of electro-optic shutter. *Optics Express* 20 (19): 21704–21712. https://doi.org/10.1364/Oe.20.021074.

29 Zhang, C., Wang, D., Cao, H., Song, P., Yang, C., Yang, H., and Hu, G.-H. (2013). Preparation and electro-optical properties of polymer dispersed liquid crystal films with relatively low liquid crystal content. *Polymers for Advanced Technologies* V 24 (5): 453–459. https://doi.org/10.1002/pat.3103.

30 Cho, J.-D., Lee, S.-S., Park, S.-C., Kim, Y.-B., and Hong, J.-W. (2013). Optimization of LC droplet size and electro-optical properties of acrylate-based polymer-dispersed liquid crystal by controlling photocure rate. *Journal of Applied Polymer Science* 130 (5): 3098–3104. https://doi.org/10.1002/app.39558.

31 Ellahi, M., Gao, Y., and Rafique, M.Y. (2013). Influence of enhanced curing temperature of epoxy monomers structure on the electro-optical properties and morphology of polymer-dispersed liquid crystal films. *American Journal of Engineering Research and Reviews* 2 (3): 1.

32 Marin, L., Ailincai, D., and Paslaru, E. (2014). Monodisperse PDLC composites generated by use of polyvinyl alcohol boric acid as matrix. *RSC Advances* 4: 38397–38404. https://doi.org/10.1039/C4RA06426G.

33 Kim, A. and Han, J.I. (2014). Effect of liquid crystal concentration on electro-optical properties of polymer dispersed liquid crystal lens for smart electronic glasses with auto-shading and auto-focusing function. *Electronic Materials Letters* 10 (3): 607–610. https://doi.org/10.1007/s13391-013-3223-y.

34 Ailincai, D., Farcau, C., Paslaru, E., and Marin, L. (2016). PDLC composites based on polyvinyl boric acid matrix – a promising pathway towards biomedical engineering. *Liquid Crystals* 43 (13–15): 1973. https://doi.org/10.1080/02678292.2016.1172353.

35 Hakemi, H. (2017). Polymer-dispersed liquid crystal technology 'industrial evolution and current market situation'. *Liquid Crystal Today* 26 (3): 70–73. https://doi.org/10.1080/1358314X.2017.1359143.

36 Chen, M., Hu, W., Liang, X., Zhang, H., Zhang, C., Song, P., Zhang, L., Li, F., Chen, F., and Yang, H. (2017). The regulation of polymer structures and electro-optical properties of epoxy-mercaptan-based phase separated liquid crystals/polymer composites. *Polymer* 127 (3): 1–7, https://doi.org/10.1016/j.polymer.2017.07.003.

37 Zhang, H., Cao, H., Chen, M., Zhang, L., Jiang, T., and Chen, H. (2017). Effects of the fluorinated liquid crystal molecules on the electro-optical properties of polymer dispersed liquid crystal films. *Liquid Crystals* 44 (14–15): 2301–2310. https://doi.org/10.1080/02678292.2017.1376715.

38 Hakemi, H. (2019). The effect of thickness on morphology and electro-optics of plastic thermoset polymer dispersed liquid crystal (PDLC). *Molecular Crystals and Liquid Crystals* 681 (1): 12; *689* (1): 34.

39 Hakemi, H. (2020). The effect of curing temperature on the morphology and electro-optical properties of a flexible UV-cured polymer dispersed liquid crystal (PDLC). *Recent Progress in Materials* 2 (4): 1–11.

40 Hakemi, H. (2020). The effect of curing temperature and time on morphology and electro-optical properties of flexible thermoset polymer dispersed liquid crystal (PDLC) films. *Molecular Crystals and Liquid Crystals* 703 (1): 1.

41 Hakemi, H. (2021). The effect of liquid crystal concentration on themorphology of few flexible polymer dispersed liquid crystals. *Materials Science and Engineering* 5 (5): 153.

42 Tursi, A., Baratta, M., De Filpo, G., Nezhdanov, A., Mashin, A., and Nicoletta, F.P. (2022). The morphology and the electro-optical properties of PDLCs cured in the presence of electric fields. *Liquid Crystals* 49 (12): 1663–1675, https://doi.org/10.1080/02678292.2022.2055179.

43 Hakemi, H. (2022). The effect of process parameters on morphology and electro-optics of a polymer dispersed liquid crystal formulation. *International Journal of Modern Research in Engineering and Technology* 7 (2): 1–13. Hakemi, H. (2022). The switching voltages in few industrial-scale polymer dispersed liquid crystal formulations, *International Journal of Modern Research in Engineering and Technology*. 7 (2): 26–29.

44 Crawford, G.P. and Zumer, S. (ed.) (1996). *Liquid Crystals in Complex Geometries*. London, UK: Taylor & Francis.

45 Hakemi, H. (1998). Industrial development of plastic PDLC: is there a future? *Liquid Crystal Today* 8 (3): 7.

46 Drzaic, P. (2006). Putting liquid crystal droplets to work: a short history of polymer dispersed liquid crystals. *Liquid Crystals* 33 (11–12): 1281–1296. https://doi.org/10.1080/02678290601140563.

47 Hakemi, H. (2018). Gauzy Ltd. a new industrial development approach in PDLC technology. *Liquid Crystal Today* 27 (3): 84–87. https://doi.org/10.1080/1358314X.2018.1525065.

48 Hakemi, H. (1998). Metallo organo liquid crystals in a polymer matrix, IT1264903; US5/843333 (granted 1998-12-01); EP0656927; JP144319/89.

49 Hakemi, H. and Carfagna, C. (1998). Chiral anisotropic thermosetting resin in liquid crystals stabilized with polymers, IT1282574 (granted 1998-03-31), EP0803525A2; US08/787,199.

50 Hakemi, H., Pane, S., and Carfagna, C. (1997). Epoxy resins liquid crystals in composites of liquid crystals dispersed in a polymer, IT1271338 (granted 1997-05-27); US08/576,196; EP96/0724006.

51 Hakemi, H., Santangelo, M., and Caporusso, M. (2000). Polymer stabilized liquid crystals and flexible devices thereof, IT1296692; EP0747461; US6/049366 (granted 2000-04-11). Assignee: Sniarcerche.

52 Huang, Z., Chidichimo, G., Golemme, A., Hakemi, H., and Santangelo, M. (1998). Process and materials for liquid crystals dispersed in highly transparent polymers, IT1282575; EP97200221.6, US08/785,132.

53 Mashen, A., De Filpo, G., Chidichimo, G., Golemme, A., Hakemi, H., and Santangelo, M. (1998). Reduced nebulosity degree PDLC film device and method for its production, IT1283618 (granted 1998-04-22); EP961030704.9.

54 Hakemi, H. and Santangelo, M. (1999). Electrooptic films with polymers dispersed and liquid crystal with reverse morphology, IT1297025 (granted 1999-08-03), EP981245836.8.

55 Hakemi, H. and Ruggieri, S. Polymer stabilized liquid crystals and flexible devices thereof, IT97A/001127; EP97103704.9.

56 Hakemi, H. and Tasselli, F. (2001). Electrooptic liquid crystal film on conductive polymer electrodes and procedure for its preparation, IT2001A/011793 (published 2001-08-14).

57 Hakemi, H. and Tasselli, F. (1999). Liquid crystal dispersed polymer electrooptical films having reverse morphology, T2002A/001307 (pub: 1999-07-07); EP03131163.

58 Hakemi, H., Lofer, A., and Peso, E. Metal Organic Liquid Crystal Dyes, US62/065,805 (granted 2020-12-01, assignee: Gauzy); PCT47459 WO2016/063273.

59 Hakemi, H., Lofer, A., and Gal-Fuss, D. (2020). Solar-Controlled Reflective & Absorbing Electrically-Switchable Film & Glazing, US 62/065,811 (pub. 2020-10-21, assignee: Gauzy).

60 Hakemi, H., Lofer, A., Peso, E., and Gal-Fuss,D. (2019). Bi-stable Liquid Crystal Film Comprising of Metal-Organic Mesogens & Applications Thereof, US62/076,002 (pub. 2019-12-25).

61 Hakemi, H., Lofer, A., Peso, E., and Gal-Fuss, D. (2021). Multiple & Single Layers Liquid Crystal Dispersion Devices for Common & Direct Glazing Applications & Methods Thereof, US62/182,415.

62 Hakemi, H., Lofer, A., and Peso, E. (2022). Dynamic Signage, Glazed & Patterned PDLC Devices & Methods for Creating Thereof, US62/065,805 (granted 2022/11/08).

63 Hakemi, H., Lofer, A., Peso, E., Gal-Fuss, D., and Kfir, K. (2017). Advanced Polymer Dispersed Liquid Crystal (PDLC) for Display Projection Screens, US62/186,414 (pub. 2017/01/05).

64 Hakemi, H., Pinshow, O., and Gal-Fuss, D. (2019). Evaluation of polymer dispersed liquid crystal (PDLC) for passive rear projection screen application. *Recent Progress in Materials* 1 (3): 002; https://doi.org/10.21926/rpm.1903002.

65 Hakemi, H., Khenkin, M., Tamari, R., and Gal-Fuss, D. (2020). Transient phenomenon in polymer-dispersed liquid crystals. *Recent Progress in Materials* 2 (1): 006; https://doi.org/10.21926/rpm.2001006.

66 Jeon, Y.J., Lee, G.H., Jang, J.E., Hwang, K.Y., Ahmad, F., Muhammad, J., Lee, J.W., and Jung, J.E. (2012). Applications of multidirectional reflective light-control films on reflective polymer-dispersed liquid crystal displays for enhancement in image quality at lower viewing angles. *Liquid Crystals* 39: 1314–1319.

67 Fuh, A.Y.-G. and Lin, T.-H. (2004). Electrically switchable spatial filter based on polymer-dispersed liquid crystal film. *Journal of Applied Physics* 96 (10): 5402–5404. https://doi.org/10.1063/1.1805185.

68 Liu, Y.J. and Sun, W. (2008). Holographic polymer-dispersed liquid crystals: materials, formation, and applications. *Advances in Opto-Electronics* 2008: 684349, https://doi.org/10.1155/2008/684349.

69 Shriyan, S.K. and Fontecchio, A.K. (2010). Improved electro-optic response of polymer dispersed liquid crystals doped with oxidized multiwalled carbon nanotubes. *Molecular Crystals and Liquid Crystals* 525 (158): 4. https://doi.org/10.1080/15421401003799128.

70 Riquelme, M., Ortuñoa, M., Márqueza A., Gallegoa, S., Pascuala, I., and Beléndez, A. (2012). A dynamic beam splitter using polymer dispersed liquid crystal materials. *SPIE Proceedings Vol. 8498, Optics and Photonics for Information Processing VI,* 84980I. https://doi.org/10.1117/12.929538.

71 Kotakonda, P., Naydenova, I., Jallapuram, R., Martin, S., and Toal, V. (2009). Electro-optical switching of the holographic polymer-dispersed liquid crystal diffraction gratings. *Journal of Optics A Pure and Applied Optics* 11 (2): 024023. https://doi.org/10.1088/1464-4258/11/2/024023.

72 Fuh, A.Y.-G., Chen, W.-K., Cheng, K.-T., Lin, Y-C., Liu, C-K., and Chen, Y-D. (2015). Formation of holographic gratings in polymerdispersed liquid crystals using off-resonant light. *Optical Materials Express* 5 (4): 774.

73 Jashnsaz, H., Hosain Nataj, N., Mohajerani, E., and Khabbazi, A. (2011). All-optical switchable holographic Fresnel lens based on azo-dye-doped polymer-dispersed liquid crystals. *Applied Optics* 50 (22): 4295.

74 Fuh, A.Y.-G., Chen, C.C., Liu, C.-K., and Chen, Y.-D. (2009). Polarizer-free, electrically switchable and optically rewritable displays based on dye-doped polymer-dispersed liquid crystals. *Optics Express* 17 (9): 7088–7094.

75 Bronnikov, S., Kostromin, S., and ZUEV, V. (2013). Polymer-dispersed liquid crystals: progress in preparation, investigation, and application. *Journal of Macromolecular Science, Part B* 52: 1718.

76 Lin, Y.-C., Cheng, K.-T., Chen, H.-F., and Fuh, A.Y.-G. (2014). Photo- and electro-isomerization of azobenzenes based on polymer-dispersed liquid crystals doped with azobenzenes and their applications. *Optics Express* 22 (4): 4404–4411.

77 Crossland, W.A. (1985). An electrically addressed smectic storage device, *SID'85 Digest,* 124, p. 124–126. Society for Information Display.

78 Sun, G., Davey, A.B., and Crossland, W.A. (2004). Smectic-A (SmA) colour LCDs in reflective mode, Patent, GB0406385.5.

79 Büyüktanir, A., Glushchenko, A., Wall, B., West, J.L., Mitrokhin, M., and Holter, B. (2005). 62.4: Flexible Bistable Smectic-A LCD Based on PDLC. *SID Symposium Digest of Technical Papers,* 36: 1778–1781. https://doi.org/10.1889/1.2036362.

80 Mitrokhin, M.V. and Holter, B. (2005). 62.3: Reflective Bistable Smectic-A Passive Matrix LCDs. *Symposium Digest of Technical Papers,* 36: 1774–1777. https://doi.org/10.1889/1.2036361.

81 Gheorghiu, N., West, J.L., Glushchenko, A.V., and Mitrokhin, M. (2006). Patterned field induced polymer walls for smectic A bistable flexible displays. *Applied Physics Letters* 88: 263511. https://doi.org/10.1063/1.2218274.

82 Büyüktanır, E.A. (2008). Electro-optical Characterization of Bistable Smectic A Liquid Crystal Displays, PhD Thesis, Kent State University, College of Arts & Sciences, Chemical Physics. http://rave.ohiolink.edu/etdc/view?acc_num=kent1207869606.

83 Gardiner, D.J., Morris, S.M., and Coles, H.J. (2009). High-efficiency multistable switchable glazing using smectic A liquid crystals. *Solar Energy Materials and Solar Cells* 93.3: 301–306. https://doi.org/10.1016/j.solmat.2008.10.023.

84 Huh, J.-W., Choi, T.-H., Kim, J.-H., Woo, J.-H., Seo, J.-H., and Yoon, T.-H. (2018). Bistable switching of diffractive smectic-A liquid crystal device between haze-free transparent and high-haze translucent states. *ACS Photonics* 5 (8): 3152.

85 Clapp T.V., Crossland W.A., Davey A.B., Grasmann M., Hannington J.P., King R.K., Pivnenko M., Robson S., and Xu H. (2015). Liquid crystal formulations and structures for smectic A optical devices. Dow Corning Corporation, US 8956548B2; EP 2 547 749A1; WO2011115976A8.

86 Grand View Research (2021). Polymer dispersed liquid crystal (PDLC) film global market (2021) (part of Research and Markets), www.researchandmarkets.com/ (accessed 5 June 2024).

87 Research Nester (2023). Polymer dispersed liquid crystal (PDLC) market size and share, www.researchnester.com/ (accessed 5 June 2024).

88 Research and Markets (2024). Smart Windows (2024), www.researchandmarkets.com/ (accessed 5 June 2024).

89 All Market Insight (2021). Smart Glass Market, https://www.gminsights.com/ (accessed 5 June 2024).

90 Frost & Sullivan – USA (2023). Technology growth opportunities series: PDLC technology; Global smart windows market (2023), www.frost.com/ (accessed 5 June 2024).

91 Market Expertz – USA (2024). The Polymer Disperse Liquid Crystal (PDLC) Glass Market Analysis by Application 2023–2032, www.marketexpertz.com/ (accessed 5 June 2024).

92 Absolute Reports (2024). Global polymer dispersed liquid crystal (PDLC) film market research report, www.absolutereports.com/ (accessed 5 June 2024).

93 Transparency Market Research (2024). Polymer dispersed liquid crystal (PDLC) film market 2017–2025, www.transparencymarketresearch.com/ (accessed 5 June 2024).

94 Market & Research Momentive (2024). SurveyMonkey, www.momentive.ai/ (accessed 5 June 2024).

95 Market Data Centre (2024). DataCentre- Effective marketing Plan for business, www.marketdatacentre.com/ (accessed 5 June 2024).

96 Cognitive Market Research (2024). Smart Glass Market (2023), www.cognitivemarketresearch.com/ (accessed 5 June 2024).

97 Research Nester (2024). Polymer dispersed liquid crystal (PDLC) market size & share (2023), www.researchnester.com/ (accessed 5 June 2024).

98 Market Watch (2024). Stock watch, www.marketwatch.com/ (accessed 5 June 2024).

6

Suspended Particle Devices

Philippe Lemarchand[1,3,4] *and Brian Norton*[1,2,3]

[1] *Dublin Energy Lab, Technological University Dublin, Dublin, Ireland*
[2] *International Energy Research Centre, Tyndall National Institute, University College Cork, Cork, Ireland*
[3] *MaREI – the SFI Research Centre for Energy, Climate and Marine, Energy Management, University College Cork, Cork, Ireland*
[4] *Sustainability Intelligence, Technological University Dublin, Dublin, Ireland*

Nomenclature

a, b	ellipsoid dimensions
AC	alternating current
C_s	particle torque constant
Cu-rGO	copper reduced graphene oxide
CNT	carbon nanotubes
DGU	double glazed unit (window) (also IGU)
EC	electrochromic
E_o	electric field
ε_f	fluid permittivity
EVA	ethylene-vinyl acetate
FTO	fluorine doped tin oxide
GC	gasochromic
IGU	insulated glass unit
ITO	indium tin oxide
k_B	Boltzmann constant
K_s	slope factor
L	length
LC	liquid crystal
η_a	absorptance efficiency
η	fluid viscosity
PVB	polyvinyl butyral
PBT	polybutylene terephthalate
PC	photochromic
PEC	photoelectrochromic
PEDOT	poly(3,4-ethylene-dioxythiophene)
PE	polyethylene
PET	polyethylene terephthalate

PI	polyimide
PIB	polyisobutylene
PMMA	polymethyl methacrylate
PP	polypropylene
PVCC	photovoltachromic
RH	relative humidity
SHGC	solar heat gain coefficient
SPD	suspended particle device
θ	angular displacement
θ_D	rod diffusion angle (Brownian motion)
$\theta_D{}^R$	rod diffusion angle for time t_r
t_{dec}	decay time
t_r	response duration
T_{sol}	solar transmittance
T_{vis}	visible transmittance
τ_{up}	dark-to-transparent switching duration
τ_{down}	transparent-to-dark switching duration
T_{uv}	UV transmittance
TC	thermochromic
UVF	ultraviolet light filter
ω	angular frequency

6.1 Introduction

Optically switchable devices [1, 2] are chromogenic devices that modulate light transmission by switching from being fully or partially absorptive or reflective to fully or partially transparent. Different devices alter optical properties by either changing light intensity, light spectral composition, temperature, applied electric field, injected charges, or physical composition [3, 4]. Their applications include:

- Control of solar heat gain and daylight transmission through a building and vehicle glazing thereby reducing heating and cooling energy consumption and/or preventing discomfort from glare [5, 6],
- Modulation of reflector surface reflectance for automotive mirrors,
- Electronic information displays, and
- Optical communication switches.

As shown in Figure 6.1, chromogenic materials based on particle movement exhibit either:

- Discrete mass movement including ion and localized electron motion. These switchable materials are either thermochromic (TC) activated by heat, photochromic (PC) activated by light intensity, electrochromic (EC) activated electrically, or gasochromic (GC) devices activated by injection of a gas. Recent developments have combined properties of electrochromic materials and solar cells into single photoelectrochromic (PEC) and photovoltachromic (PVCC) devices, or

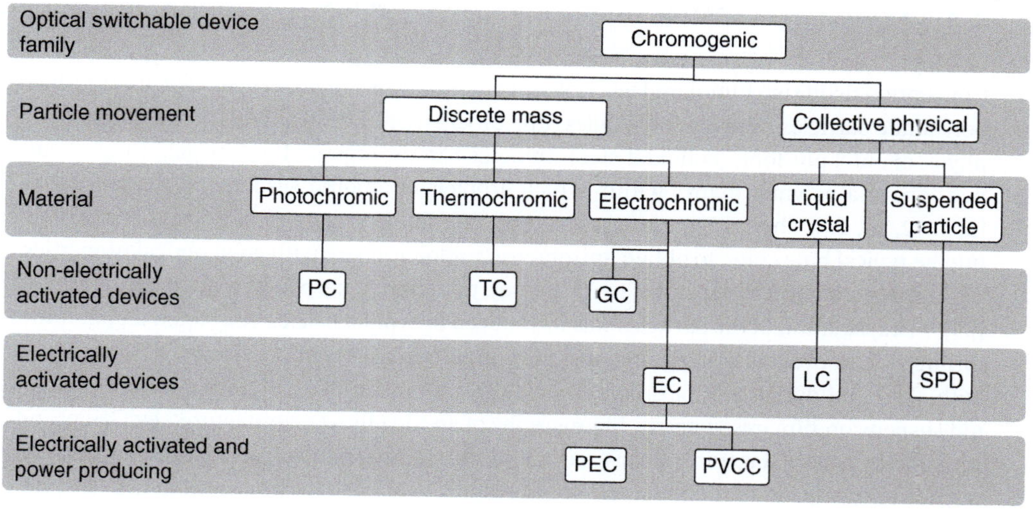

Figure 6.1 Taxonomy of chromogenic switchable technologies.

- Collective physical movement including dispersed or homogeneously distributed liquid crystal (LC) and suspended particle motion. Both LC-based and suspended particle devices (SPDs) technologies are activated electrically.

SPDs, also called light-valves or light-control particles, are electronically switchable chromogenic devices made of charged electrophoretic rod-shape particles suspended in an organic gel that change their orientation in an applied electric field, thereby enabling rapid voltage-controlled modulation of their optical transmittance. When an alternating current (AC) electric field is applied, SPDs rapidly switch from a dark bluish-black state to a clear greyish appearance. In current practical devices, a 50 Hz AC is typically applied together with voltages between 35 and 110 V_{AC} depending on the transition sought between being fully transparent and fully dark states. In a fully transparent state, low light scattering gives a non-hazy view through an SPD window. SPD technology was invented and patented by E.H. Land [7, 8] in 1934. Currently most available SPD films are developed from patents originated by Saxe in the mid-1970s [9] and in the 1980s [10, 11] and more recently in early 2000s [12, 13]. These are just examples of a larger portfolio of patents. These patents along with formulations are licensed by Research Frontiers, Inc. (New York) to over 40 companies including Hitachi Chemical (Japan), Asahi Glass Company (Japan), Cricursa (Spain), DIC/Sun Chemical (Japan), GKN Aerospace (UK), iGlass Technology, Inc. (CA, USA), Isoclima (Italy), Pittsburgh Glass Works (PA, USA), Saint–Gobain (France), Smart Glass International/Diamond Glass (Ireland), and Gauzy Ltd. (Israel), and sister companies such as Vision Systems (France) [14]. Hankuk Glass (Korea) working with RFI was granted a patent which improved the manufacturing of films and prevented dipole settling [15].

SPD-, EC-, and LC-based technologies enable large-area electrically activated switchable windows to be realized. A prominent application for SPD is to control the solar energy through "smart" building and vehicle glazing [4]. The global smart window market reached a value of US$ 1115 million in 2022 and is expected to reach US$ 2525 million by 2028 [16]. The SPD market was valued at US$ 150 million in 2018 and is expected to reach US$ 450 million by 2025 similar to the EC market [16, 17].

6.2 Materials

Conventional SPD is a film-based technology containing anisometric (needle-shape/rod-like) polyhalide particles, such as polyiodides (dihydrocinchonidine bisulfate polyiodide) or paraphyte, about 1 μm long (colloidal size) immersed in an organic fluid or gel, a liquid suspension, or a polymer matrix within which droplets of a liquid suspension are distributed [3, 4, 12, 13, 18]. There is no current practical alternative to the use of polyiodides, providing the typical blue color, to obtain anisometric submicron size with good optical properties. SPD films with particle size down to 200 nm are preferred to keep light scatter extremely low [19]. Encapsulation of the particles was introduced to avoid particles' migration and nonuniformity issues over a large glazing area [13]. Figure 6.2 [12] and Figure 6.3 [18] represent both clear and dark states of droplets of liquid suspension in a polymer matrix and a liquid suspension film, respectively. Micrographs of an SPD film (manufactured by Gauzy and supplied by SmartGlass) in transparent and dark mode are shown in Figures 6.4 and 6.5, respectively, with altered contrast and brightness to distinguish the multiple layers of droplets within the film. The droplets encapsulating the particles do not move within the polymer matrix.

An SPD glazing comprises SPD films, ethylene-vinyl acetate (EVA) copolymer or polyvinyl butyral (PVB) adhesive layers, optional infrared reflective coatings and ultraviolet light

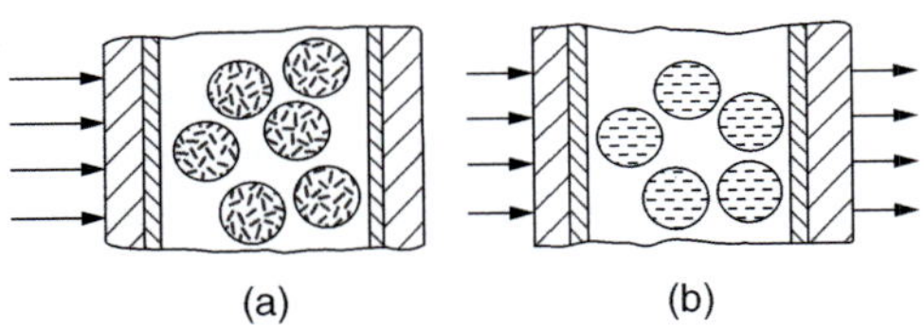

Figure 6.2 SPD film with encapsulated particles in a chaotic (a) and organized (b) order. *Source:* Adapted from US pat. 6,987,602 [13].

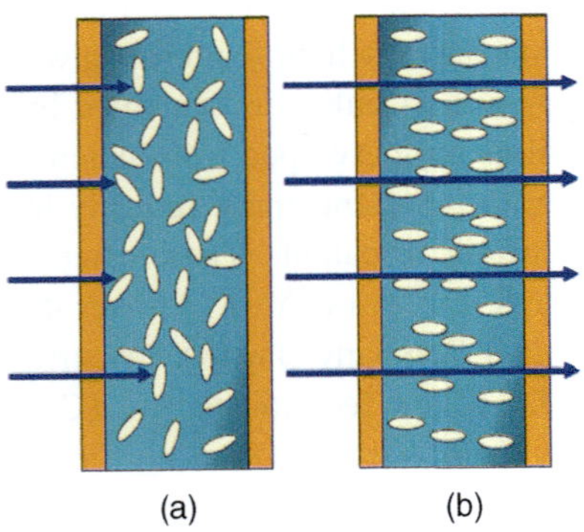

Figure 6.3 Working principle of liquid suspended particles: (a) operation with no applied voltage and (b) operation with applied voltage. *Source:* Vergaz et al. [18], Figure reproduced courtesy of Elsevier.

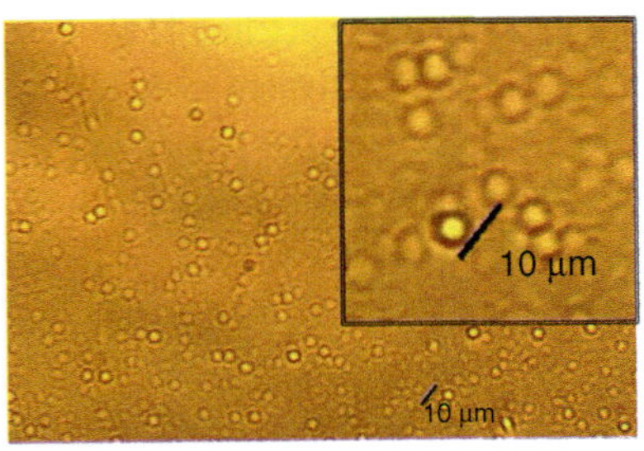

Figure 6.4 Micrograph of encapsulated particles in transparent mode and (inset) a zoom onto a few droplets (encapsulated particles).

Figure 6.5 Micrograph of encapsulated particles in the dark mode (Scale is the same as Figure 6.4).

filters (UVFs), and indium tin oxide (ITO) (i.e. In_2O_3Sn) transparent-electrode layers sandwiched between two panes. Thermoplastic polymer film preferably 0.1–0.2 mm thick containing EVA or PVB or polyurethane adhesive layers provides a smooth interface between the SPD film and the panes typically made of glass or, alternatively, polymers such as polycarbonate or polymethyl methacrylate (PMMA) (Plexiglas) [20]. The SPD film comprises the switchable material sandwiched between two flexible substrates coated or deposited with a transparent conductive layer. The transparent flexible substrate is typically fabricated from combinations of heat-stabilized polyethylene terephthalate (PET), PMMA, polyethylene (PE), polypropylene (PP), polybutylene terephthalate (PBT), and polycarbonate [21]. A substrate may incorporate UV blocker agents to enhance the lifetime of the film. A flexible substrate may have a thickness between 0.02 and 0.6 mm depending on the desired flexibility. ITO transparent conductors are expensive. Research on new transparent conducting materials aims at producing lower-cost high-quality electrodes. SnO_2:F (FTO), ZnO:Al, ZnO:Ga, poly(3,4-ethylene-dioxythiophene) (PEDOT), carbon nanotubes (CNTs), and $12CaO{\cdot}7Al_2O_3$ (C12A7) are being considered as possible replacements [6, 22].

Typical dimensions of an SPD glazing are shown in Figure 6.6; together with those for a polymer-dispersed liquid crystal (PDLC) device for comparison [23, 24]. Typical SPD film thickness is about 0.35 mm.

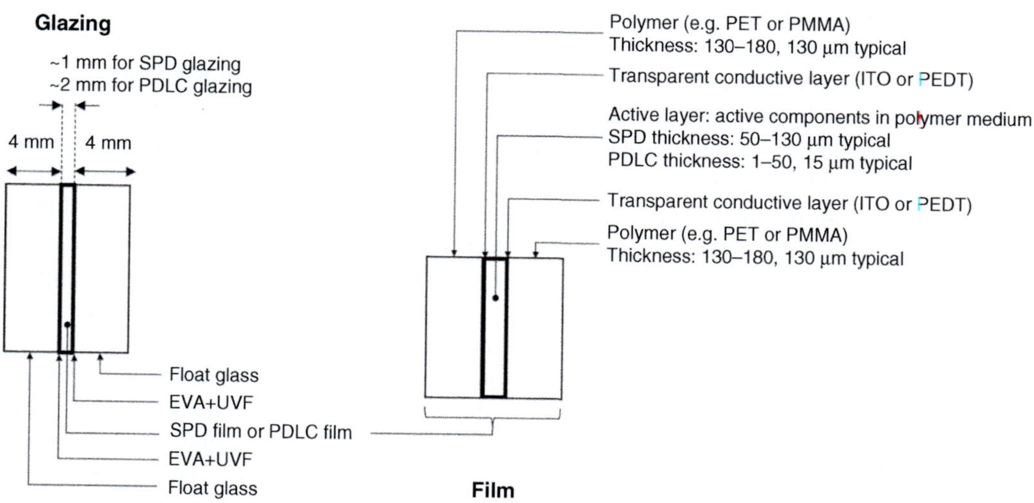

Figure 6.6 Schematic of SPD and PDLC films in laminated glazing.

6.3 Manufacture and Commercial Specifications

For cost-efficiency, SPD films are typically produced using a roll-to-roll manufacturing process [25] with a typical maximum width of 1.8 m. Potential production techniques have been presented in the US Patent 10,416,521 B2 [21]. The roll-to-roll technique uses a wet-coating process of a switchable material with a liquid-to-gel viscosity onto a flexible substrate with a transparent conductor. A second substrate with a transparent conductive layer is then applied on top to sandwich the switchable material. The viscosity of an emulsion is tuned by adapting the molecular weight of polymers, air and/or substrate temperature(s), the proportion of solvent, and the use of initiators. Incorporating solvents into the switchable material mixture resulting in a liquid viscosity can be dried to obtain the desired gel viscosity. A high-viscosity switchable material incorporating photoinitiators can be UV-cured to cross-link the mixture, forming a gel. Polymerization and/or cross-linking can be initiated by chemical-, thermal-, or photo-type initiators. A suitable polymer matrix for SPD uses siloxane with a high level of phenyl group to obtain a refractive index between 1.4630 and 1.4800 [26]. A "light valve film" composed of droplets of a liquid with a suspension of particles distributed into the plastic film can be obtained by phase separation or, more commonly, by cross-linking emulsions [13]. The liquid medium "comprises of one or more non-aqueous, electrically resistive liquids in which there is preferably dissolved at least one type of polymeric stabilizer which acts to reduce the tendency of the particles to agglomerate and to keep them dispersed and in suspension" [13]. The liquid suspension or the polymeric stabilizer used is selected to maintain the particles in gravitational equilibrium. Liquid suspending media include those disclosed in the US patent numbers 4,247,175 [10] and 4,407,565 [11], and EP1901909A4 [19]. Preferred suspension media comprises a polyalkyl (meth)acrylate and/or fluorinated (meth)acrylate and optional non-polymeric liquid(s) [26]. A polymeric stabilizer, such as nitrocellulose, bonds with particles and dissolves in the nonaqueous liquid, providing particles' dispersion and steric protection. To avoid light refraction and scattering, non-polymeric liquids and/or suspending polymers used must have a refractive index within 0.010, preferably 0.005 or less, that of the matrix polymer. Polyiodide particles degrade in terms of color and performance when exposed to UV [13]. Degradation protection is achieved by the addition of UV absorbers in the suspension and/or film or using UV filters within the SPD arrangement. Emulsions may then comprise a matrix polymer with cross-linkable polymer, a liquid suspension with coated polyiodide particles and optional non-polymeric liquids, and an initiator. Emulsions with low viscosity (<5000 centipoise at 25 °C) must be coated and cured quickly enough to avoid coalescence of small droplets into larger diameter droplets [19]. Rapid coalescence of droplets would also occur if the second electrically conductive coated substrate is applied on an uncured emulsion. Once cured, for the same operating conditions, films with droplets of 1–2.5 μm diameter present a darker off-state transmission, higher transmission range, and lower haze in the clear state than films with larger droplets (e.g. 12.5 μm). Once the sandwiched switchable material is achieved, the film is then cut, edge sealed, and electrical connections (bus bars) are added to the electrodes.

Once an SPD film has been cut to the required size and shape, a lamination process is used to assemble the layers that form an SPD glazing. This process is conducted in a climate-controlled clean room environment. Mechanical pressure is applied while heating and drying an assembled glazing in an oven at a temperature of up to 110 °C for one to four days [27]. A more recently patented lamination technique for switchable materials replaces the use of thermoplastic adhesive layers with the use of liquid resin including acrylic resin, methacrylate resin, urethane resin, silicone resin, polyester resin, epoxy resin, and polysulfide resin, which may be polymerized to form a solid [28].

Transmission ranges of SPD can be varied according to the material used and the film thickness. Typical visible light transmission (T_{vis}) ranges are <2–50% or from <5–10% to 60–65% with solar heat gain coefficients (SHGCs) ranges 0.29–0.57 and 0.06–0.25 for T_{vis} <2–50% SPD film with clear glass without and with low-e coating, respectively [29]. For an SPD with two glass panes and no low-e coating provided by SmartGlass, a 5.9 W m^{-2} K^{-1} overall heat transfer coefficient was measured experimentally [30] with SHGCs ranging from SHGC = 0.05–0.38 for visible transmittance ranging from 5% to 55% [31]. Specifications of commercially available SPD windows are shown in Table 6.1. For comparison, properties of windows with fixed properties are shown in Table 6.2.

6.3.1 Switching Duration

The ideal switching duration for an SPD glazing in a building window given in Table 6.3 is determined by occupant perceptions of the speed with which discomfort due to glare and/or solar heat gain are eliminated.

The overall duration of switching between fully clear to fully dark is the time taken for the SPD particles to complete rotation according to the voltage applied. However, spectral transmission in the transparent and dark states at particular wavelengths varies [37]. As illustrated in Figure 6.7, the time τ was defined as the time required for a window to reach 63% of its final state (transparent or dark). The defined switching duration 3τ is the time required to reach 95% of the transition and 5τ for 99% of the transition. Figure 6.8 compares the up (to transparent state) and down (to dark state) switching times of the SPD glazing with the solar spectrum. Relative spectral switching durations thereby affect the dynamic response of a switchable window to control glare, indoor lighting, solar heat absorption, and transfer.

In the unpowered (off) state, particles are randomly oriented, so the device is absorptive. In the powered (on) state, particles align with the applied electric field, reducing absorption, so the device is transparent. The orientation of suspended particles and the optical transmission of the device can be continuously controlled by varying potential differences between the range 0 to 20 V_{AC} and 150 V_{AC}. The switching speed depends on the properties of the transparent conductive electrodes and the film material properties, the film thickness and surface area, electric signal applied to the film, and spectral absorption properties of the particles together with the wavelength considered. The switching times in Table 6.1 are for simple on-off control. Obviously with predictive control, switching can be initiated before the change in conditions commences [41].

Hsu et al. [42] provided insights on the switching mechanism involved for SPD devices that enables an estimation of expected switching times. The switching mechanism from dark-to-transparent is the result of the rotation of rod-like particles when applying an AC voltage across the film. An AC electric field minimizes electrophoretic migration. This means the motion of dispersed particles is minimized relative to a fluid under the influence of a spatially uniform electric field. As an electrical field is applied (along x' in Figure 6.9), a torque force is applied to the particle. Torque depends on the electrical field strength, the particle geometry, the conductivity and permittivity of the media, and the applied frequency. Any rotation of the rod with respect to the axis parallel to the electric field is possible. Thus, rotational Brownian motion, which is the random motion of particles in a fluid, readily occurs with respect to this axis. When the x-axis of the particle coincides with the electrical

Table 6.1 Properties of commercial SPD switchable architectural windows.

Commercial SPD switchable windows									
Manufacturer	Size (cm^2)	T_{uv}%	T_{vis}(%)	T_{sol}	SHGC	Power consumption (65–110 V_{AC})	Temperature cycle tests	Switch. speed On/Off (s)	Cycles
SmartGlass [32–34]	150 × 320	<0.5	2–49 0.5–37 with low-e	—	0.29–0.57 0.06–0.25 with low-e	5 W m^{-2} to switch 0.55 W m^{-2} to maintain	−20 to 70 °C	1–3	10^5–10^6
Innovative Glass Corp. [35]	150 × 300	<1	<1–50 or 5–65	—		0.18 W m^{-2}	−20 to 71 °C	2 (on)/30 (off)	—

Table 6.2 Typical properties of static windows.

Static window type	T_{vis}(%)	SHGC
Clear single pane	71	0.72
Clear IGU	63	0.60
Low-e + Clear IGU	57	0.34
Tinted IGU	42	0.38
Low-e + Tinted IGU	43	0.27
Reflected coating IGU	10	0.17

field, maximum transparency is reached. When the electrical field is removed, Brownian diffusion ensues with consequent lower light transmission. A cluster of particles that is in chaotic disorder when the electric field is switched off thus become aligned with the electric field when it is switched on.

Assuming that incident light arrives parallel to the electric field and that the distance between the electrodes is much larger than the particle size, so that the dipole moment theory is a reasonable approximation, the application of the electrical field (E_0) causes each rod to experience an alignment torque that is balanced by the viscous resistance. This balance determines the switching response duration of the rod, which is the elapsed time taken for the rod to return to its equilibrium location after being disturbed by an angular displacement θ. The response duration t_R of an SPD device is given by Eq. (6.1) where $|C_s|$ indicates the absolute (positive) value of the real-valued torque constant C_s. K_s is a shape factor defined by Eq. (6.2). The semiaxis length of the ellipsoid in the y and z directions is "b" while that in the x direction is "a" with $a \gg b$ for a prolate ellipsoid. ε_f is the fluid permittivity, and η is the liquid viscosity.

$$t_R = \frac{12\eta a^2 K_s}{b^2 E_0^2 \varepsilon_f |C_s|} \tag{6.1}$$

$$\frac{12\eta a^2 K_s \omega}{b^2 E_0^2 \varepsilon_f |C_s|} \gg 1 \qquad \text{Condition 6.1}$$

$$K_S = \frac{4e^3(2-e^2)}{3\left(-2e+(1+e^2)\ln\left(\frac{1+e}{1-e}\right)\right)} \quad \text{and} \quad e = \sqrt{1-\frac{b^2}{a^2}} \tag{6.2}$$

$$\theta_D^2 = \frac{k_B T}{4\pi\eta a^3 K_s} t \tag{6.3}$$

$$\left(\theta_D^R\right)^2 = \frac{3k_B T}{\pi a b^2 E_0^2 \varepsilon_f |C_S|} \tag{6.4}$$

$$t_{dec} = \frac{\pi^3 \eta a^3 K_s}{k_B T} \tag{6.5}$$

In the presence of an AC field, the applied frequency is much shorter than the response time of the rods, i.e. $t_R \gg 1/\omega$ with $\omega = 2\pi f$ the angular frequency. If this condition is not satisfied, then the rods oscillate about the mean orientation. This can reduce light transmission in the On state. Eq. (6.1) together with the $t_R \gg 1/\omega$ condition implies Condition (6.1). The angle θ_D by which a rod diffuses, due to rotational Brownian motion, in time t, is given by

Table 6.3 Switching speeds and occupant comfort.

To maintain occupant comfort to an instantaneous change in	Ideal switching duration to consider		Actual switching duration of an SPD glazing. 95% of full transmission change occurs in the…	Comparison of actual and ideal switching durations
	Criteria	Duration		
Discomfort glare	Switch faster than the eye perceives discomfort from sudden glare.	Human eye responds in 13–80 ms [36] to an sudden change in illuminance.	visible within ~0.5 s and ~1 s when switching to transparent and to dark, respectively [37].	Switching times are 6–77 times slower than ideal.
Solar heat gain	Switch faster than the 0.1 °C s^{-1} [38, 39] at which thermal discomfort ensues.	1 °C mean radiant temperature change due to solar gain [40] induces discomfort in 10 s. An 8 °C variation when transiting from shade to sunlight is perceived in 1.25 s.	Near-infrared, at wavelengths above 800 nm, within ~3 s when switching to transparent but between 13 and 20 s when switching to dark.	Switching times are 3–16 times slower than ideal. Depends on the incident solar power and amplitude of solar heat that can be modulated.

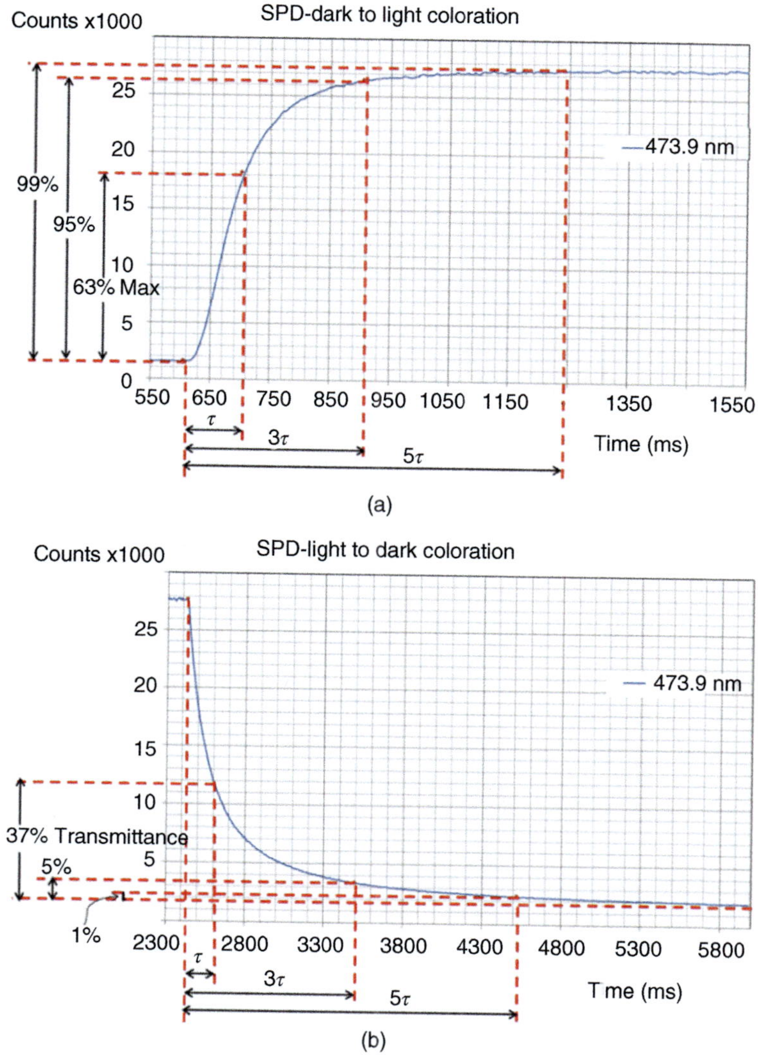

Figure 6.7 (a) Defined switching durations dark-to-transparent (τ_{up}). (b) Defined switching durations Transparent-to-dark (τ_{down}). Examples show an SPD glazing switching with 473 nm incident light.

Eq. (6.3) where the rotation is about the z-axis, T is the temperature. and k_B is the Boltzmann constant. The angle θ_D^R by which a rod will diffuse in time t_R is given in Eq. (6.4) for Brownian dispersion in the On state. A small value of θ_D^R implies less dispersion with consequently better transmission of light. Thus, the liquid permittivity, rather than its viscosity, plays a key role in determining the Brownian dispersion in the On state. Equation (6.4) implies that the strength of the electric field can control Brownian dispersion in the On state.

Once the electric field is turned off, there is no orientating torque on the rods. Hence, Brownian diffusion causes unhindered dispersion of the rods until a fully random distribution (i.e. the dark-off state) is achieved. The elapsed time it takes for the rods to go to the off state, once the electric field is turned off, is termed the decay time (t_{dec}).

When $\theta_D = \pi/2$ in Eq. (6.3), Eq. (6.5) is obtained where t_{dec} is the decay time. A low-viscosity liquid gives a short decay time. A high electric field strength or large ε_f can be chosen to

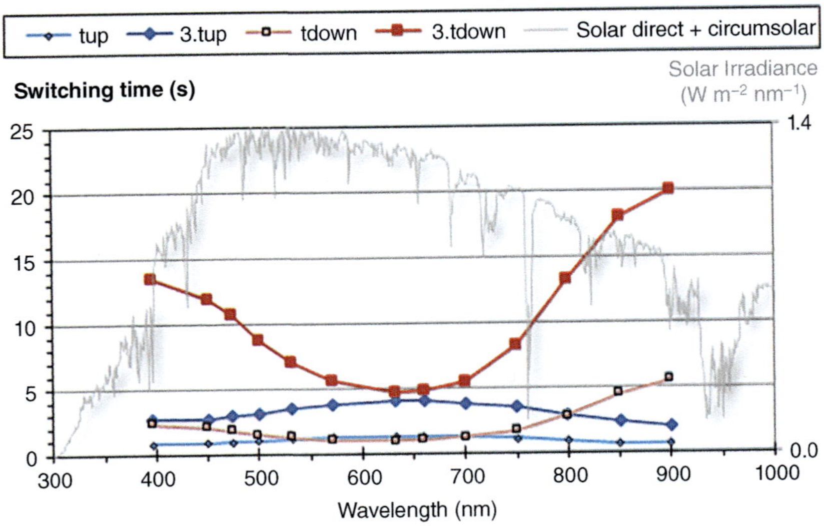

Figure 6.8 SPD window switching to transparent and to dark times for a glazing of size 280 mm × 180 mm.

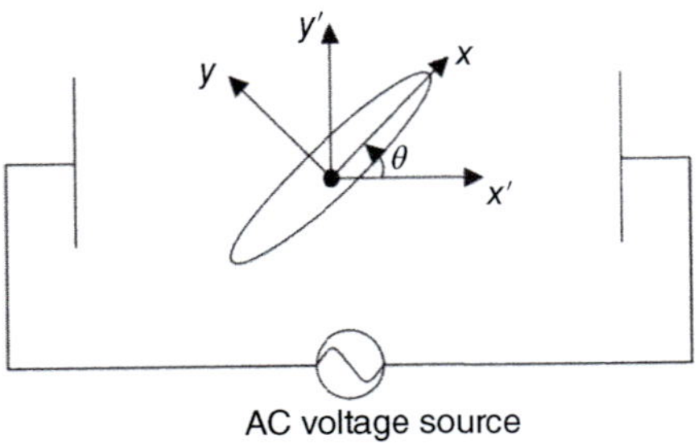

Figure 6.9 A prolate ellipsoid suspended in an AC electric field.

Table 6.4 Decay times for two organic fluids in an SPD.

	Materials properties			
Temperature T_p	***a***	***b***	**Viscosity η**	**Decay time t_{dec}**
293 K	1 μm	0.1 μm	Ethanol 1074×10^{-3} Pa s	1100 ms
			Propanol 1945×10^{-3} Pa s	1990 ms

ensure that Brownian motion is not dominant in the On state. Equation (6.5) for t_{dec} leads to criterion on the liquid viscosity that ensures hastened decay times. Table 6.4 shows examples of decay times for two organic fluids. These decay times are indicative; actual switching speeds are directly related to the glass surface area, so small windows require less elapsed time to switch than larger windows.

6.3.2 Spectral Transmittance

Using a solar lamp and integrating sphere (IS) connected to an Avantes spectrometer, for an SPD window with a total 9 mm glass thickness and 0.35 mm active layer thickness (provided

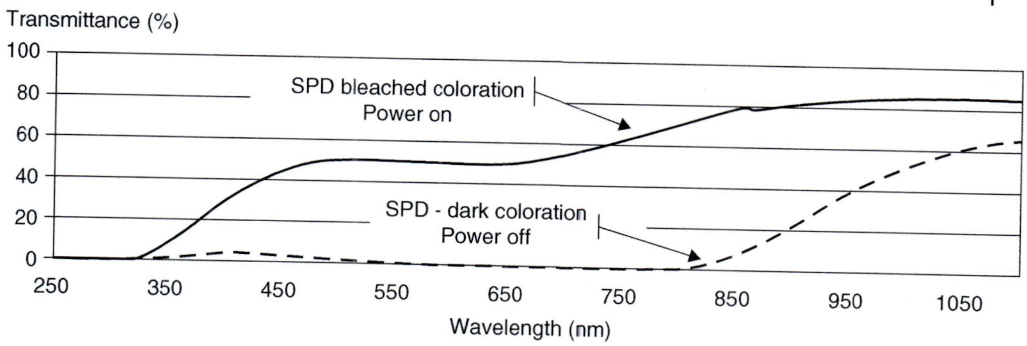

Figure 6.10 SPD spectral transmittance in the dark and light states. *Source:* Adapted from Lemarchand et al. [44].

Table 6.5 Transmittance table of an SPD window in the clear and dark (opaque) states and the associated absorbance efficiency. Averaged transmittance data was taken with a spectrophotometer.

		SPD transmittance		
		SPD Clear	SPD dark	η_a
Wavelength range	UVB (280–315 nm)	0.08%	0.01%	87.57%
	UVA (316–399 nm)	18.16%	3.05%	83.19%
	Vis (400–700 nm)	48.54%	1.27%	97.37%
	IR (701–1100 nm)	77.62%	27.43%	64.66%

by SmartGlass International), Lemarchand et al. [43] measured the transmittance spectra (recorded 35 cm away from the glazing) as shown in Figure 6.10.

A UV filter is applied between the glass and ITO layers to protect the ITO layers from yellowing under UV radiations providing a cutoff wavelength of 330 nm. The average transmittance in the UVB, UVA, visible, and infrared light range up to 1100 nm as well as the absorptance efficiency defined by Eq. (6.6) are provided in Table 6.5.

$$\eta_a = \frac{T_{clear} - T_{opaque}}{T_{clear}} \tag{6.6}$$

A transmittance of 48.5% in the clear state is however undesirably low for some applications. Changing the type of rod-like particles, reducing their density, or reducing the film thickness may further increase light transmittance in the clear state while having more light transmitted in the dark state. Light in the infrared region is absorbed less efficiently with 27.4% average transmittance of light in the solar infrared range.

SPD glazing transmittance values are usually provided in the visible (T_{vis}), at wavelengths from 380 to 780 nm. However, they are typically not provided in the solar range (T_{sol}) from 280 to 2500 nm. In an investigation of the spectral total and diffuse transmittance of an SPD, including its angular dependence [44], it was found that there was minimal infrared change. The absorption coefficient of SPD particles does not modulate from switching the glazing on and off at wavelengths above 1400 nm [45]. Short wave-infrared absorption, from 1400 to 2500 nm, has a fixed value depending on the SPD film thickness, type of particles, and their

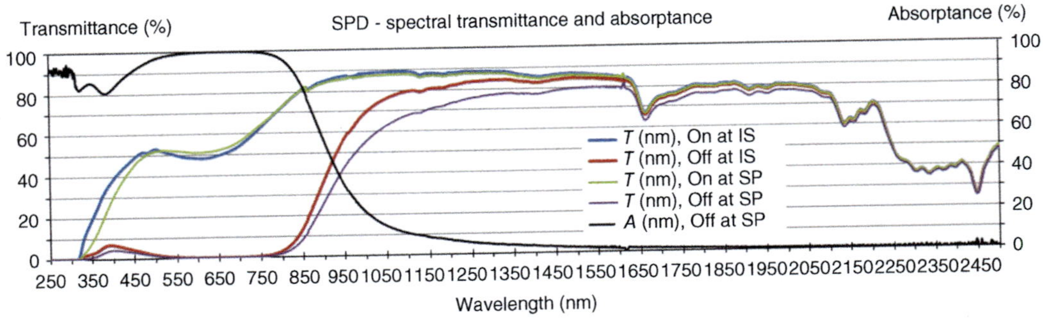

Figure 6.11 SPD film spectral transmittance and absorptance. Both integrating sphere (IS) (hemispherical) data and SP sample port data (specular) data are shown in [37].

density within the film. The solar transmission change is then mostly the result of change in the transmittance of the visible and the near-infrared (NIR) from 780 to 1400 nm.

Lemarchand et al. [37] examined a 0.35 mm thick SPD film, mainly made of a polymer (most likely PMMA) for which absorption bands occurred around 1657, 2143, and 2328–2450 nm in Figure 6.11. In the Off state, particles in random orientations strongly absorb light in the visible and NIR range. The spectral transmittance curves at the Lambda 900 IS and sample-port (SP) positions showed differences in the transmitted blue and NIR wavelength ranges. Plotting the scattered transmission (difference between the two curves), a Poisson distribution was obtained between 760 and 2250 nm with a maximum of the difference at about 1μm corresponding to the particle length [12] and confirming this difference is attributable to a Mie scattering process. In the On state, the film appears transparent as particles align with the electrical field, with their long axis along the light propagation axis. Mie scattering with the particles' long axis disappears and is privileged with the particles' width. Increased scattered transmittance is measured between 315 and 590 nm with an improved reading on the peak probability of particle width about 371 nm. The particles' width-length ratio is then approximately equal to 1/3 or aspect ratio of 3. The transmittance curves were matching at wavelength above 1500 nm; signifying reflections from all internal interfaces and absorption are the only occurring phenomena.

SPD windows show a very effective absorption modulation in the visible range but strongly decrease [46–48] modulation in the NIR region. However, considering the 278–1020 nm working range of the optical bench experiment conducted by Lemarchand et al. [37], both visible and infrared transmission switching ranges satisfied the criteria in Table 6.6. The fraction of solar energy modulated in the short wave-infrared range, from 1400 to 2500 nm, was about 1% and in line with Barios results [45]. This modulation would further diminish the absorptance from additional adhesive layers, glass layers, and coatings of windows. Scattering in the clear state was observed and implied that spectral transmittance values vary at short distances from the window before being constant in the "far field." The switching speed of the window was demonstrated to be correlated strongly to the wavelength. An SPD film with a larger absorption and a faster switching speed in the infrared wavelength range is preferable when controlling solar heat transfer through glazing.

Vergaz et al. [49] noted that an impedance analysis of commercial SPD indicated a similarity to an equivalent electric circuit comprising an active electrolyte resistance in series with a parallel combination of a capacitance and a faradaic impedance (i.e. as in a Randles circuit

Table 6.6 Requirements for switchable windows.

Solar transmission						Solar reflection						
*T*sol (300–2500 nm)		*T*vis(380–780 nm)		T_{nir}		R_{nir}(780–1400 nm)		Switching voltage (V)		Memory (h)	Cycling lifetime	Operating temperature
Clear	Dark	Clear	Dark	Clear	Dark	Clear	Dark	Small	Large	1–12	10^4–10^6 cycles 5–20 years	Unprotected
50–70%	10–20%	50–70%	≤10–20%	—	—	10–20%	≥70%	1–3	10–24			−30 to +70 °C

used in electrochemical impedance spectroscopy [50]. Such an equivalent circuit model has been developed and validated experimentally [18].

Barrios et al. [44] developed a theoretical two-flux model that provided absorption and scattering coefficients. In further work, Barrios et al. [45] used two-flux and four-flux models to derive refractive indices and extinction coefficients, as well as scattering and absorption coefficients. These coefficients were used to predict direct, total, and diffuse components of transmittance, reflectance, color, and haze, as a function of the thickness of the active layer. An optimum active layer thickness for optical contrast was found to be in the range of 200–300 nm. Ghosh Norton and Duffy [51] found that when the transmittance of the switchable window glazing increases, the CIE Colour Rendering Index also increases and the Correlated Colour Temperature decreases. However, Dangol et al. [52] suggested that findings of Ghosh and Norton [53] apply to only one particular type of switchable window glazing they examined.

Intrinsic and extrinsic scattering and absorption coefficients of an SPD smart window sample at dark and clear appearance states without and with applied electrical voltage, respectively, have been determined [54]. For the predominantly forward scattering exhibited by an SPD, extrinsic diffuse interface reflectance values were approximated by integrating all Fresnel reflection coefficients up to the critical angle for total internal reflection. A new approximation for the average light crossing parameter based on the collimated and diffuse light intensities at each interface was used by Barrios et al. [54] to calculate intrinsic scattering coefficients. Barrios et al. [55] developed a method to calculate the real and imaginary (i.e. absorption and scattering) parts of the complex refractive index of an SPD window that can be applied to determine optimum active SPD layer thickness.

An SPD using copper-reduced graphene oxide (Cu-rGO) core–shell nanowire films as a transparent conductive electrode was developed by Huang et al. [56]. The reduced graphene oxide wrapping of the copper nanowire electrodes was experimentally demonstrated to provide high optical transparency, high electrical conductivity, and improved stability under various testing conditions. A 42% change in optical transmittance between "transparent" and "dark" states was observed comparable to those of SPDs with ITO electrodes. An SPD with tuneable visual transmission using core–shell ZnS/SiO_2 colloidal nanoparticles was developed by Han et al. [57]. The measured increases in transmission as larger voltages were applied to the external electric field corresponded to those predicted by a finite-difference time-domain formulation of Maxwell's equations.

6.4 Applications in the Built Environment

There are many innovative adaptive building facade elements that alter the solar heat and daylight, air infiltration, and/or thermal insulation of buildings [41, 58]. SPD glazing alone or in combination with other systems form part of an envelope system to mediate between the internal and external environments to achieve desired internal comfort, health, and amenity. SPDs are used for daylight and glare control, solar heat transfer control, and privacy control. When introduced into supervisory control and data acquisition systems or building management systems, the tuning of SPD windows can be optimized to minimize energy consumption while improving occupant comfort. A field-programmable gate array-based electrical circuit can be customized to provide intermediate levels of transmittance [18]. For applications in building glazing, a switchable window must:

- Provide a required transmission modulation range that reduces daylight glare
- Have a fast-switching time, defined as that to reach 90% of its modulation range (this being also strongly related to the window size)
- Has an acceptable operating lifetime; defined by the number of switching cycles undergone with minimum performance degradation
- Have low transmittance of UV light
- Has an acceptably low total energy consumption and operating voltage

These requirements are summarized in Table 6.6 [6]. Desired optical transmission and reflection percentages are given in the clear and dark states.

In addition, a switchable window should:

- Change tint across a window homogeneously independently of its size
- Enable transmission to be set as desired at any point between the clearest and darkest states
- Be designed to enable integration with other coatings (e.g. with low long-wave emittance)
- Have typical commercial and residential installation costs < US$500/m^2 (463€/m^2) at 2023 prices. Switchable windows typically cost an additional ~ US$232/m^2 (215€/m^2) compared to typical IGU leading to a payback period of 26–33 years for residential buildings and 57–61 years for commercial buildings [59]. An additional cost of US$70/m^2 (€65/m^2) gives a payback time of 10–12 years and 21–22 years for residential and commercial buildings, respectively [60]
- Be available in areas sufficiently large for relevant applications (i.e. up to $3 \times 2\,m^2$)
- Have a low to high operating temperature with protective coatings and filters
- Be self-powered

The commercial SPDs shown in Table 6.1 are available in large areas and offer an excellent lifetime cycle capability, 10^6 cycles [33], with temperature cycle tests between −20 and 70 °C, close to perfectly satisfying the criteria in Table 6.6. The visible transmission modulation barely satisfies Table 6.6 criteria. Reflecting NIR as required in Table 6.6 rejects solar heat, which reduces solar heat absorption, thereby alleviate potential degradation mechanisms of the glazing, and secondary emission of energy at longer wavelengths. Converting the expected reflected modulation in terms of transmission modulation between the clear and dark state, 50% to 60%, modulation is desired. Lemarchand et al. [37] demonstrated that the SPD film alone modulates 38.8% of light transmission between 700 and 1400 nm. Further NIR modulation is desired. Although the solar transmission modulation from the two manufacturers is not mentioned, Lemarchand et al. [37] concluded a transmission modulation of 38% in the 280–2500 nm light range. With desired solar modulation targeted between 30% and 60%, manufactured SPDs fit this criterion.

Chromogenic switchable (glazing or windows) technologies vary light transmission in the visible and infrared solar range. They have the major advantage of controlling glare and solar heat transfer and, subsequently, improving the users' indoor comfort and wellbeing, and lowering the energy consumption associated with heating and cooling. SPD windows, similarly to EC windows, can potentially save up to 25% of the building energy consumption in lighting, heating, and cooling [61]. Energy saving, comfort, and wellbeing improvement potentials will vary with many parameters including the building climate, size and geometries, fabric properties, orientation, surrounding with reflection and casted shadow, window configuration, window-to-wall ratio, usage, internal heat gains, and monitoring and control.

Because of the constant power requirement of SPD windows and the apparently low transmission modulation over the solar spectral range, electrochromic switchable windows seem

Table 6.7 SPD SmartGlass transmittance with clear float glass. Values are nominal ±5%, depending on glass configuration used.

All values in %	SPD SmartGlass (11.5 mm) power off	SPD SmartGlass (11.5 mm) power on	Clear float glass (6 mm)	Frosted glass (6 mm)
Visible light transmittance	0.24	49.0	86	76
Clarity	2.90	N/A	83	18
UV transmittance	0.5	0.5	55	55

Source: Data from SPD control systems, retabulated [32].

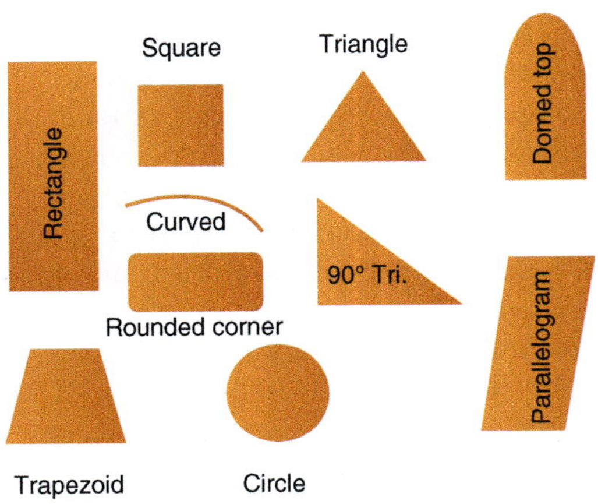

Figure 6.12 Possible shapes of SPD and PDLC windows. *Source:* Ref. [32] / SmartGlass International Ltd. 2011.

to have a significant advantage over SPD windows. Both SPD and EC windows have a specular transmittance, have an upper operating temperature limit in order of ~60 °C, and are therefore most suitable for building facades. However, the advantage of SPD over EC is its fast-switching capability (~200 ms versus 120–180 seconds for similar sizes) making SPD more appropriate when a fast visual transmission change is required.

Typical transmission ranges for SPD SmartGlass™ products are provided in Table 6.7 [32]. SmartGlass manufactures SPD windows in large sizes (about 1 × 3 m) maintaining a uniform degree of transparency across the window. The various shapes shown in Figure 6.12 are available. Manufacturer data shows an operating temperature range from −20 to +70 °C. After manufacturing, each SmartGlass panel is tested by switching approximately 57,000 times in a rapid on/off sequence.

SPDs are appropriate for façade as vision transparency is maintained through the different shading of opacity. Although the warranty of industrial products is similar, the life cycle ability of SPD appears better than EC windows. SPDs have the advantage to provide a much faster switching speed than electrochromic devices, but they suffer from the constant need of power to maintain the desired optical state else than the "off" state. The required power may be supplied by photovoltaic panels or semitransparent coating thereby forming a building-integrated photovoltaic (BIPV) system.

In a temperate climate, the lower the overall heat coefficient of a window, the larger its optimal size. Thus, highly insulated glazing can be deployed over large areas, but this often entails a correspondingly high risk of discomfort glare. To avoid this, switchable glazing is incorporated in low-emittance coated multiple-pane and vacuum glazing assemblies. For a combined SPD and evacuated glazing, the thermal performance has been measured [62], and the thermal and visual comforts have been analyzed [63].

SPD daylighting performance has been evaluated. Interior useful daylight illuminance levels for clear sunny, intermittent cloudy, and overcast cloudy days provided by an SPD and the same area of double-glazing were compared using an unfurnished outdoor south-facing test cell with glazing to wall ratio of 1 : 8 [64]. Daylight glare indexes and daylight factor were calculated for "dark" and "transparent" states. An electrically actuated SPD glazing with a dark transparency of 55% was found to control daylight glare.

When there is a high direct component of solar radiation, transmittance varies with solar incidence at incident angles greater than 60°. When the diffuse component is largest, transmittance does not vary significantly [65]. Transmittance values for different azimuthal incident angles for an SPD glazing in its "transparent" and "dark" states have been determined. In Dublin, below 0.5 clearness index, isotropic diffuse transmittance prevailed, while transmission of direct insolation was dominant above 0.5 clearness index. For a south-facing vertical plane, SPD glazing transmittance in its transparent and dark states are 0.25 and 0.025, respectively, while the clearness index is below 0.5 [51]. The clearness index is a measure of the clarity of the atmosphere. It is defined as the surface radiation divided by the extraterrestrial radiation. The difference is a measure of absorption by the Earth's atmosphere at a given time, day, and location. It is a dimensionless number between 0 and 1.

Switchable windows that operate with window-generated photovoltaic electricity are easier to install; there are no external power connections required. They also avoid operational greenhouse gas emissions incurred when the switching electricity used is generated using fossil fuels. SPD windows have been successfully powered by photovoltaics [66]. In a PV-powered SPD window, it is important to optimize the relative sizes of the PV area and battery capacity to minimize the probability of loss of power supply [67]. A 1 m^2 SPD consumes about 10.4 kWh $year^{-1}$ of electricity equivalent to 4.8 kg CO_2 emission.

6.5 Accelerated Testing

Adaptive switchable SPD glazing can be a viable component for low-energy buildings if it survives nearly 20 years without degradation of its performance. A glazing is a suitable building application component if it sustains against solar radiation exposure, operating temperature range, thermal shock, humidity, air, rain, and pollutants [68].

Generally, warranties for switchable residential architectural windows are for 10 years, following the standard for IGU windows. But in many cases, IGU windows survive much longer, more like 20–25 years. A common IGU degradation mode is seal leakage. Standards for testing architectural products are, for example, ASM G7, "Standard practice atmospheric exposure testing of non-metallic materials." For accelerated testing, standards, such as ASTM G157, G153, G155, or GB/T 14259, "Accelerated weathering test methods for accelerated weathering of building materials" are used.

So far, tens of thousands of SPD-Smart windows and sunroofs have been installed for building, automotive, aerospace, and marine applications. Some of these vehicles have been on the road or in the field for 12 years without major issues [69]. SPD sunroof panels have been installed in Mercedes, Cadillac, and other makes. The general testing standards for automotive are summarized in the Atlas Standards Guide 101 [46]. The standards for automotive weathering tests are summarized in the Atlas Testing Standards Guide 100 [47]. The popular standards are Global (ISO 16474-2) where 65C is the upper temperature. The German standard is VDA 621-429, and upper temperature is 55C. The ASTM and SAE standard cycle is 38-70C. The results of static soaking in hot climates are summarized in an article by Atkinson and Hill [48]. They correlated testing of 81 SAE cars of different types (colors and windows) in hot weather static soak. The highest upper "breath temperature" was 71C. The "breath temperature" is the air temperature at the average driver's breathing height.

Artificial accelerated weathering methods or outdoor testing methods can be used to investigate the photostability and environmental degradation of SPD glazing [21]. For example, the QUV instrument made by Q-Lab can be used over a range of temperature, ambient humidity, and a chosen UV light power density to count the number of hours for which the SPD optical degradation is less than 50%. An aging test applicable to SPD glazing is the Volkswagen durability standard PV 3929, which detects changes such as color and/or gloss of nonmetal materials, plastics, elastomers, and convertible-top fabrics that are exposed to weather and daylight. The test reproduces dry and hot climates such as in Kalahari, South Africa, and Arizona. The test conditions are an air temperature of 50 °C, 20% relative humidity, and an irradiation of 75 W m^{-2} in the 300–400 nm wavelength range. The test duration of about two months (~1600 hours) reproduces a one-year cycle. Considering glazing units must be air- and water-tight with the addition of sealants, high air temperature and solar radiation are presumably the extreme environmental conditions SPD glazing could be subject to.

In an evaluation of SPD glazing performance after three years of outdoor exposure [70], it was found that contrast ratio (ratio of minimum and maximum transmittance) of SPD glazing before and after exposures changed from 1 : 11 to 1 : 10 over the 380–780 nm wavelength range. After three years of UV exposure totaling 533.5 MJ m^{-2}, the SPD glazing appeared 2% clearer in the transparent mode and 0.5% clearer in the dark mode. This negligible change of contrast ratio is associated with the UV degradation of polymer layers, in particular the EVA. The voltage and power required to obtain the same transmission of the SPD glazing did not vary after seven year of thermal exposure with surface temperature rising from 24 to 38 °C.

Boote and Butcher [27] developed SPD glazing lamination techniques in their patent work (US 20120176656 A1). They found by accelerated testing at 100 °C of SPD laminated glazing, a major contributor to the degradation of SPD devices involved moisture present in particular components in the laminated glazing, even at very low levels of moisture. One must note that the operating temperatures for SPD architectural glazing are −20 to 70 °C, according to SmartGlass International [32]. An example given for a "typical" SPD device with unknown water content that is stored at 100 °C for 500 hours demonstrated a 20% loss of its dynamic optical transmission. Figures 6.13 and 6.14 illustrate the increase of transparency in the clear mode as water content from both adhesive and SPD film is reduced. Additionally, the SPD film is switchable between the dark and clear state for a longer period. To stabilize the film, Boote and Butcher [27] recommend keeping the water content at $\leq$0.26 wt% for PVB or $\leq$0.24 wt% and more preferably $\leq$0.16 wt% for EVA. These water levels will greatly reduce high-temperature degradation of the SPD film. This highlights the importance of the initial water content of the EVA or PVB layers depending on the storing condition before baking for the laminating process. For example, PVB stored at 20 ± 2 and 24 ± 4RH results in water content between 0.26 and 0.5 wt%. The assumption on the degradation mechanism is that under heat, water released from the adhesive layers and possibly the PET layers, forming the SPD penetrates the SPD matrix and degrades particles.

The composition and structure of the edge sealing have a great importance to prevent the degradation of the adhesive film from moisture and oxygen penetration and

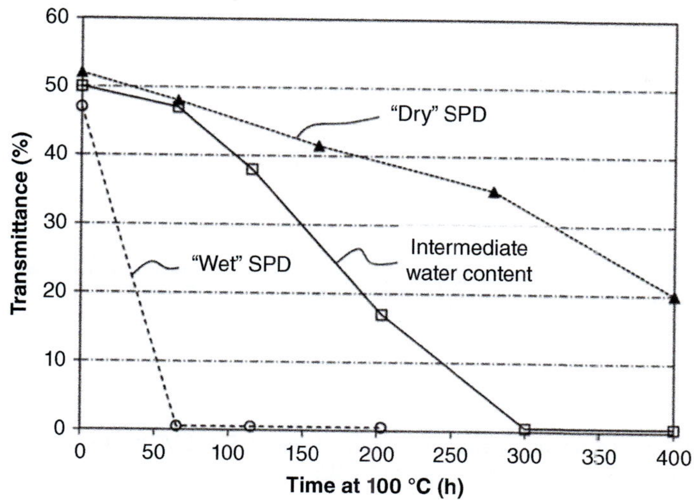

Figure 6.13 On-state light transmittance, measured using CIELAB Illuminant C 2° observer, of a SPD laminate made with standard EVA and SPD film with water content varying in time as it is baked at 100 °C. *Source:* Boote and Butcher [27] / Pilkington Group Ltd / U.S. Patent.

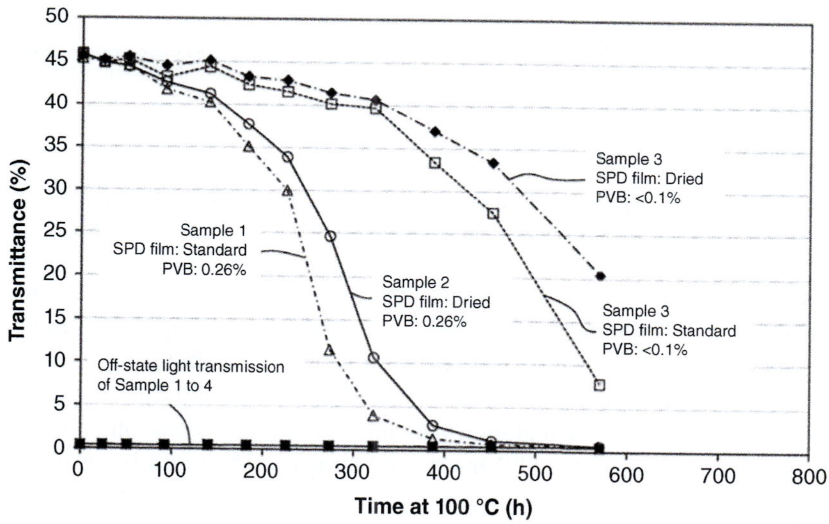

Figure 6.14 On-state light transmittance of four samples assembled with Glass/PVB/SPD/PVB/coated PET film/PVB/glass. Each SPD film has a different water content in the PVB prior to being baked at 100°C. PVB layers were all 0.38 mm thick and the coated PET was 0.05 mm thick. *Source:* Boote and Butcher [27] / Pilkington Group Ltd / U.S. Patent.

UV radiation [20]. The SPD film must be sandwiched between 2 panes, typically made of glass, with at least two adhesive layers. Figure 6.15 illustrates an edge seal preferably containing a polyimide (PI) and/or polyisobutylene (PIB) extending from the outer edge region along the horizontal side of the SPD film for a length (L) of at least 1 mm. The thermoplastic polymer film has an outer edge protrusion greater by least 2 mm than the SPD film. A third optional thermoplastic contributes to the improvement of the mechanical properties (e.g. breakage resistance) and contributes to solar protection enhanced by the optional addition of coatings and/or filters. The SPD film is cut out such that the thermoplastics are 1 mm to

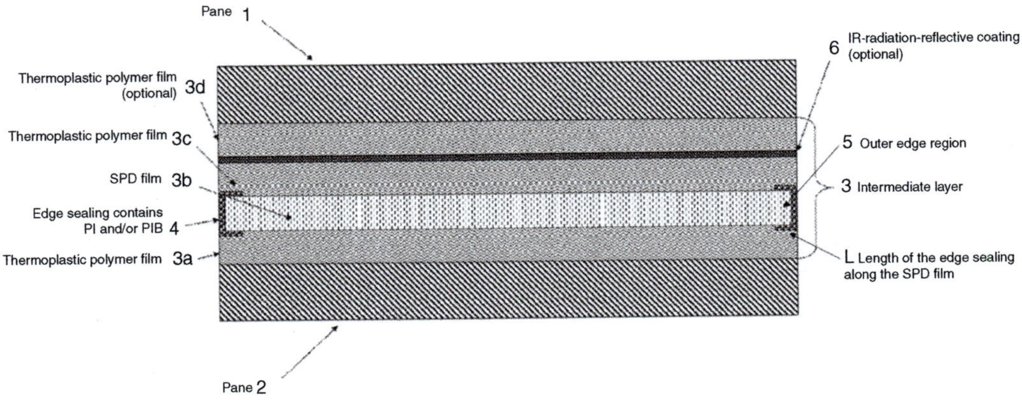

Figure 6.15 Laminated glazing cross sectional construction. *Source:* Patent US 2015/0298431 A1 (Von Der Weiden et al. 2015) [20] /Saint Gobain Glass France SAS / U.S. patent.

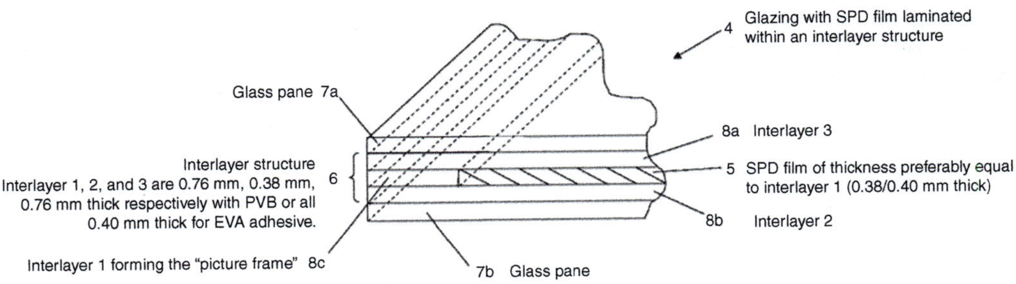

Figure 6.16 Laminated glazing cross-sectional construction. *Source:* patent WO/2007/122428 (Greenall et al. 2007) [71] / Michael Robert Greenall et al. / U.S. Patent.

4mm longer, which, in addition to the sealing edge, significantly contributes to increasing the aging resistance of the SPD film and arrangement. A second edge sealing composition with three adhesive layers is illustrated in Figure 6.16 [71]. The manufacturing process leaves a space (or "picture frame") filled with adhesive between the edge of the glazing edge and the SPD film to reduce the potential migration of mobile interlayer material components (e.g. chemical or water molecules) to the SPD matrix and particles. The result is a glazing unit with a transparent area at the edges as illustrated by Figure 6.17 where bubbles can potentially form.

SPD glazings are installed within window frames typically sealed using silicone. It is important to use only neutral cure silicones (such as Soudal Silirub 2 recommended by SmartGlass) rather than acetic silicones [32]. This is because when submitted to solar heat gain, acetic silicones release acetic acid that penetrates into the SPD interlayers. This acetic acid ingress leads to the delamination of the film and the migration of SPD particles as illustrated by Figures 6.18 and 6.19.

Additives cross-linked with raw EVA are used to cure faster, to improve UV resistance, for high transparency, additional strength, reduced shrinkage, ease storage, anti-PID (Potential Induced Degradation), and anti-snail pattern [72]. The disadvantage is that, when exposed under intense light (in particular UV), high temperature (T), and high humidity taken individually or in combination, results in the deacetylation of vinyl acetate pendant group. Interacting additives produce about 10 species including acetic acid (HAc), gases such as

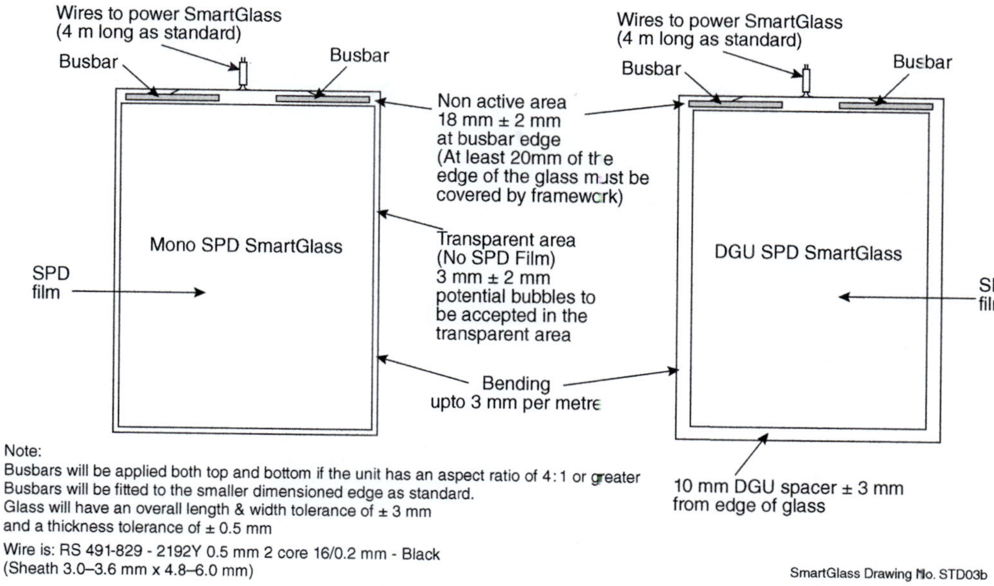

Figure 6.17 Specifications of SPD SmartGlass. *Source:* SmartGlass [32] / SmartGlass International.

Figure 6.18 View of a delaminated SPD film in the dark mode. Delamination is caused by improper sealant use, containing acetic acid. The manufacturer specifies neutral cure silicones.

acetaldehyde (CH_3CHO), butene, ethylene (C_2H_4), carbon dioxide (CO_2), monoxide (CO), methane (CH_4), and water (H_2O) that catalyze the degradation of EVA [73]. This leads to snail trail, potential discoloration or yellowing of the film, and corrosion of the metallization and electrical bus lines in modules. With the increase in temperature and UV radiation, the rate of degradation accelerates in hot, dry, and sunny climates such as in the Southwest United States, Africa, Arabia, and Asia. Special consideration on the design of glazed SPD is then needed in such climates. For example, if EVA is used, its thickness is important, as thicker films take longer to yellow.

Acetic silicone may be replaced by indium seals, as used in vacuum glazing [74] or as indium is expensive, thin indium layers between each side of a copper gasket [75] that may also form a continuous bus bar or radiation-cured adhesive resins [76]. Table 6.8 summarizes

(a)

(b)

Figure 6.19 (a) View of a delaminated SPD film in the clear mode with migration of SPD particles. (b) Magnification of the red rectangle region of the (a) picture. Delamination is caused by improper sealant use, containing acetic acid. The manufacturer recommends using neutral cure silicones.

the origins of the main SPD glazing degradation mechanisms with subjective assessments of consequence, speed of evolution, and the probability of risk level. Table 6.9 summarizes potential SPD glazing degradation mechanisms due to human errors.

6.6 Conclusion

SPDs show efficient dynamic transmission in the visible range but are still limited in the infrared range affecting solar heat transfer. Switching duration extensively varies with the light wavelength considered. SPD windows would benefit from (i) having an increased visible light transmission in the transparent state to improve indoor daylighting, (ii) a faster switching speed to avoid instantaneous glare and solar thermal discomfort, (iii) having dynamic spectral properties over a wider solar spectral range, and (iv) an improved dynamic efficiency within the infrared range. Advantages and drawbacks of SPD are summarized in Table 6.10.

SPD switchable windows are a commercially available technology. The extent of their adoption is largely dependent on manufacturing costs. More process development is required to lower the manufacturing cost. Consideration could also be given to new approaches using multiple separate active layers to extend the range of transmittance between ostensibly transparent and dark states.

Table 6.8 Summary of SPD glazing degradation mechanisms, as determined from accelerated testing at 100 °C. The recommended testing operating temperature by manufacturer is -20–70 °C. Many of these mechanisms can be eliminated by tighter quality control.

Degradation	Origin	Risk	Characteristic	Evolution	Likelihood
Transmission loss	UV degradation of polymer or ITO layer. [69]	Low	Slight variations of the window aesthetic and solar energy modulation over the windows' lifetime.	Slow but continuous degradation	High – Natural degradation
	Voltage/power change from the electrical driver.	Low	Reduction of contrast ratio.	Sudden or progressive	Low – Robust driver electronics
	Too much water content within the interlayers during glazing lamination. [23]	Medium	Lower switching repeatability, lower transmission in clear state, and potential switching failure.	Progressive over time	Low to medium – Depends on manufacturing process and operating temperature conditions
Switching repeatability	Bus bar burn from driver power surge	High	Switching failure: SPD window remains in the unpowered state. Bus bar inaccessible behind glass panes. Window replacement required. Detection: light yellow coloring appears around soldering.	Sudden	Low – Robust driver electronics
	Bus bar oxidation from acetic acid.[a]	High	Switching failure as above. Oxidation leads to a delayed failure after prolonged operation.	Delayed but sudden	Low – SPD design suited to application Medium – see below.

a) Improper sealant used.

Table 6.9 Summary of potential SPD glazing degradation mechanisms due to Human (Installation or Assembly) errors. These are unusual events.

Degradation	Origin	Risk	Characteristic	Evolution	Likelihood
Transmission loss	UV degradation of polymer or ITO layer.	High	During installation, the potential UV filter or other solar protecting layers is not facing the outdoor direction.	Fast continuous degradation	Low– Human error (installation)
Switching repeatability	Bus bar burn from bad soldering of power cables to the bus bar.	High	Switching failure: SPD window remains in the unpowered state. Bus bar inaccessible behind glass panes. Window replacement required. Detection: light yellow coloring appears around soldering.	Sudden	Medium – Human error (soldering at assembly)
Film delamination	Use of acetic silicon w/ penetration of acetic acid through the sealing tape around glazing edges.[a]	Medium	Aesthetic: Snail patterns forming from the window edge moving inward.	Slow but continuous progress over time.	Low – Human error (installation)

a) Contractor used the wrong sealant.

Table 6.10 Characteristics of SPD glazing.

Advantages	Neutral	Drawbacks
Fast switching (100 s of milliseconds) Long lifetime: ~10^6 cycles Any transmittance state possible between fully transparent and dark states Have potential for large area fabrication (>150 × 350 cm^2) Both visible and solar control possible. Operating temperature: −20 to 70 °C	Low modulation of the infrared transmission Require UV filter to protect active layer	Not Bistable (constant power required in the transparent state) Cost Limited choice of color

To decarbonize buildings, building components must enable minimal energy to be incurred both in the operation of buildings and embodied in their manufacture and installation [77]. The latter leads to imperatives for building components to be designed to (i) use recycled materials and (ii) to ease their reuse and recycling. As glass having high embodied energy, these considerations need to come to the forefront in the development of SPD switchable windows. An aspect of this is a fuller understanding of the toxicity of SPD active layer materials as well as how to minimize the wider environment impacts of their disposal.

Acknowledgment

The development of this chapter was supported by MaREI, the SFI Research Centre for Energy, Climate and Marine (Grant No. 12/RC/2302_P2).

References

1 Jelle, B.P., Hynd, A., Gustavsen, A., Arasteh, D., Goudey, H., and Hart, R. (2012). Fenestration of today and tomorrow: a state-of-the-art review and future research opportunities. *Solar Energy Materials and Solar Cells* 96: 1–28.

2 Beevor, M. (2011). Smart building envelopes. *Arab Construction World* 29: 21–22.

3 Lampert, C.M. (1991). Large-area chromogenics for smart windows. *IEEE Circuits and Devices* 16.

4 Lampert, C.M. (1993). Optical switching technology for glazings. *Thin Solid Films* 236: 6–13.

5 Karlson, J. (2001). Control system and energy saving potential for switchable windows. In: *Seventh International IBPSA Conference*, 199–206. Rio de Janeiro, Brazil: Building Simulation.

6 Beatens, R., Jelle, B.P., and Gustavsen, A. (2010). Properties, requirements and possibilities of smart windows for dynamic daylight and solar energy control in buildings: a state-of-the-art review. *Solar Energy Materials and Solar Cells* 94: 87–105.

7 Land, E.H. (1934). Colloidal suspensions and processing of making same. US Patent 1,951,664, Polaroid.

8 Land, E.H. (1934). Light valve and method of operation. US Patent 1,955,923, Polaroid.

9 Saxe, R.L., Thompson, R.I., and Forlini, M. (1977). Light valve suspension materials and suspension therefore. US Patent 4,025,163A.

10 Saxe, R.L. (1981). Light valve containing improved light valve suspension. US Patent 4,247,175A.

11 Saxe, R.L. (1983). Light valve suspension containing fluorocarbon liquid. US Patent 4,407,565.

12 Saxe, R.L. (2003). SPD films and light valves comprising liquid suspensions of heat-reflective particles of mixed metal oxides and methods of making such particles. Patent No. US 6,606,185 B2. Research Frontiers Incorporated, USA. https://patents.google.com/patent/US6606185B2/en (accessed August 2023).

13 Saxe, R.L. (2006). Anisometrically shaped carbon and/or graphite particles, liquid suspensions and films thereof and light valves comprising same. US Patent 6,987,602 B2 (Jan.

17, 2006) continuation of Patent No. US 6,529,312 B1. Research Frontiers Incorporated, USA. https://patents.google.com/patent/US6529312B1/en (accessed August 2023).

14 Research Frontiers (2015). Research Frontiers Hosts Licensee Conference at The World's Fair In Milan Under The Largest SPD-SmartGlass Roof In The World. https://www.smartglass.com/research-frontiers-hosts-licensee-conference-at-the-worlds-fair-in-milan-under-the-largest-spd-smartglass-roof-in-the-world/ (accessed August 2023).

15 Lee, J.H., Yu, B.S., and Kim, E.S., (1995). Making Liquid Suspension Type Liquid Valve Film. Patent No. US 5,409,734. Hankuk Glass Industries Incorporated, Korea. https://patents.google.com/patent/US5,409,734A/en (accessed November 2023).

16 Market Research Future (MRF) (2020). Global smart window market research report: by technology (electrochromic technology, PDLC, SPD and photochromic), by application (architectural and construction, Transportation, automotive, Aircraft, marine, consumer goods and power generation), by Region (North America, Europe, Asia-Pacific, Middle East & Africa and Central & South America) – Forecast till 2027. https://www.marketresearchfuture.com/reports/smart-window-market-9595 (accessed August 2022).

17 IMARC (2022). Smart Window Market: global Industry Trends, Share, Size, Growth, Opportunity and Forecast 2022–2028. https://www.imarcgroup.com/smart-windows-market (accessed August 2023).

18 Vergaz, R., Sanchez-Pena, J.M., Barrios, D., Vázquez, C., and Contreras-Lallana, P. (2008). Modelling and electro-optical testing of suspended particle devices. *Solar Energy Materials and Solar Cells* 92 (11): 1483–1487.

19 Chakrapani, S., Saxe, R.L., and Slovak, S.M. (2006). Materials and methods for improving properties of SPD emulsions and films. Patent No. EP1901909A4. Research Frontiers Incorporated, USA. https://patents.google.com/patent/EP1901909A4 (accessed August 2023).

20 Von Der Weiden, I., Sznerski, A., Penners, F., and Mennig, F. (2015). Switchable electrical composite pane array. Patent No. US 2015/0298431 A1. Saint-Gobain Glass France, Fance. https://www.google.com/patents/US20150298431 (accessed August 2023).

21 Lam, D. and Branda, N.R., 2019. Variable transmittance optical filter and uses thereof. Patent No. US 10,416,521 B2. https://patents.google.com/patent/US10416521B2/en (accessed August 2022).

22 Lampert, C.M. (1995). Chromogenic switchable glazing: towards the development of the smart window. In: *Window Innovations '95*. Toronto, Canada: CANMET.

23 Slovak, S. M., Chen, X.-P., Hull, W. J., and Wang, D. (2012). SPD films formed with conductive polymer-coated substrates. Patent No. US20120236393 A1. Research Frontiers Incorporated, USA.

24 Duponchel, B., Elass, A., Boussoualem, M., Buisine, J.-M., DeConinck, J., Ledauphin, V.C., Coquelet, M., and Vaillant, A. (2012). PDLC films. Patent No. US8187493 B2. Université de Mons & Université du Litteral Côte Opale, France.

25 Gauzy (2019). Company Overview. https://www.gauzy.com/wp-content/uploads/2019/07/Company-Overview-2019.pdf (accessed August 2022).

26 Chakrapani, S. and Slovak, S.M. (2005). Siloxane matrix polymers and SPD light valve films incorporating same. Patent No. US 6,900,923 B2. Research Frontiers Incorporated, USA. https://patents.google.com/patent/US6900923B2 (accessed September 2023).

27 Boote, J.J. and Butcher, R. (2011). Laminated Glazing. Patent No. WO2011/033313 A1. Pilkington Group Limited, UK. Also, US 20120176656 A1 (11/24/2015), EP 25072012, JP2013505188.

28 Wang, J. (2022). Laminated switchable panel and methods for making and using. Patent No. US 11,220,076 B2. https://patentimages.storage.googleapis.com/76/b9/d7/609216b2fffe06/US11220076.pdf (accessed August 2022).

29 Research Frontiers (2010). SPD-Smart™ Light-Control Film Technology. https://smartglass-ir.com/wp-content/uploads/2021/05/Research-Frontiers-Presentation-May-2021.pdf (accessed September 2023).

30 Ghosh, A., Norton, B., and Duffy, A. (2015). Measured overall heat transfer coefficient of a suspended particle device switchable glazing. *Applied Energy* 159: 362–369.

31 Ghosh, A., Norton, B., and Duffy, A. (2016). Behaviour of a SPD switchable glazing in an outdoor test cell with heat removal under varying weather conditions. *Applied Energy* 180: 695–706.

32 SmartGlass International Ltd. (2011). *Electronically Switchable Glass Handbook*. SmartGlass. https://www.smartglassinternational.com/wp-content/uploads/SmartGlass-Handbook-2012.pdf (accessed August 2023).

33 SmartGlass International Ltd. (2014). *Market Watch news - SmartGlass*. SmartGlass. http://www.smartglassinternational.com/market-watch-news-spd-smartglass/ (accessed August 2023).

34 SmartGlass International Ltd. (2016). *SPD-SmartGlass – Automotive Applications* SmartGlass. http://www.spdcontrolsystems.com/Docs/Automotive%20SPD-SmartGlass_FACT_SHEET.pdf (accessed August 2023).

35 Innovative Glass Corp. (2013). *SPD SmartGlass*. https //innovativeglasscorp.com/energy-efficient-exterior-smartglass/ (accessed September 2023).

36 Lewis, T. (2014). New record for human brain: fastest time to see an image. In: *Live Science*. http://www.livescience.com/42666-human-brain-sees-images-record-speed.html (accessed August 2022).

37 Lemarchand, P., McLean, E., and Norton, B. (2017). *Switchable Windows - Spectral Transmission and Switching Times*. Solar World Congress. https://arrow.tudublin.ie/cgi/viewcontent.cgi?article=1284&context=engscheleart (accessed September 2023).

38 Kenny, L.W. (1998). Part VI – General Hazards, Chapter 42 – Heat and cold: physiological responses to the thermal environment. In: *Encyclopaedia of Occupational Health and Safety*. http://www.ilocis.org/documents/chpt42e.htm (accessed August 2022). International Labour Organization.

39 Jones, L. (2009). Thermal touch. *Scholarpedia* 4 (5): 7955.

40 NWSWFO (National Weather Service Weather Forecast Office) (2004). *Heat Index*. NWSWFO. http://web.archive.org/web/20110629041320/http://www.crh.noaa.gov/pub/heat.php (Accessed August 2016).

41 Alkhatib, H., Lemarchand, P., Norton, B., and O'Sullivan, D.T.J. (2021). Deployment and control of adaptive building facades for energy generation, thermal insulation, ventilation and daylighting: a review. *Applied Thermal Engineering* 185: 116331.

42 Hsu, H.-Y., Sharma, N., Ruoff, R.S., and Patankar, N.A. (2005). Electro-orientation in particle light valves. *Nanotechnology* 16 (2): 7.

43 Lemarchand P., Doran J., Norton B., 2014. Smart switchable technologies for glazing and photovoltaic applications *Energy Procedia* 57, 1878–1887. https://doi.org/10.1016/j.egypro.2014.10.052

44 Barrios, D., Vergaz, R., Sanchez-Pena, J.M., Granqvist, C.G., and Niklasson, G.A. (2013). Toward a quantitative model for suspended particle devices: optical scattering and absorption coefficients. *Solar Energy Materials and Solar Cells* 111: 115–122.

45 Barrios, D., Vergaz, R., Sanchez-Pena, J.M., García-Cámara, B., Granqvist, C.G., and Niklasson, G.A. (2015). Simulation of the thickness dependence of the optical properties of suspended particle devices. *Solar Energy Materials and Solar Cells* 143: 613–622.

46 Atlas standards guide 101- Most Frequently Used Weathering Testing Standards. SG101, Dec. 17, 2021, https://www.atlas-mts.com/applications (accessed 11 April 2023).

47 Atlas standards guide 100 - Automotive weathering tests weathering tests guide SG100, Nov, 23, 2021, https://www.atlas-mts.com/applications/applications-overview/automotive, includes Global, German, Japanese and US standards for temperature cycling (accessed 11 April 2023).

48 Atkinson, W. and Hill, W. (2015). The heat is on…, Action Magazine, pp. 14–23 (summary of static heat soak SAE testing) www.macsw.org.

49 Vergaz, R., Pena, J., Barrios, D., Pérez, I., and Torres, J. (2007). Electrooptical behaviour and control of a suspended particle device. *Opto-Electronics Review* 15 (3): 154–158.

50 Randles, J.E.B. (1947). Kinetics of rapid electrode reactions. *Discussions of the Faraday Society* 1: 11. https://doi.org/10.1039/df9470100011. ISSN 0366-9033.

51 Ghosh, A., Norton, B., and Duffy, A. (2017). Effect of sky conditions on light transmission through a suspended particle device switchable glazing. *Solar Energy Materials and Solar Cells* 160: 134–140.

52 Dangol, R., Kruisselbrink, T., and Rosemann, A. (2017). Effect of window glazing on colour quality of transmitted daylight. *Journal of Daylighting* 4 (2): 37–47.

53 Ghosh, A. and Norton, B. (2017). Interior colour rendering of daylight transmitted through a suspended particle device switchable glazing. *Solar Energy Materials and Solar Cells* 163: 218–223.

54 Barrios, D., Alvarez, C., Miguitama, J., Gallego, D., and Niklasson, G.A. (2019). Inversion of two-flux and four-flux radiative transfer models for determining scattering and absorption coefficients for a suspended particle device. *Applied Optics* 58 (32): 8871–8881.

55 Barrios, D., Alvarez, C., and Niklasson, G.A. (2018). Determination of the optical constants of the active layer of a suspended particle device smart window with multilayer structure at the clear and dark states, with and without applied voltage. In: *LIP: 12th International Conference on Laser-light and Interaction with Particles, College Station,* March 5–9, 2018, Texas, USA.

56 Huang, S., Zhang, Q., Li, P., Ren, F., Yurtsever, A., and Ma, D. (2018). High-performance suspended particle devices based on copper-reduced graphene oxide core–shell nanowire electrodes. *Advanced Energy Materials* 8 (18): 1703658.

57 Han, J., Freyman, M.C., Feigenbaum, E., and Yong-Jin Han, T. (2018). Electro-optical device with tunable transparency using colloidal core/shell nanoparticles. *ACS Photonics* 5 (4): 1343–1350.

58 Souto, A.V., Hubschneider, C., Geisler-Moroder, D., Rounis, E., Hollick, J., Walshe, J., Bonato, P., Lemarchand, P., Fedrizzi, R., Loonen, R., Aagesen, V., Ioannidis, Z., D'Antoni, M., and Meir, M. (2018). State-of-the-art and SWOT analysis on existing solar envelope systems. In: *January 2018 Solar Heating and Cooling Technology Collaboration Programme*. Paris, France: International Energy Agency.

59 Casini, M. (2014). Smart windows for energy efficiency of buildings. In: *Proceedings of Second International Conference on Advances in Civil, Structural and Environmental Engineering–ACSEE*, 273–281.

60 Sawyer, K. (2014). *Windows and Building Envelope Research and Development: Roadmap for Emerging Technologies*. Washington, DC: US Department of Energy.

61 SmartGlass International Ltd., 2016. *SPD SmartGlass - Building Energy Management*. http://www.spdcontrolsystems.com/Docs/BEMCS%20SPD-SmartGlass_FACT_SHEET.pdf (accessed August 2023). Also, https://www.smartglassinternational.com/wp-content/uploads/2021/09/Blackout-Privacy-Solar-Smartglass-Technical-Handbook-Ver-2021.pdf.

62 Ghosh, A., Norton, B., and Duffy, A. (2016). Measured thermal performance of a combined suspended particle switchable device evacuated glazing. *Applied Energy* 169: 469–480.

63 Nundy, S. and Ghosh, A. (2020). Thermal and visual comfort analysis of adaptive vacuum integrated switchable suspended particle device window for temperate climate. *Renewable Energy* 156: 1361–1372.

64 Ghosh, A., Norton, B., and Duffy, A. (2016). Daylighting performance and glare calculation of a suspended particle device switchable glazing *Solar Energy* 132: 114–128.

65 Waide, P.A. and Norton, B. (2003). Variation of insolation transmission with glazing plane position and sky conditions. *Journal of Solar Energy Engineering* 125 (2): 182–189.

66 Ghosh, A., Norton, B., and Duffy, A. (2016). First outdoor characterisation of a PV powered suspended particle device switchable glazing. *Solar Energy Materials and Solar Cells* 157: 1–9.

67 Ghosh, A. and Norton, B. (2019). Optimization of PV powered SPD switchable glazing to minimise probability of loss of power supply. *Renewable Energy* 131: 993–1001.

68 Lampert, C.M., Agarwal, A., Baertlien, C., and Nagai, J. (1999). Durability evaluation of electrochromic devices – an industry perspective. *Solar Energy Materials & Solar Cells* 56: 449–463.

69 Correspondence with Joseph M. Harary, President, RFI; company press releases and conference and industry presentations and other sources.

70 Ghosh, A. and Norton, B. (2017). Durability of switching behaviour after outdoor exposure for a suspended particle device switchable glazing. *Solar Energy Materials and Solar Cells* 163: 178–184.

71 Greenall, M., Torr, A., and Voss, J. (2007). Glazing. Patent No. WO2007122428. Pilkington Group Limited, GB.

72 Rabindra, S. and Venkateswarlu, P. (2021). Chapter 5 – Manufacturing of crystalline silicon solar PV modules. In: *Solar PV Power* (ed. R. Satpathy and V. Pamuru), 135–241. Academic Press.

73 Czanderna, A.W. and Pern, F.J. (1996). Encapsulation of PV modules using ethylene vinyl acetate copolymer as a pottant: a critical review. *Solar Energy Materials and Solar Cells* 43: 101–181.

74 Fang, Y., Hyde, T.J., Arya, F., Hewitt, N., Eames, P.C., Norton, B., and Miller, S. (2014). Indium alloy-sealed vacuum glazing development and context. *Renewable and Sustainable Energy Reviews* 37: 480–501.

75 Zhao, J.F., Eames, P.C., Hyde, T., Fang, Y., and Wang, J. (2006). A new type of In/Cu/In seal for vacuum glazing. In: *Renewable Energy in Maritime Island Climates Conference*, Dublin, 189–194.

76 Yacobi, B.G., Martin, S., Davis, K., Hudson, A., and Hubert, M. (2002). Adhesive bonding in microelectronics and photonics. *Journal of Applied Physics* 91 (10): 6227–6262.

77 Norton, B., Gillett, W.B., and Koninx, F. (2021). Briefing: decarbonising buildings in Europe: a briefing paper. *Proceedings of the Institution of Civil Engineers: Energy* 174 (4): 147–155.

7

Inorganic Thermochromics and Photochromics

Lars Österlund, José Montero, and Gunnar A. Niklasson

Department of Materials Science and Engineering, The Ångström Laboratory, Uppsala University, Uppsala, Sweden

Nomenclature

(A)	anatase structure
$A_{lum}(\tau)$	luminous absorptance at temperature τ
$A_{sol}(\tau)$	solar absorptance at temperature τ
AR	antireflection
DCIM	double charge injection model
ESS	extended site saturation
FTO	fluorine-doped tin oxide
HiPIMS	high-power impulse magnetron sputtering
MIT	metal-to-insulator transition
M1	monoclinic 1 crystal structure
NIR	near-infrared
P(X)	crystallographic point group with "X" structure
PC	photochromic
PMMA	Polymethyl methacrylate
(R)	rutile structure
$R_{lum}(\tau)$	luminous reflectance at temperature τ
$R_{NIR}(\tau)$	NIR reflectance at temperature τ
$R_{sol}(\tau)$	solar reflectance at temperature τ
R2R	roll-to-roll
RE	rare earth
SS	site saturation
τ	temperature
τ_c	critical temperature, or transition temperature
$T_{lum}(\tau)$	luminous transmittance at temperature τ
$T_{sol}(\tau)$	solar transmittance at temperature τ
$T(\lambda,\tau)$	transmittance at a wavelength, λ, at temperature τ
ΔT_{sol}	net change in solar transmittance, before and after transition
ΔT_{NIR}	net change in NIR transmittance, before and after transition
TC	thermochromic
TCO	transparent conducting oxide

Chromogenics: Smart Switchable Optical Materials and Their Applications, First Edition. Edited by Carl M. Lampert.

UV	ultraviolet light
YHO	ytrium oxyhydride
$VO_2@MO_x$	a VO_2 core-MO_x shell nanoparticle
VIS	visible light

7.1 Introduction

Energy efficiency in the built environment is often considered secondary to that of other sectors of energy consumption, such as vehicles and industrial processes. However, the energy efficiency in the building sector is of considerable importance. Buildings are responsible for 30–40% of the worldwide use of primary energy [1], a figure that increases with global population growth, electrification, and demands for enhanced living standards. Moreover, the vast majority of that energy is consumed during the operational phase of the building, i.e. for cooling, heating, ventilation, lighting, and electronic equipment. Windows are most often the weak link for a building's energy efficiency, accounting to about 30% of its energy consumption [2]. Considerable energy savings, largely due to reduced cooling demand, can be achieved with switchable windows that consist of glazing that transmits variable amounts of solar energy and visible light; preferably in combination with reduced thermal transport through the window [3]. These glazings are often referred to as "intelligent" or "smart" and are based on so-called "chromogenic materials" [4, 5], which are able to respond to various stimuli, as described in this and other chapters in this book.

This chapter covers two types of chromogenic materials, namely, thermochromic (TC) and photochromic (PC) materials. The first one is oxide-based TC coatings, which are characterized by temperature-dependent properties. Organic TC materials are also of significant commercial interest and are covered in Chapters 1 and 8 in this book. In particular, ligand exchange thermochromic in polyvinyl butyral based materials, developed commercially by Pleotint, LLC (Georgetown twp., MI). Also, hydrogel-based materials for smart windows have been developed in recent years [6–8]. Here, we will only treat hydrogels in combination with embedded inorganic TC nanoparticles. Thin films of TC materials are able to let in more solar energy at low temperatures than at high temperatures. This means that energy-efficient glazing based on these TC films can introduce solar energy when it is needed, and reject it when it is not, e.g. during sunny days. Thus, TC coatings work as dynamically adaptive solar coatings. TC-based fenestration was suggested already three decades ago [9, 10], but inorganic oxide materials are not yet commercially available for this purpose for reasons that we discuss in this chapter. However, recent advances in inorganic TC materials indicate that practical implementation in buildings is feasible [9, 11–17], and this will be discussed in Section 7.5. Inorganic TC materials have been reviewed several times, mostly covering VO_2-based materials [13, 17–22], while research on TC perovskite materials is still at an early research stage, which also involves their applications in building integrated photovoltaics (BIPV) [23–25].

The second chromogenic material we will cover in this chapter is inorganic PC materials based on rare-earth (RE) metal hydroxides. This is a more recent discovery, first published in 2011 [26–28], and hence less mature. Nevertheless, the reported properties and the comparative simplicity of fabrication of PC films promise opportunities for broad-spectrum solar coatings with favorable esthetics. Akin to their organic PC counterparts, rare earth metal hydroxide PC coatings darken upon solar light illumination, so that less solar energy is transmitted. In the absence of light, the film reverts back to its initial high solar energy transmittance state. Organic PC coatings have been used for a long time in eyewear, specifically as

Figure 7.1 Photograph of the new Ångström Laboratory House 10, Uppsala (Sweden). The façade of House 10 is fully covered in static solar control glass with the following properties of the insulated glass units (IGU): U-value = 0.5–0.7 W K^{-1} m^{-2}, g-value between 26.4% and 26.7%, and luminous transmittance between 51.1% and 52.7%.

ophthalmic lenses, but their feasibility for large-area applications has been out of reach until today due to limitations in deposition at large areas and durability [29, 30]. Organic photochromic films are available commercially (HoHo Tint film, Shanghai). The study of organic PC materials is outside the scope of the present chapter.

Finally, a note on solar energy coatings and their use in buildings in cold climate. Glass occupies more and more areas in the façade of modern buildings. This reflects the human need and desire to have contact with the surroundings, and the technical and economic progress in glazing technology, say, since the mid-twentieth century, with the Lever House, completed in 1952 in New York City, as an iconic example. Now, it is customary to completely cover the building façades with glazing; in particular public and office buildings. Yet, solar control is necessary to achieve a good indoor climate even in cold climate, and considerable energy savings can be made by replacing existing static solar control coatings with (dynamic) chromogenic coatings when the façade is fully glass covered. An example of current practices in the Nordic countries is the Ångström Laboratory in Uppsala, Sweden, where the authors work, which is shown in Figure 7.1.

7.2 Optical Properties

The optical properties of chromogenic materials and devices are of primary importance for window applications and need to be characterized. In particular, this concerns their ability to transmit and reflect luminous (or visible light) and solar radiation. For TC and PC

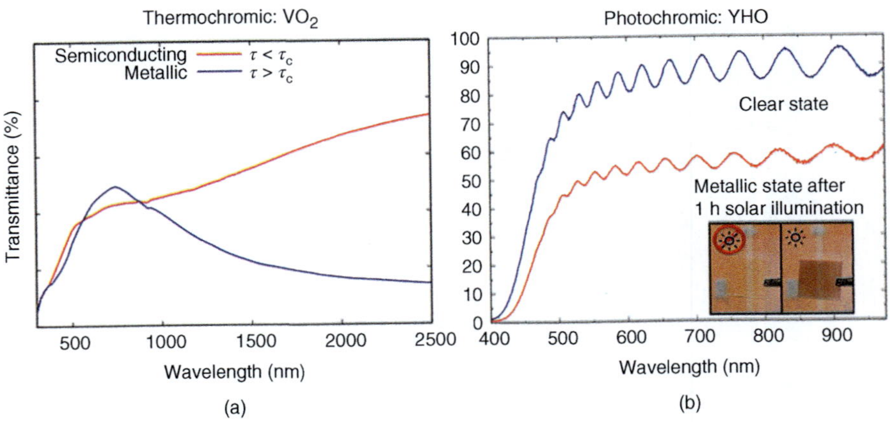

Figure 7.2 (a) Comparison of a 50 nm thick thermochromic vanadium dioxide (VO_2) thin film in semiconducting ($\tau < \tau_c$) and metallic ($\tau > \tau_c$) states, and (b) a 1000 nm thick photochromic yttrium oxyhydride (YHO) film in the clear and dark state after one hour solar light illumination. The inset in (b) shows corresponding photographs of YHO-coated glass slides in the clear and dark (metallic) state, respectively.

glazing, thermal transport is to a large extent governed by the number of glass panes in the window construction and cannot be dynamically varied to a significant extent. Solar radiation impinging upon the Earth's atmosphere can be approximated by blackbody radiation with a spectrum corresponding to the sun's surface temperature (about 5800 K) and with the majority of energy in the spectral range $0.2 < \lambda < 3\ \mu m$. Luminous radiation extends over the $0.4 < \lambda < 0.7\ \mu m$ wavelength range and has a peak at 0.55 μm (green light). Quantitative values of luminous (lum) and solar (sol) transmittance, indicated as $T_{lum}(\tau)$ and $T_{sol}(\tau)$, respectively, are obtained from

$$T_{lum,sol}(\tau) = \int d\lambda \varphi_{lum,sol}(\lambda)\, T(\lambda, \tau) / \int d\lambda \varphi_{lum,sol}(\lambda) \tag{7.1}$$

where $T(\lambda,\tau)$ is the temperature (τ) dependent spectral transmittance, φ_{lum} is the spectral sensitivity of the human eye (the daytime eye sensitivity is known as the photopic response) [31], and φ_{sol} denotes the "air mass 1.5" solar irradiance spectrum (for the sun at an angle of 37° above the horizon) [32] Analogous formulas apply to reflectance R. TC materials are predominantly modulating the solar energy throughput in the near-infrared (NIR) region, whereas the variation of the luminous properties is much less dependent on temperature (Figure 7.2a). In contrast, as we show in Section 7.4, inorganic oxyhydride lanthanide PC materials modulate both the visible and NIR region (Figure 7.2b). Hence, the solar transmittance modulation is a convenient parameter for characterizing TC performance. It is given by:

$$\Delta T_{sol} \equiv T_{sol}(\tau < \tau_c) - T_{sol}(\tau > \tau_c), \tag{7.2}$$

where τ_c indicates a "critical" temperature at which the temperature-dependent properties are changed. For PC materials, the corresponding modulation is given by:

$$\Delta T_{sol} \equiv T_{sol}(\text{light}) - T_{sol}(\text{dark}) \tag{7.3}$$

where "light" and "dark" denote the transmittance upon illumination (typically using wavelengths shorter than blue light), and in darkness, respectively.

7.3 Thermochromic Coatings

There are many inorganic TC materials [33], but only a very limited number of them can be used for glazing that operate close to ambient temperature. Thermochromism in vanadium dioxide (VO_2) was discovered more than 60 years ago [34] This material has until recently been considered to be the best available option because the switching temperature where it goes from a solar transmitting state to a more NIR reflecting state occurs at $\tau_{cc} \approx 68\,°C$, i.e. not vastly different from ordinary room temperature. The switching is accompanied by a reversible phase transition going from a transparent monoclinic (M1, $P2_1/c$) phase below τ_c to a tetragonal (rutile, R, P4/mnm) phase, above τ_c, characterized by a colossal drop of resistivity by more than 4 orders of magnitude. For the non-crystallographers, the designation M1, $P2_1/c$ signifies the monoclinic phase with a point group $P2_1/c$, which signifies the position of atoms in the crystal structure. There are 230 lattice space groups. Tables can be found at: https://en.wikipedia.org/wiki/List_of_space_groups. The nature of this metal-to-insulator transition (MIT) is still under discussion, especially whether it is driven by electron–electron (Mott–Hubbard) or electron–phonon (Peierls) interactions [35–38]. Recently, halide perovskite TC materials were shown to exhibit promising optical properties with improved luminous transparency in the clear state, and solar modulation exceeding 20%, making them viable as smart window coatings [25, 39–41]. However, VO_2-based TC materials are still considered to have more promising properties for TC glazing, and ways to combine them with other materials to enhance their properties and achieve synergistic functions is an active research area, as further discussed in Section 7.5. Most published research on TC coatings for glazing considers VO_2-based materials and their modifications by adding dopant elements, or synthesizing them as plasmonic nanoparticles with tailored shapes and overcoatings, adding the nanoparticles on plasmonic structures, embedding them in polymeric host matrices, or combining them with other light-responsive materials. Figure 7.3 presents the basic working principle and opportunities of VO_2-based smart windows for solar control.

The switching at τ_c displays significant hysteresis of the electrical and optical properties of VO_2 and takes place between two states: (i) a low-temperature semiconducting phase exhibiting low NIR absorption – and therefore large throughput of the solar NIR in a thin film, and (ii) a high-temperature metal-like phase displaying large infrared reflectance and hence a low

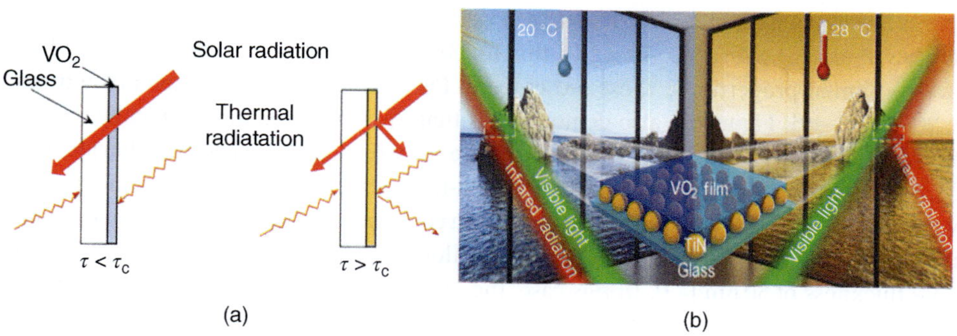

Figure 7.3 (a) Principle of VO_2 smart window for spectrally selective modulation of infrared transmission below and above the structural phase MIT transition at $\tau = \tau_c$. (b) Schematic representation of a smart window coating made of a hybrid VO_2/TiN coating with a low τ_c close to room temperature. *Source:* Reproduced with permission from Hao et al. [12] / John Wiley & Sons.

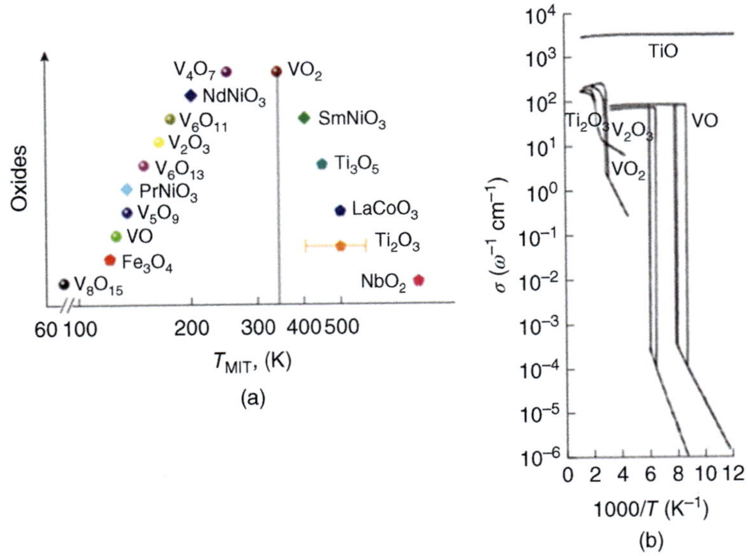

Figure 7.4 (a) Examples of insulator-to-metal transition temperatures for single crystal metal oxides. *Source:* Adapted from Yang and Ramanathan [44]. (b) Conductivity of lower oxides of vanadium and titanium single crystals (the TiO sample was polycrystalline) showing the insulator to metal transition temperature [34].

throughput of the solar NIR. Alternative TC materials have switching temperatures either below 0 °C (V_2O_3, NiS) [33], or at temperatures high above room temperature (such as the early halide perovskites [41] and V_3O_5 [33, 42, 43]). Figure 7.4a summarizes data for the MIT of different oxides, along with a historical record of early measurements of single crystals reproduced in Figure 7.4b.

Figure 7.5 presents a picture of the typical optical properties of VO_2 thin films and nanoparticle composites, computed from the optical constants [46]. The left-hand panels show spectral transmittance (upper) and reflectance (lower) for a 50-nm-thick VO_2 film on glass for wavelengths in the wavelength range pertinent to solar radiation. It is apparent that $T(\lambda,\tau)$ is much higher in the semiconducting state below τ_c than in the metallic state above τ_c at $\lambda > 1\ \mu m$. The difference in the transmittance between low and high temperature becomes larger as the wavelength increases. This type of variation in $T(\lambda,\tau)$ is clearly desired for energy efficiency: A glazing with a VO_2-based TC thin film transmits more solar energy below τ_c than above τ_c. Similar properties as those presented in Fig. 7.5 have been presented many times [15, 20–22, 47–49]. Figure 7.5 also demonstrates that $R(\lambda)$ increases monotonically toward long wavelengths for the metallic-like state above τ_c, which is the expected behavior. The reflectance seems to level off at ~40% for the 50-nm-thick film, and a higher reflectance would have been reached in a thicker film, but then with a concomitant lower ΔT_{sol}. The right-hand panels in Figure 7.5 apply to computations for a dilute suspension of VO_2 nanospheres with a mass thickness of 50 nm [42]. In this case, the maximum of the transmittance contrast is shifted toward shorter wavelengths where the solar radiation is more intense. In addition, the overall level of the luminous transmittance is increased, as compared to the thin film spectrum. Thus, coatings made of matrix-embedded VO_2 nanoparticles, when properly designed and optimized, clearly have the potential to enlarge both $T_{lum}(\tau)$ and ΔT_{sol}. In addition, $R_{lum}(\tau)$ and $R_{sol}(\tau)$ are decreased and almost temperature-independent.

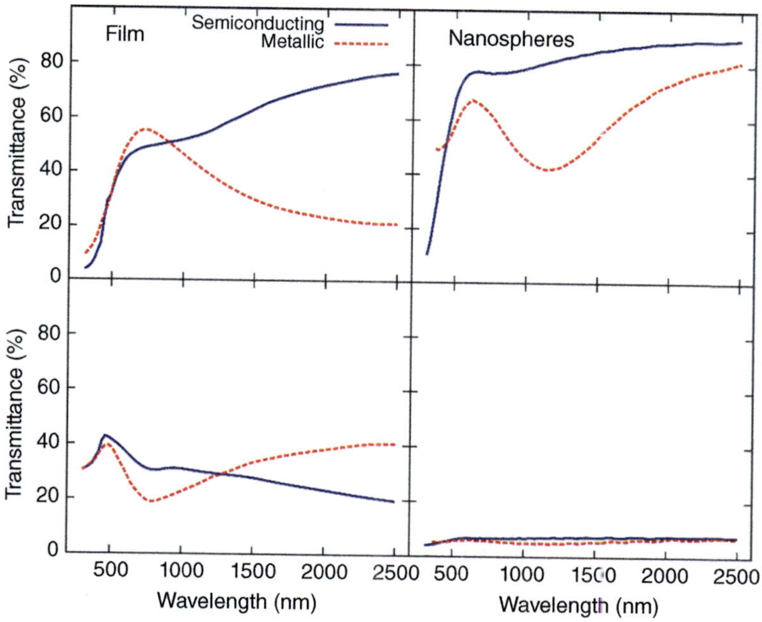

Figure 7.5 Spectral transmittance (upper panels) and reflectance (lower panels) for a 50-nm-thick thin film of VO_2 (left panels) and for a layer comprising a dilute dispersion of VO_2 nanospheres (right panels), with an equivalent VO_2 thickness of 50 nm, in a medium mimicking transparent glass or polymer.

Transmittances qualitatively similar to the nanoparticle spectrum in Figure 7.5 have been reported many times in recent years [50–53]. Studies of VO_2 nanoparticles dispersed in polymer films have recently been reviewed [54], and synergetic effects of embedding VO_2 nanoparticles in TC hydrogels have reported interesting $T_{lum}(\tau)$ and ΔT_{sol} properties [55, 56], which will be further discussed in Section 7.5.

7.3.1 Challenges for VO_2-Based Films

A number of scientific and technological challenges have hitherto prevented widespread commercialization of VO_2-based thin films for window applications. In the following, we will discuss six challenges for VO_2-based films that are currently intensively researched. A seventh challenge concerning methods to prepare high-quality M1 VO_2 materials at low processing temperatures is discussed in Sections 7.3.2 and 7.3.3.

1. The transition temperature, τ_c, occurs at ~68 °C, which is too high for most building applications. However, already in the early 1970s, it was known that addition of tungsten, as well as a number of other dopant elements, to bulk VO_2 can decrease τ_c [57–59]. The same effect can be found in thin films [60–63]. Figure 7.6 shows that ~2% of W displaces τ_c to approximately room temperature [64]. Numerous other reports give analogous results provided that the oxide film is well-crystallized and homogenous [47, 65]. An addition of Mg also gives a lowering of τ_c, but to a smaller extent [64].
2. The luminous transmittance, T_{lum}, is only ~40% for a homogeneous thin film, which is too low for most glazing (T_{lum} would be higher for a thinner film but the consequence is then that ΔT_{sol} becomes too small). For comparison, T_{lum} around 50% would be comparable to nonselective static solar control coatings (for example based on FeNiCr oxide, NiCr, and

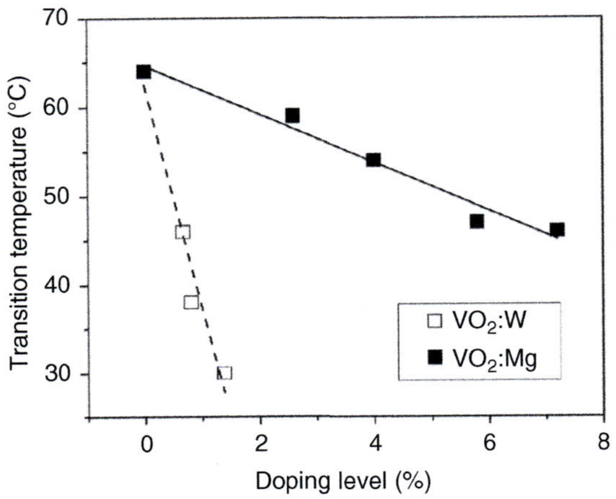

Figure 7.6 Transition temperature τ_c between semiconducting and metallic-like states in VO_2 films containing W or Mg, based on electrical measurements. *Source:* Mlyuka et al. [64] / American Institute of Physics.

others) that have been widely used on windows, especially – but not exclusively – in warm climates [66]. However, state-of-the-art selective solar control coatings typically have a luminous transmittance 70–80%, combined with a solar transmittance of around 50%.

Figure 7.5 demonstrates that strong optical absorption occurs at wavelengths below 0.5–0.6 μm, which is related to the fact that the optical band gap in VO_2 is smaller than desired. Thus, to increase T_{lum}, the absorptance of visible light should be reduced. A partial solution to this problem can be reached by replacing some of the vanadium with Mg [64, 67, 68] in order to widen the optical band gap. Hence, the dopant displaces the high-transmittance region toward shorter wavelengths, but it also has the unwanted effect of diminishing ΔT_{sol}. This effect has also been corroborated by computational results [69]. Figure 7.7 indicates that T_{lum} increases from 39% to 51% as the amount of Mg goes from 0% to 7.2%. Band gap widening can also be accomplished by adding other alkaline earth metals [67], Zn [70, 71], Al [72], Mo [73–75], Nb [75, 76], Zr [77, 78], Tb [79], or Co [80], W [81–83], or by replacing some of the oxygen with fluorine [61, 83, 84].

3. The temperature-dependent transmittance modulation is large mainly for wavelengths where the solar irradiation is weak in the NIR region. This limits ΔT_{sol} to approximately ~10%. An important step to improve on ΔT_{sol} was the discovery by means of theoretical calculations that a layer of VO_2 nanoparticles – rather than a continuous thin film of this

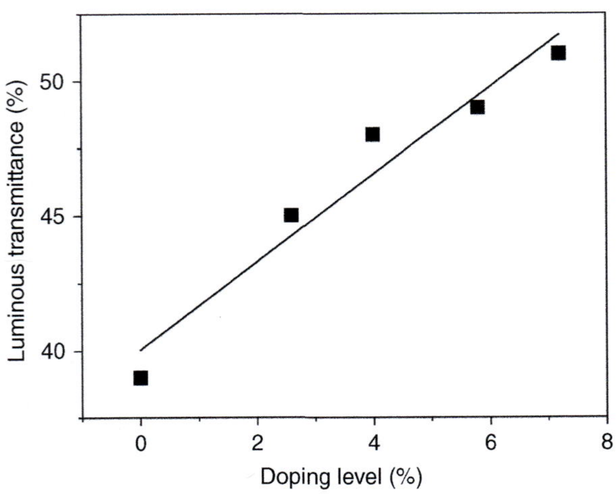

Figure 7.7 Luminous transmittance as a function of Mg doping in 50-nm-thick VO_2-based films as measured at $\tau = 22\,°C$. Line was drawn to guide the eye. *Source:* Mlyuka et al. [64] / American Institute of Physics.

material – could give improved properties [48, 85, 86]. Additional calculations showed that the particle diameter should not be larger than ~20 nm in order to prevent optical scattering ("haze") [87]. Refractive-index matching between VO_2 nanoparticles and an embedding material can also suppress scattering efficiently and therefore the small particle limit can be relaxed in practice [54, 87]. TC employing nanoparticles has been termed "nanothermochromism" [48]. The optical properties of matrix-embedded VO_2 nanospheres were already shown in the right-hand panels in Figure 7.5. There are important qualitative differences between optical data for the nanoparticle composites consisting of VO_2 nanoparticles embedded in a matrix of, e.g. a polymer or another oxide, and the corresponding thin films: (i) the nanosphere-containing film is considerably more transparent than the compact VO_2 thin film, (ii) the metallic-state nanoparticles display a pronounced minimum in the transmittance in the $0.7 < \lambda < 1.5$ μm wavelength range, which strongly depresses T_{sol}, since the solar radiation is more intense at these wavelengths than in the long-wavelength NIR range. In addition, the visible optical performance is hardly affected at all by this minimum, and (iii) the nanospheres absorb rather than reflect. The striking transmittance minimum in the NIR is seen in Figure 7.5, i.e. the absorption maximum is due to a localized plasmon resonance and is seen to occur at a wavelength of around 1200 nm. The plasmonic character implies that nano-TC films absorb rather than reflect light in the NIR, which also implies that they are heated in the metallic state. The plasmonic character of the NIR absorption has been observed and discussed several times in prior studies [88–91].

4. Environmental stability is of utmost practical importance since glazing must withstand many years without essential performance losses. VO_2 is however not thermodynamically stable. Vanadium is a multivalence element, which exhibits +2, +3, +4, +5, as well as mixed, valences. It is therefore not surprising that the phase diagram is rich and includes 20 different phases [70]. In particular, monoclinic VO_2 can be reduced to lower valence vanadium oxides, or their hydroxides, especially in the presence of moisture. Ultimately, oxidation leads irreversibly to intermediate V_6O_{13} and V_3O_7 phases and finally to V_2O_5, the highest oxidation state of vanadium, V^{5+}, which is not TC (albeit electrochromic [92]). Moreover, doping of VO_2 with various elements was discussed in points 1 and 2 mentioned previously and can cause additional unfavorable degradation characteristics. Figure 7.8

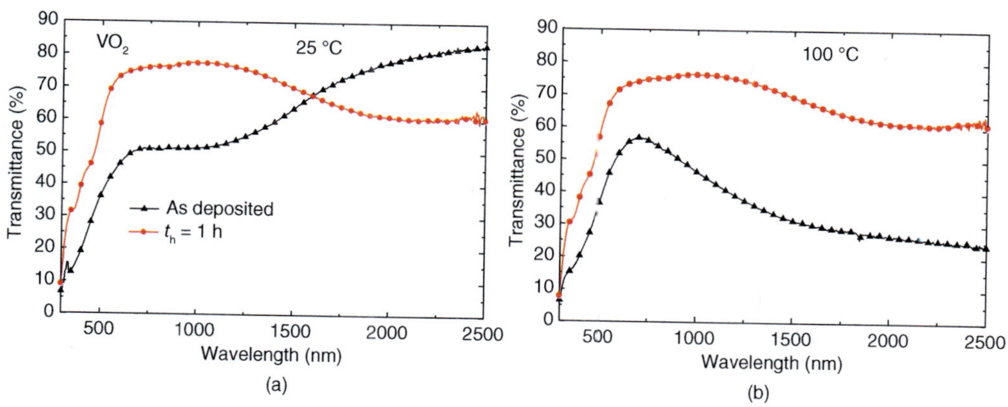

Figure 7.8 Spectral transmittance for an 80-nm-thick VO_2 film in as-deposited state and subsequent to annealing at 300 °C for one hour. Data pertain to films in (a) semiconducting ($\tau = 25$ °C) and (b) metallic-like ($\tau = 100$ °C) states. *Source:* Adapted from Piccirillo et al. [76].

reports data on 80-nm-thick sputter-deposited VO_2 thin films kept in dry air at 300 °C for one hour [93]. The as-deposited film demonstrates the expected TC properties (see Figure 7.5), whereas the annealed film displays a very different spectral transmittance indicative of non-TC V_2O_5 [94]. This gradual oxidation is expected to be much slower at room temperature, but the data in Figure 7.8 clearly show that VO_2-based materials must be protected from further oxidation. The durability of VO_2 is also dependent on the deposition technique and is improved for films with high density and large grain size [95, 96]. For example, Loquai et al. [96] demonstrated an enhanced stability of high-power impulse magnetron sputtering (HiPIMS) prepared films exhibiting higher density and larger VO_2 grain size. HiPIMS is a method for physical vapor deposition of thin films based on magnetron sputter deposition. HIPIMS utilizes deposition at extremely high-power densities of the order of 1 kW cm^{-1} in short pulses of tens of microseconds at low duty cycles.

Figure 7.9 depicts results of $T(\lambda)$ for VO_2 films, similar to those in Figure 7.8, which were over-coated with 30 nm of sputter-deposited alumina and heat treated at 300 °C for periods in the range $1 < t_h < 30$ hours [76]. Clearly, the top layer provides good protection, and only the longest heat treatment period led to a marginal decline of thermochromism. Al nitride [97] and Si nitride (SiN_x) [96] top layers have also been shown to provide similar protection for VO_2 films. The data in Figure 7.9 indicate that VO_2-based materials can be used for long times, for example in energy-efficient glazing, if appropriate environmental protection of the VO_2 film is made.

Over-coatings similar to the ones discussed above gave efficient protection also under conditions of high humidity [93]. More realistic environmental cycling stability data are rather scarce. Figure 7.10 shows recent accelerated stability tests performed in more realistic environments up to 1000 cycles [98]. The results clearly reveal the sensitivity of bare VO_2 to oxidation. Similar to the results presented in Figure 7.9, the study by Chang et al. [98] presented in Figure 7.10 also shows promising durability properties of a protected $SiO_2/VO_2/C_2O_3$ film structure, exhibiting synergistic templating (C_2O_3), antireflective, and environmental protecting (SiO_2) properties. Analogous results have also been reported for VO_2-based core–shell nanoparticles [99], which we discuss in Section 7.3.3.

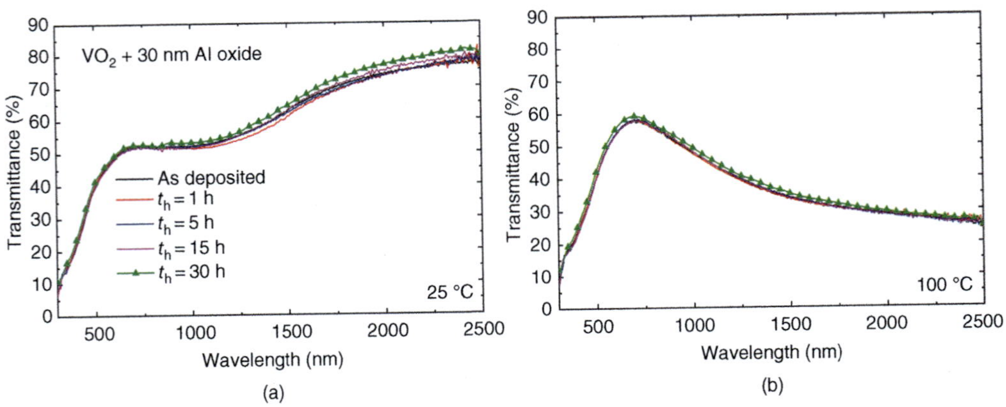

Figure 7.9 Spectral transmittance for 80-nm-thick VO_2 films, over-coated with 30 nm of Al_2O_3, in as-deposited state and after heating at 300 °C for the shown time periods t_h. Data were recorded for (a) films in semiconducting ($\tau = 25$ °C) and (b) metallic-like ($\tau = 100$ °C) states. *Source:* Reproduced with permission from Ji et al. [93] / with permission of Elsevier.

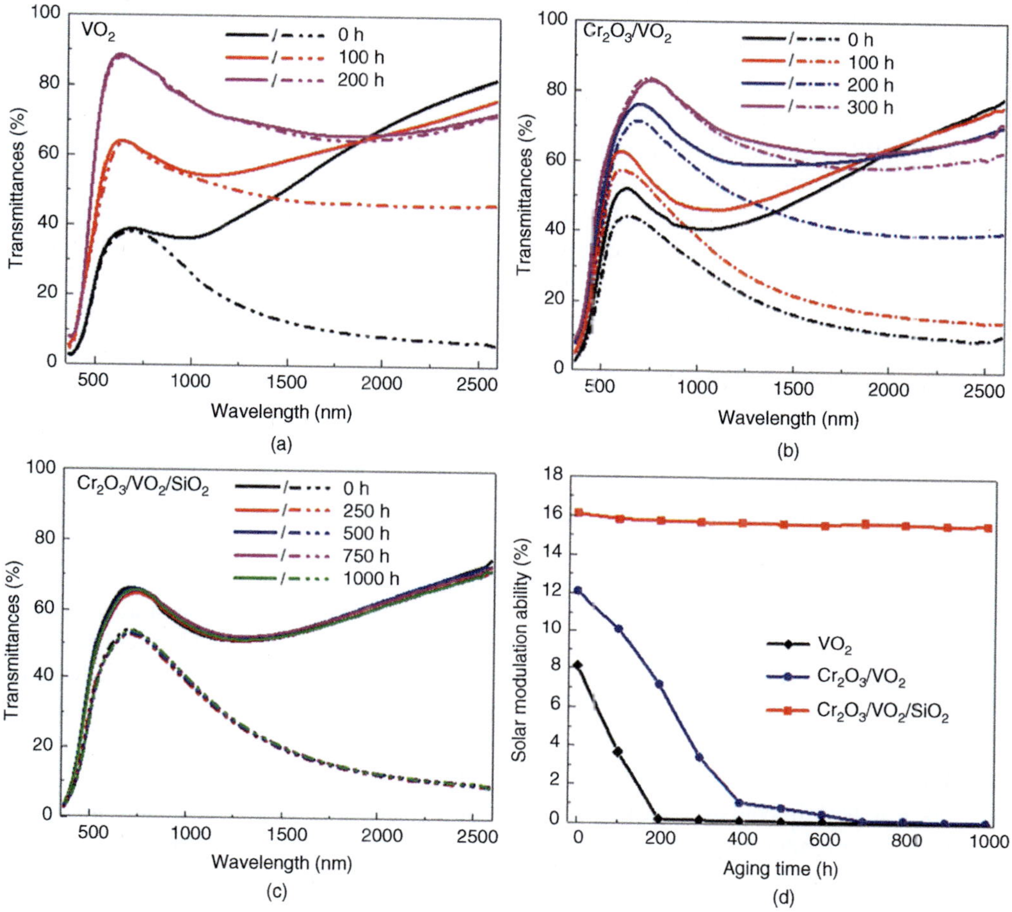

Figure 7.10 Stability tests. Transmittance spectra (350–2600 nm) at 25 °C (solid lines) and 90 °C (dashed lines) after different duration time and different fatigue cycles for: (a) VO_2, (b) Cr_2O_3/VO_2 and (c) $SiO_2/VO_2/C_2O_3$ films. (d) Corresponding ΔT_{sol} after different aging times. The accelerated aging data were obtained in a constant-temperature humidity chamber with the experimental conditions of 60 °C temperature and 90% relative humidity. *Source:* Figure reproduced with permission from Chang et al. [98], Elsevier.

5. Another possible limitation of VO_2 coatings lies in their emittance properties. In a window application, VO_2 would be coated onto a glass pane and this structure would display high emittance values close to those of the underlying glass at low temperatures below τ_c, typically, 0.6–0.7 for sputtered films [100], and higher for porous films [101]. On the other hand, at high temperatures, above τ_c, the metallic phase of VO_2 would give rise to a lower emittance, which depends on the film thickness and can be as low as 0.11–0.26 for sputtered films [100]. However, the heat transport through a window should in principle be minimized in both cold and warm climates, implying that a low emittance of a VO_2-coated pane would be beneficial in most applications. The low-temperature emittance can be decreased substantially by depositing VO_2 onto a low-emittance backing. Of course, a VO_2 coating on good conducting metal films leads to a low emittance because of the Drude behavior of metals, but it makes the coating opaque and hence unsuitable as a window coating. Overcoating the VO_2 layer with a transparent conducting oxide (TCO),

such as ZnO:Al [102], or a very thin metal film [103], leads to a significantly decreased emittance together with a good luminous transmittance. However, a better solution is obtained by instead using a glass substrate coated with a TCO. Using fluorine-doped tin oxide (FTO) as transparent conducting substrate, Zhang et al. [104] demonstrated a VO_2/FTO double-layered film that possessed both low-emissivity and TC properties. Furthermore, they showed that the rutile-structure of the FTO substrate enhanced the crystallinity of the VO_2 films and lowered the synthesis temperature to ~390 °C. The FTO layer serves as a template for the growth of VO_2. A similar strategy is employed in many template-assisted thin film growth studies, some of which we discuss in Section 7.3.2. An anti-reflected TiO_2/VO_2/FTO three-layer film on glass showed favorable properties with emittances of 0.13 (at $\tau < \tau_c$) to 0.24 ($\tau > \tau_c$), together with $T_{lum} = 0.44$ and $\Delta T_{sol} = 8.8\%$. These emittance values are approaching the emittances of TCO-based static coatings These studies show that using VO_2 in conjunction with a transparent conducting oxide on a glass substrate leads to both favorable TC and emittance properties for window applications. It should, however, be mentioned that the VO_2 film always leads to an increased emittance above that of the underlying TCO layer. As a final note, we mention that VO_2 films can be used to achieve radiative cooling at high temperatures above their phase transition. A VO_2/PMMA/ITO (PMMA is polymethyl methacrylate) structure was reported to have a 40% emittance contrast between the low- and high-temperature phases together with a $\Delta T_{sol} \approx 9\%$ [105]. The positive emittance contrast leads to temperature regulation around the phase transition temperature of the VO_2-based coating. However, the coating has to be placed in contact with the outdoor ambient in order to achieve substantial cooling, which may be problematic for durability reasons.

6. Finally, it should be noted that VO_2 absorbs in the blue region of the solar spectrum below 500 nm, both in cold and hot states, which implies that VO_2 films cannot faithfully give good color rendering performance in a window. Moreover, blue-light filtering means that it is difficult to achieve high Melanopic/Photopic ratio, which is known to suppress melatonin production and support alertness [106].

7.3.2 Synthesis of VO_2-Based Thermochromic Films

Inorganic TC thin films can be prepared by numerous techniques [20, 107–109], for example magnetron sputtering [60, 110–113], pulsed laser deposition [108], chemical vapor deposition [114], sol–gel methods [115, 116], and hydrothermal synthesis [117]. Here, we focus on synthesis methods to make polycrystalline films by sputtering techniques, which are considered promising for large-area fabrication. The main advantages of sputtering methods are that it is straightforward to make uniform coatings with well-controlled thicknesses, good substrate adhesion, high yield, and cost-effectiveness.

The complex phase diagram of vanadium oxides and the high symmetry of the monoclinic (M1)/rutile phases complicate synthesis of high-quality TC M1/rutile VO_2 films. Deposition of phase-pure thin VO_2 films typically requires high temperatures and careful control of deposition conditions. The high-temperature treatment steps are however not unique to sputtering methods, they are in general also required for chemical synthesis techniques, and therefore constitute a general synthesis challenge. In the case of sputter deposition, precise control of oxygen flow ratio, pre-sputtering of target, substrate temperature, vacuum conditions, and power is needed. In order to produce VO_2 by conventional magnetron sputtering, substrate temperatures of about 400 °C are necessary [118]. If careful control of oxygen flow is simultaneously established, the substrate temperature can be lowered below 300 °C [119]. Alternatively, sputtering of metallic V can be done without substrate heating to produce M_1

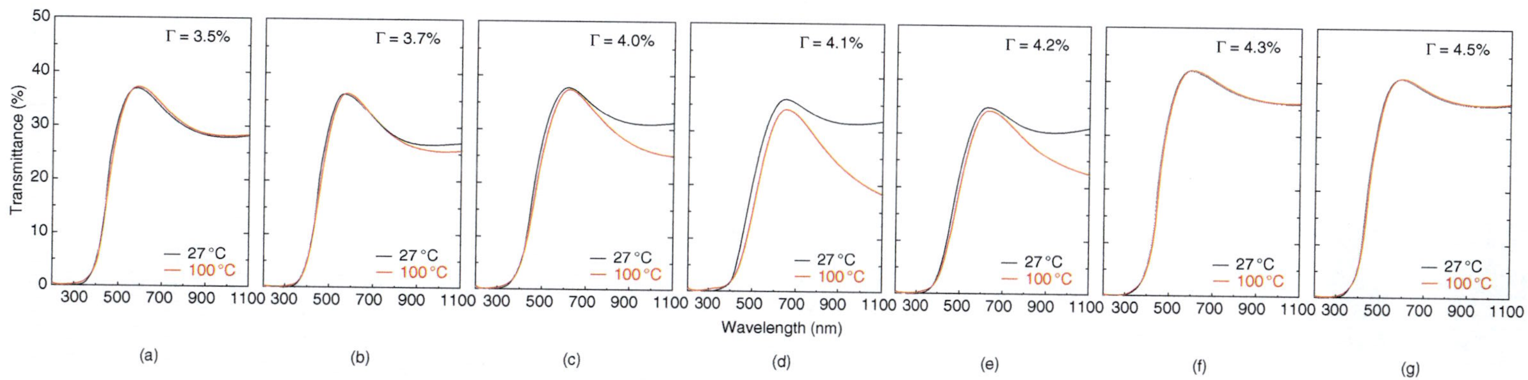

Figure 7.11 Transmittance of vanadium oxide thin films, prepared at various oxygen flow ratios Γ, measured at 25 °C (black line) and at 100 °C (red line). Γ = (a) 3.5%, (b) 3.7%, (c) 4.0%, (d) 4.1%, (e) 4.2%, (f) 4.3%, and (g) 4.5%, respectively. It is clearly shown that thermochromic properties depend on the oxygen flow ratio during the deposition. The largest thermochromic modulation is observed with Γ = 4.1%. At the lowest Γ = 3.5% and highest Γ = 4.5% ratio, no thermochromism at all is observed. In these cases; V_3O_5 and V_2O_5 structures, respectively, develop. *Source:* Figure reproduced with permission from Khairy et al. [120]. Elsevier.

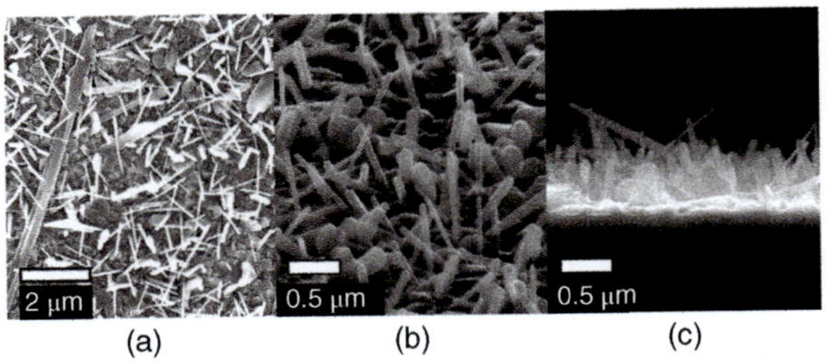

Figure 7.12 Scanning electron microscopy (SEM) images of a VO_2 layer deposited onto heated glass to an (equivalent) thickness of 102 nm. (a) and (b) images shows top-view images acquired with 70° between the electron beam and the sample's normal, and (c) shows a cross-sectional image. *Source:* Reproduced with permission from Li et al. [122] / AIP Publishing LLC.

VO_2, provided that post-annealing in air to about 400 °C is performed [120]. Figure 7.11 demonstrates the necessity of precise control of oxygen/Ar flow ratio, Γ, to obtain TC VO_2 films on glass substrates [121]. Only a narrow Γ range of about $4.1 \pm 0.1\%$ is admissible to obtain good TC properties in this case.

After heat treatment, or if deposition is done at elevated sample temperatures, rod-like structures are commonly observed as exemplified in Figure 7.12 [122], because of the preferred growth direction of the monoclinic M1 phase along the crystallographic *c*-axis [124, 125].

Due to the challenges to obtain high-quality M1/rutile(R) VO_2 phases, a viable approach is instead to start the synthesis from easily synthesized vanadium oxide phases, such as monoclinic VO_2(B) [123], V_2O_5 [119, 124], VO_2 hydrate [125], VO_2H_x hydride [126], or VO oxalate (VOC_2O_4) [127] where the deposition of VO_2 often involves post-annealing steps. In the review by Shi et al. [107], it is concluded that among post-annealing methods from precursor phases, it is beneficial to use the hydric VO_2 precursor phase, which has a similar crystal structure as rutile VO_2(R). This choice yielded better control and required only a fairly low annealing temperature (260–350 °C). In a recent study, Li et al. [127] showed that vanadyl oxalate diluted with addition of 6 wt.% film forming PVP, spin-coating, and post-annealing for 1 h at 550 °C yielded nanoporous M1 VO_2 films with very good TC properties. In addition to these methods, a V_2O_5 target can also be employed using magnetron sputtering techniques, in combination with annealing to obtain M1/R VO_2 films [107, 109, 113].

It is desirable to decrease the temperature during preparation, where hitherto only sputtering methods have shown potential for low-temperature deposition. Other strategies use complex transfer techniques in multistep approaches, where first M1/R VO_2 deposition is done on a temperature-insensitive substrate, and then etching, lift-off and transferring to a temperature-sensitive substrate, e.g. a polymer, is made [109]. Recently, work on low-temperature sputtering has been performed using the HiPIMS technique [96, 110, 111, 128]. The application of the HiPIMS technique to VO_2 films has been reviewed by Zhang et al. [110]. Loquai et al. [96] reported on the deposition of VO_2 films on polymeric Kapton substrates at substrate temperatures in the range 275–350 °C. Juan et al. [129] reported on the effect of the HiPIMS duty cycle with different on/off time ratios. They inferred that a moderate duty cycle of 2.5% yielded the best TC properties ($\Delta T_{sol} = 9.5\%$, $T_{lum} = 43\%$). It can be concluded that advanced magnetron sputtering methods such as HiPIMS, and variations of this technique, are promising for direct deposition on temperature-sensitive substrates, in particular if they are combined with protecting primary layers with the dual function of acting as physical barrier and structural template layers to assist M1/R VO_2 growth. Possibly, posttreatment with flash lamp annealing may be used to enhance crystal growth and produce higher-quality films [130]. Embedding VO_2 films in multilayer structures is also an effective way to stabilize VO_2 planar structures, and improve their optical properties [46, 131, 132]. Here, the challenge is to realize robust and up-scalable synthesis methods that can be applied to thin films and to improve material stability in realistic environments.

7.3.3 Synthesis of VO_2-Based Nanoparticles and Nanocomposites

We now focus on VO_2-based nanoparticles, which can be prepared by a variety of physical and chemical techniques. Nanothermochromism is considered a viable pathway for TC glazing applications and is currently in rapid development [13, 48]. The physical reason why TCs based on VO_2 nanoparticles are important is twofold: (i) The nanoparticle plasma frequency, $\omega_{p,NP}$, is down-shifted compared to bulk plasma frequency, ω_p, (e.g. by $\omega_{p,NP} = \omega_p/\sqrt{3}$ for spherical nanoparticles) and occurs typically around 1.2 μm (see Figure 7.5). The plasmon absorption thus modulates a more intensive part of the solar spectrum as compared to bulk VO_2 [13, 48], hence giving a higher ΔT_{sol}. (ii) Nanoparticles with appropriate sizes can be "optically diluted" and haze controlled, if they are properly dispersed in an index-matched matrix and can thus give a higher T_{lum}. Thus, it has become evident that the key to improved TC materials for glazing applications, and in particular the solution to the low luminous transparency problem, lies in the development of methods to prepare VO_2-based nanoparticles in significantly large amounts and to disperse them in a matrix, without inducing agglomeration and undesirable optical effects [54, 87]. The matrix host material may be another inorganic material, or alternatively, a polymer [12, 46].

There is a large body of work on VO_2 nanoparticle synthesis and their physicochemical properties. This work is cited, for example, in References [13, 54, 108, 117, 133–135]. Depending on precursor and reaction conditions, VO_2 nanoparticles can be made with a variety of morphologies: nanorods, round-shaped, platelets, 2D arrays, and 3D nano/micro-networks. Nanorods are commonly obtained due to the preferred growth direction of the M1 phase along the crystallographic *c*-axis if the reaction conditions do not impose kinetic crystal growth restrictions. Figure 7.13 shows examples of VO_2 nanoparticles with various morphologies that have been prepared by hydrothermal methods [117].

Hydrothermal synthesis is today considered as the most promising technique to prepare high-quality VO_2(M1) nanoparticles, although a universal one-step process that meets large-scale production requirements is still lacking. Figure 7.13 also indicates schematically how VO_2 nanoparticles are produced using hydrothermal methods [117]. State-of-the-art hydrothermal synthesis of VO_2 nanoparticles has recently been reviewed [117], and pertinent references can be found therein. Recent work has demonstrated that VO_2 nanoparticles can be prepared by direct and continuous synthesis via a two-step process incorporating hydrothermal flow synthesis followed by a short heat-treatment step [136]. In subsequent work, elaboration of the hydrothermal route was shown to yield VO_2 nanoparticles in a one-step process [134], whereby decomposition of a vanadium(IV) oxide obtained by reduction of vanadium(V) oxide was performed in a continuous hydrothermal flow synthesis yielding black VO_2 powder after washing and freeze-drying. In general, however, the growth mechanism of using either reduction–oxidation reactions starting with high valence vanadium oxide precursors, or exfoliation and self-assembly of vanadium(IV) oxides (e.g. commercial bulk VO_2) or sulfates [107] is not well-understood, and hence work is needed on hydrothermal synthesis for reliable and predicable up-scaling. Hydrothermal methods have been shown to yield unintentional hydrogen doped or nonstoichiometric VO_2

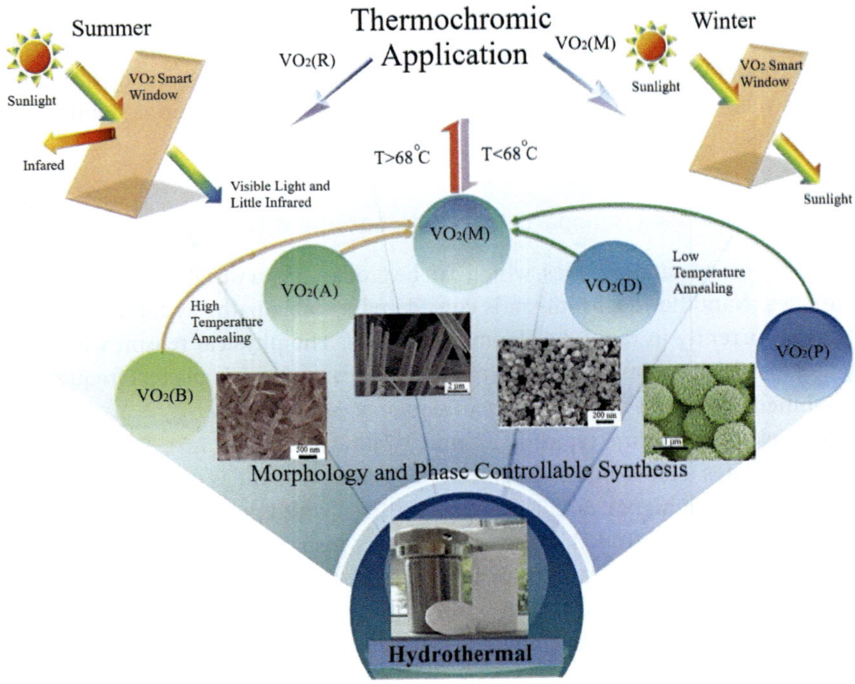

Figure 7.13 Schematic overview of processes to make monoclinic M1 VO_2 nanoparticles by hydrothermal methods, which today are considered promising to prepare nanoparticles for glazing applications. The nanoparticles can either be in the form of a pure phase or as nanocomposites, including core–shell nanoparticles. VO_2 polycrystals exhibit many polymorphic forms, including VO_2(B), VO_2(A), VO_2(M), VO_2(D), and VO_2(P), which complicates synthesis of high-quality VO_2(M) since the hydrothermal product is a metastable phase or mixture phase. Much effort has been put on one-step hydrothermal synthesis of VO_2(M), but high hydrothermal temperature and long reaction time are essential, which induce the growth of larger VO_2(M) particles. *Source:* Reproduced with permission from Li et al. [117] / John Wiley & Sons.

crystals, partly because of low synthesis temperatures. In contrast, hydrothermal synthesis routes open up the possibility to add dopants to stabilize the VO_2(M1) phase, which has the additional benefit that it lowers τ_c if e.g. Mg [86, 137], W [135, 138, 139], Mo [140], and other dopants are employed. Notwithstanding the possibilities of dedicated hydrothermal methods, simple, industrially well-known techniques such as bead-milling have also been shown to yield nanoparticles with beneficial structure and sizes [141], which may be considered in some applications. For example, Kim et al. [109] performed bead-milling of commercial V_2O_5 powder and produced VO_2 nanoparticles by a short (10 minutes) thermal treatment at 500 °C in nitrogen. Nanocomposite VO_2/PVP foil was then fabricated by using tape-casting and roll-to-roll (R2R) techniques.

Large attention has been devoted to core–shell particles. The basic idea is to protect a VO_2 core by an inert and oxidation-resistant shell to overcome the stability problems discussed in Section 7.3.1, as well as to give the particles an advantageous dispersibility in a host medium. Importantly, adding a shell that is chemically suitable for coupling to organic matrix materials allows fabrication of core–shell nanocomposites on flexible polymer substrates, which would be beneficial for glazing applications as laminated add-on options. However, it should be noted that adding a dielectric shell, X, surrounding a VO_2 core, i.e. VO_2@X, does not improve T_{lum} or ΔT_{sol} as compared to a solid VO_2 nanoparticle [85]. In contrast, the opposite configuration X@VO_2 can, depending on refractive index of the dielectric core material, improve ΔT_{sol} at the expense of a slightly lower T_{lum} [85]. Additional functionality can be achieved by arranging core–shell particles in periodic arrangements to make two-dimensional photonic crystals. Such photonic core–shell particles have been synthesized to achieve tunable T_{lum} and TC modulation of T_{sol} by changing the photonic bandgap [142]. Early reports on the synthesis of core–shell nanoparticles include the work by Gao et al. [143]. They reported on the synthesis of VO_2@SiO_2 using hydrothermal growth of M1 VO_2 nanoparticles followed by modified Stöber growth of SiO_2 on poly(vinyl-pyrrolidone) pretreated VO_2. The Stöber process is a sol–gel reaction where a suitable precursor is reacted with water in an alcoholic solution, followed by the formation of nanoparticles. The nanoparticle dispersion was subsequently cast on PET foils after adding a silane coupling agent and the polyurethane organic matrix. The VO_2@SiO_2 films with optimized hydrolysis time for the Stöber reaction exhibited $T_{lum} = 41\%$ and $\Delta T_{sol} = 15.5\%$ and showed anti-oxidation and anti-acid properties. Importantly, the synthesis of VO_2@SiO_2 enabled further modification of the silica layer that facilitated coating on flexible PET substrates. Later, Wang et al. [144] reported on a surfactant-free method to synthesize pure VO_2@SiO_2 core–shell nanoparticles at room temperature, minimizing by-products typical for the Stöber reaction. They further prepared nanocomposite films of mesoporous silica nanospheres and VO_2@SiO_2 with improved optical properties. In general, it is concluded that SiO_2 is a promising shell material yielding good optical transparency and chemical stability, as well as enabling coupling chemistries for making coatings on flexible polymer substrates. Li et al. reported on the hydrothermal synthesis of VO_2(M1) nanorods with a sol–gel prepared TiO_2 shell [145]. The VO_2@TiO_2 core–shell particles exhibited enhanced T_{lum} and color change from yellowish to blueish. At the same time, the TiO_2 shell provided photocatalytic properties as evidenced by photodegradation of RhB dye, as well as enhanced stability of VO_2 toward oxidation. Chen et al. [146] synthesized VO_2(M1)@ZnO core–shell nanoparticles by hydrothermal methods and prepared films on glass substrates by spin-coating after mixing with polyvinyl butyral. The resulting nanoparticle films demonstrated ultrahigh durability, up to 1000 hours in accelerated tests, as well as improved TC properties compared to the bare VO_2 ($\Delta T_{sol} \approx 19\%$ and $T_{lum} \approx 51\%$). Recently, Li et al. [147] reported on core–shell VO_2(M1)@SnO_2 nanoparticles synthesized by solvothermal methods that were shown to both protect the VO_2 nanoparticles from acid corrosion and oxidation, as well as improve ΔT_{sol}. Nanocomposite structures based on core–shell nanoparticles can also give new possibilities to tune the optical response by arranging them in periodic structures to make thermo-responsive metamaterials [88]. The long group prepared bioinspired VO_2@SiO_2 2D photonic crystal films, demonstrating that the antireflection (AR) structure can enhance both ΔT_{sol} and T_{lum} [142]. They also demonstrated that butterfly inspired photonic crystal structures can modulate the intrinsic yellow-brown color of VO_2. Hao et al. [12] demonstrated VO_2(M1) coated 2D TiN plasmonic nanoparticle arrays that were shown to block 70% of the IR radiation at 28 °C, while showing good IR transparency at 20 °C. The coating exhibited a good T_{lum} of about 51% and a NIR switching efficiency of $\Delta\tau_{NIR} = 48\%$ at 2000 nm wavelength.

In conclusion, despite the progress made over the past decade to synthesize VO_2-based nanoparticles and nanocomposites with various morphologies, structures, dopant additions, etc., it remains a challenge to synthesize high-quality and phase pure M1 VO_2 nanoparticles with dimensions smaller than ~20 nm, which ideally is necessary to avoid haze [87] (unless dedicated refractive-index matching is made in suitable host matrices), in amounts that are necessary for large area glazing. In order to achieve large-scale production of nanoparticle TC coatings, VO_2 particle synthesis techniques need to be upscaled. Currently, the most promising techniques for future upscaling to industrial production appear to be the continuous hydrothermal flow synthesis technique [117, 131] and high-pressure magnetron sputtering terminated cluster growth [148]. Even so, further progress is needed in order to produce core/shell particles, achieve a good particle dispersion, and avoid post-heat treatments.

7.4 Photochromic Coatings

PC materials have a historical reputation connected to their most successful commercial application as PC ophthalmic lenses. PC eyeglasses, available since the 1960s, have been a commercial success for decades and their market size continues to grow today. PC eyeglasses act as regular prescription eyeglasses but, under conditions of strong solar light – e.g. when the wearer walks outdoors on a sunny day-darken reversibly, acting in this way like sunglasses. These glasses bleach back to the original transparent state when the ambient illumination is less bright (Figure 7.14).

Despite the maturity of the PC technology for ophthalmic applications, the current rise in the number of publications containing the keyword "photochromism" demonstrates an increasing interest in the scientific community for PC materials [150]. There are several reasons for this, as we will discuss later in this section, but we note that smart PC windows are still out of reach using current state-of-the-art PC materials. Furthermore, the basic understanding of the PC mechanism is still far from clear for the well-known PC transition metal oxides (e.g. WO_3, MoO_3, and V_2O_5). But before continuing, we have to define what it means for a material to be PC, and what types of PCs can exist.

We can consider a material PC if such material is able to change its optical properties reversibly as a response to electromagnetic radiation of adequate wavelength and intensity (usually UV). Normally, this change should take place in the visible range of the electromagnetic spectrum, and hence the PC process is perceived by the human eye as a color change. Note the importance of the word *reversible* when defining photochromism, otherwise any of the many dyes that degrade under UV-light could be mistakenly be termed PC. Once the

Figure 7.14 Photochromic glass finds primary use as eyeglasses that darken when exposed to sunlight. Here only one of the lenses has been exposed to ultraviolet light. *Source:* Reproduced with permission from Trotter [149] / ITHAKA.

irradiation ceases, a PC material recovers its initial state by thermal relaxation or by irradiation with light of different spectral composition than the one that triggered the PC effect in the first place. Materials belonging to the latter case are known as *P-type* (reversed by light), while materials experiencing thermal relaxation are known as *T-type* [151]. A T-type PC material is considered to be *positive* if its final state, after irradiation, is darker than its initial state. The opposite situation is known as *negative* photochromism. Hence, PC materials of interest for glazing applications should be T-type positive. In particular, the material should be *heliochromic* T-type positive, i.e. it should darken by the action of solar light and bleach by thermal relaxation [151, 152].

The PC family, including the subgroup constituted by heliochromic T-type positive materials, comprises both organic and inorganic compounds. In this chapter, we will focus mostly on inorganic compounds because they have better UV-resilience and stability than their organic counterparts. Examples of industrial applications of PC compounds, some of them quite original, beyond eyeglasses and windows can be found in the book by M. Ferrara and M. Bengisu [29].

Historically, the first PC commercial eyeglasses were based on inorganic compounds: silver halides, which are T-type positive PC materials that are very stable and exhibit excellent resistance to cycling and UV light [153]. Unfortunately, silver halide-based PC glasses have to be prepared in bulk, by dispersing silver halide nanocrystals in a glass matrix. This is problematic because such a process is not compatible with the common float-glass fabrication approach, which is the usual method employed by the glass industry for obtaining large-area glass panes [154]. Besides, since glass materials for ophthalmic lenses in most cases have been replaced by lighter and less brittle polymer materials, silver halides have been abandoned in favor of organic PC dyes, which can easily be coated on organic substrates [151]. However, the easier fabrication process of organic dyes comes with a penalty, not only because organic PC materials are considerably more expensive than their inorganic counterparts but also because PC eyeglasses based on PC organic compounds have shorter life-span. Organic eyeglasses typically have a life span of only two or three years of continuous use. Although this is not an important drawback for applications in the ophthalmic industry – prescription eyeglasses should be regularly revised and replaced – it hinders application of such compounds in other applications that require durability of many years, even decades, such as façade glazing and windows.

In summary, there is a need for new PC materials to enable fabrication of large-area durable PC glass. Without such materials, the possibility of developing PC smart windows technology will remain out of reach. Most likely, the breakthrough will come from the inorganic side of PC materials, mostly due to their superior stability under UV radiation. However, inorganic PC compounds present their own challenges, notably small PC contrast and slow darkening/bleaching dynamics. Understanding the underlying reasons of the PC mechanism(s) for inorganic materials is crucial for improving their properties, and being able to tailor them for glazing applications.

In all T-type organic substances that are found in commercial applications, the basic PC mechanism is light-induced isomerization (photoisomerization) [151]. In contrast, in inorganic PC materials, the PC switching is generally associated with red-ox processes that change the oxidation state of the cations, which typically becomes reduced in the dark state. Following we will discuss the PC mechanism in three groups of inorganic PC materials, the well-known silver halides, transition metal oxides, and the recent development in RE oxyhydrides.

7.4.1 Silver Halides

Silver halides exhibit excellent PC properties and outstanding durability. As mentioned previously, their largest drawback, which currently remains unsolved despite considerable efforts [155], is the lack of synthesis routes that are compatible with the float glass fabrication process.

Photochromism in silver halides is explained through a well-established model [153]. A silver halide PC glass consists of a dispersion of silver halide nanocrystals (typically silver chloride or silver bromide) in a glass matrix that is rich in copper impurities. When light of sufficient energy (UV or blue) impinges on the silver halide PC glass, charge is transferred between the Ag^+ and Cl^- ions, viz.

$$Ag^+ + Cl^- \xrightarrow{h\nu} Ag^0 + Cl^0 \tag{7.4}$$

Hereinafter the term $h\nu$ signifies electromagnetic radiation (light). In silver halides, it is the formation of metallic Ag domains, according to reaction (7.4) that causes the darkening of the glass. However, the reverse of reaction (7.4) also takes place:

$$Ag^0 + Cl^0 \rightarrow Ag^+ + Cl^- \tag{7.5}$$

Reaction (7.5) causes bleaching of the film and is very fast and competes with reaction (7.4). For darkening to be observed, the process presented in reaction (7.5) has to be slowed down. This is the part where Cu^+ impurities come into play. Cu^+ impurities in the silver halide nanocrystals act as *hole traps*:

$$Cl^0 + Cu^+ \rightarrow Cl^- + Cu^{2+} \tag{7.6}$$

Therefore, in the presence of Cu^+ impurities, the bleaching process is slowed down and proceeds through the slower route:

$$Cu^{2+} + Ag^0 \longrightarrow Cu^+ + Ag^+ \tag{7.7}$$

For reaction (7.7) to take place, Cu^{2+} and Ag^0 species must be in close proximity in the lattice. For this reason, under illumination, reaction (7.4) is much faster than reaction (7.7) since diffusion of the active species in the matrix must occur, and limits the bleaching process. The film therefore becomes dark upon illumination. Once illumination ceases, the silver halide-containing glass bleaches, recovering its original transparency through reaction (7.7). The typical transmittance of a silver halide glass in the clear and darkened state is shown in Figure 7.15a and b, respectively. An approximate model of the changes that occur at the atomic scale level involves the formation of Ag^0 clusters (see Figure 7.15b) [153, 156]. Note that the decrease in transmittance occurs evenly in the visible range; therefore, silver halide glass retains color neutrality in the dark state.

7.4.2 Transition Metal Oxides

PC transition metal oxides include oxides of tungsten, molybdenum, niobium, titanium, and zinc [157]. These oxides can be synthesized by methods already utilized by the glazing industry such as by sputtering, sol–gel, and hydrothermal methods discussed in Sections 7.3.2 and 7.3.3. Figure 7.16a shows the transmittance of a typical PC WO_3 film, as reported by Bourdin et al. [158], before and after illumination. The largest change in the transmittance takes place in the NIR, spreading to the visible region. As a consequence, the human eye perceives the PC change in WO_3 as the appearance of a deep blue coloration.

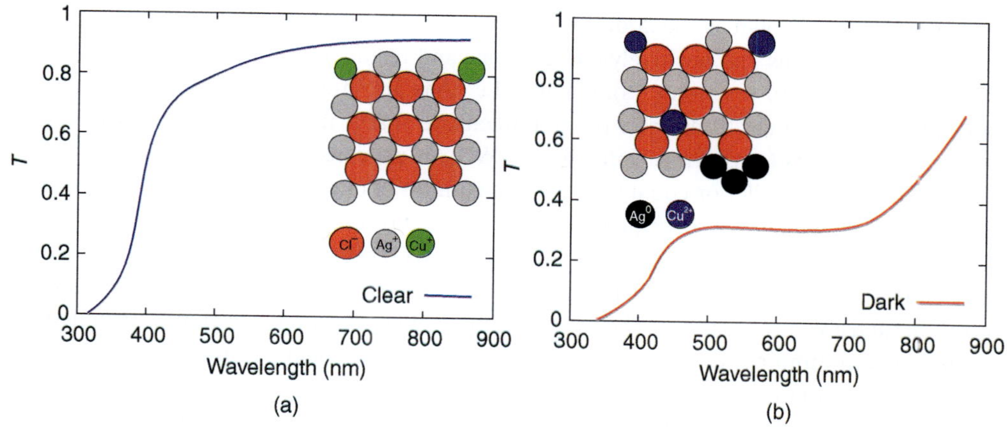

Figure 7.15 The characteristic transmittance of silver chloride glass (a) before and (b) after illumination. The insets show a simple 2D representation of the lattice in each state: transition from Cu^{+} to Cu^{2+} and the formation of Ag^{0} clusters, respectively.

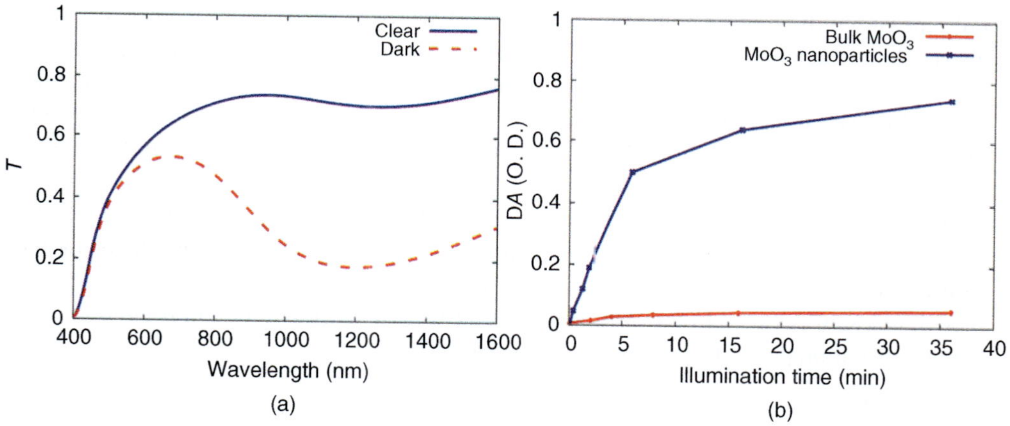

Figure 7.16 (a) Typical spectral transmittance, T, of a photochromic WO_3 film before (clear) and after illumination (dark). (b) Change in Absorbance (ΔA) (optical density O.D.) vs. Illumination time in bulk MoO_3 and MoO_3 nanoparticles. *Source:* (a) Figures adapted from a Bourdin et al. [158], Royal Society of Chemistry, and (b) Li et al. [159] / with permission of Elsevier.

In general, transition metal oxides exhibit poor PC properties in terms of contrast and switching dynamics. Although photochromism in transition metal oxides has been known for decades, the underlying physical reasons for their properties are far from being completely understood. A full understanding of their PC mechanism(s) can be a key for finding strategies to judiciously synthesize films with improved PC properties.

Here, we will use tungsten oxide, WO_3, as the model for discussing the PC mechanism in transition metal oxides. Since WO_3 is better known for its electrochromic properties, a small digression on electrochromic WO_3 thin films – which is treated elsewhere in this book – can help us understand the perspective from which researchers have historically formulated models for the PC effect in WO_3.

WO_3 exhibits a particular case of electrochromic behavior, known as *cathodic* [160], i.e. WO_3 thin films darken upon the insertion of ions, typically Li^{+} or H^{+}:

$$WO_3 + xLi^{+} + xe^{-} \rightarrow Li_xWO_3 \tag{7.8}$$

This process is known as the double charge injection model (DCIM), because it involves injection of electrons and positively charged ions and results in the formation of tungsten bronze. According to reaction (7.8), when $x \cdot Li^+$ ions are intercalated in the WO_3 lattice, an equal number of electrons enter the film to ensure charge neutrality. As a consequence, transparent WO_3 is transformed into dark Li_xWO_3. If we consider the site saturation (SS) theory [161], x is smaller or equal to 1, i.e. a maximum of one electron can be inserted per tungsten atom, which as a consequence will be reduced from W^{6+} to W^{5+}. Thus, after insertion of lithium ions, W^{6+} and W^{5+} species coexist in a WO_3 film. After absorption of a photon, an electron can jump from a W^{5+} in position i to an W^{6+} ion in position j, leaving an W^{6+} ion in position i and a W^{5+} in position j. The darkening of the WO_3 film after insertion is attributed to such "electron-jumping" exchange processes. However, experimental results show that intercalation levels up to $x \leq 2$ are possible [162]. Berggren et al. [162] introduced the extended SS (ESS) theory. In the ESS theory, up to two electrons can be intercalated per tungsten atom. Accordingly, after intercalation, W^{6+} and W^{5+} coexist with W^{4+} species. Therefore, the absorption of photons could be due to the following electron exchanges: from W^{5+} to W^{6+}, from W^{4+} to W^{6+}, and from W^{4+} and W^{5+}. The ESS theory reproduces the experimentally measured differential coloration efficiency of $WO_3{-}x$ electrochromic films very well, also for $1 \leq x \leq 2$, where the SS theory is not defined [163]. Rojas and Niklasson [163] demonstrated that it is the exchange of electrons from W^{6+} to W^{5+} that causes the coloration of the films. However, the presence of W^{4+} must be considered, because it affects the probability of electron transition, which is related to the relative number of W^{6+} to W^5 and W^{4+} sites, respectively.

The first attempts to explain the PC behavior of WO_3 were done by using the DCIM that was developed for electrochromism, i.e. assuming that the PC coloration is caused by the insertion of ions, in this case H^+, originating from adsorbed water [164]. According to this model, holes, h^+, created in the WO_3 film by inter-band absorption upon illumination have enough oxidation power to break water molecules, viz.

$$H_2O\,(\text{adsorbed}) + 2h^+ \rightarrow 2H^+ + O\uparrow \tag{7.9}$$

Oxygen atoms are either retained as interstitial or desorb (after recombination reactions to O_2), whereas the protons H^+ penetrate the WO_3 film, reacting with the oxide:

$$WO_3 + xH^+ + xe^- \rightarrow H_xWO_3 \tag{7.10}$$

Reactions (7.8) and (7.10) are essentially the same and, by analogy, would be the underlying reason for both the electrochromic and PC properties of WO_3.

According to reaction (7.10), a large increase of the hydrogen concentration in the WO_3 film is expected after illumination. However, the concentration of hydrogen after illumination in photo-darkened films is equal to the one before illumination. To remedy this discrepancy, Bechinger et al. [165, 166] proposed a modified model in which n water molecules are adsorbed on oxygen-deficient tungsten oxide. A fraction $x/2$ water molecules dissociates after illumination yielding x protons and electrons:

$$WO_{3-y} \cdot nH_2O \overset{h\nu}{\rightarrow} WO_{3-y} \cdot \left(n - \frac{x}{2}\right) H_2O + xH^+ + xe^- + \frac{x}{2}O\uparrow \tag{7.11}$$

The released protons are assumed to be captured by tungsten oxide lattice oxygen atoms:

$$WO_{3-y} \cdot \left(n - \frac{x}{2}\right) H_2O + xH^+ + xe^- \rightarrow WO_{3-y-\frac{x}{2}} \cdot nH_2O \tag{7.12}$$

Reaction (7.12) assures that the original water content in the lattice is replenished. Merging processes (7.11) and (7.12) we obtain:

$$WO_{3-y} \cdot nH_2O \overset{h\nu}{\rightarrow} WO_{3-y-\frac{x}{2}} \cdot nH_2O + \frac{x}{2}O \uparrow \tag{7.13}$$

In summary, according to reaction (7.13), illumination reduces tungsten by the desorption of O^- species, while the hydrogen concentration remains unchanged upon illumination; the role of the protons is to reduce tungsten oxide. According to reaction (7.13), photo-darkening yields coexistent W^{4+}, W^{5+}, and W^{6+} species, resulting in darkening due to electron exchange processes similar to the ones in the ESS model for electrochromism. Most oxygen atoms that are released under illumination can be retained in interstitial positions in the lattice, especially in sub-stoichiometric WO_{3-x} [166]. Considering the reverse reaction of (7.11), it follows that bleaching of photo-darkened tungsten oxide films should be faster in an oxygen-containing atmosphere, which is also observed experimentally. Bechinger et al. [166] reported the existence of two bleaching channels, one which was affected by the surrounding atmosphere, and another channel that was independent of the surrounding atmosphere. The latter was associated with the existence of interstitial oxygen. The difference in response between inert and oxygen-containing atmospheres has been reported in other related materials such as phosphotungstate-doped TiO_2 [167], and hybrid composites of nanocellulose-phosphotungstic acid [168].

PC changes in tungsten oxide (and in other transition metal oxides) are usually small and slow [157, 164, 169], with bleaching/darkening times on the order of hours or days, and small optical contrast (e.g. ΔT_{lum}). As expected from the proposed PC mechanism, contrast and speed of the PC effect strongly depends on parameters such as film density and surrounding atmosphere [169]. Low density is known to increase coloration in WO_3, a feature also observed in electrochromic films. The density of the WO_3 films is related to the deposition process and high porosity of the resulting material is beneficial. For example, in sputtering or thermal evaporation, low film density is associated with high deposition pressures and low substrate temperatures [169]. Thin films bleach faster than thick films in an atmosphere containing oxygen, but this occurs at the expense of smaller optical contrast [170].

In view of the PC mechanism, strategies that improve the PC properties of transition metal oxides include increasing the surface area-to-volume ratio for promoting oxygen permeability and charge separation processes (promoting reaction (7.11)). Reported approaches include structural modifications such as the use of nanoparticles [159], nanosheets or nanoflakes instead of films [167], decoration with noble metals [167], or fabrication of hybrid composites [168]. An illustration of the possibilities that nanotechnology poses for improving the PC properties of transition metal oxides is given in Figure 7.16b, which shows the change in absorbance A (optical density O.D.) as a function of illumination time in bulk MoO_3 and in MoO_3 nanoparticles measured at a wavelength of 750 nm [159]. It can be observed from the figure that nanoparticles exhibit a much more pronounced PC effect than bulk MoO_3.

7.4.3 Rare-Earth Oxyhydrides

In recent years, RE oxyhydrides, hydride oxides, or oxygen-containing hydrides, (RE = Y, La, Gd, Dy, Er) [171], have attracted attention due to their remarkable PC properties. An oxyhydride can be regarded as an oxide in which some of the oxygen atoms have been replaced by hydride ligands [172]. Therefore, in an oxyhydride, oxide and hydride anions coexist in a single phase. This provides extra degrees of freedom in the design of the building blocks of

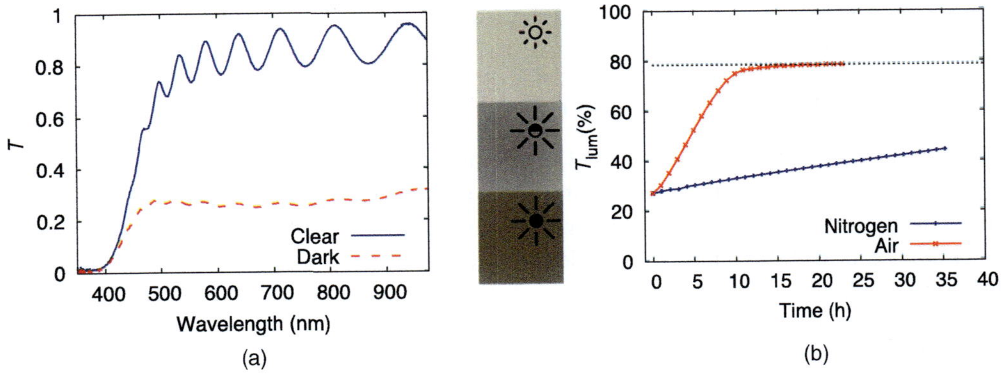

Figure 7.17 (a) Transmittance in the clear and dark state corresponding to an YHO film. (b) Pictures of an YHO sample at different stages of photo-darkening. (c) Evolution of the luminous transmittance (T_{lum}) of a photo-darkened YHO film kept in air and in nitrogen atmosphere. The dashed line at 79% indicates T_{lum} in the completely bleached state. *Source:* (b) Reproduced from Baba et al. [173] / American Physical Society / CC BY 4.0.

the material, allowing for extraordinary tunable optical, structural, and electrical properties. In comparison with single anion compounds, such as oxides, nitrides, or hydrides, mixed anion compounds constitute a true *terra ignota* in which materials with innovative functional properties can be discovered [172].

In this section, we will use yttrium oxyhydride (YHO), the archetypical PC RE oxyhydride, to exemplify physical, and in particular optical, properties of this emerging family of PC materials. Figure 7.17a shows the characteristic transmittance before and after illumination of a 1 μm-thick YHO film prepared by magnetron sputtering [173]. It can be observed that the film exhibits an excellent luminous transmittance, T_{lum}, in the clear state (80%) and a good PC contrast, $\Delta T_{lum} = 52\%$. After illumination, the decrease in transmittance is evenly distributed throughout the visible region, hence the photo-darkened film exhibits color neutrality. Figure 7.17b shows a sequence of pictures corresponding to an YHO film before illumination (top) and at different photo-darkening levels (medium and bottom). Similar to PC transition metal oxides, oxygen is reported to play a fundamental role in the bleaching mechanism of the films, at least under certain conditions. Figure 7.17c shows the evolution of T_{lum} during bleaching of a photo-darkened YHO film in air and N_2 atmosphere [173]. It is apparent that the recovery time of the darkened film in inert gas considerably increased compared to a film in ambient atmosphere (hours). Typically, photodarkening of a magnetron sputtered, 1 μm thick YHO film occurs on the time scale of minutes, while bleaching takes of the order tenth of minutes.

Reports of PC YHO films were first published by Mongstad et al. [26] in attempts to prepare YH_2 films by reactive magnetron sputtering using a metallic Y target and an argon and hydrogen plasma. The resulting films, however, turned out to be transparent, with a band gap around 3.6 eV (direct allowed inter-band transitions), very different from the metallic behavior expected for YH_2 [174]. Remarkably, such films exhibited PC properties and a large oxygen content, exhibiting a $Y_1H_1O_1$ stoichiometry [175].

As later demonstrated by Montero et al. [176], the large porosity of the precursor YH_2 film (fcc β-phase) prepared by sputtering at high pressures allowed for the fast incorporation of oxygen atoms, enabling the transformation of YH_2 into YHO. As it turned out, YH_2 constitutes an excellent scaffold to build YHO, thanks to the capacity of yttrium to retain hydrogen even at high temperatures [177]. In order for the precursor YH_2 films to be porous enough to allow for incorporation of oxygen, YH_2 has to be prepared at high deposition pressures [171], i.e. in zone 1 as defined in Thornton's growth zone model for sputter deposition (porous structure, crystallites separated by voids) [178]. The incorporation of oxygen in YH_2 has three main effects [176]: (i) expansion of the unit cell, resulting in an increased lattice constant (fcc structure) from 5.20 to 5.34 Å, (ii) opening of a band gap (of around 3.6 eV), and (iii) emergence of PC properties.

The PC mechanism in RE oxyhydrides is still open to debate. Nevertheless, we can list some important observations that have been repeated many times by different research groups: (i) under illumination/dark cycles, the YHO lattice contracts/expands reversibly [173, 179], (ii) under illumination the conductivity of the YHO films increases notably [26, 180, 181], (iii) the darkening of the films can be modeled by assuming the formation of oxygen-deficient domains of metallic character in the YHO matrix [173], which is consistent with experimental observations [182]. (iv) Similar to the PC transition metal oxides discussed above, a photo-darkened YHO requires presence of oxygen to recover efficiently [173].

Given the observations above, it is reasonable to assume that under illumination some of the Y^{3+} cations in the YHO lattice are reduced to Y^{2+} cations, but the reason why this happens is another story. Looking back to transition metal oxides, the reason could be photon-induced dissociation of adsorbed water (reaction (7.11)). Other explanations include the presence of low-lying degenerate states in the conduction band (second-order Jahn–Teller effect) [173], diffusion of hydride anions [182], or combinations of them.

As discussed above, when compared with other chromogenic materials, RE oxyhydrides exhibit distinct characteristics that are of technological relevance. First of all, they are inorganic and can be deposited on large areas by magnetron sputtering, a well-established technique in the glazing industry. Second, oxyhydrides can be deposited without intentional substrate heating, which simplifies the deposition process and allows deposition onto flexible substrates. Finally, the optical performance of PC oxyhydrides is excellent. For example, a 1 μm-thick Gd oxyhydride film exhibiting $T_{lum} = 65\%$ in the clear state has $\Delta T_{sol} = 50\%$ [183]. Considering that the technology still is in its infancy this is already competitive with other more mature chromogenic technologies. The combination of advantageous T_{lum} and ΔT_{sol} and the simplicity of the technology (only one coating is needed) makes it a promising technology.

However, the application of PC oxyhydrides as smart materials in the glazing industry is certainly not without challenges. Although the PC oxyhydrides, like other inorganic materials, are more resilient to UV radiation than organic TC materials, the durability of PC oxyhydrides – specially their resilience to "fatigue" [184], i.e. to repeated switching cycles, is still unknown. Oxyhydrides have shown promising reversibility, but this has hitherto only been shown for a very limited number of cycles [185]. In contrast, the life expectations for a window are around 30 years, often under harsh environmental conditions. This brings us to the next point: can PC oxyhydrides be protected from the ambient by lamination between glass panes, using capping layers, or by being placed on the inner pane of an IGU where the coating is only exposed to inert gas? In the literature, we can see how these materials seem to require an oxygen-containing atmosphere to bleach back to the original transparent state [173]. However, we can speculate that oxygen can be provided not only by the surrounding atmosphere but also from a source of interstitial oxygen present in the film itself, as is the case in PC WO_3 [166]. Accordingly, it is unclear how capping or an inert atmosphere will affect the recovery dynamics, although most likely it will make the recovery slower. PC oxyhydride coatings capped by compact Al_2O_3 protective layers are known to exhibit slower recovery dynamics, although it is not clear if this effect is caused by the elimination of lattice defects during the capping process, the absence of an external oxygen source, or to both [186]. In the same vein, photodarkenened oxyhydrides are T-type PC materials, i.e. bleached by a thermal process, thus under illumination a bleaching/darkening equilibrium, which condition the final value of ΔT_{sol}, is reached. When a PC oxyhydride darkens, it absorbs a large portion of the solar spectrum. This causes an increase of the temperature of the coating affecting the bleaching/darkening equilibrium and may result in lower ΔT_{sol}. In the opposite situation, cold weather can also affect the bleaching dynamics of the PC film, hindering the bleaching process when light is not impinging on the coating. Ideally, the bleaching/darkening dynamics of PC oxyhydrides should not depend to a large extent on the temperature of the ambient. All these questions, and perhaps some others, will need to be addressed to pave the way toward practical application of oxyhydrides as smart materials in smart fenestration.

7.5 Thermochromic and Photochromic Smart Windows

7.5.1 Applications of Thermochromic and Photochromic Glazing

TC and PC glazing can be used in insulated glass units, in principle using the same industrial technologies that are used for metal-based coatings with static properties [187]. Research and development of TC smart windows is much more mature than inorganic PCs, and the following examples given are therefore naturally taken from the larger body of literature on TCs. Nevertheless, as evident from Section 7.4, inorganic PC thin films can preferably be sputtered using standard sputtering methods, like VO_2 coatings, and can, similar to other chromogenic films, be transferred to flexible polymer foils [54, 109]. TC nanoparticles – perhaps with protective shells surrounding a VO_2 core – can be dispersed in polymers and used for lamination purposes, as sketched in the inset panel of Figure 7.18. A further development may be to use such TC laminates as an electrolyte in electrochromic devices, which modulate their transmittance of visible light and solar energy [188]. Here, an applied voltage leads to moving electrical charge through the electrolyte which is situated in between active tungsten oxide and nickel oxide thin films. Such a device will show combined electrochromic and TC properties. Also, straightforward addition of a VO_2 layer in an electrochromic cell configuration is possible. Lee et al. [189] reported hybrid ITO/VO_2/Ta_2O_5/WO_3/ITO multilayer structures, which showed dual electrochromic and TC properties.

Another possibility is to use a VO_2 film together with a polymer-dispersed liquid crystal (PDLC) window in order to separately regulate solar energy throughput and visibility [190]. Yet another approach is to combine TC materials with electrothermal devices, as reviewed by Ke et al. [191]. Typically, voltage is applied to highly transparent electrodes to heat up the TC materials through Joule heating. Indirect photon-assisted heating of TC materials is also a possible way to achieve a lower apparent τ_c of TC devices and also enhance the switching speed. Here, the TC material is combined with high-solar light absorbance materials (e.g. graphene oxide, antimony tin oxide (ATO), Cs_xWO_3, and plasmonic nanoparticles) [191].

Combinations of chromogenic materials with other materials have been proposed for enhanced and synergistic action to achieve multifunctional smart windows that not only modulate solar light radiation but also add other beneficial glazing functionalities. VO_2-based nanoparticles embedded in temperature-responsive hydrogels show interesting results for reversible TC smart glazing [55, 56, 192, 193]. Hydrogels, which belong to

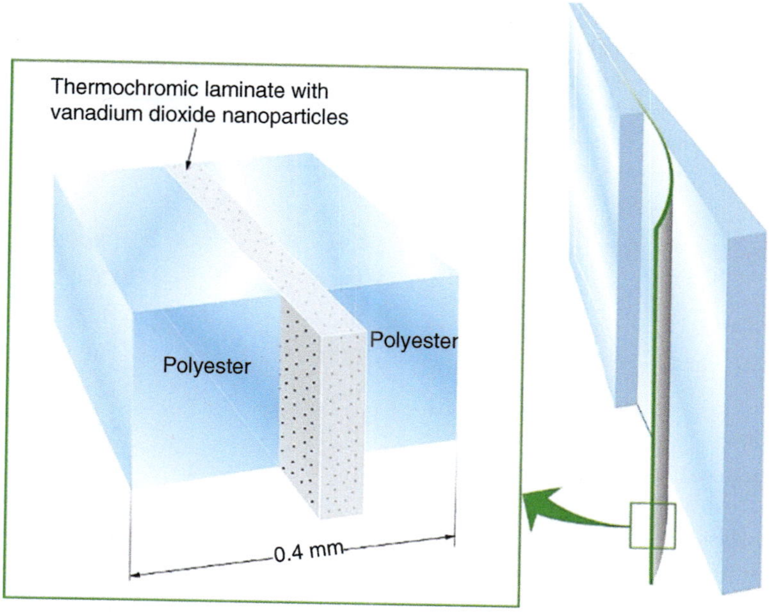

Figure 7.18 Schematic illustration of a thermochromic laminate placed between two glass panes.

the class of thermotropic materials, exhibit a reversible hydrophilic/hydrophobic phase transition of the hydrogel around the lower critical solution temperature used, which gives rise to transition from a transparent state to a translucent state. Poly(*N*-isopropylacrylmide) (PNIPAm) hydrogels with appropriate thickness have been shown to yield high modulation of visible light and moderate modulation ability in the IR range [8]. By embedding VO_2 nanoparticles [55], $VO_2@SiO_2$ core–shell nanoparticles [193] and W-doped $VO_2@SiO_2$ core–shell nanoparticles in hydrogels [192], interesting T_{lum} and ΔT_{sol} have been reported (Figure 7.19; Table 7.1), albeit the luminescent transmission in the warm state exhibits significant diffuse scattering. The addition of VO_2-based nanoparticles significantly improves the NIR modulation properties. Further challenges of these hybrid concepts include achieving an opaque state with sufficient transparency for architectural and aesthetic purposes.

(a) (b) (c)

Figure 7.19 (a) Photographs showing a 30 cm × 30 cm device based on a $V_{0.8}W_{0.2}O_2@SiO_2$/PNIPAm nanoparticle hydrogel and its response to sunlight. (b) Before the phase transition. (c) After the phase transition under outdoor sunlight. *Source:* Reproduced from Zhang et al. [192] / Royal Society of Chemistry.

Table 7.1 Typical parameters for visible light transmittance in the clear state, T_{lum}, and optical contrast calculated as integrated solar light modulation, ΔT_{sol} (or ΔT_{lum}) for different TC and PC materials.

Material	T_{lum}, clear state	Contrast, ΔT_{sol} a)	References
Thermochromic			
Nanoporous VO_2	71.6%	7.9%	[203]
TiO_2/VO_2 thin films	47.0%	15.1%	[204]
Nanoporous PVP/VO_2 films	92.7%	10.5%	[127]
$TiO_2/VO_2/TiO_2/VO_2/TiO_2$ films	44.0%	12.1%	[46]
ZnO/VO_2 bilayer films	55.9	19.3	[205]
VO_2–SiO_2 composite films	48.5	15.7	[206]
VO_2@SiO_2 core–shell polyurethane nanoparticle films	~41%	15.5%	[143]
VO_2-polyurethane nanoparticle films	45.6%	22.3%	[207]
W-doped nanoporous PVP/VO_2 films	92.0%	8.5%	[127]
W doped VO_2 thin films	52.1	6.4	[208]
V_xW_{1-x} O_2@SiO_2 core–shell nanoparticle films	51.0%	14.7%	[53]
VO_2@polydopamine/ polyurethane thin films	56.3%	14.5%	[209]
W-doped VO_2/2-hydroxy-3-butoxypropyl starch hydrogel (HBPS)	53.9	34.3	[200]
VO_2 NPs hybridized with nickel(II)-based ligand (NLETS)	73.4	18.2	[201]
VO_2/PNIPAm (*N*-isopropyl acrylamide)	62.6	34.7	[55]
Photochromic			
Silver halide photochromic glass	85.1%	53.0%	[210]
WO_3 photochromic	50.2%	$\Delta T_{lum} = 4.0$ % $\Delta T(1200\ \text{nm}) = 62.0\%$	[158]
Y oxyhydride	81.0%	$\Delta T_{lum} = 60.6\%$ $\Delta T_{sol} = 52.4$ %	Own work§
Y, Gd, Dy, Er oxyhydrides	Average: 79.8% (Y); 67.9–71.3% (Gd, Dy, Er)	Average 450-1000 nm: 33–37% (Y, Dy, Er); 45.5% (Gd)	[27]

a) 1 μm-thick YHO sputtered film onto soda-lime glass. Contrast was achieved after one hour illumination with light of 365 nm wavelength and power density of 4.5 mW cm^{-2} during 24 hours.

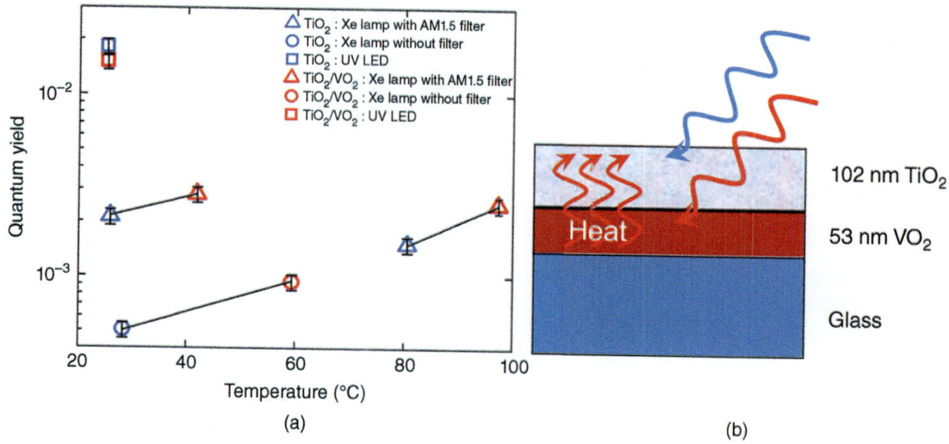

Figure 7.20 Enhanced photocatalytic properties obtained by a dual-action TiO_2(A)/VO_2(M1) bilayer coating. (a) Increased quantum yield, i.e. number of degraded molecules per absorbed photon, for degradation of stearic acid spin coated on the coating using different types of illumination (note the log-scale). (b) Schematic illustration of the photocatalytic-thermochromic bilayer coating on glass showing the heat absorption in VO_2 that heats the TiO_2 film and increased the thermally activated chemical reactions at its surface. *Source:* Data from Ji et al. [194]. Cell Press.

Progress in thermotropic polymers for smart windows is outside the scope of this chapter and the reader is instead referred to Chapter 8 in this book.

Thermochromism can be combined with photocatalytic self-cleaning in multifunctional glazing [132, 194]. Figure 7.20 shows an anatase TiO_2(A)/VO_2(M1) multilayer film on glass, which shows a significant increase in the measured quantum yield for photodegradation of stearic acid upon illumination with various light sources [194]. The results can be explained by the absorptance in the underlying VO_2 film. Notably, more than 30° heating upon illumination with simulated solar light is possible. This excess heating can be used to reach τ_c of the VO_2 film. The TiO_2/VO_2 multilayer film configuration embodies not only synergetic enhancement of the intrinsic photocatalytic and TC performance of the TiO_2 and VO_2 films, respectively, but the overlaying TiO_2 film also contributes to protect the VO_2 film from environmental degradation and acts as a $3/4\lambda$ antireflective coating.

Other multifunctional smart window configurations include antifogging/self-cleaning properties (TiO_2(R)/VO_2(M1)/TiO_2(A)) [131], oxidation resistance, and enhanced optical properties [132, 195]. Parkin et al. [132] reported a $TiO_2/SiO_2/VO_2$ multilayer coating that exhibited improved T_{lum} and ΔT_{sol}, as well as improved abrasion resistance. Their films also showed self-cleaning. Notably, the TiO_2 layer exhibits excellent ion storage ability and can be used as a counter-electrode layer in combined TC and electrochromic glazing. Yang et al. [196] fabricated TiO_2/VO_2 nanocomposite films with combined electrochromic and TC properties, reaching $\Delta T_{sol} \approx 17\%$ and $T_{lum} = 45\%$. As a practical application of large-scale glass with multifunctional properties Zheng et al. [131] devised a TiO_2(R)/VO_2(M1)/TiO_2(A) multilayer film on a glass substrate with area of 400×400 mm^2 using magnetron sputtering. Additionally, the multilayer film showed hydrophobic and photocatalytic properties in dye degradation tests using Rhodamine B dye.

Inorganic lanthanide oxyhydride PC films may be exploited in a similar way as the photothermal material combinations discussed above. The PC films absorb a large portion of the Vis–NIR light, become hot, and can therefore be used in combination with VO_2-based

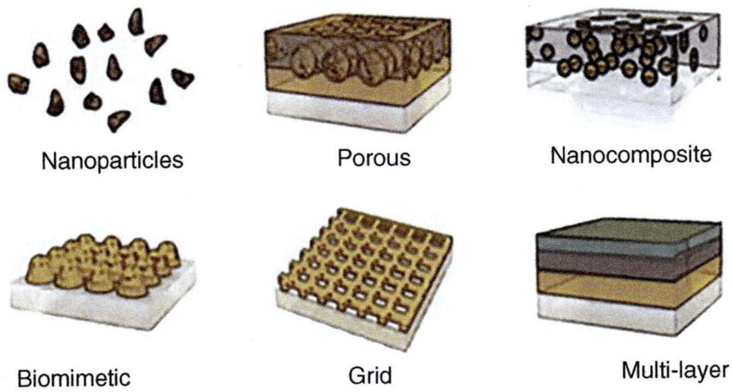

Figure 7.21 Schematic illustration of different methods to realize enhanced performance of all-inorganic thermochromic and photochromic smart windows: Nanocomposites, in particular embedding TC nanoparticles in a dielectric host matrix; meso- or nanoporous films; engineered photonic micro/nanostructures; or multilayer coatings with spectral selectivity, thermal and electrical interconnectivity. *Source:* Cui et al. [19]/with permission of Elsevier.

TC films so that the VO_2 material reaches τ_c under solar illumination. Since both the PC and the TC films act as solar control coatings, the TC film can be made thinner, allowing for an increased T_{lum}, without loss of ΔT_{sol}, while still yielding enhanced reflectance of the multilayer film, which can be beneficial in some applications.

BIPV uses PV modules as integral building components, supplying electrical energy to the building itself as well as to the local grid: PV modules become construction components, for example, in ventilated façades or curtain walls. Today's most frequent PV technologies for BIPV are based on crystalline Si PV cells. These solar cells are opaque, so the semitransparent Si PV modules made of glass laminates can only transmit light in a nonhomogeneous, partially shading way, which adversely affects indoor visual comfort and aesthetic façades, as asked for by architects and end-users. Here, glazing has enormous untapped potential for energy harvesting. This is hitherto an unresolved technical challenge, where the combination of TC and PV technologies can be fruitful. A decade ago, Zhou et al. [197] reported a novel VO_2-based smart window that combines energy-saving and electricity generation in one device using waveguiding of scattered NIR light to Si cells attached to the side of the glass pane, akin to the well-known LCD solar cells [198]. The concept of merging TC with solar harvesting for electricity generation was recently revived by the advent of perovskite-based TC materials. Lin et al. [41] reported on TC light modulating mixed halide perovskite $CsPbI_{3-x}Br_x$ ($0 \leq x \leq 3$) thin films exhibiting peak device efficiency of about 7%, and a TC phase transition temperature of 105 °C. Figure 7.21 schematically depicts different TC and PC glazing possibilities in combination with other glazing technologies discussed in this chapter. Combinations of materials as nanocomposites or multilayers can be used for multiple purposes: AR, corrosion and mechanical protection, enhanced and selective optical properties, and thermal management.

7.5.2 Performance Limits

It is evident that various combinations of T_{lum} and ΔT_{sol} can be obtained from different VO_2-based and RE oxyhydride materials and combinations of them with other materials

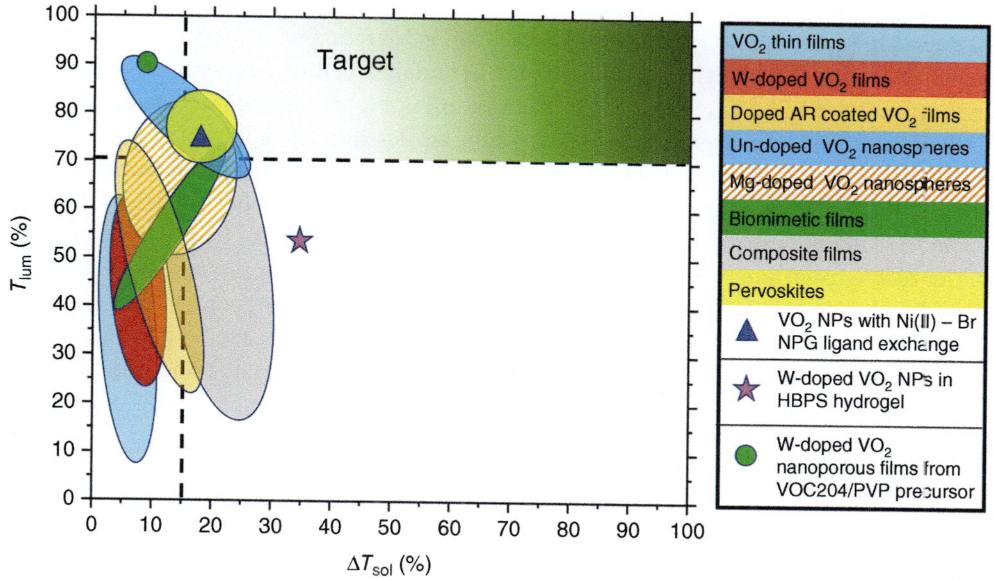

Figure 7.22 Schematic performance limits for VO_2-based thin films and nanoparticle composites with regard to TC glazing. The target limits are not definitive and merely indicate preferred performance in many window applications, but which may be case specific. Promising data indicate in green box.

(e.g. multilayer films, core–shell, and embedded VO_2 nanoparticles). Figure 7.22 illustrates the current state-of-the-art of $T_{lum}(\tau < \tau_c)$ and ΔT_{sol} for inorganic TC coatings. This figure is based on experimental data, as well as computations [86] of optical properties for thin VO_2 films with and without AR layers and thin Mg:VO_2 thin films, backed by glass substrates. In addition, calculations were performed with effective medium theory for coatings consisting of VO_2 and Mg:VO_2 nanoparticles in a matrix material having a refractive index of 1.5, deposited on glass substrates [86].

Thin films of pure VO_2 give the lowest magnitudes of $T_{lum}(\tau < \tau_c)$ and ΔT_{sol}, but the optical performance can be boosted by AR layers. Particularly good properties were obtained with overlayers of SiO_2 [199] and TiO_2 [64] on VO_2. "Bio-inspired" cone-shaped surfaces represent another possibility to increase T_{lum} [200]. Mg-containing VO_2 films have improved properties, which can be further enhanced by AR layers. However, optimum performance can be reached with VO_2 nanoparticles which, for example, can yield $T_{lum}(\tau < \tau_c) \approx 60\%$ and $\Delta T_{sol} > 20\%$. It is also seen that Mg-doped nanoparticles are not superior to nanoparticles of pure VO_2; the reason for this behavior is that the doping lowers the solar modulation due to the plasmonic effect to a degree that is not compensated by the higher short-wavelength transmittance. Embedding nanoparticles in TC hydrogels that synergistically modulate the sunlight in a large wavelength interval allows combinations of dilute concentrations of VO_2-based nanoparticles and hydrogel films to improve the T_{lum} and ΔT_{sol} properties, and with addition of W doping of the VO_2 nanoparticles also to lower TC to around 30 °C [192, 201]. Hydrogels introduce however substantial light scattering (see Figure 7.18).

Furthermore, multilayer films and nanocomposite structures allow for combinations of synergistically acting anti-reflecting, self-cleaning, antifogging, low emittance, and TC materials [131, 132, 194].

Table 7.1 shows key optical parameters for some TC and PC materials. As extensively discussed in Section 7.3, VO_2-based compounds exhibit a large optical contrast in the infrared region, while the visible region remains more or less unaltered; thus, key parameters are T_{lum} in the clear state and ΔT_{sol}. Note that, in contrast to TC, PC materials do not have a defined dark state, and different degrees of photo darkening are possible depending on various illumination parameters (wavelength, power, and time), hindering a fair comparison among different materials and studies. Nevertheless, some qualitative conclusions can be extracted from Table 7.1. After illumination, WO_3 becomes absorbed in the NIR and the longest wavelengths of the visible spectrum, resulting in darkened films with a bluish appearance to the human eye. Silver halides, on the other hand, exhibit a large and even switching contrast in the visible region, allowing for color neutrality. However, there is practically no switching in the NIR in silver halides. Lanthanide oxyhydrides, on the other hand, exhibit large and even optical contrast in the visible and a reasonable switching in the NIR, providing large ΔT_{sol} and ΔT_{lum}, combined with very good T_{lum} in the clear state.

7.6 Conclusions

TC films, particularly VO_2-based films, continue to attract a lot of attention for smart windows since the discovery of thermochromism of VO_2 more than 60 years ago. The challenge to simultaneously meet appropriate switching temperature (clear-to-dark state, and back), visible transparency (typically >60%), and solar light regulation (at least 20%), as well as stability over time in operational environment, still remains an unsolved task for all known classes of inorganic TC materials. However, as summarized in Figure 7.22 and Table 7.1, progress in synthesis of VO_2-based nanoparticles and nanocomposites, in combination with embedding strategies in host media (polymers), or as multilayer films, are promising new directions and pushing research closer to practical glazing applications. Marrying TC properties with materials having complementary and synergetic properties may be useful in glazing applications (both for the optical properties, as in the case of meta-materials, but also for adding and enhancing anti-reflection, self-cleaning, antifogging, low emittance, etc.). New materials such as perovskites, hydrogels, vanadium precursor-initiated porous materials, and ligand exchange compounds with promising optical and solar energy properties have invigorated the field to search for new materials and combinations of materials and technologies.

The recent discovery of PCRE oxyhydride materials shows that research into tertiary compounds is a fruitful way to explore and discover optical materials for smart windows with completely new properties and operating principles. Although the mechanism by which the photochromism of RE oxyhydrides works is still not understood, the optical properties, as shown in this chapter, are already very promising, showing uniform broad band shift across the visible and NIR spectral region, thus exhibiting color neutrality (see e.g. Figure 7.1 for a comparison with VO_2 TCs). Moreover, their synthesis is straightforward. The challenges that remain to solve are mainly durability pertinent to glazing applications (many thousands of cycles) and switching time. We foresee that research on inorganic PCs will be revived by this development, and at the same time old work on WO_3 PCs will be revisited. In summary, despite that neither inorganic TC nor PC materials yet have reached the tipping point for commercial applications, advances in material science research and photonic engineering over the past, say 10 years, have pushed them closer to practical applications.

References

1 Santamouris, M. and Vasilakopoulou, K. (2021). Present and future energy consumption of buildings: challenges and opportunities towards decarbonisation. *E-Prime* 100002: https://doi.org/10.1016/j.prime.2021.100002.

2 D&R International Ltd (2011). *2011 Buildings Energy Data Book*. Washington, DC: Energy Efficiency and Renewable Energy Department.

3 Hee, W.J., Alghoul, M.A., Bakhtyar, B., Elayeb, O.K., Shameri, M.A., Alrubaih, M.S., and Sopian, K. (2015). The role of window glazing on daylighting and energy saving in buildings. *Renewable and Sustainable Energy Reviews* 42: 323–343. https://doi.org/10.1016/j.rser.2014.09.020.

4 Lampert, C.M. and Granqvist, C.G. (1990). *Large-Area Chromogenics: Materials and Devices for Transmittance Control*, vol. 10304 (ed. C.M. Lampert and C.G. Granqvist). Bellingham, WA: SPIE Optical Engineering Press.

5 Granqvist, C.G. (1990). Chromogenic materials for transmittance control of large-area windows. *Critical Reviews in Solid State and Materials Sciences* 16: 291–308. https://doi.org/10.1080/10408439008242184.

6 Feng, Y.-Q., Lv, M.-L., Yang, M., Ma, W.-X., Zhang, G., Yu, Y.-Z., Wu, Y.-Q., Li, H.-B., Liu, D.-Z., and Yang, Y.-S. (2022). Application of new energy thermochromic composite thermosensitive materials of smart windows in recent years. *Molecules* 27: 1638. https://doi.org/10.3390/molecules27051638.

7 Ke, Y., Zhou, C., Zhou, Y., Wang, S., Chan, S.H., and Long, Y. (2018). Emerging thermal-responsive materials and integrated techniques targeting the energy-efficient smart window application. *Advanced Functional Materials* 28: 1800113. https://doi.org/10.1002/adfm.201800113.

8 Zhou, Y., Cai, Y., Hu, X., and Long, Y. (2014). Temperature-responsive hydrogel with ultra-large solar modulation and high luminous transmission for "smart window" applications. *Journal of Materials Chemistry A* 2: 13550–13555. https://doi.org/10.1039/C4TA02287D.

9 Jorgenson, G.V. and Lee, J.C. (1986). Doped vanadium oxide for optical switching films. *Solar Energy Materials* 14: 205–214. https://doi.org/10.1016/0165-1633(86)90047-X.

10 Babulanam, S.M., Eriksson, T.S., Niklasson, G.A., and Granqvist, C.G. (1987). Thermochromic VO_2 films for energy-efficient windows. *Solar Energy Materials* 16: 347–363. https://doi.org/10.1016/0165-1633(87)90029-3.

11 Aburas, M., Soebarto, V., Williamson, T., Liang, R., Ebendorff-Heidepriem, H., and Wu, Y. (2019). Thermochromic smart window technologies for building application: a review. *Applied Energy* 255: 113522. https://doi.org/10.1016/j.apenergy.2019.113522.

12 Hao, Q., Li, W., Xu, H., Wang, J., Yin, Y., Wang, H., Ma, L., Ma, F., Jiang, X., Schmidt, O.G., and Chu, P.K. (2018). VO_2/TiN Plasmonic Thermochromic smart coatings for room-temperature applications. *Advanced Materials* 30: 1705421. https://doi.org/10.1002/adma.201705421.

13 Granqvist, C. and Niklasson, G. (2016). Thermochromic oxide-based thin films and nanoparticle composites for energy-efficient Glazings. *Buildings* 7: 3. https://doi.org/10.3390/buildings7010003.

14 Warwick, M.E.A., Ridley, I., and Binions, R. (2016). Thermochromic vanadium dioxide thin films prepared by electric field assisted atmospheric pressure chemical vapour deposition for intelligent glazing application and their energy

demand reduction properties. *Solar Energy Materials and Solar Cells* 157: 686–694. https://doi.org/10.1016/j.solmat.2016.07.031.

15 Hoffmann, S., Lee, E.S., and Clavero, C. (2014). Examination of the technical potential of near-infrared switching thermochromic windows for commercial building applications. *Solar Energy Materials and Solar Cells* 123: 65–80. https://doi.org/10.1016/j.solmat.2013.12.017.

16 Kim, H., Kim, Y., Kim, K.S., Jeong, H.Y., Jang, A.-R., Han, S.H., Yoon, D.H., Suh, K.S., Shin, H.S., Kim, T.Y., and Yang, W.S. (2013). Flexible thermochromic window based on hybridized VO_2/graphene. *ACS Nano* 7: 5769–5776. https://doi.org/10.1021/nn400358x.

17 Xu, F., Cao, X., Luo, H., and Jin, P. (2018). Recent advances in VO_2-based thermochromic composites for smart windows. *Journal of Materials Chemistry C* 6: 1903–1919. https://doi.org/10.1039/C7TC05768G.

18 Chang, T.-C., Cao, X., Bao, S.-H., Ji, S.-D., Luo, H.-J., and Jin, P. (2018). Review on thermochromic vanadium dioxide based smart coatings: from lab to commercial application. *Advanced Manufacturing* 6: 1–19. https://doi.org/10.1007/s40436-017-0209-2.

19 Cui, Y.Y., Ke, Y.J., Liu, C., Chen, Z., Wang, N., Zhang, L., Zhou, Y., Wang, S., Gao, Y., and Long, Y. (2018). Thermochromic VO_2 for energy-efficient smart windows. *Joule* 2: 1707–1746. https://doi.org/10.1016/j.joule.2018.06.018.

20 Wang, S., Liu, M., Kong, L., Long, Y., Jiang, X., and Yu, A. (2016). Recent progress in VO_2 smart coatings: strategies to improve the thermochromic properties. *Progress in Materials Science* 81: 1–54. https://doi.org/10.1016/j.pmatsci.2016.03.001.

21 Warwick, M.E.A. and Binions, R. (2014). Advances in thermochromic vanadium dioxide films. *Journal of Materials Chemistry A* 2: 3275–3292. https://doi.org/10.1039/C3TA14124A.

22 Gao, Y.F., Luo, H.J., Zhang, Z.T., Kang, L., Chen, Z., Du, J., Kanehira, M., and Cao, C. (2012). Nanoceramic VO_2 thermochromic smart glass: a review on progress in solution processing. *Nano Energy* 1: 221–246. https://doi.org/10.1016/j.nanoen.2011.12.002.

23 Jiang, N., Chen, S., Wang, J., He, C., Fang, K., Yin, H., Liu, Y., Li, Y., and Yu, D. (2023). Smart thermally responsive perovskite materials: thermo-chromic application and density function theory calculation. *Heliyon* e12845. https://doi.org/10.1016/j.heliyon.2023.e12845.

24 Shi, B., Duan, L., Zhao, Y., Luo, J., and Zhang, X. (2020). Semitransparent perovskite solar cells: from materials and devices to applications. *Advanced Materials* 32: 1806474. https://doi.org/10.1002/adma.201806474.

25 Liu, S., Du, Y.W., Tso, C.Y., Lee, H.H., Cheng, R., Feng, S.-P., and Yu, K.M. (2021). Organic hybrid perovskite ($MAPbI_{3-x}Cl_x$) for Thermochromic smart window with strong optical regulation ability, low transition temperature, and narrow hysteresis width. *Advanced Functional Materials* 31: 2010426. https://doi.org/10.1002/adfm.202010426.

26 Mongstad, T., Platzer-Björkman, C., Maehlen, J.P., Mooij, L.P.A., Pivak, Y., Dam, B., Marstein, E.S., Hauback, B.C., and Karazhanov, S.Z. (2011). A new thin film photochromic material: oxygen-containing yttrium hydride. *Solar Energy Materials and Solar Cells* 95: 3596–3599. https://doi.org/10.1016/j.solmat.2011.08.018.

27 Nafezarefi, F., Schreuders, H., Dam, B., and Cornelius, S. (2017). Photochromism of rare-earth metal-oxy-hydrides. *Applied Physics Letters* 111: 103903. https://doi.org/10.1063/1.4995081.

28 Montero, J., Martinsen, F.A., García-Tecedor, M., Karazhanov, S.Z., Maestre, D., Hauback, B., and Marstein, E.S. (2017). Photochromic mechanism in oxygen-containing yttrium hydride thin films: an optical perspective. *Physical Review B* 95: 201301. https://doi.org/10.1103/PhysRevB.95.201301.

29 Ferrara, M. and Bengisu, M. (2014). Materials that change color. In: *Materials that Change Color*, 9–60. Switzerland: Springer International Publishing AG.

30 Nigel Corns, S., Partington, S.M., and Towns, A.D. (2009). Industrial organic photochromic dyes. *Coloration Technology* 125: 249–261. https://doi.org/10.1111/j.1478-4408.2009.00204.x.

31 Wyszecki, J. and Stiles, W.S. (2000, 2000). *Color Science: Concepts and Methods, Quantitative Data and Formulae*, 2e. New York, NY: Wiley.

32 American Society for Testing and Materials (ASTM) (2012). *G173-03 Standard Tables of Reference Solar Spectral Irradiances: Direct Normal and Hemispherical on a 37° Tilted Surface*. Philadelphia, PA: American Society for Testing and Materials.

33 Jorgenson, G.V. (1990). Thermochromic materials and devices inorganic systems. In: (ed. C.M. Lampert and C.-G. Granqvist), 1030408. https://doi.org/10.1117/12.2283611.

34 Morin, F.J. (1959). Oxides which show a metal-to-insulator transition at the Neel temperature. *Physical Review Letters* 3: 34–36. https://doi org/10.1103/PhysRevLett.3.34.

35 Shao, Z., Cao, X., Luo, H., and Jin, P. (2018). Recent progress in the phase-transition mechanism and modulation of vanadium dioxide materials. *NPG Asia Materials* 10: 581–605. https://doi.org/10.1038/s41427-018-0061-2.

36 Basu, R., Srihari, V., Sardar, M., Srivastava, S.K., Bera, S., and Dhara, S. (2020). Probing phase transition in VO_2 with the novel observation of low-frequency collective spin excitation. *Scientific Reports* 10: 1977. https://doi.org/10.1038/s41598-020-58813-x.

37 Liebsch, A., Ishida, H., and Bihlmayer, G. (2005). Coulomb correlations and orbital polarization in the metal-insulator transition of VO_2. *Physical Review B* 71: 085109. https://doi.org/10.1103/PhysRevB.71.085109.

38 Wentzcovitch, R.M., Schulz, W.W., and Allen, P.B. (1994). VO_2: peierls or Mott-Hubbard? A view from band theory. *Physical Review Letters* 72: 3389–3392. https://doi.org/10.1103/PhysRevLett.72.3389.

39 Zhang, Y., Tso, C.Y., Iñigo, J.S., Liu, S., Miyazaki, H., Chao, C.Y.H., and Yu, K.M. (2019). Perovskite thermochromic smart window: advanced optical properties and low transition temperature. *Applied Energy* 254: 113690. https://doi.org/10.1016/j.apenergy.2019.113690.

40 Rosales, B.A., Mundt, L.E., Allen, T.G., Moore, D.T., Prince, K.J., Wolden, C.A., Rumbles, G., Schelhas, L.T., and Wheeler, L.M. (2020). Reversible multicolor chromism in layered formamidinium metal halide perovskites. *Nature Communications* 11: 5234. https://doi.org/10.1038/s41467-020-19009-z.

41 Lin, J., Lai, M., Dou, L., Kley, C.S., Chen, H., Peng, F., Sun, J., Lu, D., Hawks, S.A., Xie, C., Cui, F., Alivisatos, A.P., Limmer, D.T., and Yang, P. (2018). Thermochromic halide perovskite solar cells. *Nature Materials* 17: 261–267. https://doi.org/10.1038/s41563-017-0006-0.

42 Fisher, B., Patlagan, L., Chashka, K.B., Makarov, C., and Reisner, G.M. (2016). V_3O_5: insulator-metal transition and electric-field-induced resistive-switching. *Applied Physics Letters* 109: 103501. https://doi.org/10.1063/1.4962334.

43 Rúa, A., Díaz, R.D., Kumar, N., Lysenko, S., and Fernández, F.E. (2017). Metal-insulator transition and nonlinear optical response of sputter-deposited V_3O_5 thin films. *Journal of Applied Physics* 121: 235302. https://doi.org/10.1063/1.4986486.

44 Yang, Z., Ko, C., and Ramanathan, S. (2011). Oxide electronics utilizing ultrafast metal-insulator transitions. *Annual Review of Materials Research* 41: 337–367. https://doi.org/10.1146/annurev-matsci-062910-100347.

45 Gibin, I.S. and Kotlyar, P.E. (2021). Experimental dynamics of temperature-initiated metal–insulator transition in vanadium dioxide. *Journal of Communications Technology and Electronics* 66: 340–347. https://doi.org/10.1134/S1064226921030050.

46 Mlyuka, N.R., Niklasson, G.A., and Granqvist, C.G. (2009). Thermochromic multilayer films of VO_2 and TiO_2 with enhanced transmittance. *Solar Energy Materials and Solar Cells* 93: 1685–1687. https://doi.org/10.1016/j.solmat.2009.03.021.

47 Li, S.Y., Niklasson, G.A., and Granqvist, C.G. (2012). Thermochromic fenestration with VO_2-based materials: three challenges and how they can be met. *Thin Solid Films* 520: 3823–3828. https://doi.org/10.1016/j.tsf.2011.10.053.

48 Li, S.Y., Niklasson, G.A., and Granqvist, C.G. (2010). Nanothermochromics: calculations for VO_2 nanoparticles in dielectric hosts show much improved luminous transmittance and solar energy transmittance modulation. *Journal of Applied Physics* 108: https://doi.org/10.1063/1.3487980.

49 Wang, S., Li, C., Tian, S., Liu, B., and Zhao, X. (2020). Facile synthesis of VO_2(D) and its transformation to VO_2(M) with enhanced thermochromic properties for smart windows. *Ceramics International* 46: 14739–14746. https://doi.org/10.1016/j.ceramint.2020.02.278.

50 Kim, Y., Yu, S., Park, J., Yoon, D., Dayaghi, A.M., Kim, K.J., Ahn, J.S., and Son, J. (2018). High-throughput roll-to-roll fabrication of flexible thermochromic coatings for smart windows with VO_2 nanoparticles. *Journal of Materials Chemistry C* 6: 3451–3458. https://doi.org/10.1039/C7TC05876D.

51 Xu, C., Liu, G., Li, M., Li, K., Luo, Y., Long, Y., and Li, G. (2020). Optical switching and nanothermochromic studies of VO_2(M) nanoparticles prepared by mild thermolysis method. *Materials and Design* 187: 108396. https://doi.org/10.1016/j.matdes.2019.108396.

52 Zhou, Y., Huang, A., Li, Y., Ji, S., Gao, Y., and Jin, P. (2013). Surface plasmon resonance induced excellent solar control for VO_2@SiO_2 nanorods-based thermochromic foils. *Nanoscale* 5: 9208. https://doi.org/10.1039/c3nr02221h.

53 Zhu, J., Zhou, Y., Wang, B., Zheng, J., Ji, S., Yao, H., Luo, H., and Jin, P. (2015). Vanadium dioxide nanoparticle-based Thermochromic smart coating: high luminous transmittance, excellent solar regulation efficiency, and near room temperature phase transition. *ACS Applied Materials & Interfaces* 7: 27796–27803. https://doi.org/10.1021/acsami.5b09011.

54 Murillo-Quirós, N., Vega-Garita, V., Carmona-Calvo, A., Rojas-González, E.A., Starbird-Perez, R., and Avendaño-Soto, E. (2022). Toward thermochromic VO_2 nanoparticles polymer films based smart windows designed for tropical climates. *Polymers (Basel)* 14: 4250. https://doi.org/10.3390/polym14194250.

55 Zhou, Y., Cai, Y., Hu, X., and Long, Y. (2015). VO_2/hydrogel hybrid nanothermochromic material with ultra-high solar modulation and luminous transmission. *Journal of Materials Chemistry A* 3: 1121–1126. https://doi.org/10.1039/C4TA05035E.

56 Zhou, Y., Zhao, S., Qi, S., Liu, Y., Chen, N., Wang, L., Niu, Y., and Xu, Q. (2023). VO_2 nanoparticles-based thermochromic smart windows. *ChemistrySelect* 8: e202302259. https://doi.org/https://doi.org/10.1002/slct.202302259.

57 Goodenough, J.B. (1971). The two components of the crystallographic transition in VO_2. *Journal of Solid State Chemistry* 3: 490–500. https://doi.org/10.1016/0022-4596(71)90091-0.

58 Reyes, J.M., Lynch, G.F., Sayer, M., McBride, S.L., and Hutchison, T.S. (1972). Electrical, optical, magnetic resonance and microhardness properties of tungsten-doped VO_2. *Journal of the American Ceramic Society* 69–75.

59 Hörlin, T., Niklewski, T., and Nygren, M. (1972). Electrical and magnetic properties of $V_{1-x}W_xO_2$, $0 \leq x \leq 0.060$. *Materials Research Bulletin* 7: 1515–1524. https://doi.org/https://doi.org/10.1016/0025-5408(72)90189-4.

60 Jin, P., Tazawa, M., Ikeyama, M., Tanemura, S., Macák, K., Wang, X., Olafsson, S., and Helmersson, U. (1999). Growth and characterization of epitaxial films of tungsten-doped vanadium oxides on sapphire (110) by reactive magnetron sputtering. *Journal of Vacuum Science & Technology A: Vacuum, Surfaces, and Films* 17: 1817–1821. https://doi.org/10.1116/1.581896.

61 Burkhardt, W., Christmann, T., Meyer, B.K., Niessner, W., Schalch, D., and Scharmann, A. (1999). W- and F-doped VO_2 films studied by photoelectron spectrometry. *Thin Solid Films* 345: 229–235. https://doi.org/10.1016/S0040-6090(98)01406-0.

62 Bleu, Y., Bourquard, F., Barnier, V., Loir, A.-S., Garrelie, F., and Donnet, C. (2023). Towards room temperature phase transition of W-doped VO_2 thin films deposited by pulsed laser deposition: Thermochromic, surface, and structural analysis. *Materials* 16: 461. https://doi.org/10.3390/ma16010461.

63 Greenberg, C.B. (1983). Undoped and doped VO_2 films grown from $VO(OC_3H_7)_3$. *Thin Solid Films* 110: 73–82. https://doi.org/10.1016/0040-6090(83)90175-X.

64 Mlyuka, N.R., Niklasson, G.A., and Granqvist, C.G. (2009). Mg doping of thermochromic VO_2 films enhances the optical transmittance and decreases the metal-insulator transition temperature. *Applied Physics Letters* 95: 171909. https://doi.org/10.1063/1.3229949.

65 Shen, N., Chen, S., Shi, R., Niu, S., Amini, A., and Cheng, C. (2021). Phase transition hysteresis of tungsten doped VO_2 synergistically boosts the function of smart windows in ambient conditions. *ACS Applied Electronic Materials* 3: 3648–3656. https://doi.org/10.1021/acsaelm.1c00550.

66 Gläser, H.J. (2000). *Large Area Glass Coating*. 1st English ed. Dresden: Von Ardenne Anlagentechnik.

67 Dietrich, M.K., Kramm, B.G., Becker, M., Meyer, B.K., Polity, A., and Klar, P.J. (2015). Influence of doping with alkaline earth metals on the optical properties of thermochromic VO_2. *Journal of Applied Physics* 117: 185301. https://doi.org/10.1063/1.4919433.

68 Li, S.-Y., Mlyuka, N.R., Primetzhofer, D., Hallén, A., Possnert, G., Niklasson, G.A., and Granqvist, C.G. (2013). Bandgap widening in thermochromic Mg-doped VO_2 thin films: quantitative data based on optical absorption. *Applied Physics Letters* 103: 161907. https://doi.org/10.1063/1.4826444.

69 Hu, S., Li, S.-Y., Ahuja, R., Granqvist, C.G., Hermansson, K., Niklasson, G.A., and Scheicher, R.H. (2012). Optical properties of Mg-doped VO_2: absorption measurements and hybrid functional calculations. *Applied Physics Letters* 101: 201902. https://doi.org/10.1063/1.4766167.

70 Kang, J., Liu, J., Shi, F., Dong, Y., and Jiang, S. (2021). The thermochromic characteristics of Zn-doped VO_2 that were prepared by the hydrothermal and post-annealing process and their polyurethane composite films. *Ceramics International* 47: 15631–15638. https://doi.org/10.1016/j.ceramint.2021.02.133.

71 Jiang, M., Bao, S., Cao, X., Li, Y., Li, S., Zhou, H., Luo, H., and Jin, P. (2014). Improved luminous transmittance and diminished yellow color in VO_2 energy efficient smart thin films by Zn doping. *Ceramics International* 40: 6331–6334. https://doi.org/10.1016/j.ceramint.2013.10.083.

72 Wang, S., Wei, W., Huang, T., Yuan, M., Yang, Y., Yang, W., Zhang, R., Zhang, T., Chen, Z., Chen, X., Shen, G., and Dai, N. (2019). Al-doping-induced VO_2 (B) phase in VO_2 (M) toward smart optical thin films with modulated Δ_{vis} and Δ_c. *Advanced Engineering Materials* 21: 1900947. https://doi.org/10.1002/adem.201900947.

73 Habtemariam, A.B., Simo, A., and Maaza, M. (2020). Effect of process parameters on the phase transition property of molybdenum-doped vanadium dioxide nanorods. *Applied Physics A* 126: 17. https://doi.org/10.1007/s00339-019-3190-8.

74 Hanlon, T.J., Coath, J.A., and Richardson, M.A. (2003). Molybdenum-doped vanadium dioxide coatings on glass produced by the aqueous sol–gel method. *Thin Solid Films* 436: 269–272. https://doi.org/10.1016/S0040-6090(03)00602-3.

75 Batista, C., Ribeiro, R.M., and Teixeira, V. (2011). Synthesis and characterization of VO_2-based thermochromic thin films for energy-efficient windows. *Nanoscale Research Letters* 6: 301. https://doi.org/10.1186/1556-276X-6-301.

76 Piccirillo, C., Binions, R., and Parkin, I.P. (2007). Nb-doped VO_2 thin films prepared by aerosol-assisted chemical vapour deposition. *European Journal of Inorganic Chemistry* 2007: 4050–4055. https://doi.org/10.1002/ejic.200700284.

77 Jiang, J.C., Bárta, T., Vlček, J., Houška, J., and Meletis, E.I. (2021). Microstructure of high-performance thermochromic ZrO_2/V0.984W0.016O2/ZrO_2 coating with a low transition temperature (22 °C) prepared on flexible glass. *Surface and Coating Technology* 424: 127654. https://doi.org/10.1016/j.surfcoat.2021.127654.

78 Shen, N., Chen, S., Chen, Z., Liu, X., Cao, C., Dong, B., Luo, H., Liu, J., and Gao, Y. (2014). The synthesis and performance of Zr-doped and W–Zr-co-doped VO_2 nanoparticles and derived flexible foils. *Journal of Materials Chemistry A* 2: 15087–15093. https://doi.org/10.1039/C4TA02880E.

79 Wang, N., Duchamp, M., Dunin-Borkowski, R.E., Liu, S., Zeng, X.T., Cao, X., and Long, Y. (2016). Terbium-doped VO_2 thin films: reduced phase transition temperature and largely enhanced luminous transmittance. *Langmuir* 32: 759–764. https://doi.org/10.1021/acs.langmuir.5b04212.

80 Geng, X., Chang, T., Fan, J., Wang, Y., Wang, X., Sun, Y., Selvarajan, P., Liu, C., Lin, C.-H., Wang, X., Yang, J., Cheng, Z., Kalantar-Zadeh, K., Cao, X., Wang, D., Vinu, A., Yi, J., and Wu, T. (2022). Tuning phase transition and thermochromic properties of vanadium dioxide thin films via cobalt doping. *ACS Applied Materials & Interfaces* 14: 19736–19746. https://doi.org/10.1021/acsami.2c03113.

81 Hu, L., Tao, H., Chen, G., Pan, R., Wan, M., Xiong, D., and Zhao, X. (2016). Porous W-doped VO_2 films with simultaneously enhanced visible transparency and thermochromic properties. *Journal of Sol-Gel Science and Technology* 77: 85–93. https://doi.org/10.1007/s10971-015-3832-z.

82 Zhang, Y., Zhang, X., Huang, Y., Huang, C., Niu, F., Meng, C., and Tan, X. (2014). One-step hydrothermal conversion of VO_2(B) into W-doped VO_2(M) and its phase transition and optical switching properties. *Solid State Communications* 180: 24–27. https://doi.org/10.1016/j.ssc.2013.10.028.

83 Burkhardt, W., Christmann, T., Franke, S., Kriegseis, W., Meister, D., Meyer, B.K., Niessner, W., Schalch, D., and Scharmann, A. (2002). Tungsten and fluorine co-doping of VO_2 films. *Thin Solid Films* 402: 226–231. https://doi.org/10.1016/S0040-6090(01)01603-0.

84 Khan, K.A. and Granqvist, C.G. (1989). Thermochromic sputter-deposited vanadium oxyfluoride coatings with low luminous absorptance. *Applied Physics Letters* 55: 4–6. https://doi.org/10.1063/1.102388.

85 Li, S.-Y., Niklasson, G.A., and Granqvist, C.G. (2011). Nanothermochromics with VO_2-based core-shell structures: calculated luminous and solar optical properties. *Journal of Applied Physics* 109: 113515. https://doi.org/10.1063/1.3592350.

86 Li, S.-Y., Niklasson, G.A., and Granqvist, C.G. (2014). Thermochromic undoped and mg-doped VO_2 thin films and nanoparticles: optical properties and performance limits for energy efficient windows. *Journal of Applied Physics* 115: 053513. https://doi.org/10.1063/1.4862930.

87 Laaksonen, K., Li, S.-Y., Puisto, S.R., Rostedt, N.K.J., Ala-Nissila, T., Granqvist, C.G., Nieminen, R.M., and Niklasson, G.A. (2014). Nanoparticles of TiO_2 and VO_2 in dielectric media: conditions for low optical scattering, and comparison between effective medium and four-flux theories. *Solar Energy Materials and Solar Cells* 130: 132–137. https://doi.org/10.1016/j.solmat.2014.06.036.

88 Ke, Y., Wen, X., Zhao, D., Che, R., Xiong, Q., and Long, Y. (2017). Controllable fabrication of two-dimensional patterned VO_2 nanoparticle, nanodome, and nanonet arrays with tunable temperature-dependent localized surface plasmon resonance. *ACS Nano* 11: 7542–7551. https://doi.org/10.1021/acsnano.7b02232.

89 Lopez, R., Boatner, L.A., Haynes, T.E., Feldman, L.C., and Haglund, R.F. Jr. (2002). Synthesis and characterization of size-controlled vanadium dioxide nanocrystals in a fused silica matrix. *Journal of Applied Physics* 92: 4031–4036. https://doi.org/10.1063/1.1503391.

90 Ke, Y., Zhang, Q., Wang, T., Wang, S., Li, N., Lin, G., Liu, X., Dai, Z., Yan, J., Yin, J., Magdassi, S., Zhao, D., and Long, Y. (2020). Cephalopod-inspired versatile design based on plasmonic VO_2 nanoparticle for energy-efficient mechano-thermochromic windows. *Nano Energy* 73: 104785. https://doi.org/10.1016/j.nanoen.2020.104785.

91 Gentle, A., Maaroof, A.I., and Smith, G.B. (2007). Nanograin VO_2 in the metal phase: a plasmonic system with falling dc resistivity as temperature rises. *Nanotechnology* 18: 025202. https://doi.org/10.1088/0957-4484/18/2/025202.

92 Mjejri, I., Rougier, A., and Gaudon, M. (2017). Low-cost and facile synthesis of the vanadium oxides V_2O_3, VO_2, and V_2 O_5 and their magnetic, Thermochromic and electrochromic properties. *Inorganic Chemistry* 56: 1734–1741. https://doi.org/10.1021/acs.inorgchem.6b02880.

93 Ji, Y.-X., Li, S.-Y., Niklasson, G.A., and Granqvist, C.G. (2014). Durability of thermochromic VO_2 thin films under heating and humidity: effect of Al oxide top coatings. *Thin Solid Films* 562: 568–573. https://doi.org/10.1016/j.tsf.2014.03.043.

94 Lykissa, I., Li, S.-Y., Ramzan, M., Chakraborty, S., Ahuja, R., Granqvist, C.G., and Niklasson, G.A. (2014). Electronic density-of-states of amorphous vanadium pentoxide films: electrochemical data and density functional theory calculations. *Journal of Applied Physics* 115: 183701. https://doi.org/10.1063/1.4875636.

95 Vernardou, D., Louloudakis, D., Spanakis, E., Katsarakis, N., and Koudoumas, E. (2014). Thermochromic amorphous VO_2 coatings grown by APCVD using a single-precursor. *Solar Energy Materials and Solar Cells* 128: 36–40. https://doi.org/10.1016/j.solmat.2014.04.033.

96 Loquai, S., Baloukas, B., Klemberg-Sapieha, J.E., and Martinu, L. (2017). HiPIMS-deposited thermochromic VO_2 films with high environmental stability. *Solar Energy Materials and Solar Cells* 160: 217–224. https://doi.org/10.1016/j.solmat.2016.10.038.

97 Ji, Y.-X., Niklasson, G.A., and Granqvist, C.G. (2014). Durability of VO_2-based thin films at elevated temperature: towards thermochromic fenestration. *Journal of Physics Conference Series* 559: 012005. https://doi.org/10.1088/1742-6596/559/1/012005.

98 Chang, T., Cao, X., Dedon, L.R., Long, S., Huang, A., Shao, Z., Li, N., Luo, H., and Jin, P. (2018). Optical design and stability study for ultrahigh-performance and long-lived vanadium dioxide-based thermochromic coatings. *Nano Energy* 44: 256–264. https://doi.org/10.1016/j.nanoen.2017.11.061.

99 Tong, K., Li, R., Zhu, J., Yao, H., Zhou, H., Zeng, X., Ji, S., and Jin, P. (2017). Preparation of VO_2/Al-O core-shell structure with enhanced weathering resistance for smart window. *Ceramics International* 43: 4055–4061. https://doi.org/10.1016/j.ceramint.2016.11.181.

100 Dillon, R.O., Le, K., and Ianno, N. (2001). Thermochromic VO_2 sputtered by control of a vanadium-oxygen emission ratio. *Thin Solid Films* 398–399: 10–16. https://doi.org/10.1016/S0040-6090(01)01296-2.

101 Zhang, Z., Gao, Y., Chen, Z., Du, J., Cao, C., Kang, L., and Luo, H. (2010). Thermochromic VO_2 thin films: solution-based processing, improved optical properties, and lowered phase transformation temperature. *Langmuir* 26: 10738–10744. https://doi.org/10.1021/la100515k.

102 Kang, L., Gao, Y., Luo, H., Wang, J., Zhu, B., Zhang, Z., Du, J., Kanehira, M., and Zhang, Y. (2011). Thermochromic properties and low emissivity of ZnO:Al/VO_2 double-layered films with a lowered phase transition temperature. *Solar Energy Materials and Solar Cells* 95: 3189–3194. https://doi.org/10.1016/j.solmat.2011.06.047.

103 Kang, L., Gao, Y., Chen, Z., Du, J., Zhang, Z., and Luo, H. (2010). Pt/VO_2 double-layered films combining thermochromic properties with low emissivity. *Solar Energy Materials and Solar Cells* 94: 2078–2084. https://doi.org/10.1016/j.solmat.2010.06.023.

104 Zhang, Z., Gao, Y., Luo, H., Kang, L., Chen, Z., Du, J., Kanehira, M., Zhang, Y., and Wang, Z.L. (2011). Solution-based fabrication of vanadium dioxide on F:SnO_2 substrates with largely enhanced thermochromism and low-emissivity for energy-saving applications. *Energy & Environmental Science* 4: 4290. https://doi.org/10.1039/c1ee02092g.

105 Wang, S., Jiang, T., Meng, Y., Yang, R., Tan, G., and Long, Y. (2021). Scalable thermochromic smart windows with passive radiative cooling regulation. *Science* 374: 1501–1504. https://doi.org/10.1126/science.abg0291.

106 Houser, K.W. and Esposito, T. (2021). Human-centric lighting: foundational considerations and a five-step design process. *Frontiers in Neurology* 12: https://doi.org/10.3389/fneur.2021.630553.

107 Shi, R., Shen, N., Wang, J., Wang, W., Amini, A., Wang, N., and Cheng, C. (2019). Recent advances in fabrication strategies, phase transition modulation, and advanced applications of vanadium dioxide. *Applied Physics Reviews* 6: 011312. https://doi.org/10.1063/1.5087864.

108 Zeng, W., Chen, N., and Xie, W. (2020). Research progress on the preparation methods for VO_2 nanoparticles and their application in smart windows. *CrystEngComm* 22: 851–869. https://doi.org/10.1039/C9CE01655D.

109 Kim, J. and Paik, T. (2021). Recent advances in fabrication of flexible, thermochromic vanadium dioxide films for smart windows. *Nanomaterials* 11: 2674. https://doi.org/10.3390/nano11102674.

110 Zhang, H., Cherng, J.-S., and Chen, Q. (2019). Recent progress on high power impulse magnetron sputtering (HiPIMS): the challenges and applications in fabricating VO_2 thin film. *AIP Advances* 9: 035242. https://doi.org/10.1063/1.5084031.

111 Vlček, J., Kolenatý, D., Houška, J., Kozák, T., and Čerstvý, R. (2017). Controlled reactive HiPIMS—effective technique for low-temperature (300 °C) synthesis of VO_2 films with semiconductor-to-metal transition. *Journal of Physics D: Applied Physics* 50: 38LT01. https://doi.org/10.1088/1361-6463/aa8356.

112 Shigesato, Y., Enomoto, M., and Odaka, H. (2000). Thermochromic VO_2 films deposited by RF magnetron sputtering using V_2O_3 or V_2O_5 targets. *Japanese Journal of Applied Physics* 39: 6016. https://doi.org/10.1143/JJAP.39.6016.

113 Ho, H.-C., Lai, Y.-C., Chen, K., Dao, T.D., Hsueh, C.-H., and Nagao, T. (2019). High quality thermochromic VO_2 films prepared by magnetron sputtering using V_2O_5 target with in situ annealing. *Applied Surface Science* 495: 143436. https://doi.org/10.1016/j.apsusc.2019.07.178.

114 Wang, S., Owusu, K.A., Mai, L., Ke, Y., Zhou, Y., Hu, P., Magdassi, S., and Long, Y. (2018). Vanadium dioxide for energy conservation and energy storage applications: synthesis and performance improvement. *Applied Energy* 211: 200–217. https://doi.org/10.1016/j.apenergy.2017.11.039.

115 Long, Y. and Gao, Y. (ed.) (2021). *Vanadium Dioxide-Based Thermochromic Smart Windows*. Jenny Stanford Publishing. https://doi.org/10.1201/9781003177319.

116 Seyfouri, M.M. and Binions, R. (2017). Sol-gel approaches to thermochromic vanadium dioxide coating for smart glazing application. *Solar Energy Materials and Solar Cells* 159: 52–65. https://doi.org/10.1016/j.solmat.2016.08.035.

117 Li, M., Magdassi, S., Gao, Y., and Long, Y. (2017). Hydrothermal synthesis of VO_2 polymorphs: advantages, challenges and prospects for the application of energy efficient smart windows. *Small* 13: 1701147. https://doi.org/10.1002/smll.201701147.

118 Kusano, E., Theil, J.A., and Thornton, J.A. (1988). Deposition of vanadium oxide films by direct-current magnetron reactive sputtering. *Journal of Vacuum Science & Technology A: Vacuum, Surfaces, and Films* 6: 1663–1667. https://doi.org/10.1116/1.575304.

119 Ping Jin, P.J. and Sakae Tanemura, S.T. (1994). Formation and thermochromism of VO_2 films deposited by RF magnetron sputtering at low substrate temperature. *Japanese Journal of Applied Physics* 33: 1478. https://doi.org/10.1143/JJAP.33.1478.

120 Li, B., Tian, S., Qian, J., Wu, S., Liu, B., and Zhao, X. (2023). In situ synthesis of highly dispersed VO_2(M) nanoparticles on glass surface for energy efficient smart windows. *Ceramics International* 49 (2): 2310–2318. https://doi.org/10.1016/j.ceramint.2022.09.199.

121 Khairy, K.T., Song, Y., Yoon, J.-H., Montero, J., Österlund, L., Kim, S., and Song, P. (2023). Thermochromic properties of vanadium oxide thin films prepared by reactive magnetron sputtering at different oxygen concentrations. *Vacuum* 210: 111887. https://doi.org/10.1016/j.vacuum.2023.111887.

122 Li, S.-Y., Namura, K., Suzuki, M., Niklasson, G.A., and Granqvist, C.G. (2013). Thermochromic VO_2 nanorods made by sputter deposition: growth conditions and optical modeling. *Journal of Applied Physics* 114: 033516. https://doi.org/10.1063/1.4813876.

123 Popuri, S.R., Miclau, M., Artemenko, A., Labrugere, C., Villesuzanne, A., and Pollet, M. (2013). Rapid hydrothermal synthesis of VO_2 (B) and its conversion to thermochromic VO_2 (M1). *Inorganic Chemistry* 52: 4780–4785. https://doi.org/10.1021/ic301201k.

124 Son, J.-H., Wei, J., Cobden, D., Cao, G., and Xia, Y. (2010). Hydrothermal synthesis of monoclinic VO_2 micro- and nanocrystals in one step and their use in fabricating inverse opals. *Chemistry of Materials* 22: 3043–3050. https://doi.org/10.1021/cm903727u.

125 Munozrojas, D. and Baudrin, E. (2007). Synthesis and electroactivity of hydrated and monoclinic rutile-type nanosized VO_2. *Solid State Ionics* 178: 1268–1273. https://doi.org/10.1016/j.ssi.2007.07.003.

126 Shi, R., Wang, J., Cai, X., Zhang, L., Chen, P., Liu, S., Zhang, L., Ouyang, W., Wang, N., and Cheng, C. (2017). Axial modulation of metal–insulator phase transition of VO_2 nanowires by graded doping engineering for optically readable thermometers. *The Journal of Physical Chemistry C* 121: 24877–24885. https://doi.org/10.1021/acs.jpcc.7b08946.

127 Li. B., Tian, S., Zhou, L., Wu, S., Ma, T., He, G., Liu, B., and Zhao, X. (2023). Fully Discrete VO2 Particulate Film with Ultra-High Transmittance and Excellent Thermochromic Performance. *Adv. Optical Mater.* 12: 2302042. https://doi.org/10.1002/adom.202302042.

128 Aijaz, A., Ji, Y.-X., Montero, J., Niklasson, G.A., Granqvist, C.G., and Kubart, T. (2016). Low-temperature synthesis of thermochromic vanadium dioxide thin films by reactive high power impulse magnetron sputtering. *Solar Energy Materials and Solar Cells* 149: 137–144. https://doi.org/10.1016/j.solmat.2016.01.009.

129 Juan, P.-C., Lin, K.-C., Lin, C.-L., and Lin, W.-F. (2020). Effect of duty cycle on the thermochromic properties of VO_2 thin-film fabricated by high power impulse magnetron sputtering. *Journal of Vacuum Science & Technology A* 38: 063403. https://doi.org/10.1116/6.0000409.

130 Rebohle, L., Prucnal, S., and Reichel, D. (2019). *Flash Lamp Annealing*, vol. 288. Cham: Springer International Publishing https://doi.org/10.1007/978-3-030-23299-3.

131 Zheng, J., Bao, S., and Jin, P. (2015). TiO_2(R)/VO_2(M)/TiO_2(a) multilayer film as smart window: combination of energy-saving, antifogging and self-cleaning functions. *Nano Energy* 11: 136–145. https://doi.org/10.1016/j.nanoen.2014.09.023.

132 Powell, M.J., Quesada-Cabrera, R., Taylor, A., and Teixeira, D. (2016). Intelligent multifunctional $VO_2/SiO_2/TiO_2$ coatings for self-cleaning, energy-saving window panels. *Chemistry of Materials* 28: 1369–1376. https://doi.org/10.1021/acs.chemmater.5b04419.

133 Choi, Y., Sim, D.M., Hur, Y.H., Han, H.J., and Jung, Y.S. (2018). Synthesis of colloidal VO_2 nanoparticles for thermochromic applications. *Solar Energy Materials and Solar Cells* 176: 266–272. https://doi.org/10.1016/j.solmat.2017.12.008.

134 Malarde, D., Johnson, I.D., Godfrey, I.J., Powell, M.J., Cibin, G., Quesada-Cabrera, R., Darr, J.A., Carmalt, C.J., Sankar, G., Parkin, I.P., and Palgrave, R.G. (2018). Direct and continuous hydrothermal flow synthesis of thermochromic phase pure monoclinic VO_2 nanoparticles. *Journal of Materials Chemistry C* 6: 11731–11739. https://doi.org/10.1039/C8TC00859K.

135 Zhang, L., Xia, F., Yao, J., Zhu, T., Xia, H., Yang, G., Liu, B., and Gao, Y. (2020). Facile synthesis, formation mechanism and thermochromic properties of W-doped VO_2 (M) nanoparticles for smart window applications. *Journal of Materials Chemistry C* 8: 13396–13404. https://doi.org/10.1039/D0TC03436C.

136 Powell, M.J., Marchand, P., Denis, C.J., Bear, J.C., Darr, J.A., and Parkin, I.P. (2015). Direct and continuous synthesis of VO_2 nanoparticles. *Nanoscale* 7: 18686–18693. https://doi.org/10.1039/C5NR04444H.

137 Zhou, J., Gao, Y., Liu, X., Chen, Z., Dai, L., Cao, C., Luo, H., Kanahira, M., Sunc, C., and Yan, L. (2013). Mg-doped VO_2 nanoparticles: hydrothermal synthesis, enhanced visible transmittance and decreased metal–insulator transition temperature. *Physical Chemistry Chemical Physics* 15: 7505. https://doi.org/10.1039/c3cp50638j.

138 Zhang, L., Wang, A., Zhu, T., Chen, Z., Wu, Y, and Gao, Y. (2020). Transparent wood composites fabricated by impregnation of epoxy resin and W-doped VO_2 nanoparticles for application in energy-saving windows. *ACS Applied Materials & Interfaces* 12: 34777–34783. https://doi.org/10.1021/acsami.0c06494.

139 Wang, N., Duchamp, M., Xue, C., Dunin-Borkowski, R.E., Liu, G., and Long, Y. (2016). Single-crystalline W-doped VO_2 Nanobeams with highly reversible electrical and Plasmonic responses near room temperature. *Advanced Materials Interfaces* 3: 1600164. https://doi.org/10.1002/admi.201600164.

140 Li, D., Li, M., Pan, J., Luo, Y., Wu, H., Zhang, Y., and Li, G. (2014). Hydrothermal synthesis of Mo-doped VO_2/TiO_2 composite nanocrystals with enhanced thermochromic performance. *ACS Applied Materials & Interfaces* 6: 6555–6561. https://doi.org/10.1021/am500135d.

141 Calvi, L., van Geijn, R., Leufkens, L., Habets, R., Gurunatha, K.L., Stout, K., Mann, D., Papakonstantinou, I., Parkin, I.P., Elen, K., Hardy, A., van Bael, M.K., and Buskens, P. (2022). The impact of bead milling on the thermodynamics and kinetics of the structural phase transition of VO_2 particulate materials and their potential for use in thermochromic glazing. *Solar Energy Materials and Solar Cells* 242: 111783. https://doi.org/10.1016/j.solmat.2022.111783.

142 Ke, Y., Balin, I., Wang, N., Lu, Q., Tok, A.I.Y., White, T.J., Magdassi, S., Abdulhalim, I., and Long, Y. (2016). Two-dimensional SiO_2/VO_2 photonic crystals with statically visible and dynamically infrared modulated for smart window deployment. *ACS Applied Materials & Interfaces* 8: 33112–33120. https://doi.org/10.1021/acsami.6b12175.

143 Gao, Y., Wang, S., Luo, H., Dai, L., Cao, C., Liu, Y., Chen, Z., and Kanehira, M. (2012). Enhanced chemical stability of VO_2 nanoparticles by the formation of SiO_2/VO_2 core/shell structures and the application to transparent and flexible VO_2-based composite foils with excellent thermochromic properties for solar heat control. *Energy & Environmental Science* 5: 6104. https://doi.org/10.1039/c2ee02803d.

144 Wang, M., Tian, J., Zhang, H., Shi, X., Chen, Z., Wang, Y., Ji, A., and Gao, Y. (2016). Novel synthesis of pure VO_2@SiO_2 core@shell nanoparticles to improve the optical and anti-oxidant properties of a VO_2 film. *RSC Advances* 6: 108286–108289. https://doi.org/10.1039/C6RA20636K.

145 Li, Y., Ji, S., Gao, Y., Luo, H., and Kanehira, M. (2013). Core-shell VO_2@TiO_2 nanorods that combine thermochromic and photocatalytic properties for application as energy-saving smart coatings. *Scientific Reports* 3: 1370. https://doi.org/10.1038/srep01370.

146 Chen, Y., Zeng, X., Zhu, J., Li, R., Yao, H., Cao, X., Ji, S., and Jin, P. (2017). High performance and enhanced durability of thermochromic films using VO_2 @ZnO core–shell nanoparticles. *ACS Applied Materials & Interfaces* 9: 27784–27791. https://doi.org/10.1021/acsami.7b08889.

147 Li, D., Deng, S., Zhao, Z., Yang, J., Wang, B., Li, J., and Jin, H. (2022). VO_2(M)@SnO_2 core–shell nanoparticles: improved chemical stability and thermochromic property rendered by SnO_2 shell. *Applied Surface Science* 598: 153741. https://doi.org/10.1016/j.apsusc.2022.153741.

148 Clavero, C., Slack, J.L., and Anders, A. (2013). Size and composition-controlled fabrication of thermochromic metal oxide nanocrystals. *Journal of Physics D: Applied Physics* 46: 362001. https://doi.org/10.1088/0022-3727/46/36/362001.

149 Trotter, D.M. (1991). Photochromic and photosensitive glass. *Scientific American* 264: 124–129.

150 De Sousa, F.B., Alexis, F., and Giordani, S. (2021). Editorial: photochromic materials: design and applications. *Frontiers in Materials* 8: https://doi.org/10.3389/fmats.2021.720172.

151 Towns, A.D. (2016). Colorant, photochromic. *Encyclopedia of Color Science and Technology* 447–455.

152 Towns, A.D. (2016). Industrial photochromism. In: *Applied Photochemistry: When Light Meets Molecules*, 227–279. Springer. https://doi.org/10.1007/978-3-319-31671-0_5.

153 Araujo, R.J. (2003). Photochromic glasses. In: *Encyclopedia of Physical Science and Technology*, 49–56. Elsevier. https://doi.org/10.1016/B0-12-227410-5/00567-6.

154 Fries, K., Fink-Straube, C., Mennig, M., and Schmidt, H. (2004). Photochromic coatings. In: *Sol-Gel Technologies for Glass Producers and Users*, 251–259. Boston, MA: Springer. https://doi.org/10.1007/978-0-387-88953-5_33.

155 Landa, K.A., Landa, L.M., Longobardo, A.V., and Thomsen, S. (2004). Photochromic float glasses and methods of making the same. US 6711917B2.

156 Hoffmann, H.J. (1990). *Photochromic Glass* (ed. C.M. Lampert and C.-G. Granqvist), 1030405. https://doi.org/10.1117/12.2283608.

157 Vollath, D. (2013). *Nanomaterials: An Introduction to Synthesis, Properties and Applications*, 2e. Wiley-VCH Verlag GmbH & Co: Weinheim, Germany.

158 Bourdin, M., Salek, G., Fargues, A., Messaddeq, S., Messaddeq, Y., Cardinala, T., and Gaudon, M. (2020). Investigation on the coloring and bleaching processes of WO_{3-x} photochromic thin films. *Journal of Materials Chemistry C* 8: 9410–9421. https://doi.org/10.1039/D0TC02170A.

159 Li, S. and Samy, E.-S.M. (1999). Synthesis and characterization of photochromic molybdenum and tungsten oxide nanoparticles. *Nanostructured Materials* 12: 215–219. https://doi.org/10.1016/S0965-9773(99)00102-6.

160 Granqvist, C.G. (1995). *Handbook of Inorganic Electrochromic Materials*. Elsevier. https://doi.org/10.1016/B978-0-444-89930-9.X5000-4.

161 Denesuk, M. and Uhlmann, D.R. (1996). Site-saturation model for the optical efficiency of tungsten oxide-based devices. *Journal of the Electrochemical Society* 143: L186–L188. https://doi.org/10.1149/1.1837080.

162 Berggren, L., Jonsson, J.C., and Niklasson, G.A. (2007). Optical absorption in lithiated tungsten oxide thin films: experiment and theory. *Journal of Applied Physics* 102: 083538. https://doi.org/10.1063/1.2800838.

163 Rojas-González, E.A. and Niklasson, G.A. (2020). Differential coloration efficiency of electrochromic amorphous tungsten oxide as a function of intercalation level:

comparison between theory and experiment. *Journal of Applied Physics* 127: 205101. https://doi.org/10.1063/5.0010044.

164 He, T. and Yao, J.-N. (2004). Photochromism in transition-metal oxides. *Research on Chemical Intermediates* 30: 459–488. https://doi.org/10.1163/1568567041280890.

165 Zhang, J., Benson, D.K., Tracy, C.E., Deb, S.K., Czanderna, A.W., and Bechinger, C. (1997). Chromic mechanism in amorphous WO_3 films. *Journal of the Electrochemical Society* 144: 2022–2026. https://doi.org/10.1149/1.1837737.

166 Bechinger, C., Oefinger, G., Herminghaus, S., and Leiderer, P. (1993). On the fundamental role of oxygen for the photochromic effect of WO_3. *Journal of Applied Physics* 74: 4527–4533. https://doi.org/10.1063/1.354370.

167 Wang, S., Fan, W., Liu, Z., Yu, A., and Jiang, X. (2018). Advances on tungsten oxide based photochromic materials: strategies to improve their photochromic properties. *Journal of Materials Chemistry C* 6: 191–212. https://doi.org/10.1039/C7TC04189F.

168 Santos, M.V., Barud, H.S., Alencar, M.A.S., Nalin, M., Toma, S.H., Araki, K., Benedetti, A.V., Maciel, I.O., Fragneaud, B., Legnani, C., Molina, C., and Cremona, M. (2021). Self-supported smart bacterial nanocellulose–phosphotungstic acid nanocomposites for photochromic applications. *Frontiers in Materials* 8: https://doi.org/10.3389/fmats.2021.668835.

169 Bange, K. (1999). Colouration of tungsten oxide films: a model for optically active coatings. *Solar Energy Materials and Solar Cells* 58: 1–131. https://doi.org/10.1016/S0927-0248(98)00196-2.

170 Miyazaki, H., Ishigaki, T., Suzuki, H., and Ota, T. (2016). Effect of film thickness and air atmosphere on photochromic properties of WO_3-based composite films. *Bulletin of the Chemical Society of Japan* 89: 20–23. https://doi.org/10.1246/bcsj.20150298.

171 Cornelius, S., Colombi, G., Nafezarefi, F., Schreuders, H., Heller, R., Munnik, F., and Dam, B. (2019). Oxyhydride nature of rare-earth-based photochromic thin films. *Journal of Physical Chemistry Letters* 10: 1342–1348. https://doi.org/10.1021/acs.jpclett.9b00088.

172 Kageyama, H., Hayashi, K., Maeda, K., Attfield, J.P., Hiroi, Z., Rondinelli, J.M., and Poeppelmeier, K.R. (2018). Expanding frontiers in materials chemistry and physics with multiple anions. *Nature Communications* 9: 772. https://doi.org/10.1038/s41467-018-02838-4.

173 Baba, E.M., Montero, J., Strugovshchikov, E., Özkan Zayim, E., and Karazhanov, S. (2020). Light-induced breathing in photochromic yttrium oxyhydrides. *Physical Review Materials* 4: 025201. https://doi.org/10.1103/PhysRevMaterials.4.025201.

174 Gschneider, K.A., Bunzli, J.C., and Pecharsky, V.K. (1999). *Handbook on the Physics and Chemistry of Rare Earths*. North-Holland, Amsterdam: Elsevier.

175 Moldarev, D., Primetzhofer, D., You, C.C., Karazhanov, S.Z., Montero, J., Martinsen, F., Mongstad, T., Marstein, E.S., and Wolff, M. (2018). Composition of photochromic oxygen-containing yttrium hydride films. *Solar Energy Materials and Solar Cells* 177: 66–69. https://doi.org/10.1016/j.solmat.2017.05.052.

176 Montero, J., Martinsen, F.A., Lelis, M., Karazhanov, S.Z., Hauback, B.C., and Marstein, E.S. (2018). Preparation of yttrium hydride-based photochromic films by reactive magnetron sputtering. *Solar Energy Materials and Solar Cells* 177: 106–109. https://doi.org/10.1016/j.solmat.2017.02.001.

177 Zapp, N., Auer, H., and Kohlmann, H. (2019). YHO, an air-stable ionic hydride. *Inorganic Chemistry* 58: 14635–14641. https://doi.org/10.1021/acs.inorgchem.9b02308.

178 Thornton, J.A. (1974). Influence of apparatus geometry and deposition conditions on the structure and topography of thick sputtered coatings. *Journal of Vacuum Science and Technology* 11: 666–670. https://doi.org/10.1116/1.1312732.

179 Maehlen, J.P., Mongstad, T.T., You, C.C., and Karazhanov, S. (2013). Lattice contraction in photochromic yttrium hydride. *Journal of Alloys and Compounds* 580: S119–S121. https://doi.org/10.1016/j.jallcom.2013.03.151.

180 Colombi, G. (2022). Photochromism and photoconductivity in rare-earth oxyhydride thin films. PhD thesis. Delft University of Technology.

181 Komatsu, Y., Shimizu, R., Sato, R., Wilde, M., Nishio, K., Katase, T., Matsumura, D., Saitoh, H., Miyauchi, M., Adelman, J.R., McFadden, R.M.L., Fujimoto, D., Ticknor, J.O., Stachura, M., McKenzie, I., Morris, G.D., MacFarlane, W. A., Sugiyama, J., Fukutani, K., Tsuneyuki, S., and Hitosugi, T. (2022). Repeatable Photoinduced insulator-to-metal transition in yttrium oxyhydride epitaxial thin films. *Chemistry of Materials* 34: 3616–3623. https://doi.org/10.1021/acs.chemmater.1c03450.

182 Wu, Z., de Krom, T., Colombi, G., Chaykina, D., van Hattem, G., Schut, H., Dickmann, M., Egger, W., Hugenschmidt, C., Brück, E., Dam, B., and Eijt, S.W.H. (2022). Formation of vacancies and metallic-like domains in photochromic rare-earth oxyhydride thin films studied by *in-situ* illumination positron annihilation spectroscopy. *Physical Review Materials* 6: 065201. https://doi.org/10.1103/PhysRevMaterials.6.065201.

183 Montero, J., Svedlindh, P., and Österlund, L. (2023). Photo-induced reversible modification of the Curie-Weiss temperature in paramagnetic gadolinium compounds. *ArXiv* 2309 (07978).

184 Towns, A.D. (2016). Industrial photochromism. In: *Applied Photochemistry: When Light Meets Molecules* (ed. G. Bergamini and S. Silvi), 227–279. Cham: Springer International Publishing. https://doi.org/10.1007/978-3-319-31671-0_5.

185 Montero, J. and Karazhanov, S.Z. (2018). Spectroscopic Ellipsometry and microstructure characterization of photochromic oxygen-containing yttrium hydride thin films. *Physica Status Solidi* 215: 1701039. https://doi.org/https://doi.org/10.1002/pssa.201701039.

186 Chaykina, D., Nafezarefi, F., Colombi, G., Cornelius, S., Bannenberg, L.J., Schreuders, H., and Dam, B. (2022). Influence of crystal structure, encapsulation, and annealing on photochromism in Nd oxyhydride thin films. *The Journal of Physical Chemistry C* 126: 2276–2284. https://doi.org/10.1021/acs.jpcc.1c10521.

187 Smith GB, Granqvist C-GS. Green Nanotechnology. CRC Press; 2010. https://doi.org/10.1201/EBK1420085327.

188 Granqvist, C.G. (2014). Electrochromics for smart windows: oxide-based thin films and devices. *Thin Solid Films* 564: 1–38. https://doi.org/10.1016/j.tsf.2014.02.002.

189 Lee, S.J., Choi, D.S., Kang, S.H., Yang, W.S., Nahm, S., Han, S.H., and Kim, T.Y. (2019). VO_2/WO_3-based hybrid smart windows with thermochromic and electrochromic properties. *ACS Sustainable Chemistry & Engineering* 7: 7111–7117. https://doi.org/10.1021/acssuschemeng.9b00052.

190 Madhuri, P.L., Bhupathi, S., Shuddhodana, S., Judeh, Z.M.A., Yang, S.-H., Long, Y., and Abdulhalim, I. (2021). Hybrid vanadium dioxide-liquid crystal tunable non-reciprocal scattering metamaterial smart window for visible and infrared radiation control. *Optical Materials Express* 11: 3023. https://doi.org/10.1364/OME.432784.

191 Ke, Y., Chen, J., Lin, G., Wang, S., Zhou, Y., Yin, J., Lee, P.S., and Long, Y. (2019). Smart windows: electro-, thermo-, mechano-, photochromics, and beyond. *Advanced Energy Materials* 9: 1902066. https://doi.org/10.1002/aenm.201902066.

192 Zhang, R., Xiang, B., Shen, Y., Xia, L., Xu, L., Guan, Q., and Tang, S. (2021). Energy-efficient smart window based on a thermochromic microgel with ultrahigh visible transparency and infrared transmittance modulation. *Journal of Materials Chemistry A* 9: 17481–17491. https://doi.org/10.1039/D1TA03917B.

193 Wang, Y., Zhao, F., Wang, J., Khan, A.R., Shi, Y., Chen, Z., Zhang, K., Li, L., Gao, Y., and Guo, X. (2018). VO_2@SiO_2/poly(*N*-isopropylacrylamide) hybrid nanothermochromic microgels for smart window. *Industrial and Engineering Chemistry Research* 57: 12801–12808. https://doi.org/10.1021/acs.iecr.8b02692.

194 Ji, Y., Mattsson, A., Niklasson, G.A., Granqvist, C.G., and Österlund, L. (2019). Synergistic TiO_2/VO_2 window coating with Thermochromism, enhanced luminous transmittance, and photocatalytic activity. *Joule* 3: 2457–2471. https://doi.org/10.1016/j.joule.2019.06.024.

195 Liu, C., Wang, N., and Long, Y. (2013). Multifunctional overcoats on vanadium dioxide thermochromic thin films with enhanced luminous transmission and solar modulation, hydrophobicity and anti-oxidation. *Applied Surface Science* 283: 222–226. https://doi.org/10.1016/j.apsusc.2013.06.085.

196 Yang, F., Shi, X., Chen, Z., Ma, D., Wu, Y., Luo, H., and Gao, Y. (2016). Crystallized TiO_2(a)–VO_2(M/R) nanocomposite films with electrochromism–thermochromism dual-response properties. *RSC Advances* 6: 32176–32182. https://doi.org/10.1039/C6RA05700D.

197 Zhou, J.D., Gao, Y.F., Zhang, Z.T., Luo, H., Cao, C., Chen, Z., Dai, L., and Liu, X. (2013). VO_2 thermochromic smart window for energy savings and generation. *Scientific Reports* 3: https://doi.org/10.1038/srep03029.

198 Needell, D.R., Phelan, M.E., Hartlove, J.T., and Atwater, H.A. (2021). Solar power windows: connecting scientific advances to market signals. *Energy* 219: 119567. https://doi.org/10.1016/j.energy.2020.119567.

199 Zhang, J., Wang, J., Yang, C., Jia, H., Cui, X., Zhao, S., and Xu, Y. (2017). Mesoporous SiO_2/VO_2 double-layer thermochromic coating with improved visible transmittance for smart window. *Solar Energy Materials and Solar Cells* 162: 134–141. https://doi.org/10.1016/j.solmat.2016.12.048.

200 Taylor, A., Parkin, I., Noor, N., Tummeltshammer, C., Brown, M.S., and Papakonstantinou, I. (2013). A bioinspired solution for spectrally selective thermochromic VO_2 coated intelligent glazing. *Optics Express* 21: A750. https://doi.org/10.1364/OE.21.00A750.

201 Wang, Y., Zhao, F., Wang, J., Li, L., Zhang, K., Shi, Y., Gao, Y., and Guo, X. (2019). Tungsten-doped VO_2/starch derivative hybrid nanothermochromic hydrogel for smart window. *Nanomaterials* 9: 970. https://doi.org/10.3390/nano9070970.

202 Zhu, J., Huang, A., Ma, H., Chen, Y., Zhang, S., Ji, S., Bao, S., and Jin, P. (2017). Hybrid films of VO2 nanoparticles and a nickel(ii)-based ligand exchange thermochromic system: excellent optical performance with a temperature responsive colour change. *New Journal of Chemistry* 41: 830–835. https://doi.org/10.1039/C6NJ03369E.

203 Cao, X., Wang, N., Law, J.Y., Loo, S.C.J., Magdassi, S., and Long, Y. (2014). Nanoporous thermochromic VO_2(M) thin films: controlled porosity, largely enhanced luminous transmittance and solar modulating ability. *Langmuir* 30: 1710–1715. https://doi.org/10.1021/la404666n.

204 Chen, Z., Gao, Y., Kang, L., Du, J., Zhang, Z., Luo, H., Miao, H., and Tan, G. (2011). VO_2-based double-layered films for smart windows: optical design, all-solution preparation and improved properties. *Solar Energy Materials and Solar Cells* 95: 2677–2684. https://doi.org/10.1016/j.solmat.2011.05.041.

205 Fang, Z., Tian, S., Li, B., Liu, Q., Liu, B., Zhao, X., and Sankar, G. (2021). VO_2/ZnO bilayer films with enhanced thermochromic property and durability for smart windows. *Applied Surface Science* 540: 148414. https://doi.org/10.1016/j.apsusc.2020.148414.

206 Zhao, L., Miao, L., Liu, C., Li, C., Asaka, T., Kang, Y., Iwamoto, Y., Tanemura, S., Gu, H., and Su, H. (2014). Solution-processed VO_2–SiO_2 composite films with simultaneously enhanced luminous transmittance, solar modulation ability and anti-oxidation property. *Scientific Reports* 4: 7000. https://doi.org/10.1038/srep07000.

207 Chen, Z., Gao, Y., Kang, L., Cao, C., Chen, S., and Luo, H. (2014). Fine crystalline VO_2 nanoparticles: synthesis, abnormal phase transition temperatures and excellent optical properties of a derived VO_2 nanocomposite foil. *Journal of Materials Chemistry A* 2: 2718. https://doi.org/10.1039/c3ta14612j.

208 Guo, H., Wang, Y.G., Fu, H.R., Jain, A., and Chen, F.G. (2021). Influence of dopant valence on the thermochromic properties of VO_2 nanoparticles. *Ceramics International* 47: 21873–21881. https://doi.org/10.1016/j.ceramint.2021.04.205.

209 Guo, X., Xu, H., Ma, X., Yang, S., Liu, T., Nie, Y., Wang, C., Wang, C., and Jiang, X. (2022). Photothermal polydopamine coated VO_2 nanoparticle thin film with enhanced optical property and stability. *Vacuum* 196: 110776. https://doi.org/https://doi.org/10.1016/j.vacuum.2021.110776.

210 Karmakar, B. (2016). *Functionality of Reversible Glass Nanocomposites and their Applications*. Glass Nanocomposites, p. 265–278. Elsevier. https://doi.org/10.1016/B978-0-323-39309-6.00011-0.

8

Overview of Organic Thermochromic Materials

Gunnar A. Niklasson[1], *José Montero*[1], *and Carl M. Lampert*[2,3,4]

[1] *Department of Materials Science and Engineering, The Ångström Laboratory, Uppsala University, Uppsala, Sweden*
[2] *Star Science, Cotati, CA, USA*
[3] *Formerly, Technical Director, Society of Vacuum Coaters (non-profit), Albuquerque, NM, USA*
[4] *Staff Scientist, University of California, LBNL, Energy Technologies, Berkeley, CA, USA*

Nomenclature

ATO	Sb:SnO_2
Bdmin	1-butyl-2,3-dimethylimidazolium
Bmim	1-butyl-3-methylimidazolium
FTO	F:SnO_2
HEMA	2-hydroxyethyl methacrylate
HεL	high-epsilon ligand
IL	ionic liquid
IGU	insulating glass unit (double glazing)
ITO	Sn:In_2O_3
λ_p	plasma frequency
L	liter (1 dm^3)
LBNL	Lawrence Berkeley National Laboratory
LCA	long-chain alcohol
LCST	lower critical solution temperature
LεL	low-epsilon ligand
Me	transition metal ion
n_c	index of refraction, core
n_m	index of refraction, matrix material
n_s	index of refraction, shell
n_{sd}	index of refraction, scattering domain
NIR	near-infrared
OTC	organic thermochromic
PS/HEMA:PPO	polystyrene/2-hydroxyethyl methacrylate:polypropylene oxide
PNIPAM	poly(*N*-isopropyl acrylamide)
PNIPAM-AEMA	poly(*N*-isopropylacrylamide)-2-aminoethylmethacrylate hydrochloride
PPO	polypropylene oxide
PU	poly(urethane urea)
R	reflectance

Chromogenics: Smart Switchable Optical Materials and Their Applications, First Edition. Edited by Carl M. Lampert.

SDG	Suntuitive Dynamic Glass™
SHGC	solar heat gain coefficient (0–1)
TALD	Temperatur-abhängige Lichtdurchlässigkeit
TT	thermotropic
TC	thermochromic
t_c	critical temperature, transition temperature
$t_{comfort}$	human comfort temperature
T_{lum}	integrated photopic luminous transmittance
T_{sol}	integrated solar transmittance, also called total solar transmittance.
T_{vis}	integrated visible transmittance, equal to T_{lum} and VLT.
ΔT	change in transmittance
TFSI	bis(trifluoromethylsulfonyl) imide
UCST	upper critical solution temperature
UV	ultraviolet
VIG	vacuum-insulated glazing
VLT	visible light transmittance, a term used by the glass industry equal to T_{vis} and T_{lum}

8.1 Introduction

Organic thermochromics (OTCs) have a fairly long history, and companies continue to create new types of thermochromics (TCs). Before we discuss large-area applications, it is important to understand a brief history of OTCs. The railroad, engine, and electric motor industries needed thermal indicators to show when turning components were overheating. Some of the first materials came about in 1909 with studies of bianthrone, which changes color from yellow to green upon heating [1]. In 1926, spiropyrans were reported to change from colorless to blue-violet upon heating [2]. This was followed in the 1930s with work on metal–organic temperature indicators, iron(III)-tris-dithiocarbamates, by Cambi and Szegö [3]. Since 1938, the Tempil Company, now part of LA-CO (Elk Grove, IL), has produced temperature-indicating strips, paints, as well as sterilization-indicating dyes and pigments, for a wide variety of applications. Some compounds are reversible, and some stay colored once switched, depending on their technology. In the 1960s, commercial interest increased in TC liquid crystals (e.g. mood rings), with an associated development of liquid crystals for displays. Multicolor cholesteric liquid crystals are still being studied for TC photonic applications [4]. Liquid crystals in paints and as encapsulated suspensions are also used to map heat flow. Some of the early work on OTC materials has been reviewed by Day [5]. In the 1970s, the Toray Company (Japan) developed OTC inks based on leuco dyes for color-changing fabrics. Chowdhury et al. [6] and Wang et al. [7] have reviewed TCs for silks and textiles. In the 1980–1990s, TC "Hypercolor" T-shirts were popular. Cloth products are today made by several companies, including ChroMorphous (CREOL Labs, Melbourne, FL). Paints and indicator strips are very useful for thermal mapping and are used in aerospace and satellite design [8, 9]. Thermal indicators are used to detect heat damage and fires in electrical generating equipment, motors, and electronics. Thamrin et al. [10] reviewed TC inks used as thermal indicators for cold storage of foods. Smart TC labels are still being developed. TCs is about a 2 billion (USD) industry. Some of the TC industry leaders are SpotSee (Dallas, TX) owners of TMC& LCR Hallcrest, QCR Solutions Corp. (Port Lucie, FL),

Chromatic Technologies Inc. (Colorado Springs, CO), Matsui Indust. Co. (Gardena, CA), Olikrom (Pessac, France), New Prismatic Enterprises (Taipei, Taiwan), Hali Pigments Co. (Jiangsu, China), Smarol Ind. (Hangzhou City, China), and Kolorjet (India).

In this chapter, we focus mainly on OTC for energy control of buildings, which involves large-area applications such as glazing. Many other commercial applications exist, as also noted above, for example, thermometers, temperature-indicating labels, thermo-sensitive paper, and TC inks; the inks can be added in microencapsulated form to, e.g. thermoplastic and thermosetting polymers [11]. These side applications fall outside the scope of this overview. We also do not consider the large number of TC materials that exhibit a transition from one color to another [11]. An excellent book reviews the research work and emergence of commercial products in OTCs [11].

OTC materials offer interesting alternatives to the inorganic ones that we discussed in Chapter 7, especially for applications in energy-efficient and solar-control windows. Many of these materials are based on polymers and gels, and their manufacture can often be upscaled to large sizes, employing methods used in the polymer industry [12]. On the other hand, the synthesis of complex organic compounds and mixtures may offer practical challenges, and their stability under ambient conditions constitutes another difficult issue. For window applications, materials that exhibit a reversible transition from a low-temperature transparent state to a high-temperature absorbing or scattering state are of greatest interest. OTC materials were applied to architectural glazing in the 1970–1980s in the form of a hydrogel named "Cloud Gel" (by Suntek, later Southwall), which was a clear-to-diffuse optical switching product. The transition temperature could be changed by altering the gel copolymer composition. Hydrogels were prototyped until the 1990s, but their development was hampered by long-term stability issues. Later, dynamic tinting, ligand-exchange, metal–organic materials were commercialized by Pleotint, LLC (West Olive, MI) and have been successfully installed into a number of buildings. These commercial smart windows are discussed in Section 8.2.1.

OTC materials can be divided into two classes: "clear organic thermochromics" and thermotropics (TTs) [13]. The first class is the organic TC materials proper, which are characterized by the property of exhibiting no significant light scattering, both in the low- and high-temperature states. When used in a window, the visual contact between indoors and outdoors is hence maintained irrespective of temperature. In this case, it is the optical absorption of the material that varies with temperature. These materials exhibit a low absorptance at low temperatures and a higher absorptance at high temperatures. Hence, layers of these materials exhibit a transition from a high to a low solar transmittance (T_{sol}) as the temperature increases. The optical properties vary rather gradually around a transition temperature (t_c), which is typically somewhat higher than room temperature. Room temperature is typically 20–22 °C (68–72 °F). This type of TC phenomenon can be present in different classes of polymers, either as a property of the polymer itself or by the interaction of the polymer with special additives [13]. To this class of materials belong the ligand-exchange, metal–organic materials briefly mentioned above, as well as leuco-dye composites.

The second class of materials that we will consider are the TT materials [14]. They exhibit low absorptance irrespective of temperature, and when they transit from a low-temperature to a high-temperature state, the amount of light scattering varies. The low-temperature state exhibits no significant light scattering, giving an excellent glass-like visual contact between indoors and outdoors. The materials become highly diffusely scattering at high temperatures, giving a privacy effect. There exist several mechanisms, depending on the actual material, that may give rise to a change in optical scattering around t_c. Some examples include a phase separation, a phase transition, a change of domain ("particle") size, or aggregation of "particles" into larger clusters [13]. TT materials are basically composites consisting of at least two constituents. At temperatures below t_c, the refractive indices of the constituents of a homogeneous mixture are close to equal, and/or the size of inhomogeneities is sufficiently small so that light scattering is negligible. Above t_c, on the other hand, changes in structure lead to a refractive index contrast between "particles" and matrix. In addition, domains can grow and aggregate into larger particulate structures. In either of these cases, light scattering becomes stronger, and this leads to a temperature variation in the transmittance of this type of material. At high temperatures, the appearance of the material becomes milky, and there is no clear view through the layer. TT materials include hydrogels, ionogels, polymer blends, casting resins, and polyolefins [13].

In the context of TT materials, it is important to distinguish between two types of transmittance contrast. The direct (normal–normal) transmittance gives the amount of light passing directly through the material without experiencing any light scattering. This parameter is relevant when assessing the use of TT materials as privacy windows. However, the direct transmittance tells us very little about the energy efficiency of a TT window. The visible light or solar radiation passing through a TT window is the sum of the transmitted unscattered light and the light scattered in the forward direction. The pertinent parameter to measure is directional-hemispherical transmittance (often called total transmittance). Unfortunately, the parameter of choice in the literature is often the change in direct transmittance, i.e. many studies only consider the change in the radiation flux that passes through the TT material without experiencing any scattering [15]. In the case of windows for energy efficiency, it should be noted that the diffuse reflectance is another key parameter.

8.2 Clear Organic Thermochromics

The "clear organic thermochromics" include ligand-exchange materials and leuco-dye systems. Most of the open literature on ligand-exchange thermochromic (LETC) materials actually comes from literature and journals on inorganic chemistry. In fact, while these systems are included in the chapter on OTC materials, the first book detailing these TC materials is entitled "Inorganic Thermochromism" [16]. On the other hand, organic ligands (along with inorganic ligands and the metal ions themselves) have a key role in the performance of these materials, and the LETC systems are normally contained in a polymeric matrix. Of the materials treated in this section, the ligand-exchange materials have also been the most commercially successful.

8.2.1 Ligand-Exchange, Metal–Organic Materials

In this section, we describe developments in LETC materials. Among metal–organic TC materials, polymers with TC additives are of major interest. A very large number of thermochromic (LETC) systems have been described by Byker et al. [17, 18].

The ligand-exchange mechanism is illustrated in Figure 8.1. A metal ion (Me) (e.g. Ni^{2+}, Co^{2+}, Fe^{3+}) is surrounded by ligand molecules. The continuously variable tinting of LETC systems results from the temperature-based shift in thermodynamic equilibrium of a reversible reaction, which involves the exchange of ligands around transition metal ions. The low-temperature metal–ligand complex with one set of ligands has low absorption.

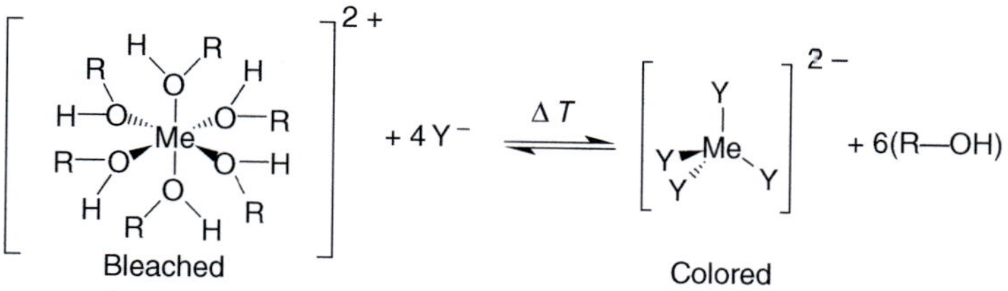

Figure 8.1 Pictorial example of a ligand-exchange reaction. This reaction supports continuous darkening as the temperature rises and continuous bleaching as the temperature falls. Me is a transition metal ion such as Ni^{2+}, Co^{2+}, or Fe^{3+}. *Source:* [13] / MDPI / CC BY 4.0.

As the temperature increases, there is a reversible thermodynamic shift of equilibrium to form another metal–ligand complex or complexes. The high-temperature metal–ligand complexes have higher absorption. The LETC systems do not switch at a specific threshold temperature but have a continuously variable tinting over a range of temperatures from about 15 °C to about 75 °C. For systems with a good thermodynamic equilibrium shift, about 3% of the metal ions are in the high-absorption form at 15 °C and over 50% of the metal ions are in the high-absorption form at 75 °C. Thus, as the temperature increases, the absorption increases, and as the temperature decreases, the absorption decreases. The LETC system is put into a polymer matrix, for example, polyvinyl butyral. The ligand-exchange process is complex, but it is generally well understood [16]. A detailed study of Ni^{2+} complexes using X-ray absorption spectroscopy in combination with theoretical calculations has been carried out by Queen et al. [19, 20].

A simple version of the chemical equilibrium process that takes place in the LETC film over a range of temperatures is described by Eq. (8.1) and pictorially in Figure 8.1. Typically, as the temperature rises, the metal ions in the LETC systems go from a six-coordinate configuration with low visible and near-infrared (NIR) light absorption to a four-coordinate configuration with much higher visible light absorption. On cooling, this reaction reverses completely, and the LETC systems go back to their high-transmission states. The reaction can be written as:

$$[\mathrm{Me(L\varepsilon L)_6}] + 4\,\mathrm{H\varepsilon L} \leftrightarrow [\mathrm{Me(H\varepsilon L)_4}] + 6\,\mathrm{L\varepsilon L}, \tag{8.1}$$

where Me is the metal ion at the center of the coordination shell. The identities of the low-epsilon ligand (LεL) and the high-epsilon ligand (HεL) define the color of the coordination complexes at ambient and elevated temperatures, respectively. The $[Me(L\varepsilon L)_6]$ corresponds to the less colored species that dominates at ambient temperatures, while the $[Me(H\varepsilon L)_4]$ is the highly colored species that dominates at elevated temperatures. For example, the low-temperature species can have a molar extinction coefficient of <10 L $(mol\ cm)^{-1}$ and the high-temperature species an extinction coefficient >250 L $(mol\ cm)^{-1}$, both in the visible wavelength region. The position of the equilibrium is determined by the composition, the thermodynamic properties, and the temperature [20].

The LETC systems are incorporated into a polymer layer, as stated above. This polymer layer, also called the interlayer, is used to laminate two sheets of glass together, as disclosed in patents [21, 22] of Pleotint, LLC. The product is called Suntuitive Dynamic Glass™ (SDG). Transmittance spectra of this laminate in the visible and NIR wavelength ranges and at several temperatures are shown in Figure 8.2. It is seen that the transmittance switches mainly at visible wavelengths, and the switching takes place gradually from below 25 to 75 °C. The LETC is different from many other TCs that have a sharp transition at a particular design

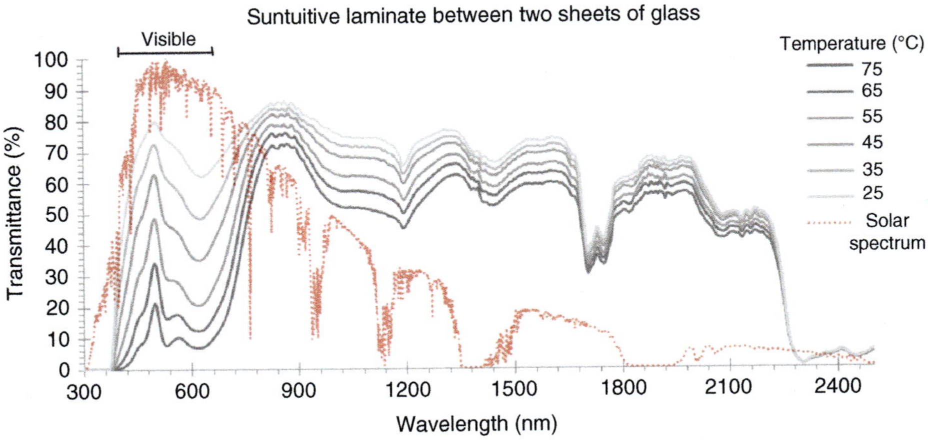

Figure 8.2 Transmittance spectra of Suntuitive® interlayer laminate (example of one formulation) sandwiched between two 3 mm glass substrates. Shown is a range of transmittance values for temperatures 25–75 °C. Similar data for the Suntuitive interlayer appear in the International Glazing Database. *Source:* Courtesy of H. Byker, Pleotint, LLC (West Olive, MI).

temperature. In the NIR, switching is small, and distinct absorption bands of the polymer matrix are visible, although these wavelengths are often blocked by high-performance low-emittance (low-e) coatings anyway. The integrated visible transmittance (or photopic luminous transmittance) goes from T_{Lum} = 0.71 at 15 °C to T_{Lum} = 0.07 at 75 °C. The corresponding switching of the solar transmittance is smaller, from T_{sol} = 0.63–0.23. The reflectance is almost independent of wavelength and temperature, displaying values of R = 0.05–0.07. Very similar results have been obtained from optical measurements in an independent study by Giovannini et al. [23].The SDG tints by the LETC mechanism via sunlight responsiveness. The windows use the light of the sun converted to heat to tint them. The level of direct sunlight and to a lesser extent the ambient temperature determine the level of tinting. Since excess solar heat gain from windows only occurs with direct sunlight exposure, SDG successfully uses the energy of the sun to block the excess heat from the sun. For practical window applications, the SDG laminate becomes the outer pane of a double-glazed window, which may involve a third pane of glass. Almost any commercially available glass, including any color of body-tinted glass, may be used in these glazing units. A high-performance low-*e* coating is provided in the double-pane window structure produced by Pleotint to ensure that most of the solar heat gain accumulated in the TC laminate is rejected to the outside of the building. An example of a Suntuitive IGU (insulating glass unit) is shown in Figure 8.3. Several different window configurations for IGU and triple-glazed window units, including the integration of an evacuated panel, are outlined in US patent 20080092456 [24].

SDG units have been incorporated into windows on about 500 buildings throughout the world from 2010 to 2020. This amounts to over 500,000 ft^2 of windows with SDG installed, with projects ranging from a few 100 to over 30,000 ft^2. The switching range of temperatures from about 15 °C to about 75 °C are typical temperatures experienced by any tinted window units when exposed to direct sunlight. This includes almost all other dynamic window glass units. The SDG is always incorporated into windows that are sunlight responsive so that direct sunlight exposure to the SDG causes most of the heating and thus most of the tinting. Window glass units or IGUs with SDG are installed in the same way as normal units.

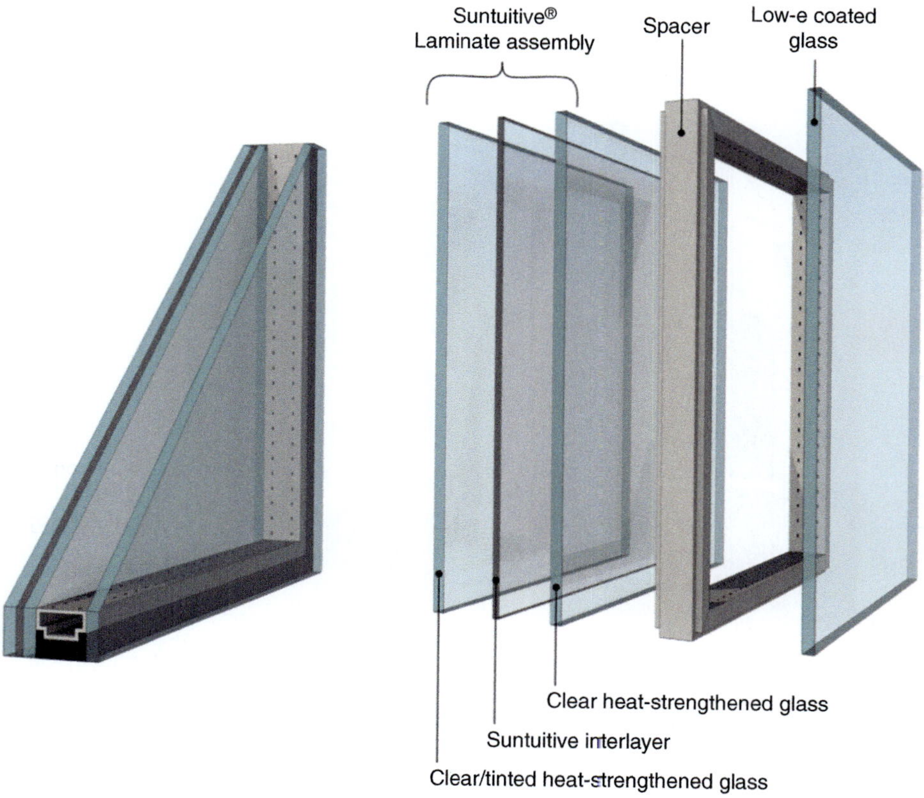

Figure 8.3 Schematic of Suntuitive IGU and its components. This is one of several window designs. *Source:* Courtesy of H. Byker, Pleotint, LLC (West Olive, MI).

This leads to fewer complications for glazing contractors as compared to dynamic windows that require electrical power and wiring. The SDG material is stable at a temperature of 85 °C and during prolonged sun exposure, which satisfies the requirements of many architectural glazing applications. An advantage of the SDG system is that it is compatible with well-known glass lamination techniques and existing equipment. Also, the LETC lamination interlayer material can be shipped directly to glass laminators throughout the world. An example of an actual building is shown in Figures 8.4 and 8.5, namely a building equipped with SGD units at NDIS Australia (National Disability Insurance Scheme) in Geelong, Australia. The facade includes cylindrically bent dynamic window glass.

Most TC systems considered for dynamic windows in the past switched completely from less tinted to a more tinted condition at a certain threshold temperature. That temperature was largely chosen based on warm outdoor ambient temperatures, and hence the windows on all sides of the building could tint on hot days regardless of whether there was direct sunlight exposure. Unlike past TC systems, the sunlight responsiveness of SDG allows windows to tint strongly only during the time of day and on the sides of buildings that are exposed to sunlight, which of course is the only time when there is excess solar heat gain. At times when windows are not exposed to sunlight and the ambient temperature is low, they remain in their high-transmission state to allow for daylighting. Thus, sunlight-responsive SDG uses all wavelengths of sunlight that are converted to heating energy in the interlayer to block the excess solar heat gain of the sun.

Figure 8.4 Suntuitive SDG glass installed at the NDIS Australia building in Geelong, Australia. The glazing is shown in lower ambient lighting conditions, hence lower darkening of the windows. *Source:* Courtesy of H. Byker, Pleotint, LLC (West Olive, MI).

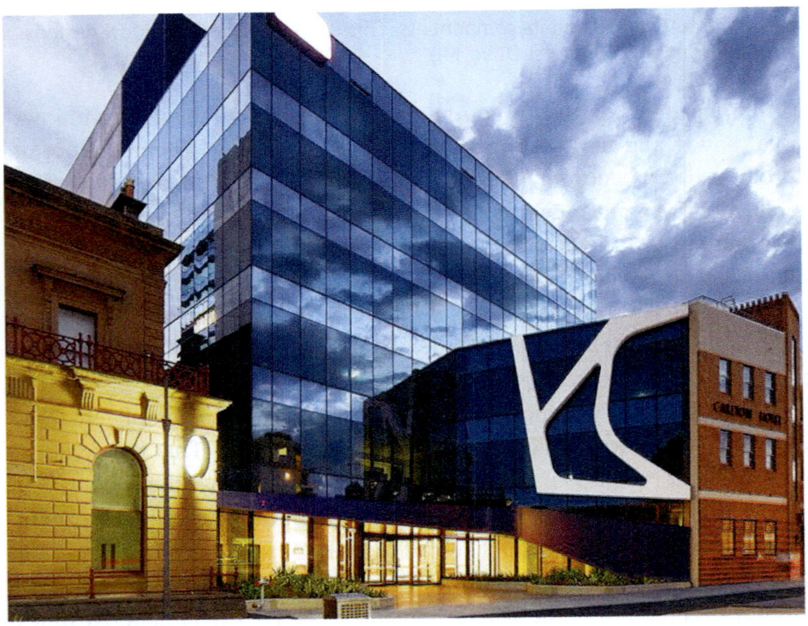

Figure 8.5 Suntuitive SDG glass installed at the NDIS Australia building in Geelong, Australia. The building is shown in high ambient lighting conditions, hence deeper darkening of the windows. *Source:* Courtesy of H. Byker, Pleotint, LLC (West Olive, MI).

The SDG has excellent durability, as demonstrated by its more than 10 years of actual use on numerous buildings. The cost is higher than conventional window glass units, but the increase is modest, and the lack of power supplies, wiring, electronic controls, and complicated installation make SDG attractive compared to other dynamic window glass technologies. The level of comfort and energy-saving performance is highly appreciated by building occupants and owners.

Several different window configurations for IGU and triple-glazed window units are produced, including the integration with vacuum-insulated glazing or VIGs. The glazing configurations are outlined in US patent 20080092456 by Millett et al. [24]. The visible and solar performance data as a function of temperature for an SDG IGU with excellent properties are shown in Figure 8.6. The actual window construction is shown in Figure 8.7. The figure includes the solar heat gain coefficient (SHGC), which is the sum of the solar transmittance and the absorbed solar radiation that is transported as heat through the window to the interior of the building. The SHGC, or "solar factor" is a measure of how well a window blocks solar radiation and solar heat from getting through the window. It is a number between 0 and 1. The lower the SHGC, the better the window prevents solar heating of the interior. Performance data for several designs of IGU units are displayed in Table 8.1. The Table provides data for different SDG IGU that differs in the tinting of the exterior glass pane and in the low-*e* coating applied on the interior glass pane. The low-*e* coatings control the amount of heat that passes through a window. Typical designations are used in Table 8.1: 2× low-*e* and 3× low-*e*. The 2× low-*e* is a two-metal layer D/M/D/M/D optical interference coating, where M is a metal and D is a dielectric layer. The 3× low-*e* is a three-metal layer D/M/D/M/D/M/D

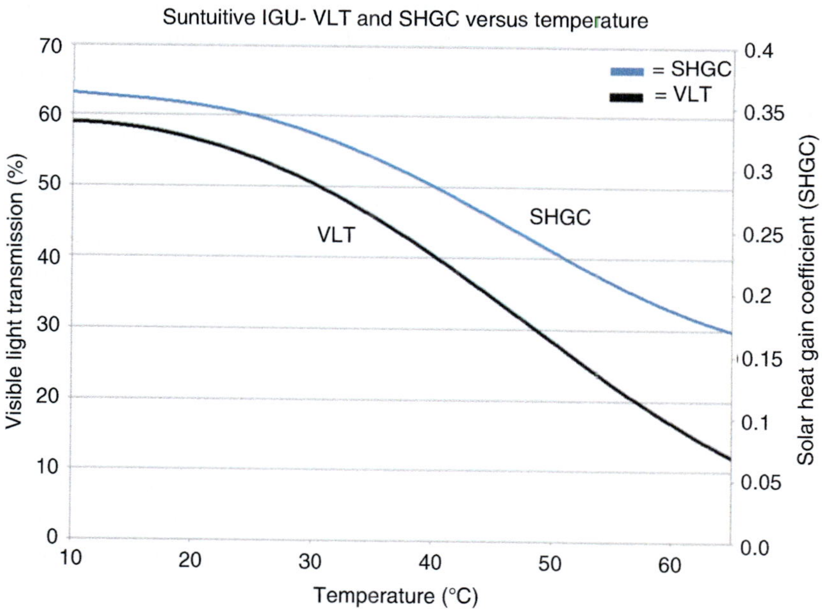

Figure 8.6 Characteristics of Suntuitive SDG in terms of visible light transmittance (VLT), which is the same as T_{lum}, and solar heat gain coefficient (SHGC), both as a function of temperature. The smooth transition from light to dark can easily be seen. The SDG construction is: exterior: 6 mm clear glass/1.2 mm Suntuitive/5 mm clear glass/11 mm air gap (90% argon + 10% air); interior: 6 mm Solarban 60 on clear glass. The data differ slightly from Table 8.1 due to the use of a different low-*e* coating. *Source:* Courtesy of H. Byker, Pleotint, LLC (West Olive, MI).

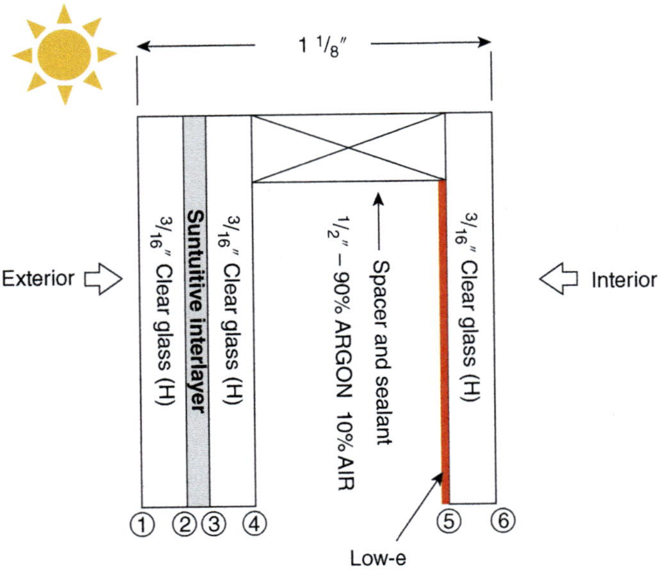

Figure 8.7 Example cross-sectional SDG clear glass construction of the IGU used to obtain the performance data displayed in Table 8.1. *Source:* Courtesy of H. Byker, Pleotint, LLC (West Olive, MI).

tighter bandpass optical filter that only allows visible light to pass through it. Other protective and transition layers are used in well-designed low-*e* coatings but are not shown here. Low SHGC ratings are significant in climates with very hot summers. The designation visible light transmission, VLT, is used by the glass and glazing industry and is equivalent to T_{lum} and T_{vis}, used elsewhere in this book, where the illuminant is the AM1.5 solar spectrum and the weighting function is the photopic response.

Pleotint, LLC has recently discovered a very high-performance, lower-cost, second-generation LETC system that it hopes to introduce into SDG for dynamic windows soon. This development may create a whole new cost-performance dynamic in the smart window arena [18]. It should also be mentioned that improved transmittance switching in the NIR can be obtained by adding VO_2 nanoparticles to the LETC system [26]. Promising results were obtained, but the LETC material used by these authors appears not to have been optimized.

8.2.2 Leuco Dyes in a Matrix

Leuco dyes are able to change their color reversibly as a function of the temperature. The TC material consists of the leuco dye which is an electron donor, and an electron-accepting "color developer" molecule (often a dissociable salt). A solvent can function as a decolorization agent. The details of the reaction mechanism are under discussion [11]. In some systems, there occurs a solid-liquid transition at t_c, which changes the pH. At the melting point, a reaction between the leuco dye and the color developer takes place. Coloration occurs at a low pH and decoloration at a higher pH, but this mechanism does not appear to be general [11].

For solar-control applications, ternary mixtures are of most interest. They consist of the dye (e.g. crystal violet lactone), a color developer (e.g. lauryl gallate), and a solvent (an alcohol with a long chain [LCA], for example, 1-Tetradecanol). The chemistry of this system has

Table 8.1 Calculated energy performance of Suntuitive glazing using different external tinted glass units and double (2×) or D/M/D/M/D and triple (3×) or D/M/D/M/D/M/D low-*e* coating types on the inner pane. VLT is the same as T_{lum}. The total solar transmittance was calculated using AM1.5 solar spectrum. The tabular values were calculated using the "Windows 7.8.55" modeling program developed by LBNL (Berkeley, CA) [25].

Suntuitive glass performance		Visible light transmission (VLT)		Solar heat gain coefficient (SHGC)		Total solar transmission	
Exterior tint	Low-*e*	Clear ⟷ Tinted		Clear ⟷ Tinted		Clear ⟷ Tinted	
Low-iron	2 × Low-*e*	0.55	0.08	0.39	0.18	0.28	0.08
	3 × Low-*e*	0.50	0.08	0.32	0.13	0.19	0.03
Clear	2 × Low-*e*	0.54	0.08	0.37	0.17	0.25	0.07
	3 × Low-*e*	0.49	0.07	0.31	0.13	0.18	0.03
Light green	2 × Low-*e*	0.45	0.07	0.28	0.13	0.18	0.04
	3 × Low-*e*	0.41	0.06	0.26	0.11	0.14	0.02
Medium gray	2 × Low-*e*	0.38	0.06	0.29	0.15	0.18	0.05
	3 × Low-*e*	0.35	0.05	0.25	0.12	0.13	0.02
Blue	2 × Low-*e*	0.35	0.05	0.29	0.15	0.18	0.05
	3 × Low-*e*	0.31	0.05	0.24	0.12	0.12	0.02
Bronze	2 × Low-*e*	0.35	0.05	0.28	0.14	0.17	0.05
	3 × Low-*e*	0.32	0.05	0.24	0.12	0.12	0.02
Gray	2 × Low-*e*	0.30	0.05	0.25	0.13	0.15	0.04
	3 × Low-*e*	0.27	0.04	0.22	0.11	0.10	0.02

- Clear values represent 5 °C and tinted values 65 °C
- Reflectance In is 0.10 (clear) and 0.09 (tinted).
- Reflectance Out is 0.10–0.06 (clear, varies with tint of glass) and 0.05 (tinted)

Source: Courtesy of H. Byker, Pleotint, LLC (West Olive, MI).

been studied in detail [27]. At low temperatures, the color developer forms a colorless complex with the solvent, and at high temperatures, the color developer forms a colored complex with the dye [13], as shown in Figure 8.8. The transition temperature for a solid-to-liquid transition is in the range of 60–80 °C [13]. These ternary TC mixtures can be dispersed in a polymer matrix, which makes possible the production of TC polymer foils by extrusion. The resulting materials can be based on different dyes and, hence, be produced in different colors. The normal-hemispherical transmittance of a green-colored system is shown in Figure 8.9. The materials exhibit a significant change in visible transmittance (up to 0.27) but a quite low modulation of solar transmittance, of the order of $\Delta T_{SOL} = 0.05–0.06$ [13].

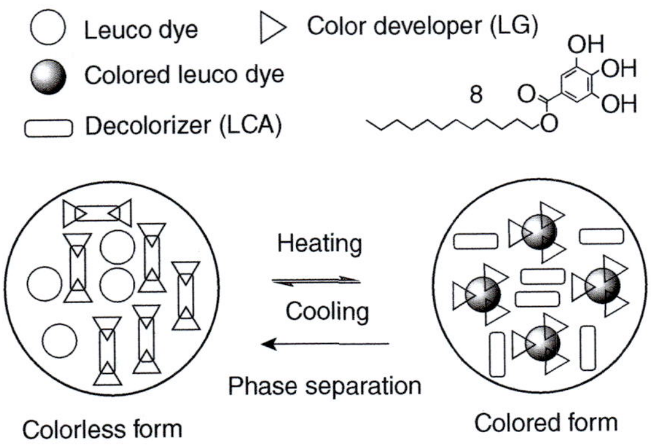

Figure 8.8 Schematic picture of thermochromism of a ternary leuco-dye system. *Source:* [13] / MDPI / CC BY 4.0.

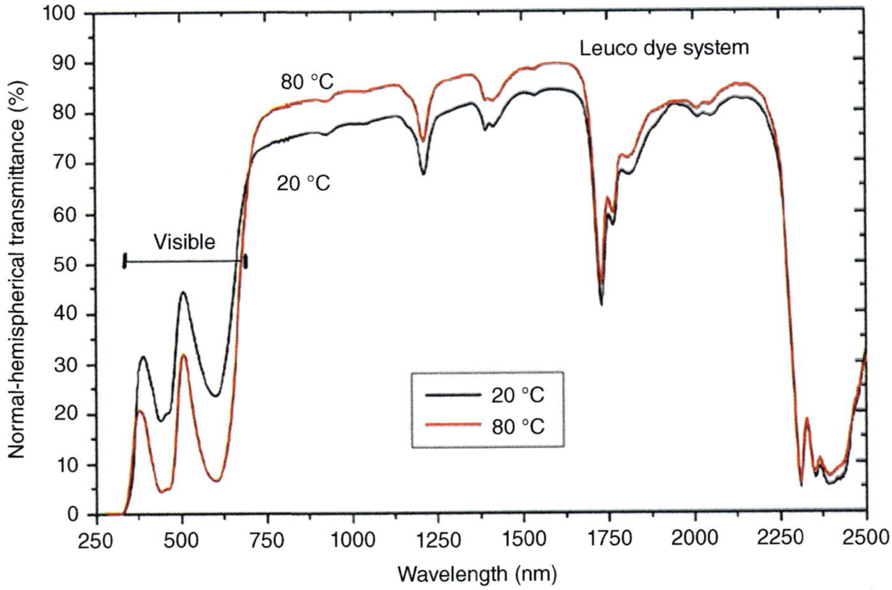

Figure 8.9 Normal-hemispherical transmittance of a ternary leuco-dye system, based on a green lactone dye, embedded in a polyethylene matrix. *Source:* [13] / MDPI / CC BY 4.0.

A serious drawback is that the leuco-dye systems experience problems with stability and durability.

In order to alleviate the stability problems, leuco dyes can be encapsulated in microspheres and used as a TC pigment in an organic binder. These materials have been considered for a rather different application, namely, as coatings on, or pigments in, facades and pavements to alleviate the so-called urban heat island problem [28]. The microspheres had a diameter of 5–10 μm and were encapsulated in the binder. In some cases, TiO_2 pigments were added to enhance light scattering [29]. The resulting paint coatings switch from a colored state at low temperatures to a white state at high temperatures. The diffuse reflectance increases considerably. Despite the encapsulation of the dye, photodegradation is a problem upon exposure to outdoor conditions [29].

Another side application concerns TC labels for detecting too high or too low storage temperatures for food. Here, the leuco dyes, together with other compounds, form the basis for inks that irreversibly or reversibly color at a preset temperature t_c [30].

8.3 Thermotropic Materials

A TT material exhibits different light-scattering properties below and above a critical temperature, t_c. The typical TT material with potential applications in solar-control smart glazing exhibits a transparent state with minimal light scattering below t_c. Above t_c, on the other hand, the scattering increases, and the material shows a milky appearance. There exist also TT materials exhibiting the opposite behavior (a transparent state above t_c and a milky state below), and they can find possible applications in privacy smart glazing. TT materials for solar control applications must necessarily be adequately characterized by the change in their directional-hemispherical transmittance (the transmitted/incident ratio of radiant flux) or by the change in their directional-hemispherical reflectance.

For a TT material to be useful in solar transmittance control, t_c must be close to the human comfort temperature ($t_{comfort} \sim 25\,°C$) [31], and the transmittance in the milky state must be lower than the transmittance in the clear state. This situation occurs (assuming that the optical absorption remains low) if a portion of the scattered solar light is directed backward [32]. Hence, to optimize the transmittance contrast between low- and high-temperature states, the backscattering should be maximized at high temperatures. This is a difficult problem in light-scattering theory, considering the possibility of multiple scattering events as light traverses a layer of TT material. It seems possible to develop design criteria for TT materials following the basic procedures outlined by Rogers et al. [33] for a related optimization problem. It seems that both high forward scattering and high backscattering are necessary for achieving a completely diffuse low transmittance [33]. Hence, to achieve a high-transmission contrast between the two states in a TT material, backscattering should be sufficiently strong [32]. We can obtain a simplified view of the conditions by considering the case of single scattering. Here, very small domains or particles do not scatter much, and particles with sizes significantly larger than the wavelength of light exhibit predominantly forward scattering. Hence there exists an optimal particle size region where scattering is strong and where backscattering exhibits a maximum. Multiple scattering modeling gives optimal domain or particle diameter of 300 nm, but the optimum is very broad toward larger particle sizes, and diameters up to 1000 nm give acceptable optical properties [32].

For window applications, basic demands of TT materials include color neutrality and minimal turbidity in the clear state as well as homogeneously distributed haze in the scattering state, good reversibility, long-term stability, and a switching temperature slightly above room temperature [15, 31, 34, 35]. Some systems, such as hydrogels, have been commercialized, but long-term stability has been an issue. TTs are somewhat opaque and scatter light in the hot switched states, which makes them useful for overhangs or clerestory windows but not necessarily for view windows, unless privacy is desired. Ionogels are the newest class of OTC materials and are still being developed along with other ionic polymer systems for applications such as batteries.

8.3.1 Hydrogels

Hydrogels consist of a mixture of a polymer with long chains and different amounts of an aqueous phase. The aqueous phase consists of water, which is thickened by a gelling agent or a second polymer [14]. The TT properties in thermally responsive hydrogels are caused by a phase separation process occurring at a critical temperature, t_c. Below t_c, we have a medium composed of gel particles consisting of a hydrophilic polymer/water mixture, which appears essentially homogeneous and transparent to the eye. In this situation, the optical properties can be described with a single effective refractive index. At t_c, there is a transition to a hydrophobic state of the polymer. This leads to the formation of polymer micro- or nanodomains surrounded by water. The difference in refractive index between the two components leads to light scattering. In particular, the refractive index of water (1.33) is much lower than the refractive index of most polymers (1.5–1.6), and this difference leads to strong light scattering in the high-temperature state. It is established that the difference in refractive index between the particles and the medium must be at least 0.03 in order to have sufficient scattering [31]. The optical switching between two states at t_c is illustrated schematically in Figure 8.10a. The polymer gel network can be cross-linked either chemically or physically. Chemical cross-linking leads to a significant volume change at the transition temperature, which is not desirable for many applications. On the other hand, physical crosslinking avoids this problem and improves the stability of the material [11]. Typically, the concentration of the polymer in the hydrogel affects t_c, which reaches a minimum at a certain concentration. Adding salt also affects t_c [11]. This minimum value of t_c, referred to as the lower critical solution temperature (LCST), depends on multiple factors [37]. Additionally, the miscibility diagram of a hydrogel usually presents an upper critical solution temperature (UCST) curve [38], which is defined as the maximum critical temperature above which the constituents of the hydrogel become miscible irrespective of their concentration. Therefore, the transparent state is, in principle, confined between the UCST and LCST curves, as shown in Figure 8.10b. The LCST transition is the one that is used in solar-control applications, while the UCST transition often takes place outside the typical operation temperature range of hydrogels [37].

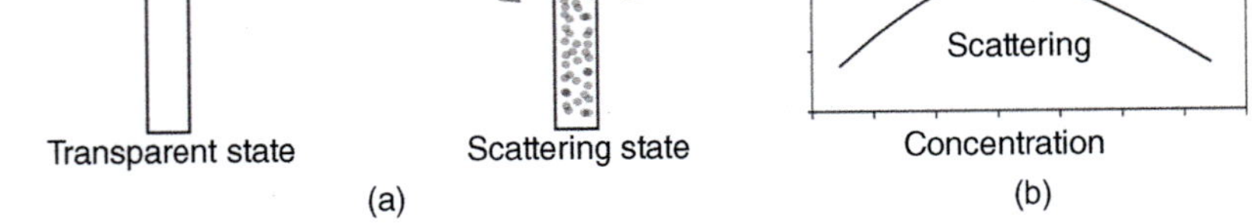

Figure 8.10 Schematic drawing of the optical switching of a thermochromic hydrogel system. Both the transparent and light-scattering states are shown in panel (a). Polymer blends display analogous properties. *Source:* [36]/with permission of ELSEVIER. Panel (b) shows an idealized miscibility diagram for a hydrogel showing upper critical solution temperature (UCST) and lower critical solution temperature (LCST) transitions.

The implementation of TT hydrogels in smart glazing presents several technological challenges. For example, the TT optical switching is usually accompanied by a significant contraction of the volume of the gel, which is challenging for window applications and often results in nonreversible deformations [13, 35]. Different approaches to hydrogel systems for keeping a constant volume of the gel during the transition have been reviewed by Seeboth et al. [39]. Additions of salt to the hydrogel are of interest both because the transition temperature can be easily controlled and because it leads to increased crosslinking [39, 40]. The presence of water in hydrogels limits the operation temperature unless anti-freezing agents are added. In addition, the gel must be rigorously sealed so that water does not diffuse into the ambient, which will decrease the optical switching performance. In fact, when the hydrogel is intended to be encapsulated between two window panes, high viscosity is preferred [34].

Achieving a significant modulation of the solar transmittance in TT hydrogels involves modulating not only the visible part of the solar spectrum (400–700 nm) but also the NIR part, from 700 to 2500 nm. In this sense, the size of the hydrogel nano- or microparticles can have a considerable effect on the scattering at longer wavelengths, significantly impacting the modulation of the solar radiation [35] and ultimately affecting the efficiency of hydrogels as solar transmittance control devices. In addition, particle size has a large impact on the fraction of backward/forward scattering. To achieve a good modulation, it is not enough that the solar light is scattered, but it has to be scattered in the backward direction, so the light is prevented from passing through the coating. Achieving an optimal solar backscattering requires optimization of the particle size [32].

Interestingly, hydrogels were used for solar protection in the Munich Zoo as early as the 1950s [13]. Later, the company Suntek (later Southwall, Mountain View, CA) attempted to commercialize hydrogels for window applications from the 1980s to the early 2000s. Their gel is reported to be based on poly(methyl vinyl ether) and methylene-bis-acrylamide [13]. Representatives of this work are patents US3953110 [12], US4307942 [41], and WO1993002380A1 [42]. In the early 2000s, Pleotint (West Olive, MI) manufactured a product known as ThermoSEE™ [22, 43]. Currently, Pleotint makes the successful ligand-exchange TC material known as Suntuitive™, discussed earlier in this chapter. The TALD (*Temperaturabhängige Lichtdurchlässigkeit*) hydrogel systems were developed in the 1990s [44]. The TALD gel consists of colloidal particles of polyether/water mixture surrounded by a gel of carboxyvinyl polymer and water [45]. The normal-hemispherical transmittance and reflectance properties have been measured [44]. Another system "Affinity Intelligent Window" was introduced to the market in 1998 [40]. The hydrogel consisted of a cellulose derivative, an amphipathic compound, NaCl, and water. All these systems showed strong TT switching at a temperature that could be varied within wide limits by changing the concentration of the constituents of the hydrogel.

The normal-hemispherical transmittance spectra of the Suntek product, commercially known as Cloud Gel™, are shown at several temperatures in Figure 8.11 [14]. The diffuse reflectance increases as the transmittance decreases since the material is essentially non-absorbing. The visible transmittance contrast is excellent, and the optical switching occurs over a relatively narrow temperature range, as compared to other organic TC and TT materials. The switching temperature could be adjusted between 10 and 70 °C by varying the composition of the hydrogel. The domains in the high-temperature state were of the order of 100 and 400 nm in size for two different formulations of Cloud Gel™ [46].

During the last decade, interest in research on hydrogels has been revived. The research field has been reviewed in detail by Maiorov [15] and also more briefly by Ke et al. [47] and

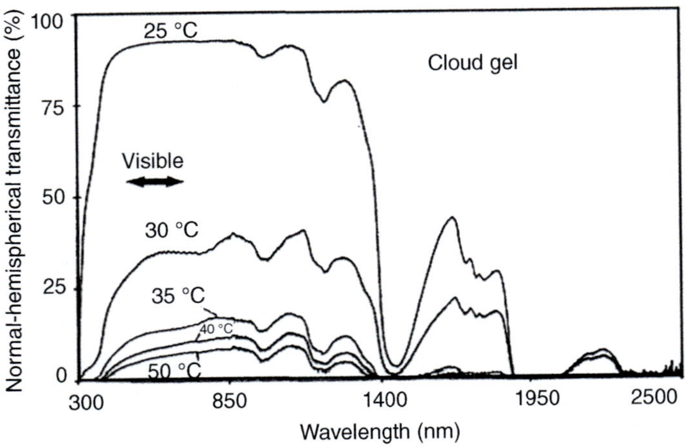

Figure 8.11 Normal-hemispherical transmittance of Cloud Gel layer with thickness of 1 mm between two polymer films (Suntek Inc.). Dips in NIR spectra at ~1450 and 1930 nm correspond to water absorption. Shown are measurements using an integrating sphere at temperatures ranging from 25 to 50 °C. As noted in [14], light scattered at high angles did not enter the integrating sphere. This effect leads to too low transmittance values for the light-scattering (high temperature) states, especially in the short-wavelength range. Model calculations show that errors may be substantial [32], although difficult to quantify. *Source:* [14] / Society of Photo-Optical Instrumentation Engineers.

Feng et al. [48]. Many different formulations of the hydrogel system have been studied [15], but hydrogels based on poly(*N*-isopropyl acrylamide) (PNIPAM) polymer have received especially great interest. This is because PNIPAM exhibits a steep TT transition at slightly above room temperature (30–35 °C) and is easy to fabricate [15]. In addition, PNIPAM hydrogel particles allow for large adjustability of their scattering properties as a function of particle size, water concentration, and internal structure [35]. A problem with recent work is the lack of normal-hemispherical optical measurements in most cases [15]. Therefore, the performance of the hydrogels in smart window applications usually cannot be assessed. In 2019, Li et al. [35] reported a TT hydrogel-based device with a luminous transmittance T_{lum} in the clear state above 80% and a very large change in direct–direct solar transmittance. The optical properties were optimized by tailoring the particle sizes in the light-scattering state. The device, which was based on poly(*N*-isopropylacrylamide)-2-aminoethylmethacrylate hydrochloride (PNIPAM-AEMA) microparticles, had a t_c slightly above room temperature. In this case, an estimate of total solar modulation was indeed possible since the authors put their sample on top of a Si solar cell illuminated with a solar simulator. It was found that the change in solar energy transmittance was almost 70% [35], which means that this material shows properties at almost the same level as the Cloud Gel. Another interesting recent development concerns a hydrogel material that is able to switch between colors, in addition to the TT switching [49].

There appears to be little open data on the thermal stability of hydrogels, but the durability of the systems attempted for commercialization has been studied. The manufacturer of Cloud Gel has carried out various studies including solar spectrum irradiation, heat aging, and thermal cycling for 200,000 cycles [14]. Data from the Fraunhofer Institute for Solar Energy Systems demonstrated excellent stability of the optical properties over at least 4000 thermal cycles [14]. The TALD hydrogels have been studied over 4000 cycles, also showing good reversibility [44]. The PNIPAM-based device of Li et al. [35] showed good stability over 1000 cycles. The Affinity window was subjected to different stability tests, including outdoor exposure for two years [40]. It was found that the hydrogel must be protected against UV irradiation with wavelengths shorter than 330 nm.

The TT switching is usually achieved in a passive way by increasing the temperature of the coating by sunlight illumination. Techniques for triggering the TT transition earlier in time or when the available solar radiation is not intense enough include the use of transparent electrodes to increase the temperature of the hydrogel by Joule heating. This approach, however, requires a source of electricity to function and transparent electrodes that typically involve oxides, such as $Sn{:}In_2O_3$ (ITO) [50], $F{:}SnO_2$ (FTO), or silver particles [51]. Alternatives to promoting an early TT switching in a passive way include the dispersion of plasmonic nanoparticles in the hydrogel matrix such as VO_2 [52] or $Sb{:}SnO_2$ (ATO) [53]. These plasmonic nanoparticles absorb solar radiation in the spectral region around their plasmon resonance wavelength (λ_p), which can be modified by doping, e.g. in the case of ATO, by changing the Sb content [54]. The value of λ_p is normally chosen to be in the NIR region. Hence, the plasmonic particles help to reduce the transmission of NIR solar radiation by increasing absorption, without reducing the transparency in the visible region of the clear state. At the same time, the plasmonic electron resonance acts as a nano-heater, raising the temperature of the surrounding hydrogel for faster TT switching. In the case of VO_2/hydrogel systems, it can also provide modulation of the NIR radiation [52].

By adding sodium dodecyl sulfate (SDS)/potassium tartrate (PTH) micelles into PNIPAM, Yu et al. [55] developed a composite hydrogel able to retain transparency only in a narrow range of temperatures. This is achieved by exploiting both the LCST and UCST transitions. On this basis, such hydrogels behave as follows (see Figure 8.10b as a guide): Starting at a low temperature, the system presents a milky appearance. As the temperature increases, the hydrogel reaches the first critical point at which a UCST-type transition takes place, causing a notable increase in transparency. Then, the hydrogel remains transparent in a certain temperature range until the temperature is high enough so that the LCST-type transition takes place and the hydrogel becomes milky again. The chemistry of the system can be adjusted so that the transparent state takes place in a narrow temperature range around $t_{comfort}$. The LCST transition can be exploited for applications in solar control, whereas the UCST transition can be used for privacy purposes (at night, when the temperature is low, the hydrogel remains in the turbid state blocking the view).

8.3.2 Polymer Blends

Polymer blends avoid some of the problems alluded to above that are associated with using water as a component in TT materials. The miscibility of polymers often depends on temperature, and the mixtures of two polymers can exhibit an LCST, which is of interest for TT materials. Below the LCST, the polymers form a homogeneous mixture with a homogeneous effective refractive index. At the LCST, a phase separation occurs, and at higher temperatures, one polymer is distributed as domains in a matrix of the other. The materials are chosen so that the refractive indices are different in this case, and light scattering is exhibited. The transition between the low- and high-temperature states around the LCST is rather wide (30–50°C, for example, in Figure 8.12).

Blends consisting of thermoplastic material in a cross-linked matrix display good reversibility of the phase separation and good mechanical properties; structures with a semi-interpenetrating network are especially beneficial [13]. One example is a material consisting of a cross-linked matrix of polystyrol and hydroxyethyl methacrylate mixed with

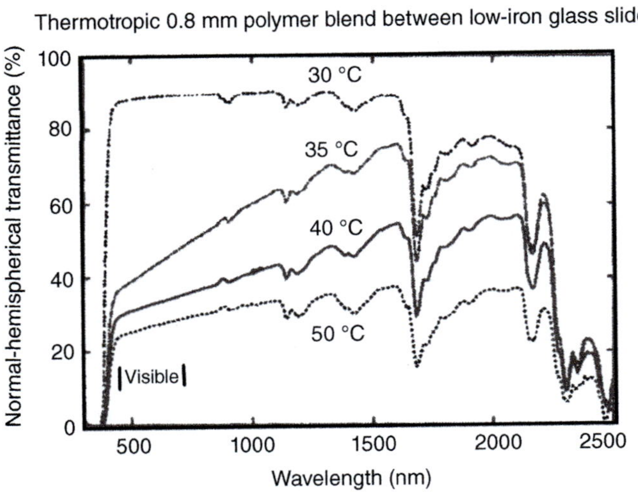

Figure 8.12 Normal-hemispherical transmittance of PS/HEMA:PPO polymer between two glass slides. The volume fraction of the two components was 50%. The thickness of the layer was 800 μm. Measurements were carried out with large integrating sphere, (620 mm diameter) with a 100 mm diameter transmittance aperture, thereby minimizing errors [56]. *Source:* [56] / John Wiley & Sons.

polypropylene oxide (PPO) [14]. The crosslinking can be achieved either thermally or by photo-initiation. The matrix has a refractive index of $n = 1.582$, while it is $n = 1.45$ in pure PPO. At low temperatures, PPO is bound to the matrix with hydrogen bonds. These bonds break at higher temperatures, and the homogeneous structure dissociates into a mixture of two components [14]. The polymer matrix network improves the stability of the high-temperature phase and the reversibility of the switching. Pronounced optical switching is observed in the visible and part of the NIR wavelength range, as shown in Figure 8.12. The refractive index difference is not as large as for the hydrogels, and this leads to a higher diffuse transmittance at high temperatures. The light-scattering properties of this material have been analyzed theoretically in detail [57].

The polymer blends showed good stability over 2000 thermal cycles, but the switching range decreased by 10% after 43 days of simulated solar irradiation (1 kW m^{-2}) at 90 °C [14]. Outdoor tests over more than six months did not show any major difference in switching performance, although the transition back to the low-temperature clear state became progressively slower [58].

These types of polymer blends were developed for use in smart windows in the 1990s and early 2000s at the Fraunhofer Institute for Solar Energy Systems in Freiburg, the BASF company, and by the company Interpane, among others [36], and prototype windows were produced. However, due to technological and marketing problems, it appears that the products were never commercialized.

Another type of polymeric TT material consists of an immiscible blend of two polymers with different temperature-dependent refractive indices. They are close to equal below a certain transition temperature. Above the transition temperature, the differences between the refractive indices gradually increase, and the material becomes light scattering [13]. Figure 8.13 gives a schematic picture of the transition. However, also in this case, the transition region is very broad (35–95 °C), but the optical switching is quite promising. For a

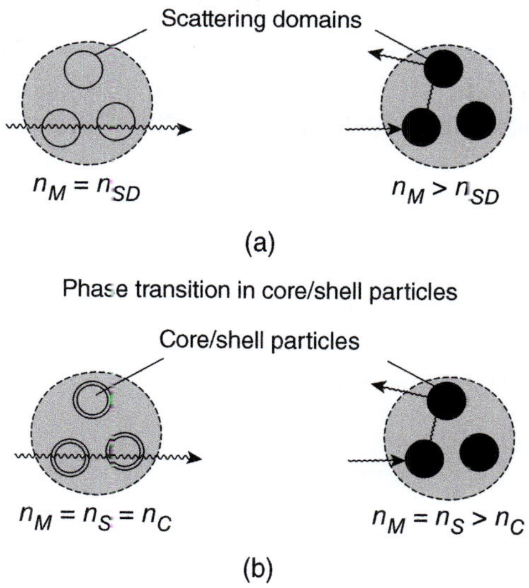

Figure 8.13 Schematic picture of the switching mechanism of TT materials with a phase transition, for example (a) casting resins and some polymer blends and (b) polyethylene with TC core–shell particles. *Source:* Image [13] / MDPI / CC BY 4.0.

layer with thickness of 2 mm, the normal-hemispherical solar transmittance switched from 0.82 at 25 °C to 0.57 at 95 °C [13].

8.3.3 Ionogels

Ionogels are an interesting alternative to hydrogels in which, instead of water, an organic molten salt, a so-called ionic liquid (IL), is immobilized in a polymer matrix. When compared to TT hydrogels, the substitution of water for an IL comes with an extra degree of freedom for manipulating the liquid/polymer miscibility. Other advantages are a much lower vapor pressure (lower volatility) and stability in a wide range of environmental conditions [59, 60]. On the other hand, the refractive index of ILs displays values above 1.4, i.e. significantly higher than that of water [61, 62]. This indicates that light scattering in a TT ionogel will not reach the same high level found in TT hydrogels. There also exist ILs with much higher refractive indices, but they show significant absorption in part of the visible wavelength range [62].

Ionogels can find applications in diverse fields, including fuel cells, electrolytic membranes, solar cells, and wearable electronics, among others [63]. Usually, these applications exploit the intrinsic properties of the IL, and the polymeric matrix only plays a scaffolding role [59, 64]. By contrast, applications of ionogels in glazing must exploit the interaction between the polymeric matrix and the IL. The polymeric matrix/ionic liquid interaction is, in some cases, altered by temperature changes, leading to TT ionogels [65–67]. A brief review of recent developments in the fields of TC IL-based materials and TT ionogels was given by Feng et al. [47].

The operating principle of TT ionogels is the same as in TT hydrogels, Figure 8.10a: TT ionogels present a turbid (milky phase) and a transparent state, and the switching between states can take place through an UCST and/or a LCST phase transition. The transition of interest for solar-control purposes is the LCST one (clear state below t_c, milky state above t_c), as for hydrogels [64], where the UCST phase transition (milky state below t_c, clear state above t_c) could find uses in glazing for privacy purposes [60], or other applications outside the scope of the present chapter, e.g. information encryption [59].

Examples of ionogels exhibiting LCST phase transitions, and hence with applications in solar control, are imidazolium-based ILs in a cross-linked poly(propylene oxide) (PPO) matrix [64], or binary IL mixtures consisting of 1-butyl-2,3-dimethylimidazolium bis(trifluoromethylsulfonyl)imide ([Bdmim][TFSI]) and 1-butyl-3-methylimidazolium bis(trifluoromethylsulfonyl)imide ([Bmim][TFSI]), in poly(urethaneurea) (PU) matrices with acylsemicarbazide (ASCZ) moieties [68].

The stability of PPO/IL and PU/IL ionogels has been proven to be excellent over 5000 [64] and 1000 cycles [68], respectively. Unfortunately, the solar modulation achieved by TT ionogels is not possible to quantify from literature data, since, to our knowledge, most works present results on changes in direct–direct transmittance instead of the direct-hemispherical (or total) transmittance.

8.3.4 Casting Resins

TT casting resins have been reviewed by Seeboth et al. [11, 13]. The casting resins are mixed with long-chain polymers (alkanes or fatty acid esters) with a phase transition in the temperature range between 10 and 50 °C. The additive is present as domains in the resin matrix. If the materials are chosen in a suitable way, their refractive indices are equal at low temperatures. The additive exhibits a phase transition, or structural rearrangements, with an associated change in the refractive index, while the resin matrix does not exhibit any significant change in refractive index. At higher temperatures, the mixture becomes light scattering (Figure 8.13).

The interest in TT casting resins originates from considerations relating to ease of production. Some types of laminated glass are produced by filling a casting resin into a space between two glass sheets and curing the resin with UV light. Producing a TT material by introducing an additive into the resin poses no technical problems, and production lines for laminated glass can be used directly [11]. Various attempts at commercializing TT casting resins have been made. In 2000, a layer of the kind outlined above called T-Opal was developed [11]. The transition temperature at which the additive experiences a discontinuous change in the refractive index could be varied from 10 to 50 °C.

The additive can also be encapsulated before being mixed with the resin. This has advantages for production reasons. The microcapsules can be produced in different sizes, and their light-scattering properties above the transition temperature are therefore easier to optimize [11]. Systems employing microencapsulating paraffin in a curable matrix can be filled into the space between two glass panes before curing the resin. Prototype windows using this concept were produced by Okalux GmbH in the early 2000s [36] but were never commercialized. In addition, the optical switching appears moderate with the best reported results of the normal-hemispherical solar transmittance difference being about 0.18 [36].

Later, a material consisting of polyacrylate core–shell microcapsules in a UV-curable acrylate-based resin was developed [69]. The normal-hemispherical transmittance was measured, and the switching was significant both in the visible and part of the NIR ranges (Figure 8.14). It was found that microcapsules with a core size of 380 nm exhibited the best properties. Production of this glass laminate was undertaken by Tilse Formglas GmbH (Germany), and the product SOLARDIM-Eco was released to the market around 2014 [11]. The glass laminate has a resin layer with thickness of 1.7 mm and exhibits a difference in solar transmittance between low- and high-temperature states of the order of 0.2–0.3 [11]. The company also commercialized IGU windows using a laminate pane and an additional

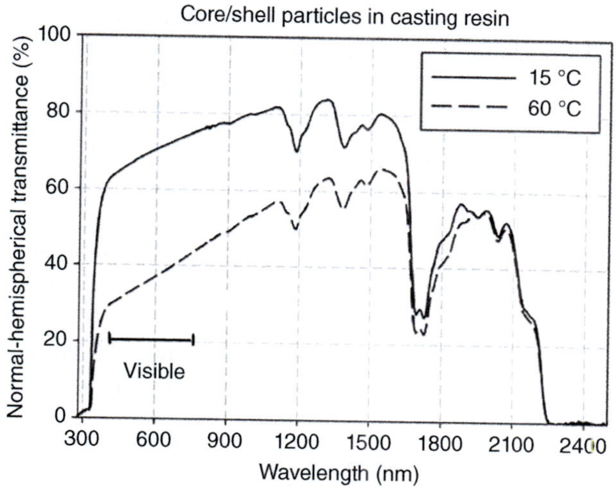

Figure 8.14 Normal-hemispherical transmittance of a TT material based on a casting resin containing 6 wt% core–shell particles in acryloyl morpholine (15 wt%) and an acrylate-based resin. *Source:* [69] / with permission of ELSEVIER.

glass pane facing the inside. SHGC differences of 0.18 and 0.16 were obtained, when the second glass pane was clear glass and low-*e*-coated glass, respectively [70]. Tilse Formglas specializes in curved laminated glazing (for marine applications), and the TT layers can be incorporated into those as well.

8.3.5 Additional Materials

Polyolefin films have been studied as a TT material, partially because of the possibility of using commercial extrusion processes to produce them. Polyethylene with additions of TC core–shell particles was produced, but the content of TC particles had to be limited due to technical problems (Figure 8.13). The resulting mixtures showed a moderate direct transmittance switching in the visible wavelength range, however, only direct transmittance data were reported [13].

In addition, hydrogels can be prepared to include nanoparticles as a component. This would in theory allow more freedom in achieving high light scattering. In one approach, core–shell nanoparticles of PNIPAM on polystyrene cores were dispersed in water. At low temperatures, the shell is hydrophilic, and a gel state was obtained. Above the LCST, the shell becomes more hydrophobic, and the particle size decreases considerably. Particle sizes shrink to the order of wavelengths of visible light, leading to a highly light-scattering state [13].

8.4 Discussion and Comparison

In this section, we compare the approaches to OTC and TT layers and window constructions, that were discussed above, and where sufficient performance data are available. For TT materials, we will only consider the cases where reliable normal-hemispherical transmittance data over the whole solar wavelength range are available. Unfortunately, many researchers report only normal–normal transmittance data, which results in misleadingly large changes in transmittance $\Delta T_{vis,sol}$. In the low-temperature clear state, the material does not exhibit light scattering, and the normal–normal transmittance should be close to the normal-hemispherical transmittance. However, in the high-temperature light-scattering state, the normal–normal transmittance is usually very much lower than the normal-hemispherical transmittance, so the switching range will be erroneously high if it is based on normal–normal transmittance data. Many TT materials scatter light, so these considerations must be taken into account. According to standard EN14500 2008, normal–normal transmittance for light-scattering materials is the fraction of the normally incident radiation flux that is transmitted in accordance with the laws of geometric optics without diffusion or redirection.

Tables 8.2 and 8.3 compare the results of OTC glazing materials where the OTC layers were either freestanding or laminated between two glass sheets. Although the data show

Table 8.2 Total transmittance data for different organic thermochromic and thermotropic materials. Visible (vis) and solar (sol) transmittance in the low-temperature (LT) and high-temperature (HT) states are displayed together with the modulation of visible (ΔT_{vis}) and solar (ΔT_{sol}) transmittance.

Normal-hemispherical transmittance	$T_{vis,LT}$	$T_{vis,HT}$	$T_{sol,LT}$	$T_{sol,HT}$	ΔT_{vis}	ΔT_{sol}
Organic thermochromic						
Pleotint [23]	0.71	0.07	0.63	0.23	0.64	0.40
LETC-VO_2 hybrid $d = 50\,\mu m$ [26]	0.65	0.52	0.69	0.44	0.12	0.24
Leuco dye (blue) [13]	0.61	0.34	0.68	0.62	0.27	0.06
Thermotropic materials						
Cloud Gel [14]	0.92	0.06[a)]	0.82	0.05[a)]	0.86	0.77[a)]
Polymer blend [56] PS/HEMA:PPO	0.89	0.26	0.83	0.27	0.63	0.56
Immiscible polymer blend PA/P(E-*co*-GMA) [13]			0.82	0.57		0.25
Casting resin $d = 1.7$ mm [69]	0.76	0.51	0.75	0.54	0.24	0.21
Casting resin [69, 70]; Tilse Formglas	0.69	0.35	0.69	0.41	0.34	0.28

a) See comment on Figure 8.11.

Table 8.3 Total transmittance data for two-pane IGUs, where one pane consists of organic thermochromic and thermotropic materials between two glass sheets. Visible (vis) and solar (sol) transmittance in the low-temperature (LT) and high-temperature (HT) states are displayed together with the modulation of visible (ΔT_{vis}) and solar (ΔT_{sol}) transmittance.

Normal-hemispherical transmittance	$T_{vis,LT}$	$T_{vis,HT}$	$T_{sol,LT}$	$T_{sol,HT}$	ΔT_{vis}	ΔT_{sol}
Organic thermochromic						
Pleotint (Table 8.1)[a)]	0.55	0.08	0.28	0.08	0.47	0.20
Thermotropic materials						
Hydrogel IGU (probably based on Affinity hydrogel) [36]	0.74	0.16	0.44	0.09	0.58	0.35
Polymer blend IGU PS/HEMA:PPO [36]	0.73	0.21	0.44	0.13	0.52	0.31
Casting resin IGU [36]	0.71	0.44	0.47	0.29	0.27	0.18

a) One example from Table 8.1.

the optical properties of TT laminates and windows, they do not characterize the angularity or distribution of scatter or view through the window during or after switching. For Pleotint, leuco dye, and LETC-VO_2 TCs, the image quality is preserved without a light-scattering effect after complete switching.

It is seen that all the materials in Table 8.2 are very transparent at low temperatures, with visible transmittances in excess of 0.6. The hydrogels and polymer blends stand out as being the most transparent ones with a transmittance equal to that of glass. These materials, together with the Pleotint OTC, also display the highest modulations of visible transmittance. The solar transmittance and its modulation are more interesting for evaluating energy efficiency. The modulation of solar transmittance is greater than 0.7 for the "Cloud Gel" hydrogel. This is to a large extent because it was optimized not only for the visible spectral range but also to exhibit a considerable modulation of NIR transmittance. In addition, high values around 0.4–0.5 were found for the polymer blends and the Pleotint LETC material.

The same data are shown in Figure 8.15, which shows the difference in solar transmittance between low- and high-temperature states as a function of visible transmittance. In addition some additional data points for TALD hydrogel [44] and different Cloud Gels [46] are shown. The OTC materials are seen to outperform the inorganic TC materials considered in Chapter 7, with the hydrogels showing desirable properties furthest up into the upper right hand corner. It is also observed that the non-scattering Pleotint LETC material exhibits a significant advantage over the inorganic TC coatings.

Smart window constructions based on TT materials have also been reported [36, 40, 58]. The hydrogel used in the Japanese construction is not specified in detail but is reported to contain as additives "amphipathic molecules and NaCl" [40]. The polymer blends and the

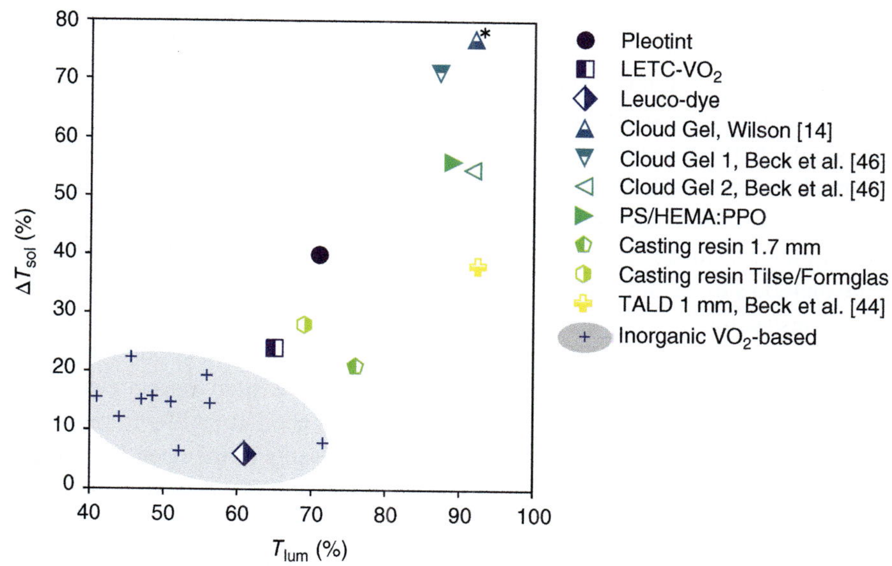

Figure 8.15 Difference in integrated solar transmittance between low and high temperature states is shown as a function of luminous transmittance for the organic thermochromic and thermotropic materials listed in Table 8.2. In addition, data from Beck et al. for the TALD hydrogel (ref. [44]) and two different Cloud Gels (ref. [46]) are displayed. The shaded area shows the performance of inorganic VO_2-based coatings (Chapter 7), for comparison. The crosses in the shaded area depict the data for VO_2-based coatings given in Table 7.1. *See comment on Figure 8.11.

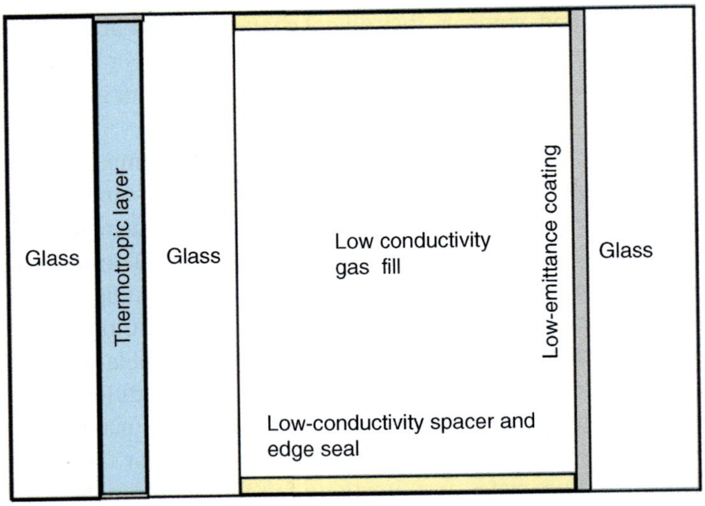

Figure 8.16 Schematic view of an insulating glass unit with a thermotropic layer and a glass with a low-e coating (not to scale).

casting resins used by Nitz and Hartwig [36] were described in Sections 8.3.2 and 8.3.4. The TT layer was laminated between two glass panes, and this construction was combined with a third glass pane coated on the inside with a low-*e* coating, as shown in Figure 8.16. The optical switching properties of three window designs of this type were studied by Nitz and Hartwig [36]. The results are included in Table 8.3, together with data for a Pleotint Suntuitive IGU from Table 8.1, for comparison.

It is observed that transmittances are generally significantly lower than in Table 8.2, because of the presence of the second glass pane and its low-emittance coating. However, the trends are similar to those of Table 8.2, with the hydrogels and polymer blends giving the highest solar modulation, closely followed by the Pleotint window. It can also be observed that the values for the different entries are closer in Table 8.3 than in Table 8.2. This is because low-emittance coatings reflect part, or in some cases, most of the NIR. Therefore, optimizing the materials for achieving large NIR modulation may not give a large advantage in this window construction.

It is evident that many of the materials covered in this review have "good enough" optical modulation and energy efficiency. Long-term durability is also an essential aspect in applications. Here, the Pleotint Suntuitive windows have experience of good durability from many installations. For the TT materials, we have found just a few results in the open literature, and it is not clear whether some of them are more durable than others, despite occasional statements in the literature.

In conclusion, it will depend on factors other than optical modulation, which materials will actually be used in smart window construction. Here, ease of manufacturing, production technology, cost, economy, and market demand will certainly play a pivotal role. OTC materials, like Pleotint, have the advantage that they allow visual contact between indoors and outdoors at both low and high temperatures. This is desirable in the majority of window applications. TT materials can be used in niche applications, for example, skylights, roof windows, and overhangs, where the see-through aspect is not essential.

References

1 Meyer, H. (1909). Über neue Derivate des Anthrachinons. *Monatshefte Für Chemie Und Verwandte Teile Anderer Wissenschaften* 30: 165–177. https://doi.org/10.1007/BF01518116.

2 Dilthey, W., Berres, C., Hölterhoff, E., and Wübken, H. (1926). Beitrag Zur Kenntnis der Spiro-di-benzopyrane (Heteropolare Kohlenstoffverbindungen. IV). *Journal für Praktische Chemie* 114: 179–198. https://doi.org/10.1002/prac.19261140108.

3 Cambi, L. and Szegö, L. (1931). Über die magnetishe susceptibilität der komplexen Verbindungen. *Berichte der Deutschen Chemischen Gesellschaft (A and B Series)* 64: 2591–2598.

4 Zhang, W., Schenning, A.P.H.J., Kragt, A.J.J., Zhou, G., and De Haanm, L.T. (2021). Reversible thermochromic photonic coatings with a protective topcoat. *ACS Applied Materials & Interfaces* 13: 3153–3160. https://doi.org/10.1021/acsami.0c19236.

5 Day, J.H. (1963). Thermochromism. *Chemical Reviews* 63: 65–80. https://doi.org/10.1021/cr60221a005.

6 Chowdhury, M.A., Butola, B.S., and Joshi, M. (2013). Application of thermochromic colorants on textiles: temperature dependence of colorimetric properties. *Coloration Technology* 129: 232–237. https://doi.org/10.1111/cote.12015.

7 Wang, Y., Ren, J., Ye, C., Pei, Y., and Ling, S. (2021). Thermochromic silks for temperature management and dynamic textile displays. *Nano Micro Letters* 13: 72. https://doi.org/10.1007/s40820-021-00591-w.

8 Smith, C.R., Sabatino, D.R., and Praisner, T.J. (2001). Temperature sensing with thermochromic liquid crystals. *Experiments in Fluids* 30: 190–201. https://doi.org/10.1007/s003480000154.

9 Roberts, G.T. and East, R.A. (1996). Liquid crystal thermography for heat transfer measurement in hypersonic flows – a review. *Journal of Spacecraft and Rockets* 33: 761–768. https://doi.org/10.2514/3.26835.

10 Thamrin, E.S., Warsiki, E., Bindar, Y., and Kartika, I.A. (2022). Thermochromic ink as a smart indicator on cold product packaging – review. *IOP Conference Series: Earth and Environmental Science* 1063: 012021. https://doi.org/10.1088/1755-1315/1063/1/012021.

11 Seeboth, A. and Lötzsch, D. (2014). *Thermochromic and Thermochromic Materials*. Singapore: Pan Stanford Publishing.

12 Chahroudi, D. (1976). Transparent thermal insulation system. Patent US3953110, granted: 1976-04-27. Assigned: Southwall.

13 Seeboth, A., Ruhmann, R., and Mühling, O. (2010). Thermotropic and thermochromic polymer based materials for adaptive solar control. *Materials (Basel, Switzerland)* 3: 5143–5168. https://doi.org/10.3390/ma3125143.

14 Wilson, H. (1994). Potential of thermotropic layers to prevent overheating: a review. In: *Proceedings of the SPIE 2255, Optical Materials Technology for Energy Efficiency and Solar Energy Conversion XIII* (ed. V. Wittwer, C.G. Granqvist, and C.M. Lampert), 214–225. SPIE.

15 Maiorov, V.A. (2020). Optical properties of thermotropic hydrogels (a review). *Optics and Spectroscopy* 128: 367–386. https://doi.org/10.1134/S0030400X20030133.

16 Sone, K. and Fukuda, Y. (1987). *Inorganic Thermochromism*. Berlin: Springer. https://doi.org/10.1007/978-3-642-51017-5.

17 Byker, H.J., Ogburn, P.H., Vander Griend, D.A., Veldkamp, B.S., and Winkle, D.D. (2011). Ligand exchange thermochromic (LETC) systems. Patent US20090283728 A1. Granted 2011-09-13.

18 Byker, H.J. (2023). Pleotint LLC, Personal communication.

19 Queen, M.S. (2014). Development of the molecular level description for Nickel(II)-based ligand-exchange thermochromism. Ph.D. Thesis. Montana State University, Bozeman, MT.

20 Queen, M.S., Towey, B.D., Murray, K.A., Veldkamp, B.S., Byker, H.J., and Szilagyi, R.K. (2013). Electronic structure of [Ni(II)S_4] complexes from S K-edge X-ray absorption spectroscopy. *Coordination Chemistry Reviews* 257: 564–578. https://doi.org/10.1016/j.ccr.2012.07.020.

21 Byker, H.J. and Ogburn, P.H. (2000). Thermochromic Devices, Patent US6084702. Granted 2000-07-04. Assigned: Pleotint LLC.

22 Byker, H.J. and Ogburn, P.H. (2002). Thermochromic devices, Patent US6446402. Granted 2002-09-10. Assigned: Pleotint LLC.

23 Giovannini, L., Favoino, F., Pellegrino, A., Lo Verso, V.R.M., Serra, V., and Zinzi, M. (2019). Thermochromic glazing performance: from component experimental characterisation to whole building performance evaluation. *Applied Energy* 251: 113335. https://doi.org/10.1016/j.apenergy.2019.113335.

24 Millett, F., Broekhuis, M., Byker, H.J., and Ogburn, P.H. (2010). Thermochromic window structures, Patent US 20080092456, Granted, 2010-10-19. Assigned: Pleotint LLC.

25 LBNL (2023). *Windows Software Downloads and Users Manuals, Windows 7.8*. Berkeley, CA, USA: LBNL. http://windows.lbl.gov/software-tools (accessed 24 April 2024).

26 Zhu, J., Huang, A., Ma, H., Chen, Y., Zhang, S., Ji, S., Bao, S., and Jin, P. (2017). Hybrid films of VO_2 nanoparticles and a nickel(II)-based ligand exchange thermochromic system: excellent optical performance with a temperature responsive colour change. *New Journal of Chemistry* 41: 830–835. https://doi.org/10.1039/C6NJ03369E.

27 Tang, H., MacLaren, D.C., and White, M.A. (2010). New insights concerning the mechanism of reversible thermochromic mixtures. *Canadian Journal of Chemistry* 88: 1063–1070. https://doi.org/10.1139/V10-069.

28 Santamouris, M., Synnefa, A., and Karlessi, T. (2011). Using advanced cool materials in the urban built environment to mitigate heat islands and improve thermal comfort conditions. *Solar Energy* 85: 3085–3102. https://doi.org/10.1016/j.solener.2010.12.023.

29 Karlessi, T., Santamouris, M., Apostolakis, K., Synnefa, A., and Livada, I. (2009). Development and testing of thermochromic coatings for buildings and urban structures. *Solar Energy* 83: 538–551. https://doi.org/10.1016/j.solener.2008.10.005.

30 Mysteria Colorum (n.d.). www.MyCol.si (accessed 7 November 2023).

31 Resch, K. and Wallner, G.M. (2009). Thermotropic layers for flat-plate collectors—a review of various concepts for overheating protection with polymeric materials. *Solar Energy Materials & Solar Cells* 93: 119–128. https://doi.org/10.1016/j.solmat.2008.09.004.

32 Nitz, P., Ferber, J., Stangl, R., Rose Wilson, H., and Wittwer, V. (1998). Simulation of multiply scattering media. *Solar Energy Materials & Solar Cells* 54: 297–307. https://doi.org/10.1016/S0927-0248(98)00081-6.

33 Benjamin Rogers, W., Corbett, M., Magkiriadou, S., Guarillof, P., and Manoharan, V.N. (2014). Breaking trade-offs between translucency and diffusion in particle-doped films. *Optical Materials Express* 4: 2621–2631. https://doi.org/10.1364/OME.4.002621.

34 Seeboth, A., Schneider, J., and Patzak, A. (2000). Materials for intelligent sun protecting glazing. *Solar Energy Materials & Solar Cells* 60: 263–277. https://doi.org/10.1016/S0927-0248(99)00087-2.

35 Li, X.-H., Liu, C., Feng, S.-P., and Fang, N.X. (2019). Broadband light management with thermochromic hydrogel microparticles for smart windows. *Joule* 3: 290–302. https://doi.org/10.1016/j.joule.2018.10.019.

36 Nitz, P. and Hartwig, H. (2005). Solar control with thermotropic layers. *Solar Energy* 79: 573–582. https://doi.org/10.1016/j.solener.2004.12.009.

37 Halperin, A., Kröger, M., and Winnik, F.M. (2015). Poly(*N*-isopropylacrylamide) phase diagrams: fifty years of research. *Angewandte Chemie, International Edition* 54: 15342–15367. https://doi.org/10.1002/anie.201506663.

38 Seuring, J. and Agarwal, S. (2013). Polymers with upper critical solution temperature in aqueous solution: unexpected properties from known building blocks. *ACS Macro Letters* 2: 597–600. https://doi.org/10.1021/mz400227y.

39 Seeboth, A., Kriwanek, J., Lötzsch, D., and Patzak, A. (2002). Chromogenic polymer gels for reversible transparency and color control with temperature at a constant volume. *Polymers for Advanced Technologies* 13: 507–512. https://doi.org/10.1002/pat.218.

40 Watanabe, H. (1998). Intelligent window using a hydrogel layer for energy efficiency. *Solar Energy Materials & Solar Cells* 54: 203–211. https://doi.org/10.1016/S0927-0248(98)00072-5.

41 Chahroudy, D. (1981). Solar control system. Patent US4307942, Granted 1981-03-10. Assigned: Southwall.

42 Chahroudi, D. (1993). Automatic light valves. Patent WD1993002380A1, pub. 1993-02-04. Assigned: Chahroudi.

43 Lampert, C.M. (2001). Progress in switching windows. In: *Solar and Switching Materials, Proceedings of the SPIE 4458* (ed. C.M. Lampert, C.-G. Granqvist, and K.L. Lewis), 95–103. San Diego, CA.

44 Beck, A., Hoffmann, T., Körner, W., and Fricke, J. (1993). Thermochromic gels for control of insolation. *Solar Energy* 50: 407–414. https://doi.org/10.1016/0038-092X(93)90061-R.

45 Eck, W., Cantow, H.-J., and Wittwer, V. (1999). Material having temperature defendant (dependent) light transmission. Patent US 5916972A, granted: 1999-06-29. Assignee: BASF, DE.

46 Beck, A., Körner, W., Scheller, H., Fricke, J., Platzer, W.J., and Wittwer, V. (1995). Control of solar insolation via thermochromic light-switching gels. *Solar Energy Materials & Solar Cells* 36: 339–347. https://doi.org/10.1016/0927-0248(94)00171-N.

47 Ke, Y., Zhou, C., Zhou, Y., Wang, S., Chan, S.H., and Long, Y. (2018). Emerging thermal-responsive materials and integrated techniques targeting the energy-efficient smart window application. *Advanced Functional Materials* 28: 1800113. https://doi.org/10.1002/adfm.201800113.

48 Feng, Y.Q., Lv, M.L., Yang, M., Ma, W.X., Zhang, G., Yu, Y.Z., Wu, Y.Q., Li, H.B., Liu, D.Z., and Yang, Y.S. (2022). Application of new energy thermochromic composite thermosensitive materials of smart windows in recent years. *Molecules* 27: 1638. https://doi.org/10.3390/molecules27051638.

49 Wang, S., Xu, Z., Wang, T., Xiao, T., Hu, X.Y., Shen, Y.Z., and Wang, L. (2018). Warm/cool-tone switchable thermochromic material for smart windows by orthogonally integrating properties of pillar[6]arene and ferrocene. *Nature Communications* 9: 1737. https://doi.org/10.1038/s41467-018-03827-3.

50 Gyenes, T., Szilágyi, A., Lohonyai, T., and Zrínyi, M. (2003). Electrically adjustable thermotropic windows based on polymer gels. *Polymers for Advanced Technologies* 14: 757–762. https://doi.org/10.1002/pat.391.

51 Zhou, Y., Layani, M., Wang, S., Hu, P., Ke, Y., Magdassi, S., and Long, Y. (2018). Fully printed flexible smart hybrid hydrogels. *Advanced Functional Materials* 28: 1705365. https://doi.org/10.1002/adfm.201705365.

52 Zhou, Y., Cai, Y., Hu, X., and Long, Y. (2015). VO_2/hydrogel hybrid nanothermochromic material with ultra-high solar modulation and luminous transmission. *Journal of Materials Chemistry A* 3: 1121–1126. https://doi.org/10.1039/C4TA05035E.

53 Xu, Z., Wang, S., Hu, X.-Y., Jiang, J., Sun, X., and Wang, L. (2018). Sunlight-induced photo-thermochromic supramolecular nanocomposite hydrogel film for energy-saving smart window. *Solar RRL* 2: 1800204. https://doi.org/10.1002/solr.201800204.

54 Xu, J.M., Li, L., Wang, S., Ding, H.L., Zhang, Y.X., and Li, G.H. (2013). Influence of Sb doping on the structural and optical properties of tin oxide nanocrystals. *CrystEngComm* 15: 3296–3300. https://doi.org/10.1039/C3CE40241J.

55 Yu, Z., Ma, Y., Mao, L., Lian, Y., Xiao, Y., Chen, Z., and Zhang, Y. (2023). Bidirectional optical response hydrogel with adjustable human comfort temperature for smart windows. *Materials Horizons* 207–216. https://doi.org/10.1039/d3mh01376f.

56 Wilson, H.R. (2000). Solar control coatings for windows. In: *Functional Materials*, EUROMAT 99, vol. 13 (ed. K. Grassie, E. Teuckhoff, G. Wegner, J. Hausselt, H., and Hanselka), 221–233. Weinheim: Wiley-VCH. https://doi.org/10.1002/3527607420.ch38.

57 Wilson, H.R., Ferber, J., and Platzer, W. (1994). Optical properties of thermotropic layers. In: *Proceedings of the SPIE 2255, Optical Materials Technology for Energy Efficiency and Solar Energy Conversion XIII* (ed. V. Wittwer, C.G. Granqvist, and C.M. Lampert), 473–484.

58 Raicu, A., Wilson, H.R., Nitz, P., Platzer, W., Wittwer, V., and Jahns, E. (2002). Facade systems with variable solar control using thermotropic polymer blends. *Solar Energy* 72: 31–42. https://doi.org/10.1016/S0038-092X(01)00093-7.

59 He, Z., Liu, Z., Liu, B., Wang, K., Dong, X., Zhang, Z., Chen, C., Wang, M., Liu, J., and Huang, W. (2024). Thermally-induced phase fusion and color switching in ionogels for multilevel information encryption. *Chemical Engineering Journal* 479: 147544. https://doi.org/10.1016/j.cej.2023.147544.

60 Dai, S., Dong, X., Jiang, Y., Ge, J., Ding, J., and Yuan, N. (2023). High adhesive force, wide temperature adaptability thermochromic ionogel smart window. *ACS Applied Polymer Materials* 5: 3398–3404. https://doi.org/10.1021/acsapm.3c00059.

61 Arosa, Y., Rodríguez Fernández, C.D., López Lago, E., Amigo, A., Varela, L.M., Cabeza, O., and de la Fuente, R. (2017). Refractive index measurement of imidazolium based ionic liquids in the Vis-NIR. *Optical Materials* 73: 647–657. https://doi.org/10.1016/j.optmat.2017.09.028.

62 Wu, X., Muntzeck, M., de los Arcos, T., Grundmeier, G., Wilhelm, R., and Wagner, T. (2018). Determination of the refractive indices of ionic liquids by ellipsometry, and their application as immersion liquids. *Applied Optics* 57: 9215–9222. https://doi.org/10.1364/ao.57.009215.

63 Le Bideau, J., Viau, L., and Vioux, A. (2011). Ionogels, ionic liquid based hybrid materials. *Chemical Society Reviews* 40: 907–925. https://doi.org/10.1039/c0cs00059k.

64 Lee, H.Y., Cai, Y., Velioglu, S., Mu, C., Chang, C.J., Chen, Y.L., Song, Y., Chew, J.W., and Hu, X.M. (2017). Thermochromic ionogel: a new class of stimuli responsive materials with super cyclic stability for solar modulation. *Chemistry of Materials* 29: 6947–6955. https://doi.org/10.1021/acs.chemmater.7b02402.

65 Chen, L., Zhao, C., Huang, J., Zhou, J., and Liu, M. (2022). Enormous-stiffness-changing polymer networks by glass transition mediated microphase separation. *Nature Communications* 13: 6821. https://doi.org/10.1038/s41467-022-34677-9.

66 Chen, L., Zhao, C., Duan, X., Zhou, J., and Liu, M. (2022). Finely tuning the lower critical solution temperature of ionogels by regulating the polarity of polymer networks and ionic liquids. *CCS Chemistry* 4: 1386–1396. https://doi.org/10.31635/ccschem.021.202100855.

67 Lee, H.N., Newell, N., Bai, Z., and Lodge, T.P. (2012). Unusual lower critical solution temperature phase behavior of poly(ethylene oxide) in ionic liquids. *Macromolecules* 45: 3627–3633. https://doi.org/10.1021/ma300335p.

68 Li, B., Xu, F., Guan, T., Li, Y., and Sun, J. (2023). Self-adhesive self-healing thermochromic ionogels for smart windows with excellent environmental and mechanical stability, solar modulation, and antifogging capabilities. *Advanced Materials* 35: 2211456. https://doi.org/10.1002/adma.202211456.

69 Muehling, O., Seeboth, A., Haeusler, T., Ruhmann, R., Potechius, E., and Vetter, R. (2009). Variable solar control using thermotropic core/shell particles. *Solar Energy Materials & Solar Cells* 93: 1510–1517. https://doi.org/10.1016/j.solmat.2009.03.029.

70 Fraunhofer Institute for Applied Polymer Research, Solardim®-ECO (2019). High energy savings through a self-regulating solar protection glazing - Technical note.

9

Other Chromogenic Technologies

Carl M. Lampert[1,2,3]

[1] *Star Science, Cotati, CA, USA*
[2] *Formerly, Technical Director, Society of Vacuum Coaters (non-profit), Albuquerque, NM, USA*
[3] *Staff Scientist, University of California, LBNL, Energy Technologies, Berkeley, CA, USA*

Nomenclature

λ	wavelength (nm)
ACeP	advanced color ePaper, E Ink
AI	artificial intelligence
CES	Consumer Electronics Show (Las Vegas, NV)
ELM	electrophoretic light modulator
EPD	electrophoretic device
HSG	heat strengthened glass
IGU	insulating or insulated glass unit, double-glazed window
LC	liquid crystal
LCD	Liquid Crystal Device
LED	light-emitting diode
MEMS	micro-electromechanical systems
NIR	near infrared
NRC	National Research Council (Canada)
NREL	National Renewable Energy Laboratory (Golden, CO)
OLED	organic light-emitting diode
PET	polyethylene terephthalate
PMMA	poly(methyl methacrylate)
PVC	polyvinyl chloride
PVD	polyvinyl butyral
R_a	color rendering index
SID	Society for Information Display
T_c	transition temperature (°C)
ΔT_c	hysteresis width (°C)
$T_{c,h}$	transition temperature in heating process (°C)
$T_{c,c}$	transition temperature in cooling process (°C)
TC	thermochromic

Chromogenics: Smart Switchable Optical Materials and Their Applications, First Edition. Edited by Carl M. Lampert.

TCO	transparent conductive oxide
TFT	thin film transistor
TGU	triple glazed unit (window)
T_{lum}	luminous transmittance (%)
ΔT_{sol}	solar modulation (%)
$T_{lum,cold}$	luminous transmittance in cold state (%)
$T_{lum,hot}$	luminous transmittance in hot state (%)
T_{sol}	solar transmittance
T_{vis}, TL	visible transmittance
UV	ultraviolet
WFG	Water-Filled-Glass
YCMW	yellow, cyan, magenta, and white (color palette)

9.1 Introduction

In this chapter, I introduce selected technologies, including encapsulated electrophoretic particles, guest-host liquid crystals, electrophoretic ink, microshutters, optofluidics, and thermochromic perovskites. These technologies are aimed at products for architectural or automotive applications. All these technologies have the potential for making large-area chromogenic switching products. Many of the products are at the prototype stage. Products like eyrise liquid crystal panels are in production and have been installed in several buildings. The biggest challenges are customer understanding, product quality and appearance, manufacturing cost, product maintenance, and lifetime. The optofluidic systems may also require routine maintenance.

There has been a considerable upswing in automotive glazing applications, especially sunroofs and visors in the last few years. Electric vehicles have presented a new challenge, to save electric energy, especially for cabin air conditioning. For electric vehicles energy-conserving glazing has become very important. The average service life of a vehicle (cars and trucks) has risen from 11 years (2012) to 12.6 years (2024), with cars alone having an average life of 14 years (2024) [1]. So, for automotive applications, one is looking for products, such as switchable sunroofs, with equivalent long life.

For architectural double-pane windows the life expectancy is 8–20 years (2024) [2]. The number varies a lot considering the construction of the framing, weather, and orientation of the window. Many window warranties are 10 years. The weakest part of the window is its sealing system. In the architectural "smart" glazing business there is an ongoing competition between several chromogenic technologies, with much of the issue situated around cost, cyclic durability, color, residual haze, speed, range, energy efficiency, and energy versatility. This is all tempered by the complete system cost, including powering and control systems. Also, we notice more smart interactive building and automotive glazing, combining vision with information display. Just as we speak to a virtual helper (like Siri, Alexa, and Google Assistant) on our cell phones, we may in the future speak to our glazing to have it perform functions. When we are not present the glazing may revert to its low energy consumption routine, powered by AI. The system learns our habits and reduces building or vehicle energy consumption, along with providing security functions.

Back in the late 1970s, a crossover of electrochromics from displays to windows occurred. At that time the display field was focused on two other technologies, liquid crystals and light-emitting diodes (LEDs). Over time liquid crystals and organic light-emitting diodes (OLEDs) dominate the large-area computer and television screen applications. Some of these displays have become as large as a window or wall and can be used to display information or to even change the ambient environment. Other display technologies, such as encapsulated electrophoretic inks continue to be improved for displays, such as electronic paper. The ePaper technologies have resulted in flexible, lightweight, high-contrast, non-emissive displays.

Today we are witnessing another crossover from electronic paper to large-area switching devices, including transparent windows. It is possible to make large-area transparent electronics. At the 2024 Society for Information Display (SID) conference (San Jose, CA). The largest transparent display window, 1.52 m (60″) diagonal, was shown by AUO (Taiwan). Another example, the E Ink electrophoretic display panels are moving toward full-color larger displays. Other, colored and transparent electrophoretic panels are being prototyped for the automotive and building markets. Also, new products are being tested in the marketplace. Examples are the liquid crystal glazing products, including eyrise B.V. (Veldhoven, NL) and electrophoretic windows from ELSTAR Dynamics (Eindhoven, NL). Some technologies, such as micro-electromechanical systems (MEMS) microshutters still need work to make the prototypes into commercial products. Using MEMS microreflectors to redirect sunlight into a building is an interesting concept. Other potential products, such as liquid film windows need more study and testing. Microfluidics is a very interesting concept which has the potential to control select bands of energy for the needs of the building. It is important to mention them here, as they are worth investigating and may direct the reader to new concepts.

9.2 E Ink – Encapsulated Electrophoretic Ink

E Ink Corporation (Billerica, MA) grew out of the MIT Media Lab in 1997 [3]. E Ink developed a bistable black and white display based on encapsulated electrophoretic ink, e.g. US 6017584 (2000) [4]. The TFT (thin film transistor) liquid crystal display company, Prime View International (Taiwan), acquired E Ink in 2009. The combined companies are now called just E Ink. Another company, SiPix Imaging, Inc. (Milpitas, CA), developed a microcup technology containing colored pigments to make flexible e-readers. E Ink acquired SiPix in 2012, combining their technologies. The SiPix patent family contains work by Robert Sprague, e.g. US 8786935B2 (2012) [5]. Their combined technology has led to the development of the ACeP® technology using multicolored pigments for e-readers, e.g. US 9921451 (2024) [6]. E Ink has shown an array of color signage, large and small displays, including flexible display sheets. The color range, depth, and brightness have improved each year, as noted in the annual SID conference presentations.

Visual contrast in E Ink products is achieved by moving ink particles within an ink capsule or microcup. The particles move when an electric field is applied. E Ink used particles that would move in a field to give contrast, with the first products comprised of blue and white particles. This display uses ambient light so no backlighting or frontlighting, as in liquid crystal displays (LCDs), is needed. Also, it is a bistable system, only needing power to change, so it is ideal for portable electronics. This performance has led to eReader applications, such as in the popular Kindle. The two-pigment system uses 17–181 μm diameter capsules made into a film that is attached to an array of pixels. Each capsule contains both negatively charged white particles and positively charged black particles. The particles are surface treated so they do not agglomerate. Both are suspended in a clear fluid with viscosity low enough to facilitate fast movement when an electric field is applied. When a field is applied the particles move either up or down, giving a viewer a color of white or black. The technology is based on a patent family including US 6982178B2 (2006) [7]. A schematic of an E Ink capsule containing black and white electrophoretic ink and fluid is shown in Figure 9.1. This configuration is for a display application. E Ink has done well commercially with their two-pigment black & white display system accounting for a wide range of display products. The magnified image of an E Ink pixelated structure, showing the ink capsules, is presented in Figure 9.2. An example of a rollable E Ink screen is shown in Figure 9.3.

The E Ink JustTint™ product is aimed at the glazing market, such as a sunroof for a car or a dimmable window for a building. It's a flexible product so it does have an advantage for curved glazing applications. The configuration of the conductors for the addressing of the particles is different from the display products. In this design the particles can be moved out of view to give transparency through the device [10]. This configuration makes an optical shutter. Depending upon the electric field and its location near the capsules, pigment particles can be moved in view to darken or sideways out of view. This effect is bistable, so power is only needed to switch the panel. This effect is shown in Figure 9.4. A prototype sunroof glazing is shown in Figure 9.5. This is a dark and clear sunroof glazing showing four segment switching. Also, it can be colored to match the car.

Color displays can be made using colored electrophoretic particles. Encapsulated pigments and electric fields are used to move colored particles in and out of view, giving images with high contrast. Different pigments are selected that require different field strengths to move, so a range of colors can be achieved with different addressing potentials. The position of a set of particles, or a mixture of particles relative to the reflective white particles determines

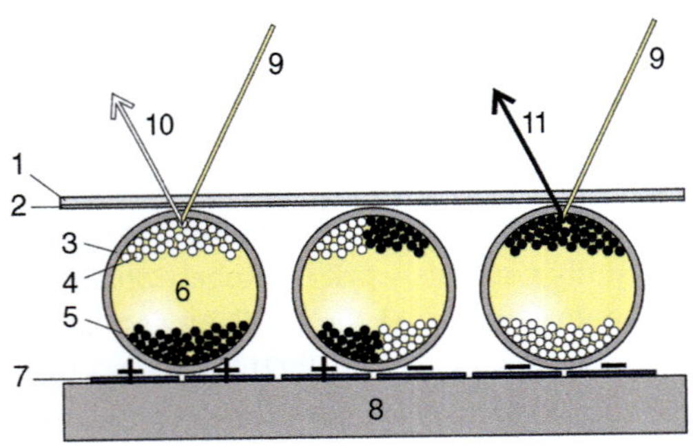

Figure 9.1 Cross section of three E Ink microcapsules showing how one can obtain contrast between black and white conditions (not to scale). The microcapsules are between 17 and 181 μm. Typical pixel density is 200–300 pixels inch^{-1}, with 16 levels of greyscale. When the applied potential (typically 15 V) is switched from negative to positive the perceived color will change from white to black [8]. Legend: (1) transparent superstrate, (2) transparent conductor, (3) microcapsule, (4) positively charged white pigment, (5) negatively charged black pigment, (6) oil, (7) electrode pixel layer, may be transparent, (8) substrate, for JustTint™ substrate is transparent, (9) incoming light, (10) reflected white image, (11) reflected black image. *Source:* FREEscan RIP, cc BY 3.0 via Wikimedia Commons.

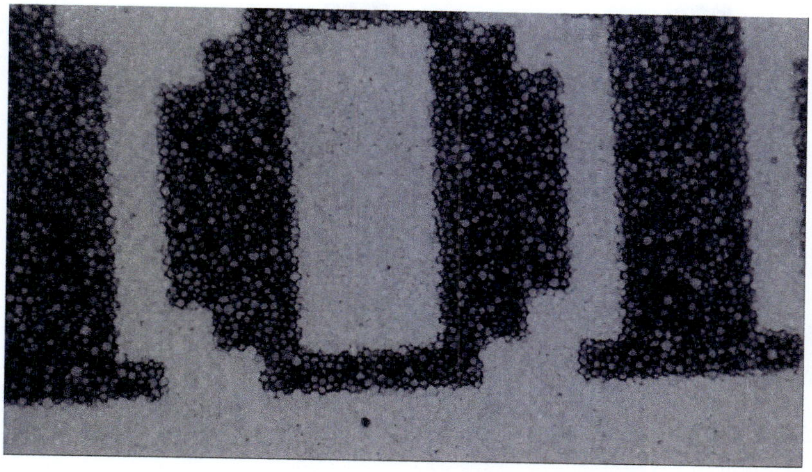

Figure 9.2 Magnified image showing E Ink pixelated microcapsules, switched to show text image [9]. Typical microcapsules are between 17 and 181 μm in size. *Source:* Alexander Gee (11 April 2021), E Ink Screen, cc BY SA4.0. This file is licensed under the Creative Commons Attribution-Share Alike 4.0 International license.

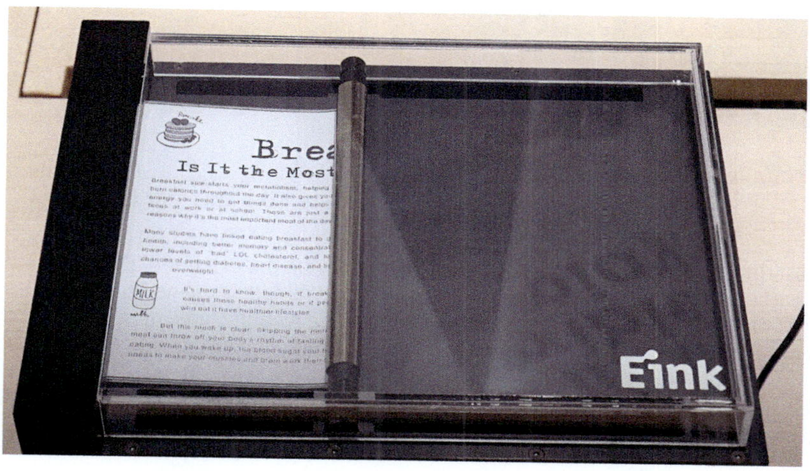

Figure 9.3 Photo of a rollable black and white E Ink screen, demonstrated at the 2024 Society for Information Display (SID) conference. *Source:* Author's photo.

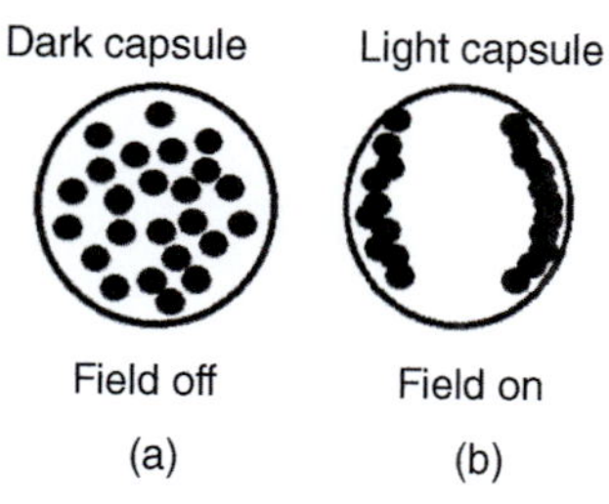

Figure 9.4 Schematic cross-section of E Ink JustTint™ capsules showing the configurations of electrophoretic particles (not to scale). (a) Dark state showing dispersed particles and how they block light. (b) Showing fully bleached and the movement of the particles to the sides of the capsule. Intermediate darkening can be achieved by reducing the potential across the capsule. *Source:* Author.

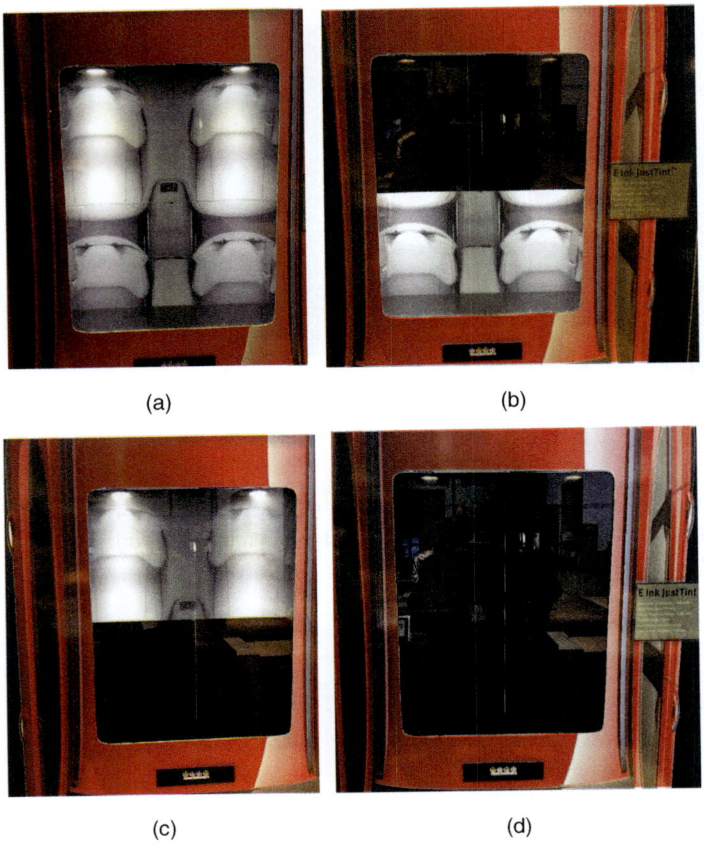

Figure 9.5 E Ink JustTint™ prototype segmented sunroof windows from the exterior roof view, showing (a) fully bleached, (b) top darkened, (c) bottom darkened, and (d) fully darkened conditions. Intermediate levels of darkening are possible. Also, other colors can be manufactured depending upon the design needs of the user. *Source:* Author.

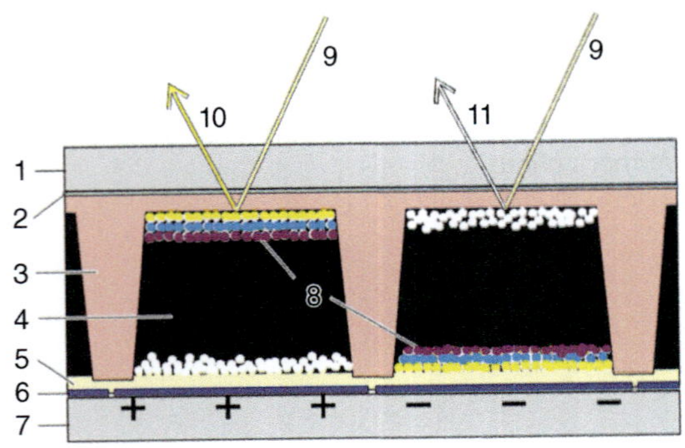

Figure 9.6 Schematic of how the four-pigment microcup system functions (not to scale). The ink is contained within microcups. By moving different colored particles in combination to the top surface, a full color gamut can be achieved. The applied potential determines the color. The different colored particles move under different applied potentials. All colors are incorporated into a pixel, eliminating the need for a color filter array. The neutral state is white, as shown in the right microcup [11]. Legend: (1) transparent superstrate, (2) transparent conductor, (3) microcup, (4) oil, (5) sealing layer, (6) electrode pixel layer, (8) colored pigment, (9) incoming light, (10) reflected yellow image, (11) reflected white image. *Source:* Modified from Tosaka (2010) cc BY 3.0 via Wikimedia Commons.

the color at a given pixel. This is called ACeP™. Figure 9.6 shows a cross-sectional schematic of the four-pigment system: yellow, cyan, magenta, and white (YCMW) display (not to scale). The colored ink is contained within a microcup structure. All colors are available at every pixel, eliminating the need for a color filter array. The neutral state is white. There are eight fundamental colors (cyan, yellow, magenta, red, blue, green, white, and black). A multicolor flexible panel is shown in Figure 9.7a. The colors can be changed by changing the applied potential.

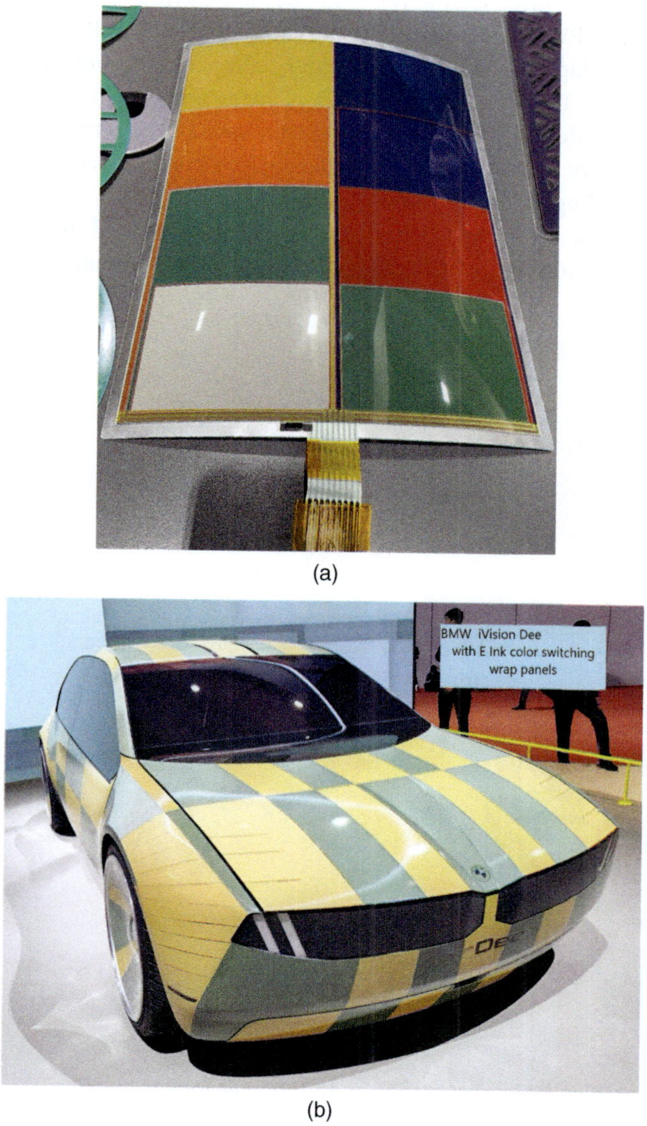

Figure 9.7 (a) Advanced flexible E Ink ACeP™ four-pigment color ePaper system. *Source:* Author. (b) A BMWi Vision Dee concept car wrapped with 240 Advanced flexible E Ink ACeP™ four-pigment color ePaper system. *Source:* Maxhdtenda / Wikimedia Commons / CC BY SA 4.0.

A BMWi Vision Dee concept car was wrapped with 240 ACeP sheet segments and shown at the 2023 Consumer Electronics Show in Las Vegas, NV [12]. Dee stands for "Digital Emotional Experience" [13]. A picture of the BMWi Vision Dee car wrapped in ACeP sheet is shown in Figure 9.7b [14]. Car wrapping is used to customize or personalize a vehicle. Wrapping is used heavily in branding commercial vehicles and for advertising on public transport, such as buses. Vehicles range from sports cars to heavy trucks. Also wrapping is used in aerospace and on trains. Companies involved in manufacturing wrapping materials (typically PVC sheeting) include 3M (St. Paul, MN), Avery Dennison (Pasadena, CA), Hexis S.A.S. (Frontignan, France), KMPF-Kay Premium Marking Films (Crumlin, Newport, Wales, UK), Orafol (Oranienburg, Germany), and Metro Restyling (Sterling Heights, MI). The wrapping materials (PVC) can be printed using equipment such as Roland DG large-scale printers (Hamamatsu, Japan). Roland has subsidiary companies throughout the world. Roland DGA is located in Irvine, CA. The largest printers are 163 cm (64″) wide-format inkjet, with 1441 dpi resolution and fastest production speed of 30.4 $m^2\ h^{-1}$ (327 $ft^2\ h^{-1}$). Some wrapping systems are for security, anti-graffiti, and anti-scratch purposes. These tend to be laminated systems using PET overlayers.

Figure 9.8 shows the same sample (a–c) of rolled sheets of flexible ACep™ films with various colors. The figures show completely different colors, switched in a matter of

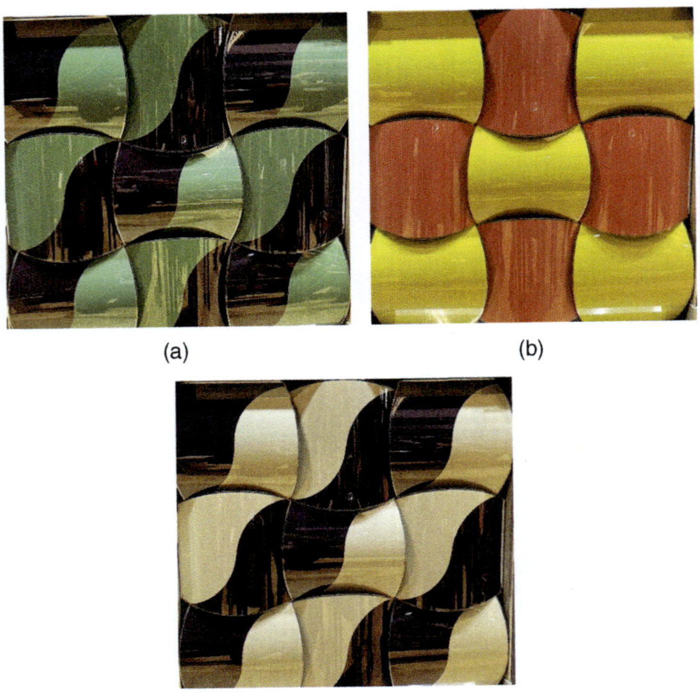

(a) (b) (c)

Figure 9.8 Examples of flexible E Ink ACep™ film shown in an array. The same sample (a–c) is shown using different coloration voltages. Any color combination can be made. Switching is very fast. Switching does pass through a white clearing state when switching to another color. The effect is barely noticeable [8]. *Source:* Author.

seconds. Switching is very fast, but it does pass through a white clearing state when switching to another color. The intermediate effect is barely noticeable. One can change 32 colors at will.

9.3 eyrise® Liquid Crystal Glazing

eyrise® is a liquid crystal (LC) glazing product developed originally by Peer+ B.V. (Veldhoven, The Netherlands) and now commercialized as eyrise®. It is now owned by Merck, KGaA (Darmstadt, Germany). Windows up to the size of 3.5 m × 1.6 m can be fabricated. See the example in Figure 9.9. The windows switch with low AC voltage and can switch as fast as one second.

The eyrise® technology is based on a liquid crystal-dye molecule combination called licrivision®, where during switching from dark to light the liquid crystal-dye alignment is retained. This is also known as a guest-host LC system. The effective visible absorption changes with applied potential. The LC director orientation and dye axis orientation are altered by the applied electric field. The rest state is oriented in-plane with the transparent conductive oxide (TCO) surface. The angle of LC molecules and dye determines the transmittance of the glazing, ranging from dark to light as voltage is increased, as shown in Figure 9.10

The eyrise® glazing element switches with 24 VAC applied potential, and power consumption is about 1 W m^{-2} (0.1 W ft^{-2}). The active element (8 mm thick) is constructed

Figure 9.9 Installation of eyrise® IGU glazing showing the bright and the dark states. The glazing shows the dynamic range of tinting, with very low scatter at a wide range of observation angles. The glazing shows good color neutrality, color rendering, and fast switching speed. *Source:* Courtesy of eyrise, B.V.

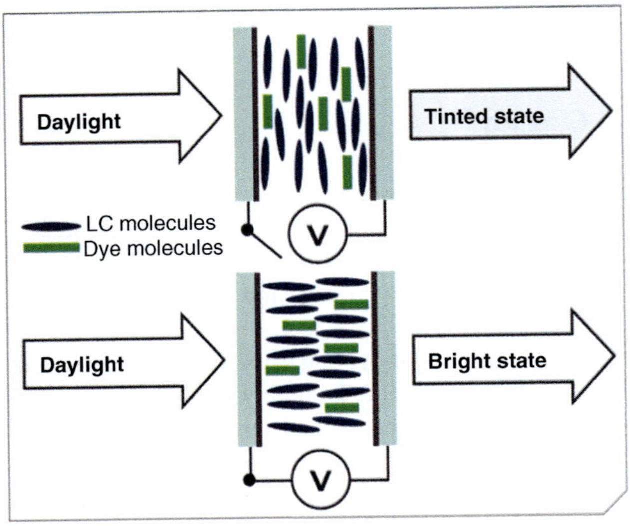

Figure 9.10 Illustration showing the off dark state (top) and the on bright state (bottom) for a liquid crystal + dye molecule combination. Intermediate states are possible. *Source:* Courtesy of eyrise, B.V.

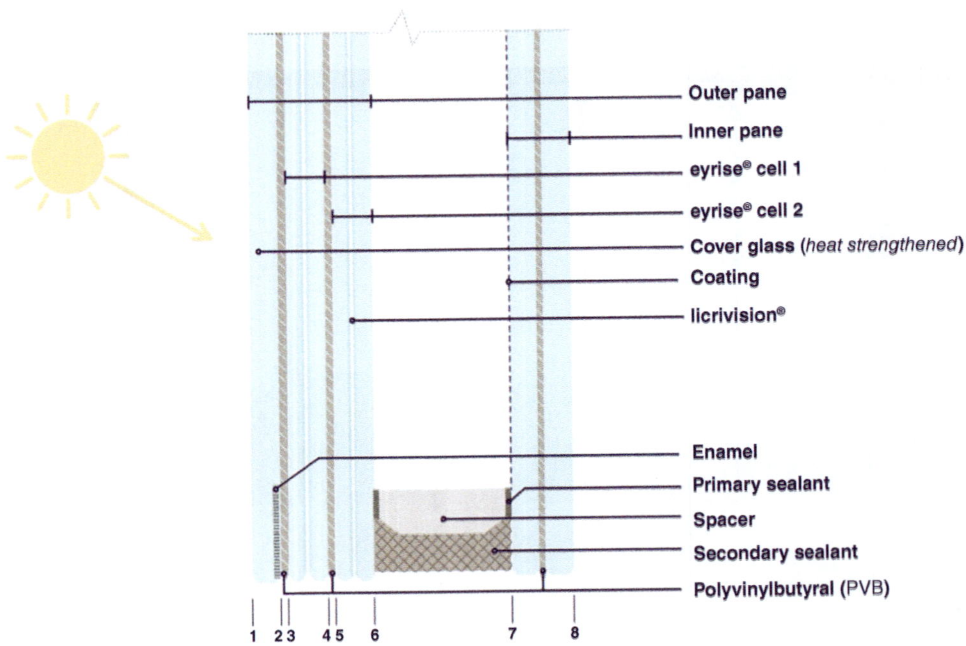

Figure 9.11 Detailed construction of an eyrise® s350 double-glazed Insulating Glass Unit (IGU). The glazing surfaces are numbered 1–8. From the outside surface, heat strengthened glass (HSG) (surface 1–2), lamination layer, eyrise® cell 1 (surfaces 3–4), lamination layer, eyrise® cell (surfaces 5–6), low-e coating (surface 7) and, laminated glass (surfaces 7–8). The laminated glass could also be HSG. Note that two eyrise® layers are laminated to the outboard pane. Liquid crystal (licrivision®) element is inside the eyrise® layers. A low-conductivity gas is used in the gap. A low-e coating on surface #7, on the inboard glass pane. *Source:* Courtesy of eyrise, B.V.

of glass/TCO/liquid crystal + dye/TCO/Glass. Some example source patents are Van Oosten: Electro-optical controllable glazing, US 9164314B2 (2015) [15], Liquid crystal dye mixture US 9249356B2 (2016) [16], and Liquid-crystalline medium European patent EP 2931839B1 (2017) [17].

The construction of an Insulating Glass Unit (IGU) is shown in Figure 9.11. Note the eyrise® layers are on the #2 surface of the outboard pane. A low-e coating is on the #3 surface of the inboard pane. The eyrise® layer (8 mm) component is shown laminated between two pieces of glass using polyvinyl butyral (PVB). In this case, two eyrise® layers are used to improve optical density range. The visible, thermal, and solar g-factor properties of the eyrise® IGU and a triple-glaze unit (TGU) are shown in Tables 9.1 and 9.2 for three liquid crystal formulations. Corresponding cross sections of the IGU and TGU glazing are shown in Figure 9.12a and b, respectively. About 50 building projects are ongoing and 50,000 m^2 of glazing have been installed. Figure 9.13 shows an installation of eyrise® windows in an office building at 16/20 Müllerstrasse, Zurich, Switzerland. Further information is available [18, 19]. Other types of LC glazing have been reviewed by Baliyan et al. [20].

Table 9.1 eyrise® double glazing performance table.

eyrise® liquid crystal mixture	Light transmittance (T_{LUM})		Solar factor/ g-value		Heat transfer coefficient U-value	General color rendering index R_a	
	Bright (%)	Dark (%)	Bright	Dark	($W\,m^{-2}K^{-1}$)	Bright	Dark
LC mixture C	50	13	0.34	0.16	1.1	93	88
LC mixture D	44	9	0.31	0.14	1.1	93	89
LC mixture E	37	4	0.28	0.11	1.1	91	89

Optical and thermal values and color rendering index for a double glazing with eyrise® s350 layers using three liquid crystal formulations. Values were calculated according to EN-410 and EN-673 standards.
Source: Courtesy of eyrise, B.V.

Table 9.2 Optical and thermal values and color rendering index for a triple glazing with two eyrise® s350 layers using three liquid crystal formulations.

eyrise® liquid crystal mixture	Light transmittance (T_{LUM})		Solar factor/ g-value		Heat transfer coefficient U-value	General color rendering index R_a	
	Bright (%)	Dark (%)	Bright	Dark	($W\,m^{-2}K^{-1}$)	Bright	Dark
LC mixture C	45	12	0.29	0.12	0.6	91	86
LC mixture D	40	8	0.26	0.10	0.6	91	88
LC mixture E	34	4	0.23	0.7	0.6	90	87

Values were calculated according to EN-410 and EN-673 standards.
Source: Courtesy of eyrise, B.V.

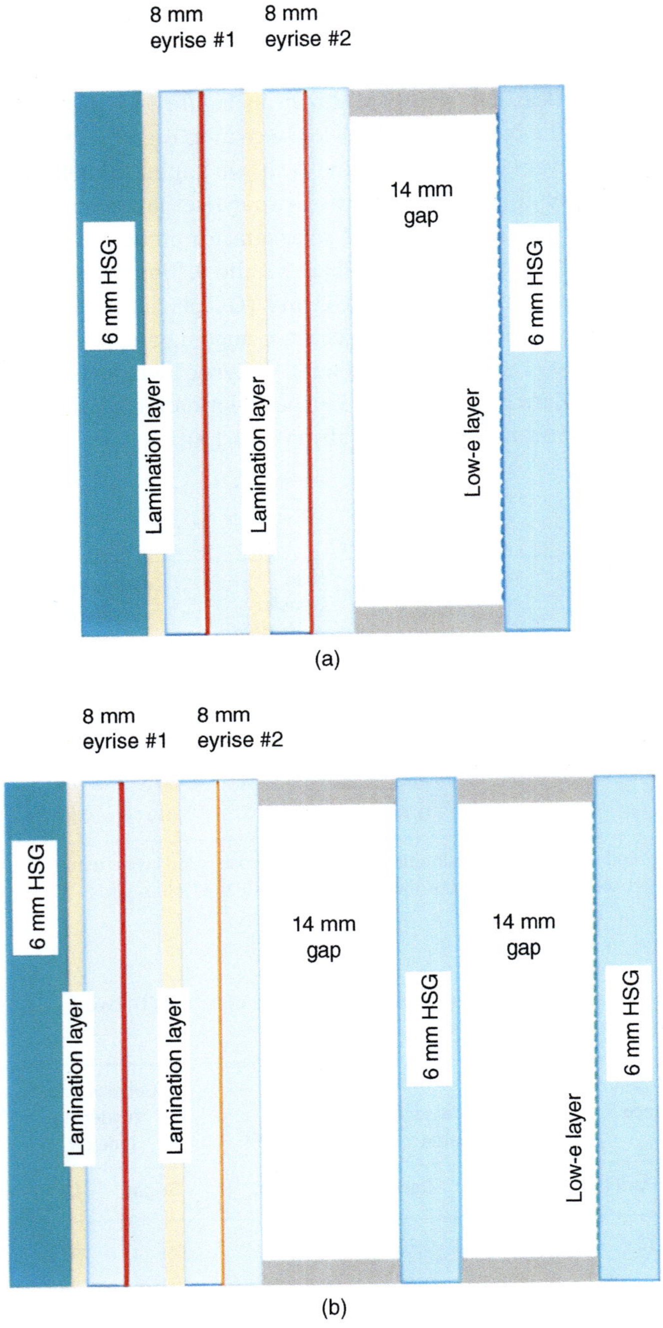

Figure 9.12 (a) Cross section of double-pane eyrise® glazing used for Table 9.1 calculations. The lamination layers are 1.52 mm thick; the gap is low-conductivity gas filled. HSG is heat-strengthened glass. (b) Cross section of triple-pane eyrise® glazing used for Table 9.2 calculations. The lamination layers are 1.52 mm thick; the gaps are filled with low-conductivity gas. *Source:* Courtesy of eyrise, B.V.

(a)

(b)

Figure 9.13 (a) A completed building façade upgrade using 3000 m^2 of dynamic liquid crystal eyrise® glass. The office building is located at 16/20 Müllerstrasse, Zurich, Switzerland. (b) Example of both bleached and darkened eyrise panes. *Source:* Courtesy of eyrise, B.V.

9.4 ELSTAR Dynamics Electrophoretic Glazing

ELSTAR Dynamics (Eindhoven, NL) showed prototype windows called eclipse™ at the SID-2024 conference in San Jose, CA. They won the 2024 I-Zone award for the best prototype. The ELSTAR Dynamics window is an electrophoretic light modulator (ELM) based on electrophoretic inks and nanoimprinted spacers with transmission switchable from $T_{vis} = 0.2–70\%$ in 20 seconds. Sizes can be made up to 2.9 m × 3.3 m. An example of a window in both lightened and darkened conditions is shown in Figure 9.14a and b, respectively. In the lightened state the color rendering index is CRI = 98%. Its operating temperature

(a)

(b)

Figure 9.14 Prototype ELSTAR Dynamics eclipse^{TM} switchable panel in its (a) bleached and (b) darkened conditions. Intermediate darkening is possible. *Source:* Courtesy of Anthony Slack, CEO, ELSTAR Dynamics, NL.

is −20 to 70 °C, with low haze. Further technical information can be found in papers by Tee et al. [21], Slack and Massard [22] and Massard and Slack [23] and patents by Massard et al. [20, 24–30].

The ELSTAR eclipse^{TM} panel is comprised of electrophoretic ink that is manipulated by a local applied electric field. When the field is on the particles migrate to microcontacts that are invisible to the naked eye, giving overall clarity to the panel. When the electric field is removed the particles migrate uniformly over the panel and the panel darkens. An expanded view of the ELM construction is shown in Figure 9.15. The construction of the nanoimprinted panel uses manufacturing equipment which is 90% compatible with standard LCD equipment, lowering engineering and development costs, according to Anthony Slack, CEO, ELSTAR Dynamics. The panel can be made using 1 mm flexible glass so it can be laminated onto curved surfaces such as automotive glazing.

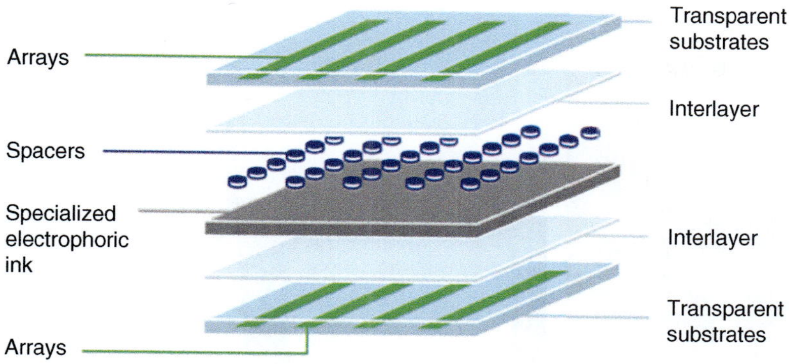

Figure 9.15 Cross-section of ELSTAR Dynamics ELM panel construction (not to scale). The electrophoretic ink moves with the applied electric field either to be hidden under the micro spacers (clear) or exposed in the transparent regions of the pane (darkened). The panel is manufactured using nanoprinting. *Source:* Courtesy of Anthony Slack, CEO, ELSTAR Dynamics, NL.

9.5 MEMS and Microshutter Materials

The microshutter or microblind systems are based on micro-electromechanical systems called MEMS, which were developed for information displays [31]. Many people are familiar with the development by Qualcomm (San Diego, CA) of capacitive MEMS displays for cell phones. The MEMS display consists of arrays of very tiny addressable shutters built onto optical cavities. The switching effect is reflective using ambient light. MEMS are made up of components between 0.001 and 0.1 mm, and MEMS devices generally range in size from 20 mm to more than 1 m^2. Ohmic MEMS are controlled by electrostatically controlled cantilevers or flaps that roll up. If all the flaps are rolled up then high transmission is achieved. If the flaps are small enough, they cannot be perceived by the eye. Ohmic MEMS switches can fail from metal fatigue of the MEMS actuator and contact wear, since curling can deform over time. MEMS Spatial light modulators were patented by Bozler and Rabe in 1999 [32]. A roll-up or microblind version of MEMS was developed by Lamontagne and Py, National Research Council (NRC) (Ottawa, Canada), for use as a switchable glazing [33]. It's called the "Louver Window." By applying a voltage, 15–25 V bipolar square waves form an electric field between the electrodes, the top electrode rolls down onto the surface, thus blocking the light transmission. Switching time is 40 ms. An example of this microshutter is shown in Figure 9.16. A typical structure of a tabular microshutter is shown in Figure 9.17. The off condition is shown in Figure 9.18, and the on condition with voltage fully applied shows fully closed microshutters in Figure 9.19.

In recent work by Donatiello et al. at the Institute of Nanostructure Technologies, Prof. Hillmer's group, University of Kassel, Germany, a fish-skin geometry MEMS array was studied as a reflective microshutter switch for a window structure. It has been studied for daylight steering into a building window. The fish skin structure has less optical aberration than the flap structure in the open condition [34]. Its interesting microstructure is shown in Figure 9.20. More details about fish skin microshutters can be found in the work of Prof. Harmut Hillmer at the University of Kassel, Germany [35] and in Hillmer and Xu (2024) Lighting module for a building façade European patent EP 410202B1.

Figure 9.16 Prototype MEMS microblind (microshutter) window made by NRC Canada. The two images show power off (right bleached image) and power on (left darkened image). The microblind is powered by an applied potential of 15–25 V. In this case the camera is focused on the tree in the background. *Source:* Courtesy of Boris Lamontagne, NRC, Canada.

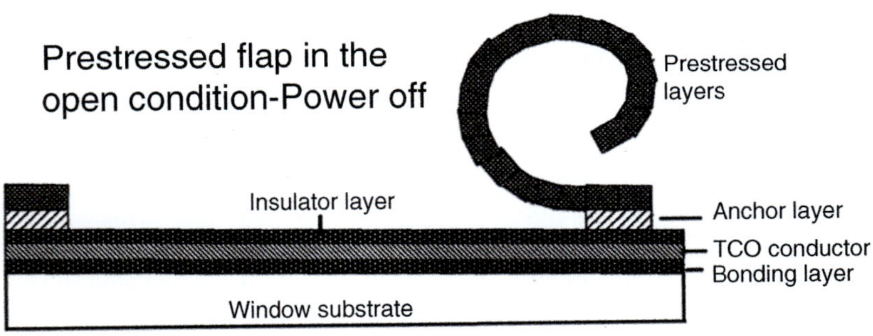

Figure 9.17 Cross-sectional structure of a microblind showing the layers. The shutter portion (top layer) shows how it rolls up when it is prestressed. This is shown in the off condition. A range of densities can be obtained by varying the voltage on the device, causing the top layer to unroll and flatten out. The potential would be applied across the prestressed layer and the TCO layer. For example, the prestressed reflective layer could be W, the insulator could be SiO_2, and the TCO could be SnO_2:F. *Source:* Modified image after Lamontagne, B., Py, C. US patent 7684105B2 (2010) [33].

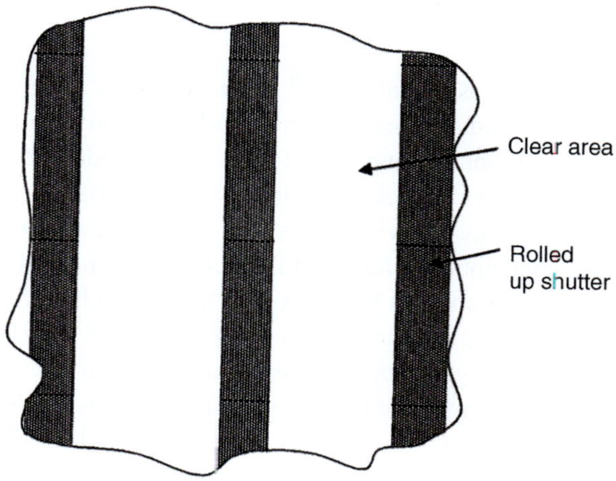

Figure 9.18 Drawing of four microblind pixels in the open, transparent condition. This is the rest position with no voltage applied. A large portion of the shutter is open, giving high visibility.
Source: Modified image after Lamontagne, B., Py, C. US patent 7684105B2 (2010) [33].

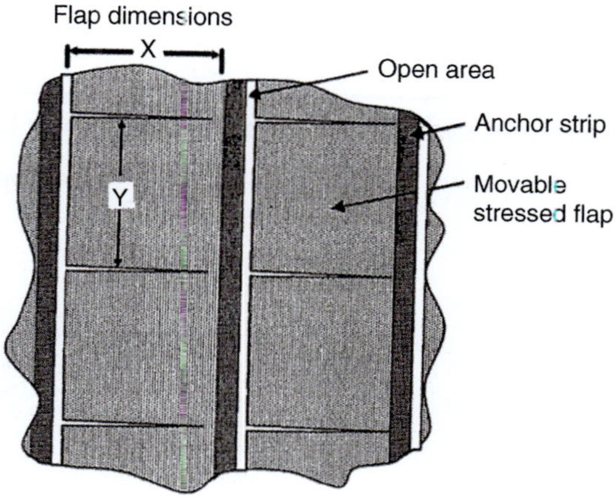

Figure 9.19 Drawing of four microblind pixels in the closed condition, with power turned on. The applied electric field has overcome the stressed flaps to cause them to flatten out. The applied voltage is 15–25 V square wave. A typical flap size might be 80 × 150 μm. *Source:* Modified image after Lamontagne, B., Py, C. US patent 7684105B2 (2010) [33].

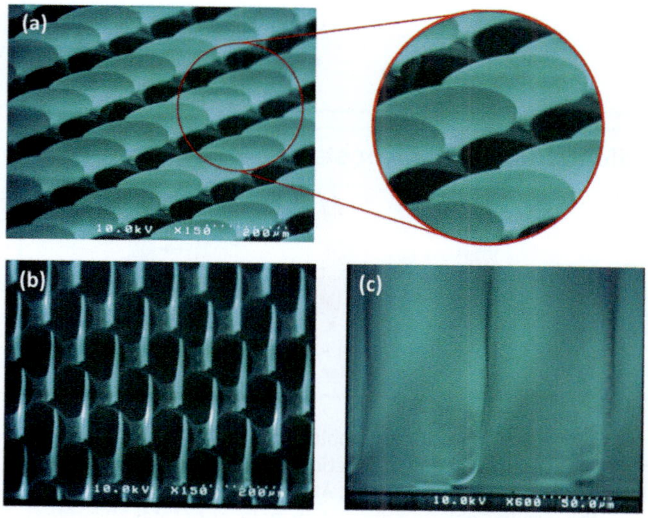

Figure 9.20 The SEM micrographs of micromirror arrays with *fish skin* geometry: (a) the array shows good homogeneity over a wide area; (b) the planarization of micromirror movable parts; (c) a fully open orientation at ~90° to the glass surface. *Source:* Roland Donatiello, Creative Commons Attribution (CC BY) license [34].

9.6 Optofluidics

Optofluidics is an interesting science used to design a dynamic building skin that uses multiple fluids to change both optical and thermal properties. For example, it can be used to regulate solar energy passing through a window. It functions by using fluids that can be pumped into channels made in the window. The window is made with multiple stacks of parallel plates, shown in Figures 9.21a,b. The window can switch into many conditions depending upon the nature of the fluid that is pumped between plates. The idea comes from the study of animals in nature and how they change their skin color structure to fit their surroundings. Raphael Kay working with the Hatton Group, Univ. of Toronto, Canada has detailed this type of building skin and simulated the energy savings of such a system for an architectural window [36]. Kay is now at Harvard University continuing his work. What was studied and evaluated was a system of parallel plates which could have a range of fluids pumped into them, giving different optical effects. This was enough to verify the engineering concept. It is still in development. Optofluidics does give a new branch to the chromogenic

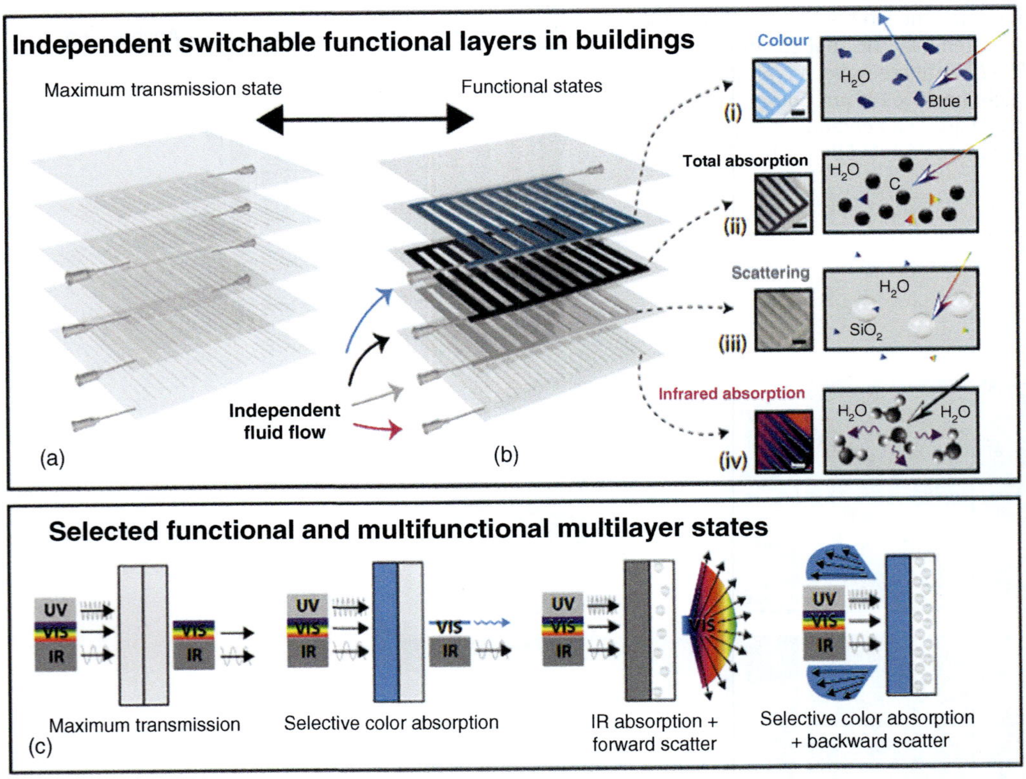

Figure 9.21 Schematic to achieve independent multilayer switchable function in glazing. Different fluids can flow within parallel layers, enabling multiple optical functions. (a) Shows the parallel layers with no flow, (b) shows specific fluids flowing into the layers. Also shown are the characteristics of each layer: (i) coloring the visible light, (ii) optical density layer to total absorption, (iii) light scattering layer (e.g. for privacy), and (iv) infrared absorption layer for the modification of the solar thermal energy. (c) Example depiction of four functional states, through coordinated fluid injections. The fluid multilayers behave as an additive light filter. *Source:* Courtesy of Raphael Kay, Harvard University, Cambridge, MA, after Kay et al. [36].

family. The pumped liquid dye system is designed with parallel poly(methyl methacrylate) (PMMA) plates with independent fluid flows for the purpose of regulating both visible absorption and near infrared (NIR) absorption, independent of each other. One can add selective scattering materials to give a frosted appearance or a privacy effect. Both backscattering and forward scattering can be enhanced depending on the concentration of titania particles, average size 20 nm, in the fluid. Strong forward scattering was shown for a concentration of 1.25 mg ml^{-1}. Both equal forward and backward scattering were obtained for 4 mg ml^{-1}. Strong backward scattering was noted for 20 mg ml^{-1} using 75 um spherical titania particles to make colloidal suspensions in water. The measurements were made on a scanning goniometer. Also, Figure 9.20a–c gives examples of how different plates can perform different functions and combined optical effects. In Figure 9.22 the transmittance properties are shown for two different fluids compared to air. Large changes in visible and near-infrared properties can be achieved with a simple change of fluids. An X, Y matrix of plates can be used to change the color and hue of the building window. It was shown that strong visible light and near-infrared modulation could be obtained.

It is estimated that the cost of the window will be about 100 US$ m^{-2}. The EnergyPlus building thermal energy modeling tool was used to determine the performance using this multilayer optofluidic window [DOE, US Government [37]]. Simulations showed better energy performance compared to electrochromic or roller shade windows. Further work needs to done to fully characterize this window technology using full-scale prototypes in actual buildings. Another structural rolled glass window system using a fluid with suspended particles has been reported. By magnetically confining the particles, the optical transmittance of the window can be adjusted. Particles of Fe_3O_4 (50–100 nm diameter) suspended in a water-glycol solution were used [38]. Also, it is possible to absorb solar energy, which can be extracted from the particles.

A company in the United Kingdom, Water-Filled-Glass (WFG), is working with water-filled windows for solar energy capture and for energy efficiency [39]. Energy studies are available on this type of structure [40–42]. Two WFG test buildings have been constructed in the

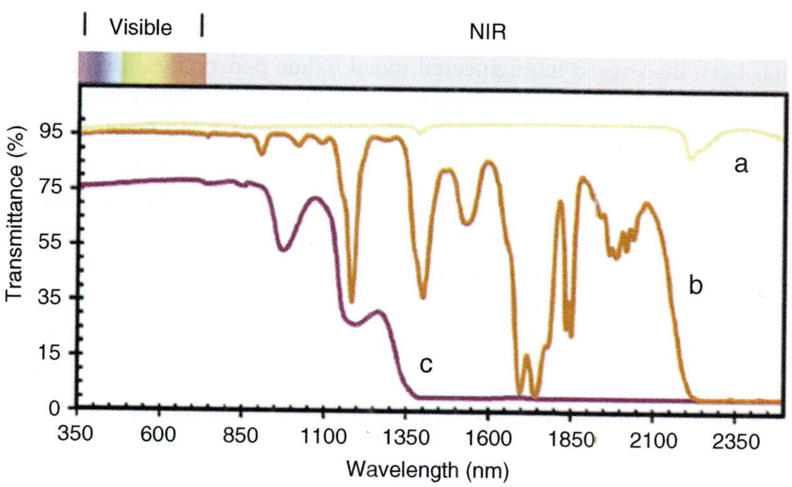

Figure 9.22 Different spectral transmittance fluid characteristics. (a) Air, no fluid, (b) glycerol, (c) 0.0156 mg carbon pigment (75 μm spherical carbon) per ml water. *Source:* Courtesy of Raphael Kay, Harvard University, after Kay et al. [36].

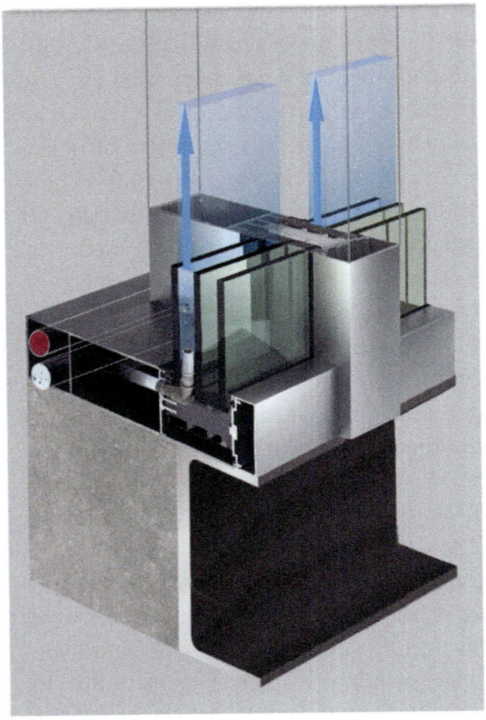

Figure 9.23 Cross-section of a triple glazing incorporating a water flow system to collect solar energy or thermal energy. Also, it works to stabilize thermal heat transport into a building or out of a building, depending upon whether heating or cooling is required in the building space. *Source:* Courtesy of Matyas Gutai, School of Architecture, Building and Civil Engineering, Loughborough University, UK and Water-filled Glass Limited, UK.

United Kingdom and Taiwan. The design of the water flow system is shown in Figure 9.23. A merger of the WFG technology with optofluidics would lead to a quite interesting energy dynamic building. Another water-filled glazing is being developed to reduce or eliminate air conditioning in buildings [43].

9.7 Thermochromic Perovskites

Thermochromism has been discovered in engineered metal halide perovskites, mentioned in Chapter 7. Much of this work is in the research stage but has potential as a new approach to thermochromic windows. Work by the Rosales et al. (NREL, Golden, CO) has shown that both hydrochromic and thermochromic materials can be engineered to make color-switchable materials [44]. These layered compounds can be used to detect changes in humidity or thermal changes. Halder et al. (IIT, Mumbai, India) explored the thermochromic behavior of hydrated lead halide hybrid perovskites $CH_3NH_3PbI_3$. They attributed the thermochromism of perovskite to the reversible hydration/dehydration process of $CH_3NH_3PbI_3$ when it encounters moisture [45]. An example of a reversible reaction is:

$$CH_3NH_3PbI_3 + 3CH_3NH_3\text{ (clear)} \leftarrow H_2O\text{, Temperature} \rightarrow 2CH_3NH_34PbI_6 \cdot 2H_2O\text{ (brown)}$$

Wheeler's group (NREL) has done work on switchable coatings for photovoltaic windows [46]. Zhang et al. in Hong Kong has made an experimental window with the spectral properties shown in Figure 9.24 [47]. The sample in Figure 9.24 is constructed of layers of

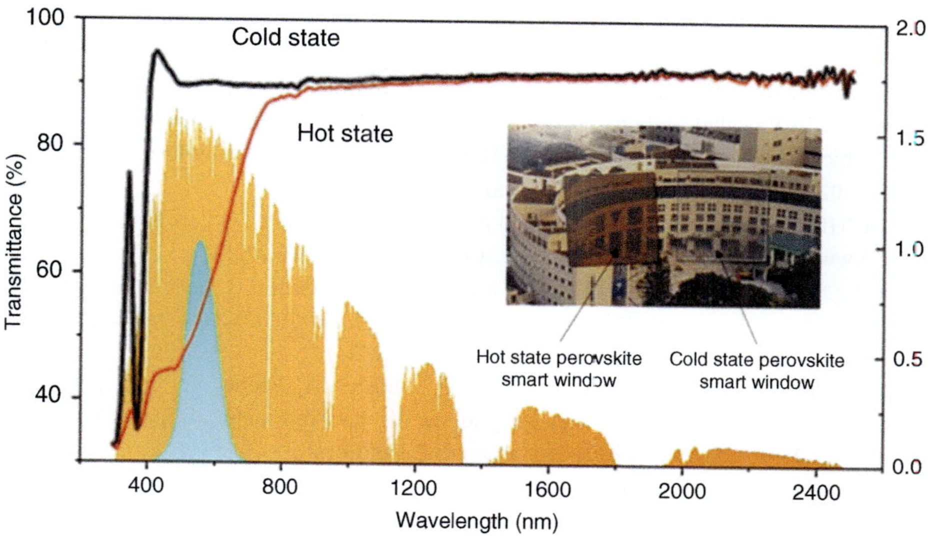

Figure 9.24 Spectral normal-hemispherical transmittance of thermochromic perovskite window. It switches with temperature but requires fairly high humidity to switch. The inset shows the color change from colorless to deep brown. Also, the solar spectrum and the photopic spectrum are shown. *Source:* Figure from Zhang et al. [47], courtesy of Elsevier.

metal halide perovskite/SiO_2 barrier layer/quartz. Spin coating is used to deposit the layers. The optimized switching parameters are: $T_{lum,cold} = 89.1\% \leftrightarrow T_{lum,hot} = 36.0\%$, $\Delta T_{sol} = 25.4\%$. At 60% RH the switching time is about 2.5 minutes; switching time is dependent on humidity level. Higher humidity gives a faster switch. There is a hysteresis in the switching effect, at 60% RH the heating switch occurs at $T_{c,h} = 54.4\,°C$, the cooling switch occurs at $T_{h,c} = 30.5\,°C$, giving a hysteresis of $\Delta T_c = 23.9\,°C$. This example does have some optical scatter in the darkened condition, possibly due to the roughness of the coating. Further work at the University of Hong Kong on metal halide perovskite windows has resulted in a mask-inspired sandwich structure with greater durability and lower haze [48]. This work used hydrated methylammonium lead chloroiodide (H-$MAPBI_{3-x}\,Cl_x$) and made a structure using nanoparticles of SiO_2/perhydropolysilazane (PHPS)/H-$MAPBI_{3-x}\,Cl_x$/glass layer system. The transmittance properties were $T_{lum,cold} = 83.4\% \leftrightarrow T_{lum,hot} = 30.4\%$, $\Delta T_{sol} = 24.4\%$.

Acknowledgments

I wish to thank Pete Valianatos and Brandon LaBelle for their help in supplying E Ink information. Also, I wish to thank Brian Bean for his review of the E Ink section. From eyrise B.V., I wish to thank Michael Grund, CEO, for all information, review of the eyrise section, and images of eyrise products. Also, I must thank Boris Lamontagne and Christophe Py of NRC Canada for pictures of their Louver window. Furthermore, I must thank Anthony Slack, CEO, ELSTAR, for his images of the ELM switchable window. I thank Dr. Raphel Kay, Harvard University for the use of his images on optofluidics. Finally, I must thank Prof. Matyas Gutai, University of Loughborough, UK, for his image of a fluid-filled window.

References

1 Culver, M. (2024). Average age of vehicles in the US continues to rise: 12.6 years in 2024, according to S&P Global Mobility. https://www.prnewswire.com/news-releases/average-age-of-vehicles-in-the-us-continues-to-rise-12-6-years-in-2024--according-to-sp-global-mobility-302152258.html (accessed 20 January 2025).

2 InterNACHI (2025). InterNACHI's standard estimated life expectancy chart for homes. https://www.nachi.org/life-expectancy.htm (accessed: 20 January 2025).

3 Hanna, J. (2009). E Ink wild ride, Harvard Business School Alumni, 1 September 2007. https://web.archive.org/web/20201012185437/https://www.alumni.hbs.edu/stories/Pages/story-bulletin.aspx?num=2760 (accessed 26 December 2024).

4 Albert, J.D., Comiskey, B., Jacobson, J.M., Zhang, L. Loxley, A., Feeney, R., Drzaic, P., and Morrison, I. (2000). Multi-color electrophoretic displays and materials for making the same, US 6017584, granted: 2000-01-25, assigned: E Ink Corp.

5 Sprague, R.A. (2014). Color electrophoretic display, US 8786935B2 publication: 2014-07-22, granted: 2012-05-3, assigned: Sipix Imaging, Inc, reassigned: E ink California, LLC.2014-07-07; E Ink Corp, 2023-09-18.

6 Telfer, S.J. Bull, S. Morrison, J.M., Slominski, L.M., Miller, D.D., Barykina-Tassa, O.V., Lee Yezek, L., MacDonald, B., Ladavac, K., and Herb, C.A. (2018). Colored electrophoretic displays, US 9921451, granted: 2018-03-20, assigned: E Ink Corp.

7 LeCain, R.D. Knaian, A.N., O'Neil, S.J., Duthaler, G.M., Danner, G.M., Zehner, R.W., Goenaga, A., Davis, B.M., Chan, R.W., Albert, J.D., and Crossley, G. (2006). Components and methods for use in electro-optical displays, US 6982178, granted: 2006-01-03, assigned: E Ink Corp.

8 FREEscan RIP (2018). Electronic paper, side view of electrophoretic display. https://en.wikipedia.org/wiki/E_Ink#/media/File:Electronic_paper_(Side_view_of_Electrophoretic_display)_in_svg.svg (accessed 26 December 2024).

9 Gee, A. (2021). E Ink Screen updating, 11 April 2021, Wikipedia. https://en.wikipedia.org/wiki/File:E_Ink_Screen_updating.webm#file (accessed 26 December 2024).

10 Page, T. (2022). New technologies series, Volume 1: E Ink JustTint™, white paper, Captains of Industry, Watertown, MA. Available from. https://go.eink.com/download-our-whitepaper-justtint-tech-series-1 (acquired 15 June 2023).

11 Tosaka-commonswiki (2010). Side view of SiPix Microcup [image heavily modified by author to represent ACep™ microcup]. https://commons.wikimedia.org/wiki/File:Electronic_paper_(Side_view_of_SiPix_Microcup).PNG (accessed 27 December 2024).

12 E Ink (2023). E Ink at CES-Consumer Electronics Show, Las Vegas, NV, January 2023. https://blog.eink.com/best-of-ces-2023-the-most-innovative-tech-we-saw (accessed 30 April 2025).

13 Woerns, S. (2023). Ultimate companion – through real and virtual worlds: BMW presents BMWi Vision Dee in Las Vegas, BMW Press Club. https://www.press.bmwgroup.com/global/article/detail/T0406898EN/ultimate-companion-%E2%80%93-through-real-and-virtual-worlds:-bmw-presents-bmw-i-vision-dee-in-las-vegas (accessed 03 November 2024).

14 Wikimedia Commons (2023). BMWi Dee. https://commons.wikimedia.org/wiki/File:BMW_Dee_IMG01.jpg (accessed 30 April 2025).

15 Van Oosten, C.L. (2015). US 9164314B2 Electro-optical controllable glazing, granted: 2015-10-20, assigned: Merck Window Technologies B.V. (2016-08-23); (same title) EP 2596403B1, granted: 2015-11-04.

16 Van Oosten, C.L. and Schenning, A.P.H.J. (2016). US 9249356B2 Liquid crystal dye mixture, granted: 2016-02-02, assigned: Merck Window Technologies B.V. (2016-08-23).

17 Junge, M., Beyer, A., Patwal, U., Kirsch, P., Beck, S., Van Oosten, C.L., Rudolf, F., and Schlosser, F.F.R. (2017). EP 2931839B1 Liquid-crystalline medium, granted 2017-11-01, assigned: Merck Patent GmbH.

18 Elementa Consulting (2019). Out of the Blue. Available from www.eyrise.com (accessed 14 September 2024).

19 Grund, M., Glipa, C., Osterodt, J., and Michiels, P.-P. (2022). LC dynamic glazing – The window becomes a pixel and the facade a giant display. *SID Symposium Digest* 53 (1): 754–756.

20 Massard, R.M., Slack, A.J., and Kate, L.T. (2019). Improved optical modulator NL 2022504B1, granted: 2019-02-04, 2020-08-19, assigned: eLstar Dynamics Patents B.V.

21 Tee, M., Galante, C.G., and Massard, R. (2024). Electrophoretic light modulator (ELM) dynamic glass technology: a daylight and energy performance study. In: *Challenging Glass Conference Proceedings*, vol. 9 (ed. C. Louter, F. Bos, and J. Belis). https://doi.org/10.47982/cgc.9.514.

22 Slack, A. and Massard, R. (2023). Smart glass an application of display technology. *Proceedings of the SID* 54 (1): 541–543. https://doi.org/10.1002/sdtp.16613.

23 Massard, R. and Slack, A. (2022). ELM: A revolutionary new smart glass technology. *Proceedings of the SID* 53 (1): 757–759. https://doi.org/10.1002/sdtp.15601.

24 Massard, R.M., Slack, A.J., and Kate, L.T. (2021). Light modulator, light modulator method and smart glazing, US 11099453B1, filed: 2020-08-03, granted: 2021-08-24, assigned: eLstar Dynamics Patents B.V.

25 Massard, R.M., Slack, A.J., and Kate, L.T. (2023). Substrate comprising electrodes and light modulator with reduced diffraction WO 2023117191A1, filed: 2021-12-24, granted: 2023-06-29, assigned: eLstar Dynamics Patents B.V.

26 Massard, R.M., Slack, A.J., and Kate, L.T. (2024). Light modulator and substrate having micro and/or nano patterned elements for optical modulation EP 4390522A1, filed: 2022-12-20, granted: 2024-06-26, assigned: eLstar Dynamics Patents B.V.

27 Massard, R.M., Slack, A.J., and Kate, L.T. (2021). Light modulator, substrate comprising electrodes and smart glazing, US 11099451B1, granted: 2021-08-31, assigned: eLstar Dynamics Patents B.V.

28 Massard, R.M., Slack, A.J., and Kate, L.T. (2024). Optical modulator, US 12001114B, filed: 22019-02-04, granted: 2024-06-04, assigned: eLstar Dynamics Patents B.V.

29 Massard, R.M., Slack, A.J., and Kate, L.T. (2024). Light modulator, substrate comprising electrodes and smart glazing, US 12007660B, filed: 22021-06-24, granted: 2024-06-11, assigned: eLstar Dynamics Patents B.V.

30 Massard, R.M., Slack, A.J., and Kate, L.T. (2024). Light modulator, light modulator method and smart glazing, US 12032263B, filed: 22020-08-03, granted: 2024-07-09, assigned: eLstar Dynamics Patents B.V.

31 Lamontagne, B., Fong, N.R., Song, I.-H., Ma, P., Barrios, P., and Poitrasa, D. (2019). Review of microshutters for switchable glass. *Journal of Micro/Nanolithography, MEMS and MOEMS*. https://doi.org/10.1117/1.JMM.18.4.040901.

32 Bozler, C.O. and Rabe, S. (1999). Spatial light modulator, EP 0706702B1, granted: 1999-01-27, assigned: MIT.

33 Lamontagne, B. and Py, C. (2010). Microblinds and a method of fabrication thereof, US 7684105B2, granted: 2010-03-23, assigned: National Research Council of Canada, Ottawa.

34 Donatiello, R., Iskhandar, M.S.Q., Hasan, M.K., Kästner, P., Qasim, M.H., Chen, J., Baby, S., Elsaka, B., Xu, G., and Hillmer, H. (2025). An analysis of arrays with irregular apertures in MEMS smart glasses for the improvement of clear view. *Micromachines* 16 (2): 176. https://doi.org/10.3390/mi16020176.

35 Hillmer, H.H., Iskhandar, M.S.Q., Hasan, M.K., Akhundzada, S., Al-Qargholi, B., and Tatzel, A. (2021). MOEMS micromirror arrays in smart windows for daylight steering. *Journal of Optical Microsystems* 1 (1): 014502. https://doi.org/10.1117/1.JOM.1.1.014502.

36 Kay, R., Jakubiec, A., Katrycz, C., and Hatton, B.D. (2023). Multilayered optofluidics for sustainable buildings. *Proceedings of the National Academy of Sciences (PNAS)* 120 (6): e2210351120.

37 U.S. Department of Energy's (DOE) Building Technologies Office (BTO). EnergyPlus building energy simulation program. energyplus.net (accessed 25 February 2024).

38 Heiz, B.P.V., Pan, Z., Su, L., Le, S.T., and Wondraczek, L. (2018). A large-area smart window with tunable shading and solar-thermal harvesting ability based on remote switching of a magneto-active liquid. *Advanced Sustainable Systems* 2: 1700140. https://doi.org/10.1002/adsu.201700140.

39 WFG (2024). Building projects using water filled windows to conserve energy. waterfilledglass.com (accessed 24 February 2024).

40 Gutai, M. and Kheybari, A.G. (2021). Energy consumption of hybrid smart water-filled glass (SWFG) building envelope. *Energy and Buildings* 230: 110508. https://doi.org/10.1016/j.enbuild.2020.110508.

41 Gutai, M. and Kheybari, A.G. (2020). Energy consumption of water-filled glass (WFG) hybrid building envelope. *Energy and Buildings* 218: 110050. https://doi.org/10.1016/j.enbuild.2020.110050.

42 Gutai, M. (2015). *Trans-structures: Fluid Architecture and Liquid Engineering*. Actar Publishers. ISBN: 978-1940291444.

43 Heiz, B.P.V., Su, L., Pan, Z., and Wondraczek, L. (2018). Fluid-integrated glass–glass laminate for sustainable hydronic cooling and indoor air conditioning. *Advanced Sustainable Systems* 2: 1800047. https://doi.org/10.1002/adsu.201800047.

44 Rosales, B.A., Mundt, L.E., Allen, T.G. Moore, D.T., Prince, K.J., Wolden, C.A., Rumbles, G., Schelhas, L.S., and Wheeler, L.M. (2020). Reversible multicolor chromism in layered formamidinium metal halide perovskites. *Nature Communications 11*: 5234. https://doi.org/10.1038/s41467-020-19009-z.

45 Halder, A., Choudhury, D., Ghosh, S., Subbiah, A.S., and Sarkar, S.K. (2015). Exploring thermochromic behavior of hydrated hybrid perovskites in solar cells. *Journal of Physical Chemistry Letters 6*: 3180–3184. https://doi.org/10.1021/acs.jpclett.5b01426.

46 Wheeler, L.M., Moore, D.M., Ihly, R., Stanton, N.J., Miller, E.M., Tenent, R.C., Blackburn, J.L., and Neale, N.R. (2017). Switchable photovoltaic windows enabled by reversible photothermal complex dissociation from methylammonium lead iodide. *Nature Communications 8*: 1–9. https://doi.org/10.1038/s41467-017-01842-4.

47 Zhang, Y., Tso, C.Y., Iñigo, J.S., Liu, S., Miyazaki, H., Chao, C.Y.H, and Yu, K.M. (2019) Perovskite thermochromic smart window: advanced optical properties and low transition temperature, *Applied Energy 254*: 113690. https://doi.org/10.1016/j.apenergy.2019.113690.

48 Liu, S., Li, Y., Wang, Y. Du, Y., Yu, K.M., Yip, H.L., Jen, A.K.Y., Huang, B., and Tso, C.Y. (2024). Mask-inspired moisture-transmitting and durable thermochromic perovskite smart windows. *Nature Communications 15*: 876. https://doi.org/10.1038/s41467-024-45047-y.

Part II

Manufacturing

10

Introduction to Manufacturing

Carl M. Lampert[1,2,3]

[1] *Star Science, Cotati, CA, USA*
[2] *Formerly, Technical Director, Society of Vacuum Coaters (non-profit), Albuquerque, NM, USA*
[3] *Staff Scientist, University of California, LBNL, Energy Technologies, Berkeley, CA, USA*

Nomenclature

AI	artificial intelligence
ALD	atomic layer deposition
BCE	before current era
B USD	billion US dollars
CIJ	continuous inkjet (inkjet)
CVD	chemical vapor deposition
DOD	drop on demand (inkjets)
EESD	electrochromic energy storage device
ρ	density
g	acceleration due to gravity
GenAI	generative artificial intelligence
IGU	insulating glass unit
ITO	indium tin oxide
γ_{LV}	liquid–vapor surface tension (sol–gel)
LHR	light heat reflecting
MAC	multilayer automatic coater
NEG	Nippon electric glass
OLED	organic light emitting diode
PDLC	polymer dispersed liquid crystals
PEN	polyethylene naphthalate
PET	polyethylene terephthalate
PI	polyimide
PVB	polyvinyl butyral
PVD	physical vapor deposition
R2R	roll-to-roll
SPD	suspended particle device
TIJ	thermal inkjet
W	plate withdrawing speed (sol–gel)
η	viscosity

Chromogenics: Smart Switchable Optical Materials and Their Applications, First Edition. Edited by Carl M. Lampert.
 Published 2026 by John Wiley & Sons, Inc.

10.1 Manufacturing Introduction

This section covers selected glass industry manufacturing subjects, including the glass float process, surface cleaning, development of flat glass sputter coating, and other flat glass processes including slot-die coating and sol–gel deposition. New processes are covered as short topics including atomic layer deposition (ALD), inkjet, and photonic processing.

The major deposition processes used by the glass industry are chemical vapor deposition (CVD) and physical vapor deposition (PVD), CVD includes pyrolytic deposition, a type of CVD used to make low-e and transparent conductors. PVD includes many large-scale processes such as vacuum deposition, magnetron sputtering, reactive sputtering, and ion plating. Sputter coating processes are used to coat float glass to make electrochromic glass at View (San Jose, CA), and SageGlass (Faribault, MN). At ChromoGenics AB (Uppsala, Sweden) they use sputter coating processes to make their laminated electrochromic layers on flexible plastic. SageGlass has a history of development using sol–gel to make their ion conductor layer. In the chapters following this introduction are deeper discussions about coatings on glass and flexible plastics by our expert authors. Sputter coating processes on glass are detailed by Wilmert Bosscher. Vacuum web (roll-to-roll) coating on flexible plastic is covered by Paul Lippens. George Neuman will discuss pyrolytic processes for the coating of float glass. These coatings are used by the glass industry to make low-e and transparent conductors for electrochromic mirrors and windows and deicing of vehicle and aircraft windows.

A factory manufacting chromogenic devices uses many other processes to make finished products. Some of these processes are surface cleaning, slot-die coating, annealing electrolyte filling or laminating, cutting, polymer laminating, bending, contact making, sealing processes, and testing. Complete manufacturing plants often include sequential coating processes using robotics, automated or artificial intelligence (AI) and generative AI (GenAI) systems. Some coating processes are incorporated into the glass-making float line process. An example is a pyrolytic coating of conductive oxide film among other layers onto hot glass surfaces emerging from a float tank.

Many of these processes are detailed subjects, which are better described by the equipment manufacturers, including Von Ardenne (Dresden, Germany), FHR Anlagenbau (Ottendorf-Okrilla, Germany), Applied Materials (Santa Clara, CA, USA), Solaras Advanced Coatings (formerly Bekaert, Belgium, US, China), Bühler Group (includes Leybold Optics, Alzenau, Germany), Mustang Vacuum Systems (Sarasota, FL), and HcVac (Guangdong, China). Specific fabrication processes and specifications of equipment are often proprietary and sometimes unique to each manufacturer. Resources for information include the various industry organizations such as the Society of Vacuum Coaters (SVC) (Albuquerque, NM) and The International Union of Vacuum Science Techniques and Applications (IUVSTA) which represents 30 nations with national vacuum organizations. Members include the American Vacuum Society (AVS) (USA), the Vacuum Society of Japan (VSJ), and the Japanese Society of Vacuum and Surface Science (JVSS).

10.2 The Glass Industry

The use of glass windows for buildings goes back to the Roman times, about 100 BCE. The Middle Ages saw the creation of the blown glass window pane. Stained glass was developed for churches. A major innovation in flat glass happened with the development of the Pilkington Float process in the 1950s, which allowed for very smooth and uniform glass to be made in large quantities. The flat glass world market in 2022 was 283.26 billion USD, and, with a CAGR growth rate of 4.41%, the estimated market will be 400 billion USD (2030) according to Statista [1]. Even with the ups and downs of the world economy, glass use has shown fairly steady growth over many years of greater than 4%. In recent times, its growth has been spurred on by new buildings and the refit of older glazing. The major growth areas are laminated glass, new housing construction, automotive and automotive aftermarket sales. Another large growth area has been the use of glass in the photovoltaic industry. In photovoltaics, both cover glass and back panel glass are flat glass. The global flat glass industry is dominated by a few major multi-billion-dollar companies including Saint-Gobain (France), Guardian Industries (USA), AGC-Asahi Glass (Japan), NSG-Pilkington (NSG) (Japan), Fuyao Glass (China), China Glass Holdings (China), PPG Industries (Glass manufacturing equipment) (USA), and Vitro Glass (Mexico), which acquired PPG Flat glass. Saint-Gobain is the world's largest producer of flat glass by tonnage. China is the largest exporter of flat glass with 13% world share (2021), (Statista) [1]. Globally, there are about 450 float plants in service or under construction in 230 locations [2]. These float plants are owned by about 40 companies according to NSG-Pilkington [3]. The average world float glass tonnage produced is 970,000 tons week^{-1}. The glass float process was invented by Sir Alastair Pilkington in 1952. The float process is shown in Figures 10.1 (side view) and 10.2 with a glass tank (top view). Prior to this invention, flat glass was drawn or cast, followed by polishing.

The float process begins on the furnace end where the glass components are mixed and melted at about 1050–1500 °C. The main components are silica, soda, lime, magnesia, and alumina. The molten glass then is floated over the molten tin and pulled over the bath to the thickness required. In the tin bath, the glass flattens and becomes very smooth. The pulling speed determines the thickness of the glass ribbon. Thickness ranges from 0.4 to 25 mm. As the glass moves it cools to about 600 °C then passes into an annealing lehr, where stresses are relieved. In this region, the glass ribbon can be pyrolytically coated. A typical electrically conductive oxide coated by this process is fluorine-doped tin oxide. Such coatings, with low haze and suitable for electrochromic and other various types of variable transmission glasses, are available in a range of 7–250 ohms/square. An example is NSG TEC™ for technical applications [5].

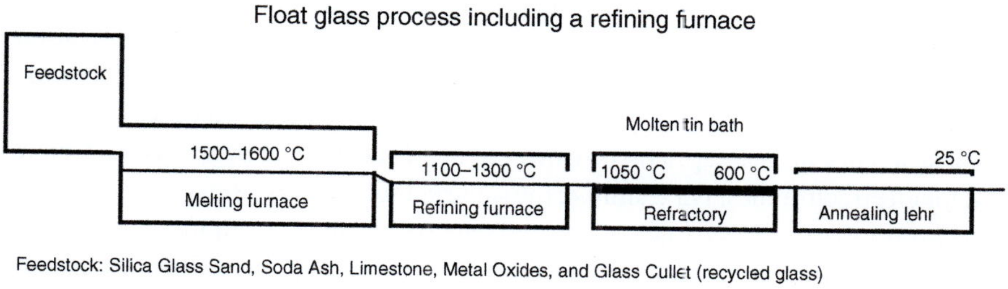

Figure 10.1 The PPG process is a modified Pilkington float process for making smooth sheet glass. The side view of the float glass process is shown. The PPG glass float process shows the method of mixing and melting raw silica, soda, lime, magnesia, and alumina at 1500–1600 °C. The glass is fed into a refining furnace to homogenize and to remove bubbles at 100–1300 °C. Then the glass flows onto a molten tin bath, creating a sheet of glass, which floats over the bath of molten tin. Next, the glass is passed into an annealing lehr to reduce stress. At this stage, pyrolytic coatings can be applied to make a doped tin oxide coating, for example. This is followed by inspection cutting and storage of the glass [4]. *Source:* Author.

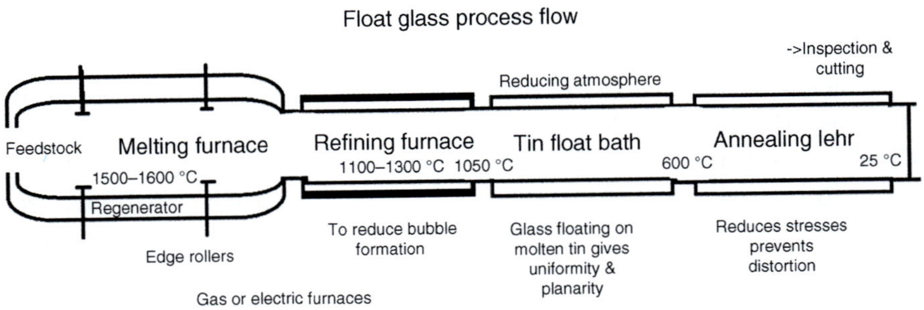

Figure 10.2 Top view of the PPG glass float line. It shows how molten glass from the furnace passes through the refining furnace, then flows over the molten tin bath. Edge rollers help the glass move over the tin bath. The speed of the glass determines the thickness. *Source:* Author.

From here, the glass is inspected for quality and then cut to the customer's specifications. Each plant runs continuously for 10–15 years until it needs rebuilding. From a sustainability standpoint of glass companies are working to use more solar electricity to melt glass instead of natural gas. New glass is being introduced with a smaller carbon footprint (about 50% less carbon compared to the old process). NSG makes Pilkington Mirai™, AGC makes Planibel, and Saint-Gobain makes their ORAÉ® products. Currently, they use glass cullet (recycled glass), in about 1/3 of the glass component mixture and this amount will increase in the future. Furnaces have improved in terms of retaining heat. In the 1970s, the author recalls barely being able to stand the intense radiated heat from the region around the glass-melting furnace. After visiting modern upgraded float plants, melting furnaces have redesigned to be more energy efficient.

Generally, the makers of glass operate several large glass float lines, coating, and processing lines. Glass companies make their money on value-added products not on basic flat glass. Specialty glass makers, including Corning, Schott, and Nippon Electric Glass (NEG) make cast and drawn glass for specific products. For example, Corning makes the strengthened "Gorilla Glass" for cell phones. Schott makes solar glass for solar cells used for space craft and low-thermal expansion glass-ceramics for telescopes and cooktops. NEG makes glass for flat panel displays among other products. Many glass companies use vacuum sputter deposition to make a variety of coated products. Solar glass has been rising in production, it is now 5% of the entire US glass market and still growing [6]. The major glass equipment investments are in the hundreds of millions of USD. Many of the major glass companies have fabrication facilities to cut shape, coat, bend, paint, and laminate glass for products such as vehicle glazing and architectural products. Normally, large vacuum and pyrolytic deposition equipment are used to make a range of coated products. However, dedicated lines at SageGlass and View are just used for electrochromics. For example, the same sputter coating line could be used to make mirrors, anti-reflected glass, low-e glass, and electrochromics. It cannot be stated too strongly that manufacturing processes that fit well with existing equipment, experience, and expertise are always preferred by the industry. The economy of each coating process and the cost of deposition equipment and source chemicals, waste chemical handling, and recycling are very important to manufacturers. Even though in research a variety of techniques are used to make switchable products, many tend to be hard to scale up to large sizes or may use too costly source chemicals. One example is spin coating.

Another important development driver in the electronics display industry, is the size of information displays. Displays are approaching that of the largest float glass sheets; wall-sized displays greater than 7.4 m diagonal size (292″) have been made. If one also looks at the information display industry, one will see several new manufacturing processes coming into the large-area fabrication arena. Thin flexible glass (200–30 μm thick), which is made by an overflow downdraw process, is now available for use in electronics. It also has the potential for use in windows. Another industry driving innovation in coating processes is the booming solar photovoltaic industry. According to Grand View Research, global photovoltaic sales were 170.25 billion USD in 2023, and are predicted to grow at 7.7% CAGR to 287.13 billion USD (2030). The US photovoltaic sales were 21.5 billion USD in 2021 and are predicted to grow 7.8% CAGR through 2030 according to Grandview Research [7].

Part of the glass industry is the glass container industry. It consists of globally of about 650 companies which produce 743 billion units (2023) and climbing to an estimated 916 billion units (2029) with a value of 96.4 billion USD (2029) globally, according to Statista [8]. The glass container industry is not a major subject of this book but is a very important industry. Chromogenic products are used for labeling of pharmaceuticals, goods and beverages in glass containers. An example is SpotSee (Dallas, TX) who manufactures environmental exposure labels to indicate if the product has been exposure to excess heat, cold or mechanical shock.

Early work on chromogenic materials started with liquid crystal and electrochromic mirrors. Automotive electrochromic mirrors have grown into a billion-dollar industry and continue to grow. In recent years, architectural applications have dominated the research and development of switchable windows. Other application areas are automotive sunroofs, marine windows, and aerospace windows. Architectural glazing is one the most attractive flat glass markets. Building energy savings have been studied by many groups with numbers of total energy savings of about 15–30% over conventional insulated glass windows. The types of switchable glazing products for buildings are windows, residential insulated glass units (IGUs), skylights, inclined glazing, security windows, commercial glazing, sunroom glazing, interior specialty glazing, and doors. The major drawbacks in the development of switchable windows have been their size, production cost, and lifetime requirements of 10–30 years. Many conventional windows are warranted for 10 years, so chromogenic glazing must satisfy this warranty requirement or must be easily replaceable. A window manufacturer study was made on the added cost of switchable glazing [9, 10]. The point at which 50% of the makers would specify a switchable panel was within the 111–250 USD m^{-2} range for the added cost. So far, many available chromogenic products are closer to 900 USD m^{-2}. However, one exception is the Pleotint Suntutive™ thermochromic window. Other big drivers in the flat glass industry are security, fire safety, and hurricane glass. More building codes are specifying tempered and laminated glass. Also, as the density of our cities increases and we live closer together, soundproof glass and privacy glass become necessary. The vacuum-insulated glazing business is estimated at 5.66 billion USD (2023) growing to 7.38 billion USD (2030) with a CAGR of 3.8%, according to MMR [11].

Many secondary glass/window companies are glass fabricators. In the United States, there are about 4,000 companies, ranging from small family operations to over 10,000 employees. Some offer specialized custom manufacturing for specific applications such as hurricane glass, bulletproof glass, fire stop glass, soundproof glass, marine and airplane glazing, and military transparency glazing. Others offer customized sizing and replacement glass for building and transportation glazing. For example, start with sheet glass, coated or uncoated, and make window product glazing comprising of a frame, fill gas, and coated products. The glass fabricators may own thin film deposition coaters, lamination autoclaves, bending and cutting equipment. For lamination, polymer-dispersed liquid crystals (PDLC), suspended particle device (SPD), and electrochromics are available on plastic films. Window makers can incorporate chromogenic film products into their window construction. Also, polyvinyl butyral (PVB) is commonly used to laminate glazing. The thermochromic Suntutive ™ product made by Pleotint is formulated into PVB as a laminating base. So, it can be used directly in the window laminating process.

The largest US window fabricators include JELD-WEN (Charlotte, NC), Marvin Windows and Doors (Warroad, MA), Milgard Andersen Windows (Bayport, MN), Pella Corp. (Pella, IA), Ply Gen Windows (Cary, NC), Alside (Akron, OH), and Velux USA (Greenwood, SC). Several of these companies have greater than one-billion-dollar annual sales. Another part of the glass fabrication industry is laminating coated plastics that are bonded onto glass. Popular window products provide increasing energy efficiency. The window efficiency guidelines (Energy Star) are provided by the US Department of Energy and the National Fenestration Rating Council. The latest Energy Star ratings for residential windows, doors, and skylights are available (Version 7.0) [12]. This has become a very successful rating system so specifiers and consumers can understand the energy efficiency of different fenestration materials.

For rating residential windows, the United States is divided into four climate zones: Northern, North-Central, South-Central, and Southern. For each of these zones, the window rating requires a certain threshold amount of the window thermal conductivity "U factor" and the solar heat gain coefficient (SHGC). Similar guidelines exist in other parts of the world. Globally, there are five major climate zones: tropical, arid, temperate, continental, and polar. Each of these climate zones also has several sub-climate zones making a total of 14 zones.

For glass industry information, there are several digital and printed sources. For example Glass Magazine (NGA), Glass Digest, Glass in Architecture, Glass in Buildings, US Glass, Glass Global, Glass International, and Window+Door magazines, among others. Every year, a meeting related to glass processors is conducted in Tampere, Finland (also in China and Brazil) called *Glass Performance Days (GPD)*, which highlights the world of glassmaking and processing on the industrial scale. Another significant conference is the *International Conference on Coatings on Glass and Plastics (ICCG)* which is the largest European scientific conference focusing on coatings on glass and plastic surfaces. In the United States, two important conferences are the National Glass Association (NGA) – *Glass Build America* show and the SVC, *TechCon*. Another conference on coatings is the *International Conference on Metallurgical Coatings and Thin Films (ICMCTF)* (San Diego, California). Yet another group of conferences incorporating thin film and their applications is organized by SPIE – The International Society for Optics and Photonics. There are several glass trade shows throughout the world, with one of the largest being Glasstec (Düsseldorf, Germany).

10.3 Surface Cleaning

Surface cleaning is extremely important before the deposition of films. Surfaces that are not clean can lead to contamination of coating films, poor adhesion, surface defects, optical aberrations, and unpredicted growth film morphologies. Float glass or soda-lime glass is widely used as glazing. In the case of a float glass process, the glass is cooled over a bath of molten tin, enriching its "float" side with tin oxide. The float glass forming processes result in flat glass sheets that are smooth on a molecular scale, requiring no further polishing. Float glass contains about 13% sodium oxide in its composition. Sodium oxide is highly soluble in water, which reacts to form sodium hydroxide. This reaction can occur in ambient air containing humidity. The water vapor in the air will generate a coating of sodium hydroxide on the surface of the glass. This layer will interfere with the adhesion of a deposited coating. The sodium hydroxide layer is best removed by rinsing in water. Another effect of forming sodium hydroxide is its reaction with carbon dioxide in the air. This reaction can lead to the formation of a white sodium carbonate on the glass surface. This is known as "blooming". Surface cleaning can be as simple as washing glass in a washing and drying station, before the pump down station in large sequential coating plants.

Other surface cleaning techniques include UV-Ozone, sputter cleaning, and reactive plasma cleaning with oxygen or hydrogen used when difficult contaminants need to be eliminated. Some cleaning processes are performed in situ, inside the vacuum chamber, or in sequential chambers. Good references on cleaning techniques are written by Don Mattox (formerly of Sandia Labs) [13, 14], David Glocker (Isoflux, Pitsford, NY) and Ismat Shah (Univ. Delaware) [15], Barbara and Edward Kanesberg (BFK Solutions, Pacific Palisades, CA) [16], and Hans Pulker (Balzers, Lichtenstein) [17].

10.4 Flat Glass Sputter Coating

The history of large-scale thin film coating starts with the development of general batch coating performed by different PVD processes [18–20]. It shows the developments needed to make the large glass coaters used by industry today. A large-scale in-line sputter coater is shown in Figure 10.3. Without the large investment in equipment by the coating industry, we would not be able to make many of the low-e or chromogenic products we have today.

The evolution of the vacuum coatings industry is very informative. The following history was set down by Russell Hill, Steve Nadel (BOC Coating Technology, Fairfield, CA), and Michael Andreasen (NxEdge, Fairfield, CA, formerly with Guardian Glass) who all worked on the design and operation of large-area glass-coating machines [22, 23].

Inline plants depositing Ag and Al coatings by chemical reduction for mirrors have existed since the late 1940s (Klopper, Dortmund, Germany), according to Joe Green [20]. Thermal

Figure 10.3 Modular sputter coating system for architectural glass (GC330H). Can be built to fit 2.54 × 3.66 m (100″ × 144″), jumbo 3.3 × 3.66 m (130″ × 144″), and super jumbo glass 3.3 × 6 m (130″ × 240″) sheet formats. *Source:* Image with permission: ©VON ARDENNE Corporate archive [21].

evaporation deposition has existed since the 1940s with equipment made by Heraeus, (Hanau, Germany). In the late 1950s, Pilkington developed a transparent conductor using a doped tin oxide coating for heating and defrosting glass. One of the first architectural coated glass products was introduced by PPG Industries, Inc. in the 1960s using a pyrolytic process. It was one of the first low-emittance coatings. The product was known as "Light Heat Reflecting" (LHR) glass. Also, in the 1960s, Saint-Gobain produced a heat-reflecting coating using resistance evaporation to deposit a thin gold layer onto glass [20]. PPG then developed a wet process called Solarban that had an improved performance compared to LHR glass. In 1968, Libbey-Owens-Ford (LOF) now part of NSG-Pilkington was the first to use an in-line vacuum process for coating [24]. The coater was based on electron beam evaporation onto vertical 3 × 3.6 m (120 × 144 in) glass (two glass sheets per load). The coating was called Vari-Tran (variable light transmission) and consisted of evaporated metals such as chromium or aluminum with a silica overcoat. Also in the 1960s, under the direction of Rolf Illsley, president of the Optical Coating Laboratory (OCLI), now MAC Films (Santa Rosa, CA) built and operated the multilayer automatic coater MAC to coat 81 × 127 cm (32 × 50 in) glass substrates with up to six-layer coatings using sequentially connected electron beam evaporation chambers. The substrates move through the coating chambers on a system of carriers in lower than 60-second cycle times. While 81 × 127 cm is not a large area by today's standards, the layer stack complexity, tolerances, and machine automation made the pathway for future larger area systems. It would be 20 years before large-area sputtering systems could match the coating uniformity and tolerances of the MAC coater.

In 1970, Leybold-Heraeus (now Applied Materials) built the first large-area coater in Europe for BSN (Danone) Glass (France). This coater was a vertical batch coater using diode sputtering. The invention of the planar magnetron in 1971 allowed sputtering at much higher film deposition rates than previously possible [25]. During that time, Al Grubb was a physicist working as the Technical and Production Manager of the LOF coater. After moving to RCA in 1973, he bought one of the first R&D planar magnetron sputtering sources from Airco Temescal (now Applied Materials) [26]. He put two innovations together: (i) for high-volume coating production, glass should be held horizontally on rollers as done in float glass manufacturing; (ii) Although it is not possible to evaporate down, it is easy to sputter in any direction. The electron beam coater operated by LOF needed special sources to evaporate sideways and OCLI's MAC coated from beneath the substrates. In 1974, Airco Temescal invented the in-line sputtering glass coater [25]. In 1975, the first sputtered glass product, for automotive sunroofs, went into mass production. The first sputtered architectural glass was made in 1977 at the Guardian-Airco joint venture coating plant in Carleton, MI [26]. This coater, built and managed by Airco, had a load size of 2 × 4 m (78 × 156 in). The coater had two coat zones having three planar magnetron cathodes in each zone. Following this success, Airco received orders for coaters from Pilkington in Sweden and SIV in Italy. During the 1970s, Airco (with Al Grubb as president) took an interesting approach. Airco created a lab-scale sputtering plant (using 0.7 m wide glass sheets) mimicking the larger area plants (with 3 m wide sheets). So, for coating development, the smaller plant could be used for experiments and then once the process was perfected, it could be transferred to the production coater. Research experimentation on large production coaters can be rather expensive, wastes substrate and ties-up production time.

Large-scale production of silver-based low-e coatings began in the United States in 1983. It allowed for a range of coated products to be developed for the residential glazing market. The first coatings on curved substrates for automotive windshields were introduced in 1983. The silver-based windshield coating was developed to provide an electrically conducting,

Figure 10.4 Dual rotatable magnetrons sputtering sources, shown with the sputtering chamber open, the glass moves on rollers shown at the bottom of the chamber. *Source:* Image with permission: ©VON ARDENNE Corporate archive [21].

optically transparent coating. This is very useful for people who have cold winters and need to eliminate ice from their windshields. Subsequently, large-aperture coaters able to accommodate deeply bent architectural glass were introduced [27]. In 1989, a significant advance was made with the development of the cylindrical rotatable magnetron cathode [28–30]. Rather than using a flat planar target, it uses target material made in the form of a rotating tube with magnetron magnets and cooling water inside the tube. The development of the rotatable magnetron source and subsequent patent battle has its own intriguing story [31, 32]. Figure 10.4 shows an open sputtering chamber with dual rotatable magnetron sources.

The continuously rotating surface offers the advantage of more target material, therefore longer production runs. This is critical to large-area production economics. The key advantage, however, is the improved performance during reactive sputtering of dielectrics (oxide and nitride compounds).

There was parallel work on the development of large-scale sputter coating plants for strip coating of metals at Manfried von Ardenne Research Institute in Dresden, Germany, according to Shiller [33]. Von Ardenne GmbH emerged in 1991 from the Manfred von Ardenne Research Institute with the fall of the German Democratic Republic. In April 2002, BOC Coating Technology sold its glass-coating business to Von Ardenne Coating Technology, Dresden. Since that time, the Von Ardenne company has become one of the world's leading companies in the development of large-scale vacuum coating systems and processes [34, 35].

Today, large-area coating is a worldwide commercial business. Since 2009, improvements in coating technology have occurred rapidly with much better mechanical systems. Specifically, improved vacuum seals and better bearings. Targets can be cantilevered for easier target changes and maintenance. Today we have many more options of target materials including alloys. Also, the target thickness is thicker so longer sputtering campaigns can be achieved. In addition, magnetic field optimization is improved, which results in better uniformity of the coated product.

As of 2024, we are in the fourth generation of large glass coaters with over three hundred large-area sputter production glass coaters in place in 22 countries worldwide. Some of the newest installations have been in Turkey, China, and Russia. Also, applications have extended from the original sputtered sunroofs to coated architectural and residential flat glass and curved automotive windshields.

10.5 Vacuum Web Coating

Commercial vacuum web coating goes back to 1935 when Hy-Sil (now Madico, Woburn, MA, owned by Lintec, Ltd.) specified a coater to deposit silver onto a flexible web. In chromogenics, vacuum web coating is used for the deposition of films onto flexible substrates, such as polymers like polyethylene terephthalate (PET), polyethylene naphthalate (PEN), and polyimide (PI), and paper to make chromogenic displays, indicators, and even windows. Vacuum web is also used to make flexible transparent conductors for plastic used in PDLCs, SPDs, and flexible displays. Further information on vacuum coating of flexible plastic is covered by Paul Lippens in the chapter on vacuum web coaters.

10.5.1 Flexible Electrochromic Coating

At ChromoGenics AB (Uppsala, Sweden), R2R coaters are used to deposit transparent conductors and electrochromic layers onto wide plastic webs. The switchable product is made by the lamination of two electrode sheets. One electrode is an anodic oxide, such as NiO_x on ITO/PET. The other is cathodic WO_x on ITO/PET. These two electrodes are laminated with a polymer electrolyte. One of the two large vacuum web coating plants at ChromoGenics AB is shown in Figure 10.5. It is an R2R vacuum sputter coater to make flexible electrochromics; in

Figure 10.5 Large roll-to-roll (R2R) vacuum web coater at ChromoGenics AB Company, Uppsala, Sweden. Shown with the chamber open. The coater can process up to a 2000 m long polymer roll with 1.6 m width in one run. *Source:* Author.

Figure 10.6 Inside the R2R vacuum coater at ChromoGenics AB showing three groups of dual rotatable sputter targets. *Source:* Author.

this view, the vacuum chamber is open. In Figure 10.6, with the interior target area open, three sets of twin rotatable sputter targets can be seen. The R2R coater can coat 1.6 m wide rolls of PET polymer with a length of 2000 m long. The length depends on the thickness of the polymer film.

10.5.2 Flexible Glass

Several glass companies are involved in the development of thin flexible glass. Corning (Willow®), Schott (UTG®), NEG (G-Leaf™), and AGC glass (Spool™). Very thin flexible glass 25-200 μm thick is made by a down-draw process. The Schott method involves a ribbon of molten glass being fed through a nozzle, then pulled down through rollers and through an annealing furnace, moving across through a cooling and inspection station. Finally, the thin glass is rolled up on a drum or cut into sheets depending upon the application. The thickness of the glass can be controlled by the speed of the process, with the resulting thickness ranging from as low as 25 μm up to 1.1 mm. A naturally fire-polished thin or ultra-thin sheet is created with <0.5 nm roughness [36]. Rolls of 1.3 m width with lengths of 300 m have been fabricated. Much of the thin glass production is aimed at display devices such as foldable cell phone screens. There is also some development for scratch resistant laminated polymer glazing applications.

Von Ardenne has developed a prototype coater to coat thin flexible glass in rolls. This coater is being evaluated at the Fraunhofer Institute for Electron Beam and Plasma Technology (FrG-FEP) in Dresden, Germany, to study transparent conductors and other films on thin glass rolls. The coater is shown in Figure 10.7 [37]. The coater was able to deposit 12 ohm/square ITO onto the flexible glass.

The first application for flexible glass is electronics, will probably be cell phone displays. A future application could be for chromogenics. From a processing viewpoint, the history

Figure 10.7 Flexible glass web roll-to-roll coating system (FOSA LAB X). It holds a 100 m (long) × 1.3 m (wide) flex glass roll, with a typical thickness of 100 μm. This was set up as a developmental coater at the Fraunhofer Institute for Electron Beam and Plasma Technology (FrG-FEP) Dresden, Germany [21]. *Source:* Image with permission: ©VON ARDENNE Corporate archive.

of R2R vacuum coating by Fenn [38] and Bishop [39] provides good industry information. Information about flexible glass can be found in a book by Garner [40].

10.6 Other Flat Glass Coating Processes

There are other techniques to deposit films. Many deposition processes are used by researchers to make small area samples. Examples of these other techniques are chemical bath deposition, electrodeposition, hydrothermal deposition, successive ionic layer deposition, ion-exchange, spin coating, dip coating, screen printing, slot-die coating, doctor blading, and sol–gel (wet-chemical) coatings [41, 42]. Some of these processes, such as slot-die, doctor blade, and dip coating, have been scaled up to large-area production machinery. Scalability is a big hurdle for many companies doing research and development. One needs to focus on processes that can be scaled to large glass sizes. For example, the largest sizes for low-e glazing are about 3.3 m × 6 m (130 × 240″) (Vitro). Some common large low-e residential window sizes in the US are for example 36 × 36″, 36 × 60″, 30 × 70″, 68 × 84″, 144 × 72″ and larger. Industry people will tell you that many times small-area issues will go away (such as gradients in thickness) in larger deposition systems with suitable control systems. They will also say new issues will pop-up which are not anticipated in small-area research studies, such as pinholes from rapidly occurring deposition processes on moving substrates. New deposition processes are developing in other areas of technology. One is inkjet coating coming from the printing industry and moving into the electronics industry. Another is atomic layer deposition coming from the electroluminescent phosphor & electronics manufacturing area.

10.7 Slot-Die Coating

Slot-die or slit-nozzle coating is used to make PDLC, SPD, and laminated electrochromics. For electrochromics, an adhesive polymer electrolyte is used as the binding agent between two electrodes. Coatema Coating Machinery, GmbH (Dormagen, Germany) makes large-area slot-die coaters. Another supplier of slot-die (slit-nozzle) coaters for smart glass is Toray (Tokyo, Japan) [43, 44]. A schematic image of the slit-nozzle coating process is shown in Figure 10.8. The image shows a stationary substrate with a moving coating gantry. Toray makes coaters to deposit liquid crystal films too. The company nTact (Dallas, TX) makes slot die coaters for the manufacture of smart windows [46]. DCN Co., Ltd. (Korea) produces slot die coaters for research up to full production size. Gradient thin films have been deposited by slot-die processes, as shown in Figure 10.9 [45]. This is an R2R coating system where the

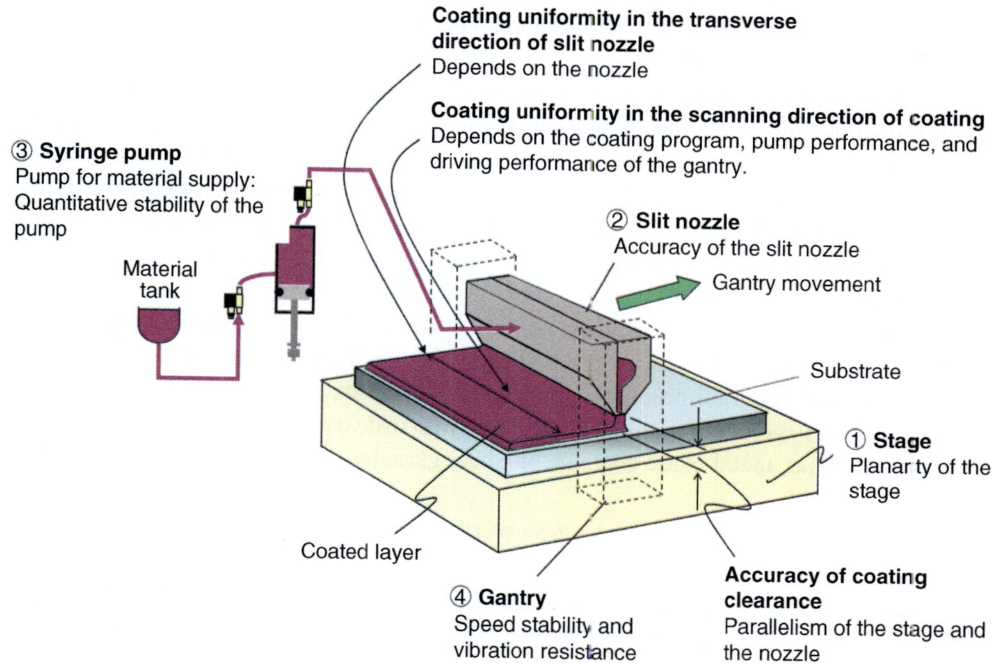

Figure 10.8 Illustration of a Toray (Japan) slit-nozzle coater for making laminated chromogenic products. Shown here, the slit-nozzle gantry moves over the substrate. The laminate mixture is pumped into the slit-nozzle coating head. Coating uniformity is the function of many variables, as noted in the illustration. In some systems, the slit nozzle is stationary and the substrate moves, as shown in Figure 10.9 [43, 44]. *Source:* [44] / With permission of Elsevier.

Figure 10.9 Gradient thin films made by slot-die coating on a roll-to-roll system. The gradient is controlled by the ratio of yellow (light) and blue (dark) feedstock pumped into the coating head. The inset shows a magnified view of the coating process at the nozzle head, Q is the flowrate, H is the coating thickness, W is the injector nozzle width, L is the effective nozzle length, and V is the web carrier speed [45]. *Source:* [45] / American Chemical Society / CC BY 4.0.

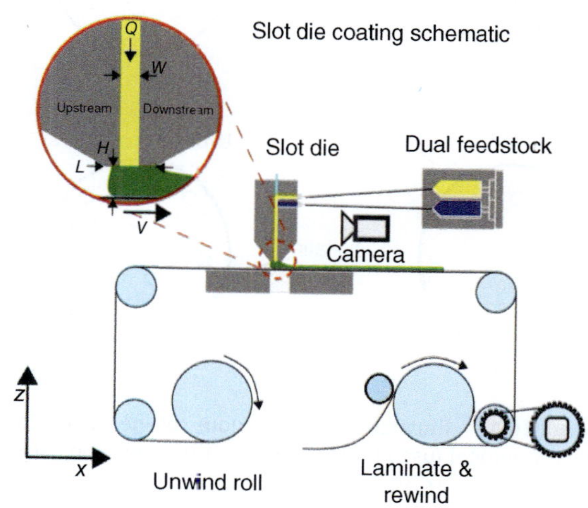

substrate moves under the stationary coating head. Gradient electrochromic films are useful for making gradient architectural windows, with a darker upper portion and a less colored lower portion of the window.

10.8 Wet-Chemical Sol–Gel Deposition

Wet-chemical deposition is a process used widely in industry to make ceramics and coatings at lower temperatures than that of traditional ceramics which require high-temperature firing. Wet-chemical deposition is composed of several related technologies. In one, it involves chemically forming a liquid and then depositing it on a surface which turns it into a gel through reactions, mainly via hydrolysis and polycondensation. The film is solidified by annealing after the sol–gel process. In another process, the sol–gel method is used to form nanoparticles or crystals of the desired compositions, which are then redispersed in a liquid matrix for the deposition of a coating. The liquid matrix itself can be a sol–gel system too. Figure 10.10 shows the process routes to make sol–gel films, ceramics, aerogels, and nanocrystals.

The sol–gel process allows for better control of the reactions involved in synthesizing solids. The chemistry of the process is adaptable to create different structures, morphology, and properties of the final product. Good sol–gel books are listed in references [47–51]. Commercially, sol–gel coating was developed for large-scale use by Schott Glaswerk (Mainz, Germany) to make metal oxide coatings on sheet glass by a low-temperature dip-coating

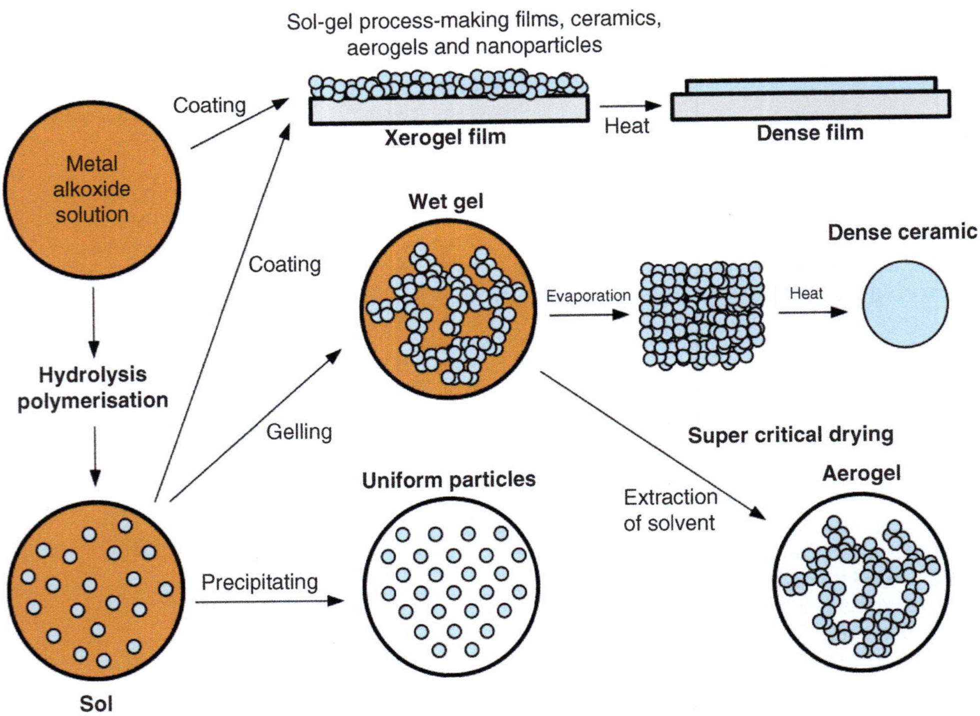

Figure 10.10 Illustration of the various sol–gel routes to make films, densified ceramics, low-density aerogel (used for transparent insulation), and nanoparticles. *Source:* Claudionico / Wikimedia Commons / CC BY SA 3.0.

process. The first product was a $TiO_2/SiO_2/TiO_2$ for an automotive mirror application, made by DESAG (Grünenplan, Germany), a subsidiary of Schott. The work at Schott is based on developments by Geffcken and Berger in 1939 [52], Schröder [53], and Dislich [54–57] for coating architectural and solar performance glazing known commercially as IROX. Schott produced coated window glazing and antireflective coatings using the sol–gel process. Schott's sol–gel draw tank could coat a 3 m × 4 m sheet of glass. The coated products showed excellent uniformity of the coating. The technique is versatile and allows multilayer coatings to be made. Sol–gel has been used to make electrochromic layers and complete devices. A range of substrates can be coated by the sol–gel process which is very useful since high firing temperatures are not used, as in conventional ceramic coatings. It is possible to coat an inorganic sol–gel coating onto a plastic substrate. The sol–gel process for coating architectural glass involves first the chemical cleaning of float glass. This is followed by a dip into a bath containing a hydrolysable metal compound. The sheet is slowly withdrawn uniformly into a humid environment which hydrolyses the film. The liquid film is formed and the effects of solvent evaporation and gelation cover both sides of the substrates with a thin film of a xerogel. Film hardening and oxidation are achieved by tempering at elevated temperatures in the air. The hardened film forms a metal oxide on the surface. The process is shown in Figure 10.11. In thicker films, sometimes residual stress caused by the large volume change during heat treatment can cause cracks in the film [58].

The sol–gel thickness during the dipping process is determined by Eqs. (10.1) and (10.2). Equation (10.1) shows the simple relationship of the thickness of the film (h) proportional to the drawing speed (V) to a power of 2/3:

$$h = k \cdot V^{2/3} \tag{10.1}$$

where h is the thickness; V is the withdrawing speed; k is the constant.

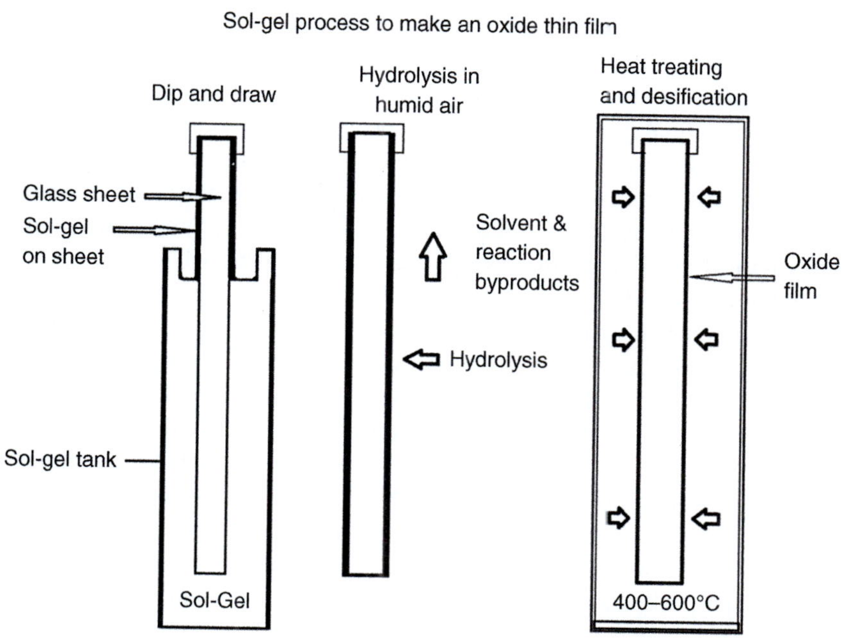

Figure 10.11 Sol–gel coating sequence (left to right) showing dipping tank, air densification, and hydrolysis, followed by baking at 400–500 °C. *Source:* Author.

Thicker coatings can be obtained by increasing the extraction speed and/or by increasing the solution viscosity. Coating thickness is mainly defined by the withdrawal speed, by the solid content, and the viscosity of the liquid. If the withdrawal speed is chosen such that the sheer rates keep the system in the Newtonian regime, the coating thickness (h) can be calculated by the Landau–Levich equation [59]:

$$h = 0.94\frac{(\eta \cdot V)^{2/3}}{\gamma_{LV}^{1/6}(\rho \cdot g)^{1/2}} \tag{10.2}$$

where η is the viscosity; V is the withdrawal speed; γ_{LV} is the liquid–vapor surface tension; ρ is the density; g is the acceleration due to gravity.

One advantage of the dip coating process is its low-temperature method to make a metal oxide coating. Another distinct advantage of this technique over others is the homogeneity and uniformity of films obtained. Coatings produced from alkoxides can also be doped with transition metals including Cu, Co, Cr, and Ni. There are several inherent advantages and disadvantages of using a solution method for coating. First, it is possible to coat all over, inside and outside the component. The solution can coat everywhere uniformly which is a distinct advantage over vacuum coating but inside and outside coating may not be desirable.

Another significant advantage is the ability to deposit coatings having several mixed metal oxides with different properties. This technique was demonstrated by Halio, Inc. in their patent "Wet-coating of thin film lithium nickel oxides for electrochromic applications", US10670936B2 [60]. An example of the formation of nanocrystalline particles (or inks) for use in electrochromic coatings was made by Heliotrope, Inc. in their patent, entitled "Electrochromic device containing metal oxide nanoparticles and ultraviolet blocking material", US9791760B2 [61]. Sol–gel deposited lithium silicate has been used as an ion conductor material for large-size windows, assigned to SageGlass [62, 63].

The capital equipment needed for wet coating is not very sophisticated although the withdrawal from the solution needs to be as smooth and vibration-free as possible. This is easily achieved using high-resolution stepping motors to drive frictionless lead screws and mounting the dip coater on a vibration-dampened table. The disadvantages depend on the coating, where the cost of chemicals can be expensive. The process can be slow compared to other deposition techniques. The final film tends to be lower density or has some porosity, compared to a densified film made by other techniques. It could be an advantage too, as it presents a large surface area for surface reaction to take place. In electrochromic films, this appears to be an advantage. The chemistry of the solutions and processing is optimized to ensure crack-free coatings at the desired thickness, while also using low-toxicity solvents.

There are several reviews and papers devoted to electrochromic films made by the sol–gel process. Anoop Agrawal reviewed sol–gel electrochromic coatings in 1993 [64]. Michel Aegerter (Institute for new Materials, Saarbrucken, Germany) reviewed sol–gel electrochromics in 1997 [65]. A special issue of *Solar Energy Materials and Solar Cells* was devoted to sol–gel films, with an emphasis on electrochromics in 2001 [66]. Also, Livage and Gangil [67] and Ozer and Lampert reviewed sol–gel deposited electrochromic coatings [68, 69]. Vo-van Troung's group Concordia University (Montreal, Canada) reviewed sol–gel for electrochromics [70]. Several cathodic electrochromics made by sol–gel include: WO_3, WO_3-TiO_2, WO_3-V_2O_5, MoO_3, Nb_2O_5 [71], $LiNbO_3$ [72, 73], TaO_x [74, 75], TiO_2, Fe_2O_3, and Fe_2O_3-TiO_2, CeO_2-TiO_2 [76]. Also, many anodic electrochromics have been made by sol–gel: NiOOH [77, 78], $NiO{\cdot}nH_2O$, $Co(OH)_2$, $IrO_2{\cdot}nH_2O$, and Co_3O_4. Cathodic/anodic materials made by sol–gel have been: V_2O_5, and V_2O-TiO_2. Also, Ta_2O_5, a popular ion conductor, used in electrochromic devices has been made by sol–gel [79].

10.9 Atomic Layer Deposition (ALD)

ALD is a form of controlled chemical surface reaction deposition. ALD has been used to make electrochromic and nanocrystalline films, including WO_3, $TiCrO_x$, V_2O_5, NiO_x, $NdNiO_3$ and highly conductive SnO_2 [80]. This technique has not been fully utilized for electrochromics but has potential. ALD is based on a binary sequence of self-limiting surface reactions that build up a film of solid material with Angstrom-level control [81]. It operates in cycles of alternating reactions with one ALD cycle depositing one "atomic layer." See Figure 10.12.

The thickness of the resulting film is directly dependent on the number of ALD cycles performed, giving high control over the film growth. While ALD has many promising features, it also suffers from slow deposition rates. Because of the long cycle times involved in pulsing and purging precursors and the layer-by-layer nature of the deposition, most ALD rates are on the order of 100–300 nm h^{-1}. During the ALD reaction, precursor molecules react with sites on the substrate surface. Once all available surface sites have become occupied, or saturated, the reactions "shut off" on their own. By driving reactions to completion every time, the resulting films are highly conformal and have a very low number of defects. This self-limiting aspect is the strength of ALD's unique process. A periodic table showing the pure materials and compounds deposited by ALD is shown in Table 10.1 [83]. The compounds include oxides, fluorides, sulfides, selenides, and tellurides, among other elements. This table is updated regularly by Prof. E. Kessels and can be viewed at his website "atomiclimits.com."

10.9.1 Spatial ALD

Spatial ALD has emerged as a promising technique for faster deposition. Spatial ALD eliminates the traditional pulse/purge chamber and replaces it with a spatially resolved deposition head, which exposes the substrate to a specific gas precursor based on location. In one design, as the head translates around the substrate, it alters the exposed precursor, resulting in film growth. Spatial ALD has also been demonstrated where the substrate moves past stationary precursor nozzles, which are arranged so that by passing by them, precursor cycling is achieved and the film is grown. In spatial ALD, the gases are delivered at different locations, so they are separated in space. In atmospheric pressure spatial-ALD, the precursor

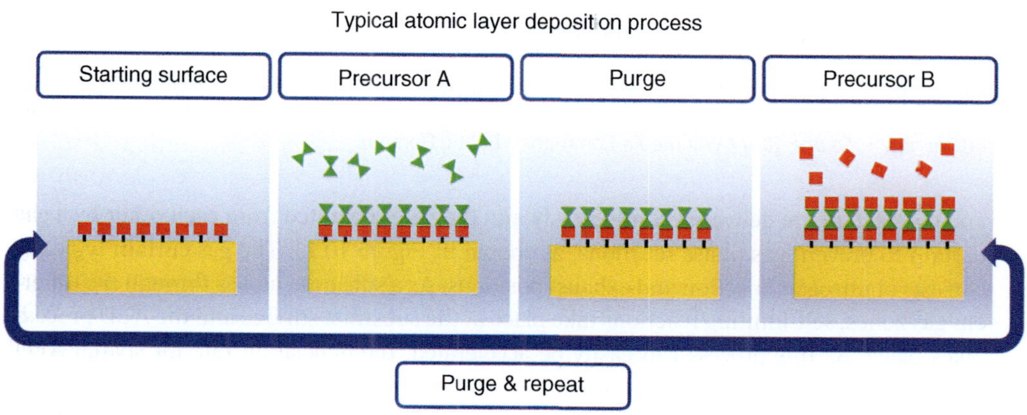

Figure 10.12 Schematic showing ALD surface reaction and purge cycle to form a film. The reactions are self-limiting and have deposition rates in the 100–300 nm h^{-1}. *Source:* [82] / Tyler Myers.

Table 10.1 Periodic table showing the pure materials and compounds deposited by ALD. Included are pure elements and compounds including F, N, O, S, Se, Te, among others.

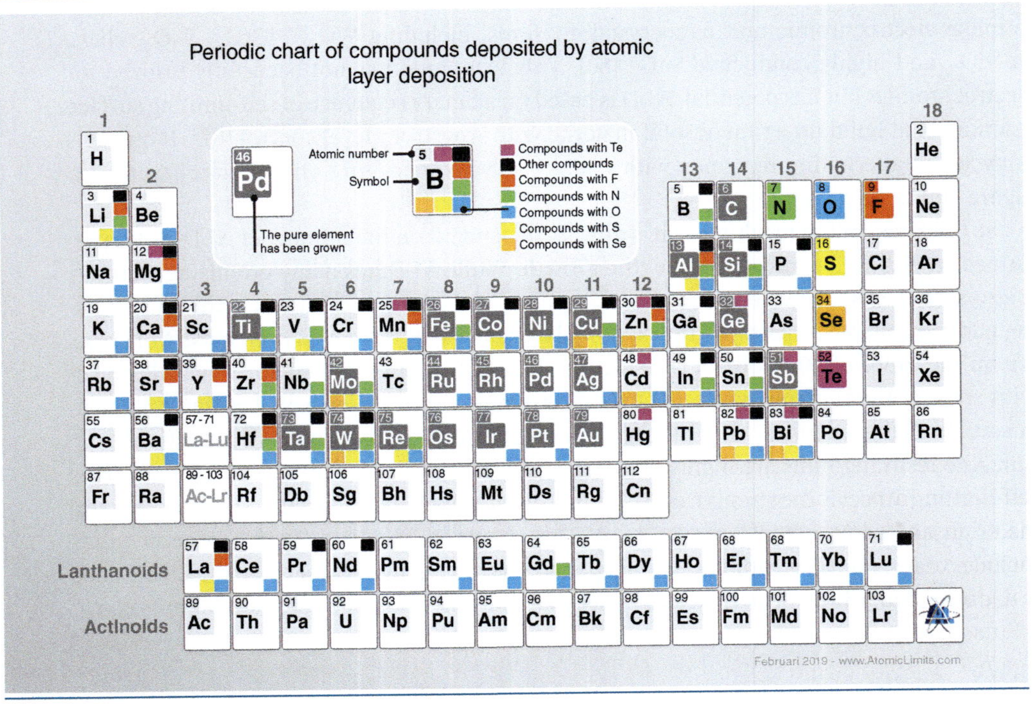

Source: Prof. E. Kessels, [83] / Erwin Kessels / CC BY 4.0.

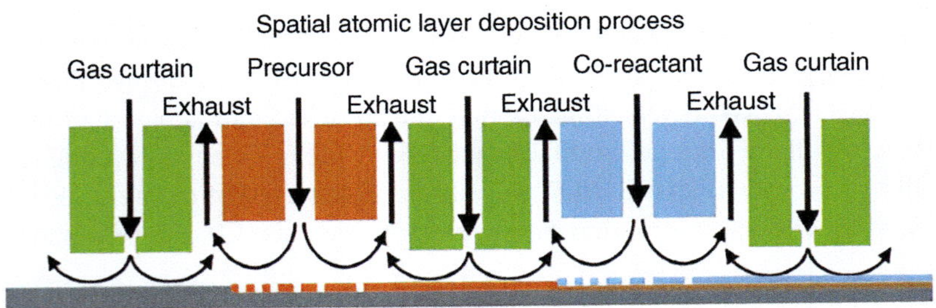

Figure 10.13 Schematic of spatial-ALD coating a substrate. Spatial ALD follows a continuous sequence, as shown above, to make thicker and faster deposition. Spatial ALD has deposition rates of 60 nm min^{-1}. *Source:* [84] / Wikimedia Commons / Public Domain.

and co-reactant are delivered continuously and they are separated from each other by a gas curtain to prevent gas phase reactions, as shown in Figure 10.13. The gas curtain typically consists of nitrogen injection and exhaust positions. As a substrate moves through the different gas zones, self-limiting reactions take place at the substrate surface and the ALD process takes place. As this process can easily be accelerated, the deposition rate for spatial ALD can be much higher than for conventional ALD. For example, for spatial ALD of Al_2O_3, the deposition rate increases to 60 nm min^{-1} compared to 2–5 nm min^{-1} [84]. For comparison, reactive sputtering can achieve about 200 nm min^{-1}. Research compounds made by ALD are listed in Table 10.1 for all applications. Electrochromism has been seen in cathodic oxides of Ti, V, Nb, Mo, Ta, and W, and anodic oxides and hydroxides of V, Cr, Mn, Fe, Co, Ni, Rh, and Ir. The glass industry has been watching the developments in the spatial-ALD area.

10.10 Inkjet Deposition

In the field of chromogenics, inkjet technology is primarily used in signage applications, but its potential for smart glass products is briefly covered here. Inks must have specific viscosities, surface tension, and rheological properties to work properly and not clog the inkjet and to deposit evenly and completely on to the substrate with fast drying. Inkjet ink typically has a viscosity range of 5–10 cP, a relative density of 1.06, and a surface tension of 28–40 mN m^{-1}. The pH ranges from 7 to 9.5. Conventional printer ink of cyan, magenta, yellow, and key (black)-CMYK colors have relative densities that run 0.95–1.9. The inkjet ink is nominally composed of seven components. (i) Pigment (can be a metal oxide or metal acetate) particle size and shape are important because the deposition velocity is 6 m s^{-1} and print nozzles (10–20-micron size) can easily be damaged from abrasive pigments; (ii) solvents for printing on glass and plastic surfaces. Solvents are typically methyl ethyl ketone (MEK), glycol esters, naphtha, water, or combinations; (iii) dispersant (to keep the pigment from agglomerating). Companies that sell dispersants are DuPont (DE, USA), Lubizol (OH, USA), and BASF (Germany); (iv) Wetting agents to help control surface tension. Agents such as Sufadol (Chongqing Acme Chem, China) and ChemPoint (US, NL, and MX) are used; (v) binders, using resins to bind the ink components together and onto the deposition surface; (vi) Humectants (glycols) to prevent dryness and improve viscosity; (vii) penetration enhancers (wetting agents) to improve ink application properties to the substrate. Depending upon formulation, inks can change with time, such as forming pigment agglomerates or changing viscosity. A company that sells ink components is Pannier Corp. (Pittsburgh, PA). Also, Saralon GmbH (Chemnitz, Germany) offers printable electrochromic ink to make display devices.

Inkjet printing is heavily used in the large fixed image and advertising industry. Inks for electronics are still under development [85]. Electronic metal oxide inks have been designed and tailored [86]. Progress in nanoparticle inks has been reviewed [87]. The development of water-based inks for organic electronics, including organic electrochromics, has been developed and studied by the Reynolds group at Georgia Institute of Technology (Atlanta, GA) [88]. Solvent-based inks have toxicity-related issues In a manufacturing environment, solvent vapor control and scrubbing are issues. Some electrochromic metal oxide inks have been developed. Tungsten oxide-based inks have been formulated [89–92] and, for vanadium oxide [93], some of the oxide ink work was done at Ynvisible (Portugal). Printed electrochromic labels for time-temperature tracking have been patented by Avery-Dennison (Pasadena, USA), see US 10282967B2 (2019). X-Card Holdings LLC. (PA, USA) has patented one-time-printed electrochromic password cards in US9004365B2 (2015). Aveso, Inc. (US) patented a printed electrochromic display label with timer US7821794 B2 (2010). Mutoh, Industries Ltd. (Tokyo, Japan) has applied for a patent for an electrochromic device made by ink deposition, JP2016065180A (pub. 2016). Other studies have been done with printable, flexible electrochromics [94], and stretchable PEDOT: PSS-based electrochromics [95].

Inkjet technology is divided into two major processes, drop-on-demand (DOD) and continuous inkjet (CIJ). In CIJ, the excess ink is recycled. In DOD inkjet, there are two types

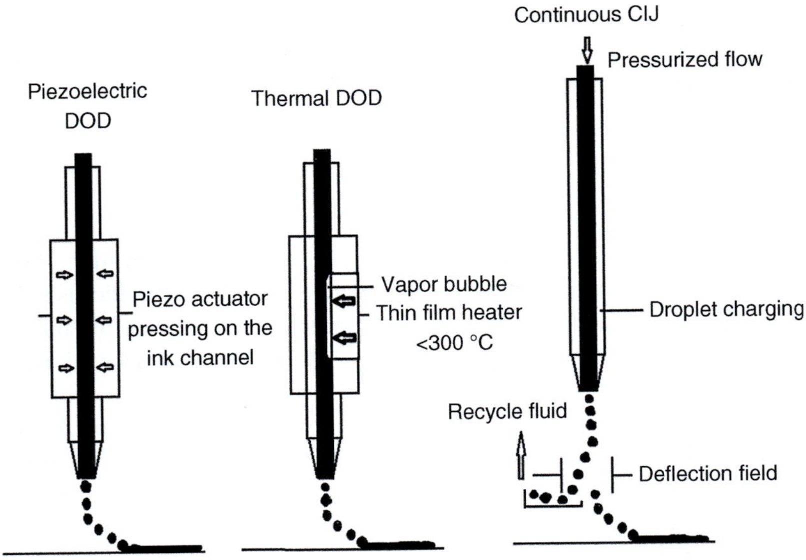

Figure 10.14 Three different inkjet print heads. Two are drop-on-demand (DOD) using a piezoelectric to pump the ink channel to make droplets or by heating forming a vapor bubble and droplets. The third is a CIJ where ink is always moving either to the substrate or deflected to a recycle ink holding tank. *Source:* Author.

of heads: thermal inkjet (TIJ) and piezoelectric or piezo DOD. Three different types of inkjet heads are shown in Figure 10.14. A quality piezo inkjet can make 50,000 drops s^{-1}. According to Epson, PreciseCore technology can go up to 40 million drops s^{-1} [96]. All the inkjet heads create drops of ink. The type of ink is important, generally, solvent-based inks are used for glass and plastics. Water-based inks are used generally for paper and porous media.

Piezoelectric industrial inkjet printers use a DOD technology that ejects ink using an electromechanical process. The DOD technology accommodates a wider range of ink types because no heating is involved. The piezoelectric crystal acts like a valve when an electrical field is applied. The valve is electrically adjustable and can produce different size ink drops. The advantage of ejecting variable-sized ink droplets is that different ink densities on the printing surface can give higher-resolution images. Piezo DOD printing systems dominate the market for printing on paper, textiles, signage, and ceramics. Their printheads can be configured to handle six or more inks giving a full-color spectrum.

The TIJ process heats the ink droplet until it is vaporized before being ejected through the printhead nozzle. High-resolution, high-speed TIJ printheads handle the toughest substrates to print using water-based or solvent-based inks. TIJ printheads are incorporated into the ink cartridge and require little or no maintenance. Some of these cartridge systems are fed from large tanks, which increases uptime for greater output and efficiency. TIJ industrial inkjet printing is often used to mark and code pharmaceutical and food packaging; offering a lot of versatility. Inks for industrial thermal printers have the lowest cost of production for shorter-run jobs.

Known for speed and long life, CIJ printers are used for large volume, large-scale industrial inkjet printing jobs. CIJ printers deliver a steady stream of ink, which is different from the DOD printheads. The ink flow is controlled by an electrical field that directs the ink stream to the substrate or diverts it into a recovery system. CIJ printing systems are generally more expensive than DOD systems because of the extra technology needed for ink recovery. The

CIJ printers remain among the fastest. Depending on various conditions, these printers can print up to 1000 ft min^{-1}. The machines can process sheets or web stock.

Inkjet manufacturing equipment has been scaled to print decorative glazing and large electronic display devices. The company Kateeva, formerly TJet Technologies (Newark, CA), has scaled up inkjet process equipment to make large-area organic light emitting diode (OLED) displays 140 cm (55″) for flat panel televisions [97]. The largest Kateeva production equipment has the upper capacity of 2.2 × 2.5 m (Gen 8.5) substrate size [98]. The inkjet printers use a movable platform to position the substrate beneath print heads that have hundreds of nozzles. The inkjet printers use a floating stage that sits on a thin cushion of nitrogen to create frictionless motion of the glass. This allows for high-stage accuracy with real-time control of ink drop position. The largest inkjet commercial printer was installed in 2020 by Tecglass (Vitrojet with Side Kinetix technology) in a joint venture with Technical Glass and Aluminum Company, UAE. The printer uses 48 print heads with 1000 nozzles each to print 1440 dpi on 3.3 × 12 m super-size glass [99].

10.11 Photonic Processing

In flexible electronic devices, we use processes that include photonic processing, which can impart high energy for a very short interval over a surface without damaging the substrate. Photonic processing can include curing, drying, reactions, annealing, and sintering of deposited films without substrate damage. In work (by the author) at Life-Space-Art (Carlsbad, CA), we successfully sintered multicolored ceramic inks produced by inkjet deposition of digital photographic images for 1 × 1.5 m sheets of glass. A rate of 1 m^2 s^{-1} sintering can be achieved. Photonic curing can be used very effectively for processing plastic and fabric substrates too.

Photonic curing is the high-energy curing or processing of thin films using a flashlamp. When this processing is performed on a low-temperature substrate, such as plastic or paper, a significantly higher temperature can be attained in the film versus the substrate. Oven curing would damage the substrate. This is because of two effects: (i) thermal processes exponentially progress with increasing temperature and (ii) thermal damage to a polymer substrate generally takes a finite amount of time. Within a range, the shorter the time, the higher the temperature a material can take without damage. In the case of a polymer or paper substrate, the ultimate temperature it can reach is right below its vaporization temperature. So as a result, an intense pulse of light can process a thin absorbing film on a low-temperature substrate in only a few hundred microseconds as effectively or better than 10 minutes in an oven. As the substrate generally absorbs less of the light energy from the flashlamp than the targeted film, the film can be selectively printed where only the film is heated. The photonic curing is very high in power for a very short duration. Over the past 20 years, photonic curing has progressed far beyond the laboratory sintering of metal on plastic and paper and now is used in many processes for the high-volume manufacturing of consumer electronics and other industries. By using shaped light pulse processing, one can stay at an overall elevated temperature point for much longer than in the simple pulse case. This increases the processing margin which is always desired. Like the simple pulse case, one can initially pulse the flashlamp to heat the film to nearly 400 °C. As the film begins to cool in about 50 μs, a smaller pulse of light is given to bring the surface temperature up. During this second pulse, the heat from the top is being conducted into the bulk of the coating, so we continue to give small pulses to move the surface temperature back up again. The effect of this effort is that the amount of time spent just below the maximum allowed temperature is a lot longer than in the single pulse case. This results in a much better cure or processing (such as crystallization of a metal ink). For this example, we have used about 70% more radiant exposure, or a total of 1.2 J cm^{-2}. However, this is well below the damage threshold for PET. The final temperature after equilibration reaches about 80 °C. (K.A. Schroder, NovaCentrix) [100].

10.12 Busbars and Electrical Connections

In electrically powered chromogenic devices, busbars and their connection to external power and internal transparent conductors are very important. Important considerations are the busbar location and manufacturing technique to make the busbar. In large segmented glazing, making transparent connections is very important to the customer. Good distributed low-resistance contacts with resistance much lower than the transparent conductors can help a large electrochromic glazing switch rapidly and uniformly. Large electrochromic glazing may be judged solely on its ability to switch in a timely and uniform manner. This is critical for electrochromics because they are low-voltage (a few volts DC), current-controlled devices and do not do well with overvoltage or high-current concentrations. Devices like liquid crystals and dispersed particle devices are electric field-controlled devices. They are activated by a large voltage (e.g. 100VAC or square wave) potential which creates an electric field across large parallel electrodes. With these devices, the concern is more over electrical leakage across parallel ITO layers. If the resistive losses are too great, one can see variations in opacity. Connections to transparent conductors can be made by several methods using strong metal clip strips directly to the transparent conductor, conductive tape with conductive adhesive, silver frit, silver-filled epoxy, printed metal inks, and ultrasonic soldering. The use of ultrasonic energy in ultrasonic soldering creates cavitation in a special zinc-containing solder. This results in the creation of zinc-oxygen bonds between the solder and the surface oxide of the glass, thus creating a stable bond. For example, Komura Technology, Ltd. (Yokohama, Japan) makes ultrasonic bonding equipment [101].

Electrical connections from the outside of the glazing to the inside of a gap or along a glass edge in an IGU glazing present some challenges. Spacers and connectors have been patented, just a few examples: (i) US 12025900 on connectors through frames to EC windows, assigned to View Corp. [102]; (ii) Ajjer patented techniques to make corrosion-resistant connections to automotive electrochromics in US 8947758 [103]. Ajjer also used nanoparticles to make superior busbars and sealants for electrochromic devices in US 8988756 [104]; (iii) SageGlass patented a series connection technique for EC glazing in US8493646B2 [105]. There are quite a few patent families that cover practices on making connections, including related electronics, smart windows, touch screens, and photovoltaic panels.

Acknowledgments

I wish to thank Frederik Fränding, CEO, Greger Gregard, CTO, and colleague, Claes Granqvist, for their hospitality at ChromoGenics AB (Uppsala, Sweden). I wish to thank Anoop Agrawal (Dyenamics) and Ric Shimshock (MLD Technologies) for their review and timely input to this chapter.

References

1 Statista (2024). World flat glass markets. Statista.com (accessed 8 Jan 2024).
2 Dick, N. (2022). World of Glass 2022 report. https://www.glassmagazine.com/article/world-glass-2022-report (accessed 18 Aug 2024).
3 Pilkington (2023). The float process. https://www.pilkington.com/en/global/knowledge-base/glass-technology/the-float-process/the-float-process# (accessed 12 Dec 2023).
4 Kent, R. (2024). Float process, Tangram Technology, UK. https://tangram.co.uk/contact/ (accessed 12 April 2024).
5 Pilkington (2024). NSG **TEC™** for technical applications. https://www.pilkington.com/en/us/products/product-categories/special-applications/nsg-tec-for-technical-applications#overview (accessed 23 Aug 2024).
6 Dick, N. (2024). World Flat Glass Report 2023. https://www.glassmagazine.com/article/world-glass-2023-report (accessed 7 June 2024).
7 Grand View Research (2024). Solar PV panels market size, share & trends analysis report by technology (thin film, crystalline silicon), by grid type (on grid, off grid), by application (residential, commercial, industrial), by region, and segment forecasts, 2024–2030. https://www.grandviewresearch.com/industry-analysis/solar-panels-market? (accessed 20 Aug 2024).
8 Jaganmohan M. (2024). Glass containers and bottles global market value 2022–2029. https://www.statista.com/statistics/700364/glass-bottles-and-containers-market-value-worldwide (released 19 Apr 2024).
9 Sottile, G.M. (2002). Assessment of attitudes and expectations of switchable glass among United States Window manufacturers. Proceedings of the 45th Annual Technical Conference Proceedings (2002), ISSN 0737-5921.
10 LaPointe, M.R. and Sottile, G.M. (2001). 2000 survey of window manufacturers on the subject of switchable glass. In: *Proceedings of the SPIE 4458, Solar and Switching Materials, (13 November 2001)* (ed. C.M. Lampert). https://doi.org/10.1117/12.448238.
11 MMR (2023). Vacuum Insulated Glass market 2023. https://www.maximize-marketresearch.com/market-report/vacuum-insulated-glass-market/13272 (accessed 12 June 2024).
12 US Department of Energy and National Fenestration Rating Council.The latest Energy Star ratings for residential windows, door, and skylights is available (Version 7.0). https://www.energystar.gov/sites/default/files/asset/document/ES_Residential_WDS_V7_Final%20Specification%202022.pdf (accessed 12 April 2024).
13 Mattox, D.M. (2007). Surface preparation for vacuum surfaces and vacuum coating (Chapter 21). In: *50 Years of Vacuum Coating Technology*, 151–153. Alburquerque, NM: Society of Vacuum Coaters, Society of Vacuum Coaters. ISBN: 978-1-878068279.
14 Mattox, D.M. (2017). Cleaning (Chapter 12). In: *Handbook of Physical Vapor (PVD) Processing*, 644–743. Cambridge University Press. ISBN: 978-0080946580.
15 Glocker, D.B. and Shah, S.I. (ed.) (2017). Processes for substrate cleaning, section E1.0, V. 2. In: *Handbook of Thin Film Process Technologies*. CRC Press. ISBN: 978-1315893586.
16 Kanesburg, B. and Kanesburg, E. (ed.) (2017). *Handbook for Critical Cleaning*, 2e. CRC Press. ISBN: 9781138074576.
17 Pulker, H. and Pulker, H.K. (1999). Cleaning of substrate surfaces (Chapter 4). In: *Coatings on Glass*, 2e. NY: Elsevier. ISBN: 978-0444501035.

18 Holland, L. (1966). *Vacuum Deposition of Thin Films*. London: Chapman and Hall.

19 Macload, H.A. (2007). The history of optical coatings. In: *50 Years of Vacuum Coating Technology and the Growth of the Society of Vacuum Coaters* (ed. D.M. Mattox and V.H. Mattox). Albuquerque, NM: Society of Vacuum Coaters. ISBN: 978-1-878068-27-9.

20 Greene, J.E. (2017). Review Article: Tracing the recorded history of thin-film sputter deposition: from the 1800s to 2017. *Journal of Vacuum Science and Technology A* 35 (5): 05C204. https://doi.org/10.1116/1.4998940.

21 Von Ardenne, GmbH (2023). Modular sputter coating system and dual rotatable cathode, vonardenne.com, vonardenne.us.

22 Hill, R.J. and Nadel, S.J. (1999). *Coated Glass Applications and Markets*. Fairfield, CA: BOC Coating Technology, BOC Group. ISBN: 0-914289-01-2.

23 Hill, R.J., Nadel, S., and Andreasen, M. (2007). History of large area coatings (Chapter 4). In: *50 Years of Vacuum Coating Technology and the growth of the Society of Vacuum CoatersSociety of Vacuum Coaters* (ed. D.M. Mattox and V.H. Mattox). Albuquerque, NM. ISBN: 978-1-878068-27-9.

24 Grubb, A.D. (1969). The LOF semi-continuous thermal evaporation plant. *R&D Magazine* 20 (6): 42–46.

25 Chapin, J.S. (1979). US Patent 4166018, Sputtering Source and Apparatus. Granted: 1979-08-28, Assigned: BOC Group/Linde LLC Expired: 1996-08-28.

26 Grubb, A.D., Mosakowski, T.S., and Overacker, W.G. (1982) Production techniques for high volume sputtered films. Proceedings of the SPIE 325 Optical Thin Films, pp. 74–80.

27 Young, P. and Bernardi, R. (1988). Auto solar control. SAE Technical Papers Series #880050.

28 McKelvey, H.E. (1982). US 4365073, Magnetron cathode sputtering apparatus (rotatable), Granted:1982-02-12 (assigned Shatterproof Glass; BOC group (1988)).

29 McKelvey, H.E. (1983). US 4422916, Magnetron cathode sputtering apparatus (rotatable), Granted:1983-12-27 (assigned Shatterproof Glass; BOC group (1988)).

30 Gläser, H.J. and Szyszka, B. (2007). History of glass coating for architectural glazing. In: *Proceedings of SVC 50*, 216–229. Albuquerque, NM: Society of Vacuum Coaters ISSN: 0737-5921.

31 Justia (1987). Libbey-Owens-Ford Co. v. BOC Group, Inc., 655 F. Supp. 897 (D.N.J. 1987). https://law.justia.com/cases/federal/district-courts/FSupp/655/897/1423646/.

32 De Gryse, R., Haemers, J., Leroy, W.P., and Depla, D. (2012). Thirty years of rotatable magnetrons. *Thin Solid Films* 520: 5833. https://doi.org/10.1016/j.tsf.2012.04.065.

33 Schiller, S., Beister, G., Neumann, M., and Jaesch, G. (1982). Vacuum coating of large areas. *Thin Solid Films* 96 (3): 199–216. https://doi.org/10.1016/0040-6090(82)90244-9.

34 Gläser, H.J. (1999). *Dünnfilmtechnologie auf Flachglas*. Verlag Karl Hofmann, Schorndorf Germany. ISBN: 13: 9783778010419.

35 Gläser, H.J. (2000). *Large area glass coating*. Dresden, Germany: Von Ardenne Anlagen-Technik GmbH. ISBN: 3000049533, 9783000049538.

36 Schott Glasswerke (2024). Glass manufacture. https://www.schott.com/en-hu/expertise/glass-melting-and-hot-forming?tab=e0366e25 (accessed 8 Sep 2024).

37 Fraunhofer FEP (2024). Functionalized ultrathin flexible glass. https://www.fep.fraunhofer.de/content/dam/fep/en/documents/Produktflyer/H10 Functionalized ultrathin flexible glass EN net.pdf (accessed 19 August 2024).

38 Fenn, J.B. (2007). The history of vacuum roll coating. In: *50 Years of Vacuum Coating Technology and the Growth of the Society of Vacuum CoatersSociety of Vacuum Coaters* (ed. D.M. Mattox and V.H. Mattox). Albuquerque, NM. ISBN: 978-1-878068-27-9.

39 Bishop, C.A. (2015). *Vacuum Deposition onto Webs, Film and Foils*, 3e. Oxford, UK: Elsevier. ISBN: 978-0-323-29644-1.

40 Garner, S.M. (ed.) (2017). *Flexible Glass: Enabling Thin, Lightweight, and Flexible Electronics*. Beverly, MA: Scrivener Publishing LLC. ISBN: 978-1-118-94636-7.

41 Sankapal, B.R., Ennaoui, A., and Gupta, R.B. (2023). Lokhande. In: *Simple Chemical Methods for Thin Film Deposition: Synthesis and Applications* (ed. C.D.). Springer. ISBN: 978-9819909605.

42 Park, J., Shin, K., and Lee, C. (2016). Roll-to-Roll coating technology and its applications: a review. *International Journal of Precision Engineering and Manufacturing* 17 (4): 1–14. https://doi.org/10.1007/s1254-016-0067-z.

43 Toray (2024). Slot-nozzle coater for smart glass. https://www.toray-eng.com/products/fpd/fpd_017.html (accessed 22 Aug 2024).

44 Tajima, K., Chan, C.Y., Kubota, T., Ito, T., Araki, K. Kamei, T., and Fukui, M. (2021). Mass-producible slit coating for large-area electrochromic devices. *Solar Energy Materials and Solar Cells* 232: 111361. https://doi.org/10.1016/j.solmat.2021.111361.

45 Jeong, T.J., Yu, X., and Harris, T.K.L. (2024). Scaled production of functionally gradient thin films using slot die coating on a roll-to-roll system. *ACS Applied Materials & Interfaces* 16 (7): 9264–9274. https://doi.org/10.1021/acsami.3c17558.

46 nTact (2024). Smart glass and architectural glass. https://ntact.com/applications/smart-glass/, Dallas, TX (accessed 22 Aug 2024).

47 Brinker, C.J. and Scherer, G.W. (2013, 1993). *Sol-Gel Science: The Physics and Chemistry of Sol-Gel Processing*. NY: Elsevier. https://doi.org/10.1016/C2009-0-22386-5.

48 Zayat, M., Almendro, D., Vadillo, V., and Levy, D. (2017). Sol-gel materials for optics and electrooptics. In: *Handbook of Sol-Gel Science and Technology* (ed. L. Klein, M. Aparicio, and A. Jitianu). Germany: Springer. https://doi.org/10.1007/978-3-319-19454-7_145-1.

49 Klein, L.C. (1989). *Sol-Gel Technology for Thin films, Fibers, Preforms, Electronics and Specialty Shapes*. Park Ridge, NJ, USA: Noyes Publications.

50 Aegerter, M.A. and Mennig, M. (ed.) (2004). *Sol-Gel Technologies for Glass Producers and Users*. New York, USA: Springer Science + Business Media.

51 Heusing, S. and Aegerter, M.A. (2018). Sol-gel coatings for electrochromic devices. In: *Handbook of Sol-Gel Science and Technology* (ed. L. Klein, M. Aparicio, and A. Jitianu). Springer. https://doi.org/10.1007/978-3-319-32101-1_36.

52 Geffcken, W. and Berger, E. (1939). GDR Patent 736,411, assigned Jenaer Glaswerk Schott, Jena, GDR.

53 Schröder, H. (1969). Oxide layers deposited from organic solutions. In: *Physics of Thin Films*, vol. 5 (ed. G. Hass and R.E. Thun), 87–141. New York: Academic Press.

54 Dislich, H. (1971, 1971). New routes to multicomponent glasses. *Angewandte Chemie, International Edition* 6: 363.

55 Dislich, H. and Hussmann, E. (1981). Amorphous and crystalline dip coatings obtained from organometallic solutions: procedures, chemical processes and products. *Thin Solid Films* 77 (1–3): 129–140. https://doi.org/10.1016/0040-6090(81)90369-2.

56 Dislich, H. (1986). Sol-gel: science processes and products. *Journal of Non-Crystalline Solids* 80 (1–3): 115–121.

57 Dislich, H., Hinz, P., Arfsten, N.-J., and Hußmann, E.K. (1989). Sol-gel yesterday, today and tomorrow. *Glastechnische Berichte* 62: 46–51.

58 Kozuka, H. (2016). Stress evolution and cracking in sol–gel derived thin films. In: *Handbook of Sol-Gel Science and Technology* (ed. L. Klein, M. Aparicio, and A. Jitianu). Switzerland: Springer International. https://doi.org/10.1007/978-3-319-19454-7_12-1.

59 Landau, L.D. and Levich, B.G. (1942). Dragging of a liquid by a moving plate. *Acta Physicochimica U.R.S.S.* 17: 45–54.

60 Choi, H.J., Bass, J.D., Lachman, E. Giaquinta, D.M., Turner, H.W., and Murphy, E. (2020). US10670936B2, Wet-coating of thin film lithium nickel oxides for electrochromic applications, granted 2020-07-13, assigned to Halio, Inc.

61 Garcia, G., Koo, B., Gregoratto, I., Basu, S., Rosen, E., Holt, J., and Thomsen, S. (2017). US9791760B2 - Electrochromic device containing metal oxide nanoparticles and ultraviolet blocking material, granted 2017-10-17, assigned Heliotrope, reassigned 202 to: Hivisq Technologies Sociedad Limitada, Spain.

62 Van Dine, J.E., Parkhe, V.D., Klein, L.A., and Trumbore, F.A. (1995). US 5,404,244A, Electrochromic structures and methods, granted: 1995-04-04, assignee: SAGE Electrochromics.

63 Gillaspie, D.T. and Weir, D.G. (2019). US 10,514,581 B2, Electrochromic nickel oxide simultaneously doped with lithium and a metal dopant, granted: 2019-12-24, assigned: Alliance for Sustainable Energy, SAGE Electrochromics.

64 Agrawal, A., Cronin, J.P., and Zhang, R. (1993). Review of solid state electrochromic coatings produced using sol-gel techniques. *Solar Energy Materials and Solar Cells* 31-1: 9–21. https://doi.org/10.1016/0927-0248(93)90003-L.

65 Aegerter, M.A., Avellaneda, C.O., Pawlicka, A., and Atik, M. (1997). Electrochromism in materials prepared by the sol-gel process. *Journal of Sol-Gel Science and Technology* 8: 689–696. https://doi.org/10.1007/BF02436924.

66 Özer, N. (2001). Sol-gel coatings special issue. *Solar Energy Materials and Solar Cells* 68 (3–4): 237–428. https://www.sciencedirect.com/journal/solar-energy-materials-and-solar-cells/.

67 Livage, J. and Ganguli, D. (2001). Sol–gel electrochromic coatings and devices: a review. *Solar Energy Materials and Solar Cells* 68: 365–381. https://doi.org/10.1016/S0927-0248(00)00369-X.

68 Özer, N. and Lampert, C.M. (1998). Electrochromic characterization of sol-gel deposited coatings. *Solar Energy Materials & Solar Cells* 54 (1–4): 147–156. https://doi.org/10.1016/S0927-0248(98)00065-8.

69 Özer, N. and Lampert, C.M. (1995). Sol-gel deposited electrochromic coatings. *Proceedings of the American Optical Society, Technical Digest* 17: 344–347.

70 Jittiarporn, P., Badilescu, S., Al Sawafta, M.N., Sikong, L., and Truong, V.-V. (2017). Electrochromic properties of sol–gel prepared hybrid transition metal oxides – a short review. *Journal of Science: Advanced Materials and Devices* 2 (3): 286–300. https://doi.org/10.1016/j.jsamd.2017.08.005.

71 Özer, N., Barreto, T., Buyuklimanli, T., and Lampert, C.M. (1995). Characterization of sol-gel deposited niobium pentoxide films for electrochromic devices. *Solar Energy Materials and Solar Cells* 36: 433–443.

72 Özer, N., Rubin, M., and Lampert, C.M. (1996). Optical and electrochemical characteristics of niobium oxide films prepared by sol-gel process and magnetron sputtering: a comparison. *Solar Energy Materials & Solar Cells* 40 (1996): 285–296.

73 Özer, N. and Lampert, C.M. (1995). Electrochemical lithium insertion in sol-gel deposited $LiNbO_3$ films (Proceedings of the 1st International Conference on Electrochromism). *Solar Energy Materials & Solar Cells* 39 (2–4): 367–375. https://doi.org/10.1016/0927-0248(96)80002-X.

74 Özer, N., He, Y., and Lampert, C.M. (1994). Ionic conductivity of tantalum oxide films prepared by sol-gel process for electrochromic devices. Proceedings of the SPIE 2255, Optical Materials Technology for Energy Efficiency and Solar Energy Conversion XIII (9 September 1994), https://doi.org/10.1117/12.185388.

75 Özer, N. and Lampert, C.M. (1995). Sol-gel deposited amorphous tantalum oxide and niobium oxide films as protonic conductors. Proceedings of the Window Innovation Conference, June 1995, Toronto, Canada, p. 3372–3382. Also, LBL-38526 report, https://eta.lbl.gov/publications/sol-gel-deposited-amorphous-tantalum.

76 Özer, N., DeSouza, S., and Lampert, C.M. (1995). Optical and electrochemical properties of sol-gel spin coated CeO_2-TiO_2 films. *Proceedings of SPIE* 2531: 143–151.

77 Šurca, A., Orel, B., and Pihlar, B. (1997). Sol-gel derived hydrated nickel oxide electrochromic films: optical, spectroelectrochemical and structural properties. *Journal of Sol-Gel Science and Technology* 8: 743–749. https://doi.org/10.1007/BF02436933.

78 Zayim, E.O., Turhan, I., Tepehan, F.Z., and Ozer, N. (2008). Sol–gel deposited nickel oxide films for electrochromic applications. *Solar Energy Materials and Solar Cells* 92 (2): 164–169. https://doi.org/10.1016/j.solmat.2007.03.034.

79 Tepehan, F.Z., Ghodsi, F.E., Özer, N., and Tepehan, G.G. (1999). Optical properties of sol–gel dip-coated Ta_2O_5 films for electrochromic applications. *Solar Energy Materials and Solar Cells* 59 (3): 265–275. https://doi.org/10.1016/S0927-0248(99)00041-0.

80 Song, Y., Zhou, L., Liu, X., and Zhang, J. (2025). Atomic layer deposition of SnO_2 enables fast-dynamics in electrochromic vanadium oxide/polyaniline films. *Next Materials* 6: 100328. https://doi.org/10.1016/j.nxmate.2024.100328.

81 Johnson, R.W., Hultqvist, A., and Bent, S.F. (2014). A brief review of atomic layer deposition: from fundamentals to applications. *Materials Today* 17 (5): 236. https://doi.org/10.1016/j.mattod.2014.04.026.

82 Myers, T. (2021). Beneq Group, (Espoo, Finland), Atomic and Molecular Layer Processes for Industrial Applications in Semiconductors, Pharmaceuticals and Optics. Ph.D. thesis, Dept. of Chemistry, University of Colorado, Boulder, CO, https://scholar.colorado.edu/concern/graduate_thesis_or_dissertations/47429b084.

83 Kessels, E. (2019). Periodic table of compounds prepared by ALD. https://www.atomiclimits.com/2019/01/28/overview-of-all-materials-prepared-by-atomic-layer-deposition-ald-an-up-to-date-and-colorful-periodic-table-to-download/ (accessed 6 June 2024).

84 JB9573321 (2024). Spatial ALD concept (Image). https://commons.wikimedia.org/wiki/File:Spatial_ALD_concept.png (accessed 6 June 2024).

85 Calver, P. (2021). Inkjet printing for materials and devices. *Chemistry of Materials* 13 (10): 3299–3305. https://doi.org/10.1021/cm0101632.

86 Fang, M., Li, T., Zhang, S., Rao, K.V., and Belova, L. (2020). Design and tailoring of inks for inkjet patterning of metal oxides. *Royal Society Open Science* 7 (4): 200242. https://doi.org/10.1098/rsos.200242.

87 Nayak, L., Mohanty, S., Nayak, S.K., and Ramadoss, A. (2019). A review on inkjet printing of nanoparticle inks for flexible electronics. *Journal of Materials Chemistry C* 7 (29): 8771–8795.

88 Schmatz, B., Yuan, Z., Lang, A.W., Hernandez, J.L. Reichmanis, E., and Reynolds, J.R. (2017). Aqueous processing for printed organic electronics: conjugated polymers with multistage cleavable side chains. *ACS Central Science* 3 (9): 961–967. https://doi.org/10.1021/acscentsci.7b00232.

89 Chang-Jian, C.-W., Cho, E.-C., Yen, S.-C., Ho, B.-C., Lee, K.-C., Huang, J.-H., and Hsiao, Y.-S. (2018). Facile preparation of WO_3/PEDOT:PSS composite for inkjet printed electrochromic window and its performance for heat shielding. *Dyes and Pigments* 148: 465–473. https://doi.org/10.1016/j.dyepig.2017.09.026.

90 Vidmar, T., Topič, M., Dzik, P., and Krašovec, U.O. (2014). Inkjet printing of sol–gel derived tungsten oxide inks. *Solar Energy Materials and Solar Cells* 125: 87–95. https://doi.org/10.1016/j.solmat.2014.02.023.

91 Jeong, C.Y., Watanabe, H., and Tajima, K. (2021). Adhesive electrochromic WO_3 thin films fabricated using a WO_3 nanoparticle-based ink. *Electrochimica Acta* 389: 138764. https://doi.org/10.1016/j.electacta.2021.138764.

92 Costa, C., Pinheiro, C., Henriques, I., and Laia, C.A.T. (2012). Inkjet printing of sol–gel synthesized hydrated tungsten oxide nanoparticles for flexible electrochromic devices. *ACS Applied Materials & Interfaces* 4 (3): 1330–1340. https://doi.org/10.1021/am201606m.

93 Costa, C., Pinheiro, C., Henriques, I., and Laia, C.A.T. (2012). Electrochromic properties of inkjet printed vanadium oxide gel on flexible polyethylene terephthalate/indium tin oxide electrodes. *ACS Applied Materials & Interfaces* 4 (10): 5266–5275.

94 Cai, G., Cheng, X., Layani, M., Tan, A.W.M., Li, S., Eh, A. L-S., Gao, D., Magdassi, S., and Lee, P.S. (2018). Direct inkjet-patterning of energy efficient flexible electrochromics. *Nano Energy* 49: 147–154. https://doi.org/10.1016/j.nanoen.2018.04.017.

95 Wang, Z. and Liu, R. (2023). PEDOT:PSS-based electrochromic materials for flexible and stretchable devices. *Materials Today Electronics* 4: 100036. https://doi.org/10.1016/j.mtelec.2023.100036.

96 Epson (2024). Printer technologies. https://corporate.epson/en/technology/search-by-products/printer-inkjet/ (accessed July 7 2024).

97 Orcutt, M. (2013). Ink-jet printing could be the key to next-generation OLED displays. *MIT Technology Review*, November 20, 2013.

98 Ahn, P. (2020). Inkjet printing: built to scale. https://kateeva.com/inkjet-printing-built-to-scale/ (accessed 18 Aug 2024).

99 Lewis, G. (2020). TecGlass installs the largest digital printer for glass in the middle-east. https://www.glass-international.com/news/tecglass-installs-largest-digital-printer-for-glass-in-the-middle-east (accessed 20 Aug 2024).

100 Schroder, K.A. (2020). NovaCentrix (20 Sept 2020), Photonic Curing. https://blog.novacentrix.com/the-evolution-of-photonic-curing (accessed 30 Oct 2023).

101 Komura-Tech (2024). Ultrasonic bonding to glass. https://www.komura-tech.co.jp/en/technology/principle/, Yokohama, Japan (accessed 22 Aug 2024).

102 Parker, R.M., Bhatnagar, Y., Frank, T., Wilbur, T.D., and Brown, S.D. (2024). US 12025900 Spacers and connectors for insulated glass units, granted: 2 July 2024, Assignee: View, Inc.

103 Agrawal, A., Tonazzi, J.C.L., LeCompte, R.S., and Cronin, J.P. (2015). US 8947758, Electrochromic mirrors and other electrooptic devices, Granted: 3 February 2015, Assignee: Ajjer, LLC.

104 Agrawal, A. Cronin, J.P., and Tonazzi, J.C.L. (2015). Conductive busbars and sealants for chromogenic devices US 8988756, Date of Patent: 24 March 2015, Assignee: Ajjer, LLC.

105 Ash, K.L., Kloeppner, L.J., Carroll, M.M., and Stephenson, M.T. (2020). US 20200393731, Electrochromic devices with separately controllable zones, Publication: 17 December 2020, Applicant: Gentex, Corp.

11

Sputter Coating Processes and Industrial Approaches

Wilmert C.S. De Bosscher

Soleras Advanced Coatings, Deinze, Belgium

Nomenclature

AC	alternating current
AR	antireflection
AZO	aluminum-doped zinc oxide
DC	direct current
DR	deposition rate
EC	electrochromic
EMI	electromagnetic interference
IC	ion conductor
ITO	indium tin oxide
PDC	power density compensated
PDCDR	power density compensated deposition rate
PDCDDR	power density compensated dynamic deposition rate
RF	radio frequency
SEEY	secondary electron emission yield
TCO	transparent conductive oxide
ZTO	zinc-doped tin oxide

11.1 Large-Area Magnetron Sputtering Basics

Over the last few decades, magnetron sputtering has become a major coating technology for depositing high-quality thin films. Magnetron sputtering has been used to make a range of flat glass glazing products. The process is responsible for the coating of many millions of square meters of low-emittance (low-E) architectural glass alone. Also, sputtering is used to coat non-flat glass. The ease of scalability, while maintaining excellent control of all crucial layer parameters, made it appropriate for large-area substrates and high mass productivity. The introduction of the rotating cylindrical magnetron concept in the 1990s has further contributed to the success of sputtering in large-area coating applications. This novelty provided a significant reduction in cost-of-ownership (higher target material consumption, more

Chromogenics: Smart Switchable Optical Materials and Their Applications, First Edition. Edited by Carl M. Lampert.

system throughput and up-time, etc.) [1, 2], while opening new perspectives for producing advanced coating stacks where tolerances are narrow and requirements are high. The move from single to triple silver low-E stacks in high volumes and with tighter control on reflection color is a good proof of the rotatable magnetron's capabilities.

Without going into the details of explaining the basics and an in-depth explanation of the magnetron sputtering processes as described in many good articles [3–5] and books [6, 7], we will start reviewing the most common types of industrial magnetrons and their field of use. For R&D applications, in semiconductor processing and storage media (magnetic or optical), circular planar magnetrons are most often used where circular disk-shaped target materials can be mounted. Target disk diameters typically range from about 25 to 150 mm; however, sizes up to 450 mm are used industrially. For applications where larger targets are desired, people most frequently use rectangular planar or rotating cylindrical magnetrons as shown in Figure 11.1 and in a coater setup where the substrates are moved past the magnetrons. The substrate movement speed drives the uniformity along the transportation direction, while the magnetic configuration along the longitudinal direction of the magnetron provides the layer uniformity across the moving substrate. Industrial implementation of chromogenic coatings fully relies on this configuration, and this will be the sole focus of this chapter. Static cylindrical cathodes are only used in some limited and specific cases, such as for decorative coatings.

Since most rectangular magnetrons have a fixed magnetic setup consisting of several arrays of permanent magnets, this type may suffer from some important shortcomings. The magnetic confinement forces the high energetic electrons to move in a well-defined racetrack-shaped zone just in front of the target surface. It is in this spatial area where the electrons interact with the available atoms forming positive ions for bombarding the target surface and realizing the sputtering effect. At the same time, there is a similar probability that these energetic electrons will excite the available atoms, generating emission of radiation (light) when the atoms return to their ground state. While looking at the sputtering plasma, the light of the excitation gives us a good understanding of where ionization and sputtering take place as depicted by the glowing zone in Figure 11.2. The limited plasma zone on the planar cathode surface will result in sputtering only at these positions and a localized erosion groove will start forming over time. Some planar cathodes may only give as low as 20% material consumption, while better static magnetic designs may provide up to 40% utilization on a full-face planar target. While the (copper) backing plate behind the

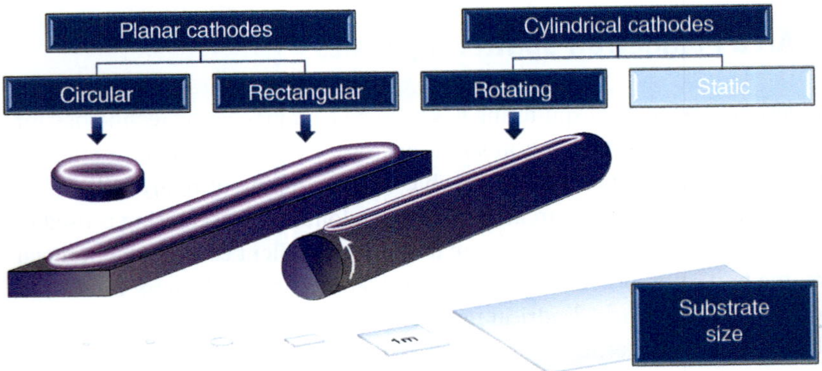

Figure 11.1 Overview of magnetron types and their relation to substrate size.

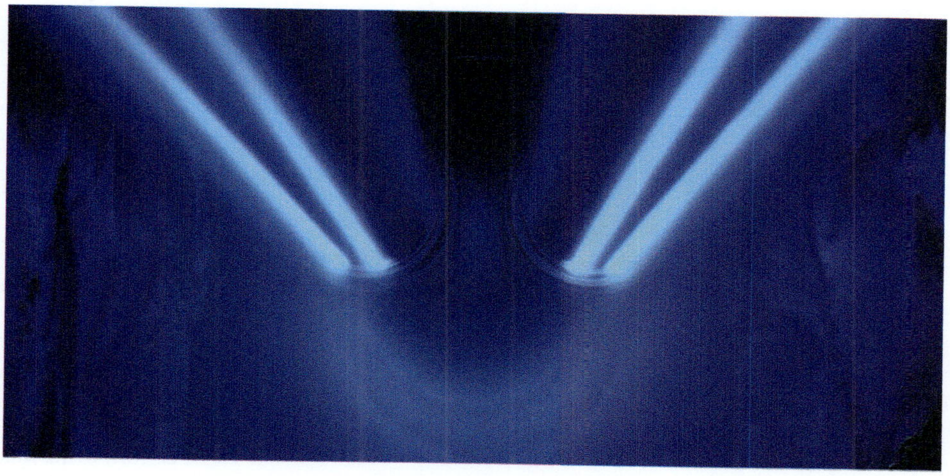

Figure 11.2 Glow discharge image of sputter target racetracks on a large area dual rotating cylindrical target cathode in AC sputtering mode, showing additionally:

- A plasma arc below the targets from a coupled magnetic field configuration.
- Slight glow on the inside and between the targets caused by the anode functionality.

target is water cooled, the heat load of sputtering on the target surface resembles a Gaussian curve and is maximal in the center of the racetrack. As such, the maximum allowed power load to the whole target, and therefore sputter rate, will be limited to what only the central zone may withstand before degrading (e.g. melting and cracking).

Rotating cylindrical cathodes are an elegant implementation of the same magnetron principles and overcome these shortcomings. In this case, the target material is placed on the outer surface of a cylindrical tube that is rotating at a fixed speed (e.g. 10–15 rpm). The target tube has cooling water running on the inside, sustaining uniform and efficient cooling. Furthermore, the magnetic system is positioned inside the rotating target type, though is not rotating with the target tube. The magnetics may generate a plasma racetrack on the target surface pointing at a desired direction toward the passing substrate. A picture of a typical plasma configuration on an industrial dual rotating cylindrical cathode is shown in Figure 11.2. Since the target material is passing under the plasma zone, no localized racetrack groove is being formed and the complete tube may be uniformly eroded: providing 75% and up to 95% of material consumption. In addition, each position on the target surface is getting a heat load while moving under the racetrack, followed by a significant amount of cooling time while completing a revolution of the target. As a result, higher power levels and deposition rates are achievable for the same target material relative to its planar counterpart. Having circumferential presence of target material increases the available inventory (about three times for a diameter like a planar target width) and allows much longer production runs (campaigns) before the targets need replacement (end-of-life). For delicate processes, it is worth mentioning that process drift over a campaign length is further reduced while using cylindrical target tubes. Indeed, target thickness reduction goes slower, and is more spatially even, having less impact on increasing the target-to-substrate distance and increasing the magnetic field strength. As the rotating target is used, it retains a tube-shaped geometry. Because of this effect, its use does not significantly impact the particle ejection angle distribution over the target lifetime. In comparison, planar targets are much more susceptible to coating nonuniformity as the target ages.

Figure 11.3 Industrial inline sputter coating line with transport of a vertical substrate, mounted into carriers. Doors may be equipped with pumps, heaters, analytical equipment, or any type of sources, such as magnetrons.

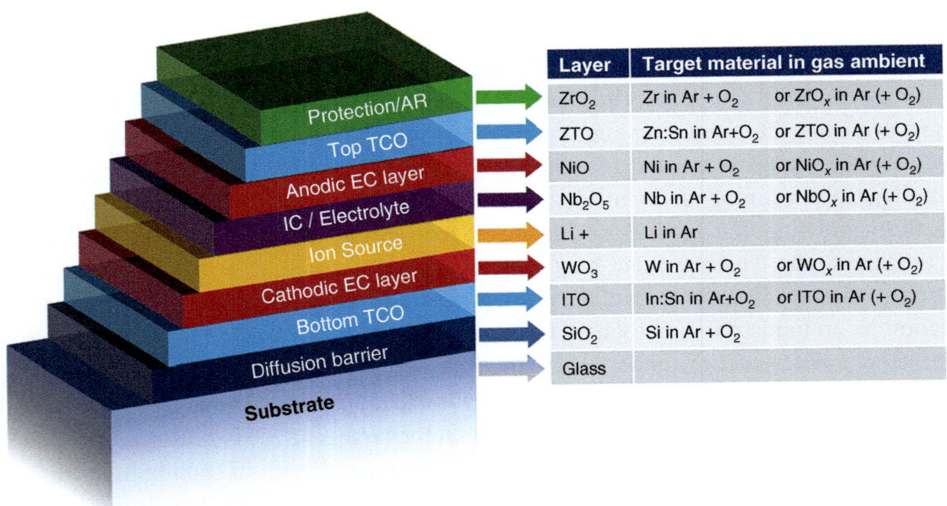

Figure 11.4 Schematic representation of an electrochromic coating stack; layer functionalities (left) and materials (right) with appropriate process possibilities.

It is only through these advances in magnetron sputtering that many types of chromogenic coating stacks may be realized on large and industrial scale. Part of such a production coater is depicted in Figure 11.3. Some chromogenic technologies, such as thermochromic devices, require less layers and have a limited complexity [8]. Other technologies are hybrid and combine sputtered layers with organic materials. Electrochromic devices, however, might be an interesting and representative showcase demonstrating most of the challenges encountered in any type of transmittance/color switching devices. In this chapter, we will focus on the various layers, each with their desired functionality and related complexity. We will hint and highlight how magnetron sputtering may need to be configured for materializing the required layer properties.

As a guideline for this journey, we will follow a generic coating stack architecture of an electrochromic (EC) device as presented in Figure 11.4.

11.2 Magnetron and Process Concepts for Sputter Deposition of Metals

While using or validating a magnetron for the first time, it may be interesting in understanding its behavior of current and voltage for a given power level and preferably at various working pressure levels. This should be done preferably on a conductive target in (pulsed) DC or (mid-frequency) AC mode and in a pure argon gas environment, to avoid any disturbance from a reactive process. Generating a couple of curves by changing the power density and the overall argon pressure in the sputtering compartment, as shown in Figure 11.5a, proves to be informative.

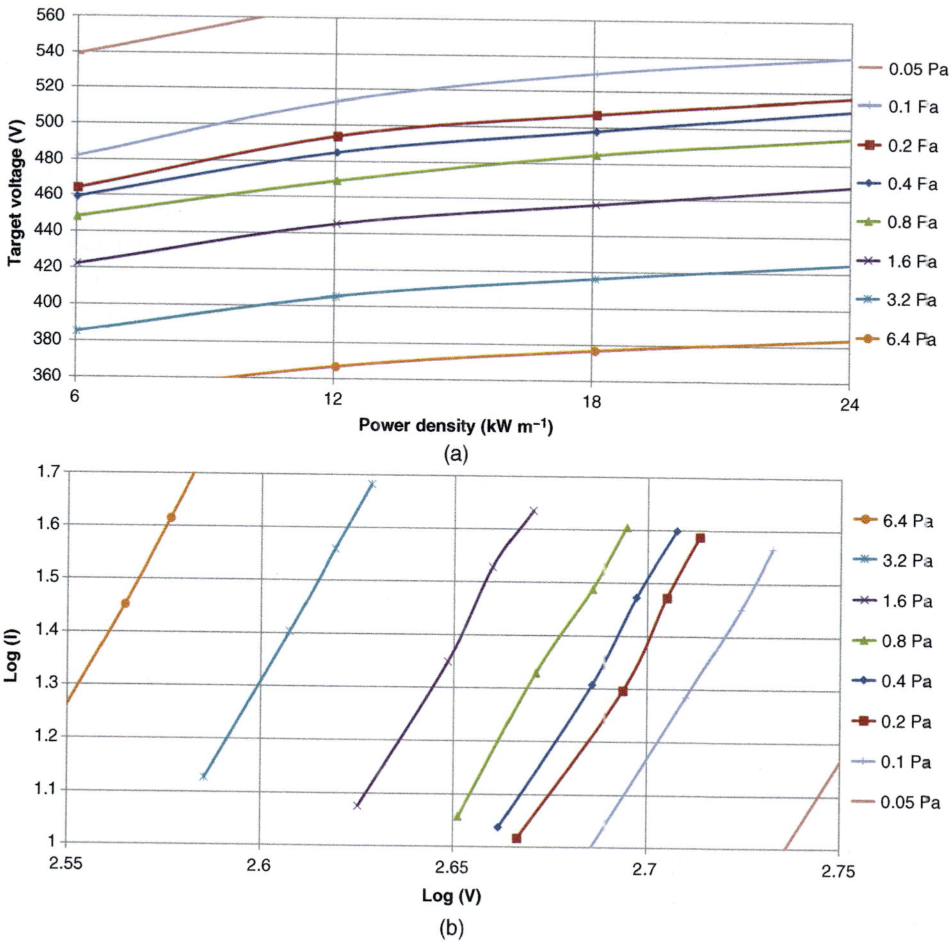

Figure 11.5 (a) Target voltage for a metallic sputtering process for various power density levels and (b) Log(*I*) versus Log(*V*) data of the same measurements at various pressure levels.

We prefer using power density (kW m^{-1}) which is the power delivered to the magnetron relative to the target length. As such, it is target length independent and can easily be scaled to any size of operation. As was expected, a higher power level will give a higher target voltage: however, the sensitivity of gas pressure on target voltage appears to be at least equally important. As a rule of thumb, the I–V (current–voltage) behavior of a magnetron in a fixed pressure condition may be approximated by the empirical formula [9]:

$$I = kV^n \tag{11.1}$$

in which n is a measure of the steepness of the I–V characteristic and is generally related to the magnetron efficiency; meaning the ability of the magnetic field configuration to confine the electrons [10]. k relates to plasma density and will be mainly defined by the secondary electron emission yield (SEEY) of the target material surface, the gas partial pressures, and the chamber geometry. To visualize the behavior of the magnetron used for these tests, Figure 11.5b shows log(I) versus log(V) of the data from Figure 11.5a. The plot shows clearly that Eq. (11.1) suits well to describe the electrical behavior for the various pressure levels. Because of the straightness of the curves, the value of n can be calculated with high confidence from the slope of the segments. Since all the curves are quite parallel, an n-value close to 12 is obtained at all pressure levels. This high n-value indicates efficient magnetic confinement of the electrons, thereby having a high-quality magnetron over the complete pressure range. Indeed, changing the pressure does not affect the magnetic bottling of the electrons significantly and hence the steady performance. An n-value below 4 may indicate having a lower quality magnetron, because of a too low magnetic field strength or some individual defective magnets. In a diode sputter configuration (no magnetic confinement), an n-value close to 1 may be expected.

On a microscopic scale, the measured sputter current I is linearly proportional with the number of incident ions hitting the target surface per second. Each single-ionized ion will collect one electron to become neutral again. The ion has a fixed probability of emitting an additional electron from the target surface, depending on the SEEY. The SEEY is defined by the target material and the bombarding species and having a value typically close to 0.1 [11]. The sputter voltage V is a measure of the energy with which the ions collide with the target surface. The amount of dislodged target atoms per single incident ion, being the sputter yield, will be proportional to the target voltage. The sputter yield is well described and tabulated for most target materials and for most ions arriving with a certain energy perpendicular to the target surface [12]. The sputter yield for Ar^+ ions ranges from 0.2 to 0.3 to about 1–2 for most materials and in the voltage range commonly used. For the lowest voltages (less than 300 V), the sputter yield bends down because an ion energy of several tens of eV is required before atoms can be dislodged from the target surface. The sputter yield for voltages above 300 V can thus be approximated by a linear relationship $aV + b$ (in which both a and b are small positive numbers).

Working with an inline or roll-to-roll web coater requires a better definition of deposition rate (DR) than just a thickness per unit of time (e.g., nm min^{-1}). For a given and fixed sputter configuration and condition, it may be expected that increasing the transport speed of the substrate will result in reduced exposure time to the magnetron source and less thickness. In fact, it is clear that the resulting thickness will be inversely proportional to the speed, e.g., doubling the speed will give half the thickness and vice versa. The dynamic deposition rate (DDR) takes this into account and is expressed as a thickness, multiplied by transport speed (e.g., nm.m min^{-1}).

The final major driver for sputter rate is the applied power level on the target. In a first-order approximation, it may be assumed that sputter rate from a target is proportional to the applied power density. Let us dive into the atomic scale again for a better in-depth understanding. The deposition rate or flux of target atoms arriving at a substrate in a coating system is linearly proportional to the target sputter rate and can be described by the following relationship:

$$\text{DR} = f(\vec{p}) \ \#\text{of ejected particles} \tag{11.2}$$

in which $f(\vec{p})$ is a scaling factor being function of position, pressure, and geometry of the coating chamber. Incorporating the amount of bombarding and incident gas ions, Eq. (11.2) can be written as:

$$\text{DR} = f(\vec{p}) \ \#\text{of incident ions} \ \frac{\#\text{of ejected particles}}{\#\text{of incident ions}} \tag{11.3}$$

in which we recognize the sputter yield. As discussed earlier while introducing SEEY, Eq. (11.3) can be approximated by:

$$\text{DR} = f(\vec{p}) \ \frac{I}{1 + \text{SEEY}} \ (aV + b) \tag{11.4}$$

in which we have introduced the sputter current I and voltage V. Dividing both sides of Eq. (11.4) by the power density allows us to get a first-order approximation of the power density compensated deposition rate (PDC DR) given by:

$$\text{PDC DR} = \frac{f(\vec{p})}{c} \left(a + \frac{b}{V}\right) \tag{11.5}$$

Since a, b, and c are constants and $f(\vec{p})$ can be taken as constant for a well-defined coating configuration as well, we conclude that the PDC DR is indeed constant in the first-order approximation (defined by a/c) but may vary slightly by a second term (with a smaller coefficient b/c) being inversely proportional to the voltage. This is indeed in rather good agreement with experimental results: sputtering at higher power levels or having a decreased magnetron field strength and confinement efficiency both lead to a higher target voltage and thus to a slightly reduced PDC DR.

We use power density and not power level to eliminate the effect of target length. Furthermore, we do not need to consider the target width for a rectangular planar cathode or the diameter and racetrack angle for a rotating cylindrical cathode in view of our power density approximation: a moving substrate under the magnetron will automatically integrate for the width. This means that we come to a power density compensated dynamic deposition rate (PDC DDR) that is a fixed value for a given target material under a given sputter configuration and being independent of power level and transport speed. PDC DDR is an easy and flexible parameter for a first-order calculation of layer thickness, substrate speed, and/or power level for a given target composition and process condition. For example, if you know that a certain material has a PDC DDR of 6 nm.m min^{-1} $(\text{kW m})^{-1}$ and you are using a power level of about 20 kW on a single target of 1 m long and having a 3 m min^{-1} substrate transport speed, then you may expect a layer thickness of about $6 \times 20/(1 \times 3) = 40$ nm. More fundamentally, the PDC DDR value is inversely proportional with the average binding energy of atoms to the target surface, also referred to as the heat of sublimation [13].

Sputtering of pure metals often requires careful handling and preparation of the coater chamber and target to avoid contamination from exposure with reactive species in the atmosphere. This is particularly critical for the proposed stack of Figure 11.4 containing a single metallic lithium layer. Li metal is known for its strong reactivity [14]. When being exposed to air, it will form LiOH (lithium hydroxide) due to its reaction with the available moisture, Li_3N (lithium nitride) from the nitrogen gas, as well as Li_2O (lithium oxide) from the available oxygen and some Li_2CO_3 from reaction with carbon dioxide; all being strongly exothermic reactions. A shiny metallic Li target surface can change to a dull black appearance in the time frame of a couple of minutes when exposed to air while forming a thick passivation layer [15]. As such, some precautions should be taken while handling and using Li targets. Preferably, the target surface should be protected while at atmosphere and only exposed during pumping of the vacuum system. Furthermore, reaching a sufficiently low base pressure (order of 10^{-5} Pa) before backfilling with pure Argon gas (e.g., 0.2–0.6 Pa), plus carrying out a thorough conditioning of the target is essential for reducing the traces of reactive gas and contamination of the target surface before its use in production.

11.3 Tweaking Deposition Rate for Reactive Sputtering of SiO_2

Besides being the material of choice for low refractive index layer optical coatings, SiO_2 may act as an efficient diffusion barrier and is compatible with many substrate materials. SiO_2 layers for large-area applications are typically deposited starting from conductive Si-targets in a gas ambient composed of a mixture of argon and oxygen. This is called a reactive sputtering process because the desired stoichiometric oxide compound formation happens on the substrate while the target remains relatively conductive. It is preferred in large-area coating applications to always keep the target conductive to avoid the need for using radio-frequency (RF) power supplies, lower sputtering efficiencies at high costs, limited power levels with complex EMI shielding, and the challenge of distributing the RF power uniformly across long targets. Conductivity in Si targets originates typically from doping with IIIa elements, such as B or Al. Although reactive sputter deposition has been well described before [16, 17], we would like to review the well-known hysteresis process behavior for two different types of Si targets, labeled as Si(A) and Si(B). Figure 11.6 represents the variation of DR, target voltage, and chamber pressure with the amount of reactive oxygen gas flow introduced in the process.

Si(B) refers to the typical Si-target that has been used over the last few decades in large-area architectural coatings and being optimized mainly for a nitride process, depositing Si_3N_4. Si(A) refers to a newer generation of Si-targets, with tuned morphology and composition. In addition, the Si microstructure may tweak the ratio of boundary to bulk atoms and impacting the average bonding strength of the atoms in the material matrix. As a result, higher DR for the same power density level may be achieved, allowing energy savings or achieving higher deposition rates and eventually higher coater throughput at similar power densities.

The configuration of Figure 11.6 was tested at a fixed power density, and the fixed base pressure of Ar-gas has been chosen at 0.26 Pa, corresponding to no oxygen flow. Adding oxygen flow while no power (no plasma) is being applied results in a straight line (black), showing that the oxygen partial pressure is linearly proportional to its flow, while the slope is inversely proportional to the pumping capacity of this sputter compartment. Adding oxygen while sputtering is active shows initially no partial pressure increase because all oxygen gas is being consumed by the sputtered Si in the formation of SiO_2 on the substrate and chamber

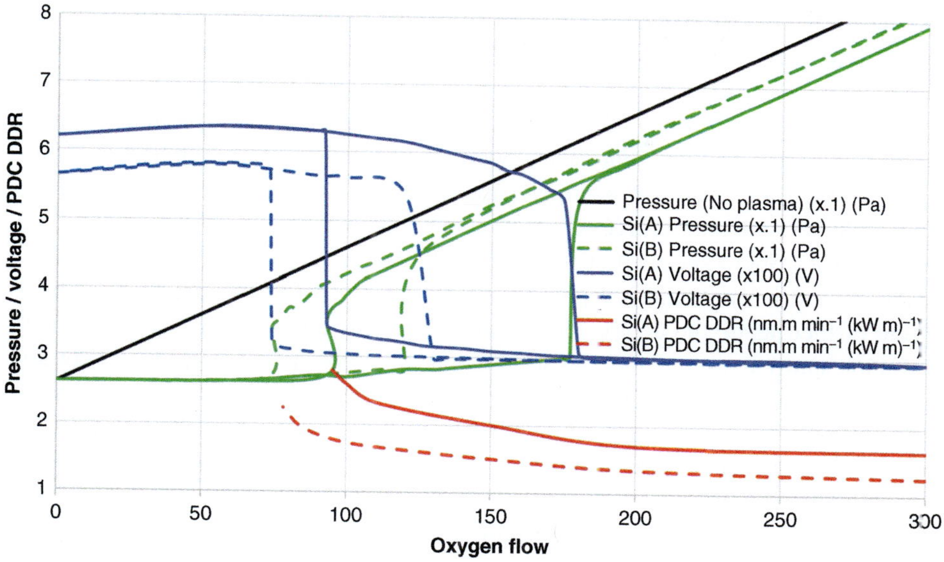

Figure 11.6 Hysteresis behaviors of deposition rate, voltage, and pressure in a reactive process for two types of Si-targets in varying oxygen flow rate. PDC DDR is power density compensated dynamic deposition rate.

walls (getter effect). When the getter pumping saturates at higher oxygen flows, the subsequent higher oxygen partial pressure will increase target oxidation and reduce sputter rate (typically threefold) until full poisoning of the target occurs. The Si(B) target falls into poisoning at about 120 sccm, while the Si(A) target requires about 175 sccm, showing that in metallic mode, the Si(A) target is sputtering significantly faster at the same power level. Further increasing the oxygen flow in poisoned mode, while the pumping capacity of Si being brought into the chamber is now saturated, gives a pressure curve below and parallel to the pumping curve. The difference in pressure between both curves is an indication of the amount of oxygen being consumed by the target in poisoned mode and therefore of its sputter rate. Accordingly, since the pressure difference between the no-plasma curve and the Si(A) target is larger than for the Si(B) target, it shows that also in the poisoned mode more Si material is being sputtered for Si(A). The red curves at the bottom show the PDC DDR of SiO_2 deposited from both Si target types in poisoned mode obtained from layer thickness measurements, which confirm our findings from the pressure behavior. The SEEY of the Si(B) target in metallic mode is slightly lower and gives a higher target voltage, but once poisoned, both targets have an oxidized surface and similar SEEY, seen from comparable voltage levels.

Returning to Figure 11.4, SiO_2 is the very first layer of our EC stack deposited on the substrate. For this application, the substrate is most commonly glass or may be a sheet or foil of polymer material. Since it is important to decouple substrate effects from the performance of the electrochromic stack, depositing a barrier layer may be essential. Polymer films may exhibit some porosity resulting in significant water vapor or oxygen transmission rate. On the other hand, soda lime glass may have Na^+ ions diffusing out and trying to participate in the coloring and bleaching cycles of the EC device, adversely affecting product performance.

11.4 Uniform Deposition of a Transparent Conductive Oxide (TCO) of Indium Tin Oxide (ITO)

Transparent Conductive Oxides (TCO) are a special breed of layer materials exhibiting at the same time the unique combined properties of visual transparency and electrical conductivity [16]. These materials are often based on doped oxides of Indium (In_2O_3), Zinc (ZnO), Tin (SnO_2), and Cadmium (CdO). Most commonly used TCOs for sputtering are ITO (In_2O_3:Sn or In_2O_3:SnO_2) and AZO (Al:ZnO) [18–20]; however, ZTO (Zn:SnO_2 or ZnO:SnO_2) is gaining interest recently and is represented as a top electrode in Figure 11.4. Without going into further detail, the doping level of TCOs may be adjusted to the desired carrier concentration and charge mobility. Most common ratios for In_2O_3: SnO_2 vary between 90 : 10 for heated ITO and 97 : 3 for non-heated ITO. Indeed, dopants require energy for activation bringing them into substitutional positions of the oxide lattice. Best-performing ITO may be achieved after a heat treatment up to 400 °C [20–22].

While those TCO materials act as outer electrodes for the electrochromic active layers in between, they must fulfill several stringent requirements. For large-area chromogenic devices (including Suspended Particle Devices or Polymer Dispersed Liquid Crystals), sufficient conductivity of these layers is essential for fast, uniform, and controlled switching performance. High transparency of the layers will also contribute to maximum transmittance in the bleached state. Furthermore, good layer uniformity across large substrate panes will have a significant impact on reflectance color uniformity, since small variations in thickness may generate a significant drift in color coordinates.

From the previous paragraph, it can be understood that reliably achieving high transmittance, low resistivity, and good uniformity across the substrate puts significant requirements on process control. In a reactive process, even while using a ceramic target, adjusting the reactive gas flow has proven to be an efficient and valuable method for tuning the DR. This can be seen in Figure 11.6 from the variation of DDR (red lines) with oxygen flow. Furthermore, reactive gas may be introduced locally in the sputter zone, impacting only a well-defined zone as to tune the uniformity across the substrate and along the target. Adding more reactive gas will bring the target surface in a more poisoned state, typically resulting in a lower sputter rate. Reducing the reactive gas flow will lower the level of poisoning and give a higher sputter rate. In the latter case, attention must be paid that the metallic material from the target gets its fully stoichiometric compound composition on the substrate as to avoid lower transmittance of the deposited layer.

In the case of ITO, however, we are being confronted with an additional complexity. Increasing the oxygen flow during sputtering from an ITO will decrease the carrier density in the deposited layer while the mobility of these free charges is increasing [23, 24]. As a result, the resistivity of the coating (being inversely proportional with mobility and charge carrier density) will exhibit a minimum for a specific oxygen flow as clearly indicated in Figure 11.7. Preferably, this minimum in resistivity is not too sharp, allowing a window of operation for tuning the optical performance and allowing for process control. Transmittance (both in the visible and near-infrared region) tends to increase with oxygen flow and is in fact related to the electrical properties. As a result, the oxygen flow is getting pinned to a specific narrow region for satisfying both electrical and optical performance and is indicated by the green zone. The control of oxygen flow in specific sputter zones is no longer available as a tool for controlling uniformity.

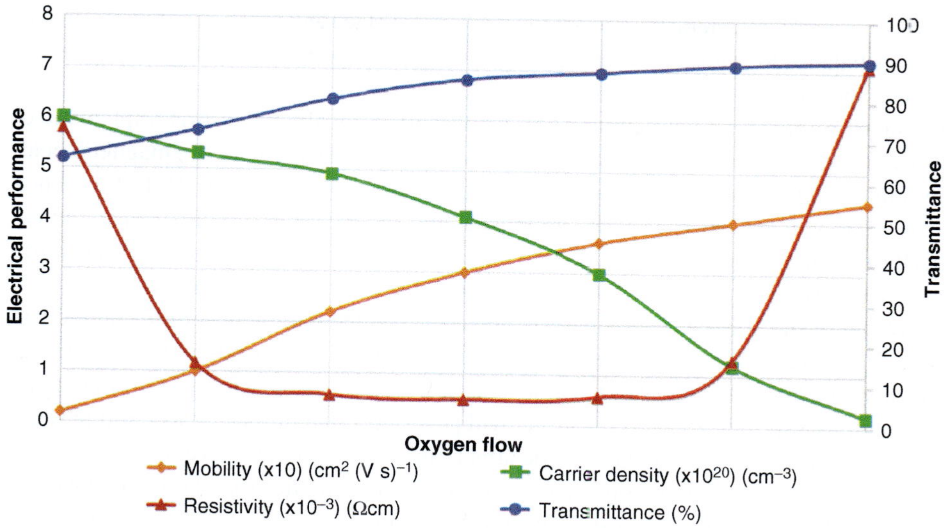

Figure 11.7 Electrical and optical performance of an ITO layer deposited for a varying flow rate of oxygen.

A new solution allowing the control of layer uniformity in addition to maintaining optimal conductivity and transparency may be found in the use of a tunable magnetic field distribution along the length of the magnetron and across the substrate [25]. In magnetron sputtering, as the name suggests, a magnetic field configuration, typically positioned below the target material, forces high energetic free electrons in the plasma into a closed-loop drift on top of the target surface. In this zone, the energetic electrons may interact with the available gas atoms (typically argon and possibly a reactive gas such as oxygen or nitrogen) to form positive ions and excited gas particles (atoms and molecules). The position of this closed-loop racetrack is indicated as the glowing zones in Figure 11.1; representing the excited species in the gas surrounding the target. The positive ions being formed close to the target surface are subjected to the electrical field **E** from the negative cathode potential and are accelerated toward the "target" for a high energy impact for dislodging target material atoms, what we call sputtering.

Tuning the strength of the magnetic field **B** along the length of the target is a powerful technique for adjusting local sputter rate without the need for adjusting gas partial pressure. For a fixed electrical and magnetic field distribution, the drift velocity of electrons in the racetrack may be described by [26]:

$$v_{E\times B} = \frac{\mathbf{E}\times\mathbf{B}}{B^2} \tag{11.6}$$

meaning that a higher magnetic field strength **B** will reduce the drift speed of the electrons along the racetrack since they are caught in more gyrating movements around the magnetic field lines. At the same time, the larger "local" movement of electrons increases the probability of interaction with gas atoms and the formation of ions. A higher density of charged particles (electrons and ions) will lower the impedance of the plasma. As a result, for a fixed sputter configuration (geometry and gas partial pressure), a higher magnetic field will result in a larger electrical current and reduced target voltage for a fixed power setting. We remember from Eq. (11.4) that sputter rate is in a first-order proportional relationship with the power density level and in a second-order inversely proportional relationship with voltage.

Changing the overall magnetic field strength on a target surface will result in a shift of current and voltage balance for a given power; however, the sputter rate will not change significantly (first-order approximation).

The impact of magnetic field adjustment changes dramatically when it is implemented locally over a limited zone of the target surface. Assume that we increase the magnetic field strength in a specific region of a longitudinal rectangular or cylindrical target. In this zone, the drift velocity of the electrons will be reduced, and the local plasma impedance will be lower. Due to a larger ion sputter flux, this zone is characterized by a higher electrical current. Since we are working with constant power, the higher local current needs to be compensated by a lowering of the voltage. The voltage is however a global parameter of the conductive target and is reduced over the complete surface, reducing the sputter rate in the remaining region where the magnetic field was not adjusted. As a result, a local increase of the magnetic field strength in a specific zone on the target will generate a higher local power and plasma density while depriving the remainder of the target. Indeed, the sputter power is a measure of the complete amount of sputter atoms that can be liberated from the target lattice at any time (related to the bonding energy). As such, we may generate a very efficient way for adjusting the local target material flux from the target to the substrate without the need for tweaking the reactive gas flow. In this way, we may decouple uniformity control for ITO layers while maintaining the exact reactive gas flow for optimal electrical and optical performance. In addition, tuning of the local magnetic field strength may be done on-the-fly by using remotely controlled actuators, as shown in Figure 11.8.

The importance of the magnetic configuration for depositing high-quality ITO layers does not stop there. It is known that the electrical quality of the resulting ITO layer may be greatly negatively impacted by bombardment from negatively charged O^- ions ejected from the target surface [27]. Several magnetron configurations (e.g., facing targets, intermediate anodes, and

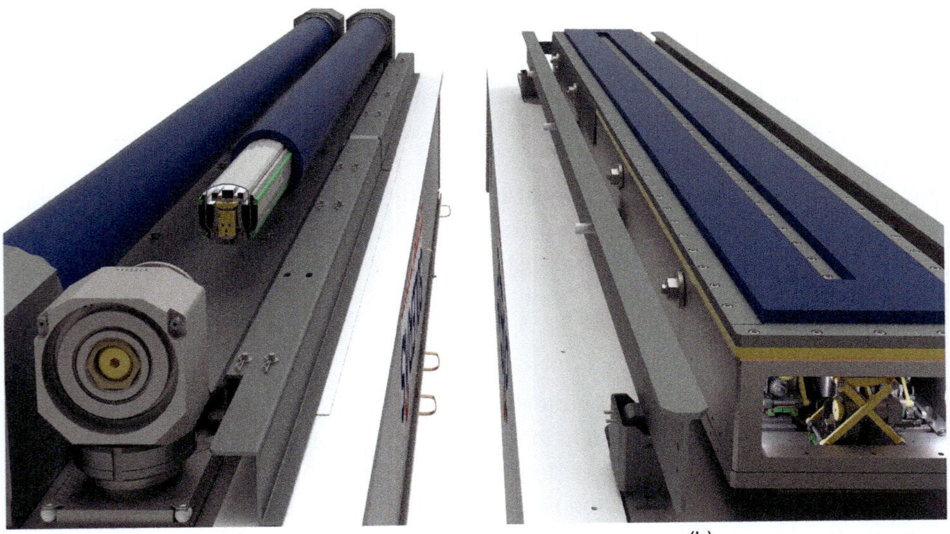

(a) (b)

Figure 11.8 Industrial sputter cathode configuration with cut-out section. (a) Dual rotating cylindrical cathode configuration showing online adjustable magnetics inside the target tube. (b) Single rectangular planar cathode configuration showing online node (scissor platform) for magnet tuning behind the target.

RF-power) have been presented to overcome this problem, though this is often leading to other shortcomings such as a significantly lower DR. As discussed earlier, increasing the magnetic field strength may be helpful, since this lowers the target voltage for a given power rating. Consequently, the damaging energy of the negative ions may be reduced. Another approach may be found by tweaking the racetrack angle of the ejection flux on the target surface as to shield or reduce the number of damaging ions reaching the growing ITO film. The use of rotating cylindrical magnetrons is especially beneficial in this case. Besides having a much higher target material utilization and efficiency for this expensive material relative to planar cathodes, the ejection angle and racetrack zone on the target surface may be readily adjusted. The whole magnet bar may be tilted inside the rotating target tube for orienting the sputter flux to a different position. Alternatively, the angle between both racetrack legs may be widened in a controlled way. As a result, a more homogeneously spread of target material flux along the direction of the substrate movement may be realized. This will provide an optimally controlled process with minimal amount of reactive gas flow needed. This is accomplished by better aligning the amount of arriving target atoms with impinging gas atoms and distributing the overall sputter energy more uniformly over the substrate surface [28].

11.5 Improved Process Stability and Performance for Metal Oxide Layers

Metal oxide thin films are extensively used as a coating material in a wide range of applications. Their optical properties make them ideally suited for realizing controlled spectral reflectance when applied in a multilayer stack. Both high reflectance (e.g., mirrors) and low reflectance (e.g., AR displays) products may be tuned in a specific wavelength band. In addition, these layers may exhibit good mechanical and chemical stability and are being used in wear and corrosion-resistant coating solutions. Many pure and stoichiometric metal oxide materials may exhibit low electrical conductivity and such ceramic targets are often impossible to sputter in DC- or AC-equivalent (non-RF) sputter configuration.

In this section, we would like to focus on the deposition of niobia, Nb_2O_5 and zirconia, ZrO_2 layers, exhibiting good electrolyte behavior: having poor electronic conductivity, while having acceptable ionic conductivity, e.g., for H^+-protons or Li^+-ions. This layer is indicated in Figure 11.4 as IC (ion conductor). Here, we will only discuss the sputtering of ZrO_2 layers in more detail; however, the proposed process is equally applicable to many other oxides as well, including the formation of Nb_2O_5-layers. ZrO_2 may also act as top coating, providing hardness and some reduced friction for improved mechanical stability and layer stack protection.

Reactive magnetron sputtering from metal Zr targets in an Ar/O_2 gas mixture is commonly used for depositing the ZrO_2 layer and is depicted in Figure 11.9a. At low oxygen flow, the target sputters in a metallic mode at high sputter rates (PDC DDR close to 6 nm.m min^{-1} $(kW\ m)^{-1}$) but yields mostly metallic and opaque coatings, while at higher oxygen flow, the target transitions to poisoned mode, governed by high oxygen coverage of the target surface and the formation of transparent oxide films. For zirconium targets, the transition to the fully poisoned mode can be very abrupt, accompanied by a severe reduction in sputter rate (PDC DDR of 1 nm.m min^{-1} $(kW\ m)^{-1}$ or lower), quite similar to the SiO_2 case that was presented earlier. In practice, users will look for a working point by first bringing the target in poisoned state and then choosing a balance in (increased) power level and (reduced) reactive gas flow

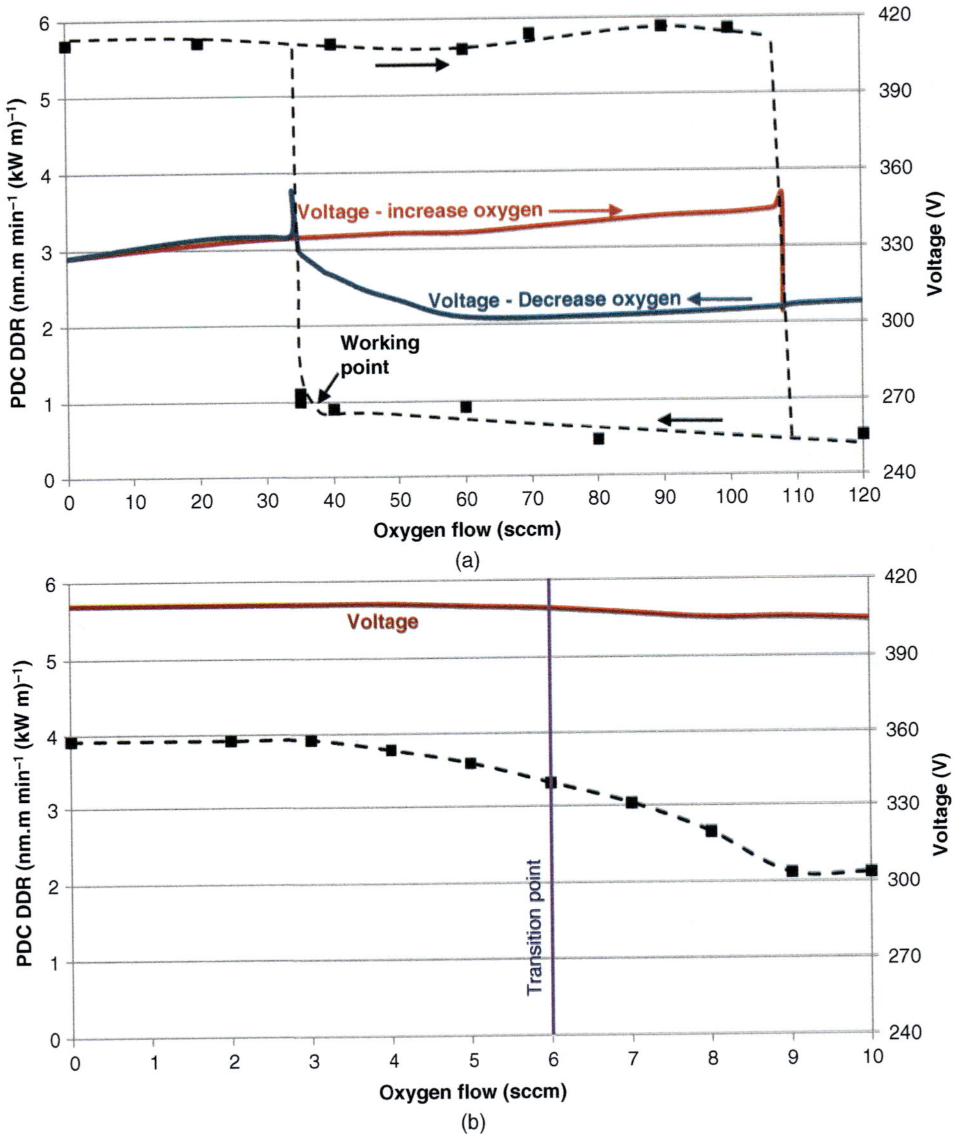

Figure 11.9 Target voltage and PDC DDR of a sputtering process starting from (a) a metallic Zr target and (b) a ceramic ZrO_x target ($x < 2$) as a function of oxygen flow rate.

just before the target surface falls back to metallic mode, as indicated in Figure 11.9a. This working point guarantees a maximal DR of the desired stoichiometric compound without the need for a fast feedback system.

An alternative and more effective method may be found for depositing ZrO_2 layers, being applicable to large area and high-volume industrial environments and significantly simplifying process control [29]. This is based on the use of sputter targets that are produced by thermal spraying of ceramic metal oxide material on a backing structure (a plate or a tube). It is known that many sub-stoichiometric transition metal oxide materials, produced in this way, show good electrical conductivity (resistivity in the order of 1 Ω-cm or lower) and allow being sputtered in a similar way as metallic targets [30]. The oxygen deficiency relative to the stable nonconducting stoichiometric metal oxide composition is typically between 5% and 20%. During sputtering, most of the oxygen required for forming the desired stoichiometric oxide layer is originating from the target material and only a small amount needs to be added through the process gas, mainly consisting of a noble gas, such as argon. The process conditions for depositing ZrO_2 layers from a ceramic target are depicted in Figure 11.9b for the same coater configuration and power level as in Figure 11.9a. Please note the difference in oxygen flow range on the abscissa while the Ar-flow rate for both experiments was close to 100 sccm, which gives an argon partial pressure close to 0.4 Pa.

While using the ceramic ZrO_x target, the hysteresis behavior completely disappears. It is interesting to note that the target voltage of over 400 V is significantly higher than in the metal target case, both in the metallic and poisoned process condition. This suggests that the SEEY on the ceramic target is lower than from the metal target in any condition and is governed by the sub-oxide species already present on the ceramic target surface.

Because of the smooth process transition with the ceramic ZrO_x target from a suboxide to fully poisoned surface, it is harder to detect at what exact oxygen point full stoichiometric ZrO_2 layers are being formed. Out of optical transmittance measurements, we can conclude that a flow of at least 6 sccm of oxygen gas is required before a completely transparent layer of ZrO_2 is obtained, as indicated with the vertical line. At lower flows, absorption in the layer originating from non-reacted Zr is observed. Looking at the PDC DDR from the ceramic target at say 7 sccm O_2, a value of 3 nm.m min^{-1} $(kW\ m)^{-1}$ may be achieved in a stable regime. This is about three times higher than what may be obtained from the metal target in a critical reactive working point. The difference in surface composition and morphology between a ceramic target and an oxidized metallic target, together with a more efficient sputter flux of Ar^+ ions for the ceramic target versus O_2^+ions dominating in the reactive process of a metallic target may account for the significant difference in sputter rate for a given power level and pressure regime. If we were to increase the oxygen flow on the ceramic target up to 30 sccm and beyond, we would achieve a similar DDR as with the metallic target, being close to 1 nm.m min^{-1} $(kW\ m)^{-1}$.

When studying and comparing layer characteristics, no real differences may be found between the metal and ceramic target. Under comparable coating conditions, comparable crystallographic structure and morphology, spectral optical constants, tribological properties, and chemical performance may be achieved. As a result, most layers in the electrochromic stack can be sputtered using either a metallic target in a reactive atmosphere or ceramic targets with no or little addition of reactive gas as indicated in the table of Figure 11.4.

Switching to ceramic targets may open opportunities for increasing the throughput of existing coating equipment or realizing the same throughput with lower energy consumption and on a smaller footprint inline coating tool, significantly reducing the coating costs. Running a reactive process close to a transition point makes layer properties very sensitive to small process variations. This complexity may be avoided with a ceramic target and better layer uniformity across larger substrate sizes may be achieved more easily. On the downside, it must be understood that a certain volume of ceramic target material may coat a smaller volume of substrates relative to its metallic counterpart. Indeed, the ceramic target already contains the oxygen compound material, while for a metallic target, all oxygen is added in the process for growing the oxide layer, having an expanded volume versus its metallic content.

11.6 Controlling Stoichiometry of Electrochromic WO_x Layers

Without any doubt, WO_3 is one of the dominant inorganic electrochromic materials exhibiting very high coloration efficiency upon introducing or extracting positive ions [31, 32]. As such, sputter deposition of this layer with the optimal morphology (e.g., structure and density) and stoichiometry is the key to high performance. The compound layer may be realized by using a metallic W-target in a reactive process containing argon and a significant amount of oxygen or by using a ceramic conductive substoichiometric WO_x-target (e.g., with $2 < x < 3$) in a process having argon and a smaller amount of oxygen.

Due to the importance of controlling the exact composition of the deposited layer, we will be describing the use of ceramic WO_x targets. It is found from literature that producing a slight sub- ($x < 3$) or superstoichiometric ($x > 3$) layer composition may be beneficial in achieving the desired coloration behavior of electrochromic devices in operation [33]. We have seen earlier that sputtering from a metallic target in a reactive oxygen process shows a clear binary behavior. Coming from the metallic regime, we may incorporate a limited amount of oxygen till we suddenly fall in poisoned mode and deposit the full stoichiometric composition. We may further increase the oxygen flow hoping to enrich the deposited layer with a superstoichiometric fraction. While returning from the poisoned mode by lowering the oxygen flow, we will most likely keep a stoichiometric layer composition till suddenly the target returns to metallic mode having a metal layer with some oxide incorporation. Controlling a reactive process to deposit a slight substoichiometric layer is quite complex. It will require sensors in the plasma process for monitoring the exact reactive gas partial pressure and possibly the target material flux as well. In addition, a fast feedback control loop system will need to be implemented, and this allows adjustment of gas flow levels or electrical power at high speed as to maintain the desired working point. Scaling to large-area deposition makes this even more difficult because of crosstalk effects in the coating compartment along the target length.

Figure 11.10 shows spectral transmittance curves of tungsten oxide layers deposited on clear glass from a $WO_{2.8}$ ceramic target and by changing only the oxygen gas flow. These optical measurements happen preferably inside an inline coater and can be positioned directly behind the coat zone to be studied, for example, in a pumping zone separating the next coating compartment. Placing inline optical measurements at various positions inside an inline coater may help understanding and correcting the process of critical sputter coatings. Furthermore, measurement heads may be placed across the substrate to fine-tune layer performance uniformity. From the transmittance spectrum before (upstream) and behind (downstream) a critical coat zone, properties of the layer(s) deposited in this critical coat zone may be defined [34]. From the results of Figure 11.10, we will provide a qualitative interpretation of the changes in the WO_x layer properties and mathematical algorithms will generate accurate quantitative layer data. Within a production environment, these results will not only allow monitoring the steady quality of a coating product, but it may also provide tuning parameters for the sputtering process as to maintain accurate control of the desired layer performance.

The dotted line on top represents the upstream measurement of the uncoated glass substrate. This glass has a refractive index n just above 1.5 in the visible (VIS: from just below 400 to just above 700 nm) and near infrared (NIR: from above 700 to 2500 nm) wavelength

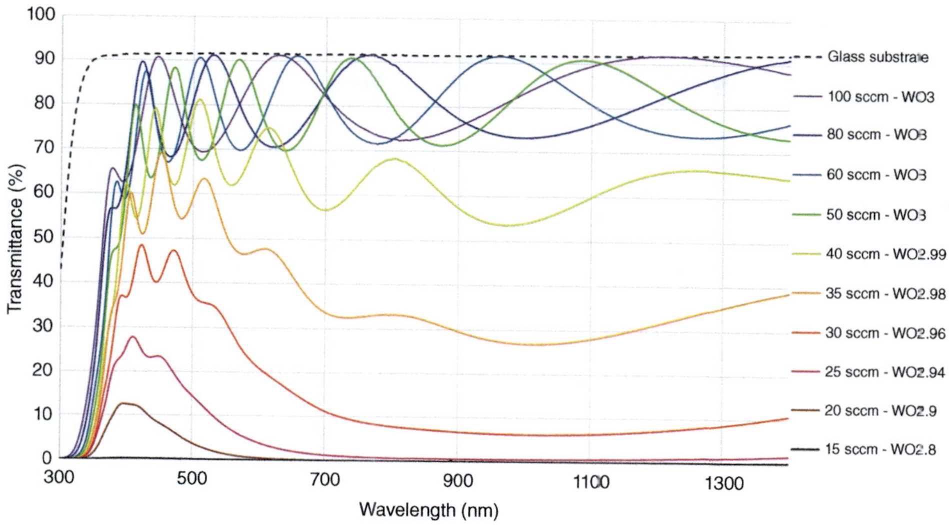

Figure 11.10 Spectral transmittance of a WO_x layer on clear glass for different oxygen flow rate conditions.

region, giving a reflectance just above 4% at each of its surfaces. Since negligible absorption occurs for this ultra-clear glass in the VIS and NIR, a transmittance close to 92% is found.

Adding a single layer to the substrate with a deviating refractive index and sufficient thickness will generate an interference pattern. This is a similar effect as can be seen from the colors in a soap bubble or oil drop on a water puddle. While sputtering from a ceramic WO_x target in high oxygen flows (60 sccm and above), a perfectly stoichiometric WO_3 layer is being deposited having virtually no absorption. This can be seen by the maxima in the interference pattern "touching" the curve of the uncoated glass. The amplitude of the interference pattern is directly proportional with the refractive index difference between the substrate and the deposited layer. We see that the amplitude is slightly diminishing when moving from the VIS to the NIR, pointing to a slight decrease of the refractive index of WO_3 with increasing wavelength. This is a typical behavior for most dielectric materials. The exact wavelength position of the maxima and minima in the interference pattern corresponds with a multiple of the optical layer thickness (being the physical layer thickness multiplied with the refractive index at that wavelength). Since we can extract the refractive index from the amplitude, we can calculate the layer thickness from the wavelength spacing between the pattern extrema. Going to lower oxygen flows (e.g., from 100 to 50 sccm) shows that the number of interference fringes is increasing, meaning that the layer thickness is increasing accordingly. This is in full agreement with the general tendency as discussed before for the ZrO_x case.

From 50 sccm of oxygen flow and lower, we notice that the maxima of the interference pattern no longer reach the level of the upstream measurement, and this may be attributed to absorption in the layer. The layer is getting more and more substoichiometric with decreasing oxygen flow, causing the optical absorption being dominant in the NIR portion of the graph [35]. The legend of the graph shows besides the oxygen flow also the layer composition. These values are defined by using Electron Probe Micro Analysis (EPMA) and provide a

Figure 11.11 Schematic view of three adjacent compartments of an inline coating system. The first and last compartments are equipped with inline optical measurement beams, between which the substrate is passing. These are representing the upstream and downstream metrology for controlling a critical coating process positioned in between, shown by the middle compartment.

relatively accurate elemental quantification. Considering the WO_x composition from EPMA with an accuracy on the x-value of 0.01, we notice a significant absorption sensitivity for even a minimal variation of the x-value. This proves that inline optical transmittance measurements may provide an accurate and easy method for defining and controlling the sub-stoichiometry of sputtered layers and achieving the desired performance parameters. An industrial configuration with inline optical measurements before and behind the coat zone is illustrated in Figure 11.11. At several spots across the moving substrate, spectral measurements can be executed for monitoring and controlling layer thickness and optical properties, both absolute and relative uniformity and repeatability.

It should be noted that an oxygen flow of 15 sccm provides a layer stoichiometry which is quite like the target stoichiometry. Sputtering the $WO_{2.8}$ target in pure argon may result in a layer stoichiometry close to $WO_{2.5}$ or even lower oxygen content. This is related to the pumping capacity of the specific coating compartment while some oxygen originating from the target will be pumped away as well. In this specific case of sputter power level and pumping configuration, we need to add about 15 sccm of oxygen for maintaining the same composition on target and layer.

A similar effect of the optical performance varying with oxygen flow during deposition may be found in the electrochromic behavior of a WO_3 layer in which positive ions are injected, showing an absorption pattern comparable to having a metallic tungsten fraction [36, 37].

Similar to Figure 11.9b, we can plot the sputter behavior of WO_x targets as shown in Figure 11.12.

The transition point in the graph indicates from which oxygen flow rate gives fully stoichiometric WO_3 layers may be deposited. For the chosen flow window, the PDC DDR and target voltage are quite linear with oxygen flow. As such, target voltage may provide some trend on target surface poisoning condition and the layer stoichiometry that may be expected. Since there is no hysteresis behavior, the oxygen flow may be tuned at will to dial in the desired working point. Target voltage alone is however not a reliable indicator. Increasing the sputtering pressure in the chamber will reduce the plasma impedance and for a fixed power setting decreasing target voltage (and increasing current). In addition, even without active tuning of the magnetic system, it may be expected that the magnetic field strength at the surface will increase over target life because of the thinning of the material, leading to a reduction in target voltage as indicated before.

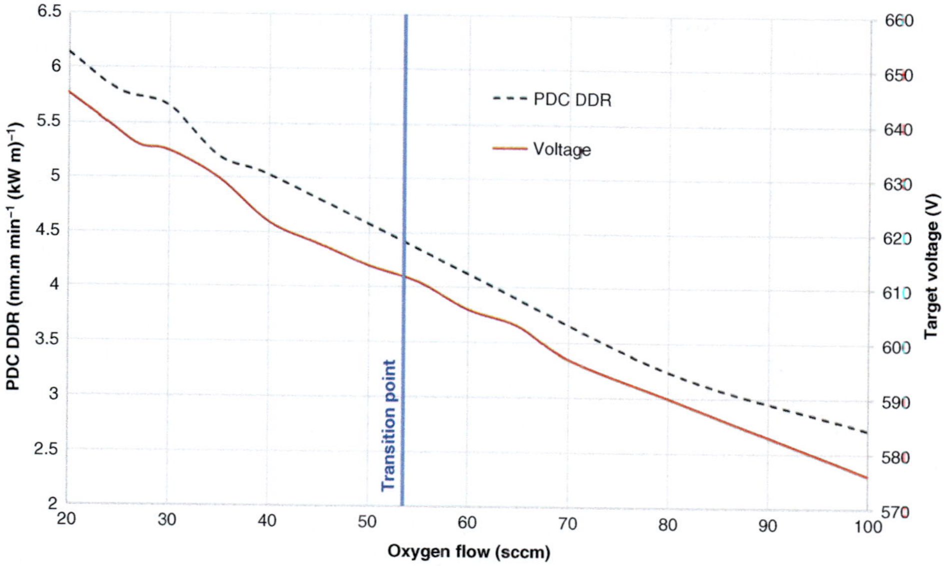

Figure 11.12 Target voltage and power density compensated dynamic deposition rate PDC DDR from a ceramic $WO_{2.8}$ target in a sputtering process for varying oxygen flow rate.

11.7 High-Pressure Sputtering Increasing Mechanical Stability of NiO Layer

Nickel oxide or a nickel compound oxide plays an essential role in an electrochromic layer stack as well. Besides its anodic coloration behavior upon positive ion injection/extraction and thus reinforcing the WO_x coloration effect in a combined stack, it is a material with a large ion storage capacity [38, 39].

From a sputter process point of view, several interesting challenges arise for the formation of an adequate layer of NiO in this application. Sputtering a Ni metal target in reactive mode proves to show some limitations. One of the main challenges is the ferromagnetic nature of the material, shorting most of the magnetic field configuration that is designed to magnetically confine the plasma zone. As such, a lower plasma density will increase its impedance and cause a high voltage for a given power level [40]. As a result, the sputtering may be limited by the maximum voltage that the power supply may provide. Limiting the target thickness is an option, though reducing the overall target life because of the lower material inventory. In planar magnetron configurations with a static magnetic configuration, we may additionally encounter significantly reduced target utilization because of the formation of a narrow and sharp erosion groove in the target. Since the target sputters primarily where the tangential field strength is strongest, that specific zone will be eroded first. Having thinner magnetic target material in this narrow zone will further enhance the local magnetic field strength and reinforce local sputtering. This effect will further reduce target utilization and thus overall target lifetime. The latter effect may be circumvented by using rotating cylindrical magnetrons. Even a sharp high-field magnetic field distribution will be distributed evenly over the complete circumference of the target tube and causing full face erosion. A high-field magnet bar can at least double the maximum allowed thickness for the sputter target. Only some attention needs to be paid to the racetrack turns at the tube ends to match the same average erosion rate relative to the middle portion of the target for sustaining optimal target utilization. Using a NiO_x ceramic substoichiometric target is a valid alternative, but will not solve the magnetic problem due to the presence of free nonbonded Ni and the antiferromagnetic material characteristic of NiO. The advantage in this case is the fact that the SEEY of the NiO is larger than for the Ni metal, lowering target voltage and allowing to go to higher power levels before the power supply reaches its voltage limit. Further advantages of substoichiometric target materials remain as presented earlier.

In this part, we would like to focus on the challenge of increasing the positive ion mobility and on generating a layer with low internal stress. When a high density of positive ions is driven into this material layer (e.g., during NiO bleaching), a significant volume expansion might occur. The solution may be found in sputtering at a higher pressure point relative to the normal 0.2–0.5 Pa regime. For the sake of clarity, we will review the sputtering of an aluminum metal target in a pure argon pressure range p from 0.1 to 10 Pa and for a target to substrate spacing d of 100 mm. In this way, we may avoid any interference in the results from material reactivity with the reactive gas partial pressure. Furthermore, any perturbations on the plasma confinement originating from magnetic interactions with the target material are eliminated. The measured and simulated PDC DDR values for Al are shown in Figure 11.13.

At each pressure point, many samples were produced with multiple power densities and at various glass transport speeds so as to generate a range of thicknesses from 10 to 600 nm. The error flags on the measurement curve cover all the samples. The PDC DDR is slightly lower at the lowest pressure point of 0.1 Pa because the probability of interaction between gas atoms and high energetic electrons in the racetrack is reduced allowing some high energetic electrons to be recaptured by the target without effectively generating ions [41]. It is surprising that in a pressure range from 0.2 to 1.6 Pa, a relatively constant PDC DDR close to 10 nm.m min^{-1} $(kW\ m)^{-1}$ was consistently found. When taking a PDC DDR of 10 at 0.2 Pa as a reference value, simulations indicate that the amount of target atoms arriving at the substrate would steadily go down with increasing sputter pressure [42]. In fact, it is the $p \times d$

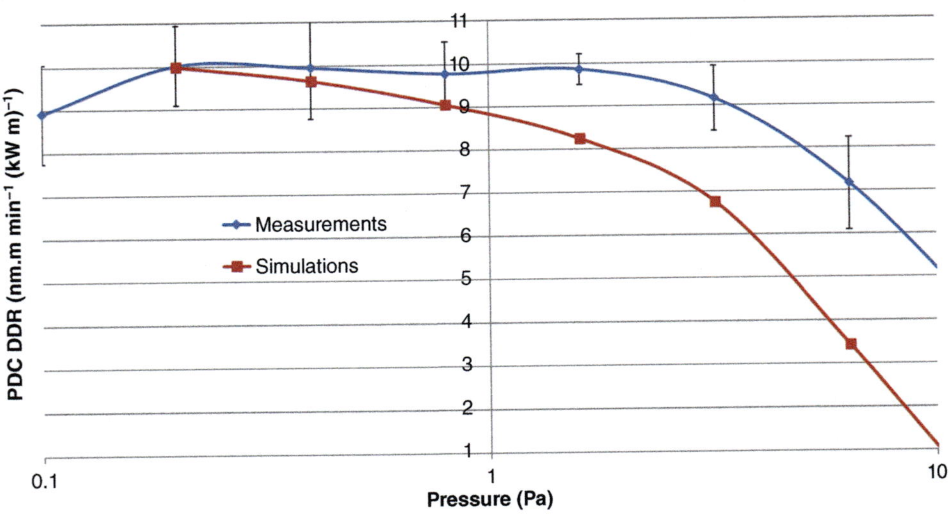

Figure 11.13 *Measured and simulated* power density compensated dynamic deposition rate, *PDC DDR for a variation of sputtering pressure.*

(pressure multiplied by target–substrate distance) product defining how many interactions a sputtered target particle undergoes with the gas ambient on its path to the substrate [43]. Each interaction alters the sputtered particle's trajectory and reduces the average probability that it reaches the substrate and reduces the DR. This is however not noticed on the thickness measurements of the samples.

Furthermore, each interaction will reduce the particle energy [44]. When target atoms leave the target surface, they have an energy distribution with an average energy of a few eV [45]. Having some tens of collisions (mainly depending on mass and energy of the target and gas atoms) is sufficient to fully thermalize these particles, reducing their energy to about 0.03 eV. When a target atom arrives on the substrate surface with insufficient energy to move to its optimal low energetic position, then the structure zone diagram predicts the formation of a porous structure of small crystallites separated by voids [46, 47]. Layer growth simulations of Al sputtered in Ar gas at 0.4 and 1.6 Pa are shown in Figure 11.14 [48].

Although at 0.4 Pa already some point defects appear in the simulations, results from 1.6 Pa show an average density of 84%, while the overall layer thickness remains quite similar for both pressure configurations. This is in agreement with the sputtering trajectory simulations from Figure 11.8, showing that indeed about 16% of the atoms no longer arrive to the substrate relative to the 0.2 Pa case. The same could be experimentally confirmed by doing optical transmittance measurements. We noticed that the amplitude of the interference pattern from high index dielectric layers on glass (like NiO) decreased while sputtering at very high pressure levels. Indeed, the vacancies in the material may be approximated as a fraction having refractive index equal to 1 and according to the Bruggeman effective medium approximation, this results in an overall reduction of average n-value [49]. The same amount of porosity could be calculated from the reduction in interference height, while the layer thickness could still be derived from the spacing between interference extrema. As a result, inline optical transmittance measurements may provide information on optical constants, thickness, and density of deposited layers.

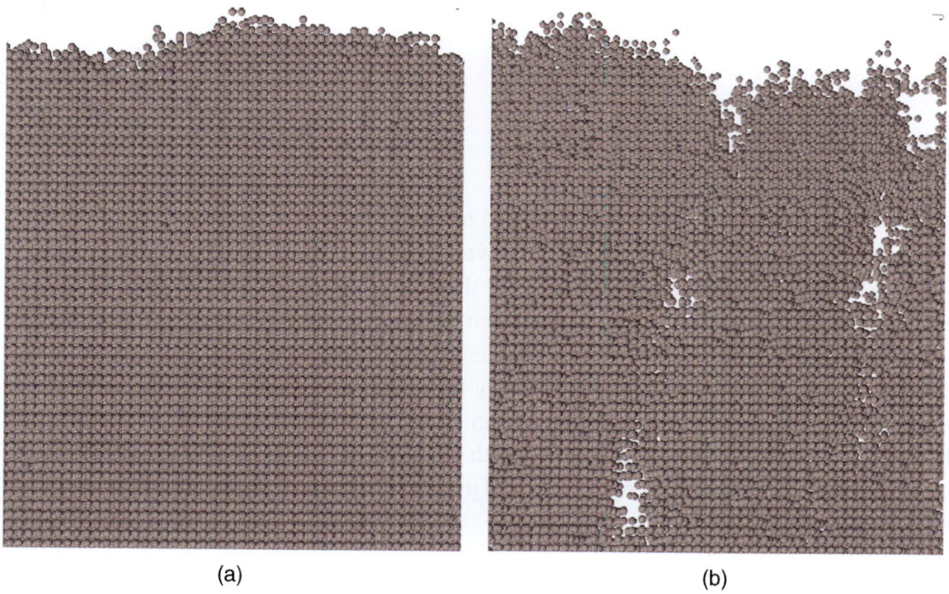

Figure 11.14 Cross section of layer growth simulations of Al-film at (a) 0.4 Pa and (b) 1.6 Pa.

Beyond 1.6 Pa, scattering of sputtered atoms becomes even larger and the number of sputtered atoms reaching the substrate is further reduced. The layer becomes physically thinner, and its porosity is further increased. Measured data and calculated data from particle trajectory and layer growth simulations prove to be in good agreement. At high $p \times d$ products (>0.1 Pa m) and depending on the atom energy and mass, a significant portion of the sputtered atoms may return to the target surface and may be resputtered. This may reduce the overall target utilization because material is removed primarily from the plasma racetrack zone while it returns more uniformly over the complete target surface. Target lifetime may still appear to be extended because the same material can be sputtered multiple times.

It has been proven that generating a porous layer structure by using high-pressure sputtering is not only an effective way for controlling the stress in the layer, it allows for material in-diffusion/injection and extraction while sustaining mechanical integrity of the layer.

11.8 Conclusions

Bringing chromogenic devices, such as electrochromic windows, into production environments with sufficiently high yield has proven to be quite challenging. Each of the thin film layers in the active stack must fulfill specific requirements that are often a combination of optical, electrical, and mechanical properties. As such, some in-depth understanding and interlinking between magnetron performance, sputter process behavior, and layer properties are key for realizing appropriate thin film stacks. In this chapter, we take the individual layers of an electrochromic layer stack as a guideline for highlighting specific facets of the sputtering process. Furthermore, we highlight recent advances in magnetron sputtering technology allowing to realize controlled and reproducible coating quality. The presented industrial approaches are readily scalable to large area and high-volume production and may serve as enabling technologies for bringing chromogenic products to mass market introduction.

Acknowledgments

I would like to express my sincere gratitude to my good Soleras colleagues that have supported many and extensive research projects over the last decade(s), leading to the overall insights as described in this chapter. Realizing the many and advanced sputter equipment developments and solutions would not have been possible without the many knowledgeable and valuable contributions from the hardware engineering team. Sincere appreciation goes to the target materials development team. They have realized leading edge target development, enabling new thin film possibilities on an industrial scale. Finally, special thanks go to the process and applications and analytical lab group members for conducting most of the sputter experiments with attention to detail and providing excellent support in the search for better and deeper understanding of the results.

References

1 Blondeel, A. (2009). Rotatable magnetron sputter technology for large area glass and web coating. *Vakuum in Forschung und Praxis* 21 (3): 6–13.

2 De Gryse, R. (2012). Thirty years of rotatable magnetrons. *Thin Solid Films* 520 (18): 5833–5845.

3 Greene, J.E. (2017). Tracing the recorded history of thin-film sputter deposition. *Journal of Vacuum Science and Technology A* 35 (5): 05C204.

4 Gudmundsson, J.T. (2020). Physics and technology of magnetron sputtering discharges. *Plasma Sources Science and Technology* 29: 113001.

5 Kelly, P.J. (2000). Magnetron sputtering: a review of recent developments and applications. *Vacuum* 56: 159–172.

6 Mattox, D.M. (2010). *Physical Vapor Deposition (PVD) Processing*, 237–286. Oxford: Elsevier Inc.

7 Wasa, K. (2012). *Handbook of Sputter Deposition Technology*. Amsterdam: William Andrew Inc.

8 Warwick, M.E.A. (2014). Advances in thermochromic vanadium dioxide films. *Journal of Materials Chemistry A* 2: 3275–3292.

9 Thornton, J.A. (1978). Magnetron sputtering: basic physics and application to cylindrical magnetrons. *Journal of Vacuum Science and Technology* 15: 171–177.

10 Waits, R.K. (1978). Planar magnetron sputtering. *Journal of Vacuum Science and Technology* 15: 179–187.

11 Depla, D. (2007). Understanding the discharge voltage behavior during reactive sputtering of oxides. *Journal of Applied Physics* 101 (1): 013301.

12 Andersen, H.H. (1981). *Sputtering by Particle Bombardment I*, 145–218. Berlin: Springer Verlag.

13 Sigmund, P. (1969). Theory of sputtering I sputtering yield of amorphous and polycrystalline targets. *Physics Review* 184: 383.

14 Schiemann, M. (2016). A review on lithium combustion. *Applied Energy* 162: 948–965.

15 Neumann, M.J. (2008). *Lithium Sputtering, Deposition and Evaporation*. Saarbrücken: VDM Verlag.

16 Depla, D. (2017). *Magnetrons, Reactive Gases, and Sputtering*. Lulu Press Inc., ISBN 978-1-304-34781-7.

17 Strijckmans, K. (2018). Tutorial: hysteresis during the reactive magnetron sputtering process. *Journal of Applied Physics* 124 (24): 241101.

18 Afre, R.A. (2018). Transparent conducting oxide films for various applications: a review. *Reviews on Advanced Materials Science* 53: 79–89.

19 Bright, C.I. (2007). Review of transparent conductive oxides, Chapter 7. In: *SVC 50th Anniversary*, 43–50.

20 Stadler, A. (2012). Transparent conducting oxides: an up-to-date overview. *Materials* 5 (12): 661–683.

21 Heinz, B. (1998). Effect of process parameters on the amorphous and the crystalline structure of Sn-doped indium oxide films. In: *Optical Interference Coatings*. Optica Publishing Group, TuC.1a.

22 Martinez, M.A. (1995). Optimization of indium tin oxide thin films for photovoltaic applications. *Thin Solid Films* 269 (1-2): 80–84.

23 Lippens, P. (2012). Indium tin oxide: sputter deposition processes. In: *Handbook of Visual Display Technology*, 779–794. Berlin: Springer-Verlag.

24 Txintxurreta, J. (2021). ITO thin film deposition by magnetron sputtering at room temperature for the manufacturing of efficient transparent heaters. *Coatings* 2021 (11): 92.

25 De Bosscher, W. (2015). On-line adjustable magnetics in rotatables for improved layer uniformity. In: *SVC 58th Technical Conference Proceedings*, vol. 2015.

26 Rauch, A. (2013). Estimating electron drift velocities in magnetron discharges. *Vacuum* 89: 53–56.

27 Ellmer, K. (2012). Reactive magnetron sputtering of transparent conductive oxide thin films: role of energetic particle (ion) bombardment. *Journal of Materials Research* 27 (5): 765–779.

28 De Bosscher, W. (2016). New rotatable magnetrons and target materials. *Converting Quarterly*, AIMCAL 2016 (Q1): 29–34.

29 Oberste-Berghaus, J. (2015). Film properties of zirconium oxide top layers from rotatable targets. In: *SVC 58th Technical Conference Proceedings*, vol. 2015.

30 Kida, O. (1998). Target, process for production thereof, and method of forming highly refractive film, EP0852566B1, *European Patent Office*.

31 Wu, W. (2018). Electrochromic metal oxides: recent progress and prospect. *Advanced Electronic Materials* 4 (8): 1800185.

32 Mehmood, A. (2020). Trends in sputter deposited tungsten oxide structures for electrochromic applications: a review. *Ceramics International* 46 (15): 23295.

33 Gillaspie, D. (2022). Counter Electrode for Electrochromic Devices, *United States Patent*, US11422426B2.

34 De Bosscher, W. (2018). Efficiency and yield enhancing optical monitoring system for inline coaters. In: *SVC 61st Technical Conference Proceedings*, vol. 2018.

35 Gies, M. (2021). Electrochromic switching of tungsten oxide films grown by reactive ion-beam sputter deposition. *Journal of Materials Science* 56: 615–628.

36 Dong, D. (2021). Colorimetric properties and structural evolution of cathodic electrochromic WO_3 thin films. *Canadian Journal of Chemistry* 99 (6): 549–556.

37 Cai, G. (2016). Ultra-large optical modulation of electrochromic porous WO_3 film and the local monitoring of redox activity. *Chemical Science* 7: 1373–1382.

38 Avendano, A. (2005). Sputter deposited electrochromic films and devices based on these: Progress on nickel-oxide-based films. *Materials Science and Engineering B* 138 (2007): 112–117.

39 Qu, H.Y. (2021). Multicolored absorbing nickel oxide films based on anodic electrochromism and structural coloration. *Journal of Applied Physics* 129: 123105.

40 Chang, S.A. (1986). High rate sputtering deposition of nickel using DC magnetron mode. *Journal of Vacuum Science and Technology A* 4 (3): 413–416.

41 Buyle, G. (2003). Recapture of secondary electrons by the target in a DC planar magnetron discharge. *Vacuum* 70 (1): 29–35.

42 Van Aeken, K. (2008). The metal flux from a rotating cylindrical magnetron: a Monte Carlo simulation. *Journal of Physics D: Applied Physics* 41: 205307.

43 Wasa, K. (1992). *Handbook of Sputter Deposition Technology*. Westwood: Noyes Publications.

44 De Bosscher, W. (2020). Rotatable magnetron sputtering at the right pressure point. *SVC Bulletin* Spring 2020: 40–46.

45 Thompson, M. (2002). Atomic collision cascades in solids. *Vacuum* 66 (2): 99–114.

46 Thornton, J.A. (1987). Structure-zone models of thin films. *Proceedings of SPIE* 821: 95–103.

47 Anders, A. (2010). A structure zone diagram including plasma-based deposition and ion etching. *Thin Solid Films* 518 (15): 4087–4090.

48 Lucas, S. (2010). Simulation at high temperature of atomic deposition, islands coalescence, Ostwald and inverse Ostwald ripening with a general simple kinetic Monte Carlo code. *Thin Solid Films* 518 (18): 5355–5361.

49 Bruggeman, D.A.G. (1935). Berechnung verschiedener physikalischer Konstanten von heterogenen Substanzen. *Annalen der Physik* 416 (7): 636–664.

12

Vacuum Web Coaters and Their Large-Area Coatings Used in Chromogenic Products: Technology and Applications of Transparent Electronic Conductors

Paul Lippens

Vacuum & Thin Film Technologies/Materials, Marciano (PG), Italy

12.1 Vacuum Web or Roll Coaters

12.1.1 Definition, Basic Configuration

Vacuum web coaters or roll coaters (in this chapter, the term 'web coater' will be used) [1, 2] are large-sized batch vacuum systems which typically consist of a vacuum vessel with pumping aggregates to evacuate the vessel. Inside the vessel a web-winding system, a coating drum, often one or more web pre-treatment systems and one or more coating sources are placed around the drum, Figure 12.1. The flexible substrate is guided by the web-winding system over the drum where it is coated with one or more thin films. The winding system can have varied complexity, ranging from three-motor systems (unwind, drum and rewind) up to industrial-grade five-motor systems. The latter have a capstan between unwind and drum, respectively rewind and drum, in order to decouple film tension between unwind respectively rewind and the drum. For thin substrates (say <50 μm), a lay-on arm containing at least one spreader roller (rubberized 'banana-roller') allows unwinding substrates onto the drum without wrinkles. The drum is typically held at temperatures between −20 and +50 °C in order to be able to coat temperature-sensitive substrates without burning or crease formation. Around the drum, in one or more coating chambers, are located one or more coating sources. Usually they are magnetron sputter sources – either planar or rotary (the term 'rotatable' is also used, but in this text the term 'rotary' will be used) counterparts as discussed in the chapter on glass coaters. Most often, sputter sources are of the cantilever type; which are supported on one side only, i.e. the cathodes are mounted on a flange or cart which is detachable from the rest of the vacuum vessel. Of course, thermal or electron-beam evaporation sources or chemical vapour deposition (CVD) sources can also be used. Usually, however, only one coating technology is installed on a single system. In the high-end chromogenics industry, most web coaters use magnetron-sputtering technology.

Provided it is rollable, the web can be any flexible thin substrate – engineered polymer films such as poly-ethylene-terephthalate (PET), poly-ethylene-naphthalate (PEN), poly-propylene (PP), poly-ethylene (PE), and poly-imide (PI); paper; thin metallic foils such as aluminium or copper foil; non-woven substrates; etc. Further in this chapter, the

Chromogenics: Smart Switchable Optical Materials and Their Applications, First Edition. Edited by Carl M. Lampert.

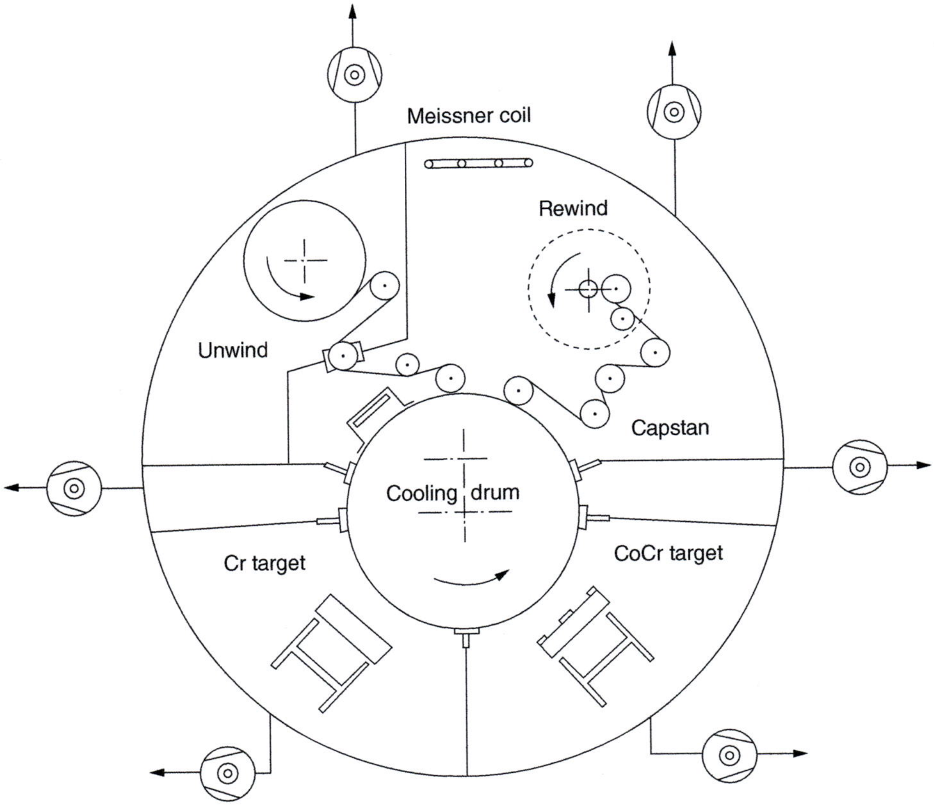

Figure 12.1 Principal cross section through a web coater with six differentially pumped vacuum zones, a cooling drum, two coating chambers with planar cathodes (here a Cr-target respectively CoCr target have been shown). In the rewind chamber is a Meissner cold trap (in the upper part of the chamber) and a glow discharge pre-treatment cathode.

discussion is limited to engineering polymer films such as PET, PEN and PI which are the major substrate families and are used in chromogenic systems.

Engineering polymer films usually contain a certain percentage of water. In PET, for instance, typically 7 wt% of water is present. This water is released partially when the film is unwound in vacuum; part of the water is also released as soon as the substrate is exposed to the condensation heat load from one or more coating sources. A web coater therefore must be able to cope with such amounts of water: typically, in the unwind section of the vacuum vessel, a Meissner trap (= cold trap, typically held at −120 °C) is used to freeze the water molecules to a cold surface. Also, certain web coaters have Meissner traps in each of the coating chambers in order to remove most of the released water from the substrate during coating. Such water release also has positive effects in the coating chambers, at the backside of the substrate, water molecules are released when the substrate is heated by the source (i.e. by the coating condensation process). As those molecules form a thin layer between substrate and drum, they help in cooling the substrate: indeed water vapour has very good heat conductivity and this improves the heat transfer coefficient from substrate to drum. The drum temperature, the power loads in the various chambers and the web speed must be chosen in such a way that the substrate is coated without the formation of creases or – worse – burning spots. This is a complex exercise every coating engineer must undertake for all new products.

Most web coaters are pumped with high-vacuum pumps such as turbomolecular pumps. However, some older systems have been constructed with oil diffusion pumps, and sometimes one or more cryo-pumps are combined with turbomolecular pumps. In the latter case, the cryopumps also help remove part of the water released from polymer film substrates.

Industrial magnetron sputter web coaters usually have more than one coating source. Some systems have several sources located in one large, differentially pumped coating chamber (covering, say, 50% or more of the perimeter of the drum). But most systems have several such chambers (for the various coating sources) around the drum and separated from each other with side chamber walls. Each chamber usually contains one coating source (single- or double-magnetron sputter source), but sometimes two or more independent coating sources are located in one of the chambers.

12.1.2 Industrial Systems Available on the Market

A web coater, having a series of coating chambers that are differentially pumped and isolated from each other, enables the production of products with complex coating stacks. Frequently in such stacks the deposition of dielectrics (oxides or nitrides) is sequentially combined with a deposition of metals or semiconducting coatings. Examples of such coating stacks are described in Section 12.2 that follows. Most suppliers of web coaters offer systems in which five coating chambers are located around one drum. The systems mentioned below can even be equipped with a maximum of six coating chambers around a single drum (there is a practical limitation on the drum diameter). Each of these coating chambers can be equipped with a single- or double (twin)-magnetron-sputtering source. Typical examples include the SmartWeb XP product line from Applied Materials Inc. (San Jose, USA, and Alzenau, Germany), Figure 12.2, and the FOSA MX product line from Von Ardenne (Dresden, Germany), Figure 12.3. Both of those systems have unwind and rewind sections located in separate side chambers, which are connected with a slit-type slot in which gate valves are mounted to the main coating chamber with the drum. This allows the latter part of the vacuum system to be kept under vacuum during roll changes (which significantly reduces the time needed to pump down to base pressure). Note that there are many more producers of web coaters all over the world, but the discussed systems are the most advanced.

The industrial web-coating systems described above use direct-current (DC), middle-frequency alternating current (MF-AC, typically 40–60 kHz) or bipolar pulsed direct-current (b-pDC, 0.5–50 kHz) as powering technologies for the sputter cathodes. Radio-frequency (RF) powering at 13.56 or 40.68 MHz is not suitable for large-area depositions, because it is difficult to match impedances automatically for large deposition sources. Large-sized RF power supplies (>12 kW) are not available on the market. It is also difficult to protect delicate web-winding aggregates such as stepper motors from RF interference.

Pump-down times for these systems (depending on the kind and amount of substrate loaded) are typically around one hour. Both manufacturers also provide double-drum versions of their web coaters, which typically have 10 individual coating chambers (FOSA MX with a double drum can be configured with up to 12 dual magnetrons), allowing the manufacture of complex stacks in a single pass through the equipment. Some web

Figure 12.2 Overview of the SmartWeb XP web coater from Applied Materials Inc. in a single-drum configuration. *Source:* Courtesy of Applied Materials, Inc.

Figure 12.3 Overview of FOSA MX vacuum web coating system in a double-drum configuration. *Source:* © Von Ardenne Corporate Archives.

coaters have complete lay-on arms on both sides of the coating drum, enabling reverse winding.

A typical web width for industrial systems is around 1600 mm, and web speeds for fairly complex products (such as index-matched indium-tin oxide [ITO], discussed further below) reach 10 m min^{-1}. This allows for production volumes in the range of 1–3 million m^2 $year^{-1}$. In sputter web coating, coating deposition rates are usually expressed as dynamic deposition

rates (DDRs) in nm.m min^{-1}. A DDR of 10 nm.m min^{-1} means that one obtains a coating thickness of 10 nm with a web speed of 1 m min^{-1}. With industrial web coaters, most coating depositions today achieve a DDR of over 100 nm.m min^{-1}.

Industrial systems also have one or more substrate pre-treatment systems, often in the unwind section of the machine: infra-red heating, DC glow discharge or any other kind of plasma pre-treatment [3]. The first technique improves the moisture removal from the substrate before it enters the coating section. The second technique improves adhesion [4, 5] between the coating (stack) and the substrate. A reactive plasma pre-treatment, often involving oxygen (O_2)-containing gases, can chemically modify the surface of polymers. In the case of PET, for instance, the density of carbonyl groups at the surface is increased. Such groups have been proven to improve adhesion of metal coatings to the polymer.

The web coaters shown in Figures 12.2 and 12.3 are typically produced with rotary (or rotatable) sputter sources. Such sources improve the stability of the sputtering processes (a reduced non-eroding target area gives rise to less arcing [6]), strongly increase target utilization (minimum 70%, often even up to 90%, depending on the design of the magnet arrays) and reduce target change times. Rotary cathodes have evolved into a standard on web coaters over the last five years, largely caused by the trend of producing coating stacks with dielectric coatings. Most dielectric coatings require the use of twin rotary cathodes powered by MF-AC or b-pDC; examples are SiO_2, Si_3N_4, Al_2O_3.

In situ measurement systems such as sheet resistance measurement and optical measurements such as reflection and transmission are usually implemented. For such optical measurements, typically a number of traces distributed over the width of the web are monitored. Industrial systems therefore have single-layer uniformities (over the web width and length) of up to ±1.5% [7]. Most of these coaters have an extensive process database which can be used for retrospective trend analysis or traceability of finished product quality issues.

The reactive processes for producing dielectrics are mostly controlled by voltage; however, there are still many coaters that use plasma emission monitoring to keep the working point in the intermediate range between metal-mode and poisoned-mode sputtering. Today many oxide coatings (e.g. titanium oxide [TiO2], niobium oxide [Nb2O5], zinc oxide [ZnO], aluminium-doped zinc oxide Al:ZnO, zinc-tin oxide [ZTO], and ITO) can be produced by quasi-reactive processes using substoichiometric sputtering targets (of respectively: TiO_x, NbO_x, ZnO, AZO, ZTO, ITO, etc.). Such targets have enough conductivity to allow MF-AC sputtering on twin cathodes without voltage or plasma-emission control. As discussed later in this text, sometimes DC sputtering with single rotatable cathodes is even possible (TiO_x) or recommended (ITO), because these quasi-reactive processes show no or almost no hysteresis.

The typical defects found in many sputter-deposited thin films such as pin holes, either caused by arcing during one or more sputter deposition processes [8] or by dust particles on the substrate, can also occur on sputter-coated webs. In addition, scratches on either the back side of the substrate or on the coated side might appear if there is some slip between one or more rollers and the web. Especially for products with important optical characteristics (such as chromogenic products), each of these defects reduces the final product quality. It is therefore important to work with clean webs, avoid slip on the web-winding system and avoid arcing as much as possible in all plasma pre-treatment and deposition chambers.

12.2 Coating Systems Produced on Web Coaters

12.2.1 Overview of a Few Important Coating Systems

A basic building block of many chromogenic devices is a PET substrate coated with a transparent conducting oxide (TCO) such as ITO. More details about this semiconductor are described below (Section 12.2.2). In order to protect the ITO and, often more importantly, the rest of the coating stack from moisture ingression through the PET, in many cases a barrier coating is first deposited between the PET and ITO. The simplest barrier is a thin 10–20 nm SiO_2 coating [9] which can be deposited in line with the ITO using a twin rotatable magnetron-sputtering source and metallic silicon (Si)-sprayed sputtering targets in fully reactive mode. Of course, more complex (and better) barrier systems consisting of multiple inorganic and inorganic/organic bi-layers are also being used. In most cases, complex barriers consisting of both inorganic oxides and organic layers are prepared off-line (e.g. by plasma-enhanced chemical vapour deposition, PE-CVD) before the product is further coated with ITO (or other coatings).

ITO is a semiconductor with a refractive index of around 2.0 in the middle of the visible part of the solar spectrum (510 nm). If such coating is deposited on PET, with a refractive index of around 1.65 (@510 nm) and such ITO coating is patterned, that pattern becomes clearly visible in reflection. If however a simple two-layer coating stack of inorganic oxides is first deposited on the substrate, the ITO pattern can be made invisible in reflection. This requires 'lifting' the refractive index of PET to about 2.0 (@510 nm, this is the design wavelength in this case, indicated by λ_0). A two-layer coating stack of approximately $\lambda_0/8$ optical thickness of a high refractive index material (such as Nb_2O_5, refractive index of 2.5 @510 nm) and approximately $\lambda_0/8$ thickness of a low refractive index material (such as SiO_2, refractive index of 1.45 @510 nm) can deliver such functionality. Such products, PET-Nb_2O_5/SiO_2/ITO, are often simply called 'invisible ITO'. Notice that the PET might have an acrylic type of hard coat and a thin SiO*x* barrier layer underneath the Nb_2O_5 coating. Also, note that the choice of Nb_2O_5 derives mainly from its high refractive index, enabling deposition of a thinner (and cheaper) coating. For the low refractive index material, no other real choice than SiO_2 exists at least where magnetron sputter deposition is used (for vacuum evaporation, MgF_2 could be an alternative).

As we will see below, ITO has its limitations in terms of conductivity, especially for temperature-sensitive substrates such as most polymer films. In order to reduce the sheet resistance further, certain products use a thin silver layer (<12 nm) interlaced between two ITO coatings. This improves conductivity considerably, but the transparency of such stack is of course slightly compromised (typically additional 10% of transmittance in the visible area is lost due to absorption in the silver layer).

Depending on the exact nature of the chromogenic product, other coatings can be deposited by magnetron sputtering in line with the ITO on a web coater. In electrochromic coating systems, for instance, tungsten tri-oxide (WO_3) is typically used as the electrochromic material. It can be sputtered reactively using tungsten (W)-sputtering targets. Also in such systems, nickel oxide (NiO) can be used as the counter-electrode material; it is also magnetron sputtered in line with ITO using nickel-sputtering targets [10].

12.2.2 An Important Building Module: Transparent Conducting Electrodes

12.2.2.1 Semiconducting ITO

The best known and most used (at least in high-end chromogenic products) transparent conducting electrode is ITO. ITO is a semiconducting oxide consisting of a matrix of indium-sesqui oxide (In_2O_3) with tin (Sn) doping. The lowest resistivities for such coatings on glass, i.e. around 1.2E-04 ohm.cm (or 120 μohm.cm), are reached at about 10 wt% Sn but require high deposition or annealing temperatures (>400 °C) or post-annealing at such temperatures. This ITO composition is usually referred to as ITO-90:10. In such cases, the coatings are fully crystalline (bixbyite unit cell of In_2O_3) and quasi all of the Sn is in substitutional lattice sites where each of those atoms contributes one free electron (Sn has four valence electrons, whereas In has only three). Free carrier densities are around 1.5E + 21 cm^{-3} with carrier mobilities in the 25–30 cm^2 $(V\ s)^{-1}$ range.

In the 1980s, ITO thin films were usually produced by reactive sputtering of In:Sn metallic alloy-sputtering targets. In:Sn alloys have a low melting temperature (<156 °C) though, this limits the usable sputtering power load (and the coating deposition rate). Moreover, this process is very prone to nodule formation [11] (see further Section 12.2.2.2) on the surface of the sputtering target, causing arcing and a very unstable deposition process. Combined with strong hysteresis of this reactive process, which requires plasma emission control [12, 13], it gave rise to very low productivity and low yield. With the arrival of high-quality ceramic ITO-sputtering targets in the 1990s (i.e. high physical density), quasi-reactive processes without the need for plasma emission (or voltage) control became possible and are today the standard for this coating deposition. The sputtering target typically consists of one or more planar ITO tiles of 6–12 mm thick metal bonded to a backing plate. In the LCD industry, very large single tiles are used (dimensions over 1 m are possible) in order to reduce the amount of gaps between tiles. Such gaps can be a source of dust formation and arcing.

The first rotary-sputtering targets of this material came on the market around 2006. They consist of one or more cylindrical segments, metal bonded to a backing tube. The latter is almost always made from pure titanium (Ti) as this metal has a coefficient of thermal expansion (8.6 μm $(m.K)^{-1}$) close to ceramic ITO (7–7.5 μm $(m.K)^{-1}$ at 20–200 °C [14]). Some ITO-sputtering target manufacturers today offer cylindrical segment lengths up to 900 mm in order to reduce the number of gaps per segment.

The conductivity of ceramic ITO-sputtering targets is close to that of the corresponding thin films and is therefore more than sufficient to allow for DC powering of the sputtering source. DC single-magnetron sputtering has established itself as the standard deposition technology in the industry.

High deposition annealing temperatures are however not compatible with sensitive polymer substrates such as PET. ITO-90/10 coatings deposited at room temperature (RT) on PET are typically amorphous and have much higher resistivities (500–800 μohm.cm) with carrier densities in the 2-3E + 20 cm^{-3} range. The conductivity mechanism now mainly relies on oxygen vacancies (oxygen deficiency). The higher the Sn concentration in such semiconductors, the more difficult their crystallization. As will be explained, other compositions of ITO, with lower Sn concentration than in ITO-90:10, become more favourable for applications on polymer substrates.

12.2.2.2 Nodule Formation

The formation of target surface asperities over the lifetime of the sputtering target due to re-deposition of sputtered matter on the target surface is a problem that exists with many target materials such as sprayed silicon, sintered or sprayed ceramic AZO and also ceramic ITO [15, 16]. As the problem is essentially linked to re-deposition, planar cathodes suffer much more than rotary cathodes because there is a lot of non-eroding surface area (all of the target area outside of the "race track"). The sputter target erosion profile is in the shape of an oval racetrack, as the result of circular magnetic fields. Nevertheless, even with rotary cathodes, there is re-deposition over the entire surface area of the target because the racetrack is continuously displacing itself over the target perimeter due to the target rotation [17].

In the specific case of ITO, at least part of the re-deposition has been shown to occur in a different crystallographic form than the In_2O_3 (the matrix of the sputtering target). Indeed, as indium oxide has three allotropic crystal forms (In_2O_3, InO and In_2O), a considerable part of the re-deposition is in the indium sub-oxide or In_2O form [18, 19]. This crystal has a different erosion rate than In_2O_3 (i.e. the surrounding target matrix), and therefore such re-deposition can grow. As such very hard cone-like surface asperities (Figure 12.4) can develop over time during long sputter runs [15, 21]. These gradually give rise to increased arc events and need to be mechanically removed after a certain time.

Research of several groups has shown [20, 22] that ITO targets are less prone to nodule development if their physical density is higher. Most ITO-90:10 targets used today in critical applications (such as in the display industry) have a density of 7.14 g cc^{-1}, which is >99.8% of theoretical density of this material. Certainly in case of rotary formats, such sputtering targets show quasi no nodule formation anymore.

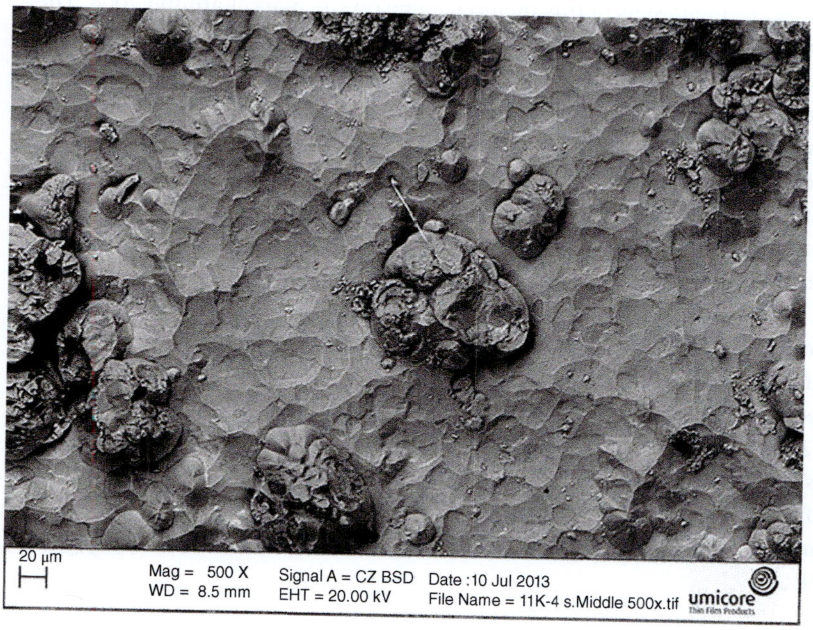

Figure 12.4 SEM micrograph of nodules on the surface of a ceramic ITO-90/10 target. *Source:* Lippens et al. [22] / reproduced with the permission from SVC.

12.2.2.3 Typical ITO Compositions for Web Coating

The deposition process of ITO coatings from ceramic ITO targets requires some oxygen addition to the inert sputter gas (and is therefore often called quasi-reactive). This is to compensate for lost oxygen in the atomic transfer process from target to substrate through the plasma zone; the lost oxygen is pumped off or reacts with other aggregates in the chamber (such as chamber walls). Typically 1.5% O_2 (in % of total gas flow) is needed to obtain the lowest resistivity of the coating, see Figure 12.5. Notice however that this oxygen amount for lowest resistivity very much depends on the coater (installed pumping capacity, base pressure, composition of the residual gas), the substrate temperature (higher temperatures will remove adsorbed water molecules from the substrate more efficiently), the kind of substrate (glass versus PET) and the ITO composition. Nevertheless, a more or less pronounced U-shaped resistivity curve as a function of oxygen concentration in the plasma is almost always found. Observe that for lowest optical absorption in the visible region, a slightly higher O_2 concentration than the location of the resistivity minimum is required (typically 0.5% more, refer to Figure 12.5). Many ITO-coating processes from ceramic targets therefore use about 2% of oxygen in the gas flow.

The graphs in Figure 12.5 also show that the minimum coating resistivity drops significantly with increased substrate (glass) temperature during deposition. Unfortunately, when using temperature-sensitive substrates such as PET, it is not possible to use substrate temperatures above 150 °C.

Other compositions of ITO with reduced Sn-doping concentration, which more easily crystallize at low deposition temperatures, then come into play.

Figures 12.6 and 12.7 show resistivities and Hall carrier densities for various ITO coatings (ITO-95:5, i.e. 5 wt% SnO_2 in the target and ITO-92.5:7.5, i.e. 7.5 wt% SnO_2 in the target) sputtered on different substrates (bare PET, hard-coated [HC] PET and double-side HC PET) from a rotary cathode with ceramic targets produced in four passages back and forth under the same cathode. Coatings (approximately 60 nm in thickness) were deposited at RT and then post-annealed at 150 °C for 30 minutes in air. Those graphs show the lowest resistivities

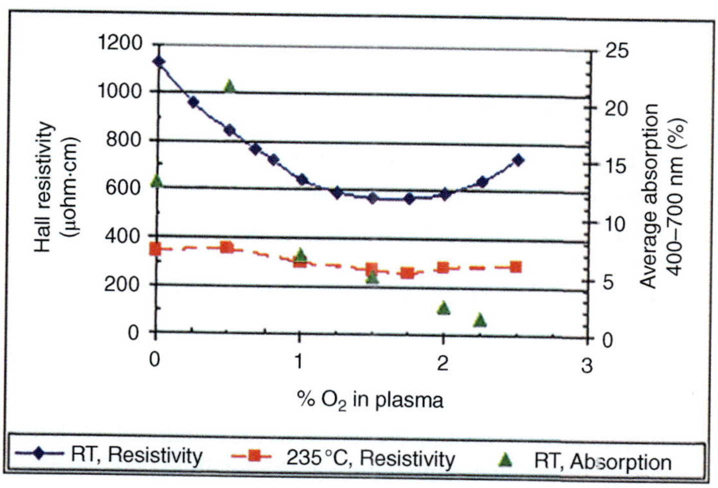

Figure 12.5 Hall resistivity of 80 nm (respectively 130–140 nm) ITO coatings deposited on glass at RT (respectively 235 °C) and 2.7E-03 mbar pressure at 10 kW m^{-1} rotary magnetron power (1 pass at 1 m min^{-1}, respectively 2 passes at 1.2 m min^{-1}) for varying O_2 gas flow. *Source:* Lippens and Haldemann [23] / Paul Lippens.

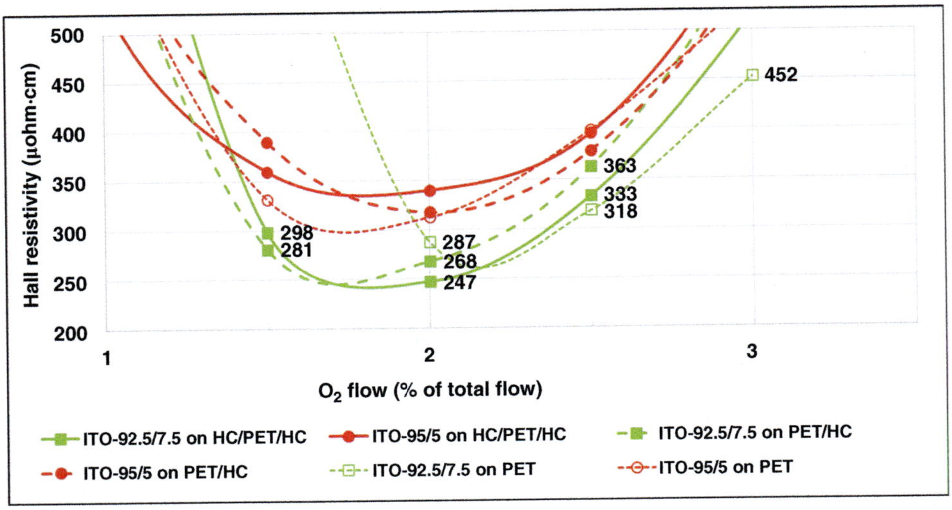

Figure 12.6 Hall resistivity for several 60 nm RT-deposited and post-annealed (150 °C, 30 minutes, air) ITO coatings on HC/PET/HC, PET/HC and bare PET. *Source:* Lippens and Frei [24] / John Wiley & Sons.

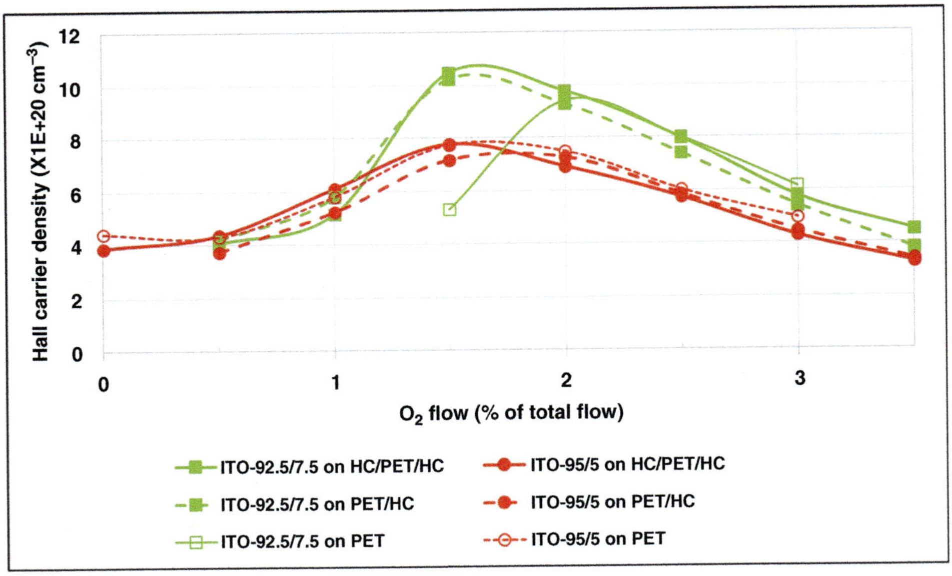

Figure 12.7 Carrier densities for the coatings of Figure 12.6. *Source:* Lippens and Frei [24] / John Wiley & Sons.

for ITO-92.5:7.5 grades when deposited on HC/PET/HC. This illustrates two things. First of all, the lowering of Sn concentration in the coating compared to the standard ITO-90:10 makes it easier for the coating to crystallize at low temperature (more details in [24]), leading to very high carrier densities (above 1.E + 21 cm^{-3}), comparable to or even better than that for ITO-90:10 on glass at this temperature. Secondly, substrates which release less water during deposition, i.e. HC PET substrates, yield better results. Indeed water has been shown to suppress the crystallization of ITO coatings.

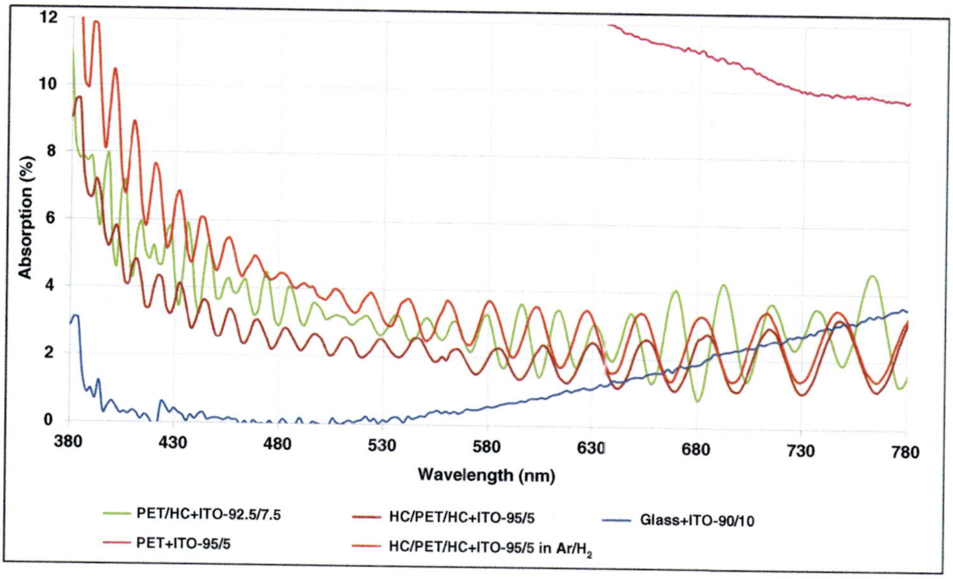

Figure 12.8 Calculated absorption graphs starting from measured transmission and reflection data in the visible part of the solar spectrum for some of the coatings from Figures 12.6 and 12.7. *Source:* Lippens and Frei [25] / Paul Lippens.

The optical properties in the visible region of some of the coatings from Figures 12.6 and 12.7 are illustrated in Figure 12.8. The absorptance curves are calculated from spectrophotometric transmission and reflection measurements on the indicated samples (the lowest resistivity condition was chosen for each of the combinations). The fringes in the curves for PET/HC- and HC/PET/HC-related data are caused by the HC (it has typical thicknesses of a few micrometers). The absorption data include the absorption of all the uncoated or hard-coated substrates and had less than 1% average absorption in the visible region. The ITO-95:5 coating on bare PET shows much more absorption due to a blooming effect of the PET (free oligomers are 'sweating out' from the PET during deposition); this was also the case for the other ITO grade. The 60 nm ITO-92.5:7.5 coated PET/HC substrate has an average absorption of around 3% in the visible. As in touch-screen applications, such coatings are often only used with a 21 nm thickness, the average absorption in such coatings drops well under 1%.

12.2.2.4 Other Transparent Electronic Conductors

Especially in thin-film photovoltaic (PV) products on glass and on flexible substrates, other TCOs have been used for which the base metals are cheaper. The most common are fluorine-doped tin oxide (F:SnO_2 or FTO) on glass for cadmium-telluride (CdTe) or thin-film silicon types of PV cells or either AZO or boron-doped zinc oxide (B:ZnO) for copper-indium-gallium-selenide (CIGS) types of PV products (both on glass and on flexible substrate). AZO has also been used for certain thin-film silicon products on glass. FTO is typically deposited by spray pyrolysis or atmospheric chemical vapour deposition (AP-CVD) starting with precursers of chlorides of tin sprayed onto hot glass (400–600 °C), e.g. directly on the float line where the glass is produced. Typical thicknesses of over 500 nm are needed in order to achieve a sheet resistance of 10 ohms/square or less and transparency in the visible is limited to the 80% range, whereas ITO achieves 90% or better. AZO is typically sputter deposited from sprayed ceramic targets. Coating thicknesses of 800–1000 nm are not uncommon as coating resistivities are typically in the 600–700 µohm.cm range for depositions on glass at 200 °C (2 wt% Al_2O_3 in the ZnO target). B:ZnO is typically produced by CVD on glass. As all those coatings are not used on web, no further details are given here. Refer to more literature on the topic [26, 27].

Indium, the primary metal in ITO (74.4 wt% In for ITO-90:10), is a minor metal whose market price fluctuates strongly. At the end of November 2022 its price was relatively low at around 208 US$ kg^{-1} (In, 99.99% or 4N purity). The price varied between a maximum of more than 270 US$ kg^{-1} in April 2022 and less than 190 US$ kg^{-1} in early November 2022. This is however insignificant compared to the recent past when the Indium price fluctuated between less than 100 US$ kg^{-1} in summer of 2002 and several peaks of more than 1000 US$ kg^{-1} in the period 2005–2006 (www.seltenerdmetalle24.de/chartcenter/). It is a typical metal which is stockpiled by speculative investors. This characteristic strongly influences the cost of an ITO coating. Indeed, the material costs, i.e. sputtering target costs, for such sputter-deposited ITO coatings are typically about 50% of the total coating cost in web-coating applications and can be around 75% of the total cost for certain glass-coating configurations in the display industry [28]. As a consequence, over the last 15 years, there has been a constant search to replace ITO by other materials.

Such newer materials are, for instance, carbon-based nanoparticles such as fullerenes. They include single- and multi-walled carbon nano-tubes (SW-CNT, MW-CNT), graphene, etc. The CNTs have a length which is typically some 3–4 orders of magnitude larger than their nanoscale diameter (of only some 0.5–2 nm). Masterbatches are typically prepared from dispersions of these nanoparticles. They can be used to prepare wet coatings on temperature-sensitive and other substrates. The nanosize of the particles makes absorption of visible light impossible (at least at low concentrations in a clear coating), and this combined with the high thermal and electrical conductivity of the CNTs (and graphene) means that transparent electrical conductors can be produced. In most applications, however, classical sputter-deposited ITO coatings remain considerably cheaper than their counterparts produced from nanoparticles, especially at Indium prices below 400 US$ kg^{-1}. Also, whereas the transparency and colour neutrality of CNT-containing coatings is usually better than that of ITO counterparts, the former still have lower conductivity for a certain optical transmittance than the latter. Refer to [29] for further reading. Nevertheless, the mechanical flexibility of CNT-containing clear coatings is considerably better than that of ITO coatings.

Other replacements for ITO are silver nanowires [30] or transparent metal mesh conductors [31]. Also these alternatives still suffer from high costs compared to sputter-deposited ITO coatings. Refer to the above-mentioned references for further reading.

12.3 Outlook

12.3.1 Web Coating on Thin Rollable Glass

It was shown above that thin-film depositions on flexible polymer substrates have their constraints due to the temperature sensitivity of those substrates. Therefore, a long-time aspiration of the flexible electronics industry has been to make such products on flexible thin glass. For several years, thin glass substrates, flexible enough to be rolled like film, are available from manufacturers such as Corning (Corning® Willow® Glass [32]), Schott (Xensation® Flex or Schott UTG® [33]) and Nippon Electric Glass, NEG (G-Leaf™, [34]). Those substrates usually have thicknesses around 100 μm and processing up to 500 °C is possible [32]. This enables producing ITO coatings with much better properties than on polymer films [35]. However, such substrates are usually only available on small width (typically 500 mm or below, but Willow glass has become available in 1300 mm width recently) and short lengths (typically only a few 100 m rolls), which make their processing on web coaters not really economical. On top of that, the thin glass substrates are in general still much more expensive than films like PET. As a consequence, sputter-coated products based on thin glass haven't really broken through in industry yet.

12.3.2 Machine-Related Developments

12.3.2.1 Combination of Several Deposition Technologies on the Same Web Coater

Not taking into account the plasma pre-treatment source (usually a DC glow discharge source), only one kind of deposition technology is usually used on the same web coater. This is usually magnetron sputtering, but it might also be thermal or electron-beam evaporation or PE-CVD. The main reason for this can be found in the different pressure ranges under which those various vacuum technologies work. It is indeed fairly easy to separate the gas atmosphere from two neighbouring sputtering chambers operating at only slightly different pressures, but things become more complicated when there is one or more orders of magnitude pressure difference between neighbouring chambers in which different technologies are applied.

In the past, several efforts were undertaken to develop web coaters on which thermal evaporation or magnetron sputtering was combined with flash evaporation of organic acrylates followed by electron-beam curing [36, 37]). None of these machines made it to series production though. In order to develop ultra-barriers, multilayers of metal oxides and polymers have proven to be an excellent solution. The latter could, for instance, be deposited by CVD. This is certainly an area where there is still quite some development needed for web coater manufacturers.

12.3.2.2 Air-to-Air Coaters

Glass-coating systems typically are air-to-air vacuum systems, i.e. the glass substrates are loaded in the machine under atmospheric pressure, fed through a washing system and via a series of load locks with different base pressure levels enter into the coating zone. After coating, those substrates exit the coater again through a series of load locks. This is the case for both horizontal architectural glass-coating systems and for vertical-display-coating systems. Consequently, the uptime of those machines is typically several days (for display-coating systems) to weeks (for architectural glass-coating systems). Only when targets need to be replaced and/or when shields need to be cleaned will such a coater be vented.

Unfortunately, air-to-air web-coating systems continue to remain a distant dream. Because of the continuous substrate ribbon, it is impossible to create load locks which are temporarily completely closed. The most recent industrial web coaters, which have been discussed in Section 12.1.2, however are a big step in the right direction: they allow to close a pinch valve on the substrate in its standstill phase at both unwind and rewind modules so that only the latter two need to be vented for roll change. Compared to older web coaters, where the pump-down time was typically several hours, this is already a big improvement.

References

1 Hebble Jr, H.B. (1966). An introduction to the continuous coating of plastic films and papers. In: *Proceedings of the 9th Annual Technical Conference of the Society of Vacuum Coater (SVC)*, Society of Vacuum Coaters, Albuquerque, NM, p. 33.

2 Bishop, C.A. (2015). *Vacuum Deposition onto Webs, Films and Foils*, 3e. Elsevier. ISBN: 978032329644.

3 Finson, E., Kaplan, S., and Wood, L. (1995). Plasma treatment of webs and films. In: *Proceedings of 38th Annual Technical Conference of the Society of Vacuum Coaters (SVC)*, Society of Vacuum Coaters, Albuquerque, NM, pp. 52–57.

4 Silvain, J.F. and Ehrhardt, J.J. (1993). An overview on metal/PET adhesion. *Thin Solid Films* 236 (1, 2): 230–235.

5 Cueff, R., Baud, G., Benmalek, M., Besse, J.P., Butruille, J.R., Dunlop, H.M., and Jacquet, M. (1995). Characterization and adhesion study of thin alumina coatings sputtered on PET. *Thin Solid Films* 270 (1, 2): 230–236.

6 Heintze, M. (2015). *Arc Management*. TRUMPF Hüttinger White Paper.

7 Von Ardenne. product brochure 'FOSA MX Vacuum Web Coating System'. See www.vonardenne.com.

8 Mitterer, C., Heuzè, O., and Derflinger, V.-H. (1997). Substrate and coating damage by arcing during sputtering. *Surface and Coatings Technology* 89 (3): 233–238.

9 De Gryse, R. and Lievens, H. (1997). Transparent Barrier Coatings: Review and State-of-the-art. In: *Proceedings of the 11th International Vacuum Web Coating Conference*, Miami, USA (9–11 November).

10 Granqvist, C.G., Bayrak Pehlivan, I., and Niklasson, G.A. (2018). Electrochromics on a roll: web coating and lamination for smart windows. *Surface and Coatings Technology* 336: 133–138.

11 Mattox, D. (1996). *Society of Vacuum Coaters (SVC)*, 8. News Bulletin, Spring.

12 Schiller, S., Heisig, U., Komdörfer, C., Beisler, G., Reschke, J., Steinfelder, K., and Strümpfel, J. (1987). Reactive DC high-rate sputtering as production technology. *Surface and Coatings Technology* 33: 405–423.

13 Lippens, P. and Verheyen, P. (1994). Electrochromic half-cells on organic substrate produced by subsequent deposition of ITO and WO3 in a roll coater using optical plasma emission monitoring. In: *Proceedings of 37th Annual Technical Conference of the Society of Vacuum Coaters (SVC)* (8–13 May), pp. 254–260. Boston, USA, Society of Vacuum Coaters, Albuquerque, NM.

14 Medvedovski, E., Alvarez, N.A., Szepesi, C.J., and Lippens, P. (2013). Advanced indium tin oxide ceramic sputtering targets (rotary and planar) for transparent conductive nano-sized films. *Advances in Applied Ceramics* 112 (5): 243–256.

15 Schlott, M., Degussa, M.K., and Hill, M. (1996). P-31: nodule formation on indium-oxide tin-oxide sputtering targets. SID96 DIGEST 31.

16 Lippens, P., Chiu, D., and Shin, J.-W. (2013). Rotary bonded multi-segment cast silicon targets as a replacement for sprayed silicon. In: *Proceedings of the 56th Annual Technical Conference of the Society of Vacuum Coaters (SVC)*, Society of Vacuum Coaters, Albuquerque, NM.

17 De Gryse, R., Haemers, J., Leroy, W.P., and Depla, D. (2012). Thirty years of rotatable magnetrons. *Thin Solid Films* 520: 5833–5845.

18 Lippens, P., Segers, A., Haemers, J., and De Gryse, R. (1998). Chemical instability of the target surface during DC-magnetron sputtering of ITO-coatings. *Thin Solid Films* 317: 405–408.

19 Wade, K. and Banister, A.J. (1975). *The Chemistry of Aluminium, Gallium, Indium and Thallium*. Oxford: Pergamon Press.

20 Omata, T., Kita, M., Okada, H., Otsuka-Yao-Matsuo, S., Ono, N., and Ikawa, H. (2006). Characterization of indium-tin oxide sputtering targets showing various densities of nodule formation. *Thin Solid Films* 503: 22–28.

21 Lewis, B.G., Mohanty, R., and Paine, D.C. (1994). Applications and processing of transparent conductive oxides. In: *Proceedings of the 37th Annual Technical Conference of the Society of Vacuum Coaters (SVC)*, pp. 432–439. Boston, USA

22 Lippens, P. and Frei, C. (1996). In: *Proceedings of the 59th Annual Technical Conference of the Society of Vacuum Coaters (SVC)*, Society of Vacuum Coaters, Albuquerque, NM.

23 Lippens, P. and Haldemann, M. (2011). ITO revisited: a study of coating properties obtained with sintered ceramic rotary targets aiming at display applications. In: *Proceedings of the 54th Annual Technical Conference of the Society of Vacuum Coaters (SVC)*, Society of Vacuum Coaters, Albuquerque, NM, pp. 210–214.

24 Lippens, P. and Frei, C. P91-What is the ideal ITO composition for touch sensor depositions on flexible substrate made with rotary cathodes? SID2014 DIGEST, 1328–1331.

25 Lippens, P. and Frei, C. (2014). Influence of substrate type and sputter gas composition on properties of touch sensor ITO coatings with various doping grades. In: *Proceedings of 10th International Conference on Coatings on Glass and Plastics (ICCG10)*, ICCG, invited, pp. 143–147.

26 Zhu, H., Bunte, E., Hüpkes, J., Siekmann, H., and Huang, S.M. (2008). Aluminium doped zinc oxide sputtered from rotatable dual magnetrons for thin film silicon solar cells. In: *Proceedings of the 7th International Conference on Coatings on Glass and Plastics (ICCG)*, ICCG, pp. 343–347.

27 Lippens, P., Leitner, K., Margadant, N., Kaitz, M., Broenstrup, A., and Sutter, S. (2010). Aluminium doped zinc oxide coatings sputtered on large area with rotary AZO targets. In: *Proceedings of the 53rd Annual Technical Conference of the Society of Vacuum Coaters (SVC)*, Society of Vacuum Coaters, Albuquerque, NM, pp. 290–295.

28 Lippens, P. and Muehlfeld, U. (2012). Indium tin oxide (ITO): sputter deposition processes. In: *Handbook of Visual Display Technology*, 2e, vol. 2, Section V, Part XVIII, 1215–1234. Springer. ISBN: 978-3-319-14345-3.

29 Schindler, A. (2016). ITO replacements: carbon nanotubes. In: *Handbook of Visual Display Technology*, (ed. J. Chen, M. Fihn, and W. Cranton) 2e, vol. 2. Springer. ISBN: 978-3-319-14345-3 Section V, Part XVIII, 1235–1255.

30 Preston, C. and Hu, L. (2016). Silver nanowires. In: *Handbook of Visual Display Technology*, (ed. J. Chen, M. Fihn, and W. Cranton) 2e, vol. 2. Springer. ISBN: 978-3-319-14345-3 Section V, Part XVIII, 1187–1203.

31 Geddes, J.B. III, Seitz, O., and Kobrin, B. (2016). Transparent metal mesh conductors. In: *Handbook of Visual Display Technology*, (ed. J. Chen, M. Fihn, and W. Cranton) 2e, vol. 2. Springer. ISBN: 978-3-319-14345-3 Section V, Part XVIII, 1205–1214.

32 See for information: Corning Willow Glass | Ultra-thin, Bendable, Flexible Glass Sheet | Corning. https://www.corning.com/worldwide/en/innovation/corning-emerging-innovations/corning-willow-glass.html.

33 See for a datasheet www.schott.com/en-gb/products/Xensation-p1000336/downloads.

34 See for datasheet. https://www.neg.co.jp/en/RD/Topics/product-g-leaf/.

35 Junghähnel, M., Westphalen, J., Naumann, F., Lorenz, G., Preussner, T., and Fahland, M. (2017). Recent developments of optimized ITO coatings on ultra-thin flexible glass in S2S and R2R processes. In: *Proceedings of the 60th Annual Technical Conference of the Society of Vacuum Coaters (SVC)*. Providence, RI, Society of Vacuum Coaters, Albuquerque, NM.

36 Shaw, D.G. and Langlois, M.G. (1994). Use of vapour deposited acrylate coatings to improve the barrier poperties of metallized film. In: *Proceedings of 37th Annual Technical Conference of the Society of Vacuum Coaters (SVC)*, Society of Vacuum Coaters, Albuquerque, NM, pp. 240–244.

37 Yalizis, A. (1995). High oxygen barrier polypropylene films using transparent acrylate-Al_2O_3 and opaque Al-acrylate coatings. In: *Proceedings of 38th Annual Technical Conference of the Society of Vacuum Coaters (SVC)*, SVC, pp. 95–102.

13

Pyrolytic Fluorine-Doped Tin Oxide on Glass for Chromogenic Products

George Neuman

Iridescence Thin Film Consulting, Michigan, USA

13.1 Introduction

Many chromogenic products use doped tin oxide on glass as an optically transparent and electrically conductive surface. Many well-known products such as electrochromic windows and mirrors use what is known as "conductive glass" as their starting substrate material. Also, doped tin oxide is used for its low thermal emittance ($e_{ir} \sim 0.1$–0.25). It is termed as "Low-*e*" coating and is used to effectively lower the radiative thermal loss in an insulating glass double pane architectural window (as discussed in Chapter 1). It has been highly refined commercially to be coated directly on the float glass line, making the added cost of the coat quite economical. The following discussion details its evolution to a viable product. Other coatings have been made using pyrolysis, but none have been refined as doped tin oxide. The wide range of applications will be shown later in Section 13.8.

13.2 The Development of Online Pyrolytic Deposition of Conductive Tin Oxide

Conductive tin oxide (SnO_2) has a rich history and continues to find new applications which benefit from its unique niche in the family of transparent conductive oxides. Roy Gordon described "The History of Processes for Making Transparent Conductors" in Criteria for Choosing Transparent Conductors [1]. The first patent filed, US 2,429,420 [2], does not describe the doping method but likely relies on a combination of oxygen vacancies and chlorine as the dopants to provide electrical conduction. This patent was filed in 1942 and was considered secret for the war effort and was not originally published (Personal communication with Kyle E. Sword of NSG). The US 2,429,420 patent covers an aircraft transparency wherein, "the conducting glass can be used for efficiently de-icing windows of high-altitude bombers in which interior icing of windows is a major problem."

The US 2,429,420 patent covers several methods for forming the conductive coatings. Stannic chloride vapors, with or without moist air, produced nonconductive and conductive tin oxide coatings, respectively. Carbon dioxide was also used as a carrier gas, thus foreshadowing modern chemical vapor deposition (CVD) methods. Spray methods, including stable

Chromogenics: Smart Switchable Optical Materials and Their Applications, First Edition. Edited by Carl M. Lampert.

formulations, were taught and can be considered foundational also. Chemical dipping, where the hot glass is immersed in a liquid formulation for a few seconds, was an additional deposition method described. "Fogging" of the glass with certain deposition conditions, in particular, excess water, describes a common problem with tin oxide coatings, haze. A significant number of patents describe inventions directed toward this particular attribute (current section). In general, the deposition of thin film coatings such as tin oxide was referred to as "iridizing" because of the iridescent colors of the coatings on the substrates. Typically, iridescence refers to multiple colors being present which indicates thickness variation of the deposited films. For early applications such as aircraft transparency de-fogging, the iridescent colors were not an issue. As the technology and applications developed, efforts to improve the deposition methods to attain uniform coatings became more important.

United States patents, in the late 1940s and early 1950s, showed the rapid advancement of tin oxide coatings, deposition methods, and insights into market applications of coated articles. Three main companies, Libby Owens Ford (LOF), Pittsburg Plate Glass (PPG), and Corning, were all active in the field during this time frame. As Gordon noted in US 2,564,706 [3] and US 2,566,346 [4], assigned to Corning and PPG, respectively, show us methods to control the sheet resistance of the tin oxide coatings using doping. The US 2,566,346 patent shows that foundational work on conductive tin oxide leads to coatings with relatively high conductivity or low resistivities. There exists a need to improve the conductivity of tin oxide so that it can be useful for more applications. The US 2,566,346 patent uses fluorine as a dopant and US 2,564,706 teaches the use of antimony to improve the conductivity of tin oxide.

US 2,564,706 shows several other attributes related to tin oxide coatings, the deposition process, and applications using this coating on vitreous (glassy) substrates. In US 2,564,706, a number of objectives are set out including: "...electrically conductive coatings which are permanently incorporated with the glass or ceramic surface, which have high chemical and thermal stability and which have sufficiently low electrical resistivities to permit the use of such articles for electrical heating", "...lower the electrical resistivities of tin iridized coatings", "...coatings which have zero or positive temperature coefficients of resistance (T_{CR})", "...control the electrical resistivities of tin iridized coatings", among others. The physical and chemical durability of tin oxide is taught as a key feature of these coatings. The improved conductivity that enables products for electrical heating with high conductivities is shown. Furthermore, the intrinsic property, T_{CR}, is developed with regard to thermal expansion, stress, and device properties.

The deposition of the antimony-doped tin oxide ($Sb:SnO_2$) coatings is a chemical process and the energy of reaction is provided by the temperature of the glass. The US 2,564,706 patent states the glass should be as hot as possible (to 500C or above) without deforming the glass. The use of borosilicate glass as an alternative to soda-lime glass allows the deposition temperature to be increased without deforming the glass and 700C is recommended for optimal properties.

The conductivity of $Sb:SnO_2$ varies with the amount of antimony in the original deposition chemistry. Figure 13.1 shows the effect of antimony percentage on the sheet resistance. Optimal sheet resistance is attained at approximately 1% of antimony relative to the tin concentration in the deposition chemistry. Figure 13.1 shows that the antimony concentration may be altered as needed, for a fixed film thickness, to change the sheet resistance of the coating. In this figure, the thickness of the coating is approximately 540 nm. Also, US 2,564,706 notes that the transmittance of $Sb:SnO_2$ coatings changes with the antimony content. At optimal antimony content for maximum conductivity, the coatings are essentially colorless.

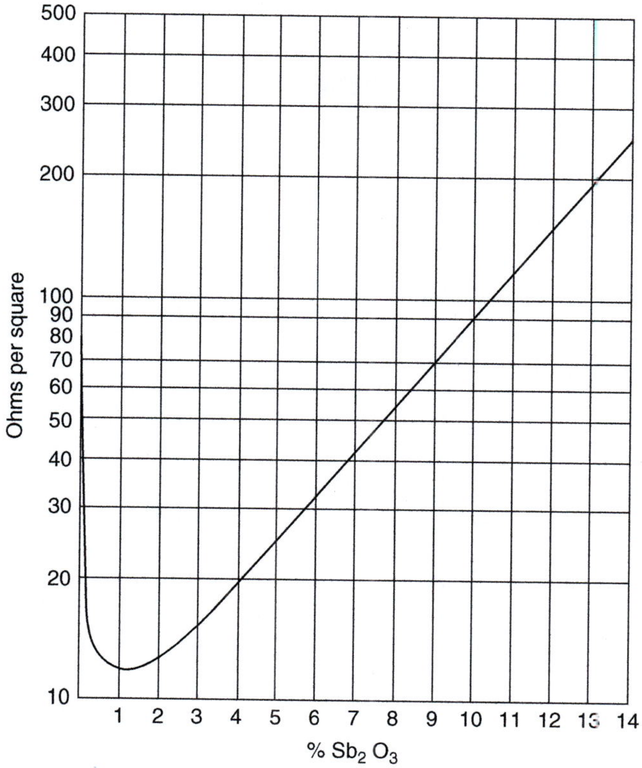

Figure 13.1 Sheet resistance of $Sb:SnO_2$ with varying percentage of antimony in the deposition chemistry. The coating is approximately 540 nm thick. *Source:* Adapted from Mochel [3].

As the antimony content is increased, the coatings develop a blue transmitted color which increases with increasing antimony content.

Figure 13.2 shows that the film thickness may also be altered to change the sheet resistance of the coating. The antimony percentage in the deposition chemistry is 1.5% in the formulation used to deposit these coatings.

The mechanisms for the variations in the T_{CR} of $Sb:SnO_2$ coatings are theorized and US '706 describes how the T_{CR} changes from an empirical perspective. US 2,564,706 notes that SnO_2 (undoped) from only tin salts demonstrate negative T_{CR}. This is most likely due to electrons being thermally elevated into the conduction band. The $Sb:SnO_2$ coating's T_{CR} changes with antimony doping level, differences in thermal expansion, and thickness of the coating. Small negative T_{CR} occur when the antimony doping is high or other metal oxides are doped into the $Sb:SnO_2$ coating. The T_{CR} values are small in this situation and do not affect practical use of the coating for various applications. Differences in the thermal expansion coefficient (T_{EC}) also affect the T_{CR} and may be further influenced by the thickness of the coating. From US 2,564,706, "…on a high silica glass having a thermal expansion coefficient of about 8×10^{-7} cm. per cm. per degree C a tin-antimony iridized film of the 4th order containing about 1% Sb_2O_3 has a negative temperature coefficient of resistance. The latter value will be positive if the film is thin, say of the 1st order, or if the antimony content is somewhat higher, say about 5% Sb_2O_3." The variations are proposed to be due to differences in the T_{EC} and that the variations in antimony content may "toughen the film."

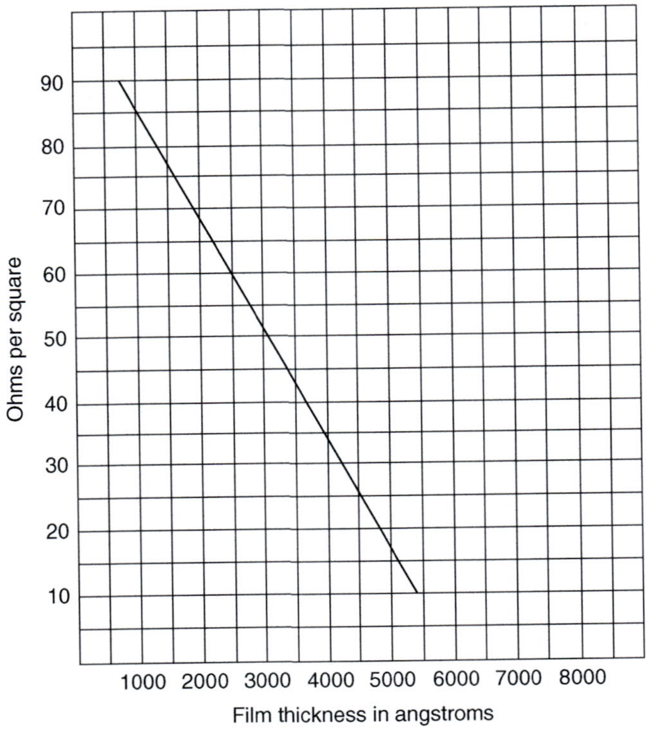

Figure 13.2 Sheet resistance (ohm/sq.) of Sb:SnO_2 with varying coating thickness (in angstroms). *Source:* Adapted from Mochel [3].

The use of additional salts of metal oxides can be used to further modify the Sb:SnO_2 film as shown in US 2,564,706. Two main effects are considered – modifying the conductivity and contact resistance. Manganese is particularly good as a conductivity modifier as the change in conductivity with concentration is low. Doping with ZnO changes the coatings and results in little to no contact resistance and enables good electrical contact for platinum metalizing.

The examples in US 2,564,706 show further practical examples and attributes of the Sb:SnO_2 films that cement this material as a practical and useful transparent conducting oxide. For example, the Sb:SnO_2 coatings may be tempered, the sheet resistance is stable over a wide temperature range, and returns to its original value after heating and cooling. The coating is inert to organic materials (Crisco) and that the sheet resistance is stable after many cycles of heating and "burning" the Crisco, and can be used to toast bread. Additionally, this patent notes that "...they emit more radiant heat from the back, that is, through the glass, than from the outer face of the film." This indicates that the inventor recognized that the coating reduced the thermal emissivity of the surface, essentially forecasting a key attribute for insulating windows. Uncoated soda-lime window glass has a thermal hemispherical emissivity of 0.84, while a Sb:SnO_2 coating can lower the emissivity to 0.15. An ideal blackbody has an emissivity of 1.0.

Like many patents, the work of John Mochel [3] gives a glimpse into what was likely very well conducted and thorough research. It is likely that the scientific community invested considerable resources to replicate what was learned as part of the commercial development of these early SnO_2 films. The vast number of coating properties and attributes taught in this single patent were foundational for future researchers.

Figure Photo 13.1 Image of water boiling on a plate glass sheet electrically heated. The coating was made by LOF Glass (c. 1940s), courtesy of NSG.

Photo 13.1 is a visible demonstration of the ability to heat objects using conductive tin oxide as a heater.[1] The glass is heated using tin oxide as a resistor. Bus bars are located on parallel edges of the coated plate glass. Electrical power is applied via the bus bars. The resistively heated glass in turn heats the water to the boiling point.

Similarly, Photo 13.2 shows an egg being cooked on the heated plate glass. Photos 13.1 and 13.4 are from the 1940s.

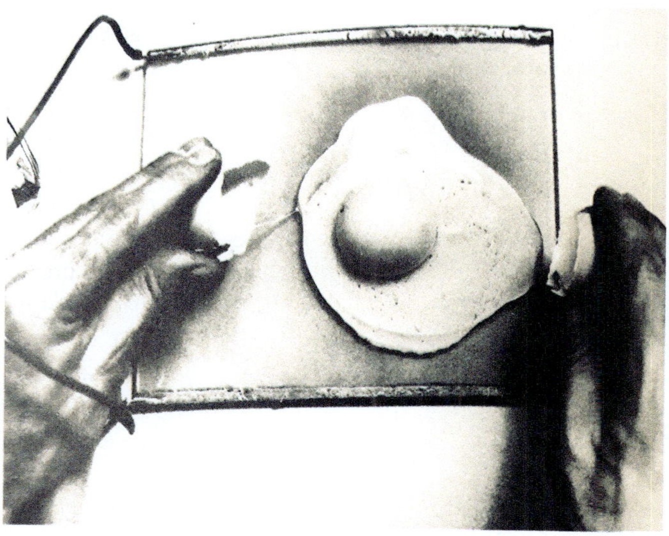

Figure Photo 13.2 Image of an egg cooking on a plate glass sheet, which is electrically heated (c. 1940s), courtesy of NSG.

1 Photo provided by NSG archives, Kyle E. Sword of NSG.

William Lytle in US 2,566,346 teaches doping SnO_2 with fluorine. This is the first US patent claiming fluorine as the doping element to attain conductivity. The development of an alternative to antimony doping allowed highly conductive SnO_2 to be produced which did not infringe the claims of US '706. With US '346, we start to see narrower claims targeting specific improvements or alterations to the deposition chemistry and/or the coating.

As noted in McMaster's US 2,429,420 patent, haze is a potential risk associated with SnO_2 coatings. Stannous chloride ($SnCl_4$) was a common precursor for the coatings. The reaction of liberated chlorine with sodium from the soda-lime silicate glass would lead to the formation of NaCl crystals. SEM images of holes left behind in the SnO_2 coating from the NaCl crystals are common in the literature [5]. Figure 13.3 shows variations in the crystal structure with processing [5]. Holes in the coatings are shown after the samples are washed. The relative chlorine content is reduced in the SnO_2 during heat treating, presumably due to the reduction of NaCl crystals in the coating. This provides support for the NaCl co-formation mechanism within the formation of SnO_2 coating. Methods were explored to minimize the interaction of the sodium from the glass with the chlorine to improve the clarity of the SnO_2 coatings deposited using stannous chloride. Several different "under layers" were considered [6–9] such as a SnO_2 coating produced from an organotin compound which was chlorine-free [6]. Other layers including a silicate [7], silica [8], or a metal oxide [9] were studied. Alternatively, the glass could be dealkalized [10] or the water vapor in the deposition carrier gas could be controlled [11]. Shifting entirely away from halide-based precursors [12]

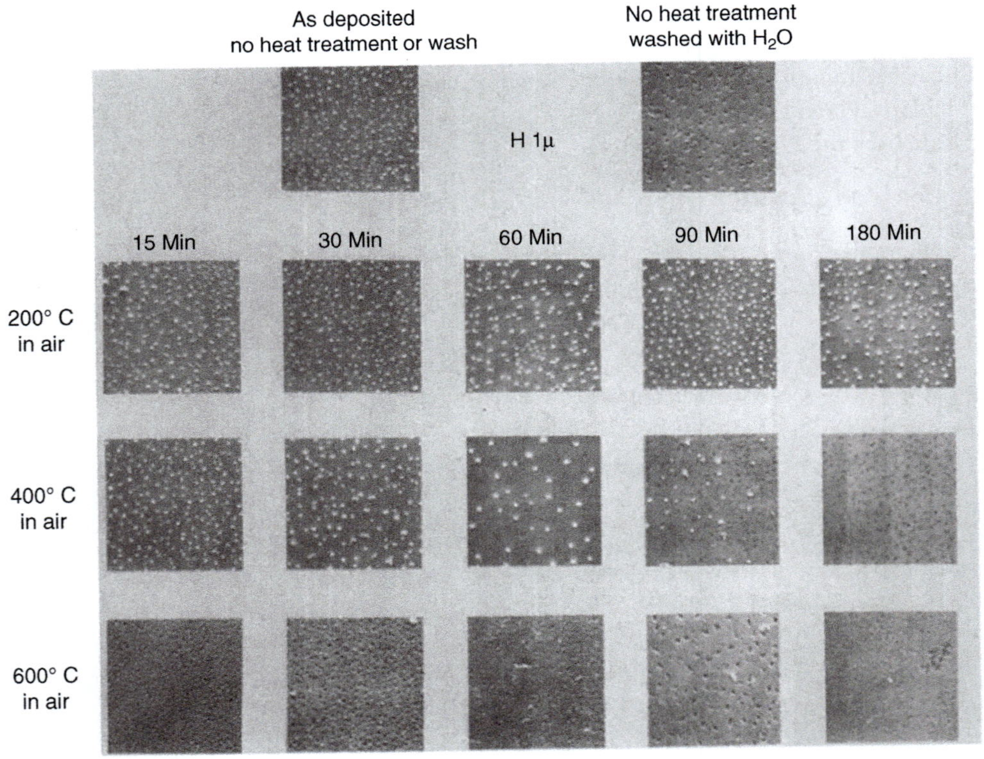

Figure 13.3 Micrographs of 50 nm thick tin oxide coatings on soda-lime silica glass. Variations in the SnO_2 crystal structure are shown after various heat treatments. *Source:* Smay [6] / Reproduced with permission from Elsevier.

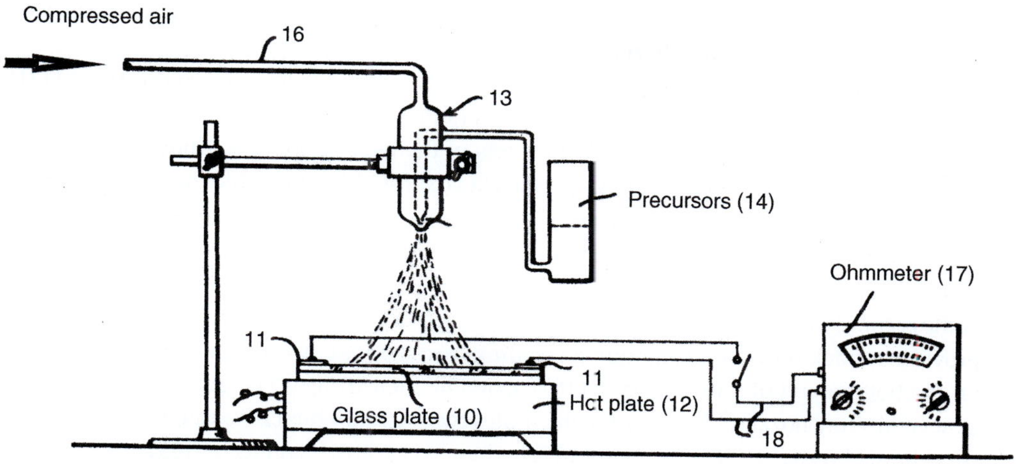

Figure 13.4 Early spray tin oxide apparatus showing chemicals spraying on to a heated glass substrate. *Source:* Adapted from Mochel, US 2,564,706 [6].

to avoid halide interaction with the hot glass surface was also considered. These methods are frequently revisited in various forms or with different processes to minimize haze in SnO_2 coatings.

As noted above, Harold McMaster [2] shows that spraying a chemical solution may be used to form conductive SnO_2 coatings and a spray apparatus is shown in reference [3] (see Figure 13.4). From work done by John Mochel [3], a glass plate (10), provided with platinized stripes (11) (shown on an exaggerated scale) on two opposite edges, is about three inches square between the stripes. It is heated uniformly on an electric hot plate (12). A glass atomizer (13) comprises a cup (14) for containing the solution of tin and antimony salts to be atomized, an atomizing nozzle (15), and a tube (16) for the introduction of compressed air to the nozzle. The atomizer is supported so that the nozzle is about one foot above the glass plate. An ohmmeter (17) with two contact leads (18) may be brought into electrical contact with the stripes before atomization. This technique is commonly referred to as "pyrolytic spray" due to elevated substrate temperature used in the process. A stationary substrate is heated via a hot plate and the precursor chemistry is combined with air, is atomized, and is directed toward the hot substrate. The earliest patents on SnO_2 do not detail the spraying process and the original solutions described are relatively simple, often comprising an alcohol, a tin chloride compound, a dopant compound (F or Sb), and water. Wide ranges are provided for components and some nuances on forming the solutions are also described. For example, the tin compound, stannic chloride, is first dissolved in the alcohol before the water is added to prevent fuming. Presumably, the fuming is related to hydrochloric acid (HCl) formation when the water reacts with the stannic chloride. Subsequent patents, such as Lytle US 2,566,346, go into more details on the composition of the spray solution to improve some aspect of the coating such as haze or electrical conductivity. The inclusion of HCl is demonstrated to improve the electrical conductivity in fluorine-doped tin oxide [13]. Work by Arnold Saunders et al. challenges the assumption that the presence of HCl in the precursor solution will always result in haze [13]. Low haze values were demonstrated that matched those of methods which employed pretreatments. The resultant films also had good contact resistance to pre-deposited bus bars. This work demonstrated that control of the chemistry and the deposition method may be used to counteract the natural tendency to form NaCl with the resultant haze.

Spraying chemicals to produce controlled resistive coatings dates back to the 1930s [14]. Other patents [11, 15, 16] discuss modifications to the spray process to improve the uniformity, quality of the coatings, or modify aspects of the deposition. T.M. Knowland et al. in US 2,588,125 [15] teach the means to produce a coating of uniform thickness on a moving substrate. Figure 13.5 shows an apparatus where the spray heads are moved in oscillating pattern perpendicular to the motion of the substrate. From Knowland's work [15], the machine shown in the drawings comprises a frame with opposing uprights (10) at opposite sides of an endless conveyor (12). A rail (14) supported at its ends on the uprights above the conveyor provides a track on which is mounted a sliding carriage (15). A spray gun (16) is mounted on the bottom end of a vertical post (18) carried by the carriage. The gun can include one or several spray nozzles. The drawing illustrates two such nozzles (20) together with air and spray liquid hoses (21). A vertical slide (22) in the carriage (16) is connected (23) to an endless chain (24) supported on two sprockets (26). The sprockets and chain are driven by a motor (27) mounted on the rail and operated through a belt (28) and reduction gearing in a box (30).

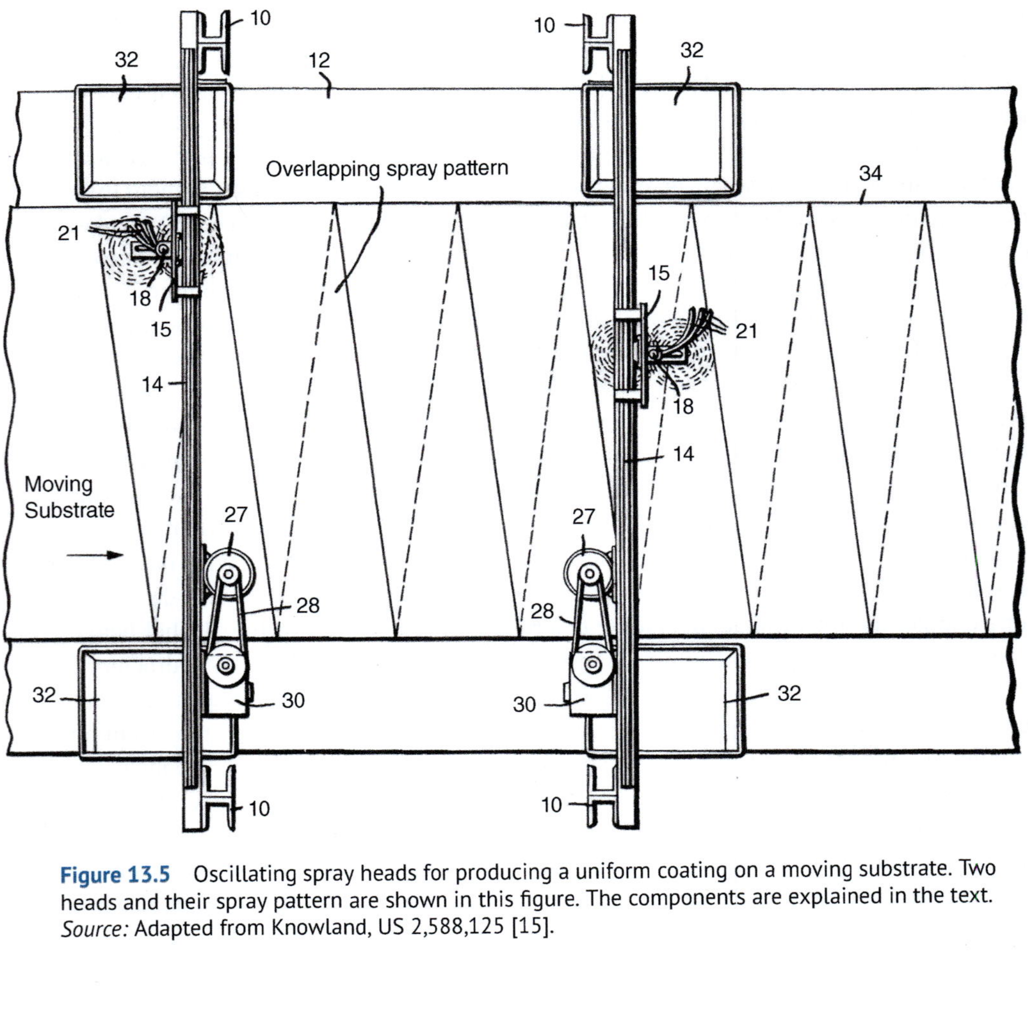

Figure 13.5 Oscillating spray heads for producing a uniform coating on a moving substrate. Two heads and their spray pattern are shown in this figure. The components are explained in the text. *Source:* Adapted from Knowland, US 2,588,125 [15].

Mounted on the two uprights at opposite margins of the conveyor (12) and in line with the two nozzles (20) are two tanks (32) located below the two sprockets (26). The substrate (34) to be sprayed is carried on the top of the conveyor (2) beneath the spraying nozzles. The conveyor is operated by a motor (not shown) and this motor and motor (27) are synchronized to operate at such relative speeds as will move the substrate longitudinally and reciprocate the carriage (15) across the substrate.

Alternatively, control of the thickness of the coating to attain varying resistance across a surface is obtained using modifications to the spray process detailed in Ref. [16]. Figure 13.6 shows a modified spray process to produce a controlled, nonuniform thickness of the SnO_2 coating. From work by Leighton Orr [16], a film of nonuniform surface resistivity is produced on substrate (20), shown on the left half of Figure 13.6, by adjusting the spray gun so that an elongated spray pattern is formed such as shown on the right half. The spray pattern is wider and more dense at the lower half (23) of the pattern than at the upper half (25). This pattern may be obtained with conventional spray guns simply by adjusting the air pressure of a spray gun (26) having two air ports so that a higher air pressure is employed at one port than at the other. The system comprises the substrate with glass base (22). The bus bars (24) are placed along nonparallel edges of the base with (28) indicating the shorter edge. The spay gun deposits the solution to the substrate.

Efficiency in the manufacturing process, such as simultaneous coating and tempering [17], helps increase speed and reduce cost. Thinner glass substrates introduce additional challenges to the deposition process. The chemical reaction which converts the precursor chemicals to the deposited film is thermally driven. As noted above, optimal properties occur at the maximum temperature allowable by the softening temperature of the substrate. Cooling occurs as the liquids are sprayed onto the substrate. Thin substrate materials do not have sufficient mass to maintain the desirable deposition temperatures. Floating the substrate on a molten metal bath [18] provides a heat source and replaces the lost thermal energy which occurs in the deposition process. Figure 13.7 shows a laboratory demonstration of this concept [18]. From John Mochel's patent [18] US 2,522,531, a thin mica sheet (10) floats on molten tin (11), contained in a shallow open container (12). The container is heated on an

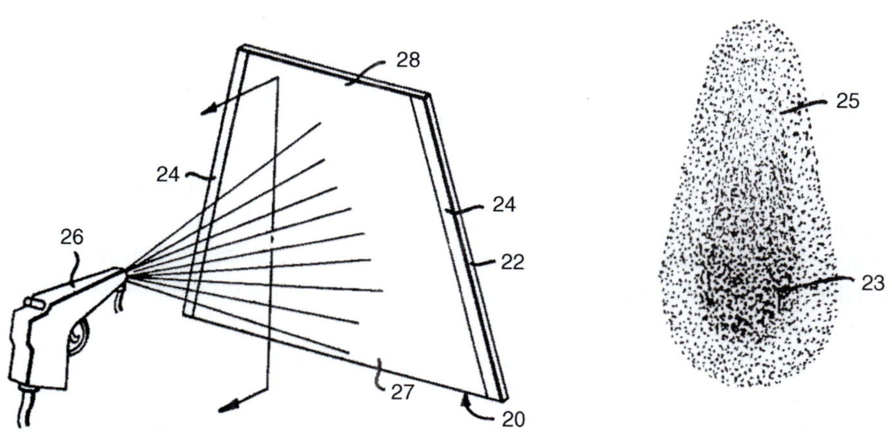

Figure 13.6 Spray process designed to produce a controlled, nonuniform thickness. On the left is the setup for spraying onto a glass substrate. On the right is the resulting spray pattern. *Source:* Adapted from Leighton Orr [16], patent US 3,019,135.

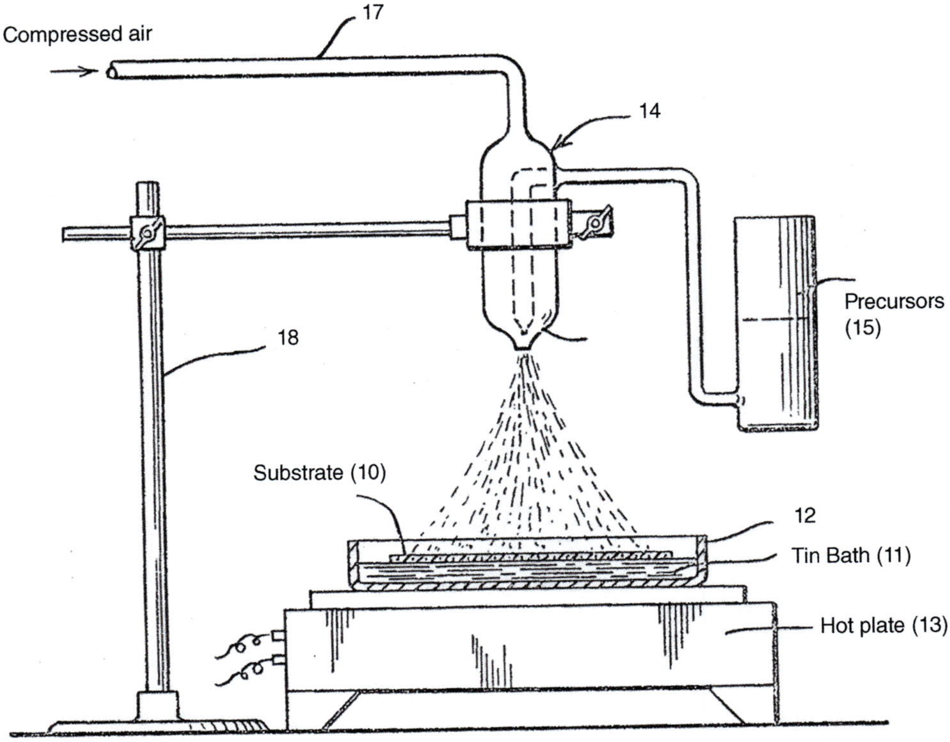

Figure 13.7 Spray coating of a conductive tin oxide coating on mica or glass floated on a molten tin bath. The molten bath supplies a constant and uniform temperature to the substrate to create a more uniform reaction and prevents substrate warping. *Source:* Adapted from John Mochel's patent [18] US 2,522,531.

electric hot plate (13). A glass atomizer (14) includes a cup (15) containing a solution of salts to be atomized, an atomizing nozzle (16), and a tube (17) for introduction of compressed air to the nozzle. The atomizer is held by a support (18) so that the nozzle (16) is about one foot above the mica sheet.

Improved uniformity of properties is a major motivator in process development in subsequent innovations. Commercial applications, such as architectural windows, require uniform appearance and iridescent coatings are not acceptable. Low emissivity, or Low-E, coatings became of interest for architectural applications after the OPEC oil embargo of 1973. As a result of the oil embargo, the United States invested in energy efficiency technologies for buildings. As noted in Ref. [3], conductive SnO_2 coatings display a reduced emittance characteristic. The use of the coatings for aircraft and other applications meant that the glass industry had decades of experience with the coating and could pivot to produce the coatings for architectural windows.

Another innovation which came of maturity in the 1970 was the float glass manufacturing process [19]. The float glass method enabled the production of high-quality glass in a continuous process. The raw materials are melted and then refined before transitioning to the tin

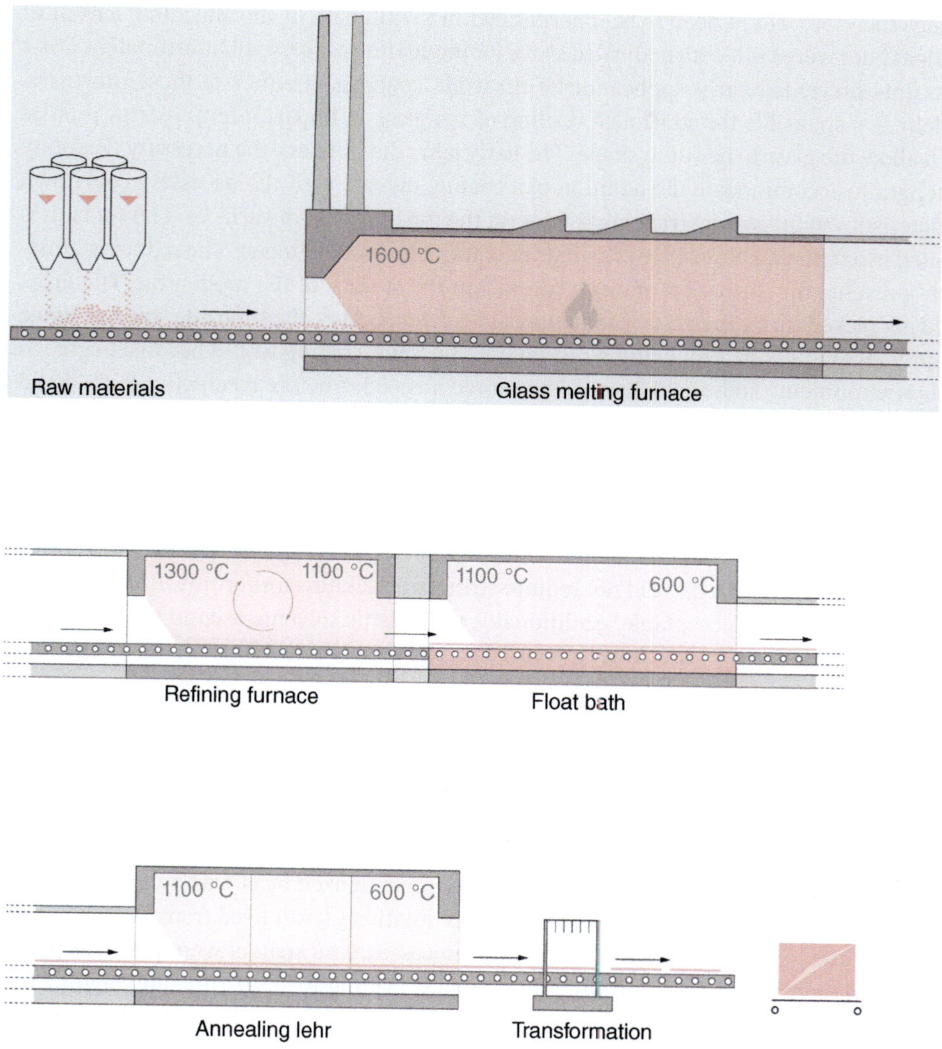

Figure 13.8 Schematic of the float glass process including temperatures of the glass in the different stages of the process [20] (for illustrative purposes, not to scale). *Source:* Used with permission from Geboes [20].

bath [20]. The steps are shown in Figure 13.8. The glass, as it cools down, goes through the optimal temperature range taught in the early patents. By applying the coatings on the float line, the cost basis could be reduced since equipment to transport and heat the glass are no longer needed. Additionally, the potential for higher throughput would improve capitalization of the coating equipment, further improving the economics of the coating manufacturing process. Unfortunately, the original float lines were not designed for online coatings. For example, an early location for the coatings was between the tin bath and the annealing lehr, but the physical space between these processes was typically less than a meter long. Online coatings therefore would need to be incorporated in the tin bath or the annealing lehr until new float lines were built with dedicated space for the coating processes. Unfortunately, other constraints prevented early application of tin oxide coatings in either of these processes. The lehr is responsible for controlled cooling of the glass to impart internal stress profiles which allow the glass to be cut to shape. The early lehrs did not have the necessary flexibility and length to accommodate the addition of a coating process with the necessary cool down characteristics with asymmetric emissivities on the top and bottom surfaces. The tin bath is operated in a reducing atmosphere, a mixture of nitrogen and hydrogen. The reducing atmosphere prevents the formation of oxide "dross" on the surface of the molten tin. The dross would get picked up by the glass leading to visible defects. Also, the tin oxide was not stable at high temperatures in a reducing atmosphere. The SnO_2 coating would become degraded in this environment. Additional innovations would be needed before conductive SnO_2 could be produced in the tin bath.

New deposition processes, with improved thickness uniformity and reduced environmental impact, were needed to be compatible with float line processes (Personal communication with Bernard Savaete, BJS Differences, France (former BSN, PPG executive)). For example, spray processes, the norm for conductive tin oxide, were historically incorporated into offline processes for applications that did not require stringent thickness uniformity. Iridescent colors in the coating were acceptable. Additionally, the organic solvents used in the spray processes were beginning to come under increased scrutiny and control from an environmental emissions perspective.

Spray technology, the standard for decades, did attain some use on the float line and is still in use today, but was no longer suitable for the new requirements, thus spurring another wave of coating innovation. Fast deposition rates, optimal emissivity per unit thickness, and uniformity became goals for innovations in new chemistry and processes. A number of companies developed powder deposition processes [21–23]. The patent art on powder processes covers approximately a decade and the technology was commercialized by at least two companies, PPG and Saint-Gobain. The powder processes were relatively short lived from a commercial perspective due to challenges with the powder processes. The coaters were prone to powder accumulation and build up in internal channels leading to short run campaigns. The cleaning of the coaters was relatively laborious and alternative deposition technologies were developed.

CVD was an alternate deposition technology which promised longer campaign lifetimes and significant development resources were committed by a number of glass companies. It should be noted that there were limited or no third-party suppliers of online deposition technology at this time in history. Each glass company independently developed their own proprietary technology in order to gain commercial advantage over their competitors.

As noted earlier, Ref. [2] described that SnO_2 could be deposited by CVD. Various aspects, improvements, and enabling technologies were patented that took the basic concept of CVD to practical development. One of the earlier patents [24] is somewhat a direct translation of spray technology to CVD to provide electrostatic dissipation coatings for TV camera tubes. The need for low haze and good clarity applies to this application. The targeted coating properties were attained by using stannic chloride in a dry air carrier gas. The suitability of using CVD technology to SnO_2 on continuous, rather than discrete substrates, such as fibers or thin glass, was recognized [25]. Figure 13.9 details the continuous substrate moved through an oven where the substrate is heated and coated by CVD. Figure 13.9 illustrates coating of a wide roll of glass fiber cloth and apparatus for that purpose. A shallow rectangular container

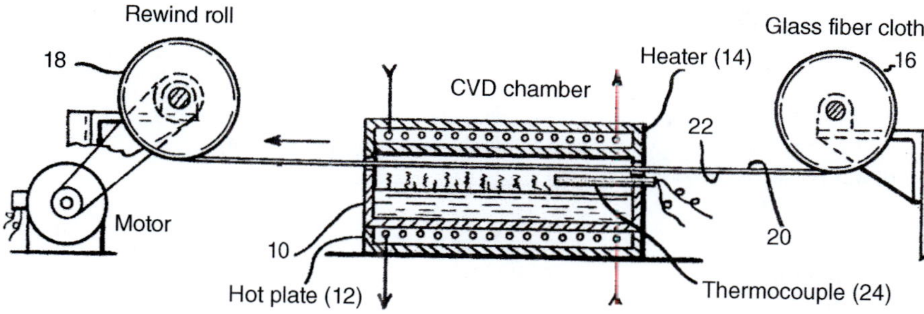

Figure 13.9 Continuous CVD deposition process for coating glass fiber cloth [25]. Shown is fiber cloth traversing through a heated cavity of chemical reactants. The components are explained in the text. *Source:* Adapted from Dates, US 3,331,702.

(10), in the nature of a trough, is partially filled with a mixture of antimony and tin chlorides. A hot plate (12) supplies heat to container (10) for the purpose of volatilizing the chloride mixture and forming within the container a confined atmosphere of quiescent fumes for coating purposes. A rectangular heater (14), which may be a conventional heater unit with wire coils, is positioned above the open mouth of the container and, as shown, slightly spaced therefrom. A roll of glass fiber cloth (16) is passed between the container and heating element and rewound on roller (18). The top surface (20) is exposed to radiant heat from the heater and the underside of fiber cloth (22) is exposed to the chemical vapors.

The support equipment for a CVD process is considerably more elaborate relative to a spray process. The complexity of the various steps and processes includes vaporization technology, gas flow metering, and temperature control among others [26]. Figure 13.10 shows a flow chart of the various processes for a SnO_2 glass container coating process. This example is for coating containers (1) which are supplied to the CVD coating station or fuming tunnel (3) via conveyor (2). The CVD gasses are supplied by the vapor distribution manifold (15) into two conduits (44) and (49) which direct the mixture of gasses to two nozzles on opposite sides of the vapor dispenser. An exhaust blower (76) provides a means to extract the reacted vapors to a stack.

The CVD precursor and carrier gasses are prepared using a vaporizer system (4). The precursor is stored in a supply tank (5) which is controlled to a particular temperature. Compressed, dry air is supplied (8) to a pressure regulator (10) and fed into flow meters (11) and (16) operated by control valves (12) and (17), respectively. The dry gas after flow control valve (12) is fed into a vaporizer (9), where it entrains the vaporized precursor. The vaporized precursor in the carrier stream is provided to trap (40') via duct (40), where the trap captures any precursor which is not fully vaporized. The gasses are then fed into vapor dilution chamber (13). The dry gas after control valve (17) is heated (18) to a desired temperature via heater control (25) and the gas is provided to dilute the gas stream via duct (41) in dilution chamber (13). The diluted air and precursor mixture is fed via duct (14) to the vapor distribution manifold (15).

Container manufacturing was one of the first significant applications of tin oxide by CVD as an anti-scuffing agent (similar to the titanate application illustration). Roger Dubble's patent US 3,438,803 also provides many equipment drawings detailing and enabling the nuances of the process needed for commercialization [26]. The requirements for the support equipment are directly transferable from container to flat glass applications. The key to producing

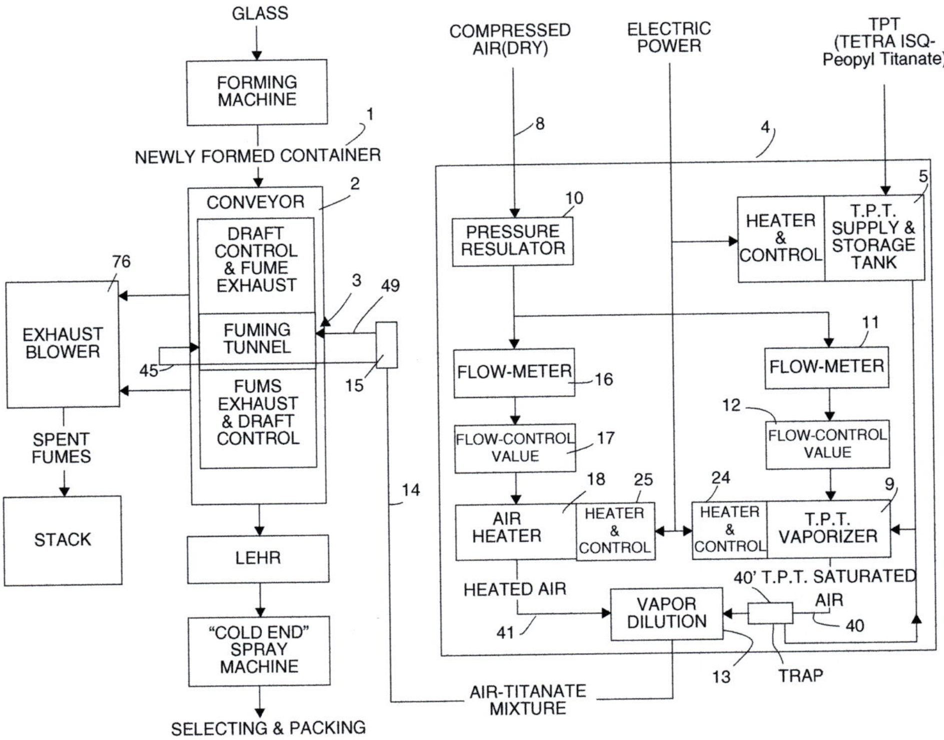

Figure 13.10 CVD process flow chart for a container tin oxide coating process [26]. *Source:* Adapted from Dubble et al., US 3,438,803.

uniform and consistent coatings lies in the stability of the support processes. The details of which will be discussed in Section 13.4.

John Mochel's patent US 2,522,531 [18] described coating a substrate floated on a molten metal bath and the concept was logically extended to coating in the tin bath on a float line by Alan Edwards et al. in patent US 3,681,042 [27]. This earliest patent on CVD coating in the float bath taught metal deposition and not SnO_2. The first tin bath CVD was a reflective silicon coating by Pilkington and Pilkington licensees called "Reflectafloat™." Equipment and materials compatible with the reducing environment of the tin bath were taught. Means to keep the deposition gases isolated from the bath environment was a key innovation. Figure 13.11 shows a cross section of an in-bath CVD system [26] which includes three deposition zones (15, 16, and 17). The coating system of Figure 13.11 is in the tin bath of a float glass line. The roof structure (6) is shown which supports refractory beams (25) via hanger bolts (26). The refractor beams support extraction exhaust ducts (44) and (45). These ducts remove the vapors through extract slot (46). The gas flows from the vapor release slots (32) in a direction indicated by (47). A hood comprising a roof (48) and side walls (49) create a reaction zone for the CVD process and prevent escape of vapors to the tin bath headspace (7). Graphite connecting pieces (29) (and 30 not shown) provide electrical connection to either end of the coating vapor release ducts (31). Each duct (31) has a slot (32) in the lower surface directed toward the upper glass surface (13) of the glass ribbon (24). Duct (31) is supported by short struts (41). The vapor comes out of slot (32), impinges on the glass surface, and then

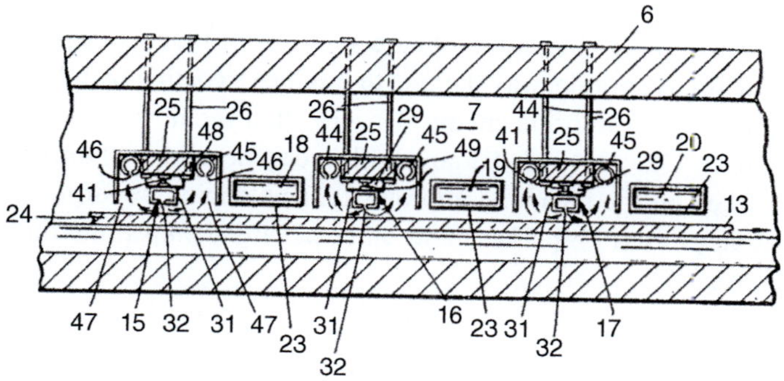

Figure 13.11 Cross section of an in-tin bath CVD deposition system [26]. The components are explained in the text. *Source:* Adapted from Dubble et al. US 3,438,803.

is extracted via exhaust ducts (44) and (45). The system is cooled via water coolers (18, 19, and 20). Each cooler has a lower surface (23) which faces the glass surface (13).

A number of patents related to CVD equipment and methods were issued to various companies detailing the development of various aspects of the technology needed for commercialization of tin oxide deposition outside the tin bath [28–32]. Two types of chemical flow methods were considered – turbulent and laminar flow. In the turbulent example, the premise is to have the vapors contact the glass with high velocity to enable faster deposition by reducing the boundary layer (to be discussed in more detail later). In the laminar flow case, the vapors are constrained to flow along the surface of the glass giving more time to contact the surface, react, and build the coating. Various aspects of each of these technologies were commercialized by the different glass companies for coating outside the tin bath.

Coating outside the tin bath was constrained by the glass temperature of the float glass process. The glass had to cool to become sufficiently solid so that it could be supported by the rollers outside the bath. This means that the glass was cooler than preferred for optimal conductivity of the tin oxide coating (as described earlier). Other processes are present at the exit of the tin bath on the float glass process, such as the application of SO_2. The SO_2 reacts with the sodium in the glass to form sodium sulphate as a solid lubricant. This is applied to the bottom of the glass, but the vapors typically can spread and affect the top surface of the glass. The presence of the SO_2 or sodium sulphate can interfere with the SnO_2 coating causing visible aesthetic defects.

The first patent covering the deposition of tin oxide in the tin bath was issued to PPG in 1989 [33]. The novelty of Ref. [33] is that the coating is deposited in a nonoxidizing atmosphere. Essentially, as the hydrogen content in the tin bath was reduced, the fluorine-doped tin oxide coating's sheet resistance was reduced. The lowest sheet resistance was attained with only nitrogen in the tin bath. Details on how to operate the tin bath without hydrogen was typically a trade secret for the different companies practicing in bath deposition of conductive tin oxide coatings.

Stannous and stannic chlorides enjoyed widespread use as precursors for tin oxide due to their low cost, fast deposition rates, and compatibility with several processes or methods of deposition. Alternatives, as noted above, were sought to minimize haze associated with NaCl formation. Glass companies and chemical companies began developing (and offering) precursors which were more aligned with the needs of the CVD processes. RCA [34] patented the use of "such as compounds having the formula (R – Sn (–OOCR), or (RSn–)–O wherein R is a lower alkyl group. Suitable compounds include dibutyl tin diacetate, bis(tri-*n*-propyl) tin oxide and bis(tin-*n*-butyl) tin oxide for example". Westinghouse patented dimethyltindichloride with dimethyltindifluoride [35]. M&T Chemicals (Arkema) patented monophenyltin trichloride which, like monobutyltin trichloride, was claimed to produce a coating with low haze [36]. As noted in Ref. [35], different fluorine chemicals were often the novel characteristic in the claims. Other patents [37, 38] teach HF and trifluoroacetic acid as fluorine precursors. Dr. Roy Gordon worked on the means to improve fluorine doping by forming Sn-F bonds during the deposition process [39]. These and other patents demonstrate the broad range of work on conductive SnO_2.

13.3 Manufacturers

Pyrolytic SnO_2 continues to be produced by glass companies who combine the process with glass manufacturing. Early companies active in the development of conductive tin oxide are either still active or have discontinued production of the coatings. For example, PPG Industries (now Vitro) no longer produces F:SnO_2. Similarly, Corning no longer produces a pyrolytic SnO_2 coating. Other companies such as LOF (which was purchased by Pilkington and then by NSG) are still active and flourishing. Other glass companies such as AGC have developed pyrolytic SnO_2 and are producing it in different locations across the globe. Saint-Gobain still offers a product. Several companies in China have entered the market. Jinjing (Group) Co. Ltd. is offering products based on PPG Industries' technology. Yaohua offered a product based on Arkema's technology and was subsequently purchased by Pilkington. Xinyi glass and Xinyi Solar also offer products. Most of these products are produced via CVD in the tin bath. Conversely, Xinyi Solar offers a product on "cast" or "rolled" glass and may employ pyrolytic spray technology (Personal communication with Bernard Savaete, BJS Differences, France (former BSN, PPG executive)). The early work on the SnO_2 deposition process was undertaken by those either selling the finished product or the precursor chemicals. Stewart Engineers (https://stewartengineers.com/), a third-party company that sells turnkey glass factories, provides CVD systems for use in the tin bath and lehr under the name "AcuraCoat®." Several of these systems are still operating today and Stewart continues to sell AcuraCoat turnkey CVD systems for new and retrofit applications.

13.4 Chemical Vapor Deposition (CVD)

CVD is the dominant deposition technology used for producing pyrolytic conductive tin oxide [40]. Reference [40] details the challenges with CVD processes for producing thin film coatings. As a chemical reaction rather than a physical vapor deposition method, the process is dependent on various reaction pathways, including parasitic reactions such as homogeneous gas-phase reactions. For a simple molecule, SiH_4, Figure 13.12 from Mark Allendorf, Sandia shows that the reactions involved in the deposition can be quite complex [40]. The investigation of new precursors suitable for CVD methods often involved empirical experimentation rather than first principal calculations. A common industrial precursor, monobutlytin trichloride, is modeled and compared to experimental deposition results [41]. The authors show the complexity of the reaction pathways and challenges of combining the fundamental reactions with geometries of specific reactors [41].

The reaction mechanisms for $SnCl_4$ and tetramethyl tin were investigated in detail [42] by Ghoshtagore. In CVD reactions, the concentration of the tin precursor and oxygen source

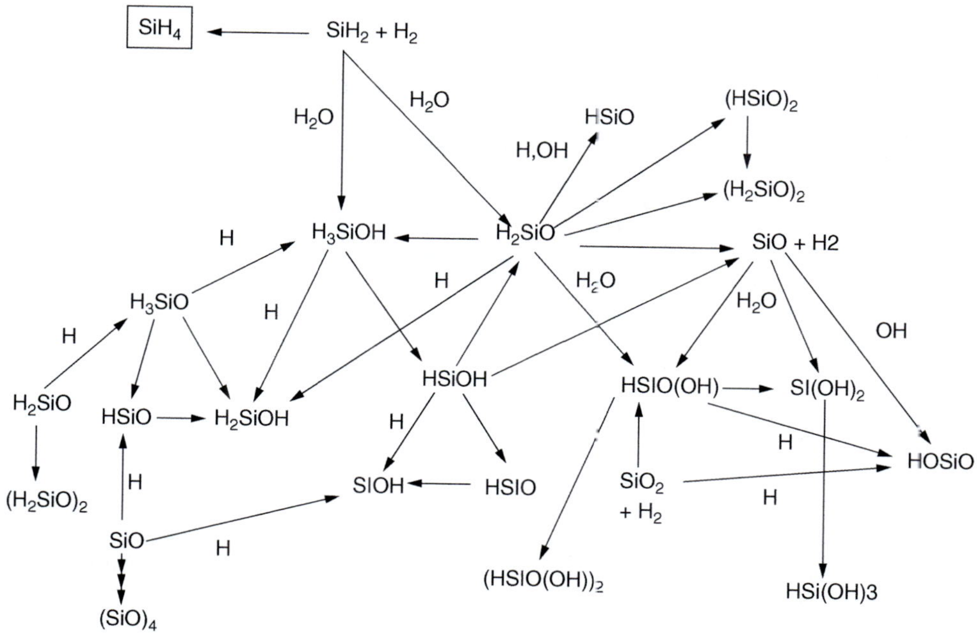

Figure 13.12 Reaction pathways for SiH_4 involved in a CVD deposition process. The direction of the arrow indicates the exothermic direction of the reaction. *Source:* Mark Allendorf [40] / reproduced with permission from IOP Publishing, Ltd.

(O_2 or H_2O, for example), and temperature are key factors which affect the growth rate of the coating. Additionally, the flow rate of the gasses in the system can affect the deposition process. Typically, "the examination of the dependence of the rate of formation of the solid reaction product on the flow rates of each of the reactants (while the others are kept constant) is the primary order of business. This information provides the fluid dynamic relationships of the particular experimental system and helps to delineate the surface reaction-controlled region from the mass-transfer controlled region." Essentially, the flow rate (gas velocity) in a given reactor needs to be high enough so that "the boundary layer thickness attains a limiting value, and the Reynolds number and the gas-phase mass-transfer coefficient increase to a value where the reaction is controlled by the chemical surface reaction rate constant. Growth rates are no longer dependent on the flow rates and the reaction becomes highly activated." If the overall flow rate is too low, then the study of the reaction rates is compromised by simple mass flow limitations. The reaction rate is limited by the amount of tin being fed into the system and not the fundamental reaction dynamics.

Figure 13.13 shows a linear increase (log/log) in deposition rate with increasing oxygen partial pressure at a fixed $SnCl_4$ concentration. A plateau is reached wherein the growth rate does not further increase with increasing oxygen partial pressure. The rate increases with increasing deposition temperature up to the plateau.

In Figure 13.14, the interaction of $SnCl_4$ partial pressure and water partial pressure is shown. In this example, the rate also plateaus at different water partial pressures and the onset of the plateau is dependent on the partial pressure of the $SnCl_4$ precursor. Interestingly, in this example, the relation is not dependent of the substrate temperature.

Figure 13.15 shows the temperature dependence of the growth rate of SnO_2 using tetramethyl tin as a precursor. The deposition rate versus temperature, in this example, has two

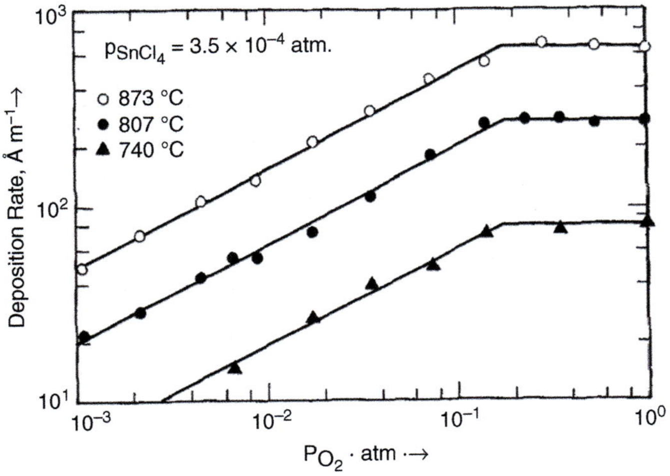

Figure 13.13 SnO_2 deposition rate dependence on oxygen partial pressure at different temperatures. This is for the $SnCl_4$ plus oxygen reaction. *Source:* Ghoshtagore [42] / reproduced with permission from IOP Publishing, Ltd.

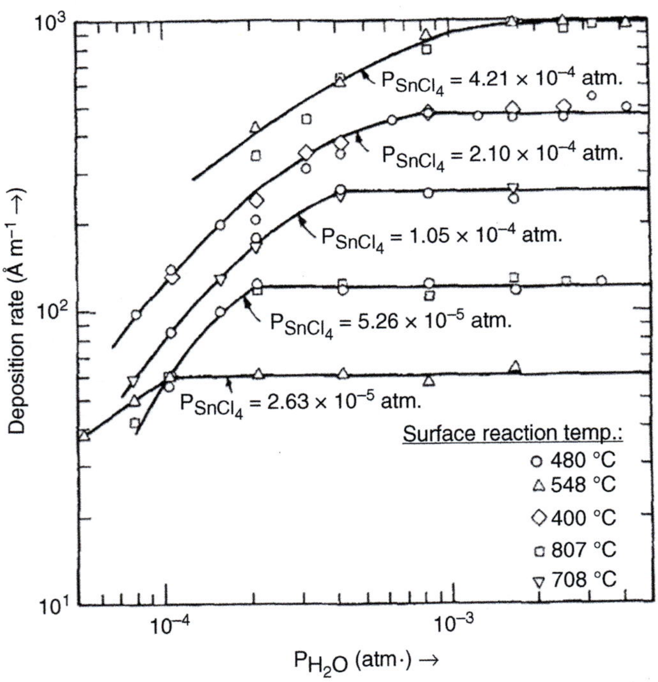

Figure 13.14 The effect of deposition rates on SnO_2 films with different partial pressures of water vapor and $SnCl_4$ concentration. Different surface reaction temperatures are shown. *Source:* Ghoshtagore [42] / reproduced with permission from IOP Publishing, Ltd.

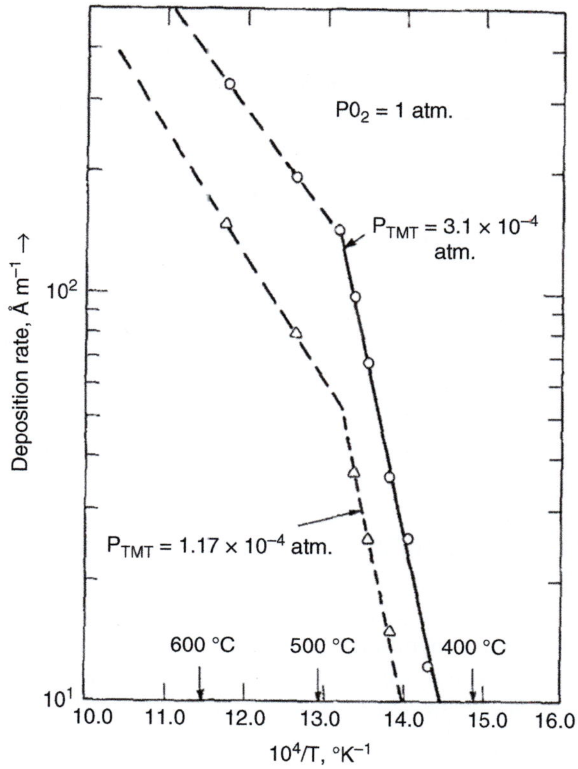

Figure 13.15 Temperature dependence of the growth rate of SnO_2 from tetramethyl tin (TMT), for two different concentrations of TMT. *Source:* Ghoshtagore [42] / reproduced with permission from IOP Publishing, Ltd.

different slopes depending on the temperature range. These examples demonstrate the complexity of CVD processes and the need to conduct empirical experiments to characterize the deposition space. The various steps in the CVD reaction, diffusion of reactants to the surface, adsorption of at least one reactant on the surface, formation of the critical nuclei by surface reactions, desorption of gaseous by-products, and diffusion of the by-products away from the surface all can have an impact on the practical growth rate. The complexity of the deposition space, with multiple compound interactions, increases as more reactants are used in a deposition process. Statistical Design of Experiments (DOE) are often employed to study a given CVD system [43, 44]. Statistical DOE approaches have the advantage that multiple responses can be simultaneously optimized for a given system.

In the Dubble et al. patent US 3.,438,803 [26], the complexity of the coating process was discussed, starting from the feed and coating side of the process. What is not covered are the environmental controls needed to prevent release of harmful chemicals into the environment. The precursor chemistry is selected based on its ability to be vaporized. The effective high vapor pressure at relatively low temperatures requires ways to condense unused precursors. This introduces a challenge for reclamation of the unused chemistry from the deposition process and any by-product. The reclamation technology to capture the various chemicals and by-products can substantially increase the capital cost of the entire CVD system.

The idealized precursor needs to react to form the appropriate coatings, but needs to be compatible with the front and back-end parts of the CVD system (https://stewartengineers.com/). The cost of the chemicals is a combination of the initial cost and the reclamation efficiency. A low-cost precursor could have an overall higher cost if the reclamation efficiency is low. This is important because the efficiency of the CVD system may be as low as 10%. In this case, 90%

of the precursor is not involved in the deposition process and reclamation of the unreacted precursor can substantially lower the overall precursor cost. The precursor should, therefore, be thermally stable through the vaporization step, react rapidly in the deposition zone but not be prone to gas-phase reactions, and then efficiently be reclaimed. Safety concerns are of important concern for production processes. The chemicals used in the CVD process need to be relatively safe from an exposure perspective, i.e., low health risk, and have a relatively low flammability level. Typically, the health concerns increase with decreasing length of organic ligands (methyl has a higher risk than butyl) and few halide constituents (tetramethyl tin has a higher risk than dimethyltindichloride) [45]. The exhaust effluents often are processed in several steps. A cyclone separator may initially be used to condense the majority of the unreacted precursors. The remaining vapors then may be altered to solids via a chemical reaction. Monobutyltin trichloride, and other metal chlorides, can form a solid compound when reacted with ammonia [46]. Russell shows a method for converting the precursor vapors to a solid which can then be captured with standard powder reclamation equipment [46].

13.5 Properties

Tin oxide coatings have a range of interesting electrical, optical, and physical properties which make them suitable for a range of industrial applications. A comparison of single crystal SnO_2 and polycrystalline SnO_2 coatings with and without doping shows the difference in electrical properties attainable for this material [47]. In Figure 13.16, the conductivity of undoped SnO_2 varies with temperature over different temperature regions (I,II,III). At low temperatures (4.2–130 K) in regions II and III, the conductivity varies by approximately nine orders of magnitude (Figure 13.16). The conductivity is a function of electron concentration and electron mobility. In this temperature range, the activation energy is sufficient to promote electrons to the conduction band. The quantity of electrons promoted to the conduction band increases with temperature leading to an increase in conductivity. In region I, from approximately 130–300 K, the electron mobility and electron carrier concentration vary differently with temperature leading to a local maximum in the conductivity as shown in Figure 13.17.

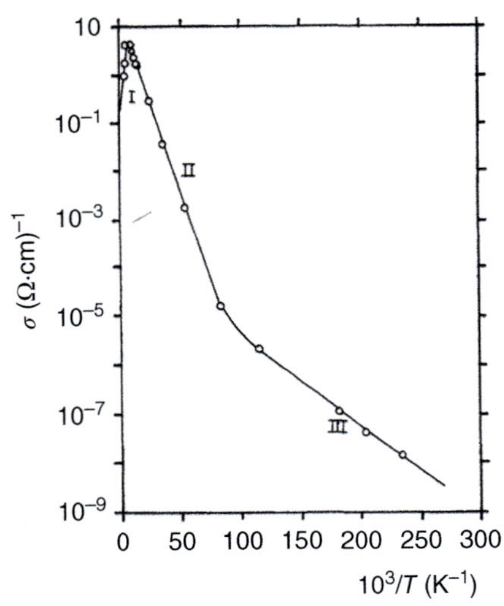

Figure 13.16 Change in conductivity (σ) of single crystal SnO_2 with temperature. *Source:* Jarzebski and Morton [47] / reproduced with permission from IOP Publishing, Ltd.

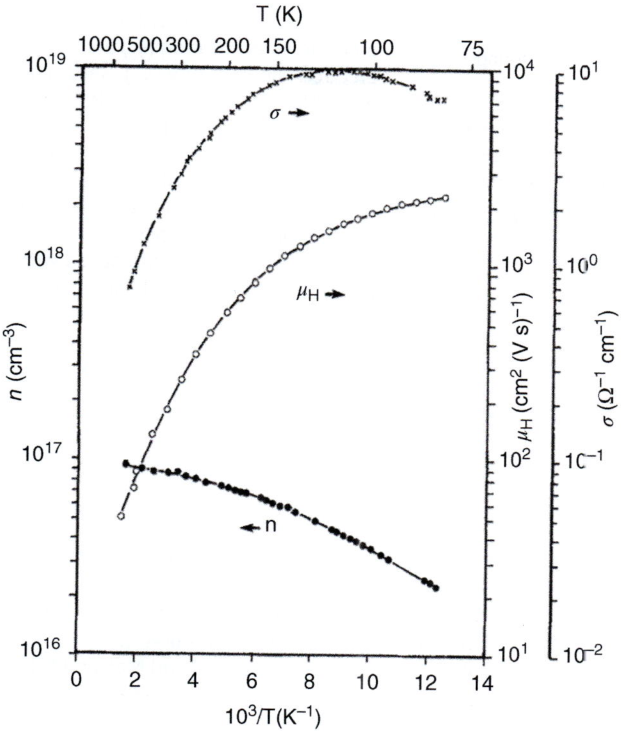

Figure 13.17 For single crystal SnO_2, the conductivity (σ), electron concentration (η_e), and electron Hall mobility (μ_H) dependence with temperature. *Source:* Jarzebski and Morton [47] / reproduced with permission from IOP Publishing, Ltd.

The change in mobility with temperature is linked to changes in the electron concentration [48]. The electron mobility is decreased with scattering of the electrons which may occur at grain boundaries or may be due to charged impurity scattering. As the electron density increases, the electron mobility decreases due to increased charged impurity scattering (see Figure 13.18). This leads to a maximum in the conductivity of the SnO_2 coatings.

In the case of thin films of SnO_2, the increase of electron concentration due to doping stabilizes the change in conductivity with temperature [47]. As the electron concentration (η_e) approaches 2×10^{20}, the change in conductivity with temperature over the range of approximately 75–300 K is essentially constant as shown in Figure 13.19.

The thermal stability of SnO_2 films is also strongly affected by doping [47]. The Sb-doped film is stable with thermal treatment while the undoped film's conductivity can change by over 3 orders of magnitude with thermal treatment. The doping of the SnO_2 films makes them more suitable for temperature dependent applications due to the stability of the electrical properties.

As discussed in Section 13.2, the electrical properties of the SnO_2 coatings are affected by the substrate. The dependence of the electron concentration and electron mobility with the substrate [49] sheds light on the root cause of the role of the substrate. Table 13.1 derived from data in Reference [49]. Table 13.1 shows the electrical properties of 5000 A thick SnO_2 films on various surfaces with the same baking condition, 500 °C for 30 minutes. The duplicate numbers represent multiple samples. The absence or reduction in sodium (silica barrier film or sodium-free glass) leads to reduced resistivity due to increases in

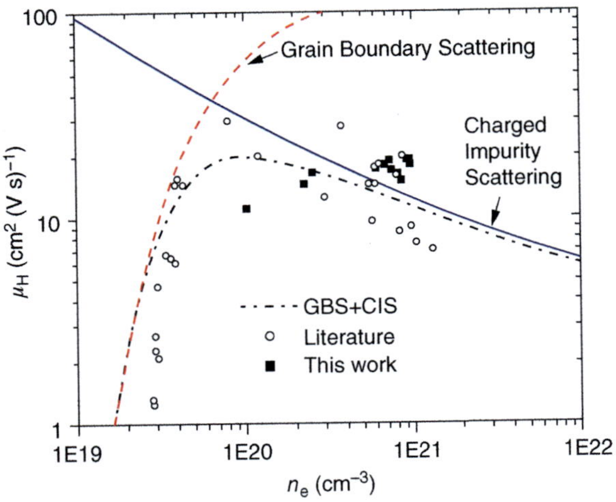

Figure 13.18 Relationship between electron concentration (η_e) and electron Hall mobility (μ_H). Shown are grain boundary scattering (GBS) and charged impurity scattering (CIS) components to Hall Mobility, both measured and calculated values. The open circles represent literature data for undoped and F-doped SnO. *Source:* Martinez et al. [48] / reproduced with permission from IOP Publishing, Ltd.

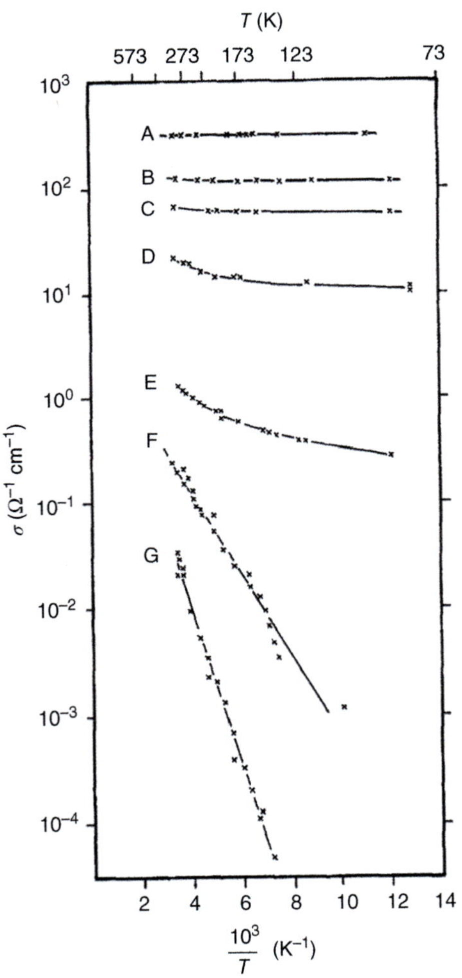

Figure 13.19 Temperature dependence of conductivity as a function of electron concentration η_e for various SnO_2 films. η_e ranges from Curve A: 2×10^{20} cm^{-3} to Curve G: 1×10^{18} cm^{-3}. *Source:* Jarzebski and Morton [47] / reproduced with permission from IOP Publishing, Ltd.

Table 13.1 Electron concentration (n_e), electron mobility (μ_H), and resistivity (ρ) for Sb:SnO_2 films on different substrates. Each coating shows a range of values.

	n (10^{20} cm^{-3})	μ (cm^2 (V s)$^{-1}$)	ρ (10^{-3} ohm cm)
SnO_2/glass	2.2	2.2	13.0
	2.1	2.0	14.0
SnO_2/SiO_2/glass	3.0	4.2	5.0
	2.4	5.6	4.5
SnO_2/fused quartz	3.2	6.5	3.0
	3.1	6.6	2.9
SnO_2/Pyrex glass	2.8	5.5	4.0
	3.1	6.0	3.0

Source: Recharted from Mitzuhashi [49], permission from Elsevier.

electron mobility. Electron concentration is slightly affected by the glass type. This indicates that sodium, and its effect on mobility, is a contributing factor in the conductivity of SnO_2 films on different substrates.

The electrical and optical properties of transparent conducting oxides are linked via the Drude equations [50, 51]. The Drude equations treat the conducting electrons as a "gas of electrons" and the core electrons do not participate in the electrical conduction. The conduction electrons lead to metallic-like behavior from an electrical perspective and high reflectance in the infrared region of the electromagnetic spectrum. The electron mobility and carrier concentration affect the refractive index of the material with consequences carrying over to the visible spectrum. For example, the absorption of a 500 nm coating is calculated [52] as a function of carrier concentration and shown in Figure 13.20. For this calculation,

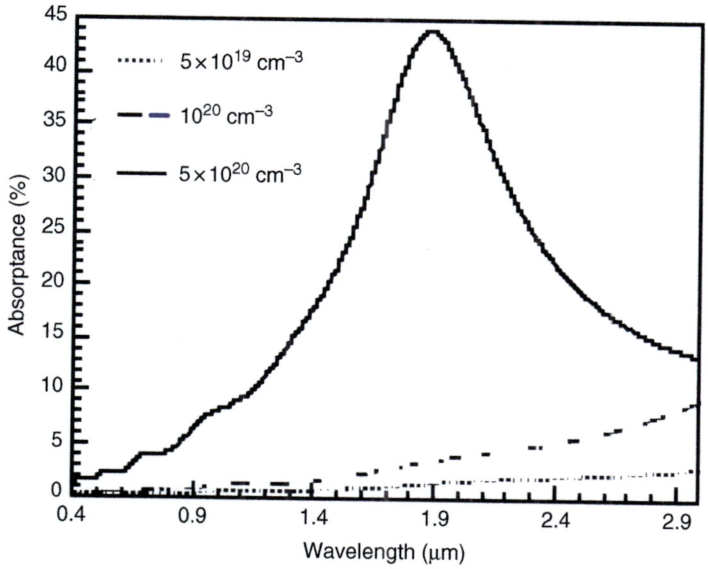

Figure 13.20 Calculated spectral absorptance of a transparent conductive oxide as a function of three carrier concentrations [52]. The carrier concentration ranges from $\eta_e = 5 \times 10^{19}$ to 5×10^{20} cm^{-3}. *Source:* Coutts, et al. [52] / U.S. Department of Energy / Public domain.

0.35 effective mass and permittivity of 5 were used. The absorption remains relatively low for low carrier concentrations but increases considerably when the concentration is higher than 10^{20} electrons cm^{-3}. The absorption tail into the visible spectrum increases substantially when the carrier concentration is high. Reference [52] indicates that for solar cell applications, high electron mobility helps mitigate the absorption. The effect of doping concentration on the absorption coefficient is shown in Figure 13.20 [52]. In this example, the absorption coefficient increases with increasing doping levels leading to increased infrared absorption.

From a practical perspective, the reflectance and transmittance are often equally important to the absorption. The Drude equations can be used to calculate the reflectance and transmittance of $F:SnO_2$ coatings at different doping levels [53]. Figure 13.21 shows the reflectance and transmittance over the solar spectrum. The doping concentration variations corresponded to different plasma frequencies for the examples in Figure 13.22. S_A, S_B, S_C, S_D, S_E, S_F, and S_G correspond to plasma frequencies of 0, 1.43, 2.02, 2.85, 3.49, 4.03, and 5.71×10^{14} S^{-1}, respectively.

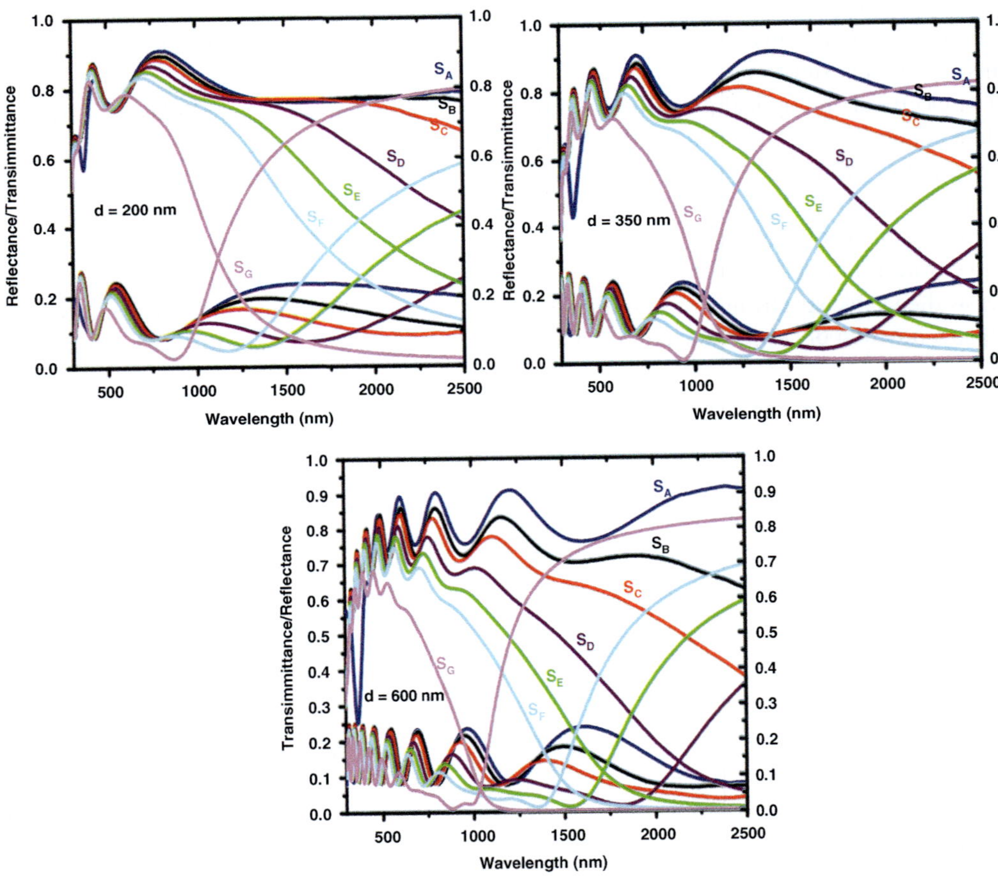

Figure 13.21 Calculated spectral reflectance and transmittance spectra for various fluorine doping levels in tin oxide coatings of different thicknesses (d). Figure 13.22a, $d = 200$ nm, Figure 13.22b, $d = 350$ nm, and Figure 13.22c, $d = 600$ nm. In each spectrum the upper curves are transmittance and the lower curves are reflectance. *Source:* Isoe et al. [54] / with permission of Kabarak Journal of Research & Innovation.

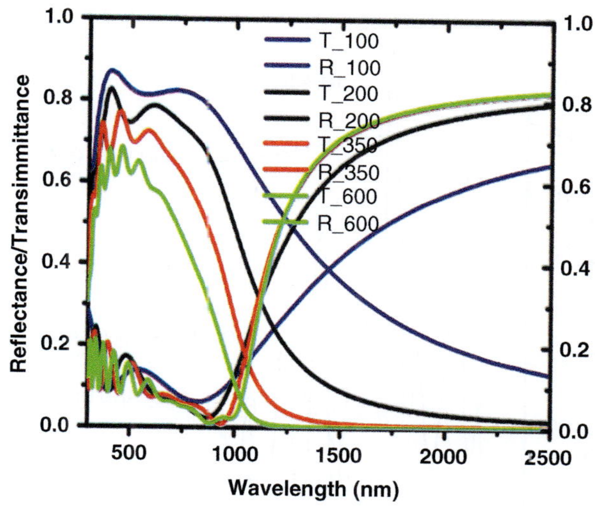

Figure 13.22 Reflectance and transmittance spectra at different film thicknesses for highly doped $S_G = 5.71 \times 10^{14}\ S^{-1}$. The film thicknesses are 100, 200, 350, and 600 nm. *Source:* Isoe et al. [54] / with permission of Kabarak Journal of Research & Innovation.

Figure 13.22 contrasts the reflectance and transmittance spectra for $S_G = 5.71 \times 10^{-4}\ S^{-1}$ doping level for different film thicknesses. The near-infrared reflectance plateaus as the thickness increases while the transmittance continues to decrease with increasing film thickness.

13.6 Color Suppression Technology (CSI)

The early conductive tin oxide coatings had an iridescent appearance. This led to the term, iridized, for the early deposition methods. As the applications for conductive tin oxide grew and expanded the need to mitigate, the interference colors became a dominant focus for innovation. An early concept to reduce the iridescent appearance of a conductive coating on a lite (sheet) of glass was to combine a reflective coating on a second lite of glass in an insulated glass unit [55]. The reflective coating provided "white" light which added to the light reflected from the conductive coating and reduced the visible intensity of the color. The use of a masking layer was logically extended to having both coatings on a single lite of glass [56].

The technology to suppress the color of a conductive coating using only layers on a single surface of a lite of glass were comprehensively developed by Dr. Roy Gordon in the late 1970s and early 1980s [57–62]. A metric was introduced, the color saturation index (CSI) ($c = (a^2 + b^2)^{1/2}$), as a means to quantify the intensity of the color. The terms "a" and "b" are from the Hunter Lab color system [63]. Figure 13.23 shows the reflected CSI for various thin film combinations as a function of the conductive tin oxide thickness. An observability threshold was proposed wherein the iridescent color was reduced in magnitude to a sufficient level as to be unobjectionable. The magnitude of the CSI for the SnO_2 layer on glass varies with the thickness of the SnO_2 coating. The CSI for single and double underlayers is also shown and lie below the observability threshold.

The anti-iridescent approaches can be explained by examining the underlayer options using thin film interference and antireflection equations. Thin film interference originates due to the combination of light waves that reflect off of different interfaces (https://eln.wikipedia.org/wiki/Thin-film_interference; https://en.wikipedia.org/wiki/Anti-reflective_coating). The conditions for antireflection between two bulk materials is $N_{AR} = (N_1 N_2)^{0.5}$, where N_1 is the refractive index of one bulk material and N_2 is the refractive index of the

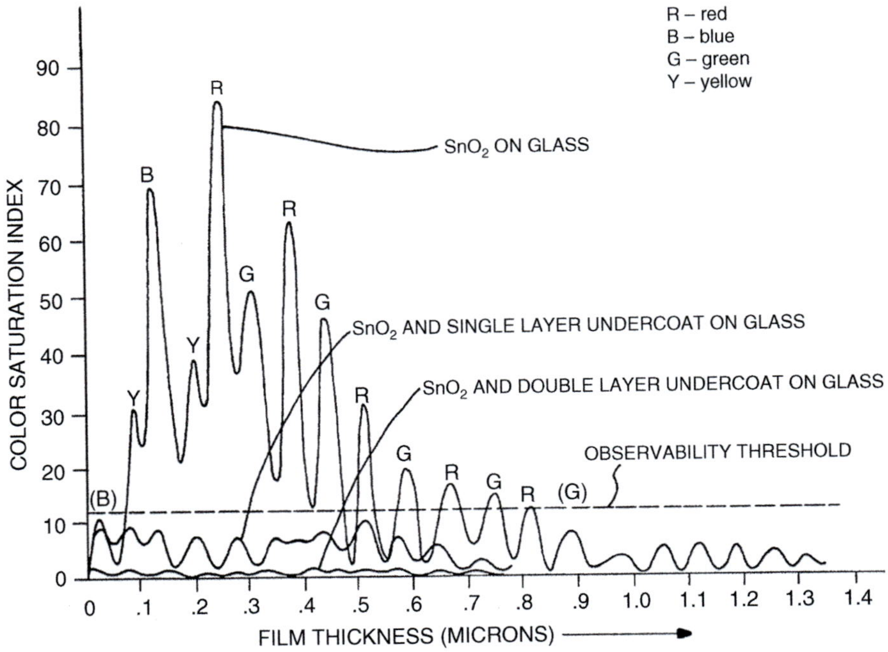

Figure 13.23 Color saturation index (CSI) as a function of film thickness demonstrating the improvements enabled by different color suppression coatings. *Source:* Adapted from Gordon (1980), US 4,187,336 [57].

other bulk material. The SnO_2 coating and substrates are the two bulk materials in this scenario. The antireflection of the interface between these materials essentially minimizes or eliminates the intensity of the light reflected from this interface, thus minimizing the intensity of the CSI. The various options developed as anti-iridescent coatings have different effectiveness at lowering the CSI and robustness with regard to thickness variations.

Alternative anti-iridescence options have been developed [64–69]. These options demonstrate alternative coating options to achieve the same optical result. Reference [69] demonstrates the use of statistical experimental design techniques to find novel solutions for an anti-iridescent design goal which led to patent claims in Ref. [68].

13.7 Physical and Chemical Properties

Fluorine-doped tin oxide is a relatively durable and chemically inert material [70] Cassiterite is the mineral form of SnO_2, and it has a Mohs hardness of 6–7 which means that it is harder than soda-lime glass (5.5), about the same as quartz (7), but not as hard as diamond (10). Despite this physical hardness, tin oxide-coated glass is not immune to damage. If the impact is forceful enough, the glass underneath the tin oxide coating can spall giving the appearance the SnO_2 has failed. Also, many metals, such as aluminum, softer than SnO_2 can give the appearance of damaging the tin oxide. But actually, it is abraded metal on the surface of the tin oxide, which can be removed with dilute acid.

Tin oxide has a long shelf life and does not dissolve in strong acids, bases, or organic solvents making it ideal for many applications. Conversely, it does not etch or pattern easily. The most common chemical etching process for SnO_2 materials consists of applying a zinc

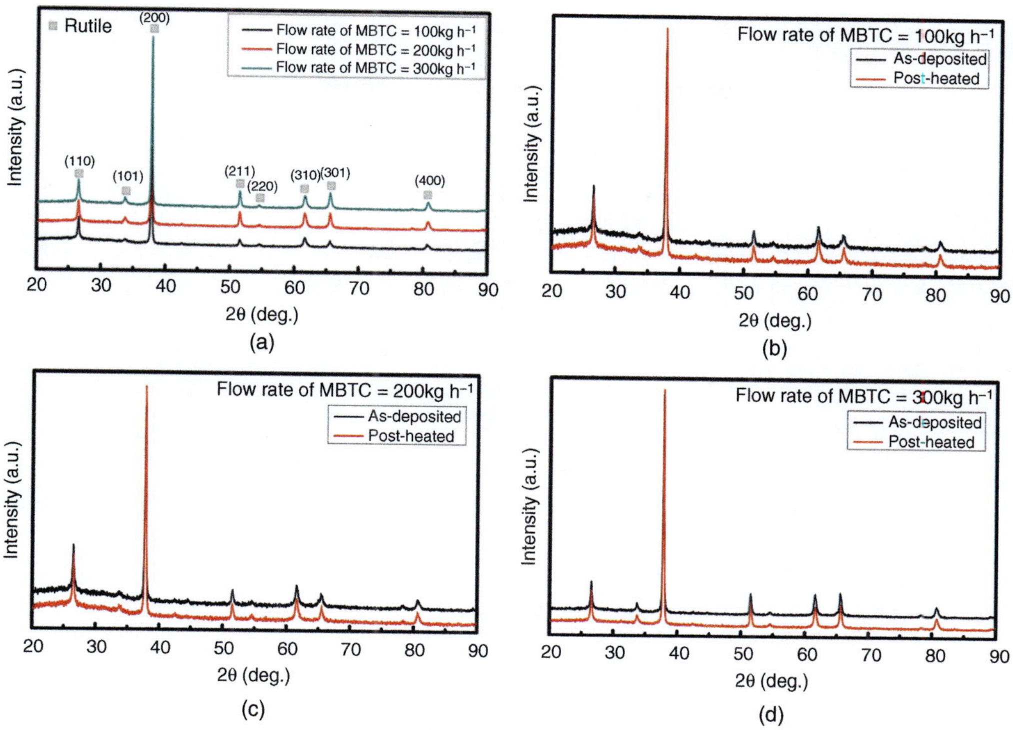

Figure 13.24 X-Ray Diffraction (XRD) patterns for F:SnO_2 produced on the float line at different monobutyltintrichloride (MBTC) flow rates. There is a strong crystal (200) orientation which grows with flow rate and post-heat treatment (as shown in Figure 13.25), both as-deposited and post-heat-treated films are shown. *Source:* Adapted from Gao, et al. [71], used with permission from Elsevier.

powder to the etch area and then contacting the zinc powder with an HCl acid solution. The rapid reaction between the Zn powder and the HCl produces hydrogen atoms in the solution which reduces the SnO_2 to metallic tin [70]. The excess HCl dissolved the remaining tin. This wet etching process can be messy and the resulting feature resolution is limited. Laser scribing is typically used for patterning fine lines. Edge deletion is done with sand blasting and an abrasive wheel.

Various physical properties of tin oxide produced by CVD using monobutyltintrichloride (MBTC) with and without post-heat treatment have been explored. The films showed a rutile crystal structure with the preferred orientation (200) becoming more crystalline with film thickness, as shown in Figure 13.24, (increasing MBTC flow rates, 100–300 kg h^{-1}). Heat treatment at 650 C for 10 minutes results in films with ~11 Ω/sq. sheet resistance and ~10^{-4} Ω cm [71]. Increasing the precursor concentration increases the growth rates and subsequently the electrical properties of the films [72]. The average grain size (crystallinity) increases with increasing MBTC flow rate and subsequent thickness increase. Figure 13.25 shows the scanning electron microscope (SEM) images for representative films at the different MBTC flow rates. The SEM images show the increased crystal sizes associated with the increasing flow rate. Figure 13.26 shows the Young's modulus and Hardness for these coatings before and after heat treatment. The properties increase with heat treatment and with increasing grain size theorized to be due to densification and reduced grain boundaries, respectively.

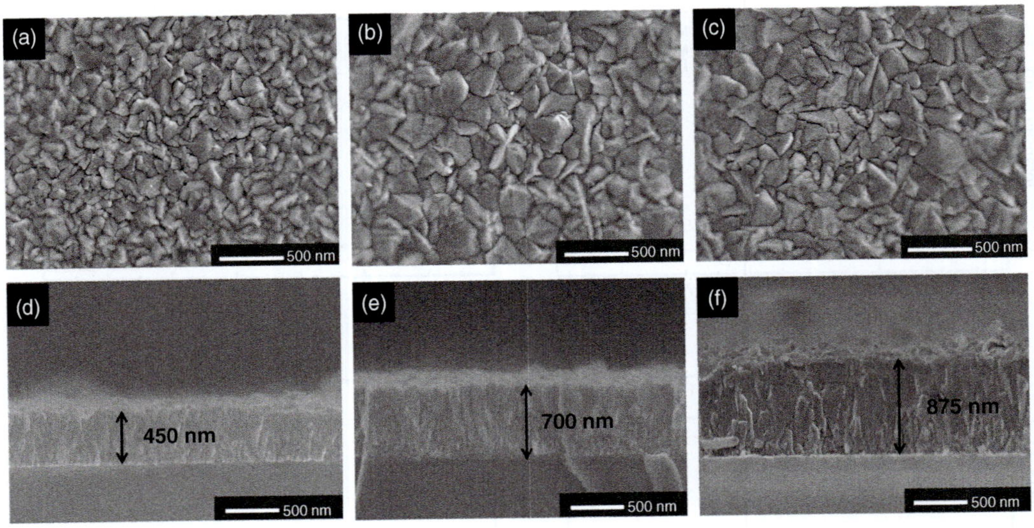

Figure 13.25 Scanning electron microscope images for F:SnO_2 produced with different monobutyltintrichloride (MBTC) flow rates. The flow rates are (a) 100 kg h^{-1}, (b) 200 kg h^{-1} and (c) 300 kg h^{-1}. In Figures (a–c) are shown the surface of the films. In Figures (d-f) are shown cross-sectional images of the coatings. *Source:* Gao, et al. [71] / reproduced with permission from Elsevier.

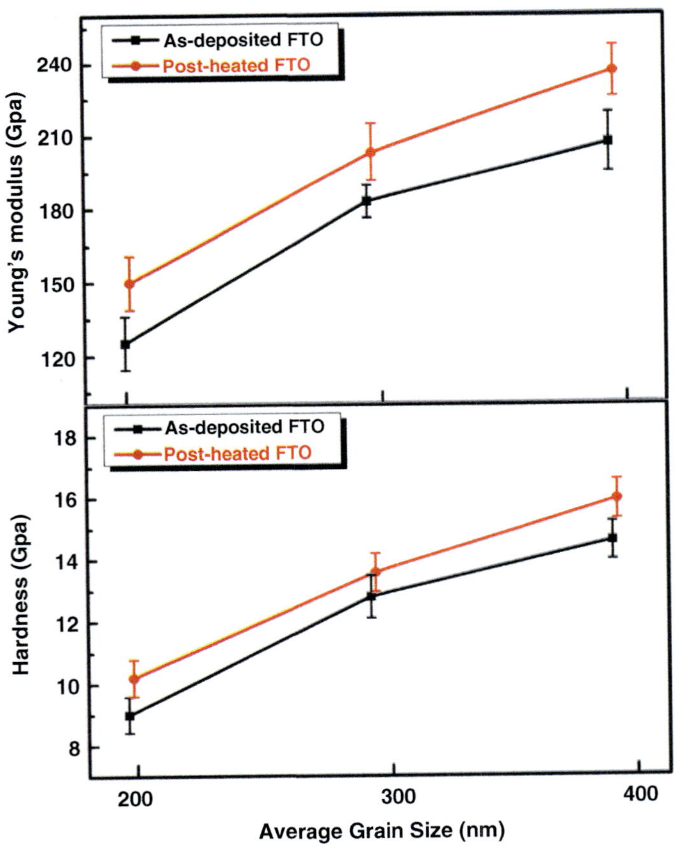

Figure 13.26 Hardness and Young's modulus for F:SnO_2 produced with different average grain sizes before and after heat treatment. *Source:* Adapted from Gao, et al. [71], used with permission from Elsevier.

Table 13.2 F:SnO_2/glass product variants offered by Nippon Sheet Glass (NSG) in their TEC™ series products [73] [only a partial list is shown here, other product variants exist, data from NSG].

NSG TEC™ product	Available thickness (mm)	Sheet resistance (Ohms/square)	Visible transmittance (%)	Haze (%)	Hemispherical emittance
TEC™ 7	2.2, 3.0, 3.2	6–8	80–82	5	0.12
TEC™ 8	2.2, 3.2	6–9	80–81.5	12	0.12
TEC™ 15	2.2, 3.0, 3.2	12–14	82–83	<0.35	0.15
	4.0, 5.0, 6.0, 8, 10		82–84.5	<0.75	0.15
TEC™ 35	3.2, 6.0	32–48	82–84	<0.65	0.35
TEC™ 1000	3.2	1000–3000	88	0.5	0.78

13.8 Commercial Products

The commercial offerings for fluorine-doped tin oxide have grown over the years, offering diverse properties for different applications. An abbreviated version of Table 13.2 was presented by Jack Brown, NSG at the TCO Workshop, Univ. of Liverpool. The table shows the optical and electrical variants available on different substrate thicknesses [73]. The sheet resistance range is large, from 6 ohms/sq. to greater than 1000 ohms/sq. The coating haze with high values is used for the solar cell industry. Rough surfaces in solar cells absorb more energy and reflect less energy compared to smooth surfaces. Low haze values are used for visual specular applications where high transmission and low scattering are desired. Relatively high visible transmittance is offered over the range of electrical properties with good aesthetics.

Figure 13.27 shows various applications for conductive tin oxide. Other applications include deicing aircraft transparencies (windows), passive freezer doors, smart panels, photocopier glass (anti-static), gaming displays (touch panels), electrochromic windows and mirrors, Low-e coatings for windows, and switchable devices [70].

13.9 Conclusion

The combination of physical, electrical, and optical properties along with a viable cost makes fluorine-doped tin oxide an excellent material for a range of commercial applications. It is used for many chromogenic technologies including electrochromic vehicle mirrors and windows. Fluorine-doped tin oxide is available on a range of glass thicknesses, electrical conductivities, and physical sizes. The material is enabling start-ups who don't have the funds to invest in a dedicated magnetron sputtering vapor deposition (MSVD) process to produce ITO for their products. The material is available from a number of suppliers who recognize that the existing offerings may not be ideal for new applications. Many suppliers are willing to adapt the properties as needed to help develop new technologies.

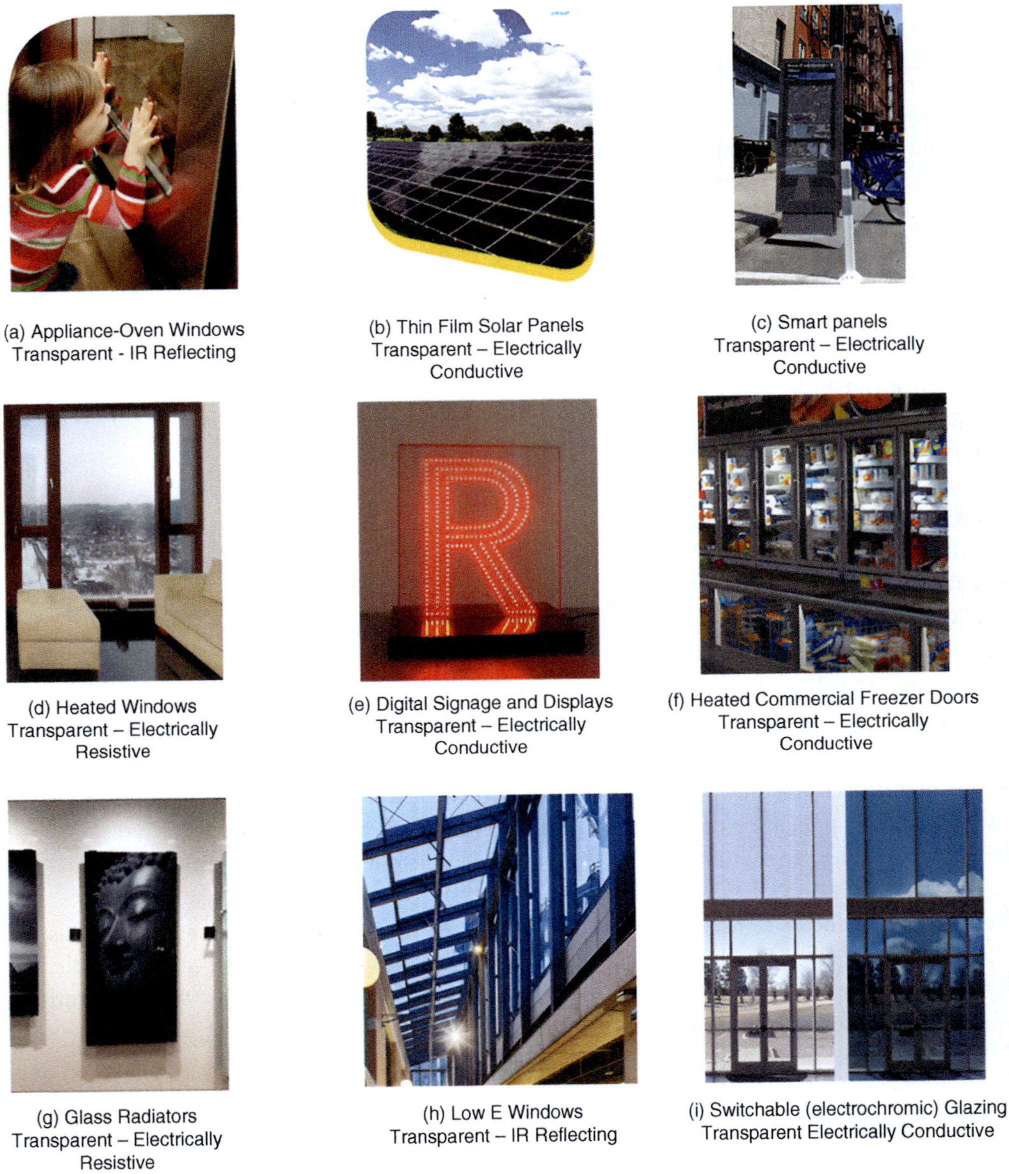

Figure 13.27 Images showing various applications for conductive tin oxide. *Source:* Credit for figures D, F, and H: with permission from NSG.

Acknowledgments

The author would like to acknowledge and thank Bernard Savaete for his time, perspective, resources, and fun discussions on the history of tin oxide. Also, I would like to thank Kyle Sword, Neil McSporran, and Kevin Sanderson from NSG for the great discussions, photographs, and historical perspectives. This work benefits greatly from these four individuals and I am grateful for the interactions I had with them.

References

1 Gordon, R. (2000). Criteria for choosing transparent conductors. *MRS Bulletin* 25 (8): 52–57. https://doi.org/10.1557/mrs2000.151.

2 Harold, A. (1947). McMaster, Conductive Coating for Glass and Method of Application, US Patent 2,429,420, Libby-Owens-Ford Glass Company.

3 Mochel, J.M. (1951). Coated Resistance, US Patent 2,564,706, Corning, Inc.

4 William, O. (1951). Lytle, Electroconductive Products and Production Thereof, US Patent 2,566,346, PPG Industries Inc.

5 Smay, G.L. (1980). Correlation between the electrical resistivity of tin oxide surface coating and their microstructure. *Journal of Non-Crystalline Solids* 38 and 39: 359–364.

6 Gaiser, R.A. and Mcauley, J.W. (1951). Method of Applying Electrically Conducting Coatings to Vitreous Bases, US 2,567,331, Libbey-Owens-Ford Glass Company.

7 Romey A. (1952). Gaiser, Electrically Conducting Surface and Method for Producing Same, US 2,602,032, Libbey-Owens-Ford Glass Company.

8 Raymond, R.L. and Dennison, B.J. (1951). Method of Producing an Electroconductive Article, US 2,617,745, PPG Industries Inc.

9 William O. (1952). Lytle, Electroconductive Article and Production Thereof, US 2,617,741, PPG Industries Inc.

10 Earl R. (1952). Olson, Electroconductive Article and Production Thereof, US 2,617,742, PPG Industries Inc.

11 Lytle, W.O. and Junge, A.E. (1953). Production of Electroconductive articles, US 2,651,585, PPG Industries Inc.

12 Gaiser, R.A. and Thomas, L.D. (1958) Electrically Conducting Glass, US 2,833,902, Libbey-Owens-Ford Glass Company.

13 Saunders, A.E. and Wagner, W.E. (1963). Method of Applying an Electroconductive Tin Oxide Film and Composition Therefor, US 3,107,177, PPG Industries Inc.

14 Schellenger, C. (1936). Method of Making Resistance Elements, US 2,061,107, Chicago Telephone Supply Co.

15 Knowland, M.T. and Marcy, G.D. (1952). Method of Uniformly Spray Coating, US 2,588,125, Boston Woven Hose and Rubber Co.

16 Orr, L.E. (1962). Method of Producing Coated Articles, US 3,019,135, PPG Industries Inc.

17 Gaiser R.A. (1955). Method of Producing Filmed and Strengthened Glass Sheets, US 2,703,949, Libbey-Owens-Ford Glass Co.

18 Mochel, J.M. (1950). Method of Producing Electrically Conducting Coatings on Glass and Mica Sheets, US 2,522,531, Corning Inc.

19 Nascimento, M.L.F. (2014). Brief history of the flat glass patent - Sixty years of the float process. *World Patent Information* 38: 50–56.

20 Geboes, E. (2020). Architectural Flat Glass: understanding the current barriers of the closed-loop alternatives to support the transition towards a circular life-cycle of flat glass, Thesis.

21 Marcault, J. (1979). Process for Coating a Glass Sheet with a Semi-Reflective Film of Tin Oxide, US 4,172,159, Saint-Gobain Industries SA.

22 Wagner, W.E. (1982). Deposition of Coatings from Fine Powder Reactants, US 4,325,988, PPG Industries Inc.

23 Hargreaves, E. and Lewis, J.E. (1988) Production of Infrared Reflecting Coatings on Glass, US 4,725,452, Feb. 16, Pilkington Group, Ltd.

24 Roe, D.W. (1970). Conductive Coatings of Tin Oxides, US 3,537,890, RCA Corp.

25 Dates, H.F. and Davis, J.K. (1967). Iridizing Method, US 3,331,702, Corning, Inc.

26 Dubble, R.D. and Alten, E.H. (1969). Method and Means for Vapor Coating, US 3,438,803, Anchor Glass Container Corp. & Manufacturers Hanover Leasing Corp.

27 Edwards, A. and Loukes, D.G. (1972). Coating of Glass on a Molten Metal Bath, US 3,681,042, Pilkington Group, Ltd.

28 Sopko, J.F. and Simhan, K. (1976). Chemical Vapor Deposition of Coatings, US 3,951,100, PPG Industries Inc.

29 Terneu, R. and Van Cauter, A. (1981). Glass Coating, US 4,293,326, Flachglas Wernberg, GmbH.

30 Chesworth, P. (1984). Coating Glass, US 4,469,045, Pilkington Group, Ltd.

31 Sleighter, G.E. (1986). Chemical Vapor Deposition of a Reflective Film on the Bottom Surface of a Float Glass Ribbon, US 4,584,206, PPG Industries Inc.

32 Villain, J.-M. and Fremaux, J. (1987). Apparatus and Method for Depositing a Metal Oxide Coating on a Glass Substrate, US 4,692,180, Saint-Gobain Vitrage SA.

33 Henery, V.A. (1989). Chemical Vapor Deposition of Tin Oxide on Float Glass in the Tin Bath, US 4,853,257, PPG Industries Ohio Inc.

34 Kane, J. and Schweizer, H. (1976). Process for Depositing Transparent Electrically Conductive Tin Oxide Coatings on a Substrate, US 3,949,146, RCA Corp.

35 Yoldas, B.E. and Mattox, D.M. (1981). Method for Forming Conductive, Transparent Coating on a Substrate, US 4,293,594, Philips North America LLC.

36 Lindner, G.H. (1986). Method of Producing Transparent, Haze-Free Tin Oxide Coatings, US 4,600,654, Arkema Inc.

37 Soubeyrand, M.J. and Halliwell, A.C. (1997). Method for Forming Tin Oxide Coating on Glass, US 5,698,262, Pilkington Group Ltd & Pilkington North America Inc.

38 Russo, D.A. and Lindner, G.H. (1988). Tin Oxide Coated Article, US 4,743,506, M&T Chemicals Inc. & Arkema Inc.

39 Gordon, R.G. (1979). Method of Depositing Electrically Conductive, Infra-Red Reflective, Transparent Coatings of Stannic Oxide, US 4,146,657, assignee Roy G. Gordon.

40 Allendorf, M.D. (2001). On-line Deposition of Oxides on Flat Glass, The Electrochemical Society Interface, Summer.

41 van Mol, A.M.B., Chae, Y., McDaniel, A.H., and Allendorf, M.D. (2006). Chemical vapor deposition of tin oxide: fundamentals and applications. *Thin Solid Films* 502: 72–78. https://doi.org/10.1016/j.tsf.205.07.247.

42 Ghoshtagore, R.N. (1978). Mechanism of CVD thin film SnO_2 formation. *Journal of the Electrochemical Society* 125: 110.

43 Athey, P.R., Dauson, D.S., Lecocq, D.E., Neuman, G.A., Sopko, J.F., and Stewart-Davis, R.L. (1994). Coating Apparatus, Method of Coating Glass, Compounds and Compositions for Coating Glass and Coated Glass Substrates, US 5,356,718, Vitro SAB de CV & Vitro Flat Glass LLC.

44 Wissmann, P.J. and Grover, M.A. (2010). Optimization of a chemical vapor deposition process using sequential experimental design. *Industrial & Engineering Chemistry Research* 49 (12): 5694–5701.

45 Piver, W.T., Organotin compounds: industrial applications and biological investigation, *Environmental Health Perspectives*, 4(1973)61–79. https://doi.org/10.1289/ehp.730461.

46 Russell, R.H. (1976). Recovery of Metal Halides, US 3,956,532, Techneglas LLC.

47 Jarzebski, Z.M. and Morton, J.P. (1976). Physical properties of SnO_2 materials: II, electrical properties. *Journal of the Electrochemical Society* 123 (9): 299C–310C. https://doi.org/10.1149/1.2133090.

48 Marinez, A.I., Huerta, L., de León, J.M.O.R., Acosta, D., Malik, O., and Aguilar, M. (2006). Physicochemical characteristics of fluorine doped tin oxide films. *Journal of Physics D: Applied Physics* 39: 5091–5096. https://doi.org/10.1088/0022-3727/39/23/029.

49 Mizuhashi, M. (1980). Electrical properties of $Sn0_2$ films on various glass substrates. *Journal of Non-Crystalline Solids* 38 and 39: 329–334.

50 Drude, P. (1900). Zur Elektronentheorie der Metalle; II. Teil. Galvanomagnetische und thermomagnetische effecte. *Ann. Phys.* 3: 369.

51 Wolf, N. (2012). Modelling the spectral reflectances of miscellaneous ITO coatings by using only the Drude theory. *Journal of Physics: Conference Series* 395: 012064.

52 Coutts, T.J., Wu, X., Sheldon, P., and Rose, D.H. (1998). Development of high-performance transparent conducting oxide and their impact on the performance of CdS/CdTe solar cells. *Presented at the 2nd World Conference and Exhibition on Photovoltaic Solar Energy Conversion*, Vienna, Austria. Also, NREL/CP-520-23880.

53 Jarzebski, Z.M. and Marton, J.P. (1976). Physical properties of SnO, materials III. Optical properties. *Journal of the Electrochemical Society* 123: 333C.

54 Isoe, W., Mageto, M.J., and Maghanga, C.M. (2020). Optical modeling of fluorine doped tin oxide films for spectrally selective applications, *Conference Proceedings of The Kabarak University International Conference on the Basic Sciences*, edited by Maghanga, C.M. 17th–18th September 2019, Nakuru, Kenya. *Journal of Research and Innovation* 9 (1): 1–17. ISSN 2410-8383. http://ojs.kabarak.ac.ke/index.php/journal1/article/view/339.

55 Stewart, J. (1973), Electrically Heated Multiple Glazed Window Having an Iridescence Masking Film, US 3,710,074, PPG Industries Inc.

56 Michelotti, J.E. (1990). Non-Iridescent Infrared-Reflecting Coated Glass, US 4,971,843, PPG Industries Ohio Inc.

57 Gordon, R.G. (1980). Non-Iridescent Glass Structures, US 4,187,336, assignee Roy. G. Gordon.

58 Gordon, R.G. (1981). Non-Iridescent Glass Structures, US 4,308,316, assignee Roy G. Gordon.

59 Gordon, R.G. (1980). Deposition Method for Coating Glass and the Like, US 4,206,252, assignee Roy G. Gordon.

60 Gordon, R.G. (1983). Non-Iridescent Glass Structures, US 4,377,613, assignee Roy G. Gordon.

61 Gordon, R.G. (1983). Non-Iridescent Glass Structures, US 4,419,386, assignee Roy G. Gordon.

62 Gordon, R.G. (1984). Non-Iridescent Glass Structures, US 4,440,822, assignee Roy G. Gordon.

63 HunterLab (2022). Measuring Color using Hunter L, a, b versus CIE 1976 L*a*b*, Hunter Lab Application Note AN 1005.00/1005b (updated 2022), Reston, VA.

64 Burns, C.G., Middleton, D.J., and Proscia, J.W. (1993). Anti-Iridescence Coatings, US 5,254,392, Ford Motor Co.

65 Proscia, J.W. (1993). Step Gradient Anti-Iridescent Coatings, US 5,271,960, Ford Motor Co.

66 Proscia, J.W. (1993). Anti-Iridescent Coatings with Gradient Refractive Index, US 5,248,545, Ford Motor Co.

67 Proscia, J.W. (1992). Step Gradient Anti-Iridescent Coatings, US 5,168,003, Ford Motor Co.

68 Neuman, G.A. and Stewart-Davis, R.L. (1995). Neutral, Low Emissivity Coated Glass Articles and Method for Making, US 5,395,698, Vitro SAB de CV & Vitro Flat Glass LLC.

69 Neuman, G.A. and Stewart-Davis, R.L. (1997). The use of experimental design techniques to meet reflected color targets in pyrolytic low emissivity coatings. *Thin Solid Films* 308: 26–30.

70 Gerhardinger, P. and Strickler, D. (2008). Fluorine doped tin oxide coatings - over 50 years and going strong. *Key Engineering Materials* 380: 169–178. https://doi.org/10.4028/www.scientific.net/KEM.380.169, ISSN: 1662-9795, Trans Tech Publications Ltd, Switzerland, 2008.

71 Gao, Q., Jiang, H., Li, M., Lu, P., Lai, X., Li, X., Liu, Y., Song, C., and Han, G. (2014). Improved mechanical properties of SnO_2:F thin film by structural modification. *Ceramics International* 40: 2557–2564.

72 Agashe, C., Takwale, M., Bhide, V., Mahamuni, S., and Kulkarni, S.K. (1991). Effect of Sn incorporation on the growth mechanism of sprayed SnO_2 films. *Journal of Applied Physics* 70: 7382–7386.

73 Brown, J. (2018). *Transparent Conducting Oxides – An Industrial Perspective, TCO Workshop*. UK: University of Liverpool. https://www.cdt-pv.org/media/resources/Presentation_Liverpool_University_-_TCO_Manufacturing.pdf (accessed July 25, 2023).

Index

Chromogenics: Smart Switchable Optical Materials and Their Applications, First Edition. Edited by Carl M. Lampert.
 Published 2025 by John Wiley & Sons, Inc.

p

r

s

Printed and bound by CPI Group (UK) Ltd, Croydon, CR0 4YY
17/09/2026
14964258-0005